Birkhäuser

Yuval Z. Flicker

# Arthur's Invariant Trace Formula and Comparison of Inner Forms

Birkhäuser

Yuval Z. Flicker
Ariel University
Ariel, Israel

The Ohio State University
Columbus, Ohio, USA

ISBN 978-3-319-31591-1 ISBN 978-3-319-31593-5 (eBook)
DOI 10.1007/978-3-319-31593-5

Library of Congress Control Number: 2016940094

Mathematics Subject Classification (2010): 11F70, 11F72, 11F41, 11M36, 22E55, 22E57



Printed on acid-free paper

This book is published under the trade name Birkhäuser, www.birkhauser-science.com
The registered company is Springer International Publishing AG, CH

# Preface

The theory of automorphic representations of the group $G(\mathbb{A})$ of the adèle points of a reductive connected group $G$ over a global field $F$, and that of admissible representations of the group $G(F_v)$ of points of a reductive connected group $G$ over a local field $F_v$, are governed by a hypothetical reciprocity law, introduced by Langlands, that relates them to representations of a variant of the Galois group of the base field, named Weil or Weil-Deligne group, into the complex Langlands dual group ${}^LG$ of $G$.

This "principle of functoriality"—not touched upon in the present tome—suggests relations between such automorphic and admissible representations of different groups $G$. These relations have been termed liftings, correspondences, transfers, and are suggested by relations amongst the underlying dual groups.

For example, establishing lifting from GL(2) to GL($n+1$) corresponding to the irreducible $n$-dimensional representation $\mathrm{Sym}^n$ from the dual group $\mathrm{GL}(2,\mathbb{C})$ to $\mathrm{GL}(n+1,\mathbb{C})$ would imply the Ramanujan conjecture for GL(2). Some of these liftings, which are analytic implications of the principle, have been established by various techniques, using various invariants of the representations. The work of Jacquet-Langlands [JL70] showed that the Selberg trace formula [Se62] could give very complete results on the correspondence of representations of GL(2) and its inner forms, the multiplicative groups of quaternion algebras.

Deligne and Kazhdan [DK] then introduced a simple form of the trace formula, which applies to test functions—and representations—with two cuspidal components ("two" was reduced later to "one" in a form of the simple trace formula developed by Flicker and Kazhdan [FK88]), and established in [DKV84] the correspondence of representations between GL($n$) and its inner forms, the multiplicative groups of simple algebras.

This work played the global trace formula against local analysis and used multiple induction arguments to prove not only the lifting of representations but also the existence of matching orbital integrals. The latter was previously considered to be a prerequisite for deriving lifting applications from the trace formula.

To remove this last constraint, and of course for other applications, Arthur developed the trace formula for a general test function, in a series of papers, over many years. In particular, Arthur put the trace formula in invariant form, namely, expressed all terms that appear in the formula as invariant distributions. This is necessary for comparison applications, as when comparing representations of different groups, only characters or orbital integrals can be related, and this is through a norm map relating conjugacy classes of elements in the two groups. In crude terms, only eigenvalues of elements can be related between two groups, not individual elements.

The present volume grew out of an attempt to study Arthur's work. We started in a course following Arthur's expository notes [Ar05], but quickly realized that to attempt to understand the theory we had to study the source articles. To make the subject more accessible, we decided to unite the main articles in one volume and rewrite them as one unit in a conventional way. Thus we cut many arguments into lemmas, propositions, and theorems, stated the claims before giving their proofs, uniformized the notation to make it easier to read (e.g., $\pi$ on $G$ but $\tau$ on $M$), and the like.

Thus, in Chapter 3, we explain Arthur's proof of the basic, noninvariant trace formula, following his early Duke and Compositio papers.

In Chapter 4, we explain Arthur's Annals and J. Funct. Anal. papers [Ar81] and [Ar89], which study the noninvariance of the terms in the basic trace formula and prepare the ingredients for setting up the invariant formula.

In Chapter 5, we explain Arthur's JAMS papers [Ar88.2, Ar88.3], where the invariant formula is finally developed. We quote some of Arthur's computations of contributions to the continuous spectrum from Amer. J. Math., [Ar82.I, Ar82.II], and of weighted orbital integrals from Duke [Ar88.1] and [Ar85, Ar86], where we felt we could not improve the exposition sufficiently to justify the increased volume. Thus Chapters 3–5 here give an almost complete attempt to develop the invariant trace formula in a form fit for applications.

To illustrate the use of this trace formula, in Chapter 6, we compare the invariant trace formulae for $G' = \mathrm{GL}(n)$ and its inner form $G$, for matching functions $f'$ on $G'(\mathbb{A})$ and $f$ on $G(\mathbb{A})$, thus functions with matching orbital integrals. This is already contained in [AC89, Chapter 2], as the secondary case, accompanying the main case of interest there: base change for $\mathrm{GL}(n)$. This marriage makes it hard to follow the inner forms case, so we decided to write it separately in our Chapter 6. This is after all the initial case of comparison; in principle it should be the simplest, and we thought consequently that it deserved its own full treatment.

The comparison of the two trace formulae is far from being simple. A key argument is a multiple induction process, reminiscent of Kazhdan's double induction argument using the simple trace formula.

Equipped with the comparison from Chapter 6, we set in Chapter 2 to prove the correspondence between $\mathrm{GL}(n)$ and its inner forms in general. This chapter is based on our course [F87.1] at Harvard 1986, where a different proof of the results of [DKV84] was given, based on using the then recent works [BDK86] on the trace Paley-Wiener theorem of Bernstein, Deligne, and Kazhdan, and on cuspidal geometry [Ka86.1] by Kazhdan.

The first section of Chapter 2 deals with the comparison of GL($n$) and the anisotropic inner forms, multiplicative groups of a division algebra, following [F90.2], using the simple trace formula. This is the only "easy" case, so we bring it first. It was once considered difficult too, but [F90.2] observed that a finiteness result known at the time permitted a relatively easy proof. This finiteness result is now known for all inner forms of GL($n$), but is not needed in the proof of the general case.

The second section of Chapter 2 discusses the asymptotic behavior of orbital integrals, following Shalika [Shal72], who used ideas of Harish-Chandra [HC70]. An ingredient here is the convergence of the orbital integrals on the unipotent orbits in $G$. There is a publication of Rao [Ra72] on this, and we sketch, toward the end of this Section 2, a proof of Deligne of this fact. This section ends with an elegant computation of Kottwitz [Ko88] of the orbital integral of the Euler-Poincaré function he introduced, which gives explicitly a pseudo-coefficient of the Steinberg representation.

In Section 24, we extend the correspondence from the *non-degenerate* case of square-integrable, tempered, standard, or relevant local representations—what follows on using purely the simple trace formula in Section 13, to a correspondence of local *unitarizable* representations from GL($n$) to its inner forms, by purely local arguments of the type of [BZ76, BZ77, Ze80, Tc90], and the Langlands classification [BW80], following Badulescu [Ba08].

Section 25 uses the comparison of the invariant trace formulae of Chapter 6 to establish the full global correspondence, for cuspidal representations without any local constraint and also for residual representations, permitting to transfer multiplicity one and rigidity theorems from the known case of the split group GL($n$) to that of the inner forms, where no global representation is generic; thus the theory of Whittaker models is not available, as well as establishing for the inner forms a description of the residual spectrum, analogous to that established by Moeglin and Waldspurger [MW89] for GL($n$).

We postpone to the final Section 26 of Chapter 2 an account of the simple trace formula of Flicker and Kazhdan [FK88] that uses one cuspidal component and a second component regular—that leads to no constraints on lifting applications. The idea of regular functions was inspired by Deligne's conjecture on the validity of the Lefschetz fixed point formula for a correspondence on a variety over a finite field, provided it is sufficiently twisted by the Frobenius; see [F13, Fu97, Va07].

Chapter 2 was used for a course at Ohio State in 2014. Chapter 1 contains a statement of the results of Chapter 2 on the correspondence between GL($n$) and its inner forms, as well as a summary of the statement of Arthur's invariant trace formula. Since the statement is so involved, it is not surprising that the proof and the development of the trace formula are so long. We hope this work would make Arthur's work more accessible. We note, however, that it, and in fact the entire theory of automorphic representations, is based on the theory of Eisenstein series, of Langlands [La66, La76]; see also [Ar79], best explained by Moeglin and Waldspurger [MW95], that remains a fundamental challenge.

This work was partially supported by grants from the Simons Foundation (#267097 and #317731 to Yuval Z. Flicker), as well as the Humboldt-Stiftung, TÜBITAK, and ERC AdG Grant 247049, while the author was a Schonbrunn professor at the Hebrew University. The author enjoyed discussions with and hospitality of Elmar Große-Klönne at HU Berlin, David Kazhdan at HU Jerusalem, Takayuki Oda at Tokyo U., Eric Opdam at Amsterdam, and Engin Özkan at Erzincan, during the preparation of this volume. This work greatly benefited from the collaboration of Michael Belfanti at Ohio State who attended the courses, took notes, improved the exposition, and was invited to coauthor this book, but declined.

The envisaged readership of this book consists of graduate students and researchers interested in the trace formula and its applications, especially to lifting problems. We hope it would simplify—in fact make it possible for such people to enter this subject.

Ariel, Israel Yuval Z. Flicker

# Contents

**Preface** ... v

**1 Introduction** ... 1
1 Motivation ... 1
2 Correspondence Between GL($n$) and Its Inner Forms ... 4
3 Arthur's Invariant Trace Formula ... 10

**2 Local Theory** ... 25
1 Case of Division Algebras ... 25
2 Orbital Integrals ... 32
3 Automorphic Forms ... 55
4 Trace Formula ... 58
5 Density ... 61
6 Characters ... 66
7 Coinvariants ... 71
8 Trace Functions ... 73
9 Stability ... 79
10 Discrete Series ... 81
11 Decay ... 82
12 Finiteness ... 84
13 Simple Algebras ... 85
14 Germs ... 90
15 Comparison ... 91
16 Existence ... 92
17 Isolation ... 94
18 Correspondence ... 96
19 Tempered ... 98
20 Irreducibility ... 101
21 Unitarity ... 103
22 Induction ... 105
23 Cuspidal Global Correspondence ... 107
24 Complements on Local Representations ... 108

25 Complete Global Correspondence ........ 116
26 One Cuspidal Place ........ 126

**3 Arthur's Noninvariant Trace Formula** ........ 139
1 Preliminary Definitions ........ 139
2 The Kernel $K_P(x,y)$ ........ 145
3 A Review of Eisenstein Series ........ 150
4 The Second Formula for the Kernel ........ 156
5 The Modified Kernel Identity ........ 163
6 Some Geometric Lemmas ........ 166
7 Integrability of $k_o^T(x,f)$ ........ 171
8 Weighted Orbital Integrals ........ 177
9 A Truncation Operator ........ 183
10 Integrability of $k_\chi^T(x,f)$ ........ 191
11 The Operator $M_P^T(\tau)_\chi$ ........ 201
12 Evaluation in a Special Case ........ 206
13 Conclusion ........ 213

**4 Study of Noninvariance** ........ 215
1 Notation ........ 215
2 A Review of the Trace Formula ........ 220
3 The Distributions $J_o$ and $J_\chi$ ........ 223
4 Noninvariance ........ 231
5 Intertwining Operators ........ 238
6 Normalization ........ 240
7 Real Groups ........ 244
8 $p$-Adic Groups ........ 251
9 Standard Representations ........ 253
10 Convex Sets and Some Related Functions ........ 257
11 Some Examples ........ 265
12 The Distributions $J_M(\tau_\lambda)$ ........ 273
13 The Distributions $J_{M,\gamma}$ and $J_{M,\pi}$ ........ 277
14 Hecke Invariance ........ 285
15 The Distributions $J_M(\tau,X)$ ........ 288
16 Residues ........ 292
17 Proof of Proposition 17.3 ........ 298
18 Changes of Contour ........ 304
19 The Spaces $\mathbb{H}_{ac}(G(F_S))$ and $\mathbb{I}_{ac}(G(F_S))$ ........ 307
20 The Map $\phi_M$ ........ 311

**5 The Invariant Trace Formula** ........ 317
1 Invariant Harmonic Analysis ........ 321
2 The Invariant Distributions $I_M(\gamma)$ ........ 325
3 The Invariant Distributions $I_M(\tau,X)$ ........ 334
4 Some Further Maps and Distributions ........ 338
5 A Contour Integral ........ 347

6 Reduction of Induction Hypotheses ..... 354
7 A Property of $(G, M)$-Families ..... 357
8 Descent ..... 365
9 Splitting ..... 374
10 Local Vanishing Properties for $\mathrm{GL}(n)$ ..... 378
11 Convex Polytopes ..... 383
12 The Invariant Trace Formula So Far ..... 390
13 The Geometric Side ..... 397
14 The Spectral Side ..... 405
15 Completion of the Induction Argument ..... 416
16 A Convergence Estimate ..... 423
17 Simpler Forms of the Trace Formula ..... 432
18 Global Vanishing Properties for $\mathrm{GL}(n)$ ..... 437

**6 Main Comparison** ..... 441
1 Notation and Conventions ..... 441
2 Statement of Parts (i) and (ii) of the Main Comparison ..... 449
3 Statement of Parts (iii) and (iv) of the Main Theorem ..... 451
4 Normalization Factors and the Trace Formula ..... 458
5 The Distributions $I_M(\gamma)$ and $I_M^{\mathcal{E}}(\gamma)$ ..... 469
6 The Distributions $I_M(\tau, X)$ and $I_M^{\mathcal{E}}(\tau, X)$ ..... 474
7 Geometric Induction Assumption ..... 476
8 The Numbers $\varepsilon_M(S)$ ..... 478
9 Comparison of Germs ..... 483
10 Comparison of $I_M^{\mathcal{E}}(\pi, X, f)$ and $I_M(\pi, X, f)$ ..... 490
11 A Formula for $I_t^{\mathcal{E}}(f)$ ..... 501
12 The Map $\varepsilon_M$ ..... 507
13 Cancellation of Singularities ..... 513
14 Separation by Infinitesimal Character ..... 526
15 Elimination of Restrictions on $f$ ..... 534
16 Completion of the Proof of the Main Theorem ..... 541

**Bibliography** ..... 549

**Symbol Index** ..... 555

**Subject Index** ..... 565

# Chapter 1
# Introduction

## 1 Motivation

### 1.1 Poisson Summation Formula

The first example of a trace formula is the classical *Poisson summation formula*. It applies to a lattice $\Gamma \subset \mathbb{R}$. If $f \in C_c^\infty(\mathbb{R})$, the formula says

$$\sum_{\gamma\in\Gamma} f(\gamma) = \frac{1}{\mathrm{vol}(\mathbb{R}/\Lambda)} \sum_{\lambda\in\Gamma^\vee} \hat{f}(\lambda),$$

where $\hat{f}$ denotes the Fourier transform $\hat{f}(x) = \int_{\mathbb{R}} f(y)e^{-2\pi iy}\,\mathrm{d}y$ of $f$ and $\Gamma^\vee$ is the dual lattice $\{a \in \mathbb{R};\ a\Gamma \subset \mathbb{Z}\}$. Set

$$a_\Gamma(\gamma) = \begin{cases} \mathrm{vol}(\mathbb{R}/\Gamma), & \gamma \in \Gamma, \\ 0, & \text{otherwise,} \end{cases}$$

and

$$\hat{a}_\Gamma(\lambda) = \begin{cases} 1, & \lambda\Gamma \subset \mathbb{Z}, \\ 0, & \text{otherwise.} \end{cases}$$

The formula then reads

$$\sum_{\gamma\in\mathbb{R}} a_\Gamma(\gamma) f(\gamma) = \sum_{\lambda\in\mathbb{R}} \hat{a}_\Gamma(\lambda)\hat{f}(\lambda). \tag{1.1.1}$$

Each summand is the product of a coefficient function and, for fixed $\gamma$ or $\lambda$, a distribution on $\mathbb{R}$. The coefficient depends on the lattice $\Gamma$ but is independent



Y.Z. Flicker, *Arthur's Invariant Trace Formula and Comparison of Inner Forms*,
DOI 10.1007/978-3-319-31593-5_1

of $f \in C_c^\infty(\mathbb{R})$. The distribution is invariant and independent of $\Gamma$. The Poisson summation formula has applications in number theory. For example, it can be used to establish the functional equations for Dirichlet $L$-functions.

## 1.2 Compact Quotient Selberg Trace Formula

A more general situation is that of a discrete cocompact subgroup $\Gamma$ of a unimodular locally compact group $G$. Let $r$ denote the *right regular representation* of $G$ on $L^2(\Gamma\backslash G)$ given by

$$(r(y)f)(x) = f(xy), \quad x, y \in G, f \in L^2(\Gamma\backslash G).$$

The representation $r$ decomposes as a direct sum of irreducible unitary representations with finite multiplicities.

Let $G_\gamma$ denote the centralizer of $\gamma$ in $G$. For a fixed invariant measure on $G_\gamma\backslash G$, the *orbital integral* is the distribution defined by setting

$$I_G(\gamma,f) = \int_{G_\gamma\backslash G} f(x^{-1}\gamma x)\, \mathrm{d}x, \quad \gamma \in G, f \in C_c^\infty(G).$$

The geometric coefficient function is defined by setting

$$a_\Gamma^G(\gamma) = \mathrm{vol}(\Gamma_\gamma\backslash G_\gamma).$$

For an irreducible unitary representation $\pi$ of $G$, its *character* is the distribution

$$I_G(\pi,f) = \mathrm{tr}\,\pi(f) = \mathrm{tr}\int_G f(y)\pi(y)\, \mathrm{d}y.$$

Since the regular representation decomposes as a direct sum of irreducible representations, the multiplicity of $\pi$ in $r$,

$$a_\Gamma^G(\pi),$$

is well defined. The *Selberg trace formula* is the equality of distributions on $G$ given by

$$\sum_\gamma a_\Gamma^G(\gamma) I_G(\gamma,f) = \sum_\pi a_\Gamma^G(\pi) I_G(\pi,f), \quad f \in C_c^\infty(G). \tag{1.2.1}$$

The left-hand side is a sum over representatives of conjugacy classes in $G$. The right-hand side is a sum over representatives of equivalence classes of irreducible unitary representations of $G$. The structure of the formula is the same as (1.1.1).

The coefficient functions $a_\Gamma^G(\gamma)$ and $a_\Gamma^G(\pi)$ depend on $\Gamma$ but not on $f \in C_c(G)$. The distributions $I_G(\gamma)$ and $I_G(\pi)$ are invariant and independent of $\Gamma$. The Selberg trace formula reduces to the Poisson summation formula when $\Gamma$ is a lattice in $\mathbb{R}$.

## 1.3 General Case Comparison

Let $G$ be a connected reductive linear algebraic group defined over a number field $F$. Let $\mathbb{A}$ be the adèle ring of $F$. Consider the subgroup $G(\mathbb{A})^1 \subset G(\mathbb{A})$ defined in 2.1. The quotient $G(F)\backslash G(\mathbb{A})^1$ is compact if and only if $G$ has no proper parabolic subgroups defined over $F$ (see [MT62, BHC62]). In particular, this observation applies when $G$ is the group of units of a quaternion algebra over $F$. In [JL70], Jacquet and Langlands stated the Selberg trace formula for the noncompact quotient $\mathrm{GL}(2,F)\backslash\mathrm{GL}(2,\mathbb{A})^1$, outlined its proof, and compared it with the trace formula for $G(F)\backslash G(\mathbb{A})^1$. From this comparison, they deduced a local injection from the set of admissible irreducible representations of the anisotropic group $G(F_v)$ to that of the split form $\mathrm{GL}(2,F_v)$, and a global injection, compatible with the local one, from the set of automorphic representations of $G$ to those of GL(2). A detailed proof of the trace formula for GL(2) is in [AFOO] when $F$ is a number field. A completely different proof in the context of a function field, still only for GL(2), is in [F14]. Its precise form is needed for the exact counting of [F15].

A generalization of the result of [JL70] is the subject of Chapter 2. This application motivates Arthur's development of the invariant trace formula, discussed in Chapters 3–5, for the (not necessarily compact) quotient $G(F)\backslash G(\mathbb{A})^1$, and the comparison in Chapter 6 of the invariant trace formulae of $\mathrm{GL}(n)$ and its inner forms.

Deligne and Kazhdan observed that for test functions with a cuspidal component and another supported on the regular set, one obtains easily a simple form of the trace formula, valid for any reductive connected group, similar to that in the compact quotient case, without weighted orbital integrals and without contributions from the continuous spectrum. Moreover they showed that the *local* correspondence of representations from the anisotropic inner form to $\mathrm{GL}(n)$ (unpublished) and then from any inner form (of $\mathrm{GL}(n)$) to $\mathrm{GL}(n)$ (in [DKV84]) can be deduced from the comparison of their simple form of the trace formula of $\mathrm{GL}(n)$ and the inner form, together with the *global* correspondence for those cuspidal representations with a cuspidal and another square-integrable components.

Another proof of this was given in [F87.1], inspired by the approach of [FK87.2], based on the work of [BDK86] and [Ka86.1], that became available. This is the approach followed in Chapter 2. In particular it is shown in [F87.1] and in Chapter 2, by a double induction argument, that matching of orbital integrals of test functions on the inner form and $\mathrm{GL}(n)$ follows from the comparison of the simple trace formulae, together with the lifting results for representations. It needs not be pre-proven by local means. The trace Paley-Wiener theorem, which became available in [BDK86], is crucial here. This type of double induction argument is the base to the comparison of the invariant trace formulae in general in Chapter 6.

The work of [F87.1] is developed in Chapter 2 from the context of square-integrable, tempered, standard, relevant local representations to that of all unitarizable local representations, that are not necessarily non-degenerate, *purely by local means*, basically of [BZ76, BZ77, Ze80, Tc90].

The invariant trace formula, whose development occupies Chapters 3–5, that is compared for a general test function on GL($n$) and its inner form in Chapter 6 using the double induction argument mentioned above and developed by Arthur, is used to compare all discrete spectrum representations on any inner form with that of GL($n$), without local cuspidality condition anywhere.

This permits extending to the context of the inner form results known for GL($n$), such as multiplicity one theorem and rigidity ("strong multiplicity one") theorem—results known for cuspidal representations with two local conditions using the simple trace formula of Deligne and Kazhdan, as well as reducing the classification of the residual spectrum of the inner form to that of GL($n$), given by [MW89].

The last section of Chapter 2 develops the simple trace formula to be applicable for the global comparison of representations with a single cuspidal component, instead of two, following [FK88], where the notion of regular functions is introduced, inspired by Deligne's conjecture on the Lefschetz fixed point formula for a correspondence twisted by a high power of the Frobenius; see [F13, Fu97, Va07]. Thus all the work of Chapters 3–5 in our context is geared to remove this last condition. The comparison of Chapter 2 is thus an application to motivate the development of the invariant trace formula in Chapters 3–5. Chapter 6 concerns then a comparison of the two invariant trace formulae in the oldest comparison case. It is discussed in [AC89, Chapter 2], as a case secondary to the main one considered there, of base change for GL($n$). We think this case of inner forms deserves a separate exposition, as the first case of comparison.

# 2 Correspondence Between GL($n$) and Its Inner Forms

## *2.1 Properties of Local Representations*

Let $F$ be a local field. Let $G$ be an inner form of GL($n$) over $F$. Thus $G$ is the multiplicative group of a central simple $F$-algebra $A$. There is a central division algebra $D$ over $F$ of rank $d$ with $A$ being the algebra $M(m, D)$ of $m \times m$-matrices with entries in $D$, $n = md$. Class field theory—see [We67]—associates with $A$ an invariant $\operatorname{inv} A$ of the form $i/d$ (modulo 1), with $i$ prime to $d$, and $\operatorname{inv} A = \operatorname{inv} D$ independent of $m$. There exists a unique central simple $F$-algebra $A$ of rank $n$ with invariant $i/d$ (modulo 1) (where $(i, d) = 1$ and $d$ divides $n$). If $F = \mathbb{C}$, then $d = 1$. If $F = \mathbb{R}$, then $d = 1$ or 2. Otherwise, $d$ can be any positive integer. We let $G' = \mathrm{GL}(n)$. Note that if $\overline{F}$ is an algebraic closure of $F$ then $G(\overline{F}) = \mathrm{GL}(n, \overline{F})$. This isomorphism is over $\overline{F}$.

A conjugacy class $\gamma$ in $G(F)$ is called (semisimple) *regular* if its characteristic polynomial $p_\gamma$ has distinct roots (in some algebraic closure of $F$). If $\gamma, \delta$ are regular and $p_\gamma = p_\delta$, then $\gamma = \delta$. There is an embedding $\gamma \mapsto \gamma'$, defined by $p_{\gamma'} = p_\gamma$, of the set of regular conjugacy classes $\gamma$ in $G(F)$ into the set of regular conjugacy classes $\gamma'$ in $G'(F)$. There is another notion of regularity for orbits, used only in Section 2.

Let $C_c^\infty(G(F))$ denote the convolution algebra of complex valued smooth compactly supported measures $f$ on $G(F)$. These are all of the form $fdg$ where $dg$ is any fixed Haar measure on $G(F)$ and $f$ is now a function. Put $R(G(F)) = R_\mathbb{Z}(G(F)) \otimes \mathbb{C}$, where $R_\mathbb{Z}(G(F))$ is the Grothendieck group of the category of smooth $G(F)$-modules. It is the free abelian group generated by the set $\mathrm{Irr}(G(F)) = \Pi(G(F))$ of equivalence classes of irreducible such $G(F)$-modules. By [BZ76, Theorem 3.25] irreducible smooth $G(F)$-modules are admissible.

If $\pi$ is an admissible $G(F)$-module, then the convolution operator $\pi(f\,\mathrm{d}g) = \int_{G(F)} f(g)\pi(g)\,\mathrm{d}g$ is of finite rank, and its trace is denoted by $\mathrm{tr}\,\pi(f)$. We often delete the measure d$g$ from the notation, namely, think of $f$ as a measure. There exists a complex valued conjugacy invariant smooth function $\chi = \chi_\pi$ on the regular set of $G(F)$ with $\mathrm{tr}\,\pi(f) = \int \chi(g)f(g)$ for any $f \in C_c^\infty(G(F))$ that is supported on the regular set of $G(F)$. It is called the *character* of $\pi$. It depends only on the image of $\pi$ in $R(G(F))$. The characters of inequivalent irreducible $G(F)$-modules are linearly independent. In particular, $\chi \neq 0$ if $\pi \neq 0$ in $R(G(F))$. Harish-Chandra [HC78] showed that $\chi$ extends to a locally integrable function on $G(F)$.

Fix a minimal parabolic subgroup $P_0$ together with its Levi decomposition $M_0N_0$ in $G$. Denote by $i_M^G$ the homomorphism $R(M(F)) \to R(G(F))$ of normalized, or unitary, induction, for any (standard) Levi subgroup $M$. Here "standard" means that $M$ is the Levi subgroup containing $M_0$ of a parabolic subgroup $P$ containing $P_0$. The map $i_M^G$ is independent of the choice of the parabolic subgroup $P$ with Levi component $M$ that is used in its definition.

An irreducible $G(F)$-module $\pi$ whose central character $\omega_\pi$ is unitary is called *square-integrable* or *discrete series* if it has a matrix coefficient that is square-integrable on $G(F)$ modulo its center $Z(F)$. In this case all of its matrix coefficients are square-integrable, and $\pi$ embeds in $L^2(G, Z, \omega_\pi)$. An alternative definition is that all of its central exponents decay.

An irreducible $G(F)$-module $\pi$ is called *tempered* if there exists a Levi subgroup $M(F)$ and a square-integrable (= discrete series) $M(F)$-module $\tau$ such that $\pi$ is a subquotient (necessarily a direct summand) of $i_M^G\tau$.

Put $\nu(x) = |x|$ for $x \in F$, where $|\cdot|$ is the normalized absolute value on $F$. Thus when $F$ is non-Archimedean with ring $R$ of integers, generator $\boldsymbol{\pi}$ of the maximal ideal and residual cardinality $q = |R/(\boldsymbol{\pi})|$, we have $|\boldsymbol{\pi}| = 1/q$. Put $\nu(g) = \nu(\det g)$, where $\det g$ is the reduced norm of $g$ in $G(F)$.

DEFINITION 2.2. *A $G(F)$-module $\pi$ is called* relevant *if there is a Levi subgroup of $G(F)$ of the form $M = \prod_{i=1}^m (M_i \times M_i)$ or $M_0 \times M$, where the $M_i$, $0 \leq i \leq m$, are multiplicative groups of central simple $F$-algebras, and tempered $M_i(F)$-modules $\tau_i$ and distinct positive numbers $s_i < 1/2$ such that $\pi$ is*

$$i_M^G\left(\prod_{i=1}^{m}(\tau_i\nu^{s_i}\times\tau_i\nu^{-s_i})\right)$$

*or*

$$i_{M_0\times M}^G\left(\tau_0\times\prod_{i=1}^{M}(\tau_i\nu^{s_i}\times\tau_i\nu^{-s_i})\right)$$

*in* $R(G(F))$.

THEOREM 2.3 (Local Theorem for Non-degenerate Representations).

(1) *Relevant* $G(F)$*-modules are unitarizable and irreducible. In particular, a* $G(F)$*-module normalizedly (= unitarily) induced from a tempered one is irreducible.*
(2) *The relation* $\chi'(\gamma')=(-1)^{n-m}\chi(\gamma)$ *for all matching* $\gamma\mapsto\gamma'$ *regular conjugacy classes* $\gamma\in G(F)$, $\gamma'\in G'(F)$, *defines a bijection between the set of equivalence classes of square-integrable (resp. tempered, relevant)* $G(F)$*-modules* $\pi$ *and the set of equivalence classes of square-integrable* $G'(F)$*-modules* $\pi'$ *(resp. tempered, relevant,* $G'(F)$*-modules* $\pi'$ *whose character* $\chi'$ *is nonzero on the set of regular* $\gamma'$ *obtained from* $\gamma$ *in* $G(F)$*).*

In particular, the character relation defines an isomorphism of $\mathrm{Irr}(G(F))$ with $\mathrm{Irr}(G_1(F))$ for inner forms $G=\mathrm{GL}(m,D)$ and $G_1=\mathrm{GL}(m,D_1)$ of $\mathrm{GL}(n)$ over $F$, where $D$ and $D_1$ are division algebras with invariants $i/d$ and $i_1/d$ modulo 1 with $(i,d)=1$ and $(i_1,d)=1$. Namely the representation theory of $\mathrm{GL}(m,D)$ depends only on the degree $d$ of $D$ (square toot of its dimension over $F$). It is independent of the numerator $i$ of $\mathrm{inv}(D)$ (which is prime to $D$).

## 2.4 Global Correspondence

Let $F$ be a global field. Let $G$ be an inner form of $G'=\mathrm{GL}(n)$ over $F$. Then $G$ is the multiplicative group of a central simple $F$-algebra $A=M(m,D)$, for $D$ a division algebra central over $F$ of rank $d$, $n=md$. Class field theory—see [We67]—associates with $A$ the sequence $\{\mathrm{inv}_v A=\mathrm{inv}\,A\otimes_F F_v\}$ of rational numbers modulo one that are almost all zero and whose sum is zero modulo one. Each such sequence $\{i_v/d_v\}$ determines, up to $F$-isomorphism, a unique division algebra $D$ central over $F$, and a unique simple algebra $A$ of rank $n$ central over $F$ with these invariants, for any $n$ that is divisible by $d_v$ for all $v$. Let $G(\mathbb{A})$ be the group of $\mathbb{A}$-points of $G$, where $\mathbb{A}$ is the ring of adèles of $F$. Let $Z$ (resp. $Z'$) denote the center of $G$ (resp. $G'$). Then $Z=Z'$ is the multiplicative group. Fix a unitary character $\omega$ of $Z(\mathbb{A})/Z(F)=\mathbb{A}^\times/F^\times$. For each place $v$ of $F$ denote by $F_v$ the completion of $F$ at $v$ and by $\omega_v$ the restriction of $\omega$ to $F_v^\times$.

Let $L(G(F)\backslash G(\mathbb{A}))$ denote the space of slowly increasing, see [BJ79], functions $\phi$ on $G(F)\backslash G(\mathbb{A})$ with $\phi(zg) = \omega(z)\phi(g)$ for $z$ in $Z(\mathbb{A})$. By $L^2(G(F)\backslash G(\mathbb{A}))$ we mean the space of $\phi$ on $G(F)\backslash G(\mathbb{A})$ with $\phi(zg) = \omega(z)\phi(g)$ for $z$ in $Z(\mathbb{A})$ that are square-integrable on $Z(\mathbb{A})G(F)\backslash G(\mathbb{A})$. The group $G(\mathbb{A})$ acts on $L = L(G(F)\backslash G(\mathbb{A}))$ and on $L^2 = L^2(G(F)\backslash G(\mathbb{A}))$ by right translation. Thus $(r(g)\phi)(h) = \phi(hg)$. Any irreducible subquotient is unitary (preserves a non-degenerate inner product) and called an *automorphic* $G(\mathbb{A})$-module.

The space $L = L(G(F)\backslash G(\mathbb{A}))$ is the direct sum of the discrete spectrum $L_d = L_d(G(F)\backslash G(\mathbb{A}))$, that is the direct sum of all irreducible *sub*representations of $L$—these $G(\mathbb{A})$-modules are called "discrete spectrum" $G(\mathbb{A})$-modules, and the continuous spectrum $L_c = L_c(G(F)\backslash G(\mathbb{A}))$, that is a "continuous sum". The space $L_d$ is also the discrete spectrum in $L^2$.

A *cuspidal* $G(\mathbb{A})$-module is an irreducible constituent of the subspace $L_0 = L_0(G(F)\backslash G(\mathbb{A}))$ that consists of the $\phi$ in $L(G(F)\backslash G(\mathbb{A}))$ with $\int_{N(F)\backslash N(\mathbb{A})} \phi(nx)\,\mathrm{d}n$ equals zero for every $x$ in $G(\mathbb{A})$ and for the unipotent radical $N$ of any proper parabolic subgroup of $G$ over $F$. Each cuspidal $\phi$ is rapidly decreasing, hence absolutely square-integrable, on $Z(\mathbb{A})G(F)\backslash G(\mathbb{A})$. The space $L_0(G(F)\backslash G(\mathbb{A}))$ is a sub-$G(\mathbb{A})$-module of $L_d(G(F)\backslash G(\mathbb{A}))$.

Any cuspidal $G'(\mathbb{A})$-module $\pi$ is non-degenerate, namely each of its local components $\pi_v$ has a Whittaker model. This means that for any additive nontrivial character $\psi_v : F_v \to \mathbb{C}^\times$, there is an embedding of the $G'(F_v)$-module $\pi_v$ in the space $W(\psi_v)$ of right smooth functions $\varphi : G'(F_v) \to \mathbb{C}$ with $\varphi(ng) = \psi_v(\sum_i n_{i,i+1})\varphi(g)$ for $n = (n_{i,j})$ upper triangular unipotent matrix and $g \in G'(F_v)$, with $G'(F_v)$ action by right shifts.

Moreover such $\pi$ is generic (has a global Whittaker model: same definition, but with a global $\psi : \mathbb{A}/F \to \mathbb{C}^\times$, $\psi \neq 1$), and it occurs with multiplicity one in $L_0(G'(F)\backslash G'(\mathbb{A}))$. See [BZ76] and [PS79] for Whittaker models and [Shal74] or [PS79] for the multiplicity one theorem.

An irreducible admissible $G(\mathbb{A})$-module $\pi$ decomposes as a restricted tensor product $\otimes_v \pi_v$ of irreducible admissible $G(F_v)$-modules $\pi_v$, almost all of which are unramified; see [Ft79].

An irreducible $\pi_v$ is call unramified if its space contains a nonzero $K_v = G'(R_v)$-fixed vector $\xi_v$, necessarily unique up to a scalar. Here $v$ is finite and $R_v$ is the ring of integers in $F_v$.

The space of the irreducible $\pi$ is the union over all sets $V$ of valuations of $F$ such that $\pi_v$ is unramified for $v \notin V$, of $(\otimes_{v\in V} V_v) \otimes (\otimes_{v\notin V}\xi_v)$, where $V_v$ is the space of $\pi_v$.

If $\pi'_1 = \otimes_v \pi'_{1,v}$ and $\pi'_2 = \otimes_v \pi'_{2,v}$ are cuspidal $G'(\mathbb{A})$-modules and $\pi'_{1,v} \simeq \pi'_{2,v}$ for almost all $v$, then $\pi'_{1,v} \simeq \pi'_{2,v}$ for all $v$ by the rigidity theorem, also called the "strong multiplicity one theorem"; see [PS79]. All components of a cuspidal $G'(\mathbb{A})$-module are relevant by [Ze80, (9.7)] and, as noted above, unitarizable. Note that an irreducible $\pi$ in $L^2$ is unitary (its space is complete in an invariant inner form). Its subspace $\pi^\infty$ of smooth vectors is admissible, and unitarizable. A representation $\pi$ is called unitarizable if its space is dense in a unitary representation $\hat{\pi}$.

Given $G$ or $D$, there is a finite set $V$ of places $v$ of $F$ such that for every $v$ outside $V$, the division algebra $D$ *splits*, namely, $D \otimes_F F_v = M(d, F_v)$. We say that $\pi_v$ *corresponds* to $\pi'_v$ if $G_v \simeq G'_v$ (thus $v \notin V$) and $\pi_v \simeq \pi'_v$, or, more generally for arbitrary $v$, if $\pi_v$ corresponds to $\pi'_v$ by the Local Theorem 2.3. In this case we also say that $\pi'_v$ comes from $\pi_v$. An irreducible $G(\mathbb{A})$-module $\pi = \otimes_v \pi_v$ *lifts*, or *corresponds*, to an irreducible $G'(\mathbb{A})$-module $\pi' = \otimes_v \pi'_v$ if $\pi_v$ *corresponds* to $\pi'_v$ for all $v$. We also say that $\pi'$ comes from $\pi$, in this case. An automorphic $G(\mathbb{A})$-module that lifts to a cuspidal $G'(\mathbb{A})$-module will be called *$G'$-cuspidal*. Using the Deligne-Kazhdan simple trace formula, one obtains the following result for representations with a cuspidal component and another square-integrable component at places outside $V$. This is reduced to a single cuspidal component in Section 26. The work of Chapters 3–6 is geared to remove the remaining constraint.

THEOREM 2.5 (Global Theorem for Generic Representations).

(1) *All local components of a $G'$-cuspidal $G(\mathbb{A})$-module are relevant.*
(2) *Each $G'$-cuspidal $G(\mathbb{A})$-module occurs in the discrete spectrum $L_d$ of $L = L(G(F)\backslash G(\mathbb{A}))$ with multiplicity one.*
(3) *If $\pi_1 = \otimes_v \pi_{1,v}$ and $\pi_2 = \otimes_v \pi_{2,v}$ are $G'$-cuspidal $G(\mathbb{A})$-modules and $\pi_{1,v} \simeq \pi_{2,v}$ for almost all $v$, then $\pi_1 \simeq \pi_2$.*
(4) *Correspondence defines a bijection from the set of $G'$-cuspidal $G(\mathbb{A})$-modules $\pi = \otimes_v \pi_v$ to the set of cuspidal $G'(\mathbb{A})$-modules $\pi' = \otimes \pi'_v$ such that $\pi'_v$ is obtained by the local correspondence for all $v$ (in $V$). Every cuspidal $G(\mathbb{A})$-module is $G'$-cuspidal.*

REMARK. Part (1) is the motivation for the definition of "relevant" representations. Part (2) is called the "multiplicity one" theorem for the $G'$-cuspidal spectrum of $G$. Part (3) is called the "rigidity" theorem for the $G'$-cuspidal spectrum. Part (4) is called the Deligne-Kazhdan correspondence. As in the local case, the representation theory of $G(\mathbb{A})$, $G = \mathrm{GL}(m, D)$, depends only on the least denominators $d_v \geq 1$ of $\mathrm{inv}(D) = \{i_v/d_v (\mathrm{mod}\ 1)\}$ at the places where $D$ ramifies, and not on the numerators $i_v$, that are prime to $d_v$ (or 0 where $D_v = D \otimes_F F_v$ splits).

## 2.6 Extended Global Correspondence

Having used the trace formula in a simple form, we were able to describe the correspondence for global cuspidal representations with some local restrictions and for local representations that may occur as their local components. These local representations, in the case of $G' = \mathrm{GL}(n)$, are non-degenerate.

Our next aim is to extend the global correspondence from the context of the cuspidal spectrum to that of the entire discrete spectrum. For this we need to use the invariant trace formula and its comparison between $G$ and $G'$. The complement of the cuspidal spectrum in the discrete spectrum is the residual spectrum. The representations that appear in the residual spectrum of $G'$ were described by Moeglin and Waldspurger [MW95]. The components of these residual spectrum representations are nontempered and degenerate, but unitarizable and of a special kind, described by [MW95].

The global correspondence will be seen to be an injection from the discrete spectrum of the inner form $G(\mathbb{A})$ to that of the quasisplit $G'(\mathbb{A})$. However it is not compatible with the local correspondence from $G(F_v)$ to $G'(F_v)$, since the correspondence is not surjective.

The problem can be seen already in the initial case, considered by Jacquet and Langlands [JL70], of $G' = \mathrm{GL}(2)$ and its inner form $G$ over $F$. In the local case, the correspondence maps the trivial representation of $G(F_v)$ to the special, or Steinberg, representation of $G'(F_v)$. The character relation $\chi_{\pi'_v}(g') = -\chi_{\pi_v}(g)$ is satisfied. The sign is $-1$ to the power $m(d-1) = 1$ ($m = 1$, $d = 2$). A non-one-dimensional representation $\pi$ corresponds to a cuspidal $\pi'$, and $\pi_v$ corresponds to $\pi'_v$ for all $v$. The one-dimensional $\pi$ corresponds to the one-dimensional $\pi'$ in the sense that $\pi_v = \pi'_v$ at all places where $G$ splits. At the finite set of places $v$ where $G$ ramifies, the components of $\pi'$ are one-dimensional. They are not the square-integrables mentioned above. We have the character relation $\chi_{\pi'_v}(g') = \chi_{\pi_v}(g)$ that differs from the previous relation by a sign. The explanation is that the trivial and the square-integrable are the constituents of an induced on $\mathrm{GL}(2, F_v)$ that does not come from the correspondence from any representation of the inner form, so its character is zero on the (elliptic) $g'$ that come from $g$ in the inner form. Consequently we need to define the correspondence from the split group to its inner form, and the sign in the character relation has to change.

This extends to the general case of GL($n$) and its inner forms. The local theory is sufficiently developed to express the values of the characters of the local components of the global residual representations in terms of those of the non-degenerate ones that appear in the correspondence, and we shall simply review and summarize these local results, in order to complete the comparison of the unrestricted trace formulae.

## 2.7 d-lifting

Let $F$ be again a local field. We say that a $G'(F)$-module $\pi'$ is a $d$-lift if $\chi_{\pi'}(g') \neq 0$ for a regular $g' \in G'(F)$ that comes from $G(F) = \mathrm{GL}(m, D)$, where $D$ is a division algebra of rank $d$ central over $F$ (and $G' = \mathrm{GL}(md)$). To extend the local theorem in the non-degenerate case above, we summarize in Section 24 the purely local—that is, using techniques of [BZ76, BZ77, Ze80, Tc90]—work of [Ba08] that extends the local correspondence theorem from the context of square-integrable, tempered, standard, or relevant representations, to that of all unitarizable representations.

THEOREM 2.8 (Unitarizable Local Correspondence). *Let $\pi'$ be an irreducible unitarizable representation of $G'(F)$ which is a d-lift. Then there exists a unique irreducible unitarizable representation $\pi$ of $G(F)$ and a unique sign $\varepsilon \in \{1, -1\}$ with $\chi_{\pi'}(g') = \varepsilon\chi_\pi(g)$ for all regular $g \in G(F)$, $g' \in G'(F)$ with equal characteristic polynomials.*

Using the non-degenerate local lifting theorem and the Langlands classification, one defines the correspondence map $C' : R(G) \to R(G')$, which injects the Grothendieck group of smooth $G$-modules of finite length, $R(G)$, into $R(G')$. The unitarizable local lifting theorem defines a map $|C| : \mathrm{Irr}^u(G') \to \mathrm{Irr}^u(G)$ from the set of equivalence classes of unitarizable irreducible $G'$-modules to such $G$-modules, and $|C|$ is not injective even for $G' = \mathrm{GL}(2)$.

THEOREM 2.9. *There exists a unique injective map, $C'$, from the discrete spectrum of $G(\mathbb{A})$, $G = \mathrm{GL}(m, D)$, $D$ a division algebra of rank $d$ central over $F$, to the discrete spectrum of $G'(\mathbb{A})$, $G' = \mathrm{GL}(md)$, such that $C'(\pi)_v = \pi_v$ for all $v \notin V$ = set of $F$-places where $D$ ramifies. We have, for all $v \in V$, that $C'(\pi)_v$ is a $d_v$-lift, and $|C_v|(C'(\pi)_v) = \pi_v$. The image of $C'$ is the set of $D$-lifts discrete spectrum $\pi'$ of $G'(\mathbb{A})$ (namely, $\pi'_v$ is a $d_v$-lift for each $v \notin V$, where $D \otimes_F F_v = M(m_v, D_v)$ and $D_v$ is a division algebra central over $F_v$ of rank $d_v$).*

Since $C'$ is injective, multiplicity one and rigidity theorems for the discrete spectrum of $G(\mathbb{A})$ follow, as well as a description of the residual spectrum of $G(\mathbb{A})$, in Proposition 25.20. See Section 25.

## 3 Arthur's Invariant Trace Formula

The goal of this section is to introduce the terms in Arthur's invariant trace formula:

$$I(f) = \sum_{M \in \mathcal{L}} |W_0^M||W_0^G|^{-1} \sum_{\gamma \in (M(F))_{M,S}} a^M(S, \gamma) I_M(\gamma, f)$$

$$= \sum_{t \geq 0} \sum_{M \in \mathcal{L}} |W_0^M||W_0^G|^{-1} \int_{\Pi(M,t)} a^M(\tau_{1,\lambda}) I_M(\tau_{1,\lambda}, f)\, d\tau_{1,\lambda}, \quad f \in \mathbb{H}_{\mathrm{ac}}(G(\mathbb{A})^1).$$

See Section 5.12 for some discussion. The distribution $I$ is a global object related to the integral kernel of the regular representation of $G(\mathbb{A})$. The trace formula is an expression for this distribution in terms of geometric and spectral data. The formula has the same structure as (1.2.1), aside from the presence of additional sums.

### 3.1 Notation

Let $G$ be a connected reductive linear algebraic group defined over a number field $F$. If $S$ is a finite set of valuations, let $F_S = \prod_{v \in S} F_v$. The $F_S$-valued points (respectively, $\mathbb{A}$-valued points) of $G$ will be denoted by $G(F_S)$ (respectively, $G(\mathbb{A})$) and similarly for other groups. All algebraic groups are assumed to be defined over $F$.

Let $M_0$ be a fixed minimal Levi subgroup of $G$. The Weyl group of $M_0$ in $G$ will be denoted by $W_0^G$. Let $\mathcal{L}$ denote the set of Levi subgroups of $G$ containing $M_0$. For $M \in \mathcal{L}$, let $\mathcal{L}(M)$ (respectively, $\mathcal{L}_0(M)$, $\mathcal{F}(M)$, $\mathcal{P}(M)$) denote the set of Levi subgroups of $G$ containing $M$ (respectively, proper Levi subgroups of $G$ containing $M$, parabolic subgroups of $G$ containing $M$, parabolic subgroups of $G$ with a Levi subgroup equal to $M$). The Levi decomposition of $P \in \mathcal{P}(M)$ is written $P = M_P N_P$ for $M_P \in \mathcal{L}(M)$. Let $A_M$ denote the maximal split torus in the center of $M$. For $P \in \mathcal{P}(M)$, set $A_P = A_M$. Fix a "good" maximal compact subgroup $K = \prod_v K_v$ of $G(\mathbb{A})$. See Section 4.1.6 for the definition and key properties. Let $G_\gamma$ denote the connected centralizer of an element $\gamma$ in $G$.

The group $X(M)_F$ of rational characters of $M$ defined over $F$ is a free abelian group of finite rank. Then

$$\mathfrak{a}_M = \mathrm{Hom}_{\mathbb{Z}}(X(M)_F, \mathbb{R})$$

and

$$\mathfrak{a}_M^* = X(M)_F \otimes_{\mathbb{Z}} \mathbb{R}$$

are finite-dimensional real vector spaces. The complexification of the latter space is denoted by

$$\mathfrak{a}_{M,\mathbb{C}}^* = X(M)_F \otimes_{\mathbb{Z}} \mathbb{C}.$$

There is a surjective homomorphism

$$H_M : M(\mathbb{A}) \to \mathfrak{a}_M$$

defined by

$$\langle H_M(m), \chi \rangle = \log |\chi(m)|, \quad m \in M(\mathbb{A}),\ \chi \in X(M)_F.$$

Here $|\cdot| = |\cdot|_{\mathbb{A}} = \prod_v |\cdot|_v$ is the product over all valuations, assumed normalized. The kernel is denoted by $M(\mathbb{A})^1$. The local version of this map, defined by restriction to $G(F_S) \subset G(\mathbb{A})$, will also be denoted by $H_M$. If $P \in \mathcal{P}(M)$, these objects are often denoted by $\mathfrak{a}_P$, $\mathfrak{a}_P^*$, and $\mathfrak{a}_{P,\mathbb{C}}^*$. The map $H_M$ extends to

$$H_P : G(\mathbb{A}) \to \mathfrak{a}_P$$

by writing $G(\mathbb{A}) = N(\mathbb{A})M(\mathbb{A})K$ and setting $H_P(nmk) = H_M(m)$.

We use both $\Pi(G)$ and $\mathrm{Irr}(G)$ to denote the set of equivalence classes of irreducible admissible representations of a group $G$.

## 3.2 Hecke Algebras

The convolution algebras defined in this subsection will be the spaces of test functions for the distributions appearing in Arthur's formula.

DEFINITION 3.2.1. Let $S$ be a finite set of valuations of $F$. Let $C_c^\infty(G(F_S))$ denote the space of compactly supported *smooth* functions on $G(F_S)$. By definition, a function $f \in C_c^\infty(G(F_S))$ is infinitely differentiable at the Archimedean places and locally constant at the non-Archimedean places. It is spanned by products $\otimes_{v\in S} f_v$, $f_v \in C_c^\infty(G(F_v))$.

The *Hecke algebra* $\mathbb{H}(G(F_S))$ is the convolution algebra (with the unique up to scalar multiple Haar measure on $G(F_S)$) of smooth compactly supported $K_S$-finite functions $f$ on $G(F_S)$. Instead of $K_S$-finite it suffices to require $f$ to be $K_{S_\infty}$-finite, where $S_\infty = \infty \cap S$ is the set of Archimedean places in $S$. A function $f_v \in C_c^\infty(G(F_v))$ is $K_v$-finite if the space spanned by its right and left $K_v$-translates is finite dimensional over $\mathbb{C}$. If $v$ is non-Archimedean, a $C_c^\infty$-function $f_v$ is $K_v$-finite.

It is possible to endow $\mathbb{H}(G(F_S))$ with a natural topology that gives it the structure of a topological vector space. See Section 4.19 for this construction.

DEFINITION 3.2.2. A finite set of valuations $S$ is said to have the *closure property* if it contains an Archimedean valuation or it contains only valuations dividing a fixed rational prime. Equivalently, for each $M \in \mathcal{L}$ the subgroup

$$\mathfrak{a}_{M,S} = \{H_M(m)\,;\, m \in M(F_S)\}$$

is a closed subgroup of $\mathfrak{a}_M$.

In this section, all finite sets of places are assumed to have the closure property.

Let $S_{\text{ram}}$ denote the (finite) set of valuations at which $G$ is ramified. Recall the $G$ is *unramified* at a place $v$ if $v$ is non-Archimedean and if $G$ is quasisplit over $F_v$ and split over an unramified extension of $v$. By convention, then, the set of Archimedean valuations $S_\infty$ is contained in $S_{\text{ram}}$. If $S$ is a finite set of valuations containing $S_{\text{ram}}$, let $\chi^S$ be the characteristic function of $\prod_{v\notin S} K_v$. The map $f \mapsto f\chi^S$ is an embedding of $\mathbb{H}(G(F_S))$ into the space of functions on $G(\mathbb{A})$. Under this embedding, if $S_1 \subset S$, then $\mathbb{H}(G(F_{S_1})) \subset \mathbb{H}(G(F_S))$.

DEFINITION 3.2.3. The *global Hecke algebra* $\mathbb{H}(G(\mathbb{A}))$ is defined as the limit $\varinjlim_S \mathbb{H}(G(F_S))$ over $S$ containing $S_{\text{ram}}$. This construction endows $\mathbb{H}(G(\mathbb{A}))$ with the structure of a topological vector space.

The space $\mathbb{H}_{\text{ac}}(G(F_S))$ of *almost compactly supported* functions can be defined as a topological vector space containing $\mathbb{H}(G(F_S))$. Its primary property is that, for all $Z \in \mathfrak{a}_{G,S}$, the restriction $f^Z$ of $f \in \mathbb{H}_{\text{ac}}(G(F_S))$ to the space

$$G(F_S)^Z = \{x \in G(F_S)\,;\, H_G(x) = Z\}$$

is compactly supported. As before, set $\mathbb{H}_{\text{ac}}(G(\mathbb{A})) = \varinjlim_S \mathbb{H}_{\text{ac}}(G(F_S))$.

The previous discussion applies to $G(F_S)^1 \subset G(\mathbb{A})^1$ and leads to analogous spaces of functions on $G(F_S)^1$ and $G(\mathbb{A})^1$. See 4.19 for further discussion.

The role of the larger spaces of test functions will be described in the next subsection. Note that any linear form on $\mathbb{H}(G(F_S))$ that depends only on $f^Z$ can be extended to $\mathbb{H}_{\mathrm{ac}}(G(F_S))$.

## 3.3 *Weighted Characters and the Map $\phi_M$*

A linear form $\theta$ acting on a convolution algebra $H$ is *invariant* if $\theta(h*f) = \theta(f*h)$ for any $f, h \in \mathbb{H}$. When comparing trace formulae on different groups, it is only possible to compare invariant distributions. This subsection introduces a linear map that will be used to systematically convert noninvariant distributions into invariant distributions.

DEFINITION 3.3.1. (i) For $\tau \in \Pi(M(F_S))$ and $\lambda \in \mathfrak{a}^*_{M,\mathbb{C}}$, define the twisted representation $\tau_\lambda$ by

$$\tau_\lambda(m) = \tau(m)e^{\lambda(H_M(m))}.$$

(ii) Let $(\tau, V_\tau)$ be a representation of $M(F_S)$. Let $P \in \mathcal{P}(M)$. The (normalized) parabolically induced representation $i_P^G(\tau_\lambda)$ is viewed as a representation of $\mathbb{H}(G(F_S))$. The representation space is

$$\begin{aligned} V_P(\tau) = \{&\phi : K \to V_\tau\,; \\ &\phi \text{ is } K\text{-finite}, \phi(nmk) = \tau(m)\phi(k), \\ &n \in N_P(F_S) \cap K, m \in M(F_S) \cap K, k \in K\}. \end{aligned} \tag{3.3.1}$$

A function $f \in \mathbb{H}(G(F_S))$ acts by the convolution operator

$$(i_P^G(\tau_\lambda, f)\phi)(k) = \int_{G(F_S)} f(y)\tau(M_P(ky))\phi(K_P(ky))e^{(\lambda+\rho_P)(H_P(ky))}\,\mathrm{dy}$$

where

$$x = N_P(x)M_P(x)K_P(x)$$

is the decomposition of $x$ according to $G(F_S) = N_P(F_S)M_P(F_S)K_S$. Note that the space $V_P(\tau)$ of $i_P^G(\tau_\lambda)$ depends only on $\tau$ and not on $\lambda \in \mathfrak{a}^*_{M,\mathbb{C}}$.

(iii) Let $P, P' \in \mathcal{P}(M)$. The intertwining operator

$$\mathcal{M}_{P'|P}(\tau_\lambda) : V_P(\tau) \to V_{P'}(\tau)$$

is defined by

$$(\mathcal{M}_{P'|P}(\tau_\lambda)\phi)(k) = \int_{N_{P'}(F_S)\cap N_P(F_S)\backslash N_{P'}(F_S)} \tau(M_P(n))\phi(K_P(n)k)e^{(\lambda+\rho_P)(H_P(n))}\, dn.$$

It converges for the real part of $\lambda$ sufficiently regular in $(\mathfrak{a}_P^*)^+$ and can be meromorphically continued to $\mathfrak{a}_{P,\mathbb{C}}^*$.

These intertwining operators have some undesirable properties. For example, they do not always satisfy

$$\mathcal{M}_{P''|P}(\tau_\lambda) = \mathcal{M}_{P''|P'}(\tau_\lambda)\mathcal{M}_{P'|P}(\tau_\lambda).$$

Moreover, the operator does not generally preserve the characteristic function of $K_S$. The following theorem provides a solution to these problems. See Sections 4.6–4.8 for a full statement and proof.

**Theorem 3.3.1.** *Let $\tau \in \Pi(M(F_S))$. For parabolic subgroups $P$ and $P'$ in $\mathcal{P}(M)$, there exists a meromorphic, scalar-valued function $m_{P'|P}(\tau_\lambda)$ of $\lambda \in \mathfrak{a}_{M,\mathbb{C}}^*$ such that the operator*

$$\mathcal{R}_{P'|P}(\tau_\lambda) = m_{P'|P}(\tau_\lambda)^{-1} \cdot \mathcal{M}_{P'|P}(\tau_\lambda)$$

*has more desirable properties.*

In particular, $\mathcal{R}_{P'|P}(\tau_\lambda)$ does not have the problems just mentioned. The functions $m_{P'|P}(\tau_\lambda)$ are the *(local) normalizing factors*, and the operators $\mathcal{R}_{P'|P}(\tau_\lambda)$ are the *normalized intertwining operators*.

The notion of a $(G, M)$-family will be discussed in more detail in Section 4.10. These families clarify the combinatorial constructions that arise in local harmonic analysis. Recall that two parabolic subgroups $P$ and $P'$ in $\mathcal{P}(M)$ are called *adjacent* if the associated chambers $\mathfrak{a}_P^+$ and $\mathfrak{a}_{P'}^+$ in $\mathfrak{a}_M$ share a wall.

DEFINITION 3.3.2. Let $M$ be a Levi subgroup. Consider a family of smooth functions

$$\{c_P(\nu)\,;\, P \in \mathcal{P}(M)\}$$

on $i\mathfrak{a}_M^*$. The collection is called a *$(G, M)$-family* if it satisfies the following condition. Whenever $P$ and $P'$ are adjacent subgroups in $\mathcal{P}(M)$ and $\nu$ belongs to the hyperplane spanned by the common wall of the chambers of $P$ and $P'$, the associated functions satisfy $c_P(\nu) = c_{P'}(\nu)$.

This is equivalent to the property that, whenever $P$ and $P'$ are contained in a parabolic subgroup $Q$ and $\nu \in i\mathfrak{a}_Q^*$, the associated functions satisfy $c_P(\nu) = c_{P'}(\nu)$.

There is a smooth function on $i\mathfrak{a}_M^*$ associated to any $(G, M)$-family. Many objects in local harmonic analysis, in particular the weight factors used to define the weighted orbital integrals and weighted characters described below, can be realized as the canonical smooth function associated to some $(G,M)$-family. The function is defined by the following proposition.

**Proposition 3.3.2.** *If* $\{c_P(\nu)\,;\, P \in \mathcal{P}(M)\}$ *is a* $(G,M)$*-family, the function*

$$c_M(\nu) = \sum_{P\in\mathcal{P}(M)} c_P(\nu)/\theta_P(\nu)$$

*can be extended to a smooth function on* $i\mathfrak{a}_M^*$.

Here $\theta_P(\nu)$ is a homogenous polynomial defined using the coroots of $(P, A_M)$, see Section 4.3.

The value of $c_M(\nu)$ at $\nu = 0$ is denoted by $c_M$.

DEFINITION 3.3.3. Fix $P_0 \in \mathcal{P}(M)$. Let $\nu \in i\mathfrak{a}_M^*$. For $P \in \mathcal{P}(M)$, define the family of operators

$$\mathcal{R}_P(\nu, \tau_\lambda, P_0) = \mathcal{R}_{P|P_0}(\tau_\lambda)^{-1}\mathcal{R}_{P|P_0}(\tau_{\lambda+\nu}).$$

This is a family of operator valued functions of $\lambda$. It is a $(G,M)$-family of functions (in $\nu$). See Section 4.11 for the proof.

The operator

$$\mathcal{R}_M(\tau_\lambda, P_0) = \lim_{\nu\to 0} \sum_{P\in\mathcal{P}(M)} \mathcal{R}_P(\nu, \tau_\lambda, P_0)/\theta_P(\nu)$$

is defined by taking the value at $\nu = 0$ of the associated smooth function $\mathcal{R}_M(\nu, \tau_\lambda, P_0)$.

The linear form $J_M(\tau_\lambda)$ on $\mathbb{H}(G(F_S))$ is defined by setting

$$J_M(\tau_\lambda, f) = \operatorname{tr}(\mathcal{R}_M(\tau_\lambda, P_0) i_{P_0}^G(\tau_\lambda, f)).$$

It is independent of the choice of $P_0$.

We will see that *weighted characters* arise naturally in Arthur's development of the preliminary noninvariant trace formula. We will prove that the globally defined spectral distributions used to expand the regular representation of $G(\mathbb{A})$ on $L^2(G(F)\backslash G(\mathbb{A}))$ can sometimes be written as a weighted character. See Section 3.12 for a precise statement and proof. The objects introduced here are a slight variant of the weighted characters of 3.12. They are discussed in more detail in Section 4.15. The weighted character is easiest to define when $J_M(\tau_\lambda)$ is regular in $\lambda \in i\mathfrak{a}_M^*$. This holds when $\tau$ is unitary. When this condition is satisfied, the weighted character is defined by

$$J_M(\tau, X, f) = \int_{i\mathfrak{a}_{M,S}^*} J_M(\tau_\lambda, f) e^{-\lambda(X)}\, d\lambda.$$

Here $X$ is a point in $\mathfrak{a}_{M,S}$. When $J_M(\tau_\lambda, f)$ is not regular in $\lambda$, one must use a sum of such distributions. Note that the weighted characters can be extended to $\mathbb{H}_{\mathrm{ac}}(G(F_S))$.

For any $f \in \mathbb{H}_{\mathrm{ac}}(G(F_S))$ and $h \in \mathbb{H}(G(F_S)^1)$, the weighted character satisfies a variance formula:

$$J_M(\tau, X, L_h f) = \sum_{Q \in \mathcal{F}(M)} J_M^{M_Q}(\tau, X, R_{Q,h} f). \tag{3.3.2}$$

Here $L_h$ denotes left convolution by $h$ and $R_{Q,h}$ denotes a modified right convolution that descends $f$ to a function in $\mathbb{H}_{\mathrm{ac}}(M_Q(F_S))$.

DEFINITION 3.3.4. (i) Let $\Pi_{\mathrm{temp}}(M(F_S))$ denote the set of equivalence classes of irreducible tempered representations of $M(F_S)$.

(ii) For $f \in \mathbb{H}(G(F_S))$, let $f_M$ denote the function on $\Pi_{\mathrm{temp}}(M(F_S))$ given by

$$f_M(\tau) = f_G(i_P^G \tau) = \operatorname{tr} i_P^G(\tau, f).$$

(iii) Let $\mathbb{I}(G(F_S))$ be the image of $\mathbb{H}(G(F_S))$ under $f \mapsto f_G$. It is a space of functions on $\Pi_{\mathrm{temp}}(G(F_S))$. It can be described more intrinsically by the trace Paley-Wiener theorems of [BDK86] in the non-Archimedean case and [CD90] in the Archimedean case. See [F95] for a different proof in the non-Archimedean case.

Note that a function $\phi \in \mathbb{I}(G(F_S))$ can be identified with the function on $\Pi_{\mathrm{temp}}(G(F_S)) \times \mathfrak{a}_{G,S}$, also denoted by $\phi$, given by the Fourier transform:

$$\phi(\pi, Z) = \int_{i\mathfrak{a}_{G,S}^*} \phi(\pi_\lambda) e^{-\lambda(Z)} \, d\lambda, \quad \pi \in \Pi_{\mathrm{temp}}(G(F_S)), \, Z \in \mathfrak{a}_{G,S}.$$

Let $\mathbb{I}_{\mathrm{ac}}(G(F_S))$ be the image of $\mathbb{H}_{\mathrm{ac}}(G(F_S))$ under $f \mapsto f_G(\pi, Z) = \operatorname{tr}(\pi(f^Z))$.

(iv) A linear form on the space $\mathbb{H}_{\mathrm{ac}}(G(F_S))$ is said to be *supported on characters* if $\theta(f) = 0$ for each $f$ such that $f_G = 0$. In this case there is a well-defined linear form $\hat{\theta}$ on $\mathbb{I}_{\mathrm{ac}}(G(F_S))$ defined by

$$\hat{\theta}(f_G) = \theta(f).$$

The weighted character can now be interpreted as a linear map. See Section 4.15 for further discussion.

DEFINITION 3.3.5. For $f \in \mathbb{H}_{\mathrm{ac}}(G(F_S))$, define $\phi_M(f)$ to be the function

$$(\tau, X) \mapsto \phi_M(f, \tau, X) = J_M(\tau, X, f)$$

on $\Pi_{\mathrm{temp}}(M(F_S)) \times \mathfrak{a}_{M,S}$.

The map $\phi_M$ satisfies a variance formula similar to (3.3.2). It is this property which is relevant for making distributions invariant.

The following theorem is one of the main results of Chapter 4; see Section 4.20.

**Theorem 3.3.3.** *The map* $f \mapsto \phi_M(f)$ *is a continuous linear transformation from* $\mathbb{H}_{\mathrm{ac}}(G(F_S))$ *to* $\mathbb{I}_{\mathrm{ac}}(M(F_S))$.

The restriction of $\phi_M$ to $\mathbb{H}(G(F_S))$ need not map into $\mathbb{I}(G(F_S))$. This is part of the reason for introducing the almost compactly supported spaces.

## 3.4 The $(M, S)$-Equivalence Classes

The next objective is to define the indexing set on the geometric side of the invariant trace formula. The equivalence classes are generally larger than a rational conjugacy class.

DEFINITION 3.4.1. Let $S$ be a finite set of valuations of $F$ containing $S_\infty$. Let $\mathcal{U}$ denote the variety of unipotent elements of $G$.

(i) Let $u_1, u_2 \in \mathcal{U}(F)$. We call $u_1$ and $u_2$ $(G, S)$-*equivalent* if they are $G(F_S)$-conjugate.
(ii) Let $\gamma, \gamma' \in G(F)$. Let $\gamma = \sigma u$ be the Jordan decomposition of $\gamma$. The elements $\gamma$ and $\gamma'$ are said to be $(G, S)$-*equivalent* if there exists $\delta \in G(F)$ satisfying the following properties:
    (1) The semisimple element $\sigma$ is the semisimple part of $\delta^{-1}\gamma'\delta$.
    (2) The elements $\sigma^{-1}\gamma$ and $\sigma^{-1}\delta^{-1}\gamma'\delta$ of $G_\sigma(F) \cap \mathcal{U}(F)$ are $(G_\sigma, S)$-equivalent in the sense of (i).

We let $(G(F))_{G,S}$ denote the collection of $(G, S)$-equivalence classes in $G(F)$. The definition applies with $G$ replaced by a Levi subgroup $M$ of $G$ and serves to define the indexing set on the geometric side. See also Section 5.13.

## 3.5 The Coefficient $a^M(S, \gamma)$

DEFINITION 3.5.1. A semisimple element $\sigma \in M(F)$ is called $F$-*elliptic* (in $M$) if the split component of the center of $M_\sigma$ is $A_M$.

DEFINITION 3.5.2. (i) Let $\sigma \in M(F)$ be semisimple. Define the characteristic function $i^M(S, \sigma)$ by setting it equal to 1 if $\sigma$ is $F$-elliptic in $M$ and if for $v \notin S$ the set

$$\{m^{-1}\sigma m;\ m \in M(F_v)\} \cap \{K_v \cap M(F_v)\}$$

is nonempty. Otherwise, $i^M(S, \sigma)$ is 0.

(ii) Let $\{u\,;\,\sigma u \sim \gamma\}$ denote the collection of $u \in (\mathcal{U}_{G_\sigma}(F))_{G_\sigma,S}$ such that $\sigma u$ is $(G,S)$-equivalent to $\gamma$.
(iii) Let $G_{\sigma,+}$ denote the full (not necessarily connected) centralizer of $\sigma$. Let $i^G(\sigma)$ be the quotient $G_{\sigma,+}(F)/G_\sigma(F)$.
(iv) The general definition requires us to define the coefficient in the unipotent case by expanding a certain noninvariant distribution associated to the unipotent class in terms of weighted orbital integrals. Once defined for unipotent elements, we set

$$a^M(S,\gamma) = i^M(S,\sigma)|i^M(\sigma)|^{-1} \sum_{\{u\,;\,\sigma u \sim \gamma\}} a^{M_\sigma}(S,u).$$

If $\sigma$ is semisimple, then

$$a^G(S,\sigma) = \varepsilon^G(\gamma)|i^G(\gamma)|^{-1} \operatorname{vol}\left(G_\gamma(F)\backslash G_\gamma(\mathbb{A})^1\right).$$

Note that the coefficient does not depend on $S$ and generalizes the coefficient from the simple trace formula discussed below.

### 3.6 Weighted Orbital Integrals and the Invariant Distribution $I_M(\gamma)$

The invariant geometric distributions are defined in terms of weighted orbital integrals. These distributions first appear in Section 3.8. They play a role similar to the weighted characters mentioned above. Namely, the global distributions of Chapter 3 can sometimes be written as weighted orbital integrals.

Let $M \in \mathcal{L}$ be a Levi subgroup. For $P \in \mathcal{P}(M)$ and a fixed $x \in G(F_S)$, the functions

$$v_P(\lambda, x) = e^{-\lambda(H_P(x))}$$

form a $(G,M)$-family. Let $v_M(x)$ be the associated smooth function. It is the volume of the convex hull of the points $\{-H_P(x)\,;\,P \in \mathcal{P}(M)\}$ [Ar05, Lemma 17.2].

For any $\gamma \in G(F_S)$, the orbital integral $I_G(\gamma)$ is a well-defined $G(F_S)$-invariant linear form on $C_c^\infty(G(F_S))$ given by

$$f \mapsto \int_{G_\gamma(F_S)\backslash G(F_S)} f(x^{-1}\gamma x)\,\mathrm{d}x.$$

When $\gamma$ is not semisimple, the convergence of the integral requires Theorem 2.32 of Deligne or Rao [Ra72]. The integral is taken with respect to a $G(F_S)$-invariant measure on the conjugacy class of $\gamma$. The *weighted orbital integrals* are linear forms on $C_c^\infty(G(F_S))$ generalizing the usual orbital integrals. They are only easy to define when $M_\gamma = G_\gamma$ for $\gamma \in M(F_S)$. When this holds, set

$$\begin{aligned} J_M(\gamma,f) &= J_M^G(\gamma,f) \\ &= |D(\gamma)|^{1/2} \int_{G_\gamma(F_S)\backslash G(F_S)} f(x^{-1}\gamma x) v_M(x)\, dx, \quad \gamma \in M(F_S), f \in C_c^\infty(G(F_S)), \end{aligned} \tag{3.6.1}$$

where $D(\gamma)$ is a normalizing factor. Since $G_\gamma \subset M$ and $v_M(mx) = v_M(x)$ for $m \in M$, the integral is well defined.

The extension of the function $J_M(f)$ to $M(F_S)$ is provided by [Ar88.1, Theorem 5.2].

**Theorem 3.6.1.** *Let $\gamma \in M(F_S)$. There are canonical functions*

$$\{m_M^L(\gamma, a)\,;\, L \in \mathcal{L}(M)\}$$

*defined for small points $a \in A_M(F_S)$ in general position such that the limit*

$$J_M(\gamma,f) = \lim_{a\to e} \sum_{L\in\mathcal{L}(M)} m_M^L(\gamma, a) J_L(a\gamma,f)$$

*exists. It is equal to the integral of $f$ with respect to a Borel measure on $\gamma^G$, a union of conjugacy classes in $G(F_S)$ depending on $\gamma$.*

The weighted orbital integrals are generally noninvariant. However, they satisfy the same variance formula (3.3.2) as the weighted characters. The following theorem defines the geometric invariant distributions.

**Theorem 3.6.2.** *For $\gamma \in M(F_S)$, there is an invariant linear form*

$$I_M(\gamma) = I_M^G(\gamma)$$

*on $\mathbb{H}_{\mathrm{ac}}(G(F_S))$ that is supported on characters and satisfies*

$$I_M(\gamma,f) = J_M(\gamma,f) - \sum_{L\in\mathcal{L}_0(M)} \widehat{I}_M^L(\gamma, \phi_L(f)).$$

The distribution $I_M(\gamma)$ is actually defined by this formula in Section 5.2. For the summands to be defined, one must assume inductively that $I_M^L(\gamma)$ is supported on characters for $L$ properly contained in $G$. The distribution is studied under this assumption in the first half of Chapter 5. One of the primary results of Chapter 5 is that $I_M(\gamma)$ is supported on characters. See Section 5.15 for the proof. The defining equation can then be written as

$$J_M(\gamma,f) = \sum_{L\in\mathcal{L}(M)} \widehat{I}_M^L(\gamma, \phi_L(f)).$$

The invariance of $I_M(\gamma)$ follows from the fact that $J_M(\gamma)$ and $\phi_L$ satisfy the same type of variance formula.

The distribution $I_M(\gamma)$ can be defined on $\mathbb{H}(G(\mathbb{A})^1)$. Indeed, for $f \in \mathbb{H}(G(\mathbb{A})^1)$ there is an $S$ such that $f \in \mathbb{H}(G(F_S)^1)$. One can also regard the geometric side of the trace formula as a limit over increasing $S$, in which case $I_M(\gamma)$ is evaluated at the restriction of $f$ to $G(F_S)^1$.

The invariant distribution satisfies two formulae that are important for applications. The first is the descent formula, Proposition 5.8.4. For $\gamma \in M(F_S)$, a Levi subgroup $M_1 \in \mathcal{L}(M)$ and $f \in \mathbb{H}_{\rm ac}(G(F_S))$, it states that

$$I_{M_1}(\gamma^{M_1}, f) = \sum_{L \in \mathcal{L}(M)} d_M^G(M_1, L) \widehat{I}_M^L(\gamma, f_L).$$

Here $d_M^G(-,-)$ is a Jacobian relating measures on different subspaces of $\mathfrak{a}_M$. It vanishes when $M_1 \supsetneq M$ and $L = G$. See Section 5.8 for the definitions.

There is also a splitting formula. It applies when $S = S_1 \cup S_2$ is a disjoint union of sets with the closure property and $\gamma = \gamma_1\gamma_2 \in M(F_S)$ and $f = f_1 f_2 \in \mathbb{H}_{\rm ac}(G(F_S))$ are the decompositions relative to $S_1 \cup S_2$. Then Proposition 5.9.1 states that

$$I_M(\gamma, f) = \sum_{L_1, L_2 \in \mathcal{L}(M)} d_M^G(L_1, L_2) \widehat{I}_M^{L_1}(\gamma_1, f_{1,L_1}) \widehat{I}_M^{L_2}(\gamma_2, f_{2,L_2}).$$

## 3.7 *The Set* $\Pi(M, t)$

The coefficient function $a^M(\tau)$ will be defined in the next subsection. The definition relies on a function $a^M_{\rm disc}$, introduced in this subsection, which is related to the usual multiplicity introduced in 1.2. The set $\Pi(M, t)$ is designed to contain the support of these generalized multiplicity functions.

DEFINITION 3.7.1. (i) There is a Lie algebra $\mathfrak{h}$ associated to $G(F_\infty)$ whose main property is that the infinitesimal character of $\tau \in \Pi(M(\mathbb{A}))$ lies in $\mathfrak{h}^*_{\mathbb{C}}$. See Section 5.14 for the definition.

(ii) Denote by $\operatorname{Im} \nu_\tau$ the imaginary part of $\nu_\tau$ relative to the real form $\mathfrak{h}^*$ of $\mathfrak{h}^*_{\mathbb{C}}$, and similarly for $\nu_{\tau^1}$. This imaginary part is a coset, and we take the representative with the smallest magnitude relative to a certain norm.

(iii) Set

$$\Pi_{\rm unit}(M(\mathbb{A})^1, t) = \{\tau \in \Pi_{\rm unit}(M(\mathbb{A})^1)\, ;\ \|\operatorname{Im}(\nu_{\tau^1})\| = t\}.$$

DEFINITION 3.7.2. (i) Consider $L^2(M(F)A_{M,\infty}\backslash M(\mathbb{A}))$ with the right regular $M(\mathbb{A})$ action. Here $A_{M,\infty}$ is the group of real points of a torus associated to $M$. Define $L^2_{\rm disc,t}(M(F)A_{M,\infty}\backslash M(\mathbb{A}))$ to be the subspace of $L^2(M(F)A_{M,\infty}\backslash M(\mathbb{A}))$ that decomposes as a direct sum of representations in $\Pi_{\rm unit}(M(\mathbb{A}), t)$.

(ii) For $M \in \mathcal{L}$, a parabolic subgroup $P \in \mathcal{P}(M)$ and $\Lambda \in \mathfrak{a}_{M,\mathbb{C}}^*$, let

$$\rho_{P,t}(\Lambda) : x \mapsto \rho_{P,t}(\Lambda, x)$$

be the representation of $G(\mathbb{A})$ induced from $L^2_{\mathrm{disc},t}(M(F)A_{M,\infty}\backslash M(\mathbb{A}))_\Lambda$.

(iii) Let $Q \in \mathcal{P}(M)$ be another parabolic subgroup. For $\Lambda$ sufficiently regular, there is an intertwining operator

$$\mathcal{M}_{Q|P}(\Lambda) : \rho_{P,t}(\Lambda) \to \rho_{Q,t}(\Lambda).$$

The Eisenstein series theory provides a meromorphic continuation of this operator to all $\Lambda \in \mathfrak{a}_{M,\mathbb{C}}^*$.

The following proposition serves to define the discrete part of the trace formula.

**Proposition 3.7.1.** (i) *Let $s \in W_0^G$ and $y \in G(\mathbb{A})$. Then there exists a linear map*

$$\rho_{P,t}(s, \Lambda, y) : \rho_{P,t}(\Lambda) \to \rho_{sP,t}(s\Lambda)$$

*satisfying*

$$\rho_{P,t}(s, \Lambda, y_1yy_2) = \rho_{sP,t}(s\Lambda, y_1)\rho_{P,t}(s, \Lambda, y)\rho_{P,t}(\Lambda, y_2)$$

*for $y_1, y_2 \in G(\mathbb{A})$.*

(ii) *For $s \in W^G(\mathfrak{a}_M)$ (= the set of linear operators on $\mathfrak{a}_M$ induced by the Weyl group), a character $\Lambda \in \mathfrak{a}_{M,\mathbb{C}}^*$ and $f \in \mathbb{H}(G(\mathbb{A}))$, set*

$$\rho_{P,t}(s, \Lambda, f) = \int_{G(\mathbb{A})^1} f(x)\rho_{P,t}(s, \Lambda, x)\, dx.$$

*The operator*

$$\mathcal{M}_{P|sP}(\Lambda)\rho_{P,t}(s, \Lambda, f^1),$$

*acting on the Hilbert space of $\rho_{P,t}(\Lambda)$, is of trace class.*

(iii) *The expression*

$$\sum_{M\in\mathcal{L}} |W_0^M||W_0^G|^{-1} \sum_s |\det(s-1)_{\mathfrak{a}_M^G}|^{-1} \operatorname{tr}\left(\mathcal{M}_{P|sP}(0)\rho_{P,t}(s, 0, f^1)\right), \qquad (3.7.1)$$

*where $P \in \mathcal{P}(M)$ and $s \in W^G(\mathfrak{a}_M)_{\mathrm{reg}}$ (the set where the determinant in the sum is nonzero), can be written as a linear combination of irreducible characters of $\pi \in \Pi_{\mathrm{unit}}(G(\mathbb{A})^1, t)$ that is finite for any $f$.*

The function $a^G_{\mathrm{disc}}(\pi)$ is defined by setting

$$(3.7.1) = \sum_{\pi\in\Pi_{\mathrm{unit}}(G(\mathbb{A})^1,t)} a^G_{\mathrm{disc}}(\pi) f_G(\pi).$$

DEFINITION 3.7.3. (i) We denote by $\Pi_{\text{disc}}(G^1, t)$ the set of representations in $\Pi_{\text{unit}}(G(\mathbb{A})^1, t)$ which are represented by irreducible summands of the induced representations $\sigma_\lambda^G$ satisfying the following properties:

(1) $\sigma \in \Pi_{\text{unit}}(M(\mathbb{A})^1, t)$ for $M \in \mathcal{L}$ and $\lambda \in i\mathfrak{a}_M^*/i\mathfrak{a}_G^*$.
(2) $a_{\text{disc}}^M(\sigma) \neq 0$.
(3) There exists $s \in W^G(\mathfrak{a}_M)_{\text{reg}}$ such that $s\sigma_\lambda = \sigma_\lambda$.

Let $M$ be a Levi subgroup. Let $M_1 \in \mathcal{L}^M$.

(ii) Set

$$\Pi_{M_1}(M, t) = \{\tau_{1,\lambda}\,;\, \tau_1^1 \in \Pi_{\text{disc}}(M_1^1, t), \lambda \in i\mathfrak{a}_{M_1}^*/i\mathfrak{a}_M^*\}.$$

(iii) Finally we set

$$\Pi(M, t) = \bigcup_{M_1 \in \mathcal{L}^M} \Pi_{M_1}(M, t).$$

## 3.8 *The Coefficient $a^M(\tau)$*

DEFINITION 3.8.1. (i) Fix $P' \in \mathcal{P}(M)$. For $P \in \mathcal{P}(M)$ and $\nu \in i\mathfrak{a}_M^*$, we define the family of functions

$$m_P(\nu, \tau_\lambda, P') = m_{P|P'}(\tau_\lambda)^{-1} m_{P|P'}(\tau_{\lambda+\nu}).$$

In fact, this is a $(G, M)$-family of functions.

(ii) Let $L \in \mathcal{L}(M)$ and $Q \in \mathcal{P}(L)$. We set

$$m_M^L(\tau_\lambda) = \lim_{\nu \to 0} \sum_{\{P \in \mathcal{P}(M)\,;\, P \subset Q\}} m_P(\nu, \pi_\lambda, P')/\theta_P^Q(\nu).$$

It is the canonical smooth function associated to a $(G, L)$-family induced from the given $(G, M)$-family.

DEFINITION 3.8.2. Let $\tau_{1,\lambda}$ with $\tau_1^1 \in \Pi_{\text{disc}}(M_1^1, t)$, $\lambda \in i\mathfrak{a}_{M_1}^*/i\mathfrak{a}_M^*$, be an element of $\Pi_{M_1}(M, t)$. We define

$$a^M(\tau_{1,\lambda}) = a_{\text{disc}}^{M_1}(\tau_1^1) m_{M_1}^M(\tau_{1,\lambda}).$$

The *global normalizing factor*, the product $\prod_v m_{P'|P}(\tau_{v,\lambda})$ of the local normalizing factors associated to a global representation $\tau \in \Pi_{\text{disc}}(M, t)$, converges for certain $\lambda \in \mathfrak{a}_{M,\mathbb{C}}^*$ and can be meromorphically continued to all $\lambda$. The global normalizing factors define a $(G, M)$-family. The resulting smooth function $m_M^L(\tau_\lambda)$, defined for $L \in \mathcal{L}(M)$, is used to define

$$a^M(\tau_{1,\lambda}) = a_{\text{disc}}^{M_1}(\tau_1^1) m_{M_1}^M(\tau_{1,\lambda})$$

when $\tau_{1,\lambda} \in \Pi(M, t)$.

## 3.9 The Invariant Distribution $I_M(\tau,f)$

The invariant spectral distributions are defined by the analogue of Theorem 3.6.2.

**Theorem 3.9.1.** *For a Levi subgroup $M \in \mathcal{L}$, a representation $\tau \in \Pi(M(F_S))$ and a point $X \in \mathfrak{a}_{M,S}$, there is an invariant linear form*

$$I_M(\tau, X, f) = I_M^G(\tau, X, f)$$

*on $\mathbb{H}_{\mathrm{ac}}(G(F_S))$ that is supported on characters and satisfies*

$$I_M(\tau, X, f) = J_M(\tau, X, f) - \sum_{L \in \mathcal{L}_0(M)} \widehat{I}_M^L(\tau, X, \phi_L(f)).$$

As on the geometric side, this formula gives an inductive definition of the linear form. One of the primary results of [Ar88.2] is an expression of $I_M(\tau, X, f)$ in terms of the geometric distributions $I_M(\gamma, f)$. See Section 5.6. In particular, the inductive hypothesis on the spectral side, stating that the distributions $I_M^L(\tau, X)$ are supported on characters for $L \subsetneq G$, is absorbed into the inductive hypothesis used to define the geometric distributions. The spectral distribution satisfies descent and splitting formulae analogous to those satisfied by $I_M(\gamma)$.

For $\tau^1 \in \Pi_{\mathrm{unit}}(M(\mathbb{A})^1)$ and $f \in \mathbb{H}(G(\mathbb{A})^1)$, the distribution $I_M(\tau^1, X, f)$ is defined by taking $S$ sufficiently large so that $\tau^1$ and $f$ are unramified outside of $S$. The distribution appearing on the spectral side of Arthur's formula is defined by setting

$$I_M(\tau^1, f) = I_M(\tau_\mu, 0, f)$$

for any $\mu \in i\mathfrak{a}_M^*$.

## 3.10 The Measure $d\tau_{1,\lambda}$

The measure $\mathrm{d}\tau_{1,\lambda}$ is defined by setting

$$\int_{\Pi(M,t)} \phi(\tau_{1,\lambda})\,\mathrm{d}\tau_{1,\lambda} = \sum_{M_1 \in \mathcal{L}^M} |W_0^{M_1}||W_0^M|^{-1} \sum_{\tau_1^1 \in \Pi_{\mathrm{disc}}(M_1,t)} \int_{i\mathfrak{a}_{M_1}^*/i\mathfrak{a}_M^*} \phi(\tau_{1,\lambda})\,\mathrm{d}\lambda$$

for any suitable function $\phi$ on $\Pi(M, t)$.

# Chapter 2
# Local Theory

## 1 Case of Division Algebras

Let $F_u$ be a local non-Archimedean field, $G(F_u)$ the multiplicative group of a division algebra $D_u$ central of rank $n$ over $F_u$, and $G'(F_u) = \mathrm{GL}(n, F_u)$. There is an embedding of the set of conjugacy classes $\gamma$ in $G(F_u)$ as the set of elliptic conjugacy classes $\gamma'$ in $G'(F_u)$, defined by $p_\gamma = p_{\gamma'}$; here $p_\gamma$ is the characteristic polynomial of $\gamma$ and $p_{\gamma'}$ is that of $\gamma'$. In a fundamental but unpublished work [DK] (see also [DKV84]) of the late 1970s, Deligne and Kazhdan proved:

THEOREM 1.1. *There is a bijection from the set of equivalence classes of irreducible $G(F_u)$-modules $\pi_u$ to the set of equivalence classes of irreducible square-integrable $G'(F_u)$-modules $\pi'_u$, defined by the character relation $\chi_{\pi'_u}(\gamma') = (-1)^{n-1}\chi_{\pi_u}(\gamma)$ for every regular $\gamma$ in $G(F_u)$ with image $\gamma'$ in $G'(F_u)$. Here $\chi_{\pi'_u}$ denotes the character* [HC78] of $\pi'_u$, and $\chi_{\pi_u}$ that of $\pi_u$.

By virtue of [Ka86.2], it suffices to prove this for $F_u$ of characteristic zero. In fact, all of our arguments hold also in the positive characteristic case, except for the reference [Ka86.1] to the orthonormality relations for characters used in the proof of Proposition 1.9. These relations follow if we use the local integrability of characters in the positive characteristic case of [Le96].

A proof of Theorem 1.1 had been sketched in [JL70] for $n = 2$. The proof of [JL70], as well as that of [DK], relies on global techniques, principally the Selberg trace formula, and on local studies of transfer of orbital integrals between $G(F_u)$ and $G'(F_u)$. There are several proofs of this local transfer; the best is the one given later in this chapter, where the relations between germs of characters and orbital integrals (due to [Ho74, HC78, Ka86.1]) are exploited. The purpose of the present section is to prove Theorem 1.1 without transferring locally the orbital integrals (except in a trivial case) and *consequently deduce this transfer* (see Corollary 1.10 below) *by global means*. These means include, in addition to the trace formula, the Hecke $L$-function theory of [GJ72]. The observation that the transfer of orbital integrals



Y.Z. Flicker, *Arthur's Invariant Trace Formula and Comparison of Inner Forms*,
DOI 10.1007/978-3-319-31593-5_2

can be deduced from the lifting Theorem 1.1 was already made in the context of [FK87.2], §27.3. The deduction relies on results of [BDK86] or [F95], and [Ka86.1].

This section is based on [F90.2]. The point in our present proof is that the theory of $L$-functions is used to show the finiteness of the set of representations which appear in the trace formulae, under some conditions. This observation was made already in [DK] (see [DKV84], pp. 78–82), which was concerned not only with Theorem 1.1, but also it contained a discussion—in the context of $\mathrm{GL}(n)$—of some fundamental ideas later developed in [BDK86] and [Ka86.1]. Here we show that this observation suffices to complete the proof of Theorem 1.1 and Corollary 1.10. We then obtain a simple proof of the Deligne-Kazhdan theorem in the division algebra case. Theorem 1.1 is used as the first step in the inductive proof of the theorem in the simple algebra case, developed later in this chapter. The wish to provide such a simple proof (essentially due to [DK]) to this first induction step used later in this chapter, following [F87.1, III], and consequently to dissipate some misconceptions concerning the difficulty of this case were a main motivation for us to write [F90.2] and this section. Namely, all of our arguments can be found in [DK], and in particular the usage of the Hecke theory, but it was assumed in some expositions that the transfer of orbital integrals had to be proven first, and this led to a proof longer and more complicated than necessary.

Our proof is not elementary. It uses deep theorems in harmonic analysis (e.g., Kazhdan's orthonormality relations for characters). Consequently the reader should have a basic knowledge of representation theory to understand the proof. However, the reduction in this section of the correspondence to "standard theorems" is considerably shorter than other reductions. The rigidity observation of [F90.2]—that using the theory of $L$-functions one can show that there are only finitely many discrete spectrum automorphic representations whose components at almost all places are fixed—was extended in [Ba05] from the context of the multiplicative group of a division algebra to that of simple algebras. Thus the $L$-function theory is sufficiently developed for our purposes in the case of $\mathrm{GL}(n)$ and its inner forms: the multiplicative group of a simple algebra. In particular, it applies also in the analogous situation of base change for $\mathrm{GL}(n)$ (proven by Arthur-Clozel [AC89]), where the local theory of base-change lifting for $\mathrm{GL}(n)$ can be established using purely global means such as the trace formula and the $L$-functions theory of [JS81]; transfer of orbital integrals is obtained as a corollary (see Corollary 1.10, following [F90.2]). It will be interesting to develop this Hecke theory for other groups, for example, to satisfy the needs of the metaplectic correspondence (see [FK87.2]), symmetric-square lifting from $\mathrm{SL}(2)$ to $\mathrm{PGL}(3)$, or base change from $\mathrm{U}(3, E/F)$ to $\mathrm{GL}(3, E)$ (see [F06]).

In the global proof one takes a number field $F$, totally imaginary for simplicity, which has a place $u$ such that the completion of $F$ at $u$ is the local field $F_u$ in Theorem 1.1. Fix a finite place $u' \neq u$ of $F$. Let $D$ be a division algebra central of rank $n$ over $F$, whose invariant $\mathrm{inv}_u D$ at $u$ is equal to the invariant $\mathrm{inv}\, D_u$ of $D_u$ (equivalently $D_u \simeq D \otimes_F F_u$) and such that $\mathrm{inv}_v D = 0$ for all $v \neq u, u'$. Then $D \otimes_F F_v$ is the matrix algebra $M(n, F_v)$ for every $v \neq u, u'$. Put $G_v = (D \otimes_F F_v)^\times$ and $G'_v = \mathrm{GL}(n, F_v)$ for every place $v$ of $F$. Note that the multiplicative group $G$ of $D$ is an inner form of $G' = \mathrm{GL}(n)$. Note that the center $Z$ of $G$ is isomorphic to that of $G'$ and to the multiplicative group. To simplify the notation, we deal here only with representations and functions which transform trivially under the center. Put $\overline{G}$ for $G/Z$.

Choose an $F$-rational invariant differential form of maximal degree on $G$. It defines Haar measures $dg_v$ on $G_v$ and $dg'_v$ on $G'_v$ for all $v$ and product measures $dg = \otimes dg_v$ on $G(\mathbb{A})$ and $dg' = \otimes dg'_v$ on $G'(\mathbb{A})$.

The trace formula is stated for a function $f = \otimes f_v$ in $C_c^\infty(\overline{G}(\mathbb{A}))$. It involves orbital integrals

$$\widetilde{\Phi}(\gamma,f) = \int_{\overline{G}(\mathbb{A})/G_\gamma(F)} f(g\gamma g^{-1})\,dg = |G_\gamma(\mathbb{A})/Z(\mathbb{A})G_\gamma(F)| \int_{G(\mathbb{A})/G_\gamma(\mathbb{A})} f(g\gamma g^{-1})\,dg,$$

(see 1.9 below for $\Phi$ without tilde) and traces

$$\operatorname{tr}\pi(f) = \prod_v \operatorname{tr}\pi_v(f_v), \qquad \text{where } \pi_v(f_v) = \int_{\overline{G}_v} f_v(g)\pi_v(g)\,dg.$$

We take the component $f_u$ to be supported on the set of $\gamma$ in $G(F_u)$ such that $\gamma^n$ is regular. Then for $\gamma$ in $G(F)$ we have $\Phi(\gamma,f) \neq 0$ only when $\gamma$ is (semisimple) regular, in which case the centralizer $G_\gamma$ of $\gamma$ in $G$ is a torus.

THEOREM 1.2 (Trace Formula for $G$). *For any $f$ as above, we have*

$$\sum_\gamma \widetilde{\Phi}(\gamma,f) = \sum_\pi m(\pi)\operatorname{tr}\pi(f). \tag{1.2.1}$$

*The sum on the left ranges over the set of conjugacy classes $\gamma$ in $\overline{G}(F)$ such that $\gamma^n$ is regular. The sum on the right ranges over the set of equivalence classes of automorphic $G(\mathbb{A})$-modules $\pi$ with trivial central character; $m(\pi)$ denotes the multiplicity of $\pi$ in the space of automorphic forms.*

The proof of this is elementary. It is Corollary 4.4 of this chapter.

The trace formula for $G'(\mathbb{A})$ will be stated for a function $f' = \otimes f'_v$ in $C_c^\infty(\overline{G}'(\mathbb{A}))$, with the following properties. Fix a finite place $u'' \neq u, u'$ of $F$. Let $f'_{u''}$ be a normalized coefficient of a (local) cuspidal $G'_{u''}$-module $\pi^0_{u''}$. Thus $\operatorname{tr}\pi_{u''}(f'_{u''}) = 0$ for any irreducible $\pi_{u''}$ inequivalent to $\pi^0_{u''}$, and $\operatorname{tr}\pi^0_{u''}(f'_{u''}) = 1$. We use the word cuspidal as in [BZ76].

Let $f'_{u'}$ be a pseudo-coefficient (its existence was proven in [Ka86.1]) of the Steinberg $G'_{u'}$-module $\mathrm{st}_{u'}$. Then $\operatorname{tr}\mathrm{st}_{u'}(f'_{u'}) = 1$, and $\operatorname{tr}\pi_{u'}(f'_{u'}) = 0$ for every irreducible tempered $G'_{u'}$-module $\pi_{u'}$ inequivalent to $\mathrm{st}_{u'}$. Moreover, the orbital integrals of $f'_{u'}$ vanish on the regular non-elliptic set, and

$$\widetilde{\Phi}(\gamma,f'_{u'}) = \int_{\overline{G}'_{u'}} f'_{u'}(g\gamma g^{-1})\,dg$$

is equal to $\chi_{\mathrm{st}_{u'}}(\gamma) = (-1)^{n-1}$ on the regular elliptic set.

Finally let $f'_u$ be a function supported on the set of $\gamma$ in $G'(F_u)$ such that $\gamma^n$ is regular.

Again by Corollary 4.4, we have

THEOREM 1.3 (Trace Formula for $G'$). *For any $f'$ as above, we have*

$$\sum_{\gamma'} \Phi(\gamma', f') = \sum_{\pi'} \operatorname{tr} \pi'(f'). \tag{1.3.1}$$

*The sum on the left ranges over the set of elliptic regular conjugacy classes $\gamma'$ in $\overline{G}'(F)$ such that $\gamma^n$ is regular. On the right the sum ranges over the set of cuspidal $G'(\mathbb{A})$-modules $\pi'$ with trivial central character.*

Note that the multiplicity of each such $\pi'$ in the cuspidal spectrum for $G'(\mathbb{A})$ is one ([Shal74]).

The trace formula for $G'$ will be used with a function $f' = \otimes f'_v$ whose components at $u'$ and $u''$ are as described above. The component $f'_u$ is taken to be supported on the set of $\gamma$ in $G(F_u)$ with regular $\gamma^n$; moreover we assume that its orbital integrals vanish on the non-elliptic set of $G'(F_u)$. In this section we call such $f'_u$ a *regular-discrete* function. The isomorphism $G_v \simeq G'_v$ for $v \neq u, u'$, can be and is used to transfer $f'_v$ to a function $f_v$ on $G_v$. Let $f_{u'}$ be a normalized matrix coefficient of the trivial $G_{u'}$-module $1_{u'}$. Then $\operatorname{tr} 1_{u'}(f_{u'}) = 1$, and $\operatorname{tr} \pi_{u'}(f_{u'}) = 0$ for any irreducible $\pi_{u'}$ inequivalent to $1_{u'}$. Moreover, $\widetilde{\Phi}(\gamma, f_{u'}) = 1$ for all $\gamma$ in $G_{u'}$. Finally, take $f_u$ to be a regular-discrete function on $G(F_u)$ (namely, $f_u$ is supported on the set of $\gamma$ in $G(F_u)$ such that $\gamma^n$ is regular), with $\widetilde{\Phi}(\gamma, f_u) = \widetilde{\Phi}(\gamma', f'_u)$ for every $\gamma$ regular in $G(F_u)$; $\gamma'$ is the image of $\gamma$ in $G'(F_u)$. We say in this case that $f_u$ and $f'_u$ have *matching orbital integrals*, and note that it is a well-known, relatively simple result of Harish-Chandra (see Proposition 2.9 below) that for every regular-discrete $f'_u$ on $G'(F_u)$ there exists such $f_u$ on $G(F_u)$, and for every regular-discrete $f_u$ on $G(F_u)$ there exists such $f'_u$ on $G'(F_u)$, with matching orbital integrals. The existence of matching functions in general is a more difficult problem, which we solve below on using Theorem 1.1; its solution is not required for the proof of Theorem 1.1.

PROPOSITION 1.4. *For the $f' = \otimes f'_v$ and $f = \otimes f_v$ related as above, we have*

$$\sum_{\pi'} \operatorname{tr} \pi'(f') = \sum_{\pi} m(\pi) \operatorname{tr} \pi(f). \tag{1.4.1}$$

*The sums are those of* (1.2.1) *and* (1.3.1).

PROOF. By the choice of $f$ and $f'$, the sums over $\gamma$ and $\gamma'$ in (1.2.1) and (1.3.1) range over isomorphic sets ($\gamma \leftrightarrow \gamma'$ iff $p_\gamma = p_{\gamma'}$), and $\widetilde{\Phi}(\gamma, f) = \widetilde{\Phi}(\gamma', f')$ for all $\gamma \leftrightarrow \gamma'$. Note that for regular $\gamma$, the centralizers $G_\gamma$ of $\gamma$ in $G$, and $G'_{\gamma'}$ of $\gamma'$ in $G'$, are isomorphic elliptic tori; this isomorphism is used to transfer measures between these groups. The proposition follows. □

Let $\pi'_{0u}$ be a square-integrable $G'(F_u)$-module. By a standard construction result, see Proposition 16.2 below, there exists a cuspidal $G'(\mathbb{A})$-module $\pi'_0$ whose component at $u$ is the chosen $\pi'_{0u}$, at $u'$ it is the Steinberg $\mathrm{st}_{u'}$, and at $u''$ it is the cuspidal $\pi^0_{u''}$. Denote by $\mathbb{A}^u$ the ring of $F$-adèles without $u$-component. Denote by $\pi^u_0 = \underset{v\neq u}{\otimes} \pi^u_{0v}$ the $G(\mathbb{A}^u)$-module $1_{u'} \otimes (\underset{v\neq u,u'}{\otimes} \pi_{0v})$. Here we identify $\pi'_{0v}$ with a $G_v$-module $\pi_{0v}$ for $v \neq u, u'$, by $G_v \simeq G'_v$. The standard-type isolation argument (see Proposition 17.1) implies the

PROPOSITION 1.5. *For the given square-integrable $\pi'_{0u}$, there exist irreducible $G(F_u)$-modules $\pi_u$, such that for any matching regular-discrete $f_u$ and $f'_u$, we have*

$$(-1)^{n-1}\,\mathrm{tr}\,\pi'_{0u}(f'_u) = \sum_{\pi_u} m(\pi_u \otimes \pi^u_0)\,\mathrm{tr}\,\pi_u(f_u). \tag{1.5.1}$$

REMARK 1.6. In the proof of (1.5.1), it is worthwhile to note that the choice of $f'_{u''}$ in (1.4.1) implies that the $\pi'$ of (1.4.1) are all cuspidal. Hence each component of $\pi'$ is non-degenerate. By [Ze80, (9.7b)], if $\mathrm{tr}\,\pi'_{u'}(f'_{u'}) \neq 0$, then $\pi'_{u'} \simeq \mathrm{st}_{u'}$ and so $\mathrm{tr}\,\mathrm{st}_{u'}(f'_{u'}) = (-1)^{n-1}$.

The usual argument, of [JL70, DK], or Section 18 below, to deduce Theorem 1.1 from (1.5.1), is based on evaluation of (1.5.1) at $f_u$ which is a normalized coefficient of some $\pi_u$ which occurs in (1.5.1) with $m(\pi_u \otimes \pi^u) \neq 0$. To do this, one has to show that there exist $f'_u$ with orbital integrals matching those of $f_u$. We shall argue differently. Using the Hecke theory of [GJ72], we prove (following [DK]) that the sum in (1.5.1) is finite *uniformly* in $f_u$. In fact, since $f_u$ is bi-invariant under some compact open subgroup $K_u$ of the compact (modulo $Z_u$) group $G(F_u)$, there are only finitely many $\pi_u$ with $\mathrm{tr}\,\pi_u(f_u) \neq 0$. However, the size of the finite set of such $\pi_u$ increases as $K_u$ decreases, and a priori the sum in (1.5.1) may be infinite (for a variable $f_u$). In order to use the orthonormality relations for characters (see the passage from Proposition 1.8 to Proposition 1.9 below), we need to know that the sum in (1.5.1) (and so in (1.8.1) below) is finite uniformly in, or independently of, $f_u$. Thus we prove

PROPOSITION 1.7. *The sum over $\pi_u$ in (1.5.1) is finite (uniformly in $f_u$).*

PROOF. Let $\psi = \prod_v \psi_v$ be a nontrivial additive character of $\mathbb{A}/F$. Denote by $L(s, \pi_v)$ the $L$-function and by $\epsilon(s, \pi_v, \psi_v)$ the $\epsilon$-factor, attached to $\pi_v$ and $\psi_v$ for every place $v$ of $F$, in [GJ72, Theorem 3.3]. Consider $\pi = \pi_u \otimes \pi^u_0$ which occurs in (1.5.1) with $m(\pi) \neq 0$. Since $\pi$ is automorphic, $\epsilon(s, \pi) = \prod_v \epsilon(s, \pi_v, \psi_v)$ is independent of $\psi$ (see [GJ72, p. 149]), and $L(s, \pi) = \prod_v L(s, \pi_v)$ satisfies the functional equation $L(s, \pi) = \epsilon(s, \pi)L(1 - s, \pi^\vee)$; here $\pi^\vee$ signifies the contragredient of $\pi$. So, if $\pi_u$ and $\pi_{0u}$ contribute to (1.5.1) (namely, $m(\pi_u \otimes \pi^u_0) \neq 0$, $m(\pi_{0u} \otimes \pi^u_0) \neq 0$), we have

$$\frac{L(1 - s, \pi^\vee_u)\epsilon(s, \pi_u, \psi_u)}{L(s, \pi_u)} = \frac{L(1 - s, \pi^\vee_{0u})\epsilon(s, \pi_{0u}, \psi_u)}{L(s, \pi_{0u})}. \tag{1.7.1}$$

Denote by $K_u$ the multiplicative group of a maximal order $M_u$ in the division algebra $D_u$ underlying $G(F_u)$. Then $K_u$ is open in $G(F_u)$, and $G(F_u)/Z_u$ is compact. Hence there are only finitely many irreducible $G(F_u)$-modules $\pi_u$ with a trivial central character which are unramified (trivial on $K_u$). If $\pi_u$ is not trivial on $K_u$, then $L(s, \pi_u) = 1 = L(s, \pi_u^\vee)$ by [GJ72, Prop. 4.4], identically in $s$. In this case (1.7.1) implies that $\epsilon(s, \pi_u, \psi_u)$ is independent of $\pi_u$ (as long as $m(\pi_u \otimes \pi_0^u) \neq 0$). Denote by $c(\pi_u)$ the positive integer ("conductor") such that $\pi_u$ is trivial on $1 + \boldsymbol{\pi}_u^{c(\pi_u)+1} M_u$ but not on $1 + \boldsymbol{\pi}_u^{c(\pi_u)} M_u$, where $\boldsymbol{\pi}_u$ is the local uniformizer in the ring $R_u$ of integers in $F_u$. Choose $\psi_u$ to be trivial on $R_u$, but not on $\boldsymbol{\pi}_u^{-1} R_u$. It is well known (see, e.g., [BF83, Theorem 3.2.11, p. 39]) that there exists a number $\alpha$ such that $\epsilon(s, \pi_u, \psi_u) = \alpha q_u^{-c(\pi_u)s}$; $q_u$ is the cardinality of the residue field $R_u/(\boldsymbol{\pi}_u)$. Consequently $c = c(\pi_u)$ is independent of $\pi_u$. Since $G(F_u)/Z_u(1 + \boldsymbol{\pi}_u^c M_u)$ is finite, there are only finitely many irreducible $G(F_u)$-modules $\pi_u$ with a trivial central character and a fixed conductor $c$. The proposition follows. □

Now that we know that the sum in (1.5.1) ranges over a finite set depending only on $\pi'_{0u}$, we can apply the Weyl integration formula

$$\int_{G/Z} f(g)dg = \int' \Phi(t,f)dt, \quad \text{where} \quad \int' \text{ signifies } \sum_{\{T\}} [W(T)]^{-1} \int_{T/Z} \Delta(t)^2.$$

Here the sum ranges over a set of representatives for the conjugacy classes of tori $T$ in $G$, $[W(T)]$ denotes the cardinality of the Weyl group of $T$, and $\Delta$ is a Jacobian. It implies

PROPOSITION 1.8. *For every regular $\gamma$ in $G(F_u)$ and $\gamma'$ in $G'(F_u)$ with $p_\gamma = p_{\gamma'}$, we have*

$$(-1)^{n-1} \chi_{\pi'_{0u}}(\gamma') = \sum_{\pi_u} m(\pi_u \otimes \pi_0^u)\, \chi_{\pi_u}(\gamma). \tag{1.8.1}$$

*The sum is the same as in* (1.5.1).

An immediate application of the orthonormality relations (due to [Ka86.1, Theorem K]), of characters of square-integrable representations, implies

PROPOSITION 1.9. *The sum in* (1.8.1) *consists of a single entry $\pi_{0u}$ with $m(\pi_{0u} \otimes \pi_0^u) \neq 0$; moreover, $m(\pi_{0u} \otimes \pi_0^u) = 1$.*

This completes the proof of one half of Theorem 1.1, asserting that for each square-integrable $\pi'_u$, there exists a corresponding $\pi_u$. To prove the opposite direction, one starts with a $G(F_u)$-module $\pi_u^0$ and constructs a cuspidal $G(\mathbb{A})$-module $\pi_0$ whose component at $u$ is $\pi_u^0$, at $u''$ it is the cuspidal $\pi_{u''}^0$, and it is $1_{u'}$ at $u'$. Then (1.5.1) is obtained and the proof proceeds as above.

Finally we use Theorem 1.1 to transfer orbital integrals. Since the following discussion is purely local, the index $u$ is omitted. Recall that for a regular $\gamma$ in $G'$, the centralizer $G'_\gamma$ is a torus, and we put

$$\Phi(\gamma,f') = \int_{G'/G'_\gamma} f'(g\gamma g^{-1})\,\mathrm{d}g.$$

Following [Ka86.1] we say that a function $f'$ is *discrete* if $\Phi(\gamma,f') = 0$ for every regular non-elliptic $\gamma$ in $G'$. The space of discrete $f'$ is denoted by $A(G')$. Theorem 1.1 has the following:

COROLLARY 1.10. *For every $f$ on $G$, there is $f'$ in $A(G')$, and for every $f'$ in $A(G')$, there is $f$ on $G$, with $\Phi(\gamma,f) = \Phi(\gamma',f')$ for all regular $\gamma$ in $G$ and $\gamma'$ in $G'$ with $p_\gamma = p_{\gamma'}$.*

The proof consists of two parts.

LEMMA 1.11. *For every $f$ there is $f'$ in $A(G')$, and for every $f'$ in $A(G')$, there is $f$, such that $(-1)^{n-1}\operatorname{tr}\pi(f) = \operatorname{tr}\pi'(f')$ for all $\pi,\pi'$ corresponding as in Theorem* 1.1.

PROOF. Given $f$, define a form $\Phi$ on the free abelian group $R(G')$ generated by the equivalence classes of irreducible tempered $G'$-modules $\pi'$ by $\Phi(\pi') = (-1)^{n-1}\operatorname{tr}\pi(f)$ if $\pi'$ is square-integrable, and it corresponds to $\pi$, and by $\Phi(\pi') = 0$ if $\pi'$ is irreducible, tempered but not square-integrable. It is clear that $\Phi$ is a good form in the terminology of [BDK86] or [F95], hence a trace form by the Theorem of [BDK86] or [F95]. Namely, there exists $f'$ on $G$ with $\Phi(\pi') = \operatorname{tr}\pi'(f')$ on $R(G')$. Since $\operatorname{tr}\pi'(f') = 0$ for every $\pi'$ in $R_I(G')$ (in the notations of [Ka86.1]), we have that $f'$ lies in $A(G')$. The proof of the opposite implication (given $f'$ in $A(G')$, there is $f$ on $G$) is analogous. □

LEMMA 1.12. *If $f'$ in $A(G')$ and $f$ on $G$ satisfy $(-1)^{n-1}\operatorname{tr}\pi(f) = \operatorname{tr}\pi'(f')$ for all $\pi,\pi'$ corresponding by Theorem* 1.1, *then $\Phi(\gamma,f) = \Phi(\gamma',f')$ for all regular $\gamma,\gamma'$ with $p_\gamma = p_{\gamma'}$.*

PROOF. The Weyl integration formula for $G$ implies that

$$\sum_{\{T\}}[W(T)]^{-1}\int_{T/Z}\Delta(t)^2[\Phi(t,f) - \Phi(t',f')]\chi_\pi(t)dt = 0$$

for every $G$-module $\pi$. Since $G/Z$ is compact, the characters $\chi_\pi$ form an orthonormal basis with respect to the inner product

$$\langle\chi,\chi'\rangle = \sum_{\{T\}}[W(T)]^{-1}\int_{T/Z}\Delta(t)^2\chi(t)\overline{\chi}'(t)\,\mathrm{d}t.$$

The lemma follows, and so does the corollary. □

REMARK 1.13. Our Theorem and Corollary are the initial, special case of the correspondence of representations of $\mathrm{GL}(n)$ and its inner forms; see Theorem 13.8 below for the general statement for the multiplicative group of any simple, not only division, algebra. Our local Theorem has a global variant, relating cuspidal representations on $\mathrm{GL}(n, \mathbb{A})$ and $G(\mathbb{A})$, for any inner form $G$ of $\mathrm{GL}(n)$; see Theorem 13.12 below, in the context of $\pi'$ with two cuspidal components, Theorem 25.2 for general representations, and Theorem 26.18 in the context of $\pi'$ with a single cuspidal component.

# 2 Orbital Integrals

## 2.1 *Points*

Let $F$ be a global field of characteristic 0. Let $\mathbb{A}$ denote its ring of adèles. The completion of $F$ at the place $v$ will be denoted $F_v$. Let $G$ be a reductive group over $F$, which is often identified with its group of $\overline{F}$-points, $G(\overline{F})$, with $\overline{F}$ a fixed algebraic closure of $F$. If $E$ is an extension of $F$, put $G(E)$ for the group of $E$-points of $G$. We write $G(\mathbb{A})$ for the group of adèle points. These conventions apply also to subgroups of $G$ defined over $F$.

## 2.2 *Parabolic Subgroups*

We usually let $P$ denote a parabolic $F$-subgroup of $G$. We let $N$ denote its unipotent radical. Fix a minimal parabolic $F$-subgroup $P_0$ with Levi decomposition $P_0 = M_0N_0$, with $M_0$ the Levi subgroup. Unless otherwise specified, we consider only *standard* $P$, the $F$-rational parabolics containing $P_0$. By a Levi subgroup of $P$, we mean the unique Levi that contains $M_0$.

## 2.3 *Rank, Semisimple Regular, Elliptic Elements*

In this subsection let $F$ be a local or global field of characteristic 0. Let $\mathfrak{g}$ be the Lie algebra of $G$. For $x \in G(F)$, consider the polynomial $\det[(t + 1 - \mathrm{Ad}(x))|_{\mathfrak{g}}]$ in $t$. Let $d$ be the degree of the first nonzero power of $t$ in the polynomial, as $x$ runs over $G$. It is called the *rank* of $G$. Let $D(x)$ be the coefficient of $t^d$. Then $x$ is called *semisimple regular* if $D(x) \neq 0$. It is then semisimple, and its centralizer $G_x$ in $G$ is a torus. A semisimple $x$ is called *elliptic* if the center of $G_x(F)/Z(F)$ ($F$ local) or $G_x(\mathbb{A})/G_x(F)Z(\mathbb{A})$ ($F$ global) is compact. If $x$ is elliptic regular, then $G_x$ is an elliptic torus of $G$. We define a general regular element in 2.9. But outside the present Section 2, we often use "regular" to mean "semisimple regular," to simplify the terminology, in the hope that no confusion would occur.

## 2.4 Orbital Integrals

Fix a Haar measure $\mathrm{d}y$ on $G(F)/Z(F)$. Let $G_x$ denote the centralizer of $x$ in $G$. Let $x \in G(F)$ and fix a Haar measure $\mathrm{d}_x$ on $G_x(F)/Z(F)$. If $G_x(F)$ and $G_{x'}(F)$ are isomorphic, we take $\mathrm{d}_x$ and $\mathrm{d}_{x'}$ to be equal under the fixed isomorphism. Define the *orbital integral* of $f$ at $x$ by

$$\Phi^G(x,f) = \Phi(x,f) = \int_{G(F)/G_x(F)} f(yxy^{-1})\frac{\mathrm{d}y}{\mathrm{d}_x}.$$

The orbital integral depends on the choice of measures. The orbit of a regular element is closed. Thus $\Phi(x,f)$ converges for all regular $x$ and $f \in C_c^\infty(G(F))$. By Theorem 2.32 or [Ra72], $\Phi(f)$ converges for $f \in C_c^\infty(G(F))$ and any $x \in G(F)$. If $x$ is regular, let $G_x^{\mathrm{sp}}$ be the split component of the center of the centralizer $G_x$, which is a torus. Define

$$\widetilde{\Phi}^G(x,f) = \widetilde{\Phi}(x,f) = \int_{G(F)/G_x^{\mathrm{sp}}(F)} f(yxy^{-1})\frac{\mathrm{d}y}{\mathrm{d}_x}.$$

## 2.5 Jacobian

Let $F$ be local. Recall that $x \in G(F)$ has a Jordan decomposition $x = su = us$ into semisimple and unipotent elements. Let $\mathfrak{g}_s$ be the Lie algebra of the centralizer $G_s$. Put

$$\Delta(x) = |\det(1 - \mathrm{Ad}(s))|_{\mathfrak{g}/\mathfrak{g}_s}|^{1/2}.$$

Put

$$I^G(x,f) = I(x,f) = \Delta(x)\Phi(x,f).$$

For regular $x$, set

$$\widetilde{I}^G(x,f) = \widetilde{I}(x,f) = \Delta(x)\widetilde{\Phi}(x,f).$$

Consider $G = \mathrm{GL}(n)$. Since $\Delta(x)$ is unchanged by field extension, assume $x = \mathrm{diag}(x_1, \ldots, x_n)$, with $x_i \neq x_j$. Then computing $\mathrm{Ad}(x)$ with the standard basis $\{e_{ij}\}_{i \neq j}$ gives $\mathrm{Ad}(x)$ as diagonal with entries $x_i x_j^{-1}$, and

$$\Delta(x) = \frac{\left|\prod_{i<j}(x_i - x_j)^2\right|^{1/2}}{|\det x|^{(n-1)/2}}.$$

## 2.6 Constant Term

Let $P = MN$ be an $F$-parabolic subgroup of $G$. Let $K \subset G(F)$ be a maximal compact subgroup such that $G(F) = KP(F) = P(F)K$. If $\mathfrak{n}$ is the Lie algebra of $N$, put $\delta_P(m) = |\det \mathrm{Ad}(m)|_{\mathfrak{n}}|$. For $f \in C_c^\infty(G(F))$, put

$$f_N(m) = \delta_P(m)^{1/2} \int_{N(F)} \int_K f(k^{-1}mnk)\, dk\, dn.$$

The function $f_N$ depends on $N$, but its orbital integral at $m \in M(F)$, regular in $G$, depends only on $M$. Indeed, a standard computation, see [FK87.2, Section7], shows $I^G(m,f) = I^M(m,f_N)$, where $I^M$ denotes the normalized orbital integral with respect to $M$. Note that $f_N$ lies in $C_c^\infty(M)$.

Denote by $J^\infty(G(F))$ the space of $f \in C_c^\infty(G(F))$ such that $I(g,f) = 0$ for all regular $g \in G(F)$. Put $\overline{C_c^\infty}(G(F)) = C_c^\infty(G(F))/J^\infty(G(F))$. The image of $f_N$ in $\overline{C_c^\infty}(M(F))$ will be denoted by $f_M$, since it depends on the Levi subgroup $M$, but not the unipotent radical $N$.

## 2.7 Germ Expansion

We recall the germ expansion of orbital integrals of $f$ in $C_c^\infty(G(F))$. Let $\overline{O}(x)$ be the closure of the conjugacy class $O(x)$ of $x$ in $G(F)$. It is the disjoint union of the conjugacy classes $O(su_i)$, $1 \leq i \leq r$, of elements $su_i$ with semisimple part $s$, so that the following properties hold:

(1) $u_1 = e$.
(2) For each $t$, $1 \leq t \leq r$, the union $O_t = \bigcup_{i=1}^t O(su_i)$ is closed.
(3) $O(su_t)$ is open in $O_t$.

The (closed) set $A_s$ of elements in $G(F)$ whose semisimple part is conjugate to $s$ is of the form $\overline{O}(x)$ for some $x$, and there are $f_i$ in $C_c^\infty(G(F))$ with $I(su_j,f_i) = \delta_{ij}$ and $f_i = 0$ on $O(su_j)$ for $j < i$. We have the following proposition, giving the germ expansion.

PROPOSITION 2.8. (1) *Given $f$ in $C_c^\infty(G(F))$ and a semisimple $s$ in $G(F)$, there exists a neighborhood $V_f$ of $s$ in $G(F)$ so that*

$$I(x,f) = \sum_i I(x,f_i) I(su_i,f)$$

*for all regular $x$ in $V_f$.*

(2) *Conversely, given a function* $I(x)$ *on* $G(F)$ *such that for each semisimple* $s$ *in* $G(F)$*, there is a neighborhood* $V$ *of* $s$ *in* $G(F)$ *with*

$$I(x) = \sum_i I(x, f_i) I(su_i)$$

*for all regular* $x \in V$*, there exists* $f$ *in* $C_c^\infty(G(F))$ *with* $I(x) = I(x, f)$.

The remainder of this section concerns a proof due to Shalika [Shal72] and Harish-Chandra [HC70], extended by Vigneras [Vi82] to the metaplectic group. We give a standard approach in Proposition 2.18, and a variant, with considerable overlap, using the uniqueness of the invariant measure on the orbit as in the proof of Corollary 4.4 below, in Proposition 2.25. We often denote $G(F)$ by $G$, $T(F)$ by $T$, etc., to simplify the notation. We start with

PROPOSITION 2.9. *Let* $f \in C_c^\infty(G)$*. Let* $T$ *be a maximal torus in* $G$*. The orbital integral* $\Phi(t, f) = \int_{T\backslash G} f(g^{-1}tg)dg$ *converges at all* $t \in T^{\text{reg}}$*. As a function of* $t$ *in* $T^{\text{reg}}$*, it is locally constant. There is a compact subset* $T(f)$ *of* $T$ *such that* $\Phi(t, f)$ *is zero for* $t \in T^{\text{reg}} - T(f) \cap T^{\text{reg}}$.

Note that the algebraic connected component $Z_G(s)^0$ of the centralizer $Z_G(s)$ of a semisimple element $s$ of a connected reductive algebraic group $G$ is reductive ([SS70, p. 197]) and connected. If $g = su = us$ is the Jordan decomposition of $g \in G$ as a product of commuting semisimple $s$ and unipotent $u$, then the centralizer $Z_G(g)$ is the centralizer $Z_{Z_G(s)}(u)$ of $u$ in $Z_G(s)$.

Let $X$ be a subset of the group $G$. A $g \in X \subset G$ is *regular in* $X$ if its conjugacy class $O(g)$ in $G$ has maximal possible dimension, thus $\dim O(g) \geq \dim O(x)$ for all $x \in X$. Put $X^{\text{reg}}$ for the set of regular elements in $X$.

The *torus* of $g \in G$ is the center $T$ of $Z_G(g)^0$. A *torus of* $G$ is a torus of a semisimple element of $G$. A *torus-unipotent pair*, or just a *pair*, is a pair $(T, u)$ consisting of a torus $T$ in $G$ and a unipotent $u$ which commutes with each element of $T$.

To prove Proposition 2.9, we use the following result of [HC70, p. 52], stated there for $u$ being the identity $e$ of $G$ and $T$ a maximal torus.

PROPOSITION 2.10. *Let* $s$ *be a semisimple element of* $G$*. Let* $T$ *be a torus containing* $s$*. Let* $u$ *be a unipotent element which commutes with every element of* $T$*. Then there is a neighborhood* $V$ *of* $s$ *in* $T$ *such that for any compact subset* $K'$ *of* $G$*, there is a compact* $C$ *in the homogeneous space* $M\backslash G$*,* $M = Z_G(s)^0$ $(= (Z_G(s)^0)(F))$ *such that if* $g^{-1}Vug \cap K'$ *is nonempty, then* $Mg \in M\backslash G$ *belongs to* $C$.

This implies the convergence of the orbital integral of $f$ at a semisimple $s$. Further, if $K'$ is open and compact, $s \in T^{\text{reg}}$, and the unipotent $u$ commutes with each element of the torus $T$, then there is a neighborhood $V$ of $s$ in $T^{\text{reg}}$ such that $g^{-1}Vug \subset K'$ if $g^{-1}sug \in K'$. This shows that the orbital integral is locally constant on the regular part of $Tu$.

PROOF. We may assume $K'$ contains the semisimple parts of its elements. Indeed, the semisimple part of $g \in G$ is a polynomial in $g$, obtained from the eigenvalues by the Chinese remainder theorem, so that the map $g = su = us \to s$ is open and closed. Then $g^{-1}Vug \cap K' \neq \emptyset$ implies $g^{-1}Vg \cap K' \neq \emptyset$. So we may assume that $u = e$. Moreover, we may take $T$ to be maximal. Indeed, if $T_1$ is a subtorus of a maximal $T$, and we found $V$ for $T$ as in the proposition, take $V_1 = T_1 \cap V$. Then $g^{-1}V_1g \cap K' \neq \emptyset$ implies $g^{-1}Vg \cap K' \neq \emptyset$, so there is $C$ in $M\backslash G$ as required.

We may assume $T$ and $G$ are split. Indeed, denote by $E$ a finite field extension of $F$ over which $G$ and $T$ split. Then $G$ is closed in $G(E)$, $T = T(E) \cap G$, and $M = M(E) \cap G$, since $M = G_s^0$ is defined over $F$. Hence $M\backslash G \hookrightarrow M(E)\backslash G(E)$. The proposition follows for $(G, T)$ provided the map is topological, in the sense that: $(a)$ $M\backslash G$ is closed in $M(E)\backslash G(E)$ and $(b)$ $M\backslash G \hookrightarrow M(E)\backslash G(E)$ is a homeomorphism onto its image. To see these, let $\{Mg_j; j \geq 1\}$ be a sequence in $M\backslash G$ which converges in $M(E)\backslash G(E)$ to $M(E)g$. We need to show that $Mg \in M\backslash G$ (namely, $M(E)g \cap G \neq \emptyset$ so that $g$ can be chosen in $G$) and $Mg_j$ converges to $Mg$ in $M\backslash G$. Choose then representatives $g_j \in G$ and $g \in G(E)$. Then $g_j^{-1}sg_j \to g^{-1}sg$ as $j \to \infty$ in $G(E)$. Hence $g_j^{-1}sg_j$ remains in a compact in $G(E)$. This implies that $Mg_j$ are all in a compact in $M\backslash G$, implying $(a)$ and $(b)$.

Suppose then that $T$ splits over $F$. Let $\Sigma$ be the set of roots of $T$ in $G$. Let $\Sigma^+$ be a choice of positive roots. Then there is a maximal compact subgroup $K$ and a unipotent subgroup $N$ such that $G = TNK$, the map $T \times N \times K \to G$, $(t, n, k) \mapsto tnk$, is continuous, $B = TN$ is a Borel subgroup ($G$ is split), and $\Sigma^+$ is the set of roots of $T$ in $N$. It suffices to show there is an open neighborhood $V$ of $s$ in $T$ such that for any compact $K' \subset G$, there is a compact $C \subset M\backslash G$ with: if $n^{-1}Vn \cap K' \neq \varnothing$ for $n \in N$, then $\overline{n} \in C$, where $\overline{n} = Mn$ is the image of $n$ in $M\backslash G$.

Write $\Sigma^+$ as a disjoint union of $\Sigma^s = \{\alpha \in \Sigma^+ ; \alpha(s) = 1\}$ and $\Sigma_s$. For each $\alpha \in \Sigma^+$, there is a unique subgroup $N_\alpha$ of $N$ and an isomorphism $\theta_\alpha : \mathbb{G}_a \to N_\alpha$, where $\mathbb{G}_a$ is the additive group, such that $\theta_\alpha(\alpha(t)a) = \mathrm{Int}(t)\theta_\alpha(a)$ for all $a \in \mathbb{G}_a(F) = F$, $t \in T$. Recall that $M = Z_G(s)^0$. Put $N^s = M \cap N$. Then $N^s = \prod_{\alpha \in \Sigma^s} N_\alpha$ in any order. Label the roots $\alpha_1, \ldots, \alpha_r$ of $\Sigma^+$ so that $i < j$ if $\alpha_i < \alpha_j$ using the basis of $\Sigma^+$. The map

$$\beta : N_{\alpha_1} \times \cdots \times N_{\alpha_r} \to N, \quad (n_1, \ldots, n_r) \mapsto n_1 \cdots n_r,$$

is an isomorphism of varieties. The image $N_k$ of $1 \times \cdots \times 1 \times N_{\alpha_k} \times \cdots \times N_{\alpha_r}$ is a subgroup with $N_k/N_{k+1} \xrightarrow{\sim} N_{\alpha_k}$. Also $[N, N_k] \subset N_{k+1}$. See [St74]. For each $\alpha \in \Sigma^+$, let $\mathrm{pr}_\alpha$ be the projection of $N_{\alpha_1} \times \cdots \times N_{\alpha_r}$ onto $N_\alpha$. Define $\varphi_\alpha(n) \in F$ by $\theta_\alpha(\varphi_\alpha(n)) = \mathrm{pr}_\alpha(\beta^{-1}(n))$, thus $\varphi_\alpha : N \to \mathbb{G}_a$.

We claim: $\varphi_{\alpha_j}(ab) = \varphi_{\alpha_j}(a) + \varphi_{\alpha_j}(b) + P_{\alpha_j}(a, b)$, where $P_{\alpha_j}(a, b)$ is a polynomial in $\varphi_{\alpha_k}(a)$ and $\varphi_{\alpha_k}(b)$ $(1 \leq k < j)$. Indeed, since $[N, N_{j+1}] \subset N_{j+2}$ and $N_j/N_{j+1} \to N_{\alpha_j}$, the restriction of $\varphi_{\alpha_j}$ to $N_j$ is a homomorphism. Since $N_j$ is a subgroup, for $b \in N_{j+1}$, we have $\varphi_{\alpha_j}(ab) = \varphi_{\alpha_j}(a)$. Since $\varphi_{\alpha_j}(ab)$ is a polynomial in the $\varphi_\alpha(a)$ and $\varphi_\alpha(b)$, the claim follows.

Let $V$ be a compact open neighborhood of $s$ in $T$ on which none of the roots $\alpha \in \Sigma_s$ is identically 1. Write $N_s = \prod_{\alpha\in\Sigma_s} N_\alpha$ with $N = N^s N_s$, according to the order on the roots. If $n = n^s n_s$, then $n^{-1}tn = n_s^{-1}tn_1^s n_s$ with $n_1^s = t^{-1}(n^s)^{-1}tn^s \in N^s \subset M$. To prove the proposition, it suffices then to show that the $\varphi_\alpha$ are bounded on the set

$$S = \{n_s \in N_s\,;\, t^{-1}n_s^{-1}tn^s n_s \in K' \text{ for some } t \in V \text{ and } n^s \in N^s\}.$$

Since $\varphi_{\alpha_j}$ is 1 on $N_k$ with $k > j$, and for $\alpha \in \Sigma^s$ we have $\varphi_\alpha = 1$ on $N_s$, we need consider only $\varphi_\alpha$ with $\alpha \in \Sigma_s$.

Suppose $n^s, n_s \in N_{\alpha_r} = N_r$. If $\alpha_r \notin \Sigma_s$, there is nothing to prove. If $\alpha_r \in \Sigma_s$, then $n^s = 1$ and $\varphi_{\alpha_r}(t^{-1}n_s^{-1}tn_s) = (1-\alpha_r(t^{-1}))\varphi_{\alpha_r}(n_s) + P_{\alpha_r}(t^{-1}n_s t, n_s)$. Since $n_s \in N_{\alpha_r}$, we have by the claim that $P_{\alpha_r}(t^{-1}n_s t, n_s)$ is 0. By assumption on $t$, we have $\alpha_r(t^{-1}) \neq 1$. Since $K'$ is compact, $\varphi_{\alpha_r}(t^{-1}n_s^{-1}tn_s)$ is bounded for $n_s$ with $t^{-1}n_s^{-1}tn_s \in K'$. This implies that $\varphi_{\alpha_r}(n_s)$ is bounded.

By induction, assume (1) the $\varphi_\alpha$ are bounded on the set of $n_s \in N_{k+1}$ such that there exists $n^s \in N_{k+1}$ with $n_s^{-1}tn^s n_s \in K'$, and (2) $n_s^{-1}tn^s n_s \in K'$ for some $n^s, n_s \in N_k$. If $\alpha_k \in \Sigma_s$, write $n_s = xy$ with $x \in N_{k+1}$ and $y \in N_{\alpha_k}$. Then $n_s \in N_{k+1}$ and $n_s^{-1}tn^s n_s = tn'_k n''_{k+1}$ where $n'_k = t^{-1}y^{-1}ty \in N_{\alpha_k}$ and $n''_{k+1} = y^{-1}t^{-1}x^{-1}tn^s xy \in N_{k+1}$. If $n_s^{-1}tn^s n_s \in K'$, then both $n'_k$ and $n''_{k+1}$ remain bounded. As before, this implies that $y$ is bounded, and $x$ is bounded by the induction assumption. The case of $\alpha_k \in \Sigma^s$ is similar, left for the reader. □

To show that the orbital integral is compactly supported on each torus, we prove

PROPOSITION 2.11. *Let $K'$ be a compact subset of $G(F)$. Put $K'^G = \{g^{-1}xg\,;\, x \in K', g \in G\}$. Then the set $S = \{t \in T\,;\, t \in$ closure of $K'^G\}$ is relatively compact.*

PROOF. Let $\rho : G \to \mathrm{GL}(n)$ be a faithful $F$-rational representation of the linear algebraic group $G$. Write $\det(x + 1 - \rho(g)) = \sum_{j=0}^{\dim\rho} p_j(g)x^j$. The $p_j$ are invariant polynomials. Hence they are bounded on the set $S$, since $K'$ is compact. For $t \in T$, the image $\rho(t)$ is semisimple. Hence $p_j$ is bounded on $S$ implies that the eigenvalues of $\rho(t)$ are bounded on $S$. But $\rho$ is faithful. Hence $S$ is relatively compact. □

Denote by $O(x)$ the conjugacy class of $x \in G$. Write $d(O(x))$ for the dimension of $O(x)$ as a variety. Put $U_d = \bigcup\{O \in U\,;\, d(O) \le d\}$, $U_\infty = \bigcup\{O\,;\, O \in U\}$. Let $U$ be the set of unipotent conjugacy classes in $G(F)$. Recall that to simplify the notation, we often write $G$ for $G(F)$.

PROPOSITION 2.12. *For all $d \ge 0$ the set $U_d$ is closed. For all $O \in U$, the union $O \bigcup U_{d(O)-1}$ is closed and $O$ is open in $U_{d(O)}$ and in the closure $\overline{O}$ of $O$. The identity $e$ of $G$ lies in $\overline{O}$. The $G$-invariant measure $d\mu_O$ on $O$ is unique up to a nonzero real multiple.*

This is well known. For $f \in C_c^\infty(G(F))$ and any $x \in O$, let $\Phi(x,f)$ or $\Phi(O,f)$ be $\int_O f\, d\mu_O$. The integral converges by Theorem 2.32 of Deligne, or by Rao [Ra72]. Hence $f \mapsto \Phi(O,f)$ is an invariant distribution on $G$. When $O = \{e\}$ write $\Phi(e)$ for $\Phi(O)$. Put $\Phi(e,f) = f(e)$.

A distribution $D$ on $G$ (a complex valued linear form on $C_c^\infty(G(F))$) is said to vanish on an open set $V$ if $D(f) = 0$ for all test functions $f$ with support in $V$. If $D$ vanishes on a family $\{V_\alpha\}$ of open sets, it vanishes on $\cup V_\alpha$, since supp($f$) is compact. The *support* of the distribution $D$ is the complement of the largest open set on which $D$ vanishes. A distribution $D$ on a subset $X$ of $G$ is called *invariant* if $D(f^g) = D(f)$ for all $g \in G(F)$ and $f \in C_c^\infty(X)$. For any subset $X$ of $G$, write $I(X)$ for the set of invariant distributions on $G$ supported in $X$.

PROPOSITION 2.13. *A basis for* $I(U_\infty)$ *is given by* $\{\Phi(O)\,;\, O \in U\}$.

PROOF. Suppose $D \in I(U_\infty)$ has supp$(D) \subset U_d$. List the $O \in U$ of dimension $d$ as $O_1, \ldots, O_m$. Each $O_j$ is open in $U_d$, and $D|_{O_j} = c_j\Phi(O_j)$ for some $c_j \in \mathbb{C}$, since $I(O_j)$ is spanned by $\Phi(O_j)$. Hence supp$(D - \sum_{j=1}^{m} c_j\Phi(O_j)) \subset U_{d-1}$, and the proposition follows by induction. □

Let $\Omega$ be an open closed subset of $\mathfrak{g} = \text{Lie}\, G$ containing 0 which is invariant under Ad($G$), such that $\exp : \Omega \to G$ is defined, the image $\omega = \exp(\Omega)$ is open and closed and $\exp(\text{Ad}(g)X) = (\text{Int}(g))(\exp X)$ for all $g \in G$ and $X \in \Omega$, and $\mathcal{O}_F\Omega \subset \Omega$ where $\mathcal{O}_F$ is the ring of integers of $F$. For example, $\Omega$ can be the image under Ad($G$) of $\mathfrak{t}(\mathcal{O}_F) + \pi\mathfrak{g}(\mathcal{O}_F)$, where $\mathfrak{t}$ is a maximally split torus in $\mathfrak{g}$, both defined over $\mathcal{O}_F$, and $\mathfrak{t}(\mathcal{O}_F)$ and $\mathfrak{g}(\mathcal{O}_F)$ indicate the $\mathcal{O}_F$ valued points, $\pi$ a generator of the maximal ideal in $\mathcal{O}_F$.

We use standard notation: $\text{Int}(g)x = gxg^{-1}$, $g, x \in G$, $X \in \mathfrak{g}$, Ad($g$) the differential of Int($g$). Fixing an embedding of $G$ in some GL($n$), we embed $\mathfrak{g}$ in $M(n, F)$, and so $\text{Ad}(g)X = gXg^{-1}$.

PROPOSITION 2.14. *We have* $U_\infty \subset \omega$.

PROOF. For every $O \in U$, we have $e \in \overline{O}$. As $e \in \omega$ and $\omega$ is open, we have $O \cap \omega \neq \varnothing$. But $\omega$ is $G$-invariant, so $O \subset \omega$. □

Each $O \in U$ is of the form $O(u)$ with $u = \exp X$ in $U_\infty$. By the Jacobson-Morozov theorem, there is a nilpotent $Y \in \mathfrak{g}$ and a semisimple $H \in \mathfrak{g}$ such that $X, Y, H$ satisfy the relations $[X, Y] = H$, $[H, X] = 2X$, $[H, Y] = -2Y$, and the Lie algebra $\mathfrak{s}$ generated by $X, Y, H$ is isomorphic to $\mathfrak{sl}(2)$. Let $S$ be the connected algebraic group obtained from $\mathfrak{s}$. Let $T_H$ be the centralizer of $H$ in $S$. There is a character $\chi : T_H \to F^\times$ with $\text{Ad}(t)X = \chi(t)X$ for all $t \in T_H$. There is a central isogeny SL(2) $\to S$. Hence $(F^\times)^2 \subset \chi(T_H)$, and for all $a \in F^\times$, the element $a^2X$ is conjugate to $X$ by some $t_a \in T$. Define, for each $a \in F^\times$, the automorphism $\alpha_a$ of $O$ by $\alpha_a(\exp X) = \exp(a^2X)$, and $\alpha_a(\text{Int}(g)\exp(X)) = (\text{Int}(g))\alpha_a(\exp X)$ for all $g \in G$. Since $\alpha_a$ commutes with the action of $G$ on $O$, we have $d\mu_O \circ \alpha_a = \lambda(a)d\mu_O$ for some $\lambda(a) \in \mathbb{R}^\times_{>0}$.

PROPOSITION 2.15.

(i) *For all* $a \in F^\times$, *we have* $d\mu_O \circ \alpha_a = |a|^{d(O)}d\mu_O$.
(ii) *For every* $f \in C_c^\infty(\omega)$, *put* $f_a(\exp Z) = f(\exp(a^2Z))$ *for all* $Z \in \Omega$. *Then* $\Phi(O, f_a) = |a|^{-d(O)}\Phi(O, f)$.

PROOF. The two claims being equivalent, we prove (ii). Denote by $G_u$ the centralizer $\{g \in G; gug^{-1} = u\}$ of $u$ in $G$ and by $\mathfrak{g}_X$ the centralizer $\{Z \in \mathfrak{g}; \mathrm{ad}(Z)X = 0\}$ of $X$ in $\mathfrak{g} = \mathrm{Lie}\, G$. Then $\Phi(O,f) = \int_{G_u\backslash G} f(g^{-1}ug)\, d\dot{g}$ for a quotient measure $d\dot{g}$ on $G_u\backslash G$. Since $X$ is an eigenvector of $\mathrm{Ad}(T_H)$, the torus $T_H$ normalizes $G_u$ and $\mathfrak{g}_X$. Then

$$\Phi(O,f_a) = \int_{G_u\backslash G} f_a(g^{-1}ug)\, d\dot{g} = \int_{G_u\backslash G} f(g^{-1}\exp(a^2X)g)\, d\dot{g}$$

$$= \int_{G_u\backslash G} f(g^{-1}t\exp(X)t^{-1}g)\, d\dot{g} \qquad (2.15.1)$$

where $t \in T_H$ has $\chi(t) = a^2$, and this is $\int_{G_u\backslash G} f((t^{-1}gt)^{-1}\exp(X)(t^{-1}gt))\, d\dot{g}$ as $\int_{G_u\backslash G} f(g^{-1}ug)\, d\dot{g}$ is invariant under $g \mapsto gx$. Changing $g \mapsto \mathrm{Int}(t)g$, we get

$$\Phi(O,f_a) = \int_{G_u\backslash G} f(g^{-1}ug)|\det(\mathrm{Ad}(t)|_{\mathfrak{g}_X\backslash\mathfrak{g}})|\, d\dot{g}.$$

To prove the proposition, we need to show that $|\det(\mathrm{Ad}(t)|_{\mathfrak{g}_X\backslash\mathfrak{g}})| = |a|^{-d(O)}$. But $\mathfrak{g}$ is unimodular, so $|\det(\mathrm{Ad}(t)|_{\mathfrak{g}})| = 1$. It remains to establish

Claim: We have $|\det(\mathrm{Ad}(t)|_{\mathfrak{g}_X})| = |\chi(t)|^{d(O)/2}$.

To prove this claim, let $\rho$ denote the restriction of the adjoint representation of $G$ to the subgroup $S$. The irreducible constituents of $\rho$ are denoted by $\mathfrak{g}_1, \ldots, \mathfrak{g}_r$, with dimensions $\dim \mathfrak{g}_j = d_j + 1$, so that $\mathfrak{g} = \oplus_{j=1}^r \mathfrak{g}_j$ and $\sum_{j=1}^r (d_j + 1) = \dim \mathfrak{g}$. Since $\mathfrak{s} \simeq \mathfrak{sl}(2)$, the representation of $S$ on $\mathfrak{g}_j$ is completely determined by the dimension of $\mathfrak{g}_j$. Let $X_j$ be the highest weight vector for $T$ in $\mathfrak{g}_j$ for $X$. Let $\chi_j$ be the character $T_H \to F^\times$ defined by $\rho(t)X_j = \chi_j(t)X_j$ for $t \in T_H$. Then $\chi_j^2 = \chi^{d_j}$. Up to scalar multiples, $X_j$ is the unique vector in $\mathfrak{g}_j$ with $\mathrm{ad}(X)X_j = 0$. Namely, $\dim(\mathfrak{g}_X \cap \mathfrak{g}_j) = 1$, $\{X_1, \ldots, X_r\}$ is a basis for $\mathfrak{g}_X$, and $\sum_{j=1}^r d_j = \dim \mathfrak{g} - \dim \mathfrak{g}_X = d(O)$. The action of $T_H$ on $\mathfrak{g}_X$ with respect to the basis $\{X_1, \ldots, X_r\}$ is diagonal: $\mathrm{Ad}(t)X_j = \chi_j(t)X_j$. Then

$$|\det(\mathrm{Ad}(t)|\mathfrak{g}_X)| = \left|\prod_{j=1}^r \chi_j(t)\right| = \left|\prod_{j=1}^r \chi(t)^{d_j}\right|^{1/2} = |\chi(t)|^{d(O)/2}.$$

The claim, and the proposition, follow. □

Recall some results of Harish-Chandra ([HC70]). In topology, a continuous map $f : M \to N$ is called *submersive* if $f$ is surjection, and for any $T \subset N$, we have that $T$ is open or closed if and only if $f^{-1}(T)$ is so. In other words, $N$ has the quotient topology relative to the surjection $M \to N$. In differential geometry submersive means moreover that the differential of $f$ is surjective everywhere. Recall: $u = \exp(X)$ is unipotent and $\{X, Y, H\}$ is an $\mathfrak{sl}(2)$-triple; we put $G_Y = \{g \in G; \mathrm{Ad}(g)Y = Y\}$.

PROPOSITION 2.16. (i) *The map* $\psi : G \times G_Y \to G$, $(g, y) \mapsto g^{-1}(yu)g$ *is everywhere submersive.*

(ii) *There is an open subset* $V(O)$ *in* $G_Y$ *containing* $e$ *such that* $\psi$ *maps* $G \times V(O)$ *to an invariant open set* $\Omega_O$ *in* $G$ *containing the unipotent orbit* $O$, *such that* $\Omega_O \cap U_{d(O)} = O$.

(iii) *Let* $\pi : M \to N$ *be a smooth submersive map of manifolds. Let* $\omega_M$ *and* $\omega_N$ *be nowhere zero differential forms of maximal degree on* $M$ *and* $N$. *Then there is a unique surjection* $C_c^\infty(M) \to C_c^\infty(N)$, $\alpha \mapsto f_\alpha$, *with* $\int_M (F \circ \pi)\alpha\omega_M = \int_N F f_\alpha \omega_N$ *for all locally integrable* $F$ *on* $N$. *Further,* supp $f_\alpha \subset \pi(\text{supp}\,\alpha)$. *If* $D$ *is a distribution on* $N$, *then the map* $\alpha \mapsto D(f_\alpha)$ *defines a distribution* $D_M$ *on* $M$; $D_M$ *determines* $D$ *uniquely.*

(iv) *Let* $D$ *be an invariant distribution on* $G$ *supported in* $\psi(G \times V(O))$. *Then there is a unique distribution* $D_O$ *on* $V(O)$ *with* $D(f_\alpha) = D_O(\beta_\alpha)$ *for all* $\alpha$ *in* $C_c^\infty(G \times V(O))$, *where* $\beta_\alpha(u) = \int_G \alpha(g, u)\, dg$.

For (ii) note that $\psi$ maps $G \times \{e\}$ onto $O$. As $O$ is open in $U_{d(O)}$, there is an open neighborhood $V(O)$ of $e$ in $G_Y$ such that $\psi(G \times V(O)) \cap U_{d(O)} = O$.

## 2.17 *Orbital Integrals*

Let $G$ be a connected reductive $F$-group and $T$ an $F$-torus. Write $\mathfrak{t}$ for the Lie algebra of $T(F)$, $t_\Omega$ for $\mathfrak{t} \cap \Omega$, and $\mathfrak{t}^{\text{reg}}$ for the set of regular elements in $\mathfrak{t}$. Then $\exp : \mathfrak{t}_\Omega \to T(F)$ is defined, and $\mathcal{O}_F \cdot \mathfrak{t}_\Omega \subset \mathfrak{t}_\Omega$. Put $T_\Omega$ for $\exp(\mathfrak{t}_\Omega)$. It is an open neighborhood of $e$ in $T(F)$. Denote the center of $G(F)$ by $Z(G)$.

Fix $s$ in $T$. Write $G_s^0$ for the connected centralizer of $s$ in $G$. It is a connected reductive $F$-group, with $\dim G_s^0 < \dim G$ if $s \notin Z(G)$. Write $M$ for $G_s^0(F)$, $T$ for $T(F)$, $G$ for $G(F)$, and $Z(M)$ for the center of $M$. We view $T$ as a torus in $M$. Write $\Sigma_M$ for the set of roots of $T$ in $M$ and $T_M^{\text{reg}}$ for $\{t \in T\,;\ \alpha(t) \neq 1 \text{ for all } \alpha \in \Sigma_M\}$. Then $T^{\text{reg}} \subset T_M^{\text{reg}}$. The latter is the set of elements of $T$ which are regular as elements of $M$. Let $\mathfrak{m}$ be the Lie algebra of $M$, and $\mathfrak{t}_M^{\text{reg}} = \{X \in \mathfrak{t}\,;\ X \text{ is regular as an element of } \mathfrak{m}\}$. Write $U^M$ for the set of unipotent conjugacy classes in $G$ which intersect $M$. Put $U_\infty^M = U_\infty \cap M$. An $O \in U$ lies in $U^M$ when $O = O(u)$ for some $u \in U_\infty^M$. If $O \subset U_\infty^M$ put $O_M = O \cap M$. The connected centralizer $G_{su}^0$ of $su$ in the $F$-group $G$ lies in $G_s^0$ for all $u \in U_\infty^M$, by the uniqueness of the Jordan decomposition. Write $M_{su}$ for the group of $F$-points of the connected centralizer $G_{su}^0$ of $su$ in the $F$-group $G$. Write $\Phi(sO, f) = \int_{M_{su}\backslash G} f(g^{-1}sug)\, d\dot{g}$ if $O = O(u)$, $u \in U_\infty^M$, for a quotient measure $d\dot{g}$ on $M_{su}\backslash G$. The integral converges by Theorem 2.32 or [Ra72]. So it defines an invariant distribution on $G$.

PROPOSITION 2.18. *For* $s$, $M$, $T$ *as above, for each* $O \in U^M$, *there exist locally constant functions* $\Gamma(O, s, T; t)$ *(called germs) in* $t \in sT_\Omega \cap T_M^{\text{reg}}$, *satisfying*

(i) $\Gamma(O, s, T; zt) = \Gamma(O, s, T; t)$ *for any* $z \in Z(M)$ *such that* $t$ *and* $zt$ *lie in* $zT_\Omega \cap T_M^{\text{reg}}$;

(ii) $\Gamma(O, s, T; s\exp(a^2H)) = |a|^{-d(O_M)}\Gamma(O, s, T; s\exp(H))$, *for all* $a \in \mathcal{O}_F$, $H \in \mathfrak{t}_M^{\mathrm{reg}}$;

(iii) *for each* $f \in C_c^\infty(G)$, *there is an open neighborhood* $V(f)$ *of* $s$ *in* $sT_\Omega \subset T$ *with*

$$\Phi(t,f) = \sum_{O \in U^M} \Gamma(O, s, T; t)\Phi(sO, f)$$

*for all* $t \in V(f) \cap T^{\mathrm{reg}}$.

The equation of (iii) is called the germ expansion of the orbital integral.

Note that if $s \in T^{\mathrm{reg}}$, then $M = T$ and the proposition asserts that $\Phi(t,f)$ is constant in an open neighborhood of $s$ in $T^{\mathrm{reg}}$. This is part of Proposition 2.9. The main case is where $s \in Z(G)$, and then $M = G$. The general case follows by a reduction argument.

Recall that $O = O(u)$, $u = \exp X$ unipotent, $\{X, Y, H\}$ is an $\mathfrak{sl}(2)$ triple, $G_Y$ and $\mathfrak{g}_Y$ are the centralizers of $Y$ in $G$ and $\mathfrak{g}$, and $\psi : G \times G_Y \to G$, $\psi(g, y) = g^{-1}yug$ is a submersion.

PROPOSITION 2.19. *There exists an open set* $V(O) \subset G_Y$ *containing the unit* $e$ *of* $G$ *and satisfying Proposition* 2.16(ii), *and also*: *if* $zu \in O(u)$ *for* $z \in V(O)$, *then* $z = e$.

Thus $V(O)$ contains representatives for the conjugacy classes which pass near $u$ but only one representative for the class of $u$.

PROOF. Since $\varphi : G_u\backslash G \to O(u)$, $g \mapsto g^{-1}ug$, is an injective analytic isomorphism, for any open neighborhood $V$ of $u$ in $G$, there is an open neighborhood $N$ of $e$ in $G_u\backslash G$ with $\varphi(N) = O(u) \cap V$. The tangent space to $O(u)$ at $u$ is $[X, \mathfrak{g}]$. The tangent space to $G_Yu$ at $u$ is $\mathfrak{g}_Y = \{Z \in \mathfrak{g}\,;\, [Z, Y] = 0\}$.

Claim: $[X, \mathfrak{g}] \cap G_Y = \{0\}$. Indeed, $[X, \mathfrak{g}] =$ image of $\mathrm{ad}(X)$, for ad as in the proof of the claim in the proof of Proposition 2.15. Also $\mathfrak{g}_Y$ is the kernel of $\mathrm{ad}(Y)$, which is the space spanned by the lowest weight vectors in the irreducible constituents $\mathfrak{g}_j$ of ad. Then $[X, \mathfrak{g}] \cap \mathfrak{g}_Y = \mathrm{Im}(\mathrm{ad}(X)) \cap \ker(\mathrm{ad}(Y)) = \{0\}$ from the well-known structure of representations of $\mathfrak{sl}(2)$. Similar arguments show that $\mathfrak{g} = [X, \mathfrak{g}] \oplus \mathfrak{g}_Y$. Hence $O(u)$ and $G_Yu$ are transverse at $u$. So we can choose the open neighborhoods $V$ of $u$ in $G$ and $N$ of $e$ in $G_u\backslash G$ to have $\varphi(N) = O(u) \cap V$ and $\varphi(N) \cap G_Yu = \{u\}$. Choose a neighborhood $V(O)$ of $e$ in $G_Y$ with $V(O)u \subset V$. Then $O(u) \cap V(O)u \subset \varphi(N) \cap G_Yu = \{u\}$, as required. □

## 2.20 *Start of Proof of Proposition* 2.18

Denote by $\rho_0$ a faithful finite dimensional $F$-rational representation of $G$, namely, an embedding of $G$ in some $\mathrm{GL}(n)$. (The symbol $\rho$ was already used in Proposition 2.15.) Put $\det(x + 1 - \rho_0(g)) = \sum_{j=0}^{\dim \rho_0} p_j(g)x^j$. For all $\epsilon > 0$ the set

$G(\epsilon) = \{g \in G; |p_j(g)| < \epsilon$ for all $j\,(0 \le j < \dim \rho_0)\}$ is an invariant open neighborhood of $U_\infty$, and $\bigcap_{\epsilon>0} G(\epsilon) = U_\infty$. In particular the sets $T(\epsilon) = G(\epsilon) \cap T$ make a basis for the set of open neighborhoods of $e$ in $T$.

We now prove proposition 2.18 when $s = e$. We use induction on $\#U$. For the initial step, suppose $f \in C_c^\infty(G)$ is supported on $G - U_\infty$. Choose $\epsilon > 0$ so that $\mathrm{supp}(f) \subset G - G(\epsilon)$. Then $\Phi(t,f) = 0$ for all $t \in T(\epsilon) \cap T^{\mathrm{reg}}$.

For the induction step note that $\#U$ is finite as $\mathrm{char}(F) = 0$. List the orbits in $U$ as $O_1, \ldots, O_m$ such that $d(O_j) \ge d(O_{j+1})$. Assume there are functions $\Gamma(O_j, e, T; t)$ in $T^{\mathrm{reg}} \cap T_\Omega$ for $j\,(0 \le j < k)$ such that

$$\text{for every } f \text{ in } C_c^\infty(G) \text{ supported in } G - \bigcup_{j=k}^{m} O_j, \text{ there is } \epsilon = \epsilon(f) > 0 \text{ with} \tag{IND}$$

$$\Phi(t,f) = \sum_{j=1}^{k-1} \Gamma(O_j, e, T; t)\Phi(O_j,f) \quad \text{for} \quad t \in T(\epsilon) \cap T^{\mathrm{reg}}.$$

By the initial step, this holds for $k = 1$. To show (iii) of Proposition 2.18, we need to show there is a function $\Gamma(O_k, e, T; t)$ in $t \in T^{\mathrm{reg}} \cap T_\Omega$ for which (IND) holds with $k$ replaced by $k+1$. Assume, as we may, that $\mathrm{supp}(f) \subset \Omega(O_k) = \psi(G \times V(O_k))$, with $V(O_k)$ as in Proposition 2.18, $\psi$ as in Proposition 2.16(i). Suppose $f = f_\alpha$ with $\alpha \in C_c^\infty(G \times V(O_k))$ as in Proposition 2.16(iii). Put

$$D_{O_k,t}(\beta_\alpha) = D_t(f_\alpha) = \Phi(t,f_\alpha) - \sum_{j=1}^{k-1} \Gamma(O_j, e, T; t)\Phi(O_j,f_\alpha), \quad t \in T^{\mathrm{reg}}.$$

Claim: If $\beta_\alpha(e) = 0$, then there is $\epsilon > 0$ with $D_t(f_\alpha) = 0$ for all $t \in T^{\mathrm{reg}} \cap T(\epsilon)$.

To verify this, put $\alpha_0 = \gamma \times \beta_\alpha \in C_c^\infty(G \times V(O_k))$, where $\gamma \in C_c^\infty(G)$ satisfies $\int_G \gamma(g)\,dg = 1$. Then $\beta_{\alpha_0} = \beta_\alpha$, hence $D_t(f_{\alpha_0}) = D_t(f)$. Also $\mathrm{supp}(f_{\alpha_0}) \subset \psi(\mathrm{supp}\,\alpha_0)$. Since $\Omega(O_k) \cap U_\infty \subset \bigcup_{j=1}^{k} O_j$, the claim will follow from the induction assumption if we show that $\psi(\mathrm{supp}(\alpha_0)) \cap O_k$ is empty. Since $\beta_\alpha$ is locally constant and $\beta_\alpha(e) = 0$, we have $e \notin \mathrm{supp}(\beta_\alpha)$. By the choice of $V(O_k)$ and Proposition 2.18, we have $\psi(\mathrm{supp}(\alpha_0)) \cap O_k = \varnothing$, so the claim follows.

## 2.21 *Induction Step*

We proceed to prove the induction step. The distribution $f_\alpha \mapsto \beta_\alpha(e)$ is invariant and supported in $O_k$. Hence there is $c \in \mathbb{C}$ with $\beta_\alpha(e) = c\Phi(O_k,f_\alpha)$. It is clear that $c \ne 0$ as we do not have $\int_G \alpha(g,e)\,dg = 0$ for all $\alpha \in C_c^\infty(G \times V(O_k))$. Choose $\beta \in C_c^\infty(V(O_k))$ with $\beta(e) = c$. Then $\beta_\alpha - \beta_\alpha(e)\beta$ satisfies the assumption of the claim. Hence there is $\epsilon > 0$ with

$$D_{O_k,t}(\beta_\alpha - \beta_\alpha(e)\beta) = 0 \quad \text{for all } t \in T^{\mathrm{reg}} \cap T(\epsilon).$$

Hence for all $t \in T(\epsilon) \cap T^{\text{reg}}$, we have

$$\Phi(t,f_\alpha) = \sum_{j=1}^{k-1} \Gamma(O_j, e, T; t)\Phi(O_j,f_\alpha) + cD_{O_k,t}(\beta)\Phi(O_k,f_\alpha).$$

Write $\Gamma(O_k, e, T; t)$ for $cD_{O_k,t}(\beta)$ and $t \in T_\Omega \cap T^{\text{reg}} \cap T(\epsilon)$. Extend it to all of $T_\Omega \cap T^{\text{reg}}$ by $\Gamma(O_k, e, T; \exp(a^2H)) = |a|^{-d(O_k)}\Gamma(O_k, e, T; \exp(H))$ for all $H \in \mathfrak{t}_\Omega \cap \mathfrak{t}^{\text{reg}}$, and $a \in \mathcal{O}_F$. By Proposition 2.15, $\Gamma(O_k, e, T; t)$ satisfies this identity on $T_\Omega \cap T^{\text{reg}} \cap T(\epsilon)$, so $\Gamma(O_k, e, T; t)$ is well defined. The induction step follows.

Note that $\Gamma(O, e, T; t)$ is locally constant in $t$ since so if $\Phi(t,f)$.

Now that we dealt with $s = e$, consider $s \in Z(G)$. Define $\Gamma(O, s, T; st)$ to be $\Gamma(O, e, T; t)$ and replace $f(g)$ by $f_s(g) = f(sg)$. Proposition 2.18 for $s \in Z(G)$ then follows from the case of $s = e$.

## 2.22 Reduction Step

Next suppose that $s \in T$ and $\dim M < \dim G$. We shall reduce Proposition 2.18 for $s$ to a germ expansion on $M$. Let $V$ be a neighborhood of $s$ in $T$ satisfying the conclusion of Proposition 2.10. Let $K' = \text{supp}(f)$. Choose $C$ as in that proposition. Then $g^{-1}Vg \cap K'$ is empty if the image of $g$ is not in $C \subset M\backslash G$. The function

$$M\backslash G \ni Mg \mapsto \int_{T\backslash M} f(g^{-1}m^{-1}tmg)\, d\dot{m}, \quad t \in V \cap T^{\text{reg}},$$

vanishes outside $C$. Choose $\alpha \in C_c^\infty(G)$ such that $\overline{\alpha}(g) = \int_M \alpha(mg)\, dg$ is 1 if $Mg \in C$ and 0 otherwise. Put $F(m) = \int_G \alpha(g)f(g^{-1}mg)\, dg$. Then $F \in C_c^\infty(M)$ as in Proposition 2.16. For $t \in V \cap T^{\text{reg}}$, we have

$$\begin{aligned}\int_{T\backslash G} f(g^{-1}tg)\, d\dot{g} &= \int_{M\backslash G}\int_{T\backslash M} \overline{\alpha}(g)f(g^{-1}m^{-1}tmg)\, d\dot{g}\, d\dot{m}\\ &= \int_G\int_{T\backslash M} \alpha(g)f(g^{-1}m^{-1}tmg)\, dg\, d\dot{m}\\ &= \int_{T\backslash M} F(m^{-1}tm)\, d\dot{m} = \Phi^M(t, F).\end{aligned}$$

Here $\Phi^M(t,f)$ is the orbital integral on $M$. A similar computation shows that $\Phi(sO,f)$, defined before Proposition 2.18, is equal to $\Phi(sO_M, F)$ for all $O \in U_M$(= the set of $O \in U$ which intersect $M$), where $O_M = O \cap M$, since $M_{su} \subset M$. Now $s$ lies in the center of $M$. Hence the germ expansion holds for $\Phi^M(t,f)$ in an open neighborhood of $s$ which intersects $T_M^{\text{reg}}$. This proves Proposition 2.18 with $\Gamma(O, s, T; t) = \Gamma(O_M, s, T; t)$. □

## 2.23 Extension of Proposition 2.8

We continue with a slight extension of Proposition 2.8, using an alternative approach, based on the uniqueness of the Haar measure.

Let $T$ be a maximal torus of $G$ containing the semisimple part $s$ of $g \in G$. Let $\Sigma$ be a root system of $T$ in $G$. Let $\Sigma_s$ be the subset of roots which do not take the value 1 at $s$. Recall that

$$\Delta(g) = \left| \prod_{\alpha \in \Sigma_s} (1 - \alpha(s)) \right|^{1/2} = |\det(1 - \mathrm{Ad}(s))|_{\mathfrak{g}/\mathfrak{g}_s}|^{1/2}$$

and

$$I(g,f) = \Delta(g) \int_{Z_G(g)\backslash G} f(x^{-1}gx)dx.$$

PROPOSITION 2.24.

(i) *Fix* $f \in C_c^\infty(G)$. *For each pair* $(T, u)$, *the normalized orbital integral* $g \mapsto I(g,f)$ *is locally constant on* $(Tu)^{\mathrm{reg}}$ *and compactly supported on* $Tu$. *For each* $s \in T$, *there is a neighborhood* $V_f$ *of* $s$ *in* $T$, *and functions* $\Gamma(Tu, su_i)$ *on* $V_f u$ *whose germs are independent of* $f$, *where the* $u_i$ *are as in* 2.7, *such that for* $t \in V_f \cap T^{\mathrm{reg}}$, *we have*

$$I(tu,f) = \sum_{i=1}^{r} I(su_i,f)\Gamma(Tu, su_i; tu).$$

(ii) *Conversely, a conjugacy invariant function on* $G$ *satisfying* (i) *with the germs* $\Gamma(Tu, su_i)$ *is an orbital integral.*

There is an obvious extension of this to the space $C_c^\infty(G, \omega)$ of locally constant complex valued functions which transform under the center $Z$ of $G$ by a character $\omega$ of $Z$. We just need to add to "compactly supported" the term "mod $Z$".

The following is used also in the proof of Corollary 4.4. It originates from [GK75], p. 100.

## 2.25 An Extension Result

If $G$ acts on a space $X$, write $C_0(X)$ for the subspace of $C_c^\infty(X)$ spanned by $g \cdot f - f$, where $f \in C_c^\infty(X)$ and $g \in G$, and $(g \cdot f)(x) = f(g^{-1}x)$. We denote by $A_s$ the set of $g \in G$ with semisimple part conjugate to $s$.

PROPOSITION. *Suppose the normalized orbital integral* $I(g,f)$ *of* $f \in C_c^\infty(G)$ *vanishes on* $A_s$. *Then there exists* $\phi \in C_0(G)$ *with* $f = \phi$ *on* $A_s$.

PROOF. This is done by induction, using the following elementary

LEMMA. *Let* $A$ *be a locally compact totally disconnected topological space. Let* $B$ *be a locally closed nonempty subspace of* $A$. *Denote by* $C_c^\infty(A,F)$ *the space of* $f \in C_c^\infty(A)$ *(the space of compactly supported locally constant functions on* $A$*) which vanish on* $F = \overline{B} - B$. *Then the restriction map* $C_c^\infty(A,F) \to C_c^\infty(B)$ *is surjective.*

We shall also use the uniqueness up to scalar of the $G$-invariant Haar measure on $O(g)$, [BZ76], which is equivalent to the statement that $C_0(O(g))$ has codimension 1 in $C_c^\infty(O(g))$. See proof of Corollary 4.4 for more details.

Since $I(g,f)$ is 0 on $A_s$, and $O(s)$ is closed in $A_s$, the restriction of $f$ to $O(s)$ is in the subspace $C_0(O(s))$ of $C_c^\infty(O(s))$, by the uniqueness of the Haar measure on $O(s)$. Hence there are finitely many pairs $(g,h)$ in $G \times C_c^\infty(O(s))$ with $f = \sum(g \cdot h - h)$ on $O(s)$. The lemma permits extending $h$ to $\overline{h} \in C_c^\infty(G)$.

The function $f_1 = f - \sum(g \cdot \overline{h} - \overline{h})$ is zero on $O(s)$ and $I(g,f_1) = 0$ for all $g \in A_s$. Note that $O(s)$ is the boundary of $A_2 = O(s) \cup O(su_2)$. Hence the restriction of $f_1$ to $O(su_2)$ lies in $C_c^\infty(O(su_2))$. Using again $[C_c^\infty(O(g)) : C_0(O(g))] = 1$, we conclude that there exist finitely many pairs $(g,h) \in G \times C_c^\infty(O(su_2))$ such that $f_1 = \sum(g \cdot h - h)$ on $O(su_2)$. By the lemma we extend $h$ to $\overline{h} \in C_c^\infty(G, O(s))$. Then $f_2 = f_1 - \sum(g \cdot \overline{h} - \overline{h})$ is zero on $A_2$ and $I(g,f_2) = 0$ for all $g \in A_s$. Continuing by induction we conclude that there is $\phi \in C_0(G)$ with $f = \phi$ on $A_s$. □

PROPOSITION 2.26. *Let* $G = \mathrm{GL}(n,F)$. *Suppose* $f \in C_c^\infty(G)$ *is zero on* $A_s$, *where* $s$ *is semisimple in* $G$. *Then there exists an open,* $G$*-invariant compactly generated neighborhood* $V_f$ *of* $s$, *depending on* $f$, *on which* $f$ *is zero.*

PROOF. The characteristic polynomial defines a continuous map $P : G \to F^n$, such that the inverse image of $P(s)$, $s$ semisimple, is the set $A_s$ of $g$ in $G$ with semisimple part conjugate to $s$. The image $P(\mathrm{supp} f)$ of the support of $f$ is compact. In the totally disconnected space $F^n$, there is an open neighborhood $N$ of $P(s)$ which does not intersect $P(\mathrm{supp} f)$. The inverse image of the open $N$ is open and it contains $s$, hence also an open compact neighborhood $sK'$ of $s$. Now $P(sK') \subset N$. Hence $\bigcup_{g \in G} gsK'g^{-1}$ does not intersect $\mathrm{supp} f$. □

COROLLARY 2.27. *Let* $G = \mathrm{GL}(n,F)$ *and* $f \in C_c^\infty(G)$. *For every semisimple* $s$ *in* $G$, *there is a neighborhood* $V_f$ *of* $f$ *such that if* $A_s = \bigcup_{i=1}^t O(su_i)$ *and* $f_i$ *are as in* 2.7, *for all* $g \in V_f$, *we have*

$$I(g,f) = \sum_{i=1}^{t} I(g,f_i) I(su_i,f).$$

PROOF. The orbital integrals of the function $f - \sum_{i=1}^t I(su_i,f) f_i$ are zero on all $g \in A_s$. Hence this function is equal to a function $\phi \in C_0(G)$ on an open $G$-invariant compactly generated neighborhood of $s$, depending on $f$, by the last two propositions. □

The formula displayed in the corollary is called the germ expansion of the orbital integral of $f$ at $s$. Note that we limited the discussion to $G = \mathrm{GL}(n)$ as we used the characteristic polynomial.

## 2.28 Extension to Reductive Groups

To extend this analysis to general reductive connected $G$ from the case of $\mathrm{GL}(n)$, we apply a sequence of reductions. Embed $G$ in $\mathrm{GL}(n, F)$.

Suppose $s = 1$. The characteristic polynomial defines a continuous map $G \to F^n$. The inverse image of $P(1)$ is the set of unipotent elements in $G$. We obtain a germ expansion of the orbital integrals in a neighborhood of $s = 1$: if $f \in C_c^\infty(G)$ and $A_1 = \bigcup_{i=1}^t O(u_i)$, and $f_i$ are as in 2.7, there is a neighborhood $V_f$ of $s = 1$ such that for $g \in V_f$ we have

$$I(g,f) = \sum_{i=1}^{t} I(g,f_i)I(u_i,f).$$

If $s$ lies in the center $Z$ of $G$, consider ${}^s f(g) = f(s^{-1}g)$ and note that $I(sg, {}^s f) = I(g,f)$. Hence in the neighborhood $sV_f$ of $s$, we have $I(g,f) = \sum_{i=1}^t I(s^{-1}g,f_i)I(su_i,f) = \sum_{i=1}^t I(g, {}^s f_i)I(su_i,f)$. The functions ${}^s f_i$ have the properties specified at the end of 2.7.

To deal with a general semisimple $s \in G$, we replace in a neighborhood of $s$ the integral on $G$ by an orbital integral on the reductive connected group $Z_G(s)^0$, in which $s$ is central.

We use Proposition 2.10 to construct a function $f_s \in C_c^\infty(Z_G(s)^0)$ such that the orbital integral of $f_s$ on $Z_G(s)^0$ is equal to that of $f$ on $G$ in a neighborhood of $su$ in $Tu$, as follows.

Choose $C$ as in Proposition 2.10 for $K' = \mathrm{supp}(f)$. We may assume that $C$ is compact and open, since $Z_G(s)^0\backslash G$ is locally compact. Choose $h \in C_c^\infty(G)$ such that the integral $\overline{h}(g) = \int_M h(xg)\, \mathrm{d}x$, $M = Z_G(s)^0$, equals the characteristic function of $C$ in $M\backslash G$. Define $f_s$ to be $f_s(g) = \int_G h(x)f(x^{-1}gx)\, \mathrm{d}x$. Then for $g \in Vu$ we have $\Phi^G(g,f) = \Phi^M(g,f_s)$ (see notation in Subsection 2.6). Indeed,

$$\Phi^G(g,f) = \int_{Z_G(g)\backslash G} f(x^{-1}gx)\, \mathrm{d}x = \int_{M\backslash G} \int_{Z_G(g)\backslash M} f(x^{-1}y^{-1}gyx)\, \mathrm{d}y\, \overline{h}(x)\, \mathrm{d}x$$

$$= \int_G \mathrm{d}x \int_{Z_G(g)\backslash M} h(x)f(x^{-1}y^{-1}gyx)\, \mathrm{d}y = \int_{Z_G(g)\backslash M} f_s(y^{-1}gy)\, y = \Phi^M(g,f_s).$$

For $g \in Vu$ we then have $I^G(g,f) = \frac{\Delta_G(g)}{\Delta_M(g)} I^M(g,f_s)$.

Suppose $A_{s,M} = \bigcup O(su_i)$ is the decomposition into orbits in $M$ of the set of elements with semisimple part conjugate in $M$ to $s$. Let $f_{s,i} \in C_c^\infty(M)$ be the functions satisfying 2.7. Then for $g \in M$ near $s$, we have

$$I^M(g,f_s) = \sum_{i=1}^{t} I^M(g,f_{s,i}) I^M(su_i,f_s).$$

A unipotent element $u$ has conjugates arbitrarily close to the identity. If $u$, $u'$ are unipotents in $M$, then $su$ and $su'$ are conjugate in $M$ if and only if they are conjugate in $G$. In $G$ we have $A_s = \bigcup_{i=1}^{t} O(su_i)$ with the same $u_i$ as for $M$. We may assume that $f_{s,i} = f_i \Delta_G(su_i)/\Delta_M(su_i)$, where $f_i \in C_c^\infty(G)$ are as in 2.7. For $g$ in a neighborhood of $su$ in $Tu$, we then have

$$I^G(g,f) = \sum_{i=1}^{t} I^G(g,f_i) I^G(su_i,f).$$

We obtain then the following analogue of Corollary 2.27 for a general connected reductive $G$, not only for $G = \mathrm{GL}(n)$.

PROPOSITION 2.29. *Let $f \in C_c^\infty(G)$. For any semisimple $s$ in $G$, torus $T$ containing $s$, and unipotent $u$ commuting with each element of $T$, there is a neighborhood $V_f$ of $s$ in $T$ such that $I(g,f) = \sum_{i=1}^{t} I(g,f_i) I(su_i,f)$ for all $g \in V_f u$. Here $A_s = \bigcup_{i=1}^{t} O(su_i)$ and the $f_i$ are as in* 2.7.

## 2.30 *Proof of Part* (i) *of Proposition* 2.24

To complete the proof of part (i) of Proposition 2.24, we introduce notation. Let $s$ be semisimple in $G$. Put $M$ for the connected centralizer $Z_G(s)^0$. Let $T$ denote the center of $M$. Let $u$ be a unipotent in $M$. Consider the normalizer $N(Tu) = \{x \in G;\, xTux^{-1} = Tu\}$, centralizer $Z(Tu) = \{x \in G;\, xtux^{-1} = tu$ for all $t$ in $T\}$, and Weyl group $W(Tu) = N(Tu)/Z(Tu)$ of $Tu$ in $G$. Then $W(Tu)$ injects into the Weyl group $W(T) = N(T)/Z(T)$ of $T$ in $G$. It is the group of inner automorphisms of $G$ mapping $Tu$ to itself. The Weyl group $W(T)$ acts continuously on the left on $T \times (Z(T)\backslash G)$ by $w(t,g) = (n_w t n_w^{-1}, n_w g)$, $t \in T$, $g \in G$, $w \in W(T)$, $w = Z(T)n_w$, $n_w \in N(T)$. Write $O(T^{\mathrm{reg}}u)$ for the orbit of $T^{\mathrm{reg}}u$ in $G$.

LEMMA. *The map $(t,g) \mapsto g^{-1}tug$ induces a canonical homeomorphism*

$$W(Tu)\backslash(T^{\mathrm{reg}} \times (Z(Tu)\backslash G)) \to O(T^{\mathrm{reg}}u).$$

PROOF. The map is bijective and a local homeomorphism. As $W(Tu)$ is finite, the map is a homeomorphism. □

It follows from the lemma that (i) for any $f \in C_c^\infty(G)$, the restriction of $I(g,f)$ to $T^{\mathrm{reg}}u$ is locally constant, and (ii) for any function $I$ in $C_c^\infty(T^{\mathrm{reg}})$ invariant under $W(Tu)$, there exists $f$ in $C_c^\infty(O(T^{\mathrm{reg}}u))$ such that $I(t) = I(tu,f)$ for all $t \in T^{\mathrm{reg}}$. Part (i) of Proposition 2.24 then follows from these and Proposition 2.29.

## 2.31 *Proof of Part* (ii) *of Proposition* 2.24

It remains to prove part (ii) of Proposition 2.24. The set of conjugacy classes of $Tu$, where $(T, u)$ is a pair as in 2.9 consisting of a torus $T$ and a unipotent $u$ which commutes with each element of $T$, is finite. We recall that it can be numbered as $(Tu)_j$, $1 \leq j \leq r$, such that $X_j = \bigcup_{i=1}^{j} O((Tu)_i)$ is closed in $G$, and $O((T^{\text{reg}}u)_j)$ is open in $X_j$, and $X_r = G$.

We proceed as follows. For a fixed $T$, arrange the $Tu$ in increasing dimensions: $Tu_1 < Tu_2$ if $\dim O(Tu_1) < \dim O(Tu_2)$. We get a finite sequence $S(T)$. Next arrange the sequences with $S(T_1) < S(T_2)$ if $T_1 \subset T_2$. Then $S(Z)$ is the first term, where $Z$ is the center of $G$. Further we have $X_j = X_{j-1} \cup O((T^{\text{reg}}u)_j)$ $(2 \leq j \leq r)$.

Now we prove (ii) of Proposition 2.24, using the lemmas in 2.25 and 2.30 and part (i) of 2.24, inductively. Let $I$ be the function of 2.24(ii). Its restriction $I|_Z$ to $Z$ lies in $C_c^\infty(Z)$. Namely, there is $f \in C_c^\infty(Z)$ with $I(z) = I(z,f) = f(z)$ for $z \in Z$. Extend $f$ to $f_0 \in C_c^\infty(G)$. Then $I_1 = I - I(\cdot, f_0)$ satisfies the requirements of (ii) of the proposition and also $I_1|_Z = 0$. By induction we assume, for $j$ $(1 \leq j < r)$, that $I_j$ satisfies the requirements of (ii), and also $I_j|_{X_j} = 0$. The restriction $I_j|_{O((T^{\text{reg}}u)_{j+1})}$ lies in $C_c^\infty(O((T^{\text{reg}}u)_{j+1}))$ and it is $W((Tu)_{j+1})$ invariant. Hence there is $f \in C_c^\infty(O((T^{\text{reg}}u)_{j+1}))$ with $I_j(x) = I(x,f)$ for $x \in (T^{\text{reg}}u)_{j+1}$. We extend $f$ to $f_j \in C_c^\infty(G, X_j)$, thus $f_j$ is zero on $X_j$, and consider $I_{j+1} = I_j - I(\cdot, f_j)$. This function satisfies the requirements of (i) of 2.24 and also $I_{j+1}|_{X_{j+1}} = 0$. At the last step of induction, we obtain our initial $I$ as an orbital integral, so (ii) of the proposition follows. □

We next study invariant measures on conjugacy classes, following a letter of P. Deligne.

THEOREM 2.32. *Let $G$ be a reductive group over a local field $F$ of characteristic zero. Fix $x \in G$ and $f \in C_c^\infty(G)$. Then the orbital integral $\int_{G/G_x} f(gxg^{-1})\, d\dot{g}$ is finite.*

Reduction to case where $x$ is unipotent. Put $x = su = us$ where $s$ is semisimple and $u$ is unipotent. Denote supp($f$) by $K'$. Then $gsug^{-1} \in K'$ implies $gsg^{-1} \in K'$ (we may assume $K'$ contains the semisimple parts of its elements since $gsg^{-1} = p(gxg^{-1})$ for some polynomial independent of $g$). Proposition 2.10 then implies that $g$ lies in a compact $C$ in $G/M$, $M = G_s^0$. Hence it suffices to replace $G$ by $G_s^0$, that is to deal with the case where $s = e$, and assume $x = u$ is unipotent. Since the orbit of $s$ is closed, its orbital integral converges so we may assume $u \neq e$ and that the center $Z(G)$ of $G$ consists of $e$. Moreover, we may assume that $f \geq 0$.

## 2.33 *Unipotent Elements*

We first discuss the structure afforded by the unipotent element $u$. Fix a faithful linear representation of $G$ on a vector space $V$ over $F$, that is an embedding $G \to \mathrm{GL}(V)$.

We claim that there exists a unique decreasing filtration $W^j$ of $V$ such that

(i) $(u-e)W^j \subset W^{j+2}$,
(ii) $(u-e)^k : W^{-k}/W^{-k+1} \xrightarrow{\sim} W^k/W^{k+1}$.

To see this, let $i : \mathrm{SL}(2) \to G$ be an embedding with $i(I+X) = u, X = \begin{pmatrix} 0 & 1 \\ 0 & 0 \end{pmatrix}$, $I = d(1)$ where $d(a) = \mathrm{diag}(a, a^{-1})$. We also put $Y = {}^tX$ for the transpose of $X$.

Consider the eigenspaces $V^j = \{v \in V;\ i(d(a))v = a^j v,\ a \in F^\times\}$ of the diagonal subgroup in $\mathrm{SL}(2, F)$, thus the vector space decomposition $V = \oplus_j V^j$. Then the filtration $W = \{W^j = \oplus_{k \geq j} V^k\}$ has the required properties.

In the special case of the adjoint representation of $G$ on $\mathfrak{g} = \mathrm{Lie}(G)$, we have

(i) $W^0(\mathfrak{g})$ is the Lie algebra of a parabolic subgroup $P$.
(ii) $W^1(\mathfrak{g})$ is the Lie algebra of the unipotent radical $N$ of $P$.
(iii) $W^0/W^1$ is the Lie algebra of the centralizer $H$ of $i(d(a))$.
(iv) If $u = \exp(U)$, then $(\mathrm{ad}(U))(W^1) = W^3$, as $\mathrm{ad}(U) : W^j/W^{j+1} \twoheadrightarrow W^{j+2}/W^{j+3}$ and $U \in W^2$.

The surjectivity follows on using $i(I + Y)$ to construct a partial inverse. Consequently the conjugacy class in $N$ of a conjugate $x$ of $u$ in $P$ depends only on $\log(x) \bmod W^3$. Let $N^j$ denote the subgroup of $N$ with Lie algebra $W^j$ $(j > 0)$. Let $N^*$ denote the set of conjugates of $u$ in $P$. Then $N^* \subset N^2$, and the map from the set $N^+$ of $N$-conjugacy classes in $N^*$, to $N^2/N^3$ (which is a commutative group), identifies $N^+$ with the orbit under $H$ of the image $\overline{u}$ of $u$ in $N^2/N^3$. This orbit is open. Thus we identify $N^+$ with an open subset of $N^2/N^3$and hence transport the measure $\mathrm{d}n$ to $N^+$.

If $x \in N^*$, then we have

(i) a duality between $W^{-k}/W^{-k+1}$ and $W^k/W^{k+1}$, using the Killing form;
(ii) an isomorphism $(x-e)^k : W^{-k}/W^{-k+1} \to W^k/W^{k+1}$.

Hence we get a bilinear form $B_x$ on $W^k/W^{k+1}$. The isomorphism (ii) and the form $B_x$ depend only on the image of $x$ in $N^+ \subset N^2/N^3$.

For any subset $A$ of $G$, write $A_u$ for $A \cap G_u$, the centralizer of $u$ in $A$. Then $G_u \subset P = HN$, and $P_u = H_u N_u$. Moreover,

$$N_u^j/N_u^{j+1} \simeq \ker(\mathrm{ad}(U) : W^j/W^{j+1} \to W^{j+2}/W^{j+3}).$$

Put (note that $(u-e)^{j+1}$ can be replaced by $U^{j+1}$)

$$F^{-j} = \ker((u-e)^{j+1} : W^{-j}/W^{-j+1} \to W^{j+2}/W^{j+3}).$$

Then

$$W^j/W^{j+1} \simeq \oplus_{k \geq 0} U^{j+k}(F^{-j-2k}),$$

where the direct sum is $B_u$-orthogonal.

The determinant of the action of $H_u$ on $F^{-j}$ is trivial, due to the existence of $B_u$.

Note that $G_u$ is unimodular, and thus $\delta_P|_{G_u} = \delta_{G_u}$ is 1, where $\delta_S(x) = \det(\mathrm{ad}(x)|\,\mathrm{Lie}\, S)$.

## 2.34 *Factorization of an Integral*

Let $\mathrm{d}\overline{g}$ denote the invariant measure on $G/G_u$, $\mathrm{d}\overline{p}$ that on $P/G_u$, $\mathrm{d}\overline{h}$ that on $H/H_u$, and $\mathrm{d}\overline{n}$ that on $N/N_u$. Denote by $\mathrm{d}g_u$ the invariant measure on $G_u$. Then

$$\mathrm{d}g = \delta_P(p)\mathrm{d}k\mathrm{d}p = \delta_P(p)\mathrm{d}k\mathrm{d}\overline{p}\mathrm{d}g_u = \mathrm{d}\overline{g}\mathrm{d}g_u$$

and

$$\mathrm{d}\overline{g} = \delta_P(p)\mathrm{d}k\mathrm{d}\overline{p}.$$

To verify the theorem, it suffices to show that

$$\int_{P/G_u} f(pup^{-1})\delta_P(p)\,\mathrm{d}\overline{p} < \infty.$$

Note that $\delta_N(H_u) = \delta_{N_u}(H_u) = 1$. Hence for $g$ on $P/G_u$, we have

$$\int_{P/G_u} g(p)\,\mathrm{d}\overline{p} = \int_{H/H_u} \mathrm{d}\overline{h} \int_{N/N_u} g(hn)\,\mathrm{d}\overline{n}.$$

Hence

$$\int_{P/G_u} f(pup^{-1})\delta_P(p)\,\mathrm{d}\overline{p} = \int_{H/H_u} \delta_N(h)\,\mathrm{d}\overline{h} \int_{N/N_u} f(hnun^{-1}h^{-1})\,\mathrm{d}\overline{n}.$$

For any function $g \geq 0$ on $N$, we have (for some $c > 0$ independent of $g$)

$$\int_{N/N_u} g(nun^{-1})\,\mathrm{d}\overline{n} \leq c \int_{N_3} g(un)\,\mathrm{d}n.$$

Hence

$$\int_{N/N_u} f(hnun^{-1}h^{-1})\,\mathrm{d}\overline{n} \leq c \int_{N_3} f(hunh^{-1})\,\mathrm{d}n$$
$$= c \int_{N_3} f(huh^{-1} \cdot hnh^{-1})\,\mathrm{d}n = \delta_{N_3}(h)^{-1} \int_{N_3} f(huh^{-1} \cdot n)\,\mathrm{d}n.$$

Consequently

$$\int_{P/G_u} f(pup^{-1})\delta_P(p)\, \mathrm{d}\overline{p} \leq c \int_{H/H_u} \delta_{N/N_3}(h)\, \mathrm{d}\overline{h} \int_{N_3} f(huh^{-1} \cdot n)\, \mathrm{d}n.$$

Since $f$ is compactly supported, the problem is to evaluate, for $F \in C_c^{\infty}(N_2/N_3)$, the integral

$$\int_{H/H_u} \delta_{N/N_3}(h) F(huh^{-1})\, \mathrm{d}\overline{h}.$$

## 2.35 Final Evaluation

By the isomorphism $H/H_u \simeq N^+ \subset N_2/N_3$, the measure $\mathrm{d}\overline{h}$ transfers to the invariant measure $\delta_{N_2/N_3}(h)^{-1}\, \mathrm{d}n$ on $N^+$, and so

$$\int_{H/H_u} \delta_{N/N_3}(h) F(huh^{-1})\, \mathrm{d}\overline{h} = \int_{N_2/N_3} \delta^*(n) F(n)\, \mathrm{d}n$$

where $\delta^*(huh^{-1}) = \delta_{N/N_2}(h)$. Every $x \in N^+ \subset N_2/N_3$ defines an isomorphism $\widetilde{x} : W^{-1}/W^0 \xrightarrow{\sim} W^1/W^2$. For $x = huh^{-1}$ we have a commutative diagram

$$\begin{array}{ccc} W^{-1}/W^0 & \xrightarrow[\widetilde{u}]{\sim} & W^1/W^2 \\ \Big\downarrow h & & \Big\downarrow h \\ W^{-1}/W^0 & \xrightarrow[\widetilde{x}]{\sim} & W^1/W^2. \end{array}$$

Then $h^{-1} : W^{-1}/W^0 \to W^{-1}/W^0$ is the transpose of $h : W^1/W^2 \to W^1/W^2$ for the Killing form. Hence $\delta_{N/N_2}(h) = c \cdot \det(\widetilde{x} : W^{-1}/W^0 \to W^1/W^2)^{1/2}$, and

$$\int_{N_2/N_3} \delta^*(n) F(n)\, \mathrm{d}n = \int_{N_2/N_3} \det(\widetilde{n})^{1/2} \cdot F(n)\, \mathrm{d}n < \infty.$$

## 2.36 Euler-Poincaré Functions

In the next subsections, we follow Kottwitz [Ko88] in computing the orbital integrals of the Euler-Poincaré function, which we now introduce.

Let $F$ be a non-Archimedean local field. Let $G$ be a connected reductive group over $F$. Assume that the connected component of 1 in the center of $G$ is anisotropic over $F$. Denote by $\mathbb{B}$ the building of $G(F)$ and by $\mathbb{F}$ the set of facets of $\mathbb{B}$. For $\sigma \in \mathbb{F}$, write $G(F)_\sigma = \{g \in G(F)\,;\, g\sigma = \sigma\}$ for the stabilizer of $\sigma$ in $G(F)$. This stabilizer is a compact open subgroup in $G(F)$. Any $g \in G(F)_\sigma$ permutes the vertices of $\sigma$.

Denote by $\mathrm{sgn}_\sigma(g)$ the sign of this permutation. We get a character $\mathrm{sgn}_\sigma : G(F)_\sigma \to \{\pm 1\}$. Extend $\mathrm{sgn}_\sigma$ to a function on $G(F)$ by 0 on $g \in G(F) - G(F)_\sigma$. Choose a Haar measure d$g$ on $G(F)$. Let $\mathbb{S}$ be a set of representatives for the orbits of $G(F)$ on $\mathbb{F}$. Define the *Euler-Poincaré function* on $G(F)$ to be

$$f_{EP} = \sum_{\sigma \in \mathbb{S}} (-1)^{\dim \sigma} |G(F)_\sigma|^{-1} \mathrm{sgn}_\sigma.$$

Here $|G(F)_\sigma|$ indicates the volume of $G(F)_\sigma$. Then $f_{EP} \in C_c^\infty(G(F))$. If d$g$ is multiplied by a scalar, then $f_{EP}$ is divided by that scalar. Changing $\mathbb{S}$ changes $f_{EP}$, but not its orbital integrals.

## 2.37 Poincaré Measures

Let $F$ be a local field of characteristic 0. Let $G$ be a connected reductive group over $F$. Let $\mu$ be an invariant measure on $G(F)$. Let $\Gamma$ be a discrete cocompact subgroup of $G(F)$. Write $\mu(\Gamma\backslash G(F))$ for the volume of $\Gamma\backslash G(F)$ with respect to the invariant measure on $\Gamma\backslash G(F)$ obtained from $\mu$. Serre [Sr71] showed that there exists an invariant measure $\mu_G$ on $G(F)$ having the following property. For every discrete, torsion-free, cocompact subgroup $\Gamma$ of $G(F)$, the Euler-Poincaré characteristic of $H^\bullet(\Gamma, \mathbb{Q})$ is equal to $\mu_G(\Gamma\backslash G(F))$. Borel and Harder [BH78] showed that $G(F)$ always has discrete, torsion-free, cocompact subgroups. Hence $\mu_G$ is uniquely determined by the property just stated. Serre names $\mu_G$ the *Euler-Poincaré measure* of $G(F)$. It can be negative or zero. It is nonzero if and only if $G$ has an anisotropic maximal $F$-torus. In the $p$-adic case, this happens if and only if the connected component of the identity in the center of $G$ is anisotropic [Kn65]. If $\mu_G \neq 0$, its sign is $(-1)^{q(G)}$, where $q(G)$ is the $F$-rank of the derived group $G^{\mathrm{ss}}$ of $G$ in the $p$-adic case, and half the dimension of the symmetric space $G^{\mathrm{ss}}/K$ attached to $G^{\mathrm{ss}}$ in the real case. See [Sr71, Propositions 23, 28].

## 2.38 Orbital Integrals of EP-Functions

Let $I$ denote the connected component $G_\gamma^0$ of the identity in the centralizer $G_\gamma$ of $\gamma$ in $G$. Write

$$O_\gamma(f) = \int_{I(F)\backslash G(F)} f(g^{-1}\gamma g)\, \mathrm{d}g/\mathrm{d}i$$

for the orbital integral of $f \in C_c^\infty(G(F))$ at $\gamma \in G(F)$.

THEOREM. *The orbital integral $O_\gamma(f_{EP})$ is* 0 *unless $\gamma$ is elliptic semisimple in $G(F)$. In the latter case, if di is the Euler-Poincaré measure on $I(F)$ (note that $di \neq 0$), then $O_\gamma(f_{EP}) = 1$. Further, the Shalika germ for the identity element e in $G(F)$ is identically* 1 *on every elliptic maximal torus of G.*

PROOF. (1) Let $\gamma$ be a semisimple element of $G(F)$. Let $\mathbb{F}(\gamma)$ be the set of facets $\tau \in \mathbb{F}$ fixed by $\gamma$; thus $\mathbb{F}(\gamma) = \{\tau \in \mathbb{F}\,;\, \gamma\tau = \tau\}$. Write $\mathbb{B}(\tau) = \{x \in \mathbb{B}\,;\, \gamma x = x\}$ for the set of points $x$ in the building $\mathbb{B}$ fixed by $\gamma$. For any $\tau$ in $\mathbb{F}(\gamma)$, put $\tau(\gamma) = \tau \cap \mathbb{B}(\gamma)$ for the set of points $x \in \tau$ fixed by $\gamma$. Then $\tau(\gamma) \neq \varnothing$, and $\mathbb{B}(\gamma)$ is a polysimplicial complex whose set of facets is $\{\tau(\gamma)\,;\, \tau \in \mathbb{F}(\gamma)\}$. One verifies that $\mathrm{sgn}_\tau(\gamma) = (-1)^{\dim(\tau)-\dim(\tau(\gamma))}$. If $f^0$ denotes the characteristic function of the maximal compact subgroup $K$ in $G(F)$, and the measure on $G(F)$ is normalized by $|K| = 1$, then

$$O_\gamma(f^0) = \sum_g |I(F)\backslash I(F)gK|; \quad g \in I(F)\backslash G(F)/K, \quad g^{-1}\gamma g \in K.$$

Put $x = gk \in \mathbb{B}$, and note that the summand is $|I(F)_x|^{-1}$, to get

$$O_\gamma(f^0) = \sum_x |I(F)_x|^{-1}, \quad x \in I(F)\backslash\mathbb{B}(\gamma).$$

To compute $O_\gamma(f_{EP}) = \sum_{\sigma\in\mathbb{S}}(-1)^{\dim\sigma}|G(F)_\sigma|^{-1}O_\gamma(\mathrm{sgn}_\sigma)$, we compute

$$O_\gamma(\mathrm{sgn}_\sigma) = \sum_g |I(F)\backslash I(F)gG(F)_\sigma| \cdot \mathrm{sgn}_\sigma(g^{-1}\gamma g), \quad g \in I(F)\backslash G(F)/G(F)_\sigma.$$

Put $\tau = g\sigma$. Then $g^{-1}\gamma g \in G(F)_\sigma$ means $g^{-1}\gamma g\sigma = \sigma$, namely $\gamma\tau = \tau$. Also, $\gamma \in gG(F)_\sigma g^{-1} = G(F)_\tau$, $\mathrm{sgn}_\sigma(g^{-1}\gamma g) = \mathrm{sgn}_\tau(\gamma)$, and $\dim\tau = \dim\sigma$. The volume factor in the summand is $|I(F)_\gamma|^{-1}|G(F)_\sigma|$, so we get

$$O_\gamma(\mathrm{sgn}_\sigma) = \sum_\tau |G(F)_\sigma||I(F)_\tau|^{-1}\mathrm{sgn}_\tau(\gamma), \quad \tau \in I(F)\backslash(I(F)\sigma) \cap \mathbb{F}(\gamma).$$

Hence

$$O_\gamma(f_{EP}) = \sum_{\sigma\in\mathbb{S}}(-1)^{\dim\sigma}\sum_\tau |I(F)_\tau|^{-1}\mathrm{sgn}_\tau(\gamma) = \sum_\rho |I(F)_\rho|^{-1}(-1)^{\dim\rho}$$

with $\rho = \tau(\gamma) \in I(F)\backslash\{\tau(\gamma) = \tau \cap \mathbb{B}(\gamma)\,;\, \tau \in \mathbb{F}(\gamma)\}$; thus $\rho$ runs through a set of representatives for the $I(F)$-orbits in the set of facets of $\mathbb{B}(\gamma)$.

If $\mathbb{B}(\gamma)$ is empty, the orbital integrals $O_\gamma(f_{EP})$ is 0. This can happen only if $\gamma$ is non-elliptic. If $\mathbb{B}(\gamma) \neq \varnothing$, then it is contractible [Ko80, Lemma 7.2], and it satisfies all the conditions of [Sr71, 3.3] relative to the group $I$. Note that Serre's condition (v) is a consequence of the semisimplicity of $\gamma$. Proposition 24 of [Sr71] asserts that $\left(\sum_\rho (-1)^{\dim \rho} |I(F)_\rho|^{-1}\right) di$ is the Euler-Poincaré measure on $I(F)$. If $\gamma$ is non-elliptic, then the Euler-Poincaré measure on $I(F)$ is 0, which implies that our orbital integral is also 0. If $\gamma$ is elliptic, we are taking $di$ to be the Euler-Poincaré measure on $I(F)$. This measure is nonzero. Hence $O_\gamma(f_{EP})$ is 1.

(2) At this point we proved the first and second sentences in the theorem for semisimple elements. We next show that the orbital integrals of $f_{EP}$ are 0 for non-semisimple elements. Let $\gamma$ be a semisimple element of $G(F)$. Let $I$ be the connected component of the identity $G_\gamma^0$ in its centralizer $G_\gamma$ in $G$. Recall the germ expansion for $f_{EP}$ about $\gamma$:

$$O_t(f_{EP}) = \sum_{u \in U} \Gamma_u(t) O_{\gamma u}(f_{EP}).$$

It holds for all regular semisimple $t$ in $G(F)$ that lie in $I(F)$ and are close enough to $\gamma$. For $u$ in the set $U$ of representatives for the conjugacy classes of unipotent elements in $I(F)$, we write $\Gamma_u$ for the germ associated with $u$. We have $O_t(f_{EP}) = 1$ if $t$ is regular elliptic, and $O_t(f_{EP}) = 0$ if $t$ is regular and not elliptic. Theorem 15 and Corollary 2 of Lemma 20 of [HC78] then imply that there is a real number $c$ with $O_t(f_{EP}) = c\Gamma_1(t)$. Hence $\sum_{u \in U} c_u \Gamma_u = 0$, where $c_u = O_{\gamma u}(f_{EP})$ if $u \neq 1$, and $c_1 = O_\gamma(f_{EP}) - c$. Then [HC78, Lemma 24] implies that $O_{\gamma u}(f_{EP}) = 0$ for all $u \neq 1$, and the first sentence of the theorem follows, as well as the third.

□

## 2.39 *Steinberg Representations*

The function $f_{EP}$ gives a pseudo-coefficient of the Steinberg representation of $G(F)$. Let $(\pi, V)$ be an irreducible admissible representation of $G(F)$. Denote by $H_e^i(G(F), V)$ the continuous cohomology groups of [BW80, X.5.1]. These are finite dimensional complex vector spaces, trivial except for finitely many values of $i$. In the $p$-adic case, we write $q(G)$ for the $F$-rank of the derived group $G^{ss}$, which is semisimple.

THEOREM. (a) *We have* $\operatorname{tr} \pi(f_{EP}) = \sum_i (-1)^i \dim H_e^i(G(F), V)$. *Moreover,* $(-1)^{q(G)} f_{EP}$ *is a pseudo-coefficient for the Steinberg representation of* $G(F)$.

(b) *Assume that* $G$ *is simple and* $\pi$ *is unitary. Then* $\operatorname{tr} \pi(f_{EP})$ *is* 0 *except in the following two cases. The trace of* $f_{EP}$ *on the trivial representation of* $G(F)$ *is* 1. *The trace of* $f_{EP}$ *on the Steinberg representation of* $G(F)$ *is* $(-1)^{q(G)}$.

PROOF. From the definition of $f_{EP}$, we have $\operatorname{tr} \pi(f_{EP}) = \sum_{\sigma \in \mathbb{S}} (-1)^{\dim \sigma} \dim V_\sigma$, where $V_\sigma$ denotes the biggest subspace of $V$ on which $G(F)_\sigma$ acts by the character $\operatorname{sgn}_\sigma$. By [BW80, X.2.4, X.5.1], the continuous cohomology of $V$ is the cohomology of a complex, denoted by $C^\bullet(Y; V)^G$ by Borel-Wallach, whose $i$th term is isomorphic to $\bigoplus_\sigma V_\sigma$. The sum is taken over the $\sigma \in \mathbb{S}$ with $\dim \sigma = i$. This gives the first statement of (a). The second statement of (a) follows from [BW80, XI.3.8], which asserts that the Steinberg representation of $G(F)$ is the only tempered representation of $G(F)$ which has nonzero continuous cohomology.

Statement (b) follows from [BW80, XI.3.9]. This says that when $G$ is simple and $\tau$ is unitary, then $H_e^i(G(F), V)$ is zero unless $i = 0$ and $V$ is the trivial representation, or $i = q(G)$ and $V$ is the Steinberg representation. In both cases $H_e^i(G(F), V)$ is one-dimensional. □

# 3 Automorphic Forms

## 3.1 Test Functions

Let $F$ be a global field. At each non-Archimedean place $v$, let $\mathcal{O}_v \subset F_v$ be the ring of integers. Fix a special maximal compact subgroup $K_v \subset G(F_v)$. At almost all $v$, we can and will take $K_v = G(\mathcal{O}_v)$ to be a hyperspecial maximal compact subgroup, see [Ti79].

Let $Z_{0\mathbb{A}}$ be a closed subgroup of $Z(\mathbb{A})$ such that $Z_{0\mathbb{A}}Z(F)$ is closed and $Z(\mathbb{A})/Z_{0\mathbb{A}}Z(F)$ is compact. Suppose that $Z_{0\mathbb{A}} = \prod_v Z_{0v}$, product over all places of $F$, with $Z_{0v}$ a closed subgroup of $Z(F_v)$. Set $Z_0 = Z_{0\mathbb{A}} \cap G(F)$. Note that $Z_{0\mathbb{A}}$, $Z_{0v}$, and $Z_0$ are *not* assumed to be the points of an algebraic group. However, any algebraic subgroup of $Z$ does satisfy these conditions.

Fix a unitary character $\omega$ of $Z_{0\mathbb{A}}/Z_0$ with local components $\omega_v$, which are characters of $Z_{0v}$. Fix also a place $v$. Let $C_c^\infty(G(F_v), Z_0(F_v), \omega)$ be the space of complex valued functions on $G(F_v)$ which transform under $Z_0(F_v)$ by $\omega_v^{-1}$, compactly supported modulo $Z_{0v}$, smooth if $v$ is Archimedean and locally constant if $v$ is non-Archimedean. To simplify the notation, we will denote this space by $C_c^\infty(G(F_v))$. We omit the place $v$ from the notation for the rest of this section, so that $F$ is a local field.

Fix a product measure $dx = \lambda \otimes_v dx_v$ on $G(\mathbb{A})/Z_{0\mathbb{A}}$ so that the product of volumes $|K_v/K_v \cap Z_{0v}|$ converges. Consider a function $f = \otimes_v f_v$ on $G(\mathbb{A})$, where $f_v \in C_c^\infty(G(F_v))$ for all $v$. Let $f_v^0$ denote the function, defined for each non-Archimedean place $v$ and supported on $Z_{0v}K_v$ with value $|K_v/K_v \cap Z_{0v}|^{-1}$ on $K_v$. We require $f_v = f_v^0$ at almost all places. Let $C_c^\infty(G(\mathbb{A}))$ be the space of functions spanned by all such $f$.

## 3.2 Kernel

Let $L = L(G(F)\backslash G(\mathbb{A}))$ be the space of complex valued functions $\phi$ on $G(F)\backslash G(\mathbb{A})$ which transform under $Z_{0\mathbb{A}}$ by the unitary character $\omega$ and are slowly increasing, see [BJ79], on $G(F)Z_{0\mathbb{A}}\backslash G(\mathbb{A})$. Define also $L^2 = L^2(G(F)\backslash G(\mathbb{A}))$ to be the space of complex valued functions $\phi$ on $G(F)\backslash G(\mathbb{A})$ which transform under $Z_{0\mathbb{A}}$ by $\omega$ and are absolutely square-integrable on $G(F)Z_{0\mathbb{A}}\backslash G(\mathbb{A})$. Let $r$ be the representation of $G(\mathbb{A})$ on $L(G(F)\backslash G(\mathbb{A}))$ or $L^2$ by right translation. Define the convolution operator $r(f\,\mathrm{d}x), f \in C_c^\infty(G(\mathbb{A}))$, on $L(G(F)\backslash G(\mathbb{A}))$ or $L^2$, by

$$(r(f\,\mathrm{d}x)\phi)(y) = \int_{G(\mathbb{A})/Z_{0\mathbb{A}}} f(x)\phi(yx)\,\mathrm{d}x.$$

This equals

$$\int_{G(\mathbb{A})/Z_{0\mathbb{A}}} f(y^{-1}x)\phi(x)\,\mathrm{d}x = \int_{G(F)\backslash G(\mathbb{A})/Z_{0\mathbb{A}}} \Big[\sum_{\gamma} f(y^{-1}\gamma x)\Big]\phi(x)\,\mathrm{d}x.$$

Hence it is an integral operator on $G(F)\backslash G(\mathbb{A})/Z_{0\mathbb{A}}$ with kernel

$$K(x,y) = K_f(x,y) = \sum_{\gamma \in G(F)/Z_0} f(x^{-1}\gamma y).$$

## 3.3 Admissible and Automorphic Representations

A complex representation of $G(F_v)$ is called *smooth* if the stabilizer $\mathrm{Stab}_{G(F_v)}(\xi) = \{g \in G(F_v);\, g\xi = \xi\}$ of any $\xi \in V$ is open in $G(F_v)$. It is called *admissible* if for all open $U \subset G(F_v)$ the subspace $V^U = \{\xi \in V;\, u\xi = \xi \text{ for all } u \in R\}$ of $U$-fixed vectors is finite dimensional. Theorem 3.25 of [BZ76] asserts that a smooth $G(F_v)$-module of finite length is admissible. Conversely, an admissible representation of $G(F_v)$ is smooth and of finite length.

An irreducible admissible $G(\mathbb{A})$-module $(\pi, V)$ is the restricted direct product $\pi = \otimes_v \pi_v$ of irreducible $G(F_v)$-modules $\pi_v$ on the spaces $V_v$, almost all of which are *unramified*. That is, almost all have a (unique up to scalar) nonzero $K_v$-fixed vector $\xi_v^0$. All $\pi_v$ are admissible and unitarizable if $\pi$ is unitarizable (a dense submodule is a unitary module). Thus $V = \cup_S \otimes_{v\in S} V_v \otimes \otimes_{v\notin S}\xi_v^0$, union over all finite sets $S$ of valuations of $F$ whose complement consists of $v$ where $\pi_v$ is unramified.

An irreducible unitary $G(\mathbb{A})$-module $(\pi, V)$ is called *automorphic* if it is a subquotient of the $G(\mathbb{A})$-module $L(G(F)\backslash G(\mathbb{A}))$.

## 3.4 Convolution Operators

For an admissible representation $(\pi, V)$, we define

$$\pi(f) = \pi(f\,\mathrm{d}x) = \int_{G(\mathbb{A})/Z_{0\mathbb{A}}} f(x)\pi(x)\,\mathrm{d}x$$

on $V$, with d$x$ the measure above. The space $V$ is spanned by vectors $\otimes_v \xi_v$ with $\xi_v$ in the space of $\pi_v$ and it is $K_v$-invariant for almost all $v$. For almost all $v$, the operator $\pi_v(f_v)$ is the projection on the space of $K_v$-fixed vectors, so has $\operatorname{tr} \pi_v(f_v) = 1$. Hence almost all factors in $\operatorname{tr} \pi(f) = \prod_v \operatorname{tr} \pi_v(f_v)$ are 1.

## 3.5 Cuspidal Spectrum

A function $\phi \in L(G(F)\backslash G(\mathbb{A}))$ is *cuspidal* if for any proper $F$-parabolic subgroup of $G$ with unipotent radical $N$, the integral

$$\int_{N(F)\backslash N(\mathbb{A})} \phi(nx)\, dn$$

is 0 for any $x \in G(\mathbb{A})$. Let $r_0$ be the restriction of $r$ to the space $L_0 = L_0(G(F)\backslash G(\mathbb{A}))$ of cuspidal functions. A standard result, see, e.g., [AFOO], is that the $G(\mathbb{A})$-module $L_0(G(F)\backslash G(\mathbb{A}))$ is the direct sum of irreducible $G(\mathbb{A})$-modules $\pi$ with finite multiplicities $m(\pi)$. The operator $r_0(f)$ is trace class, and

$$\operatorname{tr} r_0(f) = \sum_{\pi} m(\pi) \operatorname{tr} \pi(f). \tag{3.5.1}$$

The sum is over the equivalence classes of $\pi$ in $L_0(G(F)\backslash G(\mathbb{A}))$. It converges absolutely and all $\pi$ are unitary. Note that $L_0$ coincides with the analogous space of cusp forms in $L^2$.

## 3.6 Stable Conjugacy

The difference between the notions of conjugacy and stable conjugacy can be studied using Galois cohomology. See Serre [Sr94], Ribes [Ri99], Platonov-Rapinchuck [PR93], Tate [Ta66], Kottwitz [Ko86].

Elements $x, x'$ of $G(F)$ are (resp. *stably*) *conjugate* if there exists $y$ in $G(F)$ (resp. $G(\overline{F})$) such that $x' = \operatorname{Ad}(y)x = yxy^{-1}$. Here $F$ can be local or global. Let $A(x/F)$ be the set of $y \in G(\overline{F})$ with $\operatorname{Ad}(y)x$ in $G(F)$. The conjugacy classes within the stable class of $x$ are parametrized by $B(x/F) = G(F)\backslash A(x, F)/G_x$. The map

$$y \mapsto \{\tau \mapsto y_\tau = y^{-1}\tau(y)\,;\ \tau \in \operatorname{Gal}(\overline{F}/F)\}$$

is a bijection

$$B(x/F) \xrightarrow{\sim} \ker[H^1(F, G_x) \to H^1(F, G)].$$

Here $H^1(F,A) = H^1(\mathrm{Gal}(\overline{F}/F), A(\overline{F}))$. Thus, given $x$, any $x'$ stably conjugate to $x$ determines an element of $B(x,F)$, and $x'$ is conjugate to $x$ if and only if it determines the identity in $H^1(F,G_x)$. If $F$ is global, define $B(x/\mathbb{A})$ (resp. $B(x/\mathbb{A}^w)$) to be the pointed direct sum of $B(x/F_v)$ for all $v$ (resp. $v \neq w$).

# 4 Trace Formula

## 4.1 Projection to Cuspidal Spectrum

The notation will be as in Section 3. Let $u$ be a place of $F$. The function $f_u \in C_c^\infty(G(F_u))$ is called *cuspidal* if for any proper $F_u$-parabolic of $G(F_u)$ with unipotent radical $N(F_u)$, we have

$$\int_{N(F_u)} f_u(xny)\,\mathrm{d}n = 0$$

for any $x$, $y$ in $G(F_u)$.

PROPOSITION. *If $f$ has a cuspidal component at $u$, then $r(f)$ vanishes on the $G(\mathbb{A})$-invariant complement of $L_0(G(F)\backslash G(\mathbb{A}))$ in $L(G(F)\backslash G(\mathbb{A}))$. In particular, we have* $\mathrm{tr}\, r(f) = \mathrm{tr}\, r_0(f)$ *for such* $f$.

PROOF. Set $PG = G(\mathbb{A})/Z_{0\mathbb{A}}$. Then

$$\int_{N(F)\backslash N(\mathbb{A})} (r(f)\phi)(nx)\,\mathrm{d}n = \int_{N(F)\backslash N(\mathbb{A})} \int_{PG} f(y)\phi(nxy)\,\mathrm{d}y\,\mathrm{d}n$$

$$= \int_{N(F)\backslash N(\mathbb{A})} \int_{N(F)\backslash PG} \sum_{\gamma\in N(F)} f(x^{-1}n^{-1}\gamma y)\phi(y)\,\mathrm{d}y\,\mathrm{d}n$$

$$= \int_{N(F)\backslash PG} \left( \int_{N(F)\backslash N(\mathbb{A})} \sum_{\gamma\in N(F)} f(x^{-1}n^{-1}\gamma y)\,\mathrm{d}n \right) \phi(y)\,\mathrm{d}y$$

$$= \int_{N(F)\backslash PG} \left( \int_{N(\mathbb{A})} f(x^{-1}ny)\,\mathrm{d}n \right) \phi(y)\,\mathrm{d}y = 0.$$

The change in order of integration is justified by absolute convergence, since $f$ is compactly supported on $PG$ and $N(F)\backslash N(\mathbb{A})$ is compact. Now $r(f)$ preserves $L_0(G(F)\backslash G(\mathbb{A}))$ and its complement. So if $\phi$ lies in the complement, then $r(f)\phi \in L_0(G(F)\backslash G(\mathbb{A}))$ implies $r(f)\phi$ is 0. Hence $\mathrm{Im}\, r(f) = \mathrm{Im}\, r_0(f)$, and $\mathrm{tr}\, r(f) = \mathrm{tr}\, r_0(f)$. □

Let $F$ be a global field of characteristic 0.

PROPOSITION 4.2. *Let $C = \prod_v C_v$ be a compact subset of $G(\mathbb{A})$ with $C_v = G(\mathcal{O}_v)$ for almost all $v$. Then there are only finitely many regular conjugacy classes in $G(\mathbb{A})$ with a representative in $G(F)$ which intersect $C$ nontrivially.*

PROOF. Fix a faithful representation of $G$ in $\mathrm{GL}(n)$ over $F$, for some $n$. Define a map $G(\mathbb{A}) \to \mathbb{A}^{n-1} \times \mathbb{A}^\times$ by sending $x$ to the ordered set of coefficients of its characteristic polynomial. The image of $C$ is compact; that of $G(F)$ is discrete. Thus there are only finitely many semisimple conjugacy classes in $\mathrm{GL}(n, \mathbb{A})$ with a representative in $G(F)$ which intersect $C$ nontrivially. Two semisimple conjugacy classes in $G(F)$ which are conjugate in $\mathrm{GL}(n, \mathbb{A})$ are conjugate in $\mathrm{GL}(n, F)$. The theorem of [St74, p. 102] asserts that a conjugacy class of $\mathrm{GL}(n, \overline{F})$ intersects $G(\overline{F})$ in only finitely many conjugacy classes of $G(\overline{F})$. By definition, a $G(\overline{F})$-conjugacy class with a representative in $G(F)$ is a stable conjugacy class. If $\gamma_G$ is a stable conjugacy class in $G(F)$, then there exists a finite set $V$ of places of $F$ such that $\gamma_G$ intersects $G(\mathcal{O}_v)$ at most at one conjugacy class for $v$ outside $V$. This $\gamma_G$ is contained in a stable conjugacy class $\gamma_{\mathbb{A}}$, where $\gamma_{\mathbb{A}}$ is the product over all $v$ of stable conjugacy classes $\gamma_v$ in $G(F_v)$. Since $\gamma_v$ consists of finitely many classes for all $v$, the class $\gamma_G$ consists of only finitely many conjugacy classes in $G(F)$ which intersect $C$. □

## 4.3 Geometric Side

Suppose $f$ is as in Proposition 4.1. Then $r(f)$ is a trace class operator, whose trace is the integral of the kernel over the diagonal. Assume in addition $f$ vanishes on the conjugacy class in $G(\mathbb{A})$ of any $\gamma \in G(F)$ which is not elliptic regular. Note that if $\gamma$ is elliptic regular, then $\left|G_\gamma(\mathbb{A})Z(\mathbb{A})/G_\gamma(F)Z_{0\mathbb{A}}\right|$ is compact and its volume is finite. But this fails for other elements $\gamma$. Put $\widetilde{G}_\gamma$ for the centralizer of $\gamma$ in $G/Z$. Put $\{\gamma\}$ for the set of conjugacy classes of elliptic regular elements in $G(F)$. By the conditions on $f$, the sum below ranges only over $\{\gamma\}$. Thus

$$\begin{aligned}
\operatorname{tr} r(f) &= \int_{G(\mathbb{A})/Z_{0\mathbb{A}}G(F)} \sum_{\gamma \in G(F)/Z_0} f(x\gamma x^{-1})\,\mathrm{d}x \\
&= \sum_{\{\gamma\}} \int_{G(\mathbb{A})/Z_{0\mathbb{A}}\widetilde{G}_\gamma(F)} f(x\gamma x^{-1})\,\mathrm{d}x \qquad (4.3.1)\\
&= \sum_{\{\gamma\}} \left(\left|G_\gamma(\mathbb{A})/G_\gamma(F)Z_{0\mathbb{A}}\right| / [\widetilde{G}_\gamma(F) : G_\gamma(F)]\right) \Phi(\gamma, f).
\end{aligned}$$

Each integral in (4.3.1) is absolutely convergent. The sum is finite by Proposition 4.2. Note that the integral in the middle over $G(\mathbb{A})/Z_{0\mathbb{A}}\widetilde{G}_\gamma(F)$ can be written as $1/[\widetilde{G}_\gamma(F) : G_\gamma(F)]$ times the integral over $G(\mathbb{A})/Z_{0\mathbb{A}}G_\gamma(F) = G(\mathbb{A})/G_\gamma(\mathbb{A}) \times G_\gamma(\mathbb{A})/G_\gamma(F)Z_{0\mathbb{A}}$.

The following is our simple trace formula.

COROLLARY 4.4. *Suppose that $u$, $u'$, and $u''$ are places of $F$ with $u \neq u'$, with $f_u$ a cuspidal function, the orbital integral of $f_{u'}$ vanishes on the regular non-elliptic set of $G(F_{u'})$, and $f_{u''}$ vanishes on the singular set. Then* (3.5.1) *equals* (4.3.1)*, and the sum in* (4.3.1) *is finite.*

PROOF. Proposition 4.2 implies that if $f(x\gamma x^{-1}) \neq 0$ for $x \in G(\mathbb{A})$, then $\gamma$ lies in one of finitely many regular conjugacy classes. Suppose $\gamma$ lies in such a regular non-elliptic class. Then the invariant distribution $\Phi(\gamma) : h \mapsto \Phi(\gamma, h)$ on $C_c^\infty(G(F_{u'}))$ vanishes at $f_{u'}$. Let $C_0^\infty(G(F_{u'}))$ be the span of functions $h - h^g$, with $h \in C_c^\infty(G(F_{u'}))$ and $g \in G(F_{u'})$. Denote by $C_0^\infty(G(F_{u'}))_\gamma$ and $C_c^\infty(G(F_{u'}))_\gamma$ the restrictions of these sets to the orbit of $\gamma$. The uniqueness of the $G(F_{u'})$-invariant measure on the orbit of $\gamma$ means that any distribution on $C_c^\infty(G(F_{u'}))_\gamma / C_0^\infty(G(F_{u'}))_\gamma$ is a scalar multiple of $\Phi(\gamma)$. Thus $C_c^\infty(G(F_{u'}))_\gamma / C_0^\infty(G(F_{u'}))_\gamma$ is one-dimensional, and $C_0^\infty(G(F_{u'}))$ is the kernel of $\Phi(\gamma)$. Hence there are $h_i \in C_c^\infty(G(F_{u'}))$ and $g_i \in G(F_{u'})$ such that

$$f_{u'} = \sum_i (h_i - h_i^{g_i})$$

on the orbit of $\gamma$. We may choose $h_i$ to be zero outside a small neighborhood of $\gamma$.

Replacing in $f$ the component $f_{u'}$ by $f_{u'} - \sum_i (h_i - h_i^{g_i})$ will not change side (3.5.1) of the trace formula, since $\operatorname{tr} \pi(h^g) = \operatorname{tr} \pi(h)$. On the other hand, the function $f$ now vanishes on the orbit of $\gamma$, but its values on other conjugacy classes with a rational representative do not change. Consequently we may assume that if $f(x\gamma x^{-1}) \neq 0$, then $\gamma$ is elliptic regular, and the corollary follows from 4.3. □

REMARK. (1) The fact that $f_{u'}$ can be *any* function whose orbital integrals vanish on the regular non-elliptic set of $G(F_{u'})$, and it is not assumed that it is supported on the elliptic regular set, is fundamental for the primary applications of this chapter.

(2) Cuspidal functions are obtained as linear combinations of matrix coefficients of cuspidal representations of local groups.

We give next a different form of the corollary. It is used in the next section. Also it provides an example of a simple trace formula derived from the full (noninvariant) trace formula of Chapter 3 below. By the $F$-*rank* of $G$, we mean the dimension of the quotient of a maximal $F$-split torus in $G$ by the $F$-split component of a maximal $F$-torus in $Z$.

COROLLARY 4.5. *Let $f = \otimes_v f_v$, $f_v \in C_c^\infty(G(F_v))$, be a function whose components at $u_i$, $0 \leq i \leq r$ with $r \geq \operatorname{rank} G$, are supported on the elliptic regular set of $G(F_{u_i})$, and $f_{u_i}$ is zero on the $x$ in $G(F_{u_i})$ for which there are $g$ in $G(F_{u_i})$ and $z \neq 1$ in $Z(F_{u_i})$ with $gxg^{-1} = zx$. Then*

$$\sum_{\{\gamma\}} \left| G_\gamma(\mathbb{A}) Z(\mathbb{A}) / G_\gamma(F) Z_0(\mathbb{A}) \right| \Phi(\gamma, f) = \sum_\pi c_\pi \operatorname{tr} \pi(f). \tag{4.5.1}$$

*The sum over $\{\gamma\}$ is finite. It ranges over the conjugacy classes of regular $x$ in $G(F)$ which are elliptic at the $u_i$. The sum over $\pi$ is absolutely convergent. It ranges over automorphic $G(\mathbb{A})$-modules. The $c_\pi$ are complex numbers.*

PROOF. The assumption at $u_1$ alone implies that the sum $\sum J_{\mathfrak{o}}(f)$ is equal to our sum over $\{\gamma\}$. It is finite by Proposition 4.2. The sum $\sum J_\chi(f)$ consists of integrals of logarithmic derivatives of intertwining operators acting on induced representations. As the degrees of the derivatives are at most rank $G$, our $r+1$ assumptions imply the vanishing of all integrals. There remains a discrete sum of irreducible representations $\pi$ whose components at $u_i$ are elliptic. The $c_\pi$ are integral and positive for cuspidal $\pi$. □

# 5 Density

To illustrate the power and typical usage of the simple trace formula, we give here a first application, the following density theorem, due to Kazhdan [Ka86.1, Appendix]. We use various notions and facts to be used in later sections, but we prefer to give the application first, begging the reader to take on faith several standard facts, to motivate by means of this example the value of the simple trace formula. The density theorem was conjectured by Harish-Chandra, who was looking for a purely local proof. The usage of the trace formula permits playing local against global techniques, to learn more about both. We shall have two forms of the density theorem. In the first, all (irreducible, admissible) $\pi$ appear. In the second, the assumption is made only for tempered $\pi$.

Let $F$ be a local field of characteristic 0 and $G$ a reductive group over $F$, as in Section 2.

PROPOSITION 5.1. *Let $f$ be a function in $C_c^\infty(G(F))$ such that* $\operatorname{tr}\pi(f) = 0$ *for all admissible irreducible $G(F)$-modules $\pi$. Then $\Phi(x,f) = 0$ for all regular $x$ in $G(F)$.*

REMARK 5.2. Consequently $J^\infty(G(F))$ (defined in Subsection 2.6 to be the space of $f \in C_c^\infty(G(F))$ such that $I(g,f)$ is 0 for all regular $g \in G(F)$) consists of all $f$ in $C_c^\infty(G(F))$ such that $f_G(\pi) = \operatorname{tr}\pi(f)$ is 0 for every $G(F)$-module $\pi$.

PROOF. In the proof, we denote $F$, $G$, and $f$ by $F'$, $G'$, and $f'$. Due to the integration formula $I(x,f') = I^M(x,f'_N)$ (Section 2), we may assume that there exists an elliptic regular element $x_0$ in $G'(F)$ with $\Phi(x_0,f') \neq 0$ and that its centralizer $T'$ in $G'$ is an elliptic torus over $F'$ which splits over the Galois extension $F''$ over $F'$.

We prove the following lemma. It is needed to apply global methods to the local problem.

LEMMA 5.3. *Let $F'$ be a local field, $G'$ a reductive group over $F'$, $T'$ a (maximal) torus of $G'$ over $F'$, and $F''$ a Galois field extension of $F'$ such that $T'$ and $G'$ split over $F''$. Then there exists a Galois extension $E/F$ of global fields such that at a set of places $w$ of $F$ of cardinality at least two we have $F_w \simeq F'$, $E_w = E \otimes_F F_w \simeq F''$,*

*$\mathrm{Gal}(E_w/F_w) \simeq \Gamma$, where $\Gamma = \mathrm{Gal}(E/F)$, and a pair $(T, G)$ consisting of a reductive group $G$ and a torus $T$ over $F$ with $G(F_w) \simeq G'(F_w)$, $T(F_w) \simeq T'(F_w)$ for all $w$ such that $G(F)$ is dense in $G(F_w)$ and $T(F)$ is dense in $T(F_w)$.*

PROOF. By Hensel's Lemma, it is clear that there exist $E$ and $F$ with the required properties. Once $(T, G)$ is found, since the set of $w$ has cardinality at least two, it follows from [CF68, p. 361] that $(T(F), G(F))$ is dense in $(T(F_w), G(F_w))$. Now, it is well known (see [Sr94, p. III-1]) that if $K/k$ is a Galois field extension, $A$ is a torus in an algebraic group $H$, both defined over $k$, then the set of $K/k$-forms of $(A, H)$ is parametrized by the first cohomology group $H^1(\mathrm{Gal}(K/k), \mathrm{Aut}_K(A, H))$ of $\mathrm{Gal}(K/k)$ in the group $\mathrm{Aut}_K(A, H)$ of automorphisms of the pair $(A, H)$. This group $\mathrm{Aut}_K(A, H)$ consists of automorphisms of $H$ over $K$ which map $A$ to $A$. The group $A(K)$ of $K$-points injects as a normal subgroup of $\mathrm{Aut}_K(A, H)$. Denote the quotient by $W_K$.

Let $(A, H)$ be a pair consisting of a reductive group $H$ over $F$ with $H(E_w) \simeq G'(E_w)$ and a torus $A$ of $H$ over $F$ with $A(E_w) \simeq T'(E_w)$. We have

$$\begin{array}{ccccccccc} 0 & \longrightarrow & A(E) & \longrightarrow & \mathrm{Aut}_E(A, H) & \longrightarrow & W_E & \longrightarrow & 0 \\ & & \downarrow & & \downarrow & & \downarrow \wr & & \\ 0 & \longrightarrow & A(E_w) & \longrightarrow & \mathrm{Aut}_{E_w}(A, H) & \longrightarrow & W_{E_w} & \longrightarrow & 0 \end{array}$$

Since $A(E)$ is normal in $\mathrm{Aut}_E(A, H)$, by [Sr94, Prop. 38, p. I-6], we have the associated commutative diagram

$$\begin{array}{ccccccc} W_E^\Gamma & \longrightarrow & H^1(\Gamma, A(E)) & \longrightarrow & H^1(\Gamma, \mathrm{Aut}_E(A, H)) & \xrightarrow{p} & H^1(\Gamma, W_E) \\ \downarrow \wr & & \downarrow \psi & & \downarrow \phi & & \downarrow \wr \\ W_{E_w}^\Gamma & \longrightarrow & H^1(\Gamma, A(E_w)) & \longrightarrow & H^1(\Gamma, \mathrm{Aut}_{E_w}(A, H)) & \xrightarrow{p_w} & H^1(\Gamma, W_{E_w}). \end{array}$$

The Tate-Nakayama theory [Ta66] and [Ko86] implies that $\psi$ is surjective.

The pair $(T', G')$ is determined by an element $\alpha_w$ in $H^1(\Gamma, \mathrm{Aut}_{E_w}(A, H))$. To produce a pair $(T, G)$ as required, we have to find an element $\alpha$ in $H^1(\Gamma, \mathrm{Aut}_E(A, H))$ whose image under $\phi$ is $\alpha_w$. Put $\beta = p_w(\alpha_w)$. It can be regarded as an element of $H^1(\Gamma, W_E)$. As in [Sr94], denote by ${}_{\alpha_w}A$ the torus determined by the cocycle $\alpha_w$. Since it depends only on $\beta$, we denote ${}_{\alpha_w}A$ by ${}_\beta A$. For each $\gamma$ in $H^1(\Gamma, W_E)$, there exists an element $\Delta(\gamma)$ in $H^2(\Gamma, {}_\gamma A(E))$ (constructed in [Sr94, p. I-70]), such that $\gamma$ lies in the image of $p_w$ if and only if $\Delta_w(\gamma_w) = 0$ (see [Sr94, Prop. 4, p. I-70]). Also, for each $\gamma_w$ in $H^1(\Gamma, W_E)$, there is $\Delta_w(\gamma_w)$ in $H^2(\Gamma, {}_{\gamma_w} A(E_w))$ such that $\gamma_w$ lies in the image of $p_w$ if and only if $\Delta_w(\gamma_w) = 0$. The Tate-Nakayama theory

[Ta66] implies that $H^2(\Gamma,_\beta A(E))$ and $H^2(\Gamma,_\beta A(E_w))$ are isomorphic as groups. By their construction (in [Sr94, p. I-70]), $\Delta = \Delta_w$. Since $\beta = p_w(\alpha_w)$, we have $\Delta_w(\beta) = 0$; hence $\Delta(\beta) = 0$, and $\beta$ lies in the image of $p$. By [Sr94, Cor. 2, p. I-67], the inverse image by $p_w$ of $\beta$ is the quotient of $H^1(\Gamma,_\beta A(E_w))$ by Im $W^\Gamma_{E_w}$, and $p^{-1}(\beta)$ is $H^1(\Gamma,_\beta A(E))/$ Im $W^\Gamma_E$. The Tate-Nakayama theory [Ta66] implies that the map $H^1(\Gamma,_\beta A(E)) \to H^1(\Gamma,_\beta A(E_w))$ is surjective. Hence there is $\alpha$ in $H^1(\Gamma, \mathrm{Aut}_E(A,H))$ with $\phi(\alpha) = \alpha_w$. The pair $(T, G)$ determined by $\alpha$ has the required properties, and the lemma follows. □

Let $E/F$ be a global field extension and $(T, G)$ a pair defined over $F$ with the properties of the lemma. In these notations $T_w = T'$ is the centralizer $Z_{G(F_w)}(x_0)$ of $x_0$ in $G(F_w)$. We have $T = T(F)$ dense in $T(F_w)$ and similarly for $G(F)$ in $G(F_w)$. Hence the centralizer $Z_{G(F_w)}(T(F_w))$ of $T(F_w)$ in $G(F_w)$ is equal to the centralizer $Z_{G(F_w)}(T(F))$ of $T(F)$ in $G(F_w)$ and contains the centralizer $Z_{G(F)}(T(F))$ of $T(F)$ in $G(F)$ as a dense subset. Choose $x$ in $Z_{G(F)}(T(F))$ sufficiently near $x_0$ so that $T(F)$ is $Z_{G(F)}(x)$ and $\Phi(x, f_w) \neq 0$. Here we denote our local function $f'$ by $f_w$.

As in [Ka86.1, Appendix], we use an argument from Galois cohomology. Note that $T$ is a maximal torus (in the usual sense) in $G$. The set of $G(F_v)$-conjugacy classes in $G(F_v)$ which are contained in the $G(\overline{F}_v)$-conjugacy class of $\gamma$ is in bijective correspondence (see Subsection 3.6) with a subset of

$$H^1(F_v, T) = H^1(\mathrm{Gal}(\overline{F}_v/F_v), T(\overline{F}_v)).$$

A similar assertion holds for $G(F)$-conjugacy classes. Let $E/F$ be a finite Galois extension which is unramified outside a sufficiently large set $V = \{v_1, v_2, \ldots, v_k\}$ of places of $F$ and over which $T$ splits. Then $H^1(F_v, T)$ equals $H^1(\mathrm{Gal}(E_w/F_v), T(E_w))$, and Tate-Nakayama theory (see [Ta66] and [Ko86]) provides an isomorphism between this group and

$$\{\lambda^\vee \in X_*(T)\,;\, \mathrm{Norm}_{E_w/F_v}(\lambda^\vee) = 0\}/\{\lambda^\vee - \sigma\lambda^\vee\,;\, \lambda^\vee \in X_*(T), \sigma \in \mathrm{Gal}(E_w/F_v)\}, \tag{5.3.1}$$

and an isomorphism between

$$H^1(\mathrm{Gal}(E/F), T(\mathbb{A}_E)/T(E))$$

and

$$\{\lambda^\vee \in X_*(T)\,;\, \mathrm{Norm}_{E/F}(\lambda^\vee) = 0\}/\{\lambda^\vee - \sigma\lambda^\vee\,;\, \lambda^\vee \in X_*(T), \sigma \in \mathrm{Gal}(E/F)\}. \tag{5.3.2}$$

Here $w$ stands for a fixed valuation on $E$ which lies above a given $v$. Moreover, there is an exact sequence

$$H^1(\mathrm{Gal}(E/F), T(E)) \to \oplus_v H^1(\mathrm{Gal}(E_w/F_v), T(E_w)) \to H^1(\mathrm{Gal}(E/F), T(\mathbb{A}_E)/T(E)).$$

The first map is compatible with the embedding of $G(F)$-conjugacy classes into $\prod_v G(F_v)$, and the second arrow is given by the natural map

$$\oplus_v \lambda_v \mapsto \sum_v \lambda_v$$

from the direct sum of modules (5.3.1) into (5.3.2). Now, consider the conjugacy class of $x$. Any $x' \in G(F)$ with $\Phi(x',f) \neq 0$ maps to an element $\oplus_v \lambda_v$ such that $\sum_v \lambda_v = 0$. If $v$ is one of the valuations $v_2, \ldots, v_k$, then $x'$ is $G(F_v)$-conjugate to $x$, so that $\lambda_v = 0$. If $v$ lies outside $V$, then $x'$ is $G(F_v)$-conjugate to an element in $K_v^G$. Since $(G,T)$ is unramified at $v$, we again have $\lambda_v = 0$ [Ko86, Proposition 7.1]. It follows that $\lambda_{v_1} = 0$. In other words, $x'$ is $G(F_{v_1})$-conjugate to $\gamma$.

Let us rephrase this argument. The Tate-Nakayama theory [Ta66] and [Ko86] implies that the natural homomorphism from $H^1(F,T)$ to $H^1(\mathbb{A}^w,T)$ is an isomorphism, where $T = Z_G(x)$ is a torus and $H^1(\mathbb{A}^w,T)$ is the pointed direct sum of the groups $H^1(F_v,T)$ over all places $v \neq w$. If $x'$ is an element of $G(F)$ which is stably conjugate to $x$ in $G(F_v)$ for some place $v$, namely, $x$ and $x'$ are conjugate on $G(\overline{F}_v)$, then they are conjugate in $G(F')$ where $F'$ is a finite extension of $F$, and hence in $G(\overline{F})$. Consequently they are stably conjugate in $G(F)$. If $x'$ is an element of $G(F)$ which is conjugate to $x$ in $G(F_v)$ for all $v \neq w$, then it determines the identity element in $H^1(\mathbb{A}^w,T)$, hence in $H^1(F,T)$, and hence it is conjugate to $x$.

Let $V$ be a finite set of place of $F$ where $T$ is elliptic, of cardinality larger than the rank of $G$, not including the place $w$ of the proposition. At each $v$ in $V$, choose $f_v$ in $C_c^\infty(G(F_v))$ which is supported on the elliptic regular set of $G(F_v)$ and with $\Phi(x,f_v) \neq 0$. Choose $f$ in $C_c^\infty(G(\mathbb{A}))$ whose components at $v$ in $V$ are those chosen above and whose component at $w$ is the function of the proposition. As noted in Proposition 4.2 there are only finitely many conjugacy classes in $G(\mathbb{A})$ with representative $x'$ in $G(F)$, necessarily elliptic regular, with $\Phi(x',f) \neq 0$. We can replace finitely many of the components $f_v$ (for $v \neq w$) of $f$ by their product with the characteristic function of a small open and closed neighborhood of the orbit of $x$ in $G(F_v)$ to assure that if $\Phi(x',f) \neq 0$ for $x'$ in $G(F)$, then $x'$ is conjugate to $x$ in $G(F_v)$ for all $v \neq w$. Consequently, if $\Phi(y,f) \neq 0$ for $y$ in $G(F)$, then it is conjugate to $x$.

We can now apply the trace formula identity (4.5.1) of Corollary 4.5 since $f$ is chosen to satisfy the hypotheses of this corollary. The assumption of the proposition implies that the right side of (4.5.1) is equal to 0, since $\operatorname{tr} \pi(f) = 0$ for all $\pi$, while the left side of (4.5.1) is a nonzero scalar multiple of $\Phi(x,f)$. Since $\Phi(x,f_v) \neq 0$ for all $v \neq w$ by the choice of $f_v$, we conclude $\Phi(x,f_w) = 0$, as required. □

REMARK 5.4. Kazhdan proved the Proposition in [Ka86.1, Appendix] using only the simple form of the trace formula given in Corollary 4.4. To be able to use it, he produced a cuspidal function $f_u$ with $\Phi(x,f_u) \neq 0$ for the given $x$ in $G_u$, so that $\Phi(x',f_u) = 0$ for any $x'$ in $G_u$ which is stably conjugate but not conjugate to $x$ in $G_u$. This construction holds in the non-twisted case that we consider here, as well as in the case of base change, at a place which splits. However, in other twisted cases, it is more difficult to construct cuspidal functions. For example, there are no twisted-invariant cuspidal irreducible representations of $G_u = \mathrm{PGL}(3, F_u)$ if $F_u$ has odd residual characteristic and the twisting is given by $x \mapsto J\,{}^t x^{-1} J$, $J = \text{antidiagonal}(1, -1, 1)$.

DEFINITION 5.5. *An irreducible admissible representation $\pi$ of a local group G on a complex space V is called* unitarizable *if V makes a dense subspace in a unitary representation $V'$. Then $V'$ is a Hilbert space, with inner product $\langle .,. \rangle$. A* matrix coefficient *of such $\pi$ is a function on G of the form $f(g) = \langle \pi(g)\xi, \xi' \rangle$, where $\xi$, $\xi'$ are nonzero vectors in V. Then $\pi$ is called* square-integrable *if $|f|^2$ is integrable on $G/Z$. Since $\pi$ is irreducible, by Schur's Lemma* ([BZ76]), *it has a central character, $\omega_\pi$, which is unitary since $\pi$ is unitarizable. For example, cuspidal and Steinberg representations are square-integrable. A $\pi$ is called* tempered *if it is a constituent, necessarily a subrepresentation, of a representation $i_P(\sigma) = \mathrm{ind}_P^G(\delta_P^{1/2}\sigma)$, normalizedly induced (using $\delta_P^{1/2}$, see Subsection* 6.7 *below) from a square-integrable representation $\sigma$ of the Levi factor M of a parabolic subgroup $P = MN$ of G. In particular, square-integrable representations are tempered.*

PROPOSITION 5.6. *For $f$ in $C_c^\infty(G(F))$ with $\mathrm{tr}\,\pi(f) = 0$ for all tempered $G(F)$-modules $\pi$, we have $\Phi(x,f) = 0$ for all regular x in $G(F)$.*

REMARK 5.7. (i) Theorem 10 of [HC78] asserts that $I(f)$ is uniquely determined by its values on the regular set. Using this, the propositions imply that $\Phi(x,f) = 0$ for *all* $x$ in $G(F)$.

(ii) The propositions are proven for a local field $F$ of characteristic 0. They also hold for a local field $F$ of positive characteristic by virtue of Theorem A of [Ka86.1]. Moreover, [Ka86.1, Theorem B] implies that if $\mathrm{tr}\,\pi(f) = 0$ for all $G(F)$-modules $\pi$, then $f$ lies in the linear span of the commutators $[f_1, h_1] = f_1 * h_1 - h_1 * f_1$, where $f_1$ and $h_1$ lie in $C_c^\infty(G(F))$.

PROOF. As in Section 2, a minimal parabolic subgroup $P_0 = M_0N_0$ is fixed. Here $P = MN$ denotes a parabolic subgroup containing $P_0$ such that $M$ contains $M_0$. Let $A = A_M$ be the split component in the center of $M$. Let $\mathfrak{a} = \mathfrak{a}_P = \mathrm{Hom}(X(M)_\mathbb{Q}, \mathbb{R}) = \mathrm{Hom}(X(A)_\mathbb{Q}, \mathbb{R})$ where $X(M)_\mathbb{Q}$, $X(A)_\mathbb{Q}$ are the groups of rational characters of $M$, $A$ defined over $\mathbb{Q}$ and $H : M \to \mathfrak{a}$ is the homomorphism defined by $\langle H(m), \chi \rangle = \log|\chi(m)|$ for all $\chi$ in $X(M)_\mathbb{Q}$. If $\tau$ is an irreducible $M(F)$-module with central character $\omega_\tau$, define $\lambda_\tau$ in $\mathfrak{a}^* = X(M)_\mathbb{Q} \otimes \mathbb{R}$ by $\langle \lambda_\tau, H(m) \rangle = \log|\omega_\tau(m)|$. This $\tau$ is called *positive* if $\langle \lambda_\tau, \alpha \rangle$ is positive for every root $\alpha$ of $A$ in $N$. It is called *essentially tempered* if $\tau \otimes \chi$ is tempered for some $\chi$ in $X(M)_\mathbb{Q}$. The classification theorem of [BW80, XI, (2.11)] (or [La88, Si80]) used negative instead of positive, and quotient instead of subrepresentations. An equivalent formulation, established by Zelevinsky [Ze80] for $\mathrm{GL}(n)$, asserts then the following.

(i) If $\tau$ is essentially tempered and positive, then the unitarily induced $G(F)$-module $i_P(\tau)$ has a unique irreducible submodule $J_P(\tau)$.
(ii) Any irreducible $G(F)$-module is so obtained.
(iii) $J_P(\tau)$ is equivalent to $J_{P'}(\tau')$ if and only if $P = P'$ and $\tau$ is equivalent to $\tau'$.

A $G(F)$-module is called *standard* if it is equivalent to $i_P(\tau)$ with a positive $M(F)$-module $\tau$. By virtue of the relation $\operatorname{tr}(i_P(\tau))(f) = \operatorname{tr}\tau(f_N)$ (which follows from a standard computation of a character of an induced representations), the fact that $\operatorname{tr}\tau(f_N) = 0$ if and only if $\operatorname{tr}(\tau\otimes\chi)(f_N) = 0$ for any $\chi$ in $X(M)_{\mathbb{Q}}$, and the relation $I(m,f) = I^M(m,f_N)$ of orbital integrals for $m$ in $M(F)$ regular in $G$, the proposition follows at once from Proposition 5.1 and the following lemma.

Let $R_{\mathbb{Z}}(G(F))$ be the Grothendieck (free abelian) group generated by $\operatorname{Irr}(G(F))$, the set of equivalence classes of (admissible) irreducible $G(F)$-modules. Put also $R(G(F)) = R_{\mathbb{Z}}(G(F)) \otimes \mathbb{C}$.

LEMMA 5.8. *The set of standard $G(F)$-modules is a basis of $R(G(F))$ over $\mathbb{Z}$.*

PROOF. Given an irreducible $G(F)$-module $\pi$, it is equivalent to $J_P(\tau)$ for some pair $(P,\tau)$. If $\pi'$ is a submodule of $i_P(\tau)$ inequivalent to $\pi$, and $\pi' = J_{P'}(\tau')$, then $\lambda_{\tau'} < \lambda_\tau$ for the order $<$ on $\mathfrak{a}^*$ by [BW80, XI, (2.13)]. By [BZ77], $\pi'$ and $\pi$ have the same cuspidal datum $(L,\varepsilon)$, consisting of a Levi subgroup $L$ and an irreducible $L(F)$-module $\varepsilon$ whose modules $\varepsilon_U = r_{L',U}\varepsilon$ of coinvariants (see Section 7) with respect to any parabolic subgroup $L'U$ of $L$ are zero in the Grothendieck group $R(L')$ of $L'$. Hence $\pi'$ lies in a fixed finite set, and by induction on $\lambda_\pi$, we may assume that each such $\pi'$ is a linear combination over $\mathbb{Z}$ of standard $G(F)$-modules. Consequently $\pi = J_P(\tau) = i_P(\tau) - \sum \pi'$ also lies in the span of the standard $G(F)$-modules.

It remains to show that standard modules are linearly independent. Fixing a cuspidal datum $(M,\tau)$, it is shown above that all irreducible $G(F)$-modules attached to $(M,\tau)$ are linear combinations of standard $G(F)$-modules attached to $(M,\tau)$, and we obtain a (finite, square) unipotent matrix. Since irreducible $G(F)$-modules are linearly independent over $\mathbb{C}$, the standard $G(F)$-modules are linearly independent over $\mathbb{C}$, and the lemma follows. □

The proposition now follows. □

# 6 Characters

## 6.1 Invariant Distributions

Let $F$ be a non-Archimedean local field of characteristic 0. We now recall some of the fundamental results of Howe [Ho74], Harish-Chandra [HC78], and Kazhdan [Ka86.1] about Fourier transforms of invariant distributions, characters, and orbital integrals. We will not reproduce the proofs. Some of the proofs, in particular those of [Ka86.1], rely on results from the next two sections. However, their statements are independent of those results. We prefer to record these fundamental statements first, as they clarify the relationship between characters and orbital integrals, and delay the study of induction and restriction.

Fix an $F$-valued symmetric non-degenerate $G(F)$-invariant bilinear form $B$ on the Lie algebra $\mathfrak{g}$ of $G(F)$, a nontrivial additive character $\psi$ of $F$ and a Haar measure $\mathrm{d}X$ on $\mathfrak{g}$. Let $C_c^\infty(\mathfrak{g})$ denote the space of locally constant compactly supported functions on $\mathfrak{g}$. The Fourier transform $\phi \mapsto \widehat{\phi}$, where

$$\widehat{\phi}(X) = \int_{\mathfrak{g}} \psi(B(X,Y))\phi(Y)\,\mathrm{d}X,$$

is a linear bijection of $C_c^\infty(\mathfrak{g})$ onto itself. A *distribution* $T$ on $\mathfrak{g}$ is a linear complex-valued function on $C_c^\infty(\mathfrak{g})$. Its Fourier transform $\widehat{T}$ is defined by $\widehat{T}(\phi) = -T(\widehat{\phi})$. For $x$ in $G(F)$, put $\phi^x(X) = \phi(\mathrm{Ad}(x)X)$ and ${}^xT(\phi) = T(\phi^x)$. A distribution $T$ is *invariant* if ${}^xT = T$ for all $x$ in $G(F)$.

A distribution $T$ *vanishes* on an open set $U$ if it attains the value 0 at any $\phi \in C_c^\infty(\mathfrak{g})$ which is supported on $U$. If $T$ vanishes on a family of open subsets $U_\alpha$, it vanishes on $\cup_\alpha U_\alpha$, since $\phi$ is compactly supported. The *support* of $T$ is the complement of the largest open set on which $T$ vanishes. Given a set $\omega$ in $\mathfrak{g}$, let $J(\omega)$ be the space of all invariant distributions on $\mathfrak{g}$ which are supported on the closure of $\mathrm{Ad}(G)\omega$.

PROPOSITION 6.2. *If $\omega$ is compact and $T$ lies in $J(\omega)$, then there exists a locally integrable function $F$ on $\mathfrak{g}$ with*

$$T(\phi) = \int_{\mathfrak{g}} F(X)\phi(X)\,dX$$

*for all $\phi$ in $C_c^\infty(\mathfrak{g})$.*

PROOF. See [HC78, Theorem 3]. □

## 6.3 Measures on Nilpotent Orbits

Let $G_X$ be the centralizer of $X$ in $G$. Let $\mathrm{d}x$ be the unique (up to scalar) $G(F)$-invariant measure on the homogeneous space $G(F)/G_X(F)$. By Theorem 2.32 of Deligne and Rao [Ra72], the integral

$$\mu_{\mathcal{O}}(\phi) = \int_{G(F)/G_X(F)} \phi(\mathrm{Ad}(x)X)\,\mathrm{d}x$$

is well defined for $\phi \in C_c^\infty(\mathfrak{g})$. This is clear for regular $X$, whose orbit is closed, but not for unipotent orbits. It depends only on $\mathrm{d}x$ and the orbit $\mathcal{O} = \mathrm{Ad}(G)X$ of $X$. The Fourier transform $\widehat{\mu}_{\mathcal{O}}$ of the measure $\mu_{\mathcal{O}}$ is a function by Proposition 6.2.

Let $\mathfrak{N}$ be the set of nilpotent elements in $\mathfrak{g}$. It is a union of finitely many ("nilpotent") $G(F)$-orbits. Let $\omega$ be a compact subset of $\mathfrak{g}$. The local behavior of the Fourier transform of $T$ in $J(\omega)$ is described by the following:

PROPOSITION 6.4. *There exists a $G(F)$-domain $D$ (an open closed $G(F)$-invariant subset) of $\mathfrak{g}$ which contains 0 and a "nilpotent" distribution $\mu$ (a linear combination with complex coefficients, depending on $T$, of the unique up to a scalar multiple invariant measures $\mu_{\mathcal{O}}$ supported on $\mathcal{O}$, where $\mathcal{O}$ are the nilpotent orbits) so that $\widehat{T} = \widehat{\mu}$ on $D$.*

PROOF. See [HC78, Theorem 4]. □

## 6.5 Local Integrability of Characters

Fix a Haar measure $dx$ on $G(F)$. For a smooth $G(F)$-module $(\pi, V)$, we defined the endomorphism $\pi(f)$ of $V$ by

$$\pi(f) = \pi(f\,dx) = \int_{G(F)} f(x)\pi(x)\,dx.$$

Clearly $\pi(f)$ depends linearly on $dx$. If $\pi$ is admissible, then $\pi(f)$ has finite rank. In this case, we denote the trace by $\operatorname{tr}\pi(f)$. It is easy to see that if $\pi$ is admissible and irreducible, then there exists a complex-valued conjugacy invariant locally constant function $\chi$ on the *regular* set of $G(F)$ such that

$$\operatorname{tr}\pi(f) = \operatorname{tr}\pi(f\,dx) = \int_{G(F)} f(x)\chi(x)\,dx$$

for every $f$ in $C_c^\infty(G(F))$ which is supported on the *regular* set of $G(F)$. The function $\chi$ is called the *character* of $\pi$. Note that $\operatorname{tr}\pi(f)$ depends on $dx$, but $\chi(x)$ is independent of $dx$.

PROPOSITION. *The character $\chi$ of an irreducible admissible $G(F)$-module $\pi$ is a locally integrable function on $G(F)$. In particular,* $\operatorname{tr}\pi(f) = \int f(x)\chi(x)\,dx$ *for* every $f \in C_c^\infty(G(F))$.

PROOF. See [HC78, Theorem 1]. □

PROPOSITION 6.6. *Suppose that $\sigma \in G(F)$ is a semisimple element. Let $G_\sigma(F)$ and $\mathfrak{g}_\sigma$ be the centralizers of $\sigma$ in $G(F)$ and $\mathfrak{g}$. Then there exists a neighborhood $V$ of 0 in $\mathfrak{g}_\sigma$ and a $G_\sigma(F)$-invariant "nilpotent" distribution $\mu$ on $\mathfrak{g}_\sigma$, so that $\chi(\sigma \exp X) = \widehat{\mu}(X)$ for all $X$ in $V$.*

PROOF. See [HC78, Theorem 5]. □

The above results by Harish-Chandra [HC78] are based on the technique developed by Howe [Ho74] in the case of $\mathrm{GL}(n)$. Kazhdan [Ka86.1] showed that the above local behavior in fact characterizes the characters and orbital integrals, at least on the elliptic set. This characterization extends to the entire (not necessarily elliptic) set $G(F)$ by Proposition 8.2 below.

## 6.7 Induction

To describe Kazhdan's theory, let $S$ be the space of conjugation invariant functions $s$ on $G$ such that, for every semisimple $\sigma$ in $G(F)$, there is a neighborhood $V$ of 0 in the Lie algebra $\mathfrak{g}_\sigma$ of the centralizer $G_\sigma(F)$ in $G(F)$ and a $G_\sigma(F)$-invariant distribution $\mu$ on $\mathfrak{g}_\sigma$ supported on the nilpotent set of $\mathfrak{g}_\sigma$, so that $s(\sigma \exp X) = \widehat{\mu}(X)$ for all regular $X$ in $V$. Let $\eta(X)$ be the coefficient of the smallest possible power of $t$ in the polynomial $\det(t - \mathrm{ad}(X))$, $X \in \mathfrak{g}$. Then $\eta$ is a nonzero polynomial function on $\mathfrak{g}$, and $X$ is called *regular* if $\eta(X) \neq 0$.

Let $S_e(G(F))$ be the space of functions on the elliptic subset of $G(F)$ obtained by restriction of the functions in $S$.

Let $\Pi(G(F))$ be the set of equivalence classes of admissible irreducible $G(F)$-modules. We let $R_{\mathbb{Z}}(G(F))$ be the integral Grothendieck group of $G(F)$, that is of the category of smooth $G(F)$-modules, namely, the free abelian group generated by $\Pi(G(F))$. It is the group of virtual semisimplifications of the admissible representations of $G(F)$. Put $R(G(F)) = R_{\mathbb{Z}}(G(F)) \otimes_{\mathbb{Z}} \mathbb{C}$ for the Grothendieck group of $G(F)$.

Let $M$ be a Levi subgroup of a parabolic subgroup $P = MN$ with unipotent radical $N$. Denote by $i_M^G \tau$ the $G(F)$-module normalizedly induced (i.e., twisted by $\delta_P^{1/2}$) from the smooth $M(F)$-module $\tau$, which is trivially extended on $N(F)$ to $P(F)$. If $V_\tau$ denotes the space of $\tau$, then $i_M^G \tau$ acts on

$$V = \{\phi : G(F) \to V_\tau;\ \phi(mngu) = \delta_P^{1/2}(m)\tau(m)\phi(g),\ u \in K_\phi,\\ m \in M(F),\ n \in N(F),\ g \in G(F)\}$$

—here $K_\phi$ is an open compact subgroup of $G(F)$ depending on $\phi$—by right shifts: $((i_M^G \tau)(g))\phi(h) = \phi(hg)$. We write $\mathrm{ind}_M^G \tau$ for the unnormalized induced representation (thus with $\delta_P$ omitted).

Then $i_M^G$ extends to functor from the category of smooth $M(F)$-modules to the category of smooth $G(F)$-modules. Its restriction is a homomorphism from $R(M(F))$ to $R(G(F))$. We denote by $R_I(G(F))$ the space in $R(G(F))$ of the images of $i_M^G$ over all $M \neq G$. Put $\overline{R}(G(F))$ for the quotient $R(G(F))/R_I(G(F))$. Denote by $\chi$ the character of a member $\pi$ in $R(G(F))$. It is a finite linear combination with complex coefficients of characters of irreducible $G(F)$-modules.

PROPOSITION 6.8. *The map* $\overline{R}(G(F)) \to S_e(G(F))$, $\pi \mapsto \chi = \chi_\pi$, *is an isomorphism.*

PROOF. See [Ka86.1, Theorem D]. □

In particular, any function on the elliptic set of $G(F)$ whose local behavior is given by the defining property of $S$ is the restriction to the elliptic set of a character of a virtual $G(F)$-module.

## 6.9 Orbital Integrals on the Elliptic Set

Theorem C of [Ka86.1] gives another characterization of $S_e(G(F))$. Let $A_c^\infty(G(F))$ be the space of $f$ in $C_c^\infty(G(F))$ whose orbital integrals vanish on the regular non-elliptic set. As in Subsection 2.6, let $J_c^\infty(G(F))$ be the space of functions in $C_c^\infty(G(F))$ whose orbital integrals vanish on the regular set of $G(F)$. Let $\overline{A}_c^\infty(G(F))$ be the quotient $A_c^\infty(G(F))/J_c^\infty(G(F))$. We can now state Theorem C of [Ka86.1].

PROPOSITION. *The map $\overline{A}_c^\infty(G(F)) \to S_e(G(F)), f \mapsto \widetilde{\Phi}(f)$, is an isomorphism.*

See Subsection 2.4 for the definition of $\widetilde{\Phi}(x,f)$.
The isomorphism defined by Propositions 6.8 and 6.9 yields the following:

COROLLARY 6.10. *The space $\overline{A}_c^\infty(G(F))$ and $\overline{R}(G(F))$ are isomorphic.*

PROPOSITION 6.11. *Let $G$ be the multiplicative group of a simple algebra. Then $S_e(G(F))$ consists of the locally constant functions on the elliptic set of $G(F)$.*

PROOF. In this case, $G = M(m,D)^\times$, where $D$ is a division algebra of rank $d$ central over $F$. The group $G$ is an inner form of the split group $G' = \mathrm{GL}(n)$, $n = md$. A stable conjugacy class in $G(F)$ consists of a single (rational) conjugacy class. A semisimple conjugacy class $\gamma$ in $G(F)$ is determined by its characteristic polynomial $p_\gamma$ (which has coefficients in $F$). A unipotent conjugacy class determines a conjugacy class of Levi subgroups, namely, a partition $\alpha = (m_i)$ of $m$. Here the $m_i$ are positive integers and $\sum m_i = m$, $m_i \geq m_{i+1}$. There is a natural injection of the set of conjugacy classes in $G(F)$ into the set of conjugacy classes in $G'(F)$, denoted by $\gamma \mapsto \gamma'$ and defined by $p_{\gamma'} = p_\gamma$ and $(m_i) \mapsto (dm_i)$. Similarly there is an injection of the nilpotent classes in the Lie algebra $M(m,D)$ of $G$ into the set of such classes in $M(n,F)$. The nilpotent orbit $\mathcal{O}$ in $M(m,D)$ determines the partition $\alpha$ of $m$, and the corresponding standard (upper triangular) parabolic subgroup of $G$ is denoted by $P_\alpha$. Put $\mathcal{O} = \mathcal{O}_\alpha$ and $\widehat{\mu}_\alpha$ for $\widehat{\mu}_{\mathcal{O}}$. Let $\theta_\alpha$ be the character of the $G(F)$-module unitarily induced from the trivial $P_\alpha(F)$-module. Lemma 5 of [Ho74] asserts that there is a small neighborhood $V$ of zero in $M(m,D)$ such that the Fourier transform $\widehat{\mu}_\alpha$ at $X$ is equal to $\theta_\alpha(\exp X)$ for all $X$ in $V$. The result is stated there only for $\mathrm{GL}(n)$, but the proof applies to any $G$ as here. This is zero on the set of elliptic regular $\exp X$ if $\alpha$ is not the trivial partition $(m)$ of $m$. Moreover, the character $\theta_{(m)}$ is identically one. Since the centralizer of any elliptic element in $G$ is of the form $\mathrm{GL}(m', D')$, where $D'$ is a central simple algebra over a field extension $F'$ of $F$, the proposition follows. □

Combining this result (for $G(F)$ and $G'(F)$) with Proposition 6.9, we obtain the following:

COROLLARY 6.12. *For every $f$ in $A_c^\infty(G(F))$, there exists $f'$ in $A_c^\infty(G'(F))$, and for every $f'$ in $A_c^\infty(G'(F))$, there exists $f$ in $A_c^\infty(G(F))$, such that $\widetilde{\Phi}(\gamma,f) = \widetilde{\Phi}(\gamma',f')$ for all regular $\gamma$ in $G(F)$ and $\gamma'$ in $G'(F)$ with $p_\gamma = p_{\gamma'}$.*

This proves the assumptions 12.1 and 12.2 below in the special case of our $G(F)$ and $G'(F)$.

# 7 Coinvariants

## 7.1 *Module of Coinvariants*

Let $F$ be a local non-Archimedean field. Let $G$ be as in 2.1. Let $(\pi, V)$ be an admissible $G(F)$-module. Then it is of finite length. Let $P = MN$ be an $F$-parabolic subgroup with a Levi subgroup $M$ and unipotent radical $N$. Then the quotient of $V$ by the subspace generated by $\{\pi(n)\xi - \xi\,;\, \xi \in V, n \in N(F)\}$ is an $M(F)$-module $\pi'_N$, since $M(F)$ normalizes $N(F)$. Denote by $\pi''_N$ the image of $\pi'_N$ in the Grothendieck group $R(M(F))$. The (normalized) *$M(F)$-module $\pi_N$ of $N(F)$-coinvariant of $\pi$* is defined to be $\delta_P^{-1/2}\pi'_N$. It is shown in [BZ76] that if $\pi$ is admissible, then $\pi_N$ is admissible. Hence it is of finite length. The construction extends to a functor $r_M^G : \pi \mapsto \pi_N$, from the category $K(G(F))$ of smooth $G(F)$-modules to the category $K(M(F))$ of smooth $M(F)$-modules. This functor is exact. Let $i_M(\tau)$ be the $G(F)$-module $i_P^G(\tau) = \mathrm{ind}_P^G(\delta_P^{1/2}\tau)$ induced from the $P(F) = M(F)N(F)$-module $\delta_P^{1/2}\tau \otimes 1$. This construction defines a functor $i_M^G : \tau \mapsto i_M(\tau)$, from $K(M(F))$ to $K(G(F))$. It is exact. Frobenius reciprocity, see [BZ76], asserts that $\mathrm{Hom}_G(\pi, I_M(\tau)) = \mathrm{Hom}_M(\pi_N, \tau)$ for all irreducible $M(F)$-modules $\tau$ and $G(F)$-modules $\pi$. Hence $\pi_N \neq 0$ implies that there is a nonzero morphism of $G$-modules from $\pi$ to $I_M(\pi_N)$.

## 7.2 *Reduction of Characters to Levi Subgroups*

We will now introduce notation needed to state a theorem of Deligne-Casselman [De76, Cas77]. Let $A$ be a maximal $F$-split torus in $G$, $B$ a minimal parabolic subgroup of $G$ containing $A$, $\Delta$ the set of roots of $A$ in $B$. Fix a lattice $L$ in $A(F)$ so that $|\alpha(\lambda)| = 1$ if and only if $\alpha(\lambda) = 1$ for all $\lambda$ in $L$ and $\alpha$ in $\Delta$, and so that $A(F)/L$ is compact. Put $L^-$ for the set of $\lambda$ in $L$ with $|\alpha(\lambda)| \leq 1$ for all $\alpha$ in $\Delta$. For any semisimple $t$ in $G(F)$, there exists a positive integer $m$ and $y \in G(F)$ so that $yt^m y^{-1}\lambda s$, where $\lambda$ lies in $L^-$ and $s$ is a compact element in $G(F)$ (the closure of the group generated by $s$ is compact). Let $P_\lambda$ be the standard (containing $B$) parabolic subgroup of $G$ whose Levi component $M_\lambda$ is the centralizer $G_\lambda$ in $G(F)$, and put $P_t = M_t N_t$ for $y^{-1}P_\lambda y$.

The definition of the parabolic $P_t$ is the same as in [Cas77]. To recall the definition of [Cas77], put $A_\theta = \cap_{\alpha\in\theta} \ker\alpha$ for any subset $\theta$ of $\Delta$. Denote by $A^-$ the set of $x$ in $A$ with $|\alpha(x)| \leq 1$ for all $\alpha$ in $\Delta$. Given a semisimple $t$ in $G(F)$ with $yt^m y^{-1} = as$ for $a$ in $A^-(F)$ and a compact element $s$, let $\theta$ be the set of $\alpha$ in $\Delta$ with $|\alpha(a)| = 1$. Denote by $M_\theta$ the centralizer in $G$ of the torus $A_\theta$. Then $M_\theta = M_\lambda$.

We can now state the following theorem of Casselman and Deligne.

PROPOSITION 7.3. *Let $\pi$ be an admissible irreducible representation of $G(F)$. Let $t$ be a regular element of $G(F)$ so that $P = P_t$ is a parabolic subgroup. Then $\chi_\pi(t) = \chi_{\pi'_N}(t)$ $(= \chi_{\pi''_N}(t))$. Since $\Delta(t) = \Delta_M(t)\delta_P(t)^{-1/2}$, we have $(\Delta\chi_\pi)(t) = (\Delta_M\chi_{\pi_N})(t)$ for such $t$.*

Here $\chi_\pi$ denotes the character of $\pi$. Deligne discovered the result where $\pi$ is cuspidal [De76]. The extension to a general $\pi$ is in [Cas77].

## 7.4 Reduction of Traces to Levi Subgroups

We now recall Lemma 5.1 of [Cas77]. Suppose $t$ is in $A^-$. The associated $\lambda$ in $L$ then lies in $L^-$. We put $P = P_t = MN$. Let $C$ be an open compact congruence subgroup of $G(F)$ with the properties of $K_i$ in [Cas77, Lemma 2.1], and in particular

$$C = (C \cap \overline{N}(F))(C \cap M(F))(C \cap N(F)),$$

where $\overline{N}$ is the unipotent radical of the parabolic $P_{t^{-1}} = \overline{P}$ opposite to $P$. Let $f_t$ be the function in $C_c^\infty(G(F))$ supported on $Z(F)CtC$ which attains the value $|Z(F)CtC/Z(F)|^{-1}$ on $CtC$. Let $f_t^M$ be the function on $M(F)$ which is supported on $t(C \cap M(F))Z(F)$, transforms under $Z(F)$ by $\omega_\pi^{-1}$ (where $\omega_\pi$ is the central character of $\pi$), and attains the value $\delta_P^{1/2}(t)/|C \cap M(F)|$ on $t(C \cap M(F))$.

The following is Lemma 5.1 of [Cas77].

PROPOSITION 7.5. *We have* $\operatorname{tr}\pi(f_t) = \operatorname{tr}\pi_N(f_t^M)$ *for any $G(F)$-module $\pi$.*

## 7.6 Multiplicities

This proposition will be used below as follows. The $\tau$ be an irreducible constituent of the $M(F)$-module $\pi_N$. Denote its central character by $\omega_\tau$ and its character by $\chi_\tau$. We are interested in the function $f_t^M$ on $M(F)$ since

$$\operatorname{tr}\tau(f_t^M) = \int \chi_\tau(tx) f_t^M(tx)\,dx = \omega_\tau(t)\delta_P^{1/2}(t)\operatorname{tr}\tau(1_C^M)$$

for $t$ in the center of $M(F)$, where $1_C^M$ is the function in $C_c^\infty(M(F))$, transforming under $Z(F)$ by $\omega_\pi^{-1}$ and equal to $1/|C \cap M(F)|$ on $C \cap M(F)$. Then $\operatorname{tr}\tau(1_C^M)$ is the (non-negative integral) multiplicity of the trivial representation of $C \cap M(F)$ in $\tau$. On the other hand, $f_t$ is a $C$-bi-invariant function, where $C$ is independent of $t$. Thus $\operatorname{tr}\pi(f_t) \neq 0$ only for $\pi$ which have a nonzero $C$-invariant vector.

Let $W(M(F), G(F)) = N(M(F), G(F))/M(F)$ be the quotient by $M(F)$ of the normalizer $N(M(F), G(F))$ of $M(F)$ in $G(F)$, where $P(F) = M(F)N(F)$ is a parabolic subgroup of $G(F)$.

PROPOSITION 7.7. *Let $x$ be a regular element in $G(F)$. Then the orbital integral $I(x,f_t)$ vanishes unless $x$ is conjugate to an element of $M(F)$. For $x$ in $M(F)$, we have*

$$I(x,f_t) = \sum_{w \in W(M(F),G(F))} I^M(wxw^{-1},f_t^M).$$

REMARK. The proof of this proposition relies on Corollary 8.13. It is given here since the functions $f_t$ and $f_t^M$ do not appear in Section 8, and it is clear that the work of Section 8 does not depend on Proposition 7.7.

PROOF. Corollary 8.13 implies that given $f_t^M$, there exists a function $f$ on $G(F)$ such that $I(x,f) = 0$ unless $x$ is conjugate in $G(F)$ to an element of $M(F)$, and when $x$ lies in $M(F)$, then

$$I(x,f) = \sum_{w \in W(M(F),G(F))} I^M(wxw^{-1},f_t^M).$$

The Weyl integration formula and Proposition 7.3 imply that $\operatorname{tr}\pi(f) = \operatorname{tr}\pi_N(f_t^M)$ for every admissible $G(F)$-module $\pi$, since the parabolic subgroup $P_x$ associated with any element $x$ in the support of $I(x,f)$ is $P_t = P$. On the other hand, Proposition 7.5 implies that $\operatorname{tr}\pi(f_t) = \operatorname{tr}\pi_N(f_t^M)$, hence $\operatorname{tr}\pi(f_t) = \operatorname{tr}\pi(f)$, for every admissible $G(F)$-module $\pi$. But then Proposition 5.1 implies that $I(x,f) = I(x,f_t)$ for every regular $x$ in $G(F)$, and the proposition follows. □

# 8 Trace Functions

## 8.1 Orbital Integrals of Elliptic Elements

Let $F$ be a local non-Archimedean field, $G$ and $C_c^\infty(G(F))$ as before. We continue to let $J_c^\infty(G(F))$ be the space of $f$ in $C_c^\infty(G(F))$ whose orbital integrals vanish at each regular element in $G(F)$ and $A_c^\infty(G(F))$ the space of $f$ in $C_c^\infty(G(F))$ whose orbital integrals vanish on every regular non-elliptic element in $G(F)$, $\overline{C}_c^\infty(G(F)) = C_c^\infty(G(F))/J_c^\infty(G(F))$ and $\overline{A}_c^\infty(G(F)) = A_c^\infty(G(F))/J_c^\infty(G(F))$.

Our final aim in this section is to prove the following:

PROPOSITION 8.2. *Let $M$ be a Levi subgroup of $G$ and $f^M$ an element of $C_c^\infty(M(F))$ with the following property. For every $m$, $m'$ in $M(F)$ which are regular in $G(F)$ and conjugate to each other by $G(F)$, we have*

$$I^M(m,f^M) = I^M(m',f^M). \tag{8.2.1}$$

*Then there exists $f$ in $C_c^\infty(G(F))$ with $f_M = f^M$ and $f_L = 0$ for every Levi subgroup $L$ of $G$ which does not contain a conjugate of $M$. See 2.6 for $f_M$.*

This proposition, which concerns "lifting" of orbital integrals from a Levi subgroup of $G(F)$ to $G(F)$ itself, is proven below using representation theoretic techniques, in the spirit of Corollary 6.10. We will use the trace Paley-Wiener theorem of [BDK86] and [F95] and the geometric lemma of [BZ77, (2.12)], which we now proceed to state.

## 8.3 Trace Paley-Wiener Theorem

As in Section 6, let $R_{\mathbb{Z}}(G(F))$ denote the integral Grothendieck group of the category of smooth $G(F)$-modules, namely, the group of virtual representations of $G(F)$ of finite length (the free abelian group with basis $\Pi(G(F))$). Set $R(G(F)) = R_{\mathbb{Z}}(G(F)) \otimes \mathbb{C}$. Let $i_M^G : R(M(F)) \to R(G(F))$ be the induction homomorphism. As in Subsection 7.1, let $r_M^G : R(G(F)) \to R(M(F))$ be the coinvariants homomorphism. Let $X(G(F))$ be the group of unramified characters of $G(F)$. Then $X(G(F))$ acts naturally on $\Pi(G(F))$ and $R(G(F))$ by $\psi : \pi \mapsto \pi\psi$.

Recall that an unramified character of $F^\times$ has the form $x \mapsto |x|^s$, with $s \in \mathbb{C}/(2\pi i\mathbb{Z}/\ln q)$, since $|u| = 1$ for the units $u \in F^\times$, and $|\boldsymbol{\pi}| = q^{-1} = e^{-\ln q}$ for any generator $\boldsymbol{\pi}$ of the maximal ideal $(\boldsymbol{\pi})$ in the ring $R$ of integers in $F$, where $q = R/(\boldsymbol{\pi})$ is the residual cardinality. Thus $X(F^\times) \simeq \mathbb{C}/(2\pi i\mathbb{Z}/\ln q) \simeq \mathbb{C}^\times$; the last isomorphism is via $z \mapsto q^z$.

The set $X(G(F))$ has a natural structure of a complex algebraic group, isomorphic to $(\mathbb{C}^\times)^d$, where $d = d(G(F)) = \dim X(G(F))$. As usual, fix a Haar measure $dx$ on $G(F)$. We work in this section with the space $C_c^\infty(G(F))$ of smooth compactly supported functions on $G(F)$. Passing to the space $C_c^\infty(G(F), Z(F), \chi)$ as before is easy. Each function $f$ in $C_c^\infty(G(F))$ defines a linear form $\beta_f : R(G(F)) \to \mathbb{C}$ by $\beta_f(\pi) = \operatorname{tr} \pi(f)$. The form $\beta = \beta_f$ satisfies the following two conditions:

(PW(1)) For any Levi subgroup $M$ and irreducible $M(F)$-module $\tau$, the function $\psi \mapsto \beta(i_M^G(\tau\psi))$ is a regular function on the complex algebraic variety $X(M(F))$.

(PW(2)) There exists an open compact subgroup $K$ in $G(F)$ which dominates $\beta$, in the sense that $\beta$ vanishes on each $G(F)$-module $\pi$ which has no nonzero $K$-fixed vector.

Let $R^*(G(F)) = \operatorname{Hom}_{\mathbb{C}}(R(G(F)), \mathbb{C}) = \operatorname{Hom}(\Pi(G(F)), \mathbb{C})$ be the space of all linear forms on $R(G(F))$. A form $\beta : R(G(F)) \to \mathbb{C}$ is called *good* if it satisfies conditions (PW(1)) and (PW(2)). The form is called *trace* if $\beta = \beta_f$ for some $f \in C_c^\infty(G(F))$. We denote the spaces of good and trace forms by $F_{\text{good}} = F_{\text{good}}(G(F))$ and $F_{\text{tr}} = F_{\text{tr}}(G(F))$. We can now state the primary theorem of [BDK86]. See [F95] for a different proof and for a twisted analogue.

THEOREM 8.4 (Trace Paley-Wiener). *For every p-adic reductive group $G(F)$, we have $F_{\text{tr}} = F_{\text{good}}$.*

This theorem describes the image of the natural morphism $\mathrm{tr} : C_c^\infty(G(F)) \to R^*(G(F))$. As noted at the end of Section 5, Proposition 5.1 and Theorem B of [Ka86.2] imply that $\ker \mathrm{tr} = [C_c^\infty(G(F)), C_c^\infty(G(F))]$ for every local non-Archimedean field (of any characteristic).

## *8.5 Weyl Group*

We continue to assume that a Levi subgroup $M$ of $G$ contains the fixed Levi component $M_0$ of the minimal parabolic subgroup $P_0$. Denote by $W_M$ the quotient by $M_0$ of the normalizer $N(M_0, M)$ of $M_0$ in $M$. It is the Weyl group of $M_0$ in $M$. Let $L$ be a Levi subgroup of $G$ and let $W(M, L)$ denote a set of representatives in $W_G$, of minimal length, for $W_M \backslash W_G / W_L$. For every $w$ in $W(M, L)$, put $M_w = M \cap wLw^{-1}$ and $L_w = w^{-1}Mw \cap L$.

We can now state the Geometric Lemma of [BZ77, (2.12)].

PROPOSITION 8.6 (Geometric Lemma). *For every $\rho$ in $R(L(F))$, we have*

$$F(\rho) \stackrel{\text{defn}}{=} r_M^G \circ i_L^G(\rho) = \sum_{w \in W(M,L)} i_{M_w}^M \circ w \circ r_{L_w}^L(\rho).$$

COROLLARY 8.7. *For each Levi subgroup $M$ of $G$, put $T_M = i_M^G \circ r_M^G : R(G(F)) \to R(G(F))$. Then*

(1) $T_L \circ i_M^G = \sum_{w \in W(L,M)} i_{M_w}^G \circ r_{M_w}^M$
(2) $T_L \circ T_M = \sum_w T_{M_w}$.

*As before, $M_w = M \cap w^{-1}Lw$. The $w$ range over $W(L, M)$.*

PROOF. (1) We have that

$$i_L^G \circ r_L^G \circ i_M^G = \sum_w i_L^G \circ i_{L_w}^L \circ w \circ r_{M_w}^M = \sum_w i_{L_w}^G \circ w \circ r_{M_w}^M$$

is equal to the required expression, since $i_{L_w}^G \circ w = i_{M_w}^G$ by [BDK86, Lemma 5.4(iii)].

(2) We also have

$$T_L \circ i_M^G \circ r_M^G = \sum_w i_{M_w}^G \circ r_{M_w}^M \circ r_M^G = \sum_w i_{M_w}^G \circ r_{M_w}^G = \sum_w T_{M_w}.$$

□

## 8.8 Dual Maps

We now proceed to establish the Proposition 8.2. Denote the pairing $R^*(G(F)) \times R(G(F)) \to \mathbb{C}$ by $(\beta, \pi) \mapsto \langle \beta, \pi \rangle$. Let

$$i_M^{G*} : R^*(G(F)) \to R^*(M(F))$$

and

$$r_M^{G*} : R^*(M(F)) \to R^*(G(F))$$

be the morphisms adjoint to $i_M^G$ and $r_M^G$. Note that $\overline{C}_c^\infty(G(F))$ is a subspace of $R^*(G(F))$. The function $f$ defines the form $\beta = \beta_f : \pi \mapsto \langle \beta, \pi \rangle = f_G(\pi) = \operatorname{tr} \pi(f)$. Put $\langle f, \pi \rangle$ for $\langle \beta, \pi \rangle$ in this case.

LEMMA. *For every $M(F)$, $i_M^{G*}$ maps $\overline{C}_c^\infty(G(F))$ to $\overline{C}_c^\infty(M(F))$ and $r_M^{G*}$ maps $\overline{C}_c^\infty(M(F))$ to $\overline{C}_c^\infty(G(F))$.*

PROOF. For $f$ in $\overline{C}_c^\infty(G(F))$, $f_M$ satisfies $\langle f_M, \tau \rangle = \langle f, i_M^G \tau \rangle$ for every $\tau$ in $R(M(F))$ by virtue of a standard formula for the character of an induced representation. By virtue of Proposition 5.1, we have $i_M^{G*} f = f_M$, as required. For the second part of the lemma, for every $f^M$ in $\overline{C}_c^\infty(M(F))$, define a form $\beta = r_M^{G*}(f^M)$ in $R^*(G(F))$ by $\langle \beta, \pi \rangle = \langle f^M, r_M^G \pi \rangle$, for $\pi \in R(G(F))$. This is a good form, hence a trace form by the trace Paley-Wiener theorem. Namely, $r_M^{G*}(f^M)$ is a function in $\overline{C}_c^\infty(G(F))$, and the second part of the lemma follows. □

COROLLARY 8.9. *The homomorphisms $i_M^G : R(M(F)) \to R(G(F))$ and $r_M^G : R(G(F)) \to R(M(F))$ admit adjoints $i_M^{G*} : \overline{C}_c^\infty(G(F)) \to \overline{C}_c^\infty(M(F))$ and $r_M^{G*} : \overline{C}_c^\infty(M(F)) \to \overline{C}_c^\infty(G(F))$.*

A function $f$ in $\overline{C}_c^\infty(G(F))$ is called *discrete* if $i_M^{G*} f = 0$ for all Levi subgroups $M \neq G$. By Proposition 5.1 the space of discrete functions in $\overline{C}_c^\infty(G(F))$ is $\overline{A}_c^\infty(G(F))$.

PROPOSITION 8.10 (Combinatorial Lemma). *For each proper Levi subgroup $M$ of $G$, there is a rational number $c_M$ such that*

$$f^d = f - \sum_{M \neq G} c_M r_M^{G*} \circ i_M^{G*}(f)$$

*is discrete for every $f$ in $\overline{C}_c^\infty(G(F))$.*

PROOF. This is an analogue of Lemma 3.3 of [BDK86]. In [BDK86] a form $\beta$ in $R^*(G(F))$ is called discrete if $i_M^{G*} \beta = 0$ for all $M \neq G$. Lemma 3.3 [BDK86] asserts that there are $c_M$ such that for each $\beta$ in $R^*(G(F))$ the form

$$\beta^d = \beta - \sum_{M \neq G} c_M r_M^{G*} \circ i_M^{G*}(\beta)$$

is discrete. But Lemma 8.8 asserts that if $f$ lies in $\overline{C}_c^\infty(G(F))$, then $f^d$ lies in $\overline{C}_c^\infty(G(F))$, not only in $R^*(G(F))$; hence it is in $\overline{A}^\infty(G(F))$, as asserted. □

THEOREM 8.11. *The space $\overline{C}_c^\infty(G(F))$ is the direct sum over a set of representatives for the conjugacy classes of Levi subgroups in $G$ of $r_M^{G*}(\overline{A}_c^\infty(M(F)))$.*

PROOF. To show that $\overline{C}_c^\infty(G(F))$ is the sum of $r_M^{G*}(\overline{A}_c^\infty(M(F)))$, we assume by induction that this claim holds for every proper Levi subgroup $M$ of $G$. Namely, we assume that for each $M \neq G$, and for each $L \subset M$, there is a rational number $c_{M,L}$ with the following property. Given $f^M$ in $\overline{C}_c^\infty(M(F))$, there are $f^{M,L}$ in $\overline{A}_c^\infty(L(F))$ for each $L \subset M$, such that $f^M = \sum_{L \subset M} c_{M,L} r_L^{M*}(f^{M,L})$. Hence, given $f$ in $\overline{C}_c^\infty(G(F))$ there are $f^{M,L}$ in $\overline{A}_c^\infty(L(F))$ for every $M \neq G$ and $L \subset G$ with $i_M^{G*} f = \sum_{L \subset G} c_{M,L} r_L^{M*}(f^{M,L})$. Using the Combinatorial Lemma Proposition 8.10, we conclude that

$$\begin{aligned} f &= f^d + \sum_{M \neq G} c_M r_M^{G*}(i_M^{G*}(f)) \\ &= f^d + \sum_{M \neq G} c_M r_M^{G*} \sum_{L \subset M} c_{M,L} r_M^{L*} f^{M,L} \\ &= f^d + \sum_{L \neq G} r_L^{G*} \left( \sum_M c_M c_{M,L} f^{M,L} \right), \end{aligned}$$

where $M$ ranges over the $M \neq G$ which contain $L$, as required.

To prove that the sum is direct, note that if $f^M$ lies in $\overline{C}_c^\infty(M(F))$, then by the Geometric Lemma 8.6 for each Levi subgroup $L$ and $\rho$ in $R(L(F))$, we have

$$\begin{aligned} \langle i_L^{G*} r_M^{G*} f^M, \rho \rangle &= \langle f^M, r_M^G \circ i_L^G(\rho) \rangle \\ &= \sum_{w \in W(M(F), L(F))} \langle f^M, i_{M_w}^M \circ w \circ r_{L_w}^L(\rho) \rangle. \end{aligned} \qquad (8.11.1)$$

If $f^M$ lies in $\overline{A}_c^\infty(M(F))$ and $w$ contributes a nonzero term in the sum, then $M_w = M$, namely, $L \supset w^{-1}Mw$. Consequently (8.11.1) is zero if $L$ contains no conjugate of $M$. If $L$ is conjugate to $M$, say $L = s^{-1}Ms$ for some $s$ in $W(M, L)$, then (8.11.1) is equal to $\langle f^M, s\rho \rangle$. Now, if $f = \sum_M r_M^{G*}(f^M)$ is zero, where the $f^M$ lies in $\overline{A}_c^\infty(M(F))$, then choose $L$ to be a minimal Levi subgroup (up to conjugation) for which $f^L \neq 0$ in this sum. Then $i_L^{G*}(f) = f^L$, and $f = 0$ implies that $f^L = 0$. This contradiction completes the proof of the theorem. □

## 8.12 *Proof of Proposition* 8.2

Given $M \neq G$ and $f^M$ in $\overline{C}_c^\infty(M(F))$, since

$$\overline{C}_c^\infty(M(F)) = \oplus_{L \subset M} r_L^{M*}(\overline{A}_c^\infty(L(F)))$$

by Theorem 8.11, we may assume that $f^M = r_Y^{M*}(f^Y)$ for some Levi $Y$ in $M$ and $f^Y$ in $\overline{A}_c^\infty(Y(F))$. We claim that the product of $f = r_M^{G*}f^M = r_Y^{G*}(f^Y)$ by a scalar which depends only on $Y$, $M$, and $G$ has the properties required by the proposition. Indeed, as in (8.11.1), for each $\rho$ in $R(Y(F))$, we have

$$\langle i_L^{G*} r_Y^{G*}(f^Y), \rho\rangle = \sum_{w \in W(Y,L)} \langle f^Y, i_{Y_w}^Y \circ w \circ r_{L_w}^L(\rho)\rangle$$

and it suffices to consider $w$ with $L \supset w^{-1}Yw$, since $f^Y$ lies in $\overline{A}_c^\infty(Y(F))$. Hence if $L$ contains no conjugate of $M$, then the sum is empty and $i_L^{G*}(f) = f_L$ is zero, as required. If $L = M$, our sum becomes the sum over all $w$ in $W(Y, M)$ with $w^{-1}Yw \subset M$ of $\langle f^Y, w \circ r_{M_w}^M(\rho)\rangle$. The condition (8.2.1) implies that each of the summands is equal to $\langle f^Y, r_Y^M(\rho)\rangle = \langle r_Y^{M*}(f^Y), \rho\rangle = \langle f^M, \rho\rangle$; hence $i_M^{G*}(f)$ is equal to $f^M$ up to a multiple by the cardinality of the set of $w$ in $W(Y, M)$ with $w^{-1}Yw \subset M$. The proposition follows. □

Proposition 8.2 implies that a function $f^M$ in $C_c^\infty(M(F))$ can be "lifted" to a function $f$ in $C_c^\infty(G(F))$ with the "same" orbital integrals on the regular conjugacy classes of $G(F)$ which intersect $M(F)$. The orbital integrals of $f$ are not necessarily zero on $x$ in $G(F)$ whose conjugacy class does not intersect $M(F)$. However, we have the following:

COROLLARY 8.13. *Suppose that $f^M$ has the property that $I^M(m, f^M)$ is supported on the set of $m$ in $M(F)$ with $|\alpha(m)| \neq 1$ for every root of the split component of the center of $M$ in $N$ of the parabolic subgroup $P = MN$. Then $f$ can be chosen to have the property that $I(x,f)$ is zero unless $x$ is conjugate in $G(F)$ to an element of $M(F)$.*

PROOF. Let $S_M$ denote the support of $I^M(f^M)$ in $M(F)$. Put $S = (S_M)^G = \{g^{-1}sg\, ;\, g \in G(F), s \in S_M\}$. Then $S$ is open and closed in $G(F)$. Replace the $f$ obtained in the proposition by its product with the characteristic function of $S$ to obtain the function $f$ of the corollary. □

# 9 Stability

## 9.1 *Stable Conjugacy*

Continue to let $F$ be a local non-Archimedean field of characteristic 0. Let $G'$ be a quasisplit connected reductive linear algebraic group, defined over $F$. Recall that an $F$-group is called *quasisplit* if it has a Borel subgroup defined over $F$. By a Borel subgroup, we mean a minimal parabolic subgroup $B$ over an algebraic closure $\overline{F}$ of $F$. The Levi subgroup of $B$ is an $F$-torus, and $G$ is called *split* if this $F$-torus is split, namely, isomorphic over $F$ to a product $\mathbb{G}_m^r$ of $r$ copies of the multiplicative group $\mathbb{G}_m$ over $F$. Two $F$-groups $G_1$ and $G_2$ are *F-forms* if they are isomorphic over $\overline{F}$. We recall in the next subsection the definition of inner forms and simply note here that given a reductive connected linear algebraic $F$-group $G$, there is a unique quasisplit $F$-group $G'$ which is an inner form of $G$.

In the current work, we are interested mainly in the example of $G' = \mathrm{GL}(n)$ and its inner forms $G$. To put the problem in perspective, we comment first on a more general situation, where stable conjugacy does not reduce to conjugacy.

The stable conjugacy class of $x$ in $G'(F)$ is defined in Subsection 3.6. We recall that two elements are stably conjugate if they are conjugate by an element of $G'(\overline{F})$. Let $T$ be the centralizer of $x$ in $G'$. We will only be interested in regular $x$, in which case $T$ is an $F$-torus. The conjugacy classes within the stable class of $x$ are parametrized by the (finite) set

$$B(T/F) = \ker[H^1(F,T) \to H^1(F,G')],$$

see also 3.6. If $x$ and $x'$ are stably conjugate, then $G'_x = \mathrm{Ad}(y)G'_{x'}$ is isomorphic to $G'_{x'}$ over $\overline{F}$. This allows us to transfer a differential form of maximal degree, yielding compatible Haar measure on $G'_x(F)$ and $G'_{x'}(F)$.

Let $\{\mathrm{Ad}(b)x\,;\, b \in B(T/F)\}$ be a set of representatives for the conjugacy classes within the stable conjugacy class of the regular element $x$ of $G'(F)$.

DEFINITION. Let $\Xi$ be a function on the regular conjugacy classes in $G'(F)$. The *stable* function $\Xi^s$ associated with $\Xi$ is defined by

$$\Xi^s(x) = \sum_b \Xi(\mathrm{Ad}(b)x).$$

It depends only on the stable conjugacy class of the regular $x$. In particular, for any $f$ in $C_c^\infty(G'(F))$, we have the stable orbital integral $\Phi^s(\cdot,f)$ of $f$, and the normalized $I^s(\cdot,f)$.

## 9.2 *Inner Twisting*

The stable orbital integrals are introduced for purposes of comparison between the group $G'$ and a reductive connected $F$-group $G$, such that the following holds. Let $G$ and $H$ be $F$-groups. An isomorphism $\psi : G \to H$ over $\overline{F}$ is called an *inner twisting* if for every $\tau$ in $\mathrm{Gal}(\overline{F}/F)$ there is $g_\tau$ in $G(\overline{F})$ such that $(\tau\psi)^{-1} \circ \psi = \mathrm{Ad}(g_\tau)$. If such $\psi$ exists, then $G$ and $H$ are called *inner forms*. Suppose that $G$ is an inner form of $G'$ and fix an inner twisting $\psi : G \to G'$. Fix a maximal split torus $A$ in $G$. It can be identified with a torus $A'$ of $G'$ via $\psi$. Each Levi subgroup $M$ of $G$ containing $A$ corresponds by $\psi$ to a Levi subgroup $M'$ of $G'$ containing $A'$. Fix a lattice $L$ as in Subsection 7.2, so that $M(F)$ is of the form $M_\lambda(F) = G_\lambda(F)$ for some $\lambda$ in $L^-$.

## 9.3 Norm Map

In every known comparison situation (base change, symmetric square or more generally symplectic and orthogonal groups in $\mathrm{GL}(n)$, metaplectic correspondence, inner twisting), there exists a map $N$ which we call a *norm map*, with at least the following properties. The map $N$ is a bijection, from a subset $S'_0$ of the set $S'$ of stable conjugacy classes of regular elements in $G'(F)$, to a subset $S_0$ of the set $S$ of stable conjugacy classes of regular elements in $G(F)$, such that the following properties hold:

(1) $G'_x$ and $G_{Nx}$ are inner forms.
(2) $Nx = \psi^{-1}(x)$ for $x$ in $A'(F)$.
(3) $x$ has a representative in $M'(F)$ if and only if $Nx$ has a representative in $M(F)$.
(4) At least one of the subsets $S_0$, $S'_0$ is equal to $S$, $S'$.

We use (1) to relate measures on the two groups there. Fix a norm map $N$.

## 9.4 Matching Functions

We let $W(M,G) = N(M,G)/M$ be the quotient by $M$ of the normalizer $N(M,G)$ of $M$ in $G$. We define similarly $W(M',G')$. Given $f^M$ in $C_c^\infty(M(F))$, let ${}^MI(f^M)$ be the conjugacy class function on the set of regular $x$ in $G(F)$ which attains the value 0 unless (a conjugate of) $x$ lies in $M(F)$ when we put

$$^MI(x,f^M) = \sum_{w\in W(M,G)} I^M(wxw^{-1},f^M).$$

Similarly, for $\phi^M$ in $C_c^\infty(M'(F))$, we define ${}^MI(x',\phi^M)$, for regular $x' \in G'(F)$. In particular, ${}^GI = I = I^G$. Recall that ${}^MI^s$ indicates the stable function on $G(F)$ associated with ${}^MI$.

DEFINITION. The functions $\phi^M$ in $C_c^\infty(M'(F))$ and $f^M$ in $C_c^\infty(M(F))$ are called *matching* if the following hold:

(1) ${}^M I^s(s, f^M)$ is zero for any $s$ in $S \smallsetminus S_0$.
(2) ${}^M I^s(s, \phi^M)$ is zero for any $s$ in $S' \smallsetminus S'_0$.
(3) ${}^M I^s(s, \phi^M)$ is equal to ${}^M I^s(Ns, f^M)$ for all $s$ in $S'_0$.

In the comparison of $G' = \mathrm{GL}(n)$ with its inner form $G$, the norm map associates a regular $x \in G(F)$ to $x' \in G'(F)$ which has the same characteristic polynomial. In the notations of this section: $Nx' = x$ for the $x'$ coming from $x$. This relation defines not the elements but only the conjugacy classes $x$ and $x'$, and these coincide with the stable conjugacy classes.

# 10 Discrete Series

## *10.1 Central Exponents*

Let $F$ be a local non-Archimedean field of characteristic 0. Let $G$ be a connected reductive $F$-group. Let $G_s$ be the centralizer of $s$ in $G$. Let $\pi$ be a smooth $G(F)$-module of finite length. As noted in Subsection 3.3, it is admissible. By a *central exponent* of $\pi$ with respect to a Levi subgroup $M$ of $G$, we mean the central character of an irreducible constituent of the module $\pi_N$ of coinvariants (defined in 7.1) of $\pi$ with respect to any parabolic subgroup $P = MN$ with Levi component $M$.

## *10.2 Small Part*

As in Subsection 7.2, let $A$ be a maximal split torus in $G$ and let $L$ be a cocompact lattice in $A(F)$. Let $B$ be a minimal parabolic subgroup of $G$ containing $A$, and $A^-$ the set of $a$ in $A$ with $|\alpha(a)| \leq 1$ for any $\alpha$ in the set $\Delta$ of roots of $A$ in $B$, $L^- = L \cap A^-(F)$. To any semisimple $t$, we associate (using the action of the Weyl group) $a$ in $A^-(F)$ (or $\lambda$ in $L^-$) and a subset $\theta$ of $\Delta$, consisting of the $\alpha$ with $\alpha(\lambda) = 1$.

## *10.3 Decay of Central Exponents*

Given $\lambda$ in $L^-$, consider the centralizer $M_\lambda$ of $\lambda$ in $G$ and denote by $P_\lambda = M_\lambda N_\lambda$ the standard parabolic subgroup of $G$ with Levi component $M_\lambda$. The center $A_\lambda$ of $M_\lambda$ lies in $A$. We say that the central exponent $\omega$ of $\pi$ with respect to $M_\lambda$ *decays* if $|\omega(a)| < 1$ for every $a$ in $A_\lambda(F)$ with the following properties:

(1) $|\alpha(a)| \leq 1$ for any root $\alpha$ of $A_\lambda$ in $N_\lambda$.
(2) $|\alpha(a)| < 1$ for some such $\alpha$.

We define $\pi$ to be *discrete series* if its central character is unitary, and its central exponents with respect to any proper Levi subgroup $M_\lambda$, where $\lambda$ is any element in $L^-$, all decay.

### *10.4 Square Integrability*

Harish-Chandra's criterion for square integrability (see [Cas, Theorem 4.4.6] or [Si80, Theorem 4.4.4]) asserts that $\pi$ is a discrete-series representation in the sense above if and only if it is *square-integrable* $\pi$, in the sense that its matrix coefficients $f_{v,v'}(x) = \langle \pi'(x)v, v' \rangle$ are absolutely square-integrable functions on $G(F)/Z(F)$.

DEFINITION 10.5. We say that a discrete series $G'(F)$-module $\pi'$ satisfies a *trace identity* if the following hold:

(1) There is a set $\{\pi\}$ which, for any open compact subgroup $C$ in $G(F)$, contains only finitely many $G(F)$-modules with a $C$-fixed vector.
(2) There are positive integers $m(\pi)$ (depending on $\pi'$) and a complex number $c$ so that, for all matching $\phi$ in $C_c^\infty(G'(F))$ and $f$ in $C_c^\infty(G(F))$, we have

$$c \operatorname{tr} \pi'(\phi) = \sum m(\pi) \operatorname{tr} \pi(f). \tag{10.5.1}$$

**Assumption 10.6.** We make the following assumption. For any proper Levi subgroup $M$, and any open compact subgroup $C$ as in Proposition 7.5, there exists $\phi^M$ in $C_c^\infty(M'(F))$ matching the characteristic function of $C \cap M(F)$ in $C_c^\infty(M(F))$.

Our assumption is tantamount to the following. For any proper Levi subgroup $M$ with center $A_M$ contained in $A$, and any $t_0$ in $A_M(F)$, we have the following. There exists a function $\phi_{t_0}^M$ in $C_c^\infty(M'(F))$ matching the function $f_{t_0}^M$ in $C_c^\infty(M(F))$ defined in Subsection 7.4. Indeed, the function $f_{t_0}^M$ is obtained from the characteristic function of $C \cap M(F)$ on translating by the central element $t_0$ and multiplying by a scalar, so that $\phi_{t_0}^M$ can be obtained from $\phi^M$ on translating by the central element $t_0$ and multiplying by the same scalar.

## 11 Decay

PROPOSITION 11.1. *Suppose that the discrete-series $G'(F)$-module $\pi'$ satisfies a trace identity* (10.5.1), *and $G(F)$ satisfies Assumption* 10.6. *Then all $\pi$ in* (10.5.1) *are discrete series $G(F)$-modules.*

PROOF. Let $M$ be a proper Levi subgroup, $C$ a compact open subgroup of $G(F)$ as in Proposition 7.5, $t_0$ in $A_M(F)$ such that $|\alpha(t_0)| \leq 1$ for all roots $\alpha$ of $A_M$ in the unipotent radical of the standard parabolic subgroup with Levi component $M$, and $f_{t_0}^M$ the function of Proposition 7.5. Proposition 7.5 and Proposition 7.7 imply that the function $f_{t_0}$ on $G(F)$ defined in Proposition 7.5, which is $C$-bi-invariant, satisfies $I(x,f_{t_0}) = {}^M I(x,f_{t_0}^M)$; hence

$$I^s(x,f_{t_0}) = {}^M I^s(x,f_{t_0}^M)$$

for all regular $x$ in $G(F)$.

As noted following Proposition 7.5, the function $f_{t_0}$ is $C$-bi-invariant. Hence $\operatorname{tr}\pi(f_{t_0}) \neq 0$ only for $\pi$ with a nonzero $C$-invariant vector. By definition of the trace identity (10.5.1), there are only finitely many such $\pi$ in (10.5.1). On the other hand, if $\omega_\tau$ is the central character of the irreducible constituent $\tau$ of the $M(F)$-module $\pi_N$, then $\operatorname{tr}\pi(f_{t_0}) = \operatorname{tr}\pi_N(f_{t_0}^M)$ is a sum over $\tau$ of $\omega_\tau(t_0)\delta_P^{1/2}(t_0)n(\tau, C)$, where $n(\tau, C)$ is the nonnegative integral multiplicity of the trivial representation of $C$ in $\tau$ (the dimension of the space of $C$-fixed vectors in $\tau$).

Assumption 10.6 asserts that there exists a function $\phi_{t_0}^M$ in $C_c^\infty(M'(F))$ matching $f_{t_0}^M$. Proposition 7.7 asserts that there exists a function $\phi_{t_0}$ in $C_c^\infty(G'(F))$ with $I^s(x,\phi_{t_0}) = {}^M I^s(x,\phi_{t_0}^M)$ for all regular $x$ in $G'(F)$. Hence the functions $f_{t_0}$ on $G(F)$ and $\phi_{t_0}$ on $G(F)$ are matching. Since $\pi'$ appears in the trace identity, it is clear that its character $\chi_{\pi'}$ is a stable function, depending only on the stable conjugacy class of $x$ in $G'(F)$. Using the Weyl integration formula, we have

$$\operatorname{tr}\pi'(\phi) = \sum_T{}^s\, w(T)^{-1}\int_{T(F)/Z(F)} (\Delta\chi_{\pi'})(x) I^s(x,\phi_{t_0})\, \mathrm{d}(Nx).$$

The sum is over the stable conjugacy classes of $F$-tori in $G(F)$; $w(T)$ is the cardinality of the quotient $W(T) = N(T)/T$ by $T$ of the group $N(T)$ of $x$ in $G(\overline{F})$ such that $\operatorname{Ad}(x) : T(F) \to T(F)$, $t \mapsto xtx^{-1}$, is defined over $F$. Recall that

$$I^s(x,\phi_{t_0}) = {}^M I^s(x,\phi_{t_0}^M) = {}^M I^s(Nx, f_{t_0}^M).$$

Since $t_0$ lies in the center of $M(F)$, we have $M \subset M_{t_0}$. As we assumed that $|\alpha(t_0)| < 1$ for all roots $\alpha$ of $A_M$ in $N_M$, we have $M = M_{t_0}$. But $f_{t_0}^M$ is supported on a small neighborhood of $t_0$. Hence $I^s(x,\phi_{t_0}) \neq 0$ implies that $M_x^0$ is equal to $M$. Proposition 7.3 now implies that we have

$$\sum_T w(T)^{-1}\int (\Delta_M\chi_{\pi'_N})(x)\cdot {}^M I^s(x,\phi_{t_0}^M)\, \mathrm{d}(Nx).$$

As $t_0$ lies in the center of $M(F)$, changing variables $x \mapsto t_0x$, we obtain

$$\sum_{\tau' \text{ in } \pi'_N} \omega_{\tau'}(t_0)\sum_T w(T)^{-1}\int (\Delta_M\chi_{\tau'})(x)\cdot {}^M I'(x,\phi_1^M)\, \mathrm{d}(Nx) = \sum_{\tau'} \omega_{\tau'}(t_0)\operatorname{tr}\tau'(\phi_1^M),$$

with $\tau'$ ranging over the irreducible subquotients of $\pi'_N$. Here $\omega_{\tau'}$ is the central character of $\tau'$.

We conclude from the trace identity that for any $t_0$ in the center of $M(F)$ with $M_{t_0} = M$ in the notation of 7.2, we have

$$\sum_{\tau'} c(\tau')\omega_{\tau'}(t_0) = \sum_{\tau} n(\tau)\omega_{\tau}(t_0).$$

The sum on the left ranges over the constituents $\tau'$ of $\pi'_N$; hence it is finite, since $\pi'$, whence $\pi'_N$, is admissible. The $c(\tau')$ are complex numbers. On the right the sum is finite, depending on the compact open subgroup $C$, and the coefficients are positive, so that no cancelation may occur. Linear independence of characters (on the set of $t_0$ in $A_M(F)$ with $|\alpha(t_0)| < 1$ for the positive roots $\alpha$) implies that for each $\tau$ there exists $\tau'$ with $\omega_\tau(t_0) = \omega_{\tau'}(t_0)$. Consequently the character $\omega_\tau$ decays, where $\tau$ is any constituent of $\pi_N$. Here $M$ is any proper Levi subgroup of $G$, and $\pi$ is any $G(F)$-module with a nonzero $C$-fixed vector. Since any $\pi$ has a nonzero $C$-fixed vector for a sufficiently small $C$, it follows that all $\pi$ are discrete series, as required. □

REMARK 11.2. It is clear that if $\pi'$ is assumed to be only tempered (see 13.6), then the above proof implies that the $\pi$ of (10.5.1) are tempered.

## 12 Finiteness

We now continue with the situation and Assumption 10.6 of Section 11. We make two additional assumptions.

**Assumption 12.1.** Suppose that $\Phi^s$ is a stable function in $S_e(G'(F))$ (see Section 6). Then there exists $\Phi$ on $G(F)$ in $S_e(G(F))$ matching $\Phi^s$.

Namely, we suppose that $\Phi^s(x) = \Phi^s(x')$ for all stably conjugate $x, x'$ and assume the existence of a function $f \in A_c^\infty(G(F))$ with $\widetilde{\Phi}^s(Nx, f) = \Phi^s(x)$ on the regular set.

**Assumption 12.2.** For any $f$ in $A_c^\infty(G(F))$, there exists a matching function $\phi$ in $A_c^\infty(G'(F))$.

Using these assumptions, we have the following:

PROPOSITION 12.3. *Suppose that the discrete series $G'(F)$-module $\pi'$ satisfies a trace identity* (10.5.1). *Then the set of $\pi$ is finite.*

PROOF. Note that Proposition 11.1 asserts that the $\pi$ are all discrete series. To prove our proposition, note that by the trace identity (10.5.1), $\operatorname{tr} \pi'(\phi)$ depends only on $f$, namely, on the stable orbital integral of $\phi$; hence the character $\chi_{\pi'}$ of $\pi'$ on $G(F)$ is a stable function. Assumption 12.1 implies that there exists a finite linear combination of $G(F)$-modules $\pi$ with complex coefficients $c(\pi)$ so that

$$\sum_{\pi} c(\pi)\chi_\pi(Nx) = \chi_{\pi'}(x)$$

for any elliptic regular $x$ in $G'(F)$, and $\sum_\pi c(\pi)[\chi(\pi)](y) = 0$ for the elliptic regular $y$ which are not norms. We may assume that all $\pi$ here are tempered by [Ka86.1, Proposition 1.1].

Applying the Weyl integration formula, we deduce that

$$\operatorname{tr}\pi'(\phi) = \sum_T^s w(T)^{-1} \int (\Delta\chi_{\pi'})(x) I^s(x,\phi)\,\mathrm{d}x.$$

Only elliptic tori occur since we take $\phi$ in $A_c^\infty(G'(F))$. Further, we take $\phi$ so that it has a matching $f$, so that $I^s(x,\phi) = I^s(Nx,f)$. Replacing $\chi(\pi')$ by our linear combination $\sum_\pi c(\pi)\chi(\pi)$, we obtain

$$\sum_\pi c(\pi) \sum_T^s w(T)^{-1} \int (\Delta\chi_\pi)(x) I^s(x,f)\,\mathrm{d}x = \sum_\pi c(\pi)\operatorname{tr}\pi(f).$$

We deduce from (10.5.1) the identity $\sum_\pi c(\pi)\operatorname{tr}\pi(f) = \sum_\pi m(\pi)\operatorname{tr}\pi(f)$. On the left the sum is finite and consists of tempered $\pi$. On the right all $\pi$ are discrete series. The identity holds for all $f$ in $A_c^\infty(G(F))$ which have a matching function $\phi$. So fix $\pi_0$ on the right. By [Ka86.1, Theorem K], there exists a pseudo-coefficient $f_0$ in $A_c^\infty(G(F))$ with $\operatorname{tr}\pi_0(f_0) = 1$ and $\operatorname{tr}\pi(f_0) = 0$ for any tempered irreducible $\pi$ inequivalent to $\pi_0$. But Assumption 12.2 implies that $f_0$ has a matching function $\phi$. Using our identity with $f = f_0$, we conclude that $m(\pi_0) = 0$ for all $\pi_0$ on the right which are not equivalent to any of the finitely many $\pi$ on the left. Consequently, the set of $\pi$ with $m(\pi) \neq 0$ is finite, as asserted. □

# 13 Simple Algebras

## *13.1 Invariants of Simple Algebras*

Let $F$ be a local field. Let $G$ be an inner form of $\mathrm{GL}(n)$ over $F$. Thus $G$ is the multiplicative group of a central simple $F$-algebra $A$. There is a central division algebra $D$ over $F$ of rank $d$ with $A = M(m,D)$, $n = md$. Class field theory—see [We67]—associates with $A$ an invariant $\operatorname{inv} A$ of the form $i/d$ (modulo 1), with $i$ prime to $d$, and $\operatorname{inv} A = \operatorname{inv} D$ independent of $m$. There exists a unique central simple $F$-algebra $A$ of rank $n$ with invariant $i/d$ (modulo 1) (where $(i,d) = 1$ and $d$ divides $n$). If $F = \mathbb{C}$, then $d = 1$. If $F = \mathbb{R}$, then $d = 1$ or 2. Otherwise, $d$ can be any positive integer. We let $G' = \mathrm{GL}(n)$. Note that if $\overline{F}$ is an algebraic closure of $F$, then $G(\overline{F}) = \mathrm{GL}(n,\overline{F})$. This isomorphism is over $\overline{F}$.

## 13.2 Regular Conjugacy Classes

A conjugacy class $\gamma$ in $G(F)$ is called *regular* if its characteristic polynomial $p_\gamma$ has distinct roots (in some algebraic closure of $F$). If $\gamma, \delta$ are regular and $p_\gamma = p_\delta$, then $\gamma = \delta$. There is an embedding $\gamma \mapsto \gamma'$, defined by $p_{\gamma'} = p_\gamma$, of the set of regular conjugacy classes $\gamma$ in $G(F)$ into the set of regular conjugacy classes $\gamma'$ in $G'(F)$.

## 13.3 Grothendieck Group

Let $C_c^\infty(G(F))$ denote the convolution algebra of complex valued smooth compactly supported measures $f$ on $G(F)$. These are all of the form $fdg$ where $dg$ is any fixed Haar measure on $G(F)$ and $f$ is now a function. Put $R(G(F)) = R_{\mathbb{Z}}(G(F)) \otimes \mathbb{C}$, where $R_{\mathbb{Z}}(G(F))$ is the Grothendieck group of the category of smooth $G(F)$-modules. It is the free abelian group generated by the set $\Pi(G(F))$ of equivalence classes of irreducible such $G(F)$-modules. By [BZ76, Theorem 3.25] irreducible smooth $G(F)$-modules are admissible.

## 13.4 Character

If $\pi$ is an admissible $G(F)$-module, then the convolution operator $\pi(f\,dg)$, defined by $\int_{G(F)} f(g)\pi(g)\,dg$, is of finite rank, and its trace is denoted by $\operatorname{tr}\pi(f)$. We often delete the measure $dg$ from the notation, namely, think of $f$ as a measure. There exists a complex valued conjugacy invariant smooth function $\chi = \chi_\pi$ on the regular set of $G(F)$ with $\operatorname{tr}\pi(f) = \int \chi(g)f(g)$ for any $f \in C_c^\infty(G(F))$ which is supported on the regular set of $G(F)$. It is called the *character* of $\pi$. It depends only on the image of $\pi$ in $R(G(F))$. The characters of inequivalent irreducible $G(F)$-modules are linearly independent. In particular, $\chi \neq 0$ if $\pi \neq 0$ in $R(G(F))$. Harish-Chandra [HC78] showed that $\chi$ extends to a locally integrable function on $G(F)$.

## 13.5 Induction

Fix a minimal parabolic subgroup $P_0$ together with its Levi decomposition $M_0N_0$ in $G$. Denote by $i_M^G$ the homomorphism $R(M(F)) \to R(G(F))$ of normalized, or unitary, induction, for any (standard) Levi subgroup $M$. Here "standard" means that $M$ is the Levi subgroup containing $M_0$ of a parabolic subgroup $P$ containing $P_0$. The map $i_M^G$ is independent of the choice of the parabolic subgroup $P$ with Levi component $M$, which is used in its definition.

## 13.6 Tempered Representations

An irreducible $G(F)$-module $\pi$ whose central character $\omega_\pi$ is unitary is called *square-integrable* $\pi$ or *discrete series* if it has a matrix coefficient which is square-integrable on $G(F)$ modulo its center $Z(F)$. In this case all of its matrix coefficients are square-integrable, and $\pi$ embeds in $L^2(G, Z, \omega_\pi)$. An alternative definition is that all of its central exponents decay.

An irreducible $G(F)$-module $\pi$ is called *tempered* if there exists a Levi subgroup $M(F)$ and a square-integrable (= discrete series) $M(F)$-module $\tau$ such that $\pi$ is a subquotient (necessarily a direct summand) of $i_M^G \tau$.

Put $\nu(x) = |x|$ for $x \in F$, where $| \cdot |$ is the normalized valuation on $F$. Put $\nu(g) = \nu(\det g)$, where $\det g$ is the reduced norm of $g$ in $G(F)$.

DEFINITION 13.7. *A $G(F)$-module $\pi$ is called* relevant *if there is a Levi subgroup of $G(F)$ of the form $M = \prod_{i=1}^m (M_i \times M_i)$ or $M_0 \times M$, where the $M_i$, $0 \leq i \leq m$, are multiplicative groups of central simple $F$-algebras, and tempered $M_i(F)$-modules $\tau_i$ and distinct positive numbers $s_i < 1/2$ such that $\pi$ is*

$$i_M^G \left( \prod_{i=1}^m (\tau_i \nu^{s_i} \times \tau_i \nu^{-s_i}) \right)$$

*or*

$$i_{M_0 \times M}^G \left( \tau_0 \times \prod_{i=1}^M (\tau_i \nu^{s_i} \times \tau_i \nu^{-s_i}) \right)$$

*in $R(G(F))$.*

THEOREM 13.8 (Local Theorem for Non-degenerate Representations).

(1) *Relevant representations of $G(F)$ are unitarizable and irreducible. In particular, a $G(F)$-module normalizedly (= unitarily) induced from a tempered one is irreducible.*
(2) *The relation $\chi'(\gamma') = (-1)^{n-m} \chi(\gamma)$ for all matching $\gamma \mapsto \gamma'$ regular conjugacy classes $\gamma \in G(F)$, $\gamma' \in G'(F)$, defines a bijection between the set of equivalence classes of square-integrable (resp. tempered, relevant) $G(F)$-modules $\pi$ and the set of equivalence classes of square-integrable $G'(F)$-modules $\pi'$ (resp. tempered, relevant, $G'(F)$-modules $\pi'$ whose character $\chi'$ is nonzero on the set of regular $\gamma'$ obtained from $\gamma$ in $G(F)$).*

The bijection of (2) is called the Deligne-Kazhdan correspondence.

## 13.9 Global Invariants

Let $F$ be a global field. Let $G$ be an inner form of $G' = \mathrm{GL}(n)$ over $F$. Then $G$ is the multiplicative group of a central simple $F$-algebra $A = M(m, D)$, for $D$ a central division algebra over $F$ of rank $d$, $n = md$. Class field theory—see [We67]—associates with $A$ the sequence $\{\mathrm{inv}_v A = \mathrm{inv}\, A \otimes_F F_v\}$ of rational numbers modulo one which are almost all zero and whose sum is zero modulo one. Each such sequence $\{i_v/d_v\}$ determines, up to $F$-isomorphism, a unique division algebra $D$ central over $F$ and a unique simple algebra $A$ of rank $n$ central over $F$ with these invariants, for any $n$ which is divisible by $d_v$ for all $v$. Let $G(\mathbb{A})$ be the group of $\mathbb{A}$-points of $G$, where $\mathbb{A}$ is the ring of adèles of $F$. Let $Z$ (resp. $Z'$) denote the center of $G$ (resp. $G'$). Then $Z = Z'$ is the multiplicative group. Fix a unitary character $\omega$ of $Z(\mathbb{A})/Z(F) = \mathbb{A}^\times/F^\times$. For each place $v$ of $F$, denote by $F_v$ the completion of $F$ at $v$ and by $\omega_v$ the restriction of $\omega$ to $F_v^\times$.

## 13.10 Automorphic Representations

Let $L(G(F)\backslash G(\mathbb{A}))$ denote the space of slowly increasing, see [BJ79], functions $\phi$ on $G(F)\backslash G(\mathbb{A})$ with $\phi(zg) = \omega(z)\phi(g)$ for $z$ in $Z(\mathbb{A})$. By $L^2(G(F)\backslash G(\mathbb{A}))$ we mean the space of $\phi$ on $G(F)\backslash G(\mathbb{A})$ with $\phi(zg) = \omega(z)\phi(g)$ for $z$ in $Z(\mathbb{A})$ that are square-integrable on $Z(\mathbb{A})G(F)\backslash G(\mathbb{A})$. The group $G(\mathbb{A})$ acts on $L = L(G(F)\backslash G(\mathbb{A}))$ and on $L^2 = L^2(G(F)\backslash G(\mathbb{A}))$ by right translation. Thus $(r(g)\phi)(h) = \phi(hg)$. Any irreducible subquotient of $L^2$ is unitary (preserves a non-degenerate inner product). The irreducible subquotients of $L$ are called *automorphic* $G(\mathbb{A})$-modules.

The space $L = L(G(F)\backslash G(\mathbb{A}))$ is the direct sum of the discrete spectrum $L_d = L_d(G(F)\backslash G(\mathbb{A}))$, which is the direct sum of all irreducible *sub*representations of $L$—these $G(\mathbb{A})$-modules are called "discrete spectrum" $G(\mathbb{A})$-modules, and the continuous spectrum $L_c = L_c(G(F)\backslash G(\mathbb{A}))$, which is a "continuous sum." The space $L_d$ is also the discrete spectrum in $L^2$.

A *cuspidal* $G(\mathbb{A})$-module is an irreducible constituent of the subspace $L_0 = L_0(G(F)\backslash G(\mathbb{A}))$, which consists of the $\phi$ in $L(G(F)\backslash G(\mathbb{A}))$ with $\int_{N(F)\backslash N(\mathbb{A})} \phi(nx)\, dn$ equals zero for every $x$ in $G(\mathbb{A})$ and for the unipotent radical $N$ of any proper parabolic subgroup of $G$ over $F$. Each cuspidal $\phi$ is rapidly decreasing, hence absolutely square-integrable, on $Z(\mathbb{A})G(F)\backslash G(\mathbb{A})$. The space $L_0(G(F)\backslash G(\mathbb{A}))$ is a sub-$G(\mathbb{A})$-module of $L_d(G(F)\backslash G(\mathbb{A}))$.

Any cuspidal $G'(\mathbb{A})$-module $\pi$ is non-degenerate, namely, each of its local components $\pi_v$ has a Whittaker model. This means that for any additive nontrivial character $\psi_v : F_v \to \mathbb{C}^\times$, there is an embedding of the $G'(F_v)$-module $\pi_v$ in the space $W(\psi_v)$ of right smooth functions $\varphi : G'(F_v) \to \mathbb{C}$ with $\varphi(ng) = \psi_v(\sum_i n_{i,i+1})\varphi(g)$ for $n = (n_{i,j})$ upper triangular unipotent matrix and $g \in G'(F_v)$, with $G'(F_v)$ action by right shifts.

Moreover such $\pi$ is generic (has a global Whittaker model: same definition, but with global $\psi : \mathbb{A}/F \to \mathbb{C}^\times$, $\psi \neq 1$), and it occurs with multiplicity one in $L_0(G'(F)\backslash G'(\mathbb{A}))$. See [BZ76] and [PS79] for Whittaker models and [Shal74] or [PS79] for the multiplicity one theorem.

An irreducible admissible $G(\mathbb{A})$-module $\pi$ decomposes as a restricted tensor product $\otimes_v \pi_v$ of irreducible admissible $G(F_v)$-modules $\pi_v$, almost all of which are unramified; see [Ft79].

An irreducible $\pi_v$ is call unramified if its space contains a nonzero $K_v = G'(R_v)$-fixed vector $\xi_v$, necessarily unique up to a scalar. Here $v$ is finite and $R_v$ is the ring of integers in $F_v$.

The space of the irreducible $\pi$ is the union over all sets $S$ of valuations of $F$ such that $\pi_v$ is unramified for $v \notin S$, of $(\otimes_{v\in S} V_v) \otimes (\otimes_{v\notin S}\xi_v)$, where $V_v$ is the space of $\pi_v$.

If $\pi'_1 = \otimes_v \pi'_{1,v}$ and $\pi'_2 = \otimes_v \pi'_{2,v}$ are cuspidal $G'(\mathbb{A})$-modules and $\pi'_{1,v} \simeq \pi'_{2,v}$ for almost all $v$, then $\pi'_{1,v} \simeq \pi'_{2,v}$ for all $v$ by the rigidity theorem, also called the "strong multiplicity one theorem"; see [PS79]. All components of a cuspidal $G'(\mathbb{A})$-module are relevant by [Ze80, (9.7)] and, as noted above, unitarizable. Note that an irreducible $\pi$ in $L^2$ is unitary (its space is complete in an invariant inner form). Its subspace $\pi^\infty$ of smooth vectors is admissible and unitarizable. A representation $\pi$ is called unitarizable if its space is dense in a unitary representation $\widehat{\pi}$.

## 13.11 Correspondence

Given $G$ or $D$, there is a finite set $S$ of places $v$ of $F$ such that for every $v$ outside $S$ the division algebra $D$ *splits*, namely, $D \otimes_F F_v = M(d, F_v)$. We say that $\pi_v$ *corresponds* to $\pi'_v$ if $G_v \simeq G'_v$ (thus $v \notin S$) and $\pi_v \simeq \pi'_v$, or, more generally for arbitrary $v$, if $\pi_v$ corresponds to $\pi'_v$ by the Local Theorem 13.8. In this case we also say that $\pi'_v$ comes from $\pi_v$. An irreducible $G(\mathbb{A})$-module $\pi = \otimes_v \pi_v$ *lifts*, or *corresponds*, to an irreducible $G'(\mathbb{A})$-module $\pi' = \otimes_v \pi'_v$ if $\pi_v$ *corresponds* to $\pi'_v$ for all $v$. We also say that $\pi'$ comes from $\pi$ , in this case. An automorphic $G(\mathbb{A})$-module which lifts to a cuspidal $G'(\mathbb{A})$-module will be called *$G'$-cuspidal.*

THEOREM 13.12 (Global Theorem for Generic Representations).

(1) *All local components of a $G'$-cuspidal $G(\mathbb{A})$-module are relevant.*
(2) *Each $G'$-cuspidal $G(\mathbb{A})$-module occurs in the discrete spectrum $L_d$ of $L = L(G(F)\backslash G(\mathbb{A}))$ with multiplicity one.*
(3) *If $\pi_1 = \otimes_v \pi_{1,v}$ and $\pi_2 = \otimes_v \pi_{2,v}$ are $G'$-cuspidal $G(\mathbb{A})$-modules and $\pi_{1,v} \simeq \pi_{2,v}$ for almost all $v$, then $\pi_1 \simeq \pi_2$.*
(4) *Correspondence defines a bijection from the set of $G'$-cuspidal $G(\mathbb{A})$-modules $\pi = \otimes_v \pi_v$ to the set of cuspidal $G'(\mathbb{A})$-modules $\pi' = \otimes \pi'_v$ such that $\pi'_v$ is obtained by the local correspondence for all $v$ (in $S$). Every cuspidal $G(\mathbb{A})$-module is $G'$-cuspidal.*

REMARK. Part (1) is the motivation for the definition of “relevant” representations. Part (2) is called the “multiplicity one” theorem for the $G'$-cuspidal spectrum of $G$. Part (3) is called the “rigidity” theorem for the $G'$-cuspidal spectrum. Part (4) is called the Deligne-Kazhdan correspondence.

### *13.13 Remarks*

The local theorem is proven below for $F$ of characteristic 0. The positive characteristic case follows from [Ka86.2]. Theorem 13.12 is proven here only for the subset of the cuspidal $G'(\mathbb{A})$-modules $\pi'$ with two cuspidal components, using the simple form of the trace formula of Corollary 4.4. We reduce “two” to “one” using the same version of the trace formula and “regular functions” in Section 26. This Corollary 4.4 applies to any test function $f = \otimes_v f_v$ which has a cuspidal component $f_u$, and at a second place $u'$, the component $f_{u'}$ is *any* function whose orbital integrals vanish on the regular non-elliptic set. This $f_{u'}$ lies in the class $A_c^\infty(G(F_{u'}))$ of [Ka86.1]; see also 6.9, which is called the class of discrete functions in [BDK86] (see also 8.9). In particular, $f_{u'}$ can be taken to be a pseudo-coefficient of any square-integrable $G(F_{u'})$-module. Had we proven Corollary 4.4 only for $f$ such that $f_{u'}$ is supported on the elliptic regular set, we would not have been able to prove Global Theorem 13.12 except in the special, more elementary case where the simple algebra underlying $G$ is a division algebra which is a division algebra at some place. An elementary proof of this case is given in the first section.

## 14 Germs

### *14.1 Comparison*

The goal is to compare representations of and orbital integrals on $G' = \mathrm{GL}(n)$ with the analogous objects associated to its inner form $G$. Note that there is no difference between conjugacy and stable conjugacy; if two elements of $G(F)$ are conjugate in $G(\overline{F})$, then they are conjugate in $G(F)$. This property fails for a general reductive connected $F$-group.

### *14.2 Matching*

Let $F$ be a local non-Archimedean field of characteristic 0. Fix Haar measures $\mathrm{d}x$ and $\mathrm{d}x'$ on $G(F)$ and $G'(F)$. Write $\gamma \mapsto \gamma'$ if $\gamma$, $\gamma'$ are semisimple elements of $G(F)$ and $G'(F)$ with $p_\gamma = p_{\gamma'}$. If $\gamma$, $\gamma'$ are regular, that is, have distinct eigenvalues, their centralizers in $G$, $G'$ are tori $T$, $T'$. These tori are isomorphic if $\gamma \mapsto \gamma'$. In this case we take the Haar measures on $T(F)$ and $T'(F)$ to be equal. The orbital integral $\Phi(x,f)$ of a function $f$ in $C_c^\infty(G(F))$ and its analogue for $G'(F)$ was defined in Subsection 2.4.

PROPOSITION. *For every $f$ in $A_c^\infty(G(F))$, there exists $f'$ in $A_c^\infty(G'(F))$, and for every such $f'$, there is such $f$, so that $\widetilde{\Phi}(\gamma,f) = \widetilde{\Phi}(\gamma',f')$ for every elliptic regular $\gamma$ and $\gamma'$ with $p_\gamma = p_{\gamma'}$.*

PROOF. This follows from Proposition 6.11. □

This proposition proves assumptions 12.1 and 12.2 in the present case.

DEFINITION 14.3. The functions $f$ in $C_c^\infty(G(F))$ and $f'$ in $C_c^\infty(G'(F))$ are called *matching* if $\Phi(x,f) = \Phi(x',f')$ for all regular $x'$ in $G'(F)$ and $x$ in $G(F)$ with $p_x = p_{x'}$, and $\Phi(x',f') = 0$ for all regular $x'$ in $G'(F)$ which do not come from $G(F)$.

We also state the following:

THEOREM 14.4. *For every $f$ in $C_c^\infty(G(F))$, there exists $f'$ in $C_c^\infty(G'(F))$, and for every $f'$ in $C_c^\infty(G'(F))$ so that $\Phi(f')$ is zero at any regular $x$ in $G'(F)$ which does not come from $G(F)$, there exists $f$ in $C_c^\infty(G(F))$, so that $f$ and $f'$ are matching.*

This theorem will be proven by induction on the Levi subgroup of $G(F)$. Hence we now assume the validity of the theorem for every proper Levi subgroup $M$ of $G$. Consequently, we can use Assumption 10.6 in our case. The proof is based on the correspondence theorem for tempered local representations. We will complete the proof in Section 22.

# 15 Comparison

## 15.1 Measures

Let $F$ be global, $n = md$, $G = \mathrm{GL}(m, D)$ the multiplicative group of the $m \times m$ matrix algebra over the central division algebra $D$ of dimension $d^2$ over $F$, and $G' = \mathrm{GL}(n)$. Note that $G(F_v) = \mathrm{GL}(m, D_v)$. By definition of inner forms, $G$ and $G'$ are isomorphic over an algebraic closure $\overline{F}$ of $F$. Using this isomorphism, we can transfer a differential form of maximal degree on $G'$ rational over $F$ to one on $G$. These define Haar measures $\mathrm{d}x_v$ and $\mathrm{d}'x_v$ on $G(F_v)$ and $G'(F_v)$ for all $v$, which we call compatible, and consequently we can choose compatible product measures $\mathrm{d}x = \otimes \mathrm{d}x_v$ and $\mathrm{d}'x = \otimes \mathrm{d}'x_v$ on $G(\mathbb{A})$, $G'(\mathbb{A})$.

### 15.2 Conjugacy Classes

There is a bijection from the set of conjugacy classes in $D^\times$ (over a local or global field), to the set of elliptic conjugacy classes in $\mathrm{GL}(d, F)$. Similarly, there is a bijection from the set of semisimple conjugacy classes in $G(F) = \mathrm{GL}(m, D)$ to the set of semisimple conjugacy classes in $G'(F) = \mathrm{GL}(n, F)$ with an elliptic representative in the Levi subgroup $\prod_i \mathrm{GL}(da_i, F)$, $\sum_i a_i = m$. Globally, if $G$ ramifies at the finite set $V$ of places of $F$, there is a bijection from the set of conjugacy classes of tori $T$ in $G$ over $F$ into the set of conjugacy classes of tori $T'$ in $G'$ such that at each $v$ in $V$ the torus $T'(F_v)$ of $G'(F_v)$ is obtained from a $F_v$-torus $T(F_v)$. We choose compatible product measures $\mathrm{d}t = \otimes_v \mathrm{d}t_v$, $\mathrm{d}'t = \otimes_v \mathrm{d}'t_v$ on the matching tori $T(\mathbb{A})$, $T'(\mathbb{A})$, which are isomorphic over $F$.

### 15.3 Test Functions

We choose functions $f = \otimes_v f_v$ on $G(\mathbb{A})$ and $f' = \otimes_v f'_v$ on $G'(\mathbb{A})$ such that $f_v$ and $f'_v$ are matching for all $v$. In fact, for $v$ outside $V$, the groups $G(F_v)$ and $G'(F_v)$ are isomorphic over $F_v$, and we take $f_v, f'_v$ equal under this isomorphism. For almost all $v$, we take $f_v = f_v^0 = f'_v$. Corollary 14.2 and the inductive assumption of Theorem 14.4 show that there exist sufficiently many matching pairs in $C_c^\infty(G(F_v))$, $C_c^\infty(G'(F_v))$ for our purposes.

PROPOSITION 15.4. *If $f$ and $f'$ are matching and satisfy (each) the (three) requirements of Corollary 4.4, then $\sum \operatorname{tr} \pi'(f') = \sum m(\pi) \operatorname{tr} \pi(f)$. The sums range over the cuspidal spectra of $L^2(G'(F)\backslash G'(\mathbb{A}))$ and $L^2(G(F)\backslash G(\mathbb{A}))$.*

PROOF. This follows from Corollary 4.4. □

We used the multiplicity one theorem for $L_0^2(G'(F)\backslash G'(\mathbb{A}))$ to conclude that the multiplicities $m(\pi')$ on the left are equal to 1.

## 16 Existence

### 16.1 Pseudo-Coefficients

Let $G(F)$ be a reductive $p$-adic group and $\pi_0$ a square-integrable $G(F)$-module. A *pseudo-coefficient* of $\pi_0$ is a function $f$ in $A_c^\infty(G(F))$ (see Section 6) with $\operatorname{tr} \pi_0(f) = 1$ and $\operatorname{tr} \pi(f) = 0$ for every tempered (irreducible) $G(F)$-module $\pi$ inequivalent to $\pi_0$. If $\pi_0$ is cuspidal, then each of its (normalized) matrix coefficients is a pseudo-coefficient (in fact $\operatorname{tr} \pi(f) = 0$ if $\pi$ is irreducible and inequivalent to $\pi_0$). In general, the existence of a pseudo-coefficient is proven in [Ka86.1, Theorem K]. See also [BDK86].

Let $F$ be a global field. Fix a finite set $V$ of non-Archimedean places. Fix three distinct non-Archimedean places $w$, $u$ and $u'$ outside $V$. Although more general variants of the following proposition can be proven, for simplicity we now assume that $G = \mathrm{GL}(n)$.

PROPOSITION 16.2. *Fix a cuspidal $G(F_u)$-module $\pi_{0,u}$. Let $\pi_{0,w}$ be a square-integrable $G(F_w)$-module. Then there exists a cuspidal $G(\mathbb{A})$-module $\pi = \otimes_v \pi_v$, such that the following properties hold:*

(1) $\pi_w \simeq \pi_{0,w}$.
(2) $\pi_u \simeq \pi_{0,u}$.
(3) *For each $v$ in $V$ the component $\pi_v$ is Steinberg.*
(4) *$\pi_{u'}$ is square-integrable.*
(5) *$\pi_v$ is unramified for each non-Archimedean place $v \neq u, u', w$ outside $V$.*

PROOF. We use Corollary 4.4 with a function $f = \otimes_v f_v$, chosen to have the following properties.

- $f_w$ is a pseudo-coefficient of $\pi_{0,w}$.
- $f_u$ is a matrix coefficient of $\pi_{0,u}$.
- For each $v$ in $V$, the component $f_v$ is a pseudo-coefficient of the Steinberg $G(F_v)$-module.
- $f_{u'}$ is supported on the regular elliptic set in $G(F_{u'})$.
- At each finite $v \neq u, u', w$ outside $V$, we take spherical ($K_v$-bi-invariant) $f_v$, with $f_v = f_v^0$ for almost all $v$.

These components can be and are chosen so that $\Phi(x,f) \neq 0$ for some elliptic regular $x$ in $G(F)$. Since the sum of (4.3.1) is finite, we can reduce the support of $f_{u'}$ so that the sum (4.3.1) consists of a single entry; hence it is nonzero. Hence there is a cuspidal $\pi$ with $\mathrm{tr}\,\pi(f) \neq 0$. This $\pi$ is generic; hence each of its local components $\pi_v$ is non-degenerate. It is easy to check that $\pi$ has the properties required by the proposition, using the following:

REMARK. A $G(F_v)$-module is called *elliptic* if its character is not identically zero on the regular elliptic set of $G(F_v)$. Theorem 9.7(b) of [Ze80] implies that every irreducible non-degenerate elliptic $G(F_v)$-module is square-integrable (in fact of a "generalized Steinberg" type).

The proposition follows. □

## 16.3 Hilbert-Schmidt Operators

For this subsection we consider a locally compact unimodular group $H$ with center $Z$, $\omega$ a character of $Z$ of absolute value one, and set $f^*(h) = \overline{f}(h^{-1})$. Let $L(H)$ denote the convolution $*$-algebra of complex valued functions on $H$ with $f(zh) = \omega(z)^{-1}f(h)$, $h \in H$ and $z \in Z$, such that $|f(h)|^2$ is integrable on $H/Z$. For a unitary irreducible $H$-module $\pi$, put $\pi(f) = \int_{H/Z} f(h)\pi(h)\,dh$. Suppose $B$ is a dense $*$-closed subalgebra of $L(H)$, $I$ is a set, $\{\pi_i\}_{i\in I}$ is a set of irreducible unitary pairwise inequivalent $H$-modules such that $\pi(f)$, $\pi_i(f)$ are Hilbert-Schmidt operators for all $f$ in $B$, and $\|\cdot\|$ is the norm. Suppose that $\{c_i\}_{i\in I}$ is a set of nonnegative real numbers such that $\sum_i c_i\|\pi_i(f)\|^2$ is finite for all $f$ in $B$. Then the remark in the second (proven in the third) paragraph in the proof of Lemma 16.1.1 (on page 251 of recent printing and page 496 in the original draft) of [JL70] asserts the following. For each positive $\epsilon$, there exists $f$ in $B$ with $\|\pi(f)\| \neq 0$ and $\sum_i c_i\|\pi_i(f)\|^2 \leq \epsilon\|\pi(f)\|^2$.

LEMMA. *If $\{d_i\}_{i\in I}$ are complex numbers such that $\sum_i d_i \operatorname{tr}\pi_i(f*f^*)$ is absolutely convergent to zero for all $f$ in $B$, then $d_i = 0$ for all $i$.*

PROOF. Note that $\operatorname{tr}\pi_i(f*f^*) = \|\pi_i(f)\|^2$. If $d_0 \neq 0$, there is $f$ in $B$ such that

$$\sum_{i\neq 0} |d_i| \operatorname{tr}\pi_i(f*f^*)$$

is bounded by $(1/2)|d_0| \operatorname{tr}\pi_0(f*f^*) \neq 0$, and we arrive at a contradiction. □

## 17 Isolation

Let $F_w$ be a local non-Archimedean field of characteristic 0. Let $G_w$ be the multiplicative group of the matrix algebra $M(m, D_w)$, where $D_w$ is a central division algebra over $F_w$ of rank $d$ and invariant $i/d$ (modulo one), with $(i,d) = 1$. Put $G' = \mathrm{GL}(n)$, $n = md$.

Recall the following standard notation. If $S$ is any finite set of places of $F$, put $\pi^S = \otimes_v\pi_v$ and $f^S = \otimes_{v\notin S} f_v$. Put also $\pi_S = \otimes_{v\in S}\pi_v$ and $f_S = \otimes_{v\in S} f_v$. Denote by $S_\infty$ the set of Archimedean places of $F$. Put $\mathbb{A}_f$ for the ring of adèles without Archimedean components.

PROPOSITION 17.1. *For every square-integrable $G'(F_w)$-module $\pi'_w$, there exist $G_w$-modules $\pi_w$ and positive integers $m(\pi_w)$ such that, for all matching $f'_w$ and $f_w$, we have*

$$(-1)^{n-m} \operatorname{tr}\pi'_w(f'_w) = \sum m(\pi_w) \operatorname{tr}\pi_w(f_w).$$

*If $C_w$ is an open compact subgroup of $G_w$, then the sum consists only of finitely many $\pi_w$ with a nonzero $C_w$-invariant vector.*

PROOF. Let $F$ be a totally imaginary number field whose completion at some place $w$ is our local field $F_w$. Choose a set $V$ of $n - m + 1$ non-Archimedean places including $w$. We may assume that $i$ is prime to $n = md$, since there are infinitely many primes in the arithmetic progression $\{i+kd;\ k \geq 0\}$. Choose a division algebra $D$ central over $F$ with the following invariants. At $w$ it is $i/d$, it is $i/n$ at each $v \neq w$ in $V$ and 0 outside $V$. Take $G = D^\times$. Then $G(F_w)$ is our $G_w = \mathrm{GL}(m, D_w)$, where $\mathrm{inv}_w D_w = i/d$. Fix three distinct non-Archimedean places $u$, $u'$, and $u''$ of $F$ outside $V$, a cuspidal $G(F_u)$-module $\pi_u$ and a matrix coefficient $f_u$ of $\pi_u$. Choose a unitary irreducible $G(F_\infty)$-module $\pi_\infty$. Using Lemma 16.3 with the dense $*$-closed subalgebra $B = C_c^\infty(G(F_\infty))$, we conclude from Proposition 15.4 that if $f'^\infty = \otimes f'_v$ and $f^\infty = \otimes_{v \notin S_\infty} f_v$ and $f'_v, f_v$ are matching for all $v$, then

$$\sum \mathrm{tr}\, \pi'^\infty(f'^\infty) = \sum m(\pi)\, \mathrm{tr}\, \pi^\infty(f^\infty). \tag{17.1.1}$$

On the left, the sum ranges over all $G(\mathbb{A}_f)$-modules $\pi'^\infty$ such that $\pi' = \pi'^\infty \otimes \pi_\infty$ is a cuspidal $G'(\mathbb{A})$-module with the cuspidal component $\pi'_u = \pi_u$ at $u$. On the right the sum is over the $G(\mathbb{A}_f)$-modules $\pi^\infty$, whose component at $u$ is the cuspidal $\pi_u$, so that $\pi = \pi^\infty \otimes \pi_\infty$ appears with positive multiplicity $m(\pi)$ in the (cuspidal) spectrum $L_0(G(F)\backslash G(\mathbb{A}))$ of $G(\mathbb{A})$.

Recall the following theorem of Harish-Chandra (see [BJ79]). This is just an adelic translation of a classical result of Siegel and others on the finiteness of the number of normalized modular forms with fixed level and weight.

LEMMA 17.2. *Let $C$ be an open compact subgroup of $G(\mathbb{A}_f)$. Then there are only finitely many (irreducible) automorphic $G(F)$-modules $\pi$ with a nonzero $C$-fixed vector and a given infinitesimal character at each Archimedean place (in particular with the fixed component $\pi_\infty$ at $S_\infty$).*

Let $V'$ be the union of $V$ and $\{u, u', u''\}$. Fix $f_v$ and $f'_v$ for $v$ in $V'$, and let $f_v = f'_v$ be a variable spherical ($K_v = G(\mathcal{O}_v)$-bi-invariant) function for the finite $v$ outside $V'$. Lemma 17.2 implies that the sum in (17.1.1) are both finite. It is clear from the theory of the Satake transform that, given a finite set $\{\pi_{iv}\,;\, i \geq 0\}$ of irreducible unramified pairwise-inequivalent $G(F_v)$-modules, there exists a spherical function $f_v$ with $\mathrm{tr}\, \pi_{iv}(f_v) = 0$ if $i \neq 0$ and $\mathrm{tr}\, \pi_{0v}(f_v) = 1$. We conclude that, given an irreducible $G(\mathbb{A}^{V'})$-module $\pi^{V'}$, for all matching $f_v, f_{v'}$ ($v$ in $V'$), we have

$$\sum \mathrm{tr}\, \pi'_{V'}(f'_{V'}) = \sum m(\pi)\, \mathrm{tr}\, \pi_{V'}(f_{V'}). \tag{17.2.1}$$

On the left the sum is over the irreducible representations $\pi'_{V'}$ of $\prod_{v \in V'} G'(F_v)$ such that $\pi' = \pi'_{V'} \otimes \pi^{V'}$ is cuspidal. The component at $u$ is our fixed cuspidal $\pi_u$. By the rigidity theorem of [JS81], there exists at most one such $\pi'$. We choose $\pi^{V'}$ so that $\pi'$ of Proposition 16.2 appears on the left. On the right the sum is over the equivalence classes of irreducible $\pi_{V'}$ such that $\pi = \pi_{V'} \otimes \pi^{V'}$ is cuspidal, with multiplicity $m(\pi)$. The sum on the right is not finite, *a priori*.

Since $f_u$ is a normalized coefficient of a cuspidal $G(F_u)$-module $\pi_u$, we have $\mathrm{tr}\, \pi_u(f_u) = 1$ and $\mathrm{tr}\, \pi'_u(f'_u) = 1$ for the $\pi$, $\pi'$ which appear in (17.2.1). At each $v \neq w$ in $V$, let $f_v$ be the function $1_v$ and $f'_v$ a matching function on $G(F'_v)$. The function $f'_v$ exists by Corollary 14.2. At such $v$, let $\pi_v$ be the trivial $G(F_v)$-module and $\pi'_v$ the Steinberg $G'(F_v)$-module. Then $\chi'_v(x') = (-1)^{n-1}\chi_v(x)$ on the elliptic regular set and $\mathrm{tr}\, \pi'_v(f'_v) = (-1)^{n-1}$. Moreover, if $v \neq w$ in $V$ and $\pi_v$ appears on the right of (17.2.1), then $\operatorname{tr} \pi_v(f_v)$ is 0 or 1. Since $(n-1)(n-m) \equiv (n-m)$ (modulo 2), we conclude that for all matching $f_w, f'_w$, and for all $f_{u''}$ that vanish on the singular set of $G_{u''}$, we have

$$(-1)^{n-m} \operatorname{tr} \pi'_w(f'_w) \operatorname{tr} \pi'_{u''}(f'_{u''}) = \sum m(\pi) \operatorname{tr} \pi_w(f_w) \operatorname{tr} \pi_{u''}(f_{u''}). \tag{17.2.2}$$

The sum is over an easily specified set of $(\pi_w, \pi_{u''})$. Note that $G$ splits at $u''$; hence $f'_{u''} = f_{u''}$. Moreover, the place $u''$ is chosen so that $\pi'_{u''}$ is unramified. The $\pi'$ of Proposition 16.2 is cuspidal; hence it has a Whittaker model, and $\pi'_{u''}$ is nondegenerate. Consequently, $\pi'_{u''}$ is equal to an irreducible representation which is induced from an unramified character of the upper triangular subgroup, by [Ze80, Theorem 9.7(b)].

Let $f'_{u''}$ be any function such that $\Phi(f'_{u''})$ is supported on the split regular set of $G(F_{u''})$, and its restriction to $A(F_{u''})$ is $A(\mathcal{O}_{u''})$-invariant. It is clear that if $I(t, f'_{u''}) \neq 0$, then the Levi subgroup $M_t$ of Section 7 is $A$, so that $\operatorname{tr} \pi(f'_{u''}) = \operatorname{tr} \pi_N(f'_{u''N})$ for any irreducible $G'(F_{u''})$-module $\pi$, where $N$ is the upper triangular unipotent group. The support of $I(f'_{u''})$ is an open closed set. Denote by $\theta$ its characteristic function, and replace $f'_{u''}$ by its product with $\theta$. This does not change the value of the orbital integral, but assures the vanishing of the compactly (modulo center) supported $f'_{u''}$ on the singular set. Note that Theorem 4.2 of [BZ76] implies that if $\operatorname{tr} \pi_N(f'_{u''N}) \neq 0$, then $\pi$ has a nonzero vector fixed by the first congruence subgroup, as it has Iwahori decomposition. By virtue of the Lemma 17.2, the sum of (17.2.2) is then finite, uniformly in the $f'_{u''}$ considered here. Hence we can apply linear independence of (finitely many) characters on $A_{u''}$. This, together with Frobenius reciprocity, implies that we may consider on the right only $\pi_{u''}$ which are subquotients of, hence equal to the irreducible unramified $\pi'_{u''}$. The first claim of the proposition follows. The last assertion of the proposition follows from Lemma 17.2. □

## 18 Correspondence

Let $F$ be non-Archimedean. Let $G$ be an inner form of $G' = \mathrm{GL}(n)$ over $F$, with $G(F) = \mathrm{GL}(m, D)$ for a division algebra $D$. We have an injection $x \mapsto x'$ of conjugacy classes from $G(F)$ to $G'(F)$. We denote the characters of the $G(F)$-module $\pi$ and $G'(F)$-module $\pi'$ by $\chi_\pi$ and $\chi_{\pi'}$ (or $\chi'$).

THEOREM 18.1. *The relation $\chi_{\pi'}(x') = (-1)^{n-m} \chi_\pi(x)$ for all matching regular conjugacy classes $x$, $x'$ in $G(F)$, $G'(F)$ defines a bijection between the set of equivalence classes of square-integrable $G(F)$-modules $\pi$ and such $G'(F)$-modules $\pi'$ and between tempered $G(F)$-modules $\pi$ and such $G'(F)$-modules $\pi'$ whose character $\chi'$ is nonzero on the set of regular $x'$ obtained from $x$ in $G(F)$.*

We start the proof here. It is completed in the next Section 19. In this section we deal only with the square-integrable part of the theorem. It is used in the next section to study the general, tempered case.

Let $\pi'$ be a square-integrable $G'(F)$-module. Proposition 17.1 (where we now omit the subscript $w$) establishes the existence of a Trace Identity (10.5.1) for the $\pi'$. By virtue of Proposition 14.2 and the induction assumption of Theorem 14.4 for $M \neq G$, the assumptions 10.6, 12.1 and 12.2 are valid. By Proposition 11.1 the $\pi$ of the Trace Identity (10.5.1) are square-integrable, and by Proposition 12.3, there are only finitely many $\pi$ in the sum. Since $f$ is an arbitrary function on $G(F)$, we conclude an identity of characters

$$(-1)^{m-n}\chi_{\pi'}(x') = \sum_{\pi} m(\pi)\chi_{\pi}(x)$$

for regular matching classes $x \mapsto x'$. On the right the sum ranges over a finite set of square-integrable $G(F)$-modules $\pi$. Applying the orthonormality relations for square-integrable $G(F)$ and $G'(F)$-modules of [Ka86.1, Theorem K], we conclude from $1 = \sum_{\pi} m(\pi)^2$ that the sum consists of a single $\pi$ with coefficient $m(\pi) = 1$.

## *18.2 Weyl Integration Formula*

Recall the Weyl integration formula

$$\int_{G/Z} f(g)\mathrm{d}g = \int' \Phi(t,f)\mathrm{d}t, \quad \text{where} \quad \int' \text{signifies} \quad \sum_{\{T\}}[W(T)]^{-1}\int_{T/Z}\Delta(t)^2.$$

Here the sum ranges over a set of representatives for the conjugacy classes of tori $T$ in $G$, $[W(T)]$ denotes the cardinality of the Weyl group of $T$, and $\Delta$ is the Jacobian.

For conjugacy invariant functions $\chi$, $\chi_1$ on the elliptic set of $G$, let $\langle\chi,\chi_1\rangle$ be $\int' \chi(x)\chi_1(x)\mathrm{d}x$, where $\{T\}$ extends only over the conjugacy classes of the elliptic tori $T$.

REMARK 18.3. Another proof for the existence of a square-integrable $\pi$ to match such $\pi'$, without using the finiteness result of Proposition 12.3, yet using Proposition 14.2 or the assumptions 12.1, 12.2, is as follows. It is clear that some $\pi$ appears in the sum of Proposition 17.1, since we can take $f'$ to be a pseudo-coefficient of $\pi'$ by Proposition 14.2. Fixing such $\pi$ we take $f$ in $A_c^{\infty}(G(F))$ with $\widetilde{\Phi}(f) = \chi_{\pi}$ on the elliptic regular set. It exists by [Ka86.1, Theorem K]. Then the sum of Proposition 17.1 is equal to $m(\pi)$. Then if $f'$ is a matching function (which exists by Proposition 14.2), then

$$|\operatorname{tr}\pi'(f')|^2 = \left|\int' \chi'(x')\widetilde{\Phi}(x',f')\,\mathrm{d}x'\right|^2 \leq \int' |\chi'(x')|^2\,\mathrm{d}x' \int' |\chi(x)|^2\,\mathrm{d}x$$

by the Schwarz inequality. The integrals are taken only over the elliptic tori of $G(F)$ or $G'(F)$, and we use the fact that $\widetilde{\Phi}(x',f') = \widetilde{\Phi}(x,f) = \chi(x)$. Here $\chi$, $\chi'$ are the characters of $\pi$ and $\pi'$. By the orthonormality relations $\langle \chi_\pi, \chi_{\pi_1} \rangle = \delta(\pi, \pi_1)$ of [Ka86.1, Theorem K], we conclude that $m(\pi) \leq 1$. As $m(\pi)$ is a positive integer, we conclude that $m(\pi) = 1$ and that the Schwarz inequality is an equality in our case, so that $\chi'(x') = c\widetilde{\Phi}(x',f') = c\widetilde{\Phi}(x,f) = c\chi(x)$ on the elliptic regular set, where $c$ is a number with $|c| = 1$. Hence $\pi$ is the only term in the sum, and $c = (-1)^{m-n}$.

In the opposite direction, given a square-integrable $\pi$, we take a pseudo-coefficient $f$ in $A_c^\infty(G(F))$ of $\pi$, and a matching function $f'$ in $A_c^\infty(G'(F))$. By the density Proposition 5.6 and the orthonormality relations of [Ka86.1], there exists a tempered elliptic, hence by [Ze80] square-integrable, $G'(F)$-module $\pi'$, with $\operatorname{tr} \pi'(f') \neq 0$. By the orthonormality relations on $G(F)$, the $G(F)$-module matching $\pi'$, whose existence was proven above, is our $\pi$.

## *18.4 Remark on Tempered Case*

We have now completed the proof of that part of the theorem which concerns square-integrable $\pi$, $\pi'$. The extension to the case of any tempered $\pi$ and $\pi'$ follows once we establish in Section 19 below that any tempered $G(F)$-module is *equal* to an induced $G(F)$-module from a square-integrable module. This result is well known in the case of the split group $G'$. In its proof we use that part of the theorem proven above, for square-integrable modules.

REMARK 18.5. In particular, we completed the proof and hence can use the assertion of Theorem 18.1 in the case $m = 1$, namely, when $G$ is the multiplicative group $D^\times$ of a division algebra $D$ central over $F$. Indeed, all $G(F)$-modules in this case are square-integrable, and the image of the correspondence here is the set of elliptic tempered, hence square-integrable, $G'(F)$-modules. This case was independently established in Section 1.

# 19 Tempered

PROPOSITION 19.1. *Any (irreducible) elliptic tempered $G(F)$-module is square-integrable.*

See Remark in 16.2, Subsection 13.6, and Subsection 10.4 for the respective definitions.

PROOF. Suppose that the character $\chi$ of $\pi$ is nonzero on the elliptic regular element $y$. Let $f$ be the characteristic function of a small neighborhood of $y$ (modulo $Z(F)$), where $\chi$ is constant. It is clear from the Weyl integration formula that

$$\langle \chi, \widetilde{\Phi}(f) \rangle = \int \chi(x) f(x) \, dx \; (= \operatorname{tr} \pi(f))$$

for $x$ in the elliptic set of $G(F)$. For our $f$ and $\chi$, we have $\langle\chi, \widetilde{\Phi}(f)\rangle \neq 0$. Since $f$ is supported on the regular set, there is a matching $f'$, with $\widetilde{\Phi}(x,f) = \widetilde{\Phi}(x',f')$ on the elliptic set. As $f'$ lies in $A_c^\infty(G'(F))$, there is a matching function $\varphi$ on the multiplicative group $D(F)$ of a division algebra of dimension $n^2$ central over $F$. Since $D(F)$ is compact modulo its center $Z(F)$, there are only finitely many $D(F)$-modules $\sigma_i$, with characters $\chi_{\sigma_i}$, and complex numbers $c_i$, so that $\widetilde{\Phi}(x,\varphi) = \sum c_i\chi_{\sigma_i}(x)$ on the regular $x$ in $D(F)$. If $\chi_i'$ are the characters of the $G'(F)$-modules $\pi_i'$ that correspond to the $\sigma_i$, then $\widetilde{\Phi}(x',f') = \sum c_i\chi_i'(x')$ on the elliptic regular set. Since the $\pi_i'$ are square-integrable, they correspond to square-integrable $G(F)$-modules $\pi_i$ with characters $\chi_i$. Hence $\widetilde{\Phi}(x,f) = \sum c_i\chi_i(x)$ on the regular elliptic set. Then $\langle\chi, \widetilde{\Phi}(f)\rangle = \sum c_i\langle\chi, \chi_i\rangle$. Since this is nonzero, we have $\langle\chi, \chi_i\rangle \neq 0$ for some $i$. But the orthonormality relations for square-integrable $G(F)$-modules of [Ka86.1, Theorem K] imply that $\pi$ is equivalent to the square-integrable $\pi_i$, as required. □

PROPOSITION 19.2. *Suppose $I$ is a $G(F)$-module normalizedly induced from a square-integrable $M(F)$-module, where $M$ is a Levi component of a proper parabolic subgroup. Then $I$ is irreducible.*

In particular, the same conclusion holds when the representation of $M(F)$ is tempered.

PROOF. Let $I = i_M^G\tau$ where $\tau$ is a square-integrable $M(F)$-module. Suppose that $I$ contains the irreducible representation $\pi$. We shall first show that $I$ is a multiple of the irreducible representation $\pi$. By induction, we assume this assertion for all proper Levi subgroups of $G$.

Suppose that $\pi$ does not lie in the space $R_I(G)$, see Subsection 6.7, which is spanned over $\mathbb{C}$ by the properly induced representations. By [Ka86.1, Theorem D], $\pi$ is elliptic. Since $\pi$ is also tempered, Proposition 19.1 implies it is square-integrable. The uniqueness theorem [BW80, Theorem 2.10] part of the Langlands classification implies that $\pi = \tau, M = G$ an so $I = \pi$ as required.

Suppose $\pi$ does lie in $R_I(G)$. Then by [Ka86.1, Proposition 1.1], there are finitely many proper Levi subgroups $L_i$, irreducible tempered $L_i(F)$-modules $\tau_i$ and complex numbers $\alpha_i$ so that in the Grothendieck group $R(G(F))$, we have

$$\pi = \sum_i \alpha_i i_{L_i}^G\tau_i.$$

Since $\tau_i$ is tempered, there is a unique (up to conjugacy) pair $(\sigma_i, R_i)$, where $R_i$ is a Levi subgroup of $L_i$ and $\sigma_i$ is a square-integrable $R_i(F)$-module, such that $\tau_i$ is a direct summand of $i_{R_i}^{L_i}\sigma_i$. Since $L_i \neq G$, the induction assumption implies that there is a positive integer $\beta_i^{-1}$ such that $i_{R_i}^{L_i}\sigma_i = \beta_i^{-1}\tau_i$. Hence

$$\pi = \sum_i \alpha_i\beta_i i_{R_i}^G\sigma_i.$$

Since $\pi$ is a constituent of $I = i_M^G \tau$, where $\tau$ is square-integrable, the expression for $\pi$ and the uniqueness part of the Langlands classification implies that either $I = i_{R_i}^G \sigma_i$ or $I$ and $i_{R_i}^G \sigma_i$ are not relatives in the terminology of [Ka86.1]. Hence $\pi = \alpha I$, where $\alpha$ is the sum of $\alpha_i \beta_i$ over the $i$ where $I = i_{R_i}^G \sigma_i$, as required. Thus $I = \alpha^{-1}\pi$, where $\alpha^{-1}$ is a positive integer.

It remains to show that $I$ is irreducible. For that we use the work of [Si80]. First consider the case where $M$ is of rank one; thus $M = M' \times M''$, with $M' = \mathrm{GL}(a, D)$, $M'' = \mathrm{GL}(b, D)$. Then [Si80, Theorem 2.5.8, p. 99] implies that $I$ is irreducible unless $a = b$, in which case its composition series has length bounded by the order of the Weyl group $W(A)$ of [Si80, p. 100], which is two. But if $I$ is the direct sum of $k$ copies of an irreducible, its commuting algebra has dimension $k^2$, which is at least 4, unless k=1.

Next we consider the general case. We shall express the $M(F)$-module $\tau$ as a product of square-integrables on $\mathrm{GL}(n_i, D)$'s. Thus we may, upon rearranging the factors, assume that

$$\tau = (\tau_1 \times \cdots \times \tau_1) \times \cdots \times (\tau_s \times \cdots \times \tau_s)$$

where each square-integrable $M_i(F)$-module $\tau_i$, with $M_i = \mathrm{GL}(n_i, D)$, occurs $t_i$ times and $\tau_i$, $\tau_j$ are inequivalent if $i \neq j$. Then $n = \sum_i n_i t_i$. Put $t_0 = 0$. The center of

$$M = (M_1 \times \cdots \times M_1) \times \cdots \times (M_s \times \cdots \times M_s)$$

is

$$A = A_1^{t_1} \times \cdots \times A_s^{t_s},$$

where $A_i$ is the center of $M_i$. Let $W(A)$ be the product of the symmetric groups $S_{t_i}$. The Harish-Chandra commuting algebra theorem [Si80, 5.5.3] asserts that the commuting algebra of $I$ is spanned by the intertwining operators $R(w)$, $w \in W(A)$, subject to the relations $R(1) = 1$ and $R(ww') = R(w)R(w')$. Hence it is generated by the $R(s(i))$, where $s(i)$ is a reflection of the form $(i, i+1)$, $t_{j-1} < i < t_j$. However, the operator $R(s(i))$ is induced (recall the induction is a functor) from the intertwining operator of the representation induced from $\tau$ on $M = M'_1 \times \cdots \times M'_m$ (the $M'_i$ are the $M_i$, relabeled) to

$$G_i = M'_1 \times \cdots \times M'_{i-1} \times X_i \times M'_{i+2} \times \cdots \times M'_m$$

and $X_i = \mathrm{GL}(2n_j, D)$ if $M'_i = \mathrm{GL}(n_j, D) = M'_{i+1}$. It follows from the rank one case that $R(s(i))$ is a scalar. Hence the commuting algebra of $I$ consists of scalars, which proves that $I$ is irreducible, as required. □

**Remark 1.** (i) We do not discuss here the normalization of intertwining operators.
(ii) It is possible to complete the proof of irreducibility above by further analyzing the proof of [Si80, Chapter II]. This proof does not require the commuting algebra theorem of [Si80, Chapter V], but we do not give it here.

# 20 Irreducibility

## 20.1 Induction

Let $\tau_i$ be cuspidal $M_i$-modules, where $M_i = \mathrm{GL}(n_i, D)$ and $\sum_{i=1}^m n_i = n$. Let $\nu$ be the character $\nu(x) = |\det(x)|$ of $M_i$—where by det we mean determinant on $\mathrm{GL}(n_i)$ followed by reduced norm on the division algebra $D$. Let $s_i$ $(1 \leq i \leq m)$ be real numbers. Put $\mathbf{s} = (s_1, \ldots, s_m)$. Let $I = i_M^G(\tau(\mathbf{s}))$ be the $G(F)$-module obtained by induction from the $M = \prod_i M_i$-module $\tau(\mathbf{s}) = \prod_{1\leq i\leq m} \tau_i \nu^{s_i}$. If $P$ is a parabolic with Levi $M$, then the module of coinvariants $I_P$ of $I$ with respect to $P$ consists, by [BZ76, (2.12)] (see the Geometric Lemma 8.6), of composition factors of the form $\tau_\alpha(\mathbf{s}) = \prod_i \tau_{\alpha(i)} \nu^{s_{\alpha(i)}}$, where $\alpha$ ranges over the symmetric group $S_m$ on $m$ letters.

DEFINITION 20.2. (i) The (cuspidal) *support* of a subquotient $\pi$ of $I$ is the set of $M(F)$-modules $\tau_\alpha(\mathbf{s})$ which are constituents of $\pi_P$.

(ii) The representation $\pi$ is called *multiplicity free* if each $\tau_\alpha(\mathbf{s})$ occurs in $\pi_P$ at most once.

(iii) A reflection in $S_m$ of the form $r(i) = (i, i+1)$ is called *admissible* if $|s_{i+1} - s_i| \neq 1$ or $\tau_{i+1}$ is inequivalent to $\tau_i$. This term depends on $\tau(\mathbf{s})$.

PROPOSITION 20.3. (i) *If $m = 2$ and $|s_1 - s_2| \neq 1$ or the cuspidal $\tau_1$, $\tau_2$ are inequivalent, then $I = i_M^G(\tau_1\nu^{s_1} \times \tau_2\nu^{s_2})$ is irreducible.*

(ii) *The support of $\pi$ is invariant under the action of the set of admissible reflections.*

PROOF. (i) By Proposition 19.2, which deals with the tempered case $s_1 = s_2$, we may assume that $s_1 \neq s_2$, hence $s_1 > s_2$ without loss of generality. The module of coinvariants of $I$ with respect to the parabolic subgroup of type $(n_1, n_2)$ has two exponents, one increasing and one decaying. If $I$ is reducible, then its composition series has length two by [Si80, Theorem 2.5.8], since the $\tau_i$ are cuspidal. One of the constituents has the decaying exponent and hence is square-integrable by Harish-Chandra's criterion, [Si80, (4.4.4)], [Cas, (4.4.6)], quoted in Subsection 10.4. But this square-integrable should correspond by Theorem 18.1 to a square-integrable constituent of the representation $I' = i_M^G(\tau_1'\nu^{s_1} \times \tau_2'\nu^{s_2})$ corresponding to $I$. As $I'$ is irreducible by [BZ76], (i) follows.

(ii) Suppose that $\tau(\mathbf{s})$ lies in the support of $\pi$. We have to show that so does $\tau_{r(i)}(\mathbf{s}) = \tau_1\nu^{s_1} \times \cdots \times \tau_{i+1}\nu^{s_{i+1}} \times \tau_i\nu^{s_i} \times \cdots$. For that we consider the parabolic subgroup $Q$ of type $(n_1, \ldots, n_{i-1}, n_i + n_{i+1}, n_{i+2}, \ldots)$, and its standard Levi subgroup $L$. Since $\pi_P = (\pi_Q)_{L\cap P}$, there is an irreducible $L(F)$-module $\varepsilon$ in the composition series of $\pi_Q$ such that $\varepsilon_{L\cap P}$ contains $\tau(\mathbf{s})$. But part (i) implies that if $|s_{i+1} - s_i| \neq 1$ or $\tau_i$, $\tau_{i+1}$ are inequivalent, then $\varepsilon_{L\cap P}$ contains also $\tau_{r(i)}(\mathbf{s})$, and (ii) follows.

□

PROPOSITION 20.4. (i) *If $\pi$ is a multiplicity-free subquotient of $I$, and the set of admissible transpositions acts transitively on the support of $\pi$, then $\pi$ is irreducible.*

(ii) *Suppose that $\sigma_1$ and $\sigma_2$ are square-integrable and $|s| < 1/2$. Then*

$$I = i_M^G(\sigma_1 \nu^s \times \sigma_2 \nu^{-s})$$

*is irreducible.*

REMARK. Part (ii) here sharpens (i) of Proposition 20.3.

PROOF. (i) This is clear by (ii) of Proposition 20.3 and the fact that each subquotient of $\pi$ has a nonzero subquotient $\tau_\alpha(\mathbf{s})$ in its module of coinvariants $\pi_P$.

(ii) Since $\sigma_1$ and $\sigma_2$ are square-integrable on $\mathrm{GL}(n_i, D)$, they correspond by Theorem 18.1 to such representations $\sigma_1'$ and $\sigma_2'$ on the split groups $\mathrm{GL}(dn_i, F)$. The $\sigma_i'$ are classified by [BZ76] for $G'$: a square-integrable representation $\sigma'$ of $G' = \mathrm{GL}(n, F)$, $n = md$, is the unique subrepresentation of the representation induced to $G'$ from a cuspidal representation

$$\left(\left(\tau', \frac{m'-1}{2}\right), \left(\tau', \frac{m'-3}{2}\right), \ldots, \left(\tau', \frac{1-m'}{2}\right)\right)$$

—where we put $(\tau', s)$ for $\tau'\nu^s$—on a Levi subgroup of type $(d', \ldots, d')$ ($m'$ times). Thus $\tau'$ is a cuspidal representation of $\mathrm{GL}(d', F)$. Since $n = m'd' = md$, we gather these representations in segments of length least common multiply of $d$ and $d'$, which is $de$, $e = d'/(d, d')$. Denote by $\tau$ the cuspidal representation of $\mathrm{GL}(e, D)$ corresponding to the square-integrable subrepresentation of the $\mathrm{GL}(ed, F)$-module induced from

$$\left(\left(\tau', \frac{e'-1}{2}\right), \left(\tau', \frac{e'-3}{2}\right), \ldots, \left(\tau', \frac{1-e'}{2}\right)\right), \qquad e' = d/(d, d').$$

We conclude that there exist cuspidal $\tau_1$ and $\tau_2$ on $\mathrm{GL}(e_1, D)$ and $\mathrm{GL}(e_2, D)$, with $e_i$ dividing $n_i$, thus $m_i = n_i/e_i$ is an integer, and $n = dn_1 + dn_2$, such that the *support* of $I$ consists of all $m_1 + m_2$ tuples $(a_{\alpha(i)})$ obtained from

$$(a_i) =$$
$$\left(\left(\tau_1, \frac{m_1-1}{2}+s\right), \left(\tau_1, \frac{m_1-3}{2}+s\right), \ldots, \left(\tau_1, \frac{1-m_1}{2}+s\right); \left(\tau_2, \frac{m_2-1}{2}-s\right), \ldots\right) \tag{20.4.1}$$

on permuting by $\alpha$ in $S_{m_1+m_2}$ which satisfies $\alpha(i) < \alpha(j)$ if $i < j \leq m_1$ or $m_1 < i < j$. We may assume that $s \neq 0$ by Proposition 19.2. This set is multiplicity free, and the set of admissible transpositions act transitively if (i) $\tau_1$ is inequivalent to $\tau_2$; when $\tau_1 = \tau_2$, if (ii) $m_1 - m_2$ is even, as $2|s| < 1$; or

(iii) $m_1 - m_2$ is odd, unless $|s| = 1/4$. Hence the proposition follows from part (i) for $\pi$ being $I$, except that we have to deal with the case when $\tau_1 = \tau_2$ (and $m_1 - m_2$ is odd, $|s| = 1/4$). In this case we use the notation (20.4.1) for vectors in the support, omitting the reference to $\tau_1$, $\tau_2$, namely, from now on we deal with the case $\tau_1 = \tau_2$.

By a *segment* we mean a vector $(c_i)$ of real numbers with $c_i - c_{i+1} = 1$ for all $i$. The center of the segment $(c_1, \ldots, c_m)$ is $(c_1 + c_m)/2$. The vector $(c_i)$ is called an *L-vector* if it has a partition $(\mathbf{b}_j)$ into segments $\mathbf{b}_j = (b_{ij})$, whose centers are nondecreasing. The description of tempered representations of $G'$ by [BZ76], transferred to $G$ by Theorem 18.1 and Proposition 19.2, together with the classification theorems of [BW80, IV§2] asserts that each irreducible $G(F)$-module has (at least one) $L$-vector in its support. But it is easy to check that the support of our $I$, namely, the set of $(a_{\alpha(i)})$ obtained from the $(a_i)$ of (20.4.1), contains only one $L$-vector. Hence $I$ is irreducible, as required. □

COROLLARY 20.5. *Given any irreducible* $\tau_1$, $\tau_2$ *and* $\sigma_1$, $\sigma_2$, $s$ *as in* (*ii*) *of Proposition* 20.4, $i_M^G(\tau_1 \times \sigma_1\nu^s \times \sigma_2\nu^{-s} \times \tau_2)$ *is equal to* $i_M^G(\tau_1 \times \sigma_2\nu^{-s} \times \sigma_1\nu^s \times \tau_2)$. *In particular, one of them is unitarizable if and only if the other is.*

PROOF. This follows from induction in stages, since $i_M^G(\sigma_1\nu^s \times \sigma_2\nu^{-s})$ is irreducible, hence equal to $i_M^G(\sigma_2\nu^{-s} \times \sigma_1\nu^s)$. □

# 21 Unitarity

## *21.1 Relevance*

Recall (Definition 13.7) that a $G(F)$-module $\pi$ is called *relevant* is there is $M$ as in (1) and $\tau_i$ as in (2) such that $\pi$ is as in (3).

(1) A Levi subgroup $M$ of $G$ of the form $M_0 \times \prod_{i=1}^m (M_i \times M_i)$, or of the form $\prod_{i=1}^M (M_i \times M_i)$, where $M_i$ is a multiplicative group of a simple algebra for each $i$.
(2) Irreducible tempered $M_i$-modules $\tau_i$, $0 \leq i \leq m$.
(3) Distinct positive numbers $s_i < 1/2$, $1 \leq i \leq m$, such that $\pi$ is equivalent to $i_M^G(\tau)$ or $i_M^G(\tau_0 \times \tau)$ and $\tau$ is the $\prod_{i=1}^m (M_i(F) \times M_i(F))$-module $\prod_{i=1}^m (\tau_i\nu^{s_i} \times \tau_i\nu^{-s_i})$.

The motivation for this definition is the fact that each component of any cuspidal automorphic $G(\mathbb{A})$-module is unitarizable and non-degenerate, hence relevant by [Be84, Lemma 8.9] and [Ze80, Theorem 9.7(b)].

In this section we complete the proof of the Local Theorem 13.8 and show that a relevant $G(F)$-module is irreducible and unitarizable.

Let $U$ be a finite dimensional complex vector space.

LEMMA 21.2. *Let $\langle\cdot,\cdot\rangle_s$ be a family of non-degenerate Hermitian forms on $U$ depending continuously on a parameter $s$ in a connected set. If $\langle\cdot,\cdot\rangle_s$ is positive definite for some value of $s$, then it is positive definite for all $s$.*

PROOF. The set of $s$ where $\langle\cdot,\cdot\rangle_s$ is positive definite is open, and also closed, since $s$ ranges over a connected set. □

PROPOSITION 21.3. *Let $\sigma$ be square-integrable and $s \in \mathbb{R}$, $-1/2 < s < 1/2$. Then the $G(F)$-module $I(s) = i_M^G(\sigma\nu^s \times \sigma\nu^{-s})$ is unitarizable.*

PROOF. Let $V_s$ be the space of the representation $I(s)$. As a space $V_s$ is independent of $s$, but the action of $G$ does depend on $s$. Since $I(s)$ is irreducible, it is equivalent to its contragredient (see [BZ76]) $I(s)' = i_M^G(\sigma\nu^{-s}, \sigma\nu^s)$. Note that the square-integrable $\sigma$ is self-contragredient, so $\sigma' = \sigma$, $(\sigma\nu^s)' = \sigma\nu^{-s}$, and $i(\tau)' = i(\tau')$. The choice of an isomorphism $I(s) \to I(s)' = I(-s)$, which is unique up to a scalar, determines an Hermitian inner product $\langle\cdot,\cdot\rangle$ on $V_s$, which is non-degenerate. We can choose the isomorphism, or the inner product $\langle\cdot,\cdot\rangle_s$, to vary continuously with $s$. For each compact open congruence subgroup $C$ in $K$, the isomorphism $V_s \to V_{-s}$ determines an isomorphism from the space $V_s^C$ of $C$-fixed vectors in $V_s$, to the dual $(V_s^C)' = (V_s')^C = V_{-s}^C$. For each $C$ we obtain a continuous family $\langle\cdot,\cdot\rangle_{s,C}$ of non-degenerate Hermitian inner product, which varies continuously with the parameter $s$ in $-1/2 < s < 1/2$.

Now the tempered $I(0)$ is unitarizable, being normalizedly induced from a unitarizable representation $\sigma \times \sigma$. Hence $\langle\cdot,\cdot\rangle_{s,C}$ is positive definite at $s = 0$. Consequently it is positive definite for all $s$ with $-1/2 < s < 1/2$ and for all $C$ by Lemma 21.2. As $V_s$ is the union of $V_s^C$ over all $C$, we conclude that $\langle\cdot,\cdot\rangle_s$ is positive definite for all $s$; hence $I(s)$ is unitarizable. □

COROLLARY 21.4. *Let $\sigma_i$, $1 \le i \le m$, be square-integrable. Let $s_i$ be positive numbers, $1 \le i \le k$, $k \le m$, with $s_i < 1/2$. Let $\sigma$ denote the product $\prod_{i=1}^k(\sigma_i\nu^{s_i} \times \sigma_i\nu^{-s_i}) \times \prod_{j=k+1}^m \sigma_j$. Then $I = i_M^G\sigma$ is unitarizable for any choice of a parabolic subgroup or, equivalently, for any order of the factors $\sigma_i\nu^{s_i}$, $\sigma_i\nu^{-s_i}$, $\sigma_j$.*

PROOF. The representation $I$ is independent of the choice of factors by Corollary 20.5. Let $M = \prod_{i=1}^k(M_i \times M_i) \times \prod_j M_j$ be the Levi subgroup from which we induce. If $M_i = \mathrm{GL}(n_i, D)$, put $L_i = \mathrm{GL}(2n_i, D)$, and $L = \prod_i L_i \times \prod_j M_j$. Since

$$i_M^G(\sigma) = i_L^G i_M^L \sigma,$$

and $i_M^L\sigma$ is unitarizable by Proposition 21.3, we conclude that $I$ is unitarizable. □

THEOREM 21.5. *Suppose that the $\tau_i$, $0 \le i \le m$, are irreducible and tempered and $s_i$ are distinct positive numbers with $s_i < 1/2$. Put $\tau = \prod_{i=1}^m(\tau_i\nu^{s_i} \times \tau_i\nu^{-s_i})$. Then the induced representations $I = i_M^G(\tau)$ and $i_M^G(\tau_0 \times \tau)$ are irreducible.*

PROOF. We induce from the Levi $M = M_0 \times \prod(M_i \times M_i)$. Here $M_0 = \mathrm{GL}(n_0, D)$ for $n_0 \ge 0$, and $n_0 = 0$ means $M_0$ does not appear in $M$. There exists a parabolic subgroup $P$ with Levi subgroup $M$ such that the vector determined by

$(0; s_1, -s_1; s_2, -s_2; \ldots)$ lies in the positive Weyl chamber (in the Lie algebra of the diagonal subgroup) determined by $P$. Consequently $I$ has a unique quotient $J$; see [BW80, IV, (4.6)]. On the other hand, Corollary 21.4 implies that $I$ is unitarizable. As each constituent of a unitarizable representation is a direct summand of it, the unique quotient $J$ must be $I$ itself, and we conclude that $I$ is irreducible. □

## *21.6 End of Proof of Local Theorem* 13.8

It is now clear that by parabolic induction Theorem 18.1 extends to hold also for relevant, not only tempered, $G(F)$ and $G'(F)$-modules. This completes the proof of the Local Theorem 13.8. □

REMARK 21.7. The result of Proposition 19.2 and Theorem 18.1 are due to [DKV84] and that of 21.5 to [F87.1]. These theorems are proven here for local fields of characteristic zero. The positive characteristic case follows from [Ka86.2].

# 22 Induction

It remains to complete the proof of Theorem 14.4, which asserts the existence of matching orbital integrals, using the correspondence for tempered representations Theorem 18.1. In the proof of Theorem 18.1, we used the induction assumption of Theorem 14.4, namely, the statement of Theorem 14.4 for all proper Levi subgroups. Our aim is to show that for any $f$ in $C_c^\infty(G(F))$, there exists a matching $f'$ in $C_c^\infty(G'(F))$, and for any suitable $f'$ in $C_c^\infty(G'(F))$ (thus $\Phi(x', f') = 0$ for any regular $x'$ not obtained from $x$ in $G(F)$), there is a matching $f$ in $C_c^\infty(G(F))$.

Note we have the following:

LEMMA 22.1. *For every $f$ there exists $f'$ with $\operatorname{tr}\pi(f) = \operatorname{tr}\pi'(f')$ for all corresponding tempered $\pi$, $\pi'$, and $\operatorname{tr}\pi'(f') = 0$ for the tempered $\pi'$ which are not obtained by the correspondence.*

PROOF. Given $f$ we define the function $f'_{G'}$ on the space of tempered $\pi'$ by $f'_{G'}(\pi') = \operatorname{tr}\pi(f)$ if $\pi$ corresponds to $\pi'$ by Theorem 18.1 and by $f'_{G'}(\pi') = 0$ if the character of $\pi'$ is zero on the set of regular $x'$ obtained from $x$. Then $f'_{G'}$ is in the space $F_{\mathrm{good}}$ of [BDK86] or [F95]. Thus it is a trace function by the trace Paley-Wiener Theorem 8.4. Namely, there is an $f'$ with $f'_{G'}(\pi') = \operatorname{tr}\pi'(f')$ for all tempered $\pi'$, as required.

The same argument implies the existence of $f$ for a given suitable $f'$. □

PROPOSITION 22.2. *Suppose that $f$ and $f'$ satisfy $\operatorname{tr}\pi(f) = \operatorname{tr}\pi'(f')$ for all corresponding tempered $\pi$ and $\pi'$, and $\operatorname{tr}\pi'(f') = 0$ for the tempered $\pi'$ not obtained by the correspondence. Then $f, f'$ are matching.*

PROOF. By induction on the Levi subgroup $M$ of the parabolic subgroup $P = MN$ of $G$. Denote by $P'$, $M'$, $N'$ the corresponding objects of $G'$. Let $\delta_P$ be the modulus homomorphism on $P$. Thus $\mathrm{d}(ab) = \delta_P(a)\mathrm{d}b$, for $a$, $b$ in $P(F)$, for any right Haar measure $\mathrm{d}b$ on $P(F)$. For $a$ in the center $A(F)$ of $M(F)$ we have $\delta_P(a) = \prod |\alpha(a)|$. The product ranges over all roots of $A$ in $N$. As usual, we put

$$f_N(m) = \delta_P(m)^{1/2} \int_K \int_N f(k^{-1}mnk)\,\mathrm{d}n\,\mathrm{d}k.$$

Here $K$ is a maximal compact subgroup of $G(F)$ with $G(F) = KP(F)$. For any $m$ in $M(F)$ regular in $G(F)$, we have $I(m,f) = I^M(m,f_N)$, where $I(x,f) = \Delta(x)\Phi(x,f)$ and

$$\Delta(x) = |\prod_{i<j}(x_i - x_j)^2/x_ix_j|^{1/2}$$

if $x$ has distinct eigenvalues $x_i$. The orbital integral $I^M$ is the analogous object defined for $M$. Analogous notations are employed in the case of $G'$. Further, we note that if $\pi = i_M^G(\tau)$ is the $G(F)$-module normalizedly induced from the $M(F)$-module $\tau$, then $\operatorname{tr}\pi(f) = \operatorname{tr}\tau(f_N)$ by a standard evaluation of the character of an induced representation. Consequently, if $\tau$, $\tau'$ are corresponding tempered $M(F)$ and $M'(F)$, we have $\operatorname{tr}\tau(f_N) = \operatorname{tr}\tau'(f'_N)$, and $\operatorname{tr}\tau'(f'_N) = 0$ for the tempered $\tau'$ not obtained from any $\tau$. By induction we have $I(x',f') = I^M(x',f'_N) = I^M(x,f_N) = I(x,f)$ for the regular $x'$ in $M'(F)$ which come from $x$, and $I(x',f') = 0$ for the regular $x'$ in $M'(F)$ which do not come from $G(F)$. It remains to show the proposition for elliptic regular $x$, $x'$.

Choose matching elliptic regular $y$, $y'$. Let $U'$ be a sufficiently small compact neighborhood of $y'$, and $f'_0$ a function on $G'(F)$, supported near $y'Z(F)$, whose orbital integrals $\widetilde{\Phi}(f'_0)$ is the characteristic function of $Z(F)U'^{G'}$. Let $f_0$ be a matching function on $G(F)$. It exists by Corollary 6.12. Now, $\widetilde{\Phi}(f'_0)$ is a finite linear combination of the characters of square-integrable $\pi'_i$ with coefficients $c_i$, by [Ka86.1, Theorem K]. Then $\widetilde{\Phi}(f_0)$ is the corresponding combination of the characters of the $\pi_i$ which correspond to the $\pi'_i$. Since $U'$ is small, the Weyl integration formula implies that $\int_{T'(F)/Z(F)} I(t',f'_0)I(t',f')\,\mathrm{d}t'$ is equal to $\sum c_i \operatorname{tr}\pi'_i(f')$. Here $T'$ is the centralizer of $y'$ in $G'$. The assumption of our proposition implies that this is equal to $\sum c_i \operatorname{tr}\pi_i(f)$. But this is $\int_{T(F)/Z(F)} I(t,f_0)I(t,f)\,\mathrm{d}t$. We take $U'$ to be so small that both $I(t',f'_0)$ and $I(t,f_0)$ are constant on $U'$. The desired equality $I(y,f) = I(y',f')$ now follows from the choice of $f_0$ and $f'_0$, which guarantees that $I(t,f_0) = F(t,f'_0)$. □

# 23 Cuspidal Global Correspondence

## 23.1 *Injection of Conjugacy Classes*

Let $F$ be a global field. We continue to let $G$, $G'$ be as before. We have $G'(\mathbb{A}) = \mathrm{GL}(n, \mathbb{A})$, $G(\mathbb{A}) = \mathrm{GL}(m, D_{\mathbb{A}})$, where $D_{\mathbb{A}}$ denotes the adèle points of a division algebra $D$ of dimension $d^2$ central over $F$ and $n = md$. We let $D(F_v)$ denote the $F_v$-points of $D$. Then $G(F_v) = \mathrm{GL}(m_v, D(v))$, where $D(v)$ is a division algebra of dimension $d_v^2$ over $F_v$, and $n = m_v d_v$. We have $G(F_v) \simeq G'(F_v)$ outside of a finite set of places $V$. We have an injection $x \mapsto x'$ of conjugacy classes from $G(\mathbb{A})$ to $G'(\mathbb{A})$ and from $G(F_v)$ to $G'(F_v)$ for all $v$. It is a bijection for $v$ outside $V$, but it is not surjective for $v$ in $V$, the set where $D$ is ramified.

## 23.2 *Local Correspondence*

Recall that an irreducible admissible $G(F_v)$-module $\pi_v$ is said to *correspond* (or lift) to a $G'(F_v)$-module $\pi_v'$ if their characters $\chi_v$, $\chi_v'$ are related by $(-1)^{n-m}\chi_v'(x') = \chi_v(x)$ for all regular matching $x, x'$. At $v$ outside $V$ we have $m_v = n$ and this relation amounts to $\pi_v \simeq \pi_v'$. Our Local Theorem 13.8 asserts that the map $\pi_v \mapsto \pi_v'$ induces an embedding of the set of (equivalence classes of) tempered (resp. relevant) $G(F_v)$-modules as a subset of the set of tempered (resp. relevant) $G'(F_v)$-modules.

## 23.3 *Global Correspondence*

A $G(\mathbb{A})$-module $\pi = \otimes_v \pi_v$ is said to (quasi-) correspond to a $G'(\mathbb{A})$-module $\pi' = \otimes_v \pi_v'$ if $\pi_v$ corresponds to $\pi_v'$ for (almost) all $v$. Results about global lifting depend on the form of the trace formula which is available. Here we use only Corollary 4.4. It implies, on using transfer of orbital integrals (Theorem 14.4), that any discrete spectrum (automorphic) $G(\mathbb{A})$-module $\pi$, whose components at two places $u$, $u'$ lift to cuspidal $G'(F_u)$ and square-integrable $G'(F_u)$-modules, quasi-lifts to an automorphic (necessarily cuspidal) $G'(\mathbb{A})$-module with cuspidal component at $u$, an elliptic component at $u'$, and components $\pi_v'$ with characters $\chi_v'$ which are not identically 0 on the set of regular classes $\chi'$ obtained from $x$ in $G(F_v)$ for all $v$ in $V$, is a quasi-lift of a discrete spectrum $G(\mathbb{A})$-module. Note that $u$, $u'$ are not required to be in or out of $V$.

### 23.4 Multiplicity One and Rigidity Theorems

Since an automorphic $\pi'$ with a cuspidal component is cuspidal, multiplicity one and rigidity theorems for the cuspidal spectrum of $L(G'(F)\backslash G'(\mathbb{A}))$ imply that the discrete spectrum quasi-correspond $\pi'$ of $\pi$ is unique if it exists. We shall now deal with the notion of correspondence, rather than quasi-correspondence, and conclude the uniqueness of $\pi$ too, thereby obtaining multiplicity one and rigidity-type theorems for discrete spectrum $G(\mathbb{A})$-modules.

THEOREM 23.5. *Suppose that $\pi'$ is an automorphic $G'(\mathbb{A})$-module with cuspidal components at two places $u$ and $u'$, and components $\pi'_v$ whose characters are not identically zero on the set of the $x'$ which come from $G(F_v)$ for all $v$ in $V$. Then there exists a unique automorphic $G(\mathbb{A})$-module $\pi$ which quasi-corresponds to $\pi'$. Moreover, $\pi$ corresponds to $\pi'$.*

PROOF. The condition at $u$ implies that $\pi'$ is cuspidal. Hence it has a Whittaker model, and its components are all non-degenerate and unitarizable. Hence, by [Ze80, Theorem 9.7(b)], each $\pi'_v$ is relevant. The Local Theorem 13.8 implies that $\pi'_v$ corresponds to a relevant $G(F_v)$-module $\widetilde{\pi}_v$. The identity of Proposition 15.4, say in the form (17.2.1), with $\pi'$ as the only term on the left and with a sufficiently large but finite set $S$ of places of $F$ ($S$ depends on $\pi'$), implies that

$$\prod_{v\in S} \operatorname{tr} \widetilde{\pi}_v(f_v) = \prod_{v\in S} \operatorname{tr} \pi'_v(f'_v) = \sum_{\pi} m(\pi) \prod_{v\in S} \operatorname{tr} \pi_v(f_v)$$

for *all* functions $f_v$ on $G(F_v)$ ($v$ in $S$). The "generalized linear independence" of characters in Lemma 17.2 implies that the sum on the right consists of a single summand $\pi$ with $m(\pi) = 1$, and the theorem follows. □

## 24 Complements on Local Representations

### 24.1 Non-Degenerate Case

Having used the trace formula in a simple form, we were able to describe the correspondence for global cuspidal representations with some local restrictions and for local representations that may occur as their local components. These local representations, in the case of $G' = \mathrm{GL}(n)$, are non-degenerate.

Our next aim is to extend the global correspondence from the context of the cuspidal spectrum to that of the entire discrete spectrum. For this we need to use the invariant trace formula and its comparison between $G$ and $G'$. The complement of the cuspidal spectrum in the discrete spectrum is the residual spectrum. The representations that appear in the residual spectrum of $G'$ were described by Moeglin and Waldspurger [MW95]. The components of these residual spectrum representations are nontempered and degenerate. However, they are unitarizable of a special type, described by [MW95].

The global correspondence will be seen to be an injection from the discrete spectrum of the inner form $G(\mathbb{A})$ to that of the quasisplit $G'(\mathbb{A})$. However, it is not compatible with the local correspondence from $G(F_v)$ to $G'(F_v)$, since the correspondence is not surjective.

The problem can be seen already in the initial case, considered by Jacquet and Langlands [JL70], of $G' = \mathrm{GL}(2)$ and its inner form $G$ over $F$. In the local case, the correspondence maps the trivial representation of $G(F_v)$ to the special, or Steinberg, representation of $G'(F_v)$. The character relation $\chi_{\pi'_v}(g') = -\chi_{\pi_v}(g)$ is satisfied. The sign is $-1$ to the power $m(d-1) = 1$ $(m = 1, d = 2)$. A non-one-dimensional representation $\pi$ corresponds to a cuspidal $\pi'$, and $\pi_v$ corresponds to $\pi'_v$ for all $v$. The one-dimensional $\pi$ corresponds to the one-dimensional $\pi'$ in the sense that $\pi_v = \pi'_v$ at all places where $G$ splits. At the finite set of places $v$ where $G$ ramifies, the components of $\pi'$ are one-dimensional. They are not the square-integrables mentioned above. We have the character relation $\chi_{\pi'_v}(g') = \chi_{\pi_v}(g)$, which differs from the previous relation by a sign. The explanation is that the trivial and the square-integrable are the constituents of an induced on $\mathrm{GL}(2, F_v)$ which does not come from the correspondence from any representation of the inner form, so its character is zero on the (elliptic) $g'$ which come from $g$ in the inner form. Consequently we need to define the correspondence from the split group to its inner form, and the sign in the character relation has to change.

This extends to the general case of $\mathrm{GL}(n)$ and its inner forms. The local theory is sufficiently developed to express the values of the characters of the local components of the global residual representations in terms of those of the non-degenerate ones that appear in the correspondence, and we shall simply review and summarize these local results, in order to complete the comparison of the unrestricted trace formulae.

## 24.2 Langlands Classification

We start with recalling the Langlands classification in our context. Let $\pi$ be an irreducible representation of $G = \mathrm{GL}(m, D)$. Then [BZ76] asserts that there exists a standard (diagonal) Levi subgroup $M = \prod_{i=1}^{r} \mathrm{GL}(m_i, D)$ of $G$, and a cuspidal irreducible $\tau_i$ on $\mathrm{GL}(m_i, D)$ $(1 \leq i \leq r)$ such that $\pi$ is a constituent of $i_M^G(\tau_1 \times \cdots \times \tau_r)$. The unordered multiset (set with multiplicities) $\{\tau_1, \ldots, \tau_r\}$ of cuspidal representations is called the *cuspidal support* of $\pi$, or simply the *support* of $\pi$.

Next let $\sigma_i$ be quasi-square-integrable irreducible representations of $\mathrm{GL}(m_i, D)$; thus there are real exponents $e_i$ and square-integrable $\sigma_i^0$ with $\sigma_i = \nu^{e_i}\sigma_i^0$. Let $\alpha$ be a permutation in the symmetric group $S_r$ of $\{1, \ldots, r\}$ such that $e_{\alpha(i)}$ decrease with $i$. Put $M_\alpha = \prod_{i=1}^{r} \mathrm{GL}(m_{\alpha(i)}, D)$ and $\sigma_\alpha = \sigma_{\alpha(1)} \times \cdots \times \sigma_{\alpha(r)}$ and $\sigma = \sigma_1$. The Langlands classification, [BW80, Si80], asserts that $i_{M_\alpha}^G \sigma_\alpha$ has a unique irreducible quotient $\pi$, denoted $L(\sigma)$ or $L(i_M^G \sigma)$, called the *Langlands quotient* of $i_M^G \sigma$. Every irreducible $\pi$ is so obtained. The unordered multiset $\{\sigma_1, \ldots, \sigma_r\}$ is uniquely determined by $\pi$ (if $L(\sigma) = L(\sigma')$, $\sigma' = \sigma'_1 \times \cdots \sigma'_{r'}$, then $r' = r$ and there is $\beta \in S_r$ with $n'_i = n_{\beta(i)}$, $\sigma'_i = \sigma_{\beta(i)}$). It is called the *quasi-square-integrable support* of $\pi$. The representations $i_M^G \sigma$ are called *standard representations*.

The set of standard representations of $G$ makes a basis of the Grothendieck group of admissible representations of $G$. Another basis is the set (of equivalence classes) Irr$G$ of irreducible representations of $G$, and $i_M^G\sigma \mapsto L(i_M^G\sigma)$ is a bijection between these bases. Write $\pi_1 < \pi$ for $\pi_1 \in \mathrm{Irr}G$ if $\pi_1 \neq \pi$, $\pi = L(i_M^G\sigma)$ and $\pi_1$ is a constituent of $i_M^G\sigma$. This is a partial order relation on Irr$G$. Note that if $\pi_1 < \pi$, then the cuspidal support of $\pi_1$ coincides with that of $\pi$.

As in [Ze80], we often write $\tau_1 \times \cdots \times \tau_r$ for $i_M^G(\tau_1 \times \cdots \times \tau_r)$. Then $L(\sigma_1 \times \sigma_2)$ is a constituent of $L(\sigma_1) \times L(\sigma_2)$ with multiplicity one; see [Ze80, Proposition 8.4] and [Tc90, Proposition 2.3]. Since any irreducible constituent of $L(\sigma_1) \times L(\sigma_2)$ is also a constituent of $\sigma_1 \times \sigma_2$, any irreducible constituent $\pi$ of $L(\sigma_1) \times L(\sigma_2)$ other than $L(\sigma_1 \times \sigma_2)$ has $\pi < L(\sigma_1 \times \sigma_2)$. In particular, if $L(\sigma_1) \times L(\sigma_2)$ is reducible, then it has at least two different constituents.

## 24.3 Generalized Steinberg Representations

The quasi-square-integrable representations $\sigma$ of $G = \mathrm{GL}(m, D)$ can be listed as follows. Consider first the case of $G' = \mathrm{GL}(n, F)$, where $d = \mathrm{rk}\,D$ is 1, following [Ze80]. Suppose $n = kr$ for positive integers $k$, $r$ and $\tau'$ is a cuspidal representation of $\mathrm{GL}(k, F)$. Then the $G$-module $\tau' \times \tau'\nu \times \cdots \times \tau'\nu^{r-1}$ induced from $M = \mathrm{GL}(k, F) \times \cdots \times \mathrm{GL}(k, F)$ has a unique quotient $\sigma'$, denoted $Z(\tau', r)$ or $Z(\tau' \times \tau'\nu \times \cdots \times \tau'\nu^{r-1})$. It is quasi-square-integrable; every quasi-square-integrable irreducible $G'$-module is so obtained (and is sometimes called generalized Steinberg). A set $S' = \{\tau', \nu\tau', \ldots, \nu^{r-1}\tau'\}$, where $\tau'$ is a cuspidal $\mathrm{GL}(k, F)$-module, is called a *segment of length r with end* $\nu^{r-1}\tau'$.

## 24.4 Segments

To extend this to $G = \mathrm{GL}(m, D)$, note that a standard parabolic subgroup $M$ of $G$ is determined by a sequence $(m_1, \ldots, m_r)$ of positive integers whose sum is $m$. Its split form is a standard Levi subgroup $M'$ of $G' = \mathrm{GL}(n, F)$, $n = md$, associated with the sequence $(n_1, \ldots, n_r)$, $n_i = m_i d$. The standard Levis $M'$ of $G'$ of type $(n_1, \ldots, n_r)$ with $d|n_i$ for all $i$ is said to come from $G$. Given a cuspidal $\mathrm{GL}(k, D)$-module $\tau$, denote by $\sigma'$ the corresponding quasi-square-integrable $\mathrm{GL}(kd, F)$-module. Then $\sigma' = Z(\tau', r')$ for some integer $r' > 0$ and cuspidal $\mathrm{GL}(kd/r', F)$-module $\tau'$. We say that $r'$ is the *Zelevinsky length* $z(\tau)$ of $\tau$. Put $\nu_\tau = \nu^{z(\tau)}$. Then the $G = \mathrm{GL}(kr, D)$-module $i_M^G(\tau \times \nu_\tau\tau \times \cdots \times \nu_\tau^{r-1}\tau)$ induced from $M = \mathrm{GL}(k, D)^r$ has a unique irreducible quotient representation $\sigma$, denoted $Z(\tau, r)$. This $\sigma$ is quasi-square-integrable $G$-module. Every quasi-square-integrable $G$-module is so obtained. We put $\sigma = Z(\tau, r)$ and set $z(\sigma) = z(\tau)$. The numbers $k$, $r$, and the cuspidal $\tau$ are uniquely determined by $\sigma$. For this extension of [Ze80], see [Tc90]. A set $S = \{\tau, \nu_\tau\tau, \ldots, \nu_\tau^{r-1}\tau\}$ is called a *segment of length r and end* $\nu_\tau^{r-1}\tau$.

## 24.5 Multi-Segments

In summary, to a quasi-square-integrable $\mathrm{GL}(m,D)$-module $\sigma$, we can associate a segment $X$. A *multisegment* $X$ is a multiset of segments; its multiset of ends is denoted by $E(X)$.

If $\pi$ is an irreducible $G = \mathrm{GL}(m,D)$-module, the multiset of the segments of the representations of the quasi-square-integrable support of $\pi$ is a multisegment, denoted by $X(\pi)$. This $X(\pi)$ determines $\pi$. The sum of the representations in $X(\pi)$ is the cuspidal support of $\pi$.

Two segments $S_1, S_2$ are *linked* if $S_1 \cup S_2$ is a segment not equal to $S_1$ or $S_2$. Linked $S_1$, $S_2$ are *adjacent* if $S_1 \cap S_2 = \varnothing$. If $S_1$, $S_2$ are linked segments in a multisegment $X$, define a multisegment $Y$ to be obtained from $X$ by *elementary operations* if $Y$ is

$$X \cup \{S_1 \cup S_2\} \cup \{S_1 \cap S_2\} - \{S_1, S_2\}$$

if $S_1$, $S_2$ are not adjacent (thus $S_1 \cap S_2 \neq \varnothing$) or

$$X \cup \{S_1 \cup S_2\} - \{S_1, S_2\}$$

if $S_1$, $S_2$ are adjacent ($S_1 \cap S_2 = \varnothing$). We then write $Y < X$. This extends by transitivity to a partial order relation on the set of multisegments of $G$. By [Ze80, Theorem 7.1] for $G' = \mathrm{GL}(n,F)$ and [Tc90] for $G = \mathrm{GL}(m,D)$, we have

PROPOSITION 24.6. *Let $\pi_1$, $\pi_2$ be irreducible $G$-modules. Then $\pi_1 < \pi_2$ if and only if $X(\pi_1) < X(\pi_2)$.*

Define the *length* $l(X)$ of a multisegment $X$ to be the maximum of the lengths of the segments in $X$. For an irreducible $\pi$, put $l(\pi) = l(X(\pi))$.

PROPOSITION 24.7. (i) *If $Y$ is obtained from $X$ by an elementary operation, then $l(X) \leq l(Y)$ and $E(Y) \subseteq E(X)$. Thus $l$ is decreasing on the set* $\mathrm{Irr}\, G$ *of irreducible representations of $G$.*

(ii) *Let $\pi_1 \in \mathrm{Irr}\ \mathrm{GL}(m,D)$, $\pi_2 \in \mathrm{Irr}\ \mathrm{GL}(k,D)$. If for all $S_1 \in X(\pi_1)$ and $S_2 \in X(\pi_2)$ the segments $S_1$ and $S_2$ are not linked, then $\pi_1 \times \pi_2$ is irreducible.*

PROOF. Part (i) is easy. Part (ii) is a result of [Ze80] for $G'$ extended to $G$ by [Tc90]. □

If $\tau$ is a cuspidal $\mathrm{GL}(k,D)$-module, the *line* generated by $\tau$ is $\{\nu_\tau^r \tau\,;\, r \in \mathbb{Z}\}$; it is equal to the line of $\nu_\tau^r \tau$ for any integer $r$. An irreducible $\pi$ of $G = \mathrm{GL}(m,D)$ is called *rigid* if all representations in the cuspidal support of $\pi$ lie on the same line.

COROLLARY 24.8. *Let $\pi$ be an irreducible $G$-module. Let* $\mathrm{supp}(\pi)$ *be the set of irreducibles in the cuspidal support of $\pi$. Let $L_1,\ldots, L_r$ be the set of lines which have nonempty intersection with* $\mathrm{supp}(\pi)$. *Then up to permutation, $\pi$ has a unique presentations as $\pi_1 \times \cdots \times \pi_r$ where $\pi_i$ are rigid and* $\mathrm{supp}(\pi_i) \subset D_i$ *for all $i$ $(1 \leq i \leq r)$.*

We say that $\pi_1 \times \cdots \times \pi_r$ is the *standard decomposition* of $\pi$ as a product of rigid representations.

## 24.9 Opposite Correspondence

We defined a correspondence $C' : R(G) \to R(G')$ by a character relation. It maps irreducible square-integrable (resp. tempered, resp. standard, resp. relevant) $G$-modules to such $G'$-modules. Since the standard representations make a basis of the Grothendieck group, $C'$ extends to an injection on all of $R(G)$ by $C' : i_M^G \sigma \mapsto i_{M'}^{G'}(C'(\sigma))$, where $\sigma$ is a quasi-square-integrable $G$-module. The character relation commutes with induction. Let us define a map $Q : \mathrm{Irr}(G) \to \mathrm{Irr}(G')$ by $L(i_M^G \sigma) \to L(i_{M'}^{G'}(C'(\sigma)))$, using the fact that each irreducible is a Langlands quotient of a standard representation. If $\pi_1 < \pi_2$, then $Q\pi_1 < Q\pi_2$. Define $\pi_1 << \pi_2$ if $Q\pi_1 < Q\pi_2$. Then $\pi_1 < \pi_2$ implies $\pi_1 << \pi_2$, but in principle there can be irreducible $\pi_1, \pi_2$ with $Q\pi_1 < Q\pi_2$ without $\pi_1 < \pi_2$.

Define a correspondence $C : R(G') \to R(G)$ in the opposite direction to be the $\mathbb{Z}$-morphism with $C(i_{M'}^{G'}\sigma') = 0$ if $i_{M'}^{G'}\sigma'$ is not in the image of the correspondence $C'$, and $C(i_{M'}^{G'}\sigma') = i_M^G \sigma$ if $\sigma' = C'(\sigma)$, where $\sigma'$ is quasi-square-integrable. Then $C$ is a surjective group homomorphism, and it is the unique map from $R(G')$ to $R(G)$ satisfying the character relation $\chi_{\pi'}(g') = (-1)^{m(d-1)}\chi_{C(\pi')}(g)$ whenever $g' \in G'$ and $g \in G$ are regular with equal characteristic polynomials. It commutes with induction. We have

$$C(Q(\pi)) = \pi + \sum_{\{\pi_j\,;\,\pi_j << \pi\}} b_j \pi_j$$

for irreducibles $\pi_j$ and integers $b_j$.

DEFINITION. A virtual $G'$-module $\pi'$ in $R(G')$ is a *d-lift* if $C(\pi') \neq 0$; thus $\chi_{\pi'}(g') \neq 0$ for a regular $g' \in G'$ which comes from a $g \in G$. Thus the characteristic polynomial of $g'$ is a product of irreducible factors of degrees divisible by $d$. Hence $\pi'$ being a $d$-lift depends only on $d$, not on $D$. An induced $\pi' = i(\tau_1' \times \cdots \times \tau_r')$ is a $d$-lift if and only if each $\tau_j'$ is a $d$-lift.

## 24.10 Langlands Quotients

Recall that $i(\tau' \times \nu\tau' \times \cdots \times \nu^{r-1}\tau')$ has a unique irreducible quotient $Z(\tau', r)$, which is quasi-square-integrable when $\tau'$ is an irreducible cuspidal $\mathrm{GL}(k, F)$-module. It is unitarizable if and only if $\tau_u' = \nu^{\frac{r-1}{2}}\tau'$ is unitarizable. Thus $i(\nu^{-\frac{r-1}{2}}\tau_u' \times \nu^{1-\frac{r-1}{2}}\tau_u' \times \cdots \times \nu^{\frac{r-1}{2}}\tau_u')$ has a unique irreducible quotient $Z_u(\tau_u', r)$ which is square-integrable when $\tau_u'$ is unitarizable. We write $Z_u(\tau', r)$ only when $\tau'$ is unitarizable.

When $\sigma'$ is a square-integrable representation of $\mathrm{GL}(k,F)$, write $u(\sigma',r)$ for the Langlands quotient $L(\prod_{j=1}^{r}\nu^{\frac{r+1}{2}-j}\sigma')$. It is an irreducible representation of $\mathrm{GL}(rk,F)$.

For $s$ $(0<s<\frac{1}{2})$, the representation $\pi(u(\sigma',r),s)=i(\nu^{s}u(\sigma',r)\times\nu^{-s}u(\sigma',r))$ is irreducible by Proposition 24.7(ii).

PROPOSITION 24.11 (Bernstein, Tadic). *Let $U'$ be the set of all representations $u(\sigma',r)$ and $\pi(u(\sigma',r),s)$, where $\sigma'$ is any square-integrable* $\mathrm{GL}(k,F)$*-module, $r,k\in\mathbb{Z}_{>0}$. Then any product of members of $U'$ is irreducible and unitarizable, and any unitarizable $G'$-module $\pi'$ is such a product. The nonordered multiset of the factors in the product is uniquely determined by $\pi'$.*

## 24.12 Unitarizable Representations

Let $\tau$ be a cuspidal irreducible $\mathrm{GL}(k,D)$-module. It corresponds to a quasi-square-integrable $\mathrm{GL}(kd,F)$-module $\sigma'=C'(\tau)$, of the form $Z(\tau',r)$ for some cuspidal $\mathrm{GL}(kd/r,F)$-module $\tau'$ and divisor $r$ of $kd$. Put $z(\tau)=r$ and $\nu_\tau=\nu^{z(\tau)}$.

The representations $i\left(\prod_{j=0}^{r}\nu_\tau^{j}\tau\right)$ has a unique irreducible quotient $\sigma$, which is a quasi-square-integrable $\mathrm{GL}(kr,D)$-module, denoted by $Z(\tau,r)$. By [Tc90] every quasi-square-integrable $\sigma$ is so obtained, and we put $z(\sigma)$ for $z(\tau)$.

Then $\sigma=Z(\tau,r)$ is unitarizable if and only if $\tau_u=\nu_\tau^{\frac{r-1}{2}}\tau$ is unitarizable, and we write $\sigma=Z_u(\tau_u,r)$. For a square-integrable $\sigma$, write $u(\sigma,r)$ for the Langlands quotient $L\left(\prod_{j=1}^{r}\nu_\tau^{\frac{r+1}{2}-j}\tau\right)$. Also put $\overline{u}(\sigma,r)$ for the Langlands quotient $L\left(\prod_{j=1}^{r}\nu^{\frac{r+1}{2}-j}\tau\right)$. Both are irreducible unitarizable representations of $\mathrm{GL}(kr,D)$. For $s$ $(0<s<\frac{1}{2})$, put $\pi(u(\sigma,r),s)=\nu_\tau^{s}u(\tau,r)\times\nu_\tau^{-s}u(\tau,r)$. It is an irreducible representation of $\mathrm{GL}(2kr,D)$. There are the relations

$$\overline{u}(\sigma,rz(\sigma))=\prod_{j=1}^{z(\sigma)}\nu^{j-\frac{z(\sigma)+1}{2}}u(\sigma,r)$$

and for all integers $m$ $(1\le m<z(\sigma))$:

$$\overline{u}(\sigma,rz(\sigma)+m)=\left(\prod_{j=1}^{m}\nu^{j-\frac{m+1}{2}}u(\sigma,r+1)\right)\times\left(\prod_{j=1}^{z(\sigma)-m}\nu^{j-\frac{z(\sigma)-m+1}{2}}u(\sigma,r)\right).$$

The second factor is erased if $r=0$. These products are irreducible as the segments in the quasi-square-integrable support of any two factors are not linked.

## 24.13 Hermitian Representations

Let $\pi$ be an irreducible $\mathrm{GL}(m, D)$-module. Denote by $h(\pi)$ the Hermitian conjugate of $\pi$. This is an irreducible $\mathrm{GL}(m, D)$-module with a non-degenerate sesquilinear form $\langle\cdot,\cdot\rangle$ on $\pi \times h(\pi)$. Then $\pi$ is said to be *Hermitian* if it is equivalent to $h(\pi)$. Unitarizable representations are Hermitian. Let $H = \{\sigma_j\, ;\, 1 \leq j \leq k\}$ be a multiset of quasi-square-integrable representations of $G(m_j, D)$. Let $h(H)$ be the multiset $\{h(\sigma_j)\, ;\, 1 \leq j \leq k\}$. For $s \in \mathbb{R}$, $h(\nu^s\pi) = \nu^{-s}h(\pi)$.

PROPOSITION 24.14. (i) *If H is the quasi-square-integrable support of an irreducible* $\mathrm{GL}(m, D)$*-module* $\pi$*, then* $h(H)$ *is this support of* $h(\pi)$*. In particular,* $\pi$ *is Hermitian iff* $h(H) = H$.

(ii) *Let* $\pi_j$ *be irreducible unitarizable* $\mathrm{GL}(m_j, D)$*-modules such that* $i(\pi_j \times \pi_j)$ *is irreducible. Then* $i\left(\prod_{j=1}^k \pi_j\right)$ *is irreducible.*

## 24.15 Permutations

Let $\sigma$ be a square-integrable $\mathrm{GL}(m, D)$-module. It corresponds to a square-integrable $\mathrm{GL}(n, F)$-module $\sigma' = C'(\sigma)$. Then $\sigma = Z_u(\tau, k)$ for a cuspidal unitarizable $\mathrm{GL}(m/k, D)$-module $\tau$ for a $k|m$. This $\tau$ corresponds to a square-integrable $C'(\tau)$ of the form $Z_u(\tau', z(\sigma))$ for some cuspidal unitarizable $\mathrm{GL}(\frac{md}{kz(\sigma)}, F)$-module $\tau'$. Then $\sigma' \in Z_u(\tau', k')$ where $k' = kz(\sigma)$. In particular, the cuspidal support of $\sigma'$ can be found from that of $\sigma$, since the correspondence commutes with taking the module of coinvariants.

Let $W_r^k$ be the set of permutations $w$ of $\{1, \ldots, r\}$ with $w(j) + k \geq j$ for all $j$ $(1 \leq j \leq r)$. Let $R$ be the group of permutations $w$ of $\{1, \ldots, r'\}$, $r' = rz(\sigma)$, with $z(\sigma)|w(j) - j$ for all $j$ $(1 \leq j \leq r')$.

PROPOSITION 24.16. (i) *We have* $C(u(\sigma', r')) = \overline{u}(\sigma, r')$.

(ii) *The induced representation* $i(\overline{u}(\sigma, r') \times \overline{u}(\sigma, r'))$ *is irreducible.*

(iii) *We have the character formula*

$$\overline{u}(\sigma, r') = \nu^{-\frac{k'+r'}{2} + \frac{z(\sigma)-1}{2}} \sum_{w \in R \cap W_{r'}^{k'}} \mathrm{sgn}(w) \prod_{j=1}^{r'} Z\left(\nu^j\tau, \frac{w(j) - j}{z(\sigma)} + k\right).$$

(iv) *The induced* $i(u(\sigma, r) \times u(\sigma, r))$ *is irreducible, and* $\pi(u(\sigma, r), s)$ *is unitarizable for all* $s$ $(0 < s < \frac{1}{2})$.

(v) *Suppose* $\sigma \in Z_u(\tau, k)$ *for a unitarizable cuspidal representation* $\tau$*. Then*

$$u(\sigma, r) = \nu_\sigma^{-\frac{r+k}{2}} \sum_{w \in W_r^k} \mathrm{sgn}(w) \prod_{j=1}^{r} Z(\nu_\sigma^j, w(j) + k - j).$$

This is Theorem 3.2 and Corollary 3.6 of [Ba08], extending a result of Tadic when $D = F$ (and so $z(\sigma) = 1$).

## 24.17 Indices

Starting with a cuspidal unitarizable representation of $\mathrm{GL}(q, F)$, let $\sigma'_k = Z_u(\tau', k)$ be the associated square-integrable $\mathrm{GL}(kq, F)$-module, and similarly put $\sigma'_r = Z_u(\tau', r)$. Denote $d/(d, q)$ by $z$. It is the smallest natural number with $d|zq$. Assume $d|n = rkq$, thus $z|rk$. Proposition 3.7 of [Ba08] asserts that (i) if $d|kq$, thus $z|k$, then $C(\sigma'_k) = \sigma_k$ is a square-integrable $\mathrm{GL}(\frac{kq}{d}, D)$-module and $z$ is $z(\sigma_k)$ and $C(u(\sigma'_k, r)) = \overline{u}(\sigma_k, r)$. (ii) If $d$ divides neither $kq$ nor $rq$, thus $z$ does not divide $k$ and $r$, then $C(u(\sigma'_k, r)) = 0$. If $d|rq$, thus $z|r$, then $C(\sigma'_r) = \sigma_r$ is a square-integrable $\mathrm{GL}(\frac{rq}{d}, D)$-module, we have $z = z(\sigma_r)$, and $C(u(\sigma'_k, r))$ is computed in [Ba08, subsection 3.3]. Using this, [Ba08, Proposition 3.9] concludes the following. Recall that the correspondence $C : R(G') \to R(G)$ is defined by a character relation.

Denote by $U$ the set of representations $u(\sigma, r)$ and $\pi(u(\sigma, r), s)$, where $0 < s < \frac{1}{2}$, $\sigma$ is a square-integrable $\mathrm{GL}(k, F)$-module, and $k, r \geq 1$.

PROPOSITION 24.18. (i) *The representations in $U$ are all irreducible and unitarizable.*

(ii) *If $\pi_1, \ldots, \pi_r \in U$, then the induced $i(\pi_1 \times \cdots \times \pi_r)$ is irreducible and unitarizable.*

(iii) *If $\pi'$ is an irreducible unitarizable $\mathrm{GL}(md, F)$-module, then $C(\pi')$ is $0$ or it is a unitarizable irreducible representation $\pi$ of $\mathrm{GL}(m, D)$, or it is $-\pi$ for such a $\pi$.*
*Write $|C|(\pi') = \pi$ in the second case.*

(iv) *If $\pi$ is a unitarizable irreducible $\mathrm{GL}(m, D)$-module, and $i(\pi \times \pi)$ is irreducible, then $\pi$ is induced from a product of representations in $U$.*

(v) *Let $\prod U$ be the set of induced representations $i(\pi_1 \times \cdots \times \pi_r)$, $\pi_i \in U$. It is a set of irreducible unitarizable representations including the $\overline{u}(\sigma, r)$. It is stable under the Zelevinsky involution—see* [Ze80]*—$\iota : R(G) \to R(G)$.*

(vi) *If $\pi'$, a unitarizable representation of $\mathrm{GL}(md, F)$, is a lift, then $|C|(\pi') \in \prod U$.*

## 24.19 All Components

Finally we note that [Ze80] implies that a non-degenerate irreducible unitarizable $\mathrm{GL}(n, F)$-module $\pi'$ has the form $\pi' = i(\nu^{e_1}\sigma'_1 \times \cdots \times \nu^{e_q}\sigma'_q)$ where the $\sigma'_j$ are square-integrable and $e_j \in (-\frac{1}{2}, \frac{1}{2})$. Then for every $r > 0$ in $\mathbb{Z}$ the induced $i(\nu^{\frac{r+1}{2}-1}\pi' \times \cdots \times \nu^{\frac{r+1}{2}-r}\pi')$ is a standard representation. Denote its Langlands quotient by $L(\pi', r)$. Then $L(\pi', r) = i(\nu^{e_1}u(\sigma'_1, r) \times \cdots \times \nu^{e_q}u(\sigma'_q, r))$, and it is unitarizable as so was $\pi'$. Note that $\pi'$ is tempered if and only if all $e_j$ are 0.

Now all local components of any cuspidal representation are non-degenerate, so [MW95] implies that all components of a discrete spectrum representation of $\mathrm{GL}(n, \mathbb{A})$ are of this form, $L(\pi', r)$. When do they map by $C$ to a nonzero representations?

To answer this question, write $\sigma'_j = Z_u(\tau'_j, k_j) = Z(\nu^{-\frac{k_j-1}{2}}\tau'_j \times \cdots \times \nu^{\frac{k_j-1}{2}}\tau'_j)$ with cuspidal unitarizable $\mathrm{GL}(p_j, F)$-modules $\tau'_j$. Put $J = \{j \in \mathbb{Z};\ 1 \leq j \leq q, d|p_j k_j\}$. Let $z(\pi', d)$ be the least natural number $z$ such that $d|zp_j$ for all $j \in \{1, \ldots, q\} - J$. Then $C(L(\pi', r)) \neq 0$ if and only if $C(u(\sigma'_j, r)) \neq 0$ for all $j$ ($1 \leq j \leq q$) if and only if $z(\pi', d)|r$ by 24.17. In this case $C(L(\pi', r)) = i\left(\prod_{j=1}^{q} \nu^{e_j} C(u(\sigma'_j, r))\right)$.

# 25 Complete Global Correspondence

## *25.1 Local Lifting*

Using the comparison of the trace formulae for arbitrary matching test functions, achieved in Chapter 6 below, a more complete correspondence theorem can be obtained, without any conditions at any local place, and for the entire discrete spectrum, rather than only for the cuspidal representations with cuspidal components. A form of such a global theorem is as follows.

Let $F$ be a number field. Put $G = \mathrm{GL}(m, D)$ where $D$ is a central division algebra over $F$ of rank $d$. Put $G' = \mathrm{GL}(n, F)$, $n = md$, $G'_v = G'(F_v)$. Put $G_v = G(F_v) = \mathrm{GL}(m_v, D_v)$, where $D \otimes_F F_v = M(m'_v, D_v)$, $D_v$ is a division algebra of rank $d_v$ central over $F_v$ and $m'_v d_v = d$, $m_v = mm'_v$.

Recall that we defined, in Subsection 24.9, the local correspondence $C'_v : R(G_v) \to R(G'_v)$ and its partial inverse $C_v : R(G'_v) \to R(G_v)$, by means of a character relation on all regular elements, in such a way that each irreducible $G_v$-module $\pi_v$ corresponds to a $G'_v$-module $\pi'_v$. Also we defined—in Proposition 24.18(iii)—a correspondence $|C_v|$ from the set $\mathrm{Irr}_u(G'_v)$ of irreducible unitarizable representations of $G'_v$ to the union of $\{0\}$ and the analogous set $\mathrm{Irr}_u(G_v)$ for $G_v$. It coincides with the correspondence $C_v$ up to a sign. If $\pi_v$ corresponds to $\pi'_v$ up to a sign—thus $|C_v|(\pi'_v) = \pi_v$—we also say that $\pi'_v$ *comes from* $\pi_v$. If $\pi'_v$ is a unitarizable irreducible $d_v$-lift, namely, its character $\chi_{\pi'_v}$ is nonzero at a regular $g'$ in $G'_v$ whose characteristic polynomial equals that of some $g \in G_v$, then $|C_v|(\pi'_v) \neq 0$, and it is an irreducible $G_v$-module, by Proposition 24.18(vi).

For an irreducible representation $\pi'$ of $G'(\mathbb{A})$, we say that $\pi'$ comes locally from $G(\mathbb{A})$ if $\pi'_v$ comes from $G_v$ for all $v$ (namely, comes from some $\pi_v$ on $G_v$). By Proposition 24.18(vi), this is the same as saying that $\pi'$ is a $D$-lift, in the sense that $\pi'_v$ is a $d_v$-lift for all $v \in V$.

Let $V$ be the set of places of $F$ where $D$ ramifies.

THEOREM 25.2. *There exists a unique map $C'$ from the discrete spectrum* $\mathrm{DS}(G(\mathbb{A}))$ *to the discrete spectrum* $\mathrm{DS}(G'(\mathbb{A}))$ *such that* $C'(\pi) = \pi'$ *is defined* *by $\pi'_v \simeq \pi_v$ for all $v \notin V$. It is injective. If $C'(\pi) = \pi'$, then $|C_v|(\pi'_v) = \pi_v$ for all $v \in V$. If $\pi'$ is cuspidal, then $C_v(\pi'_v) = \pi_v$ for all $v \in V$. The latter means that the character relation $\chi_{\pi'_v}(g') = (-1)^{(m_v-1)d_v}\chi_{\pi_v}(g)$ holds for all regular $g \in \mathrm{GL}(m_v, D_v)$ and $g' \in \mathrm{GL}(m_v d_v, F_v)$ with equal characteristic polynomials.*

*The global map $C'$ is injective. Its image is the set of discrete spectrum $\pi'$ which come locally from $G(\mathbb{A})$, equivalently $\pi'_v$ is a $d_v$-lift for all $v \in V$, namely, $\chi_{\pi'_v}$ is nonzero at some regular $g'_v \in G'_v$ coming from $G_v$. Thus if $\pi'$ is a discrete spectrum representation of $G'(\mathbb{A})$, and $\chi_{\pi'_v}(g') \neq 0$ for a regular $g'$ coming from $G_v$, then $\pi'_v$ comes from $G_v$, for all $v \in V$.*

*Multiplicity one theorem holds for $G(\mathbb{A})$. Thus each irreducible representation in the discrete spectrum of $G(\mathbb{A})$ occurs with multiplicity one.*

*Rigidity theorem, also called "strong multiplicity one theorem", holds for $G(\mathbb{A})$. Thus if $\pi_1$, $\pi_2$ occur in the discrete spectrum of $G(\mathbb{A})$ and $\pi_{1,v} \simeq \pi_{2,v}$ for almost all $v$, then $\pi_{1,v} \simeq \pi_{2,v}$ for all $v$.*

*If $\pi$ is in the discrete spectrum of $G(\mathbb{A})$, then $\pi_v \in \prod U_v$ for all $v \in V$.*

## 25.3 Trace Formula

Let $F$ be a number field. Let $F_\infty^\times$ be $\prod_{v \in S_\infty} F_v^\times$, product over the Archimedean places of $F$, embedded in $\mathbb{A}^\times$ as the idèles with component 1 at the non-Archimedean places. Thus $F_\infty^\times \subset \mathbb{A}^\times = Z(\mathbb{A})$, where $Z$ is the center of $\mathrm{GL}(n)$. Let $\omega$ be a unitary character of $F_\infty^\times$. Let $D$ be a division algebra of rank $d$ central over $F$. Put $G = \mathrm{GL}(m, D)$. Let $\mathcal{L}$ denote the set of $F$-Levi subgroups of $G$ which are standard, namely, contain the (maximal split torus in the) center of the diagonal group of $G$. Write the expression 6.(3.4.1) in the form which appears in 6.(16.11.1) as

$$I_{t,\omega,\mathrm{disc}}(f) = \sum_{M \in \mathcal{L}} |W_0^M|/|W_0^G| \sum_{s \in W(\mathfrak{a}_M)_{\mathrm{reg}}} |\det(s-1)_{\mathfrak{a}_M^G}|^{-1} \operatorname{tr} \mathcal{M}_M^G(s,0) r_{M,t}(0,f). \tag{25.3.1}$$

Recall that $r_{M,t}(0,f)$ is $r_{Q,t}(0,f)$, where $Q$ is any parabolic with Levi $M$, the induced representation of $G(\mathbb{A})$ obtained from the subrepresentation of $M(\mathbb{A})$ in $L^2(M(F)\backslash M(\mathbb{A}))$ which decomposes into a discrete sum of irreducible representations $\tau$ of $M(\mathbb{A})$ which transform under $F_\infty^\times$ by $\omega$, and the imaginary part of the Archimedean infinitesimal character of $\tau$ has norm $t$, where $t \in \mathbb{R}_{\geq 0}$. The other terms are $|W_0^M|$, the cardinality of the Weyl group of $M$, so $|W_0^G|$ is that of $G$. Then $\mathfrak{a}_M^G = \mathfrak{a}_M/\mathfrak{a}_G$, where $\mathfrak{a}_M = \mathrm{Hom}(X(M)_F, \mathbb{R})$ and $X(M)_F$ is the lattice of $F$-rational characters of $M$. Further, $W(\mathfrak{a}_M)$ is the Weyl group of $\mathfrak{a}_M$, and $W(\mathfrak{a}_M)_{\mathrm{reg}}$ is the set of $s$ in $W(\mathfrak{a}_M)$ with $\det(s-1)_{\mathfrak{a}_M^G} \neq 0$. Finally, $\mathcal{M}_M^G(s,0) = \mathcal{M}(s,0)$ is the intertwining operator associated to $s$ at the point 0. It intertwines the representations $i_M^G\tau$ and $i_{sM}^G s\tau$, where $\tau$ is a representation of $M$.

The identity 6.(16.11.1) asserts that for matching test functions $f = \otimes_v f_v$ on $G(\mathbb{A})$ and $f' = \otimes_v f'_v$ on $G'(\mathbb{A})$, $G' = \mathrm{GL}(n)$, $n = md$, the expression $I_{t,\omega,\mathrm{disc}}(f)$ for $G$ is equal to the analogous expression $I_{t,\omega',\mathrm{disc}}(f')$.

## 25.4 Weyl Groups

The set $W(\mathfrak{a}_M)_{\mathrm{reg}}$ is empty unless $M$ is $\mathrm{GL}(r_i, D)$ along the diagonal, $1 \leq i \leq k$, with $r_i$ independent of $i$, up to conjugation. Suppose $M$ is then the Levi subgroup $\mathrm{GL}(r, D) \times \cdots \times \mathrm{GL}(r, D)$, with $k$ factors and $m = kr$. Then $W(\mathfrak{a}_M)$ is isomorphic to the symmetric group $S_k$ and $W(\mathfrak{a}_M)_{\mathrm{reg}}$ is the set of $k$-cycles. Then $W(\mathfrak{a}_M)_{\mathrm{reg}}$ consists of $(k-1)!$ elements. For every $s \in W(\mathfrak{a}_M)_{\mathrm{reg}}$, we have $|\det(s-1)_{\mathfrak{a}_L^G}| = k$. Further, $|W_0^M| = (r!)^k$ and $|W_0^G| = m!$. Thus the coefficient of the term indexed by $M$ in $I_{t,\omega,\mathrm{disc}}$ is $(r!)^k/m!k$.

If $M_1$ is conjugate to $M$, the contribution of $M_1$ to the sum is equal to that of $M$. If $M_1$ is conjugate to $M$, and it contains the diagonal torus, then the center of $M_1$ is contained in the (center of the) diagonal torus. But $M_1$ is the centralizer of its center. Hence the number of $M_1$ is the number of non-ordered partitions of $\{1, \ldots, m\}$ into $k$ subsets of cardinality $r$. This number is $\binom{m}{m-r}\binom{m-r}{m-2r}\binom{m-2r}{m-3r}\cdots\binom{2r}{r}/k!$, which is $m!/(r!)^k k!$.

## 25.5 Rewriting the Formula

We can then rewrite the formula as follows. For $k|m$, let $M_k$ be the Levi subgroup of $G$, diagonal with $k$ blocks of size $r = m/k$. Then

$$I_{t,\omega,\mathrm{disc}}(f) = \sum_{k|m} \frac{1}{k!k} \sum_{s \in W(\mathfrak{a}_{M_k})_{\mathrm{reg}}} \mathrm{tr}\, \mathcal{M}_{M_k}^G(s, 0) r_{M_k,t}(0, f).$$

For any $M_k$, the $(k-1)!$ elements $s \in W(\mathfrak{a}_{M_k})_{\mathrm{reg}}$ give the same contribution to the sum. Thus if $s$ denotes now the cycle $(1, 2, \ldots, k)$, we get

$$I_{t,\omega,\mathrm{disc}}(f) = \sum_{k|m} k^{-2}\, \mathrm{tr}\, \mathcal{M}_{M_k}^G(s, 0) r_{M_k,t}(0, f).$$

## 25.6 Intertwinging Operators

To describe the operator $\mathcal{M}_{M_k}^G(s, 0) r_{M_k,t}(0, f)$, note that a discrete spectrum representation $\tau$ of $M_k = \mathrm{GL}(r, D) \times \cdots \times \mathrm{GL}(r, D)$ has the form $\tau_1 \times \cdots \times \tau_k$, where each $\tau_i$ is a discrete spectrum representation of $\mathrm{GL}(r, D(\mathbb{A}))$. The symmetric group

$S_k$ permutes the entries of the sequence $(\tau_1, \ldots, \tau_k)$. Denote by $\mathrm{Fix}(\tau)$ the subgroup which fixes $(\tau_1, \ldots, \tau_k)$. Let $\mathrm{Rep}(\tau)$ be a set of representatives in $S_k$ for $S_k/\mathrm{Fix}(\tau)$. The subspace $V_\tau = \otimes_{\alpha \in \mathrm{Rep}(\tau)} \tau_{\alpha(1)} \times \cdots \times \tau_{\alpha(k)}$ of $r_{M_k,t}$ is invariant under the operator $\mathcal{M}^G_{M_k}(s,0)$. If $\tau_i \neq \tau_j$ for some $i \neq j$, then this operator acts on each $\tau_{\alpha(1)} \times \cdots \times \tau_{\alpha(k)}$ without fixed point, and its trace on $V_\tau$ is zero. Thus contributions to the formula are only from $\tau = \tau_1 \times \cdots \times \tau_1$, $\mathrm{Fix}(\tau) = S_k$ and $\mathrm{Rep}(\tau) = \{1\}$, so

$$I_{t,\omega,\mathrm{disc}}(f) = \sum_{\pi \in \mathrm{DS}(m,t,\omega)} m(\pi) \operatorname{tr} \pi(f) + \sum_{1<k|m} k^{-2} \sum_{\tau \in \mathrm{DS}(m/k,t/k,\omega_k)} m(\tau)^k \operatorname{tr} \mathcal{M}^G_{M_k}(s,0) i(\tau^k,0;f).$$

Here $\mathrm{DS}(m,t,\omega)$ is the set of discrete spectrum representations $\tau$ of $\mathrm{GL}(m, D(\mathbb{A}))$ such that $\tau$ transforms via $\omega$ on $F^\times_\infty$ and the norm of the imaginary part of its infinitesimal character is $t \geq 0$. The coefficient $m(\tau)$ is the multiplicity of the irreducible $\tau$ in the discrete spectrum of $\mathrm{GL}(m, D(\mathbb{A}))$. The same notation of course applies to $\mathrm{DS}(m/k, t/k, \omega_k)$, where $\omega_k$ is a character of $F^\times_\infty$ with $\omega_k^k = \omega$. By $\tau^k$ we mean $\tau \times \cdots \times \tau$ ($k$ times).

## 25.7 Discrete Part of Trace Formula

The same formula holds of course when $D^\times = \mathbb{G}_m$, the multiplicative group; thus $d = 1$, for $G' = \mathrm{GL}(n)$. In this case the multiplicities $m(\tau)$ are 1 by the multiplicity one theorem [Shal74] and its extension to the residual spectrum by [MW89]. In this case

$$I_{t,\omega,\mathrm{disc}}(f') = \sum_{\pi' \in \mathrm{DS}'(n,t,\omega)} \operatorname{tr} \pi'(f') + \sum_{1<k|n} k^{-2} \sum_{\tau \in \mathrm{DS}'(n/k,t/k,\omega_k)} \operatorname{tr} \mathcal{M}^{G'}_{M'_k}(s,0) i(\tau'^k,0;f').$$

We write $\mathrm{DS}'$ for DS of the split group.

Then 6.(16.11.1) asserts that for matching $f$ on $G(\mathbb{A})$, $G = \mathrm{GL}(m,D)$, and $f'$ on $G'(\mathbb{A})$, $G' = \mathrm{GL}(n)$, $n = md$, we have

$$I_{t,\omega,\mathrm{disc}}(f') = I_{t,\omega,\mathrm{disc}}(f).$$

PROPOSITION 25.8. *Assume that the multiplicity one theorem holds for all* $\mathrm{GL}(m', D)$, $m' < m$. *Then*

$$\sum_{\pi' \in \mathrm{DS}'_D(n,t,\omega)} \operatorname{tr} \pi'(f') + \sum_{1<k|m} k^{-2} \sum_{\tau' \in \mathrm{DS}'_D(n/k,t/k,\omega_k)} \operatorname{tr} \mathcal{M}^{G'}_{M'_k}(s,0) i(\tau'^k,0;f')$$

$$= \sum_{\pi \in \mathrm{DS}(m,t,\omega)} m(\pi) \operatorname{tr} \pi(f) + \sum_{1<k|m} k^{-2} \sum_{\tau \in \mathrm{DS}(m/k,t/k,\omega_k)} \operatorname{tr} \mathcal{M}^G_{M_k}(s,0) i(\tau^k,0;f)$$

*where* $\mathrm{DS}'_D$ *signifies the subset of* $\tau'$ *all of whose components* $\tau'_v$ *comes from* $\tau_v$ *in* $G(F_v)$.

PROOF. The $m(\tau)^k$ in the expression right of $=$ is replaced with 1 by the induction assumption. As for the first expression, left of =, associated with $G'$, let $k$ divide $n = md$, let $\tau \in \mathrm{DS}'(md/k, t/k, \omega_k)$, and let $f, f'$ be matching functions. We need to show that if $k$ does not divide $m$, or if $k|m$ and $\tau'_v$ does not come from $\tau_v$ on $G(F_v)$ for all $v$, then $\operatorname{tr} \mathcal{M}_{M'_k}^{G'}(s,0) i(\tau'^k, 0; f')$ is zero.

Suppose then that $k$ does not divide $m$. Then $d$ does not divide $md/k$. Since $d$ is the least common multiple of the integers $d_v$, there is a place $u$ such that $d_u$ does not divide $md/k$. But then $\tau'_u$ does not come from $\tau_u$, and $\tau_u'^k$ does not come from the correspondence. Now the operator $\mathcal{M}_{M'_k}^{G'}(s,0)$ acts as a scalar on the irreducible $i(\tau'^k)$. So $\operatorname{tr} i(\tau_u'^k, 0; f'_u) = 0$, as required. □

Using the proposition we prove the theorem, by induction on $m$. If $m = 1$ the identity of the proposition reduces to

$$\sum_{\pi' \in \mathrm{DS}'_D(d,t,\omega)} \operatorname{tr} \pi'(f') = \sum_{\pi \in \mathrm{DS}(1,t,\omega)} m(\pi) \operatorname{tr} \pi(f) \tag{25.8.1}$$

for all matching functions $f$, $f'$. Consider a discrete spectrum representation $\pi'_0$ which occurs on the left. Denote by $V$ the finite set of places where $D$ ramifies. Write $\pi'_0$ as $\pi'_{0V} \otimes \pi_0'^V$, accordingly. The standard argument of generalized linear independence of characters used in this chapter implies that for matching $f_v$, $f'_v$ $(v \in V)$ we have

$$\sum_{\{\pi' ; \pi'^V = \pi_0'^V\}} \operatorname{tr} \pi'_V(f'_V) = \sum_{\{\pi ; \pi^V = \pi_0'^V\}} m(\pi) \operatorname{tr} \pi_V(f_V). \tag{25.8.2}$$

The sum on the left consists of the discrete spectrum representations $\pi'$ of $\mathrm{GL}(d, \mathbb{A})$ with $\pi'^V = \pi_0'^V$.

By multiplicity one theorem and rigidity theorem for the cuspidal spectrum of $\mathrm{GL}(d)$, extended to the discrete noncuspidal spectrum by [MW89], the only contribution to the left side is $\pi'_0$. Since $\pi'_{0v}$ comes from $\pi_{0v}$ for all $v \in V$, we can write the term on the left as $\prod_{v \in V} \operatorname{tr} \pi_{0v}(f_v)$. Now on the right there are only finitely many discrete spectrum $\pi$ with component $\pi_0'^V$ outside $V$ by the argument involving $L$-functions explained in Section 1. Even without this argument, generalized linear independence of characters used in this chapter implies that the sum on the right reduces to a single term, say $\pi_0$, with $m(\pi_0) = 1$. Thus the discrete spectrum $\pi'_0$ on $\mathrm{GL}(d, \mathbb{A})$ corresponds to the discrete spectrum $\pi_0$ on $D(\mathbb{A})^\times$ with the same $t$ and character $\omega$ on $F_\infty^\times$.

## 25.9 Lifting in First Case

Conversely, starting with $\pi_0 \in \mathrm{DS}(1, t, \omega)$ which appears in the sum on the right of (25.8.1) we obtain once again the identity (25.8.2), with $\pi_0'^V$ replaced by $\pi_0^V$. If the left side is 0 (its sum is empty), since $f_V$ is arbitrary, we would obtain a contradiction (it suffices to take $f_V$ to be the characteristic function of a small neighborhood of the identity); we need not appeal to the argument that the sum on the right is finite, which was used in Section 1 since we could not use arbitrary test functions $f_V$ at that stage. Then there exists $\pi'_0$ on the left with component $\pi_0'^V = \pi_0^V$, and as in the first part of this discussion of the case $m = 1$, the discrete spectrum $\pi'_0$ exists and is unique, and it corresponds to the discrete spectrum $\pi_0$; both have the same infinitesimal character, hence $t$, and character $\omega$ on $F_\infty^\times$. Moreover, multiplicity one and rigidity theorems hold on the discrete spectrum of $D(\mathbb{A})^\times$.

## 25.10 Induction

To prove the theorem for $m > 1$, we can assume it by induction for all $m' < m$, having just established it for $m' = 1$. We then use the identity of the proposition. Note that the $\tau$ which appear on the right are discrete spectrum, in particular unitary, hence $i_{M_k}^G(\tau^k, 0)$ is irreducible, and the intertwining operator $\mathcal{M}_{M_k}^G(s, 0)$ is a scalar $\lambda(\tau)$ on the unit circle in $\mathbb{C}^\times$, as $\mathcal{M}$ is unitary. The same holds in particular for the left side; thus $\mathcal{M}_{M'_k}^{G'}(s, 0)$ is $\lambda(\tau')$ on $i_{M'_k}^{G'}(\tau'^k, 0)$ of absolute value one. By induction each discrete spectrum $\tau$ in $\mathrm{DS}(m/k, t/k, \omega_k)$, where $1 < k|m$, corresponds to a $\tau' \in \mathrm{DS}'_D(m/k, t/k, \omega_k)$, so we can write the identity of the proposition as

$$\sum_{\pi' \in \mathrm{DS}'_D(n,t,\omega)} \operatorname{tr} \pi'(f') + \sum_{1<k|m} k^{-2} \sum_{\tau' \in \mathrm{DS}'_D(n/k,t/k,\omega_k)} (\lambda(\tau') - \lambda(\tau \to \tau')) \operatorname{tr} i(\tau'^k, 0, f')$$
$$= \sum_{\pi \in \mathrm{DS}(m,t,\omega)} m(\pi) \operatorname{tr} \pi(f). \tag{25.10.1}$$

Here $\lambda(\tau \to \tau')$ is $\lambda(\tau)$ for the $\tau \in \mathrm{DS}(m/k, t/k, \omega_k)$ which corresponds to $\tau'$ by induction, as $1 < k|m$.

## 25.11 Trace Identity

As in the case $m = 1$, fix a discrete spectrum $\pi'_0$ in $\mathrm{DS}'_D(n, t, \omega)$. The standard argument of generalized linear independence of characters implies that for matching $f_v, f'_v$ $(v \in V)$, we have

$$\sum_{\{\pi' ; \pi'^V = \pi_0'^V\}} \operatorname{tr} \pi'_V(f'_V) + \sum_{1<k|m} \frac{\lambda(\tau') - \lambda(\tau \to \tau')}{k^2} \operatorname{tr} i(\tau_V'^k, 0, f'_V)$$
$$= \sum_{\{\pi ; \pi^V = \pi_0'^V\}} m(\pi) \operatorname{tr} \pi_V(f_V). \tag{25.11.1}$$

The sum on the left ranges over the $\pi'$ in $\mathrm{DS}'_D(n,t,\omega)$ with $\pi'^V = \pi_0'^V$. By the rigidity theorem for GL($n$), the only term in this sum is that of $\pi'_0$. The other sum on the left ranges over the discrete spectrum $\tau' \in \mathrm{DS}'_D(n/k, t/k, \omega_k)$, for $1 < k|m$, with $i(\tau'^k, 0)^V \simeq \pi_0'^V$.

## 25.12 Residual Spectrum

Moeglin and Waldspurger [MW89] classified the discrete spectrum of $\mathrm{GL}(n,\mathbb{A})$. Let $\tau'$ be a cuspidal representation of $\mathrm{GL}(k,\mathbb{A})$. Then the induced representation $i(\nu^{\frac{r+1}{2}-1}\tau' \times \cdots \times \nu^{\frac{r+1}{2}-r}\tau')$ from the Levi of type $(k,\ldots,k)$ ($k$ appears $r$ times, $n = kr$) has a unique irreducible constituent $\pi'$ which is a discrete spectrum of $\mathrm{GL}(n,\mathbb{A})$, denoted $\mathrm{MW}(\tau', r)$. Note that for all $v$ the $\tau'_v$ is non-degenerate, hence irreducible standard, and when written as induced from quasi-square-integrables with decreasing central exponents, $\pi'_v$ is the Langlands quotient $L(\tau'_v, r)$ of $i(\nu_v^{\frac{r+1}{2}-1}\tau'_v \times \cdots \times \nu_v^{\frac{r+1}{2}-r}\tau'_v)$. Every discrete spectrum representation $\pi'$ of $\mathrm{GL}(n,\mathbb{A})$ is of the form $\mathrm{MW}(\tau', r)$ for uniquely determined $\tau'$ and $r$, and $\pi'$ is cuspidal if and only if $r = 1$.

LEMMA 25.13. *Let $\tau_i$ $(1 \le i \le r)$ be discrete spectrum of $\mathrm{GL}(k_i,\mathbb{A})$ with $k_1 + \cdots + k_r = n$, and $\tau'_i$ $(1 \le i \le r')$ discrete spectrum of $\mathrm{GL}(k'_i,\mathbb{A})$, $k'_1+\cdots+k'_{r'}=n$, such that for almost all finite places $v$ the local components of the irreducible induced $i(\tau)$, $\tau = \tau_1 \times \cdots \times \tau_r$, and $i(\tau')$, $\tau' = \tau'_1 \times \cdots \times \tau'_{r'}$, are equivalent. Then $r' = r$ and $(\tau_1,\ldots,\tau_r)$ equals $(\tau'_1,\ldots,\tau'_{r'})$ up to permutation.*

PROOF. By [JS81, Theorem 4.4], the cuspidal support of the automorphic representations $i(\tau)$ and $i(\tau')$ is equal. If $\sigma$ is a cuspidal representation of $\mathrm{GL}(k,\mathbb{A})$ for some integer $k$, we name the set $\{\nu^j\sigma\,;\, j \in \mathbb{Z}\}$ a line, and the set $\{\nu^{j+\frac{1}{2}}\sigma\,;\, j \in \mathbb{Z}\}$ a shifted line. By [MW95], the irreducibles in the cuspidal support of $\tau_i$ or $\tau'_i$ lie in a line or a shifted line. Separating the supports, we reduce the question to that where there is a line or a shifted line $L$ such that the irreducibles in the cuspidal support of all $\tau_i$ and $\tau'_i$ lie in $L$. Namely, there is a cuspidal $\sigma$ such that $\tau_i = \mathrm{MW}(\sigma, a_i)$ for all $i$ $(1 \le i \le r)$ and $\tau'_j = \mathrm{MW}(\sigma, b_j)$ for all $j$ $(1 \le j \le r')$ where either all $a_i$, $b_j$ are even, or odd. We claim that the cuspidal support of $i(\tau)$ and $i(\tau')$ determines the $\tau_i$ (and $\tau'_i$) up to permutation. If the $a_i$ are odd this follows from the following claim. The case where the $a_i$ are even simpler.

If $X$ is a multiset of integers that can be written as a union with multiplicities of sets of the form $x = \{-r, -r+1, -r+2, \ldots, r-1, r\}$, then $X$ determines the sets $x$.

Indeed, if $m : \mathbb{Z} \to \mathbb{Z}_{\ge 0}$ is the multiplicity map, thus $m(a)$ is the multiplicity of $a$ in $X$, then $m(a)$ is also the number of the sets $x$ containing $a$. If $a \ge 1$ and $x$ contains $a$, then it contains also $a-1$, so $m$ is decreasing on $\mathbb{Z}_{\ge 0}$, and the number of the sets $\{-r, -r+1, \ldots, r-1, r\}$ in $X$ is given by $m(r) - m(r+1)$. □

## 25.14 Reduced Identity

We conclude that the sum over $1 < k|m$ and $\tau'$ in the last displayed equation, over the discrete spectrum $\tau'$ with $i(\tau'^k, 0)^V \simeq \pi_0'^V$ contains at most one element.

Now the $\pi_0'$ and the possible $\tau'$, or rather $i(\tau'^k, 0)$ which may occur on the left side of the last displayed equation come locally from representations of the inner form $G = \mathrm{GL}(m, D)$, we deduce from generalized linear independence of characters that the term $i(\tau'^k, 0)$ cannot contribute, since its coefficient $(\lambda(\tau') - \lambda(\tau \to \tau'))/k^2$ is less than 1/2 in absolute value, in particular not an integer. Hence $\lambda(\tau') = \lambda(\tau \to \tau')$, and we are left with the identity

$$\sum_{\{\pi' ; \pi'^V = \pi_0'^V\}} \mathrm{tr}\, \pi_V'(f_V') = \sum_{\{\pi ; \pi^V = \pi_0'^V\}} m(\pi)\, \mathrm{tr}\, \pi_V(f_V). \tag{25.14.1}$$

Linear independence of characters, as used earlier in this chapter, implies that the right side is not zero and consists of a single term indexed by some $\pi_0$ with $\pi_0^V = \pi_0'^V$, and its multiplicity $m(\pi_0)$ is 1.

## 25.15 Multiplicity One

Similarly, fixing a discrete spectrum $\pi_0$ in $\mathrm{DS}(m, t, \omega)$, we cannot have that both sums over $\pi'$ with $\pi'^V = \pi_0^V$ and $\tau'$ with $i(\tau'^k)^V = \pi_0^V$ are empty. By rigidity the sum over $\pi'$ has at most one entry, and the argument above shows that $\lambda(\tau') = \lambda(\tau \to \tau')$, so that the sum over $\tau'$ is zero. This implies that there is a nonzero entry in the sum over $\pi'$, denote it by $\pi_0'$, and that $\pi_0$ corresponds to $\pi_0'$ and the multiplicity $m(\pi_0)$ of $\pi_0$ in the cuspidal spectrum of $\mathrm{GL}(m, D)$ is one.

## 25.16 End of Proof of Global Theorem

In fact, we can use equation (25.11.1) with $\pi_0' = i(\tau'^k)$ for any $k > 1$, $k|m$, and a discrete spectrum $\tau'$. Using Proposition 7.5 we obtain an analogous identity for the cuspidal support of $i(\tau'^k)$. Once again, since the coefficient $(\lambda(\tau') - \lambda(\tau \to \tau'))/k^2$ is less than one in absolute value, while all other terms have integral coefficients, our generalized linear independence of characters implies that $\lambda(\tau \to \tau') = \lambda(\tau')$ for all discrete spectrum $\tau$ of $\mathrm{GL}(m/k, D(\mathbb{A}))$ which we know correspond to a discrete spectrum $\tau'$ of the split form of the group. Hence the sum over $k$ $(1 < k|m)$ and $\tau'$ can be removed from (25.11.1).

From this, namely, from (25.14.1), we conclude that for each discrete spectrum $\pi$, there is a unique discrete spectrum $\pi'$ with $\pi_v' \simeq \pi_v$ for all $v \notin V$, such that $\mathrm{tr}\, \pi_v'(f_v') \neq 0$ for some $f_v'$ matching some $f_v$, for all $v \in V$. Since (25.14.1) holds for all matching $f'_v$ and $f_v$ ($v \in V$), using again our generalized linear independence of characters, using Proposition 7.5 and the finiteness argument of Section 12, we conclude that each component $\pi'_v$ of $\pi'$ comes from $G_v$ ($v \in V$). Conversely, starting from a discrete spectrum $\pi'$ in (25.14.1), we see that $\operatorname{tr} \pi'_V(f'_V) \neq 0$ for $f'_V$ matching $f_V$ implies, by the same argument, that $\pi'$ comes from a discrete spectrum $\pi$, in particular each of its components at $v \in V$ comes from $G_v$. This completes the proof of the global theorem. □

## 25.17 *Refinement*

Let $M = \mathrm{GL}(m_1, D) \times \cdots \times \mathrm{GL}(m_r, D)$ be a standard Levi subgroup of $\mathrm{GL}(m, D)$. A quasi-cuspidal, or a quasi-discrete spectrum, representation of $M(\mathbb{A})$ is a representation $\tau = \nu^{a_1}\tau_1 \times \cdots \times \nu^{a_r}\tau_r$ where $a_j \in \mathbb{R}$ and $\tau_j$ are cuspidal, or discrete spectrum, on $\mathrm{GL}(m_j, D(\mathbb{A}))$. Such $\tau' = \nu^{a_1}\tau'_1 \times \cdots \times \nu^{a_r}\tau'_r$ on $M'(\mathbb{A})$ will be said to *come locally* from $M(\mathbb{A})$ if all $\tau'_j$ come locally from $\mathrm{GL}(m_j, D(\mathbb{A}))$; thus $\tau'_{jv}$ comes from $\mathrm{GL}(m_j, D(F_v))$ for all $v \in V$.

Let $\tau'$ be a cuspidal representation of $\mathrm{GL}(k, \mathbb{A})$. As described in Subsection 24.19, each component $\tau'_v$ is generic and unitarizable so has the form $\tau'_v = i\left(\nu_v^{e_{1,v}}\sigma'_{1,v} \times \cdots \times \nu_v^{e_{q_v,v}}\sigma'_{q_v,v}\right)$ with $e_{jv} \in (-1/2, 1/2)$ and square-integrable $\sigma'_{jv} \in Z_u(\tau'_{jv}, k_{jv})$, where the $\tau'_{jv}$ are cuspidal unitarizable $\mathrm{GL}(p_{jv}, F_v)$-modules. Put $J_v = \{j \in \mathbb{Z};\, 1 \leq j \leq q_v, d_v | p_{jv}k_{jv}\}$. Let $z(\tau'_v, d_v)$ be the smallest integer $z_v > 0$ with $d_v | z_v p_{jv}$ for all $j \in \{1, \ldots, q_v\} - J_v$. As in 24.19, $C(L(\tau'_v, r_v)) \neq 0$ if an only if $z(\tau'_v, d_v) | r_v$.

Proposition 25.18. *Let $\tau'$ be a cuspidal representation of $\mathrm{GL}(k, \mathbb{A})$. Let $z(\tau', D)$ be the least common multiple of the $z(\tau'_v, d_v)$, $v \in V$. Then $\mathrm{MW}(\tau', r)$ comes locally from $\mathrm{GL}(kr/d, D(\mathbb{A}))$ if and only if $z(\tau', D)$ divides $r$. Further, the discrete spectrum representation $\pi = C(\mathrm{MW}(\tau', z(\tau', D)))$ of $\mathrm{GL}(kz(\tau', D)/d, \mathbb{A})$, corresponding to the discrete spectrum $\mathrm{MW}(\tau', z(\tau', D))$ on $\mathrm{GL}(kz(\tau', D), \mathbb{A})$, is cuspidal. In particular, a cuspidal $G'(\mathbb{A})$-module can only be obtained from a cuspidal $G(\mathbb{A})$-module.*

Proof. The first claim follows from the definition of $z(\tau', D)$ and 24.19. For the further claim, suppose $\pi$ is not cuspidal. Then there is a quasi-cuspidal representation $\theta$ of $M(\mathbb{A})$, where $M$ is a proper standard Levi subgroup of $G = \mathrm{GL}(kz(\tau', D)/d, D)$, such that $\pi$ is a constituent of the representation induced to $\mathrm{GL}(kz(\tau', D)/d, D(\mathbb{A}))$ from $\theta$. Then $\theta$ corresponds to $\theta' = C'(\theta)$. This $\theta' = C'(\theta)$. This $\theta'$ is a quasi-square-integrable $M'(\mathbb{A})$-module; $M'$ is the proper Levi subgroup of $\mathrm{GL}(kz(\tau', D))$ corresponding to $M$, coming locally from $M(\mathbb{A})$. By [JS81, Theorem 4.4], $\theta'$ has the same cuspidal support as $\mathrm{MW}(\tau', z(\tau', D))$. Then $\theta'$ is a quasi-square-integrable representation of the smaller group $M'(\mathbb{A})$, coming from $M(\mathbb{A})$. This contradicts the minimality of $z(\tau', D)$.

The last, "in particular," claim is that where $z(\tau', D) = 1$, thus $d_v | p_{jv} k_{jv}$ for all $j$ ($1 \le j \le q_v$) and $v \in V$. Then $\mathrm{MW}(\tau', z(\tau', D))$ is the cuspidal $\tau'$, and the corresponding $\pi = C(\tau')$ is cuspidal, rather than just discrete spectrum. □

DEFINITION 25.19. The cuspidal representation $\pi = C(\mathrm{MW}(\tau', z(\tau', D)))$ will be called *basic* cuspidal. Put $z(\pi)$ for $z(\tau', D)$, and $\nu_\pi$ for $\nu^{z(\tau', D)}$. Let $M = \mathrm{GL}(m_1, D) \times \cdots \times \mathrm{GL}(m_r, D)$ be a standard Levi subgroup of $G = \mathrm{GL}(m, D)$. A *basic* quasi-cuspidal representation of $M$ is $\nu^{a_1}\tau_1 \times \cdots \times \nu^{a_r}\tau_r$ where $a_j \in \mathbb{R}$ and $\tau_j$ is a basic cuspidal representation of $\mathrm{GL}(m_j, D(\mathbb{A}))$.

PROPOSITION 25.20. (i) *Let $\tau$ be a basic cuspidal representation of $\mathrm{GL}(k, D(\mathbb{A}))$. The induced representation $i\left(\nu_\tau^{\frac{r+1}{2}-1}\tau \times \cdots \times \nu_\tau^{\frac{r+1}{2}-r}\tau\right)$ has a unique quotient $\pi$, denoted $\mathrm{MW}(\tau, r)$. It is in the discrete spectrum of $\mathrm{GL}(rk, D(\mathbb{A}))$. Every discrete spectrum representation $\pi$ of $\mathrm{GL}(m, D(\mathbb{A}))$ is of this form, where $r$ and $\tau$ are uniquely determined by $\pi$. The discrete spectrum $\pi$ is basic cuspidal if and only if $r = 1$. Put $\pi = \mathrm{MW}(\tau, r)$. Then $C'(\tau) = \mathrm{MW}(\tau', z(\tau', D))$ if and only if $C'(\pi) = \mathrm{MW}(\tau', rz(\tau', D))$.*

(ii) *Let $M_1$, $M_2$ be standard Levi subgroups of $G = \mathrm{GL}(m, D)$, and $\tau_1$, $\tau_2$ basic quasi-cuspidal representations of $M_1(\mathbb{A})$, $M_2(\mathbb{A})$. Let $V'$ be a finite set of places of $F$ including the Archimedean places and the finite places where $\tau_1$ and $\tau_2$ are not both unramified. If the unramified subquotients of the induced representations $i(\tau_{1v})$ and $i(\tau_{2v})$ of $G_v$ are equivalent for all $v \notin V'$, then $(M_1, \tau_1')$ is conjugate to $(M_2, \tau_2')$.*

(iii) *Let $\pi$ be an automorphic representation of $G(\mathbb{A})$, $G = \mathrm{GL}(m, D)$. Then there is a standard Levi subgroup $M$ of $G$ and a basic quasi-cuspidal representation $\tau$ of $M(\mathbb{A})$ such that $\pi$ is a constituent of the induced representation $i_M^G\tau$. Then the pair $(M, \tau)$ is uniquely determined by $\pi$ up to conjugation.*

Part (i) generalizes [MW89] from the case $G' = \mathrm{GL}(n)$ to that of $\mathrm{GL}(m, D)$, that is, from the case of $d = 1$ to that of $d \ge 1$. Part (ii) coincides with [JS81, Theorem 4.4] when $d = 1$.

PROOF. (i) Write $C'(\tau)$ as $\mathrm{MW}(\tau', z(\tau', D))$. The discrete spectrum $\mathrm{MW}(\tau', rz(\tau', D))$ comes locally from $\mathrm{GL}(krz(\tau', D), D(\mathbb{A}))$ by Proposition 25.18. We claim that $C(\mathrm{MW}(\tau', rz(\tau', D))$ is a constituent of $i\left(\nu_\tau^{\frac{r+1}{2}-1}\tau \times \cdots \times \nu_\tau^{\frac{r+1}{2}-r}\tau\right)$. It suffices to show that $|C|_v(\mathrm{MW}(\tau', rz(\tau', D))_v)$ in a constituent of $\nu_{\tau_v}^{\frac{r+1}{2}-1}\tau_v \times \cdots \times \nu_{\tau_v}^{\frac{r+1}{2}-r}\tau_v$ for all $v$ in $V$. It suffices to show that the quasi-square-integrable support of $|C|_v(\mathrm{MW}(\tau', rz(\tau', D))_v)$ is the union of such support of the representations $\nu_{\tau_v}^{\frac{r+1}{2}-j}\tau_v$.

As in 24.19 we write the non-degenerate $\tau_v'$ as induced $i(\nu^{e_1}\sigma_1' \times \cdots \times \nu^{e_{q_v}}\sigma_{q_v}')$, where $\sigma_j'$ are square-integrable and $e_j \in (-1/2, 1/2)$. For simplicity we write $\sigma_j'$ for $\sigma_{jv}'$. We have seen that

$$\tau_v = |C|_v(L(\tau_v', z(\tau', D))) = \prod_{j=1}^{q_v} \nu^{e_j} |C|_v(u(\sigma_j', z(\tau', D)))$$

and

$$|C|_v(L(\tau_v', rz(\tau', D))) = \prod_{j=1}^{q_v} \nu^{e_j} |C|_v(u(\sigma_j', rz(\tau', D))).$$

If $\sigma_j'$ comes from $\sigma_j$, then $|C|_v(u(\sigma_j', z(\tau', D))$ is $\overline{u}(\sigma_j, z(\tau', D))$, and $|C|_v(u(\sigma_j', rz(\tau', D))$ is $\overline{u}(\sigma_j, rz(\tau', D))$. The quasi-square-integrable support of $\overline{u}(\sigma_j, z(\tau', D))$ is the union of the quasi-square-integrable support of $\nu^{(\frac{r+1}{2}-j)z(\tau', D)}\overline{u}(\sigma_j, z(\tau', D))$ $(1 \leq i \leq r)$. If $\sigma_j'$ is not in the image of the correspondence (see reference in 24.17), a similar discussion applies. The conclusion is that $i\left(\nu_\tau^{\frac{r+1}{2}-1}\tau \times \cdots \times \nu_\tau^{\frac{r+1}{2}-r}\tau\right)$ has a subquotient which is in the discrete spectrum. Rigidity theorem for $\mathrm{GL}(m, D(\mathbb{A}))$ implies that there is no other subquotient that is in the discrete spectrum.

Now let $\pi$ be in the discrete spectrum of $G(\mathbb{A})$, $G = \mathrm{GL}(m, D)$. We claim that it is obtained as in the last paragraph. Write $C'(\pi) = \mathrm{MW}(\tau', r)$. Then $z(\tau', D)|r$ since $\mathrm{MW}(\tau', q)$ comes locally from $G(\mathbb{A})$, by Proposition 25.18. Put $\theta = C(\mathrm{MW}(\tau', z(\tau', D)))$. It is basic cuspidal. Then we have that $\pi = \mathrm{MW}(\theta, q/z(\tau', D))$. Rigidity theorem for $\mathrm{GL}(md, \mathbb{A})$ implies that $\tau'$ and $q$ are uniquely determined by $\pi$. Hence $r = q/z(\tau', D)$ and $\theta$ are determined by $\pi$. Finally, $\pi$ is basic cuspidal if and only if $q = z(\tau', D)$, if and only if $r = 1$.

(ii) Write $\tau_1' = C'(\tau_1)$ as a product $\otimes_{j=1}^{q_1} \nu^{a_j}\mathrm{MW}(\alpha_j, z(\alpha_j, D))$ and $\tau_2' = C'(\tau_2)$ as the product $\otimes_{j=1}^{q_2} \nu^{b_j}\mathrm{MW}(\beta_j, z(\beta_j, D))$, where $\alpha_j$ and $\beta_j$ are cuspidal. The representations of $\mathrm{GL}(md, \mathbb{A})$ induced from $\tau_1'$ and $\tau_2'$ have equivalent unramified components at all places outside $V' \cup V$. By [JS81, Theorem 4.4], they have the same quasi-cuspidal support. But $\alpha_j$ and $\beta_j$ are cuspidal. From the formulae for $\tau_1'$ and $\tau_2'$, it follows that the multisets $\{(a_j, \alpha_j)\}$ and $\{(b_j, \beta_j)\}$ are equal. Hence the products representing $\tau_1'$ and $\tau_2'$ are equal up to permutation.

(iii) Existence follows from part (i) and uniqueness from part (ii).

□

The appendix to [Ba08], by Neven Grbac, shows—upon further analyzing normalizing factors of intertwining operators and poles of Eisenstein series on the inner form $G = \mathrm{GL}(m, D)$—that all cuspidal representations of $G(\mathbb{A})$ are basic. Consequently in Proposition 25.20, we may erase the word “basic.”

## 26 One Cuspidal Place

The Selberg trace formula is of unquestionable value for the study of automorphic forms and related objects. In principal it is a simple and natural formula, generalizing the Poisson summation formula, relating traces of convolution operators with orbital integrals. This section is motivated by the belief that such a fundamental and natural relation should admit a *simple and short* proof in key and useful cases. This is accomplished here for test functions with a single discrete component and another component which is spherical and "sufficiently admissible" with respect to the other components. The resulting trace formula is then used to sharpen and extend the (metaplectic correspondence and the) simple algebras correspondence, of automorphic representations, to the context of automorphic forms with a *single* cuspidal component, over any global field. We deal with the extension of these theorems to the context of all automorphic representations in the following chapters. In the previous sections, a simple form of the trace formula was developed for test functions with two discrete components; this was used to establish these correspondences for automorphic forms with two discrete components. The notion of "sufficiently admissible" spherical functions has its origins in Drinfeld's study of the reciprocity law for GL(2) over a function field, and our form of the trace formula is analogous to Deligne's conjecture on the fixed point formula in étale cohomology, for a correspondence which is multiplied by a sufficiently high power of the Frobenius, on a separated scheme of finite type over a finite field, first published in [FK87.1], and proven by Fujiwara [Fu97] and Varshavsky [Va07]. Our trace formula can be used (see the announcement [FK87.1], and in more detail in [F13]), to prove the Ramanujan conjecture for automorphic forms with a cuspidal component on GL($n$) over a function field and to reduce the reciprocity law for such forms to Deligne's conjecture. Similar techniques are used in [F90.1] to establish base change for GL($n$) in the context of automorphic forms with a single cuspidal component. They can be used to give short and simple proofs of rank one lifting theorems for *arbitrary* automorphic forms; see [F90.1] for base change for GL(2), [F06] for base change for U(3), and for the symmetric square lifting from SL(2) to PGL(3).

Let $F$ be a global field, $\mathbb{A}$ its ring of adèles and $\mathbb{A}_f$ the ring of finite adèles, $G$ a connected reductive algebraic group over $F$ with center $Z$. The group $G(F)$ of $F$-rational points on $G$ is discrete in the adèle group $G(\mathbb{A})$ of $G$. Put $G' = G/Z$. The quotient $G'(F)\backslash G'(\mathbb{A})$ has finite volume with respect to the unique (up to scalar multiple) Haar measure $\mathrm{d}g$ on $G'(\mathbb{A})$. Fix a *unitary* complex-valued character $\omega$ of $Z(F)\backslash Z(\mathbb{A})$. For any place $v$ of $F$, let $F_v$ be the completion of $F$ at $v$. If $F_v$ is non-Archimedean, let $R_v$ denote its ring of integers. For almost all $v$, the group $G(F_v)$ is defined over $R_v$, quasisplit over $F_v$, split over an unramified extension of $F_v$, and $K_v = G(R_v)$ is a maximal compact subgroup. For an infinite set of places (of positive density) $u$ of $F$, the group $G(F_u)$ is split (over $F_u$). A fundamental system of open neighborhoods of 1 in $G(\mathbb{A})$ consists of the set $\prod_{v\in V} H_v \times \prod_{v\notin V} K_v$, where $V$ is a finite set of places of $F$ and $H_v$ is an open subset of $G(F_v)$ containing 1.

Let $L^2(G(F)\backslash G(\mathbb{A}))$ denote the space of all complex-valued functions $\phi$ on $G(F)\backslash G(\mathbb{A})$ which satisfy $\phi(zg) = \omega(z)\phi(g)$ ($z \in Z(\mathbb{A})$, $g \in G(\mathbb{A})$) and are square-integrable on $G'(F)\backslash G'(\mathbb{A})$. The group $G(\mathbb{A})$ acts on $L^2(G(F)\backslash G(\mathbb{A}))$ by right translation: $(r(g)\phi)(h) = \phi(hg)$. The representation is unitary since $\omega$ is unitary. The function $\phi$ in $L^2(G(F)\backslash G(\mathbb{A}))$ is called *cuspidal* if for each proper parabolic subgroup $P$ of $G$ over $F$ with unipotent radical $N$, we have

$$\int_{N(F)\backslash N(\mathbb{A})} \phi(ng)\,\mathrm{d}n$$

for any $g$ in $G(\mathbb{A})$. Let $L_0^2(G(F)\backslash G(\mathbb{A}))$ denote the space of cuspidal functions in $L^2(G(F)\backslash G(\mathbb{A}))$, and $r_0$ the restriction of $r$ to $L_0^2(G(F)\backslash G(\mathbb{A}))$. The space $L_0^2(G(F)\backslash G(\mathbb{A}))$ decomposes as a direct sum with finite multiplicities of invariant irreducible unitary $G(\mathbb{A})$-modules called *cuspidal* $G$-modules.

Let $f$ be a complex-valued function on $G(\mathbb{A})$ with $f(g) = \omega(z)f(zg)$ for $z$ in $Z(\mathbb{A})$, which is supported on the product of $Z(\mathbb{A})$ and a compact open neighborhood of 1 in $G(\mathbb{A})$, smooth as a function on the Archimedean part $G(F_\infty)$ of $G(\mathbb{A})$, and bi-invariant by an open compact subgroup of $G(\mathbb{A}_f)$. Fix Haar measures $dg$ on $G'(F_v)$ for all $v$, such that the product of the volumes $|K_v/Z_v \cap K_v|$ converges. Then $dg = \otimes_v dg_v$ is a measure on $G'(\mathbb{A})$. The convolution operator

$$r_0(f) = \int_{G'(\mathbb{A})} f(g) r_0(g)\, dg$$

is of trace class. Its trace is denoted by $\operatorname{tr} r_0(f)$. Then

$$\operatorname{tr} r_0(f) = \sum_\pi m(\pi) \operatorname{tr} \pi(f), \tag{26.0.1}$$

where the sum is over all equivalence classes of cuspidal representations $\pi$ of $G(\mathbb{A})$, and $m(\pi)$ denotes the multiplicity of $\pi$ in $L_0^2(G(F)\backslash G(\mathbb{A}))$. Each $\pi$ here is unitary, and the sum is absolutely convergent.

The Selberg trace formula is an alternative expression for (26.0.1). To introduce it we recall the following:

DEFINITION 26.1. Denote by $H_\gamma = \{h \in H\,;\, h\gamma = \gamma h\}$ the centralizer of an element $\gamma$ in a group $H$. A semisimple element $\gamma$ in $G(F)$ is called *elliptic* if $G'(\mathbb{A})_\gamma/G'(F)_\gamma$ has finite volume. It is called *regular* if $G'(\mathbb{A})_\gamma$ is a torus, and *singular* otherwise. Let $\gamma$ be an elliptic element of $G(F)$. The *orbital integral* of $f$ at $\gamma$ is defined to be

$$\Phi(\gamma, f) = \int_{G'(\mathbb{A})/G'(F)_\gamma} f(g\gamma g^{-1})\, dg.$$

Similarly, for any place $v$ of $F$, the element $\gamma$ of $G(F_v)$ is called *elliptic* if $G'(F_v)_\gamma$ has finite volume, and *regular* if $G(F_v)_\gamma$ is a torus. If $\gamma$ is an element of $G(F)$ and there is a place $v$ of $F$ such that $\gamma$ is elliptic (resp. regular) in $G(F_v)$, then $\gamma$ is elliptic (resp. regular). The orbital integral of $f_v$ at $\gamma$ in $G(F_v)$ is defined to be

$$\Phi(\gamma, f_v) = \Phi(\gamma, f_v\, d_\gamma) = \int_{G'(F_v)/G'(F_v)_\gamma} f_v(g\gamma g^{-1}) \frac{dg}{d_\gamma}.$$

It depends on the choice of a Haar measure $d_\gamma$ on $G'(F_v)_\gamma$.

Let $\{\phi_\alpha\}$ be an orthonormal basis for the space $L_0^2(G(F)\backslash G(\mathbb{A}))$. The operator $r_0(f)$ is an integral operator on $G'(\mathbb{A})$ with kernel $K_f^0(x,y) = \sum_{\alpha,\beta}(r(f)\phi_\alpha(x))\overline{\phi}_\beta(y)$. The operator $r(f)$ is an integral operator on $G'(\mathbb{A})$ with kernel $K_f(x,y) = \sum_\gamma f(x^{-1}\gamma y)$ with $\gamma$ in $G'(F)$. If $G$ is *anisotropic*, namely, $G'(F)\backslash G'(\mathbb{A})$ is compact, then $L_0^2(G(F)\backslash G(\mathbb{A})) = L^2(G(F)\backslash G(\mathbb{A}))$ and $r = r_0$. Since $K_f^0(x,y) = K_f(x,y)$ is smooth in both $x$ and $y$, we integrate over the diagonal $x = y$ in $G'(\mathbb{A})$, change the order of summation and integration as usual, and obtain the Selberg trace formula in the case of compact quotient, as follows:

PROPOSITION 26.2. *If $G$ is anisotropic, then for every function $f$ on $G(\mathbb{A})$ as above, we have*

$$\sum_\pi m(\pi)\,\mathrm{tr}\,\pi(f) = \sum_{\{\gamma\}} \Phi(\gamma,f). \tag{26.2.1}$$

*The sum on the left is as in* (26.0.1). *The sum on the right is finite. It ranges over the conjugacy classes of elements in $G'(F)$.*

REMARK 26.3. If $G$ is anisotropic, then each element $\gamma$ in $G(F)$ is elliptic.

For a general group $G$, we introduce the following:

DEFINITION 26.4. The function $f$ is called *discrete* if for every $x$ in $G(\mathbb{A})$ and $\gamma$ in $G(F)$, we have $f(x^{-1}\gamma x) = 0$ unless $\gamma$ is elliptic regular.

Changing again the order of summation and integration as usual, we obtain the

PROPOSITION 26.5. *If $f$ is discrete, then*

$$\int_{G'(\mathbb{A})} \left( \sum_{\gamma\in G'(F)} f(x^{-1}\gamma x) \right) dx = \sum_{\{\gamma\}} \Phi(\gamma,f). \tag{26.5.1}$$

*The sum on the right is finite. It ranges over the set of conjugacy classes of elliptic regular elements in $G'(F)$.*

REMARK 26.6. It is well known that the sum on the right is finite. For a proof see [FK87.2] when $G = \mathrm{GL}(n)$ and Proposition 4.2 in general.

DEFINITION 26.7. The function $f$ is called *cuspidal* if for every $x$, $y$ in $G(\mathbb{A})$ and every proper $F$-parabolic subgroup $P$ of $G$, we have

$$\int_{N(\mathbb{A})} f(xny)\,dn = 0,$$

where $N$ is the unipotent radical of $P$.

When $f$ is cuspidal, the convolution operator $r(f)$ factorizes through the projection to the cuspidal spectrum $L_0^2(G(F)\backslash G(\mathbb{A}))$; it is of trace class with $\operatorname{tr} r(f) = \operatorname{tr} r_0(f)$, and $K_f(x,y) = K_f^0(x,y)$. We obtain the

COROLLARY 26.8. *If $f$ is cuspidal and discrete, then the equality* (26.2.1) *holds. The sum on the left is as in* (26.0.1). *The sum on the right is as in* (26.5.1).

For some applications we need to replace the requirement that $f$ be discrete by a requirement on the orbital integrals of $f$ (but not on $f$ itself). The purpose of this section is to present such a requirement and apply the resulting trace formula to extend some global lifting theorems, such as those of [FK87.2].

Fix a non-Archimedean place $u$ of $F$ such that $G(F_u)$ is split, and the component $\omega_u$ of $\omega$ at $u$ is unramified (namely trivial on the multiplicative group $R_u^\times$ of $R_u$).

DEFINITION 26.9. A complex-valued compactly supported modulo center function $f_u$ on $G(F_u)$ is called *spherical* function if it is $K_u$-bi-invariant. Let $\mathbb{H}_u$ be the convolution algebra of such functions. Of course $\mathbb{H}_u$ is empty unless the central character $\omega_u$ is unramified.

For any maximal (proper) $F_u$-parabolic subgroup $P(F_u) = M(F_u)N(F_u)$ of $G(F_u)$, where $N(F_u)$ is the unipotent radical of $P(F_u)$ and $M(F_u)$ is a Levi subgroup, define an $F_u^\times$-valued character $\alpha_{P(F_u)}$ of $M(F_u)$ by $\alpha_{P(F_u)}(m) = \det(\operatorname{ad}(m)|\mathfrak{n})$, where $\mathfrak{n}$ is the Lie algebra of $N(F_u)$ and $\operatorname{ad}(m)|\mathfrak{n}$ denotes the adjoint action of $m$ in $M(F_u)$ on $\mathfrak{n}$. Let $\nu_u : F_u^\times \to \mathbb{Z}$ be the normalized additive valuation. Let $A(F_u)$ be a maximal split torus in $G(F_u)$. For any nonnegative integer $n$, let $A(F_u)^{(n)}$ be the set of $a$ in $A(F_u)$ such that $|\nu_u(\alpha_{P(F_u)}(a))| < n$ for some maximal $F_u$-parabolic subgroup $P(F_u)$ containing $A(F_u) \subset G(F_u)$.

DEFINITION 26.10. A spherical function $f_u$ is called *$n$-admissible* if the orbital integral $\Phi(a,f_u)$ is zero for every regular $a$ in $A(F_u)^{(n)}$.

Let $\mathbb{A}^u$ denote the ring of $F$-adèles without $u$-component. Write $f = f_u f^u$ if $f$ is a function on $G(\mathbb{A})$, $f_u$ is a function on $G(F_u)$, $f^u$ is a function on $G(\mathbb{A}^u)$, and $f(x,y) = f_u(x)f^u(y)$ for $x$ in $G(F_u)$ and $y$ in $G(\mathbb{A}^u)$. We choose the place $u$ such that the central character $\omega$ is unramified at $u$.

THEOREM 26.11. *Let $f^u$ be a function on $G(\mathbb{A}^u)$ which is compactly supported modulo $Z(\mathbb{A}^u)$ and vanishes on the $G(\mathbb{A}^u)$-orbit of any singular $\gamma$ in $G(F)$. Then there exists a positive integer $n_0 = n_0(f^u)$ such that for every spherical $n_0$-admissible function $f_u$, there is a function $f_u'$ on $G(F_u)$ such that $\Phi(x,f_u') = \Phi(x,f_u)$ for all regular $x$ in $G(F_u)$ and $f' = f_u' f^u$ is discrete.*

PROOF. For every maximal $f$-parabolic subgroup $P$ of $G$ and every place $v \neq u$ of $F$, there exists a nonnegative integer $C_{v,P}$ which depends on $f^u$, with $C_{v,P} = 0$ for almost all $v$, such that if $\gamma$ lies in a Levi subgroup $M$ of $P$ and $f^u(x^{-1}\gamma x) \neq 0$ for some $x$ in $G(\mathbb{A}^u)$, then

$$|\nu_v(\alpha_P(\gamma))| \leq C_{v,P}. \tag{26.11.1$_v$}$$

Put $C_{u,P} = \sum_{v \neq u} C_{v,P}$. Since $\gamma$ is rational in $G(F)$, the product formula $\sum_v \nu(\alpha_P(\gamma)) = 0$ on $F^\times$ implies that the inequality (26.11.1$_v$) remains valid also for $v = u$. Choose $n_0 > C_{u,P}$ for all (of the finitely many conjugacy classes of) $P$. Let $f_u$ be any spherical $n_0$-admissible function. Put $f = f_u f^u$. Since a discrete and compact set is finite (see the proof of Proposition 4.2), there are only finitely many rational conjugacy classes $\gamma$ in $G'(F)$ such that $f$ is not zero on the $G'(\mathbb{A})$-orbit of $\gamma$. Note that $f$ is zero on the $G(\mathbb{A})$-orbits of all singular $\gamma$ in $G(F)$ by assumption. Let $\gamma_i$ $(1 \leq i \leq m)$ be a set of representatives for the regular non-elliptic rational conjugacy classes in $G(F)$ such that $f$ is nonzero on their $G(\mathbb{A})$-orbits. Since $\gamma_i$ is non-elliptic, it lies in a Levi subgroup $M_i$ of a maximal parabolic subgroup $P_i$ of $G$. Since $f_u$ is $n_0$-admissible, the relation $\Phi(\gamma_i, f_u) \neq 0$ implies that $|\nu_u(\alpha_{P_i}(\gamma_i))| > n_0$. This contradicts (26.11.1)$_u$. Hence $\Phi(\gamma_i, f_u) = 0$ for all $i$. Let $S_i$ denote the characteristic function of the complement in $G(F_u)$ of a sufficiently small open-closed neighborhood of the orbit of $\gamma_i$ in $G(F_u)$. Since $\gamma_i$ is regular non-elliptic, we may and do take $S_i$ to be one on the elliptic set of $G(F_u)$. Put $f'_u = f_u \prod_{i=1}^m S_i$. Then $f'_u$ is zero on the orbit of $\gamma_i$ $(1 \leq i \leq m)$, and $\Phi(\gamma, f'_u) = \Phi(\gamma, f_u)$ for every regular $\gamma$ in $G(F_u)$. Since $f' = f'_u f^u$ vanishes on the $G(\mathbb{A})$-orbit of each rational $\gamma$ in $G(F)$ which is not elliptic regular, the theorem follows. □

Since both sides of (26.2.1) are invariant distributions, we conclude the immediate

COROLLARY 26.12. *Suppose that $f = f_u f^u$ is a cuspidal function which vanishes on the $G(\mathbb{A})$-orbit of every singular $\gamma$ in $G(F)$, and $f_u$ is a spherical $n_0$-admissible function with $n_0 = n_0(f^u)$. Then the equality* (26.2.1) *holds, where the sum on the left is as in* (26.0.1)*, while the sum on the right is as in* (26.5.1).

DEFINITION 26.13. A $G(F_u)$-module $\pi_u$ is called *unramified* if it has a nonzero $K_u$-fixed vector.

For applications such as those given in Theorem 26.18 below, we need to show that the set of $n$-admissible functions is sufficiently large in the following sense.

THEOREM 26.14. *Let $\{\pi_i\,;\, i \geq 0\}$ be a sequence of inequivalent unitarizable unramified $G(F_u)$-modules, and $c_i$ complex numbers, such that $\sum_i c_i \operatorname{tr} \pi_i(f_u)$ is absolutely convergent for every spherical $f_u$. Suppose that there is a positive integer $n_0$ such that $\sum_i \operatorname{tr} \pi_i(f_u) = 0$ for all $n_0$-admissible $f_u$. Then $c_i = 0$ for all $i$.*

PROOF. Delayed to the end of this section. □

REMARK 26.15. The notion of $n$-admissible functions is suggested by Drinfeld [Dr77], at least in the case of $G = \mathrm{GL}(2)$. For a general $G$, the corollary is a representation theoretic analogue of Deligne's conjecture on the Grothendieck-Lefschetz fixed-point formula for the trace of a finite flat correspondence on a separated scheme of finite type over a finite field, which is multiplied by a sufficiently high power of the Frobenius morphism, proven by Fujiwara [Fu97] and Varshavsky [Va07]. This analogy in more detail in the work [F13] on the geometric Ramanujan conjecture for $\mathrm{GL}(n)$ (see also [FK87.1]).

In the proofs of Theorem 26.14 and Theorem 26.18 below, we shall use some results concerning unramified representations and spherical functions (see [Car79]) and regular functions. These will be recalled now in order to be able to give an uninterrupted exposition of the proof of Theorem 26.18.

Let $G$ be a split $p$-adic reductive group with minimal parabolic subgroup $B = AN$, where $N$ is the unipotent radical of $B$ and the Levi subgroup $A$ is a maximal (split) torus. Let $X^* = X^*(A)$ be the lattice of rational characters on $A$ and let $X_* = X_*(A)$ be the dual lattice. If $A^0$ is the maximal compact subgroup of $A$, then $X_* \simeq A/A^0$. Let $T = X^*(\mathbb{C})$ denote the complex torus $\mathrm{Hom}(X_*, \mathbb{C}^\times)$. The Weyl group $W$ of $A$ in $G$ acts on $A$, $X^*$, $X_*$, and $T$. Each $t$ in $T$ defines a unique $\mathbb{C}^\times$-valued character of $B$ which is trivial on $N$ and on $A^0$. The $G$-module $I(t) = i_B^G(\delta^{1/2}t)$ normalizedly induced from the character $t$ of $B$ is unramified and has a unique unramified irreducible constituent $\pi(t)$. We have $\pi(t) \simeq \pi(t')$ if and only if $t' = wt$ for some $w$ in $W$. The map $t \mapsto \pi(t)$ is a bijection from the variety $T/W$ to the set of unramified irreducible $G$-modules. Put $t(\pi)$ for the $t$ associated with such a $\pi$. Let $\alpha_i$ $(1 \le i \le m)$ be a set of simple (with respect to $N$) roots in the vector space $X^* \otimes \mathbb{R} = \mathrm{Hom}(X_*, \mathbb{R})$ and $\alpha_i$ the corresponding character of $A$, defined as usual by $\alpha_i(a) = \mathrm{ad}(a)|\mathfrak{n}_i$, where $\mathrm{ad}(a)$ denotes the adjoint action of $A$ on the Lie algebra $\mathfrak{n}_i$ of the root subgroup $N_i$ of $\alpha_i$ in $N$. Denote by $\alpha_i^\vee$ $(1 \le i \le m)$ the corresponding set of coroots in the dual space $X_* \otimes \mathbb{R}$, and by $\alpha_i^\vee$ the corresponding set of characters of the torus $T = X^*(\mathbb{C}) = \mathrm{Hom}(X_*, \mathbb{C}^\times)$, defined as usual by $\alpha_i^\vee(\exp T) = \exp\langle \alpha_i^\vee, T\rangle$ for all $T$ in $X^* \otimes \mathbb{C} = \mathrm{Hom}(X_*, \mathbb{C})$. Here $\langle \cdot, \cdot \rangle$ is the pairing between $X_*$ and $X^*$. There exists $q = q(G) > 1$ such that if $\pi$ is (irreducible, unramified and) unitarizable, then $q^{-1} < |\alpha_i^\vee(t)| < q$ for all $1 \le i \le m$ and the complex conjugate $\bar{t}$ of $t$ is equal to $wt^{-1}$ for some $w$ in $W$.

If $f$ is a spherical function, the value of the normalized orbital integral $F(a,f) = \Delta(a)\Phi(a,f)$ at a regular $a$ in $A$ depends only on the $W$-orbit of the image $x$ of $a$ in $X_*$; it is denoted by $F(x,f)$. Let $\mathbb{C}[X_*]^W$ be the algebra of $W$-invariant elements in the group ring $\mathbb{C}[X_*]$. The Satake transform $f \mapsto f^\vee = \sum_{x \in X_*} F(x,f)x$ defines an algebra isomorphism from the convolution algebra $\mathbb{H}$ of spherical functions, to $\mathbb{C}[X_*]^W$. For each $x$ in $X_*$, let $f(x)$ be the element of $\mathbb{H}$ with $f(x)^\vee = \sum_{w \in W} wx$. Then $f(x)$ is $n_0$-admissible if $|\nu\alpha_P(w(a(x)))| \ge n_0$ for every $w$ in $W$ and parabolic subgroup $P$ containing $A$; $a(x)$ is an element of $A$ which corresponds to $x$ under the isomorphism of $A/A^0$ with $X_*$ fixed above. We have $\mathrm{tr}(\pi(t))(f) = \mathrm{tr}(i(t))(f) = f^\vee(t)$ for every $f$ in $\mathbb{H}$ and $t$ in $T$, where $f^\vee(t) = \sum_{x \in X_*} F(x,f)t(x)$.

DEFINITION 26.16. Consider $x$ in $X_*$ with $\nu(a(x)) \neq 0$ for each root $\alpha$ of $A$ on $N$. A complex-valued locally constant function $f$ with $f(zg)\omega(z) = f(g)$ for all $g$ in $G$ and $z$ in $Z$ which is compactly supported modulo $Z$ is called *x-regular* if $f(g)$ is zero unless there is $z$ in $Z$ such that $zg$ is conjugate to an element $a$ in $A$ whose image in $X_*$ is $x$, in which case the normalized orbital integral $F(g,f)$ is equal to $\omega(z)^{-1}$. If $f$ is $x$-regular, then we denote it by $f_x$. A *regular* function is a linear combination with complex coefficients of $x$-regular functions.

REMARK 26.17. (1) Any regular function vanishes on the singular set. In fact, it is supported on the regular split set by definition.

(2) If $\pi$ is an admissible $G$-module with central character $\omega$, then the normalized module $\pi_N$ of coinvariants [BZ76] is an $A$-module. Its character is denoted by $\chi_{\pi_N}$. If $f_x$ is an $x$-regular function, then a simple application of the Weyl integration formula and the theorem of Deligne-Casselman [Cas77] implies that

$$\operatorname{tr}\pi(f_x) = [W]^{-1}\int_{A/Z}(\Delta\chi_{\pi_N})(a)F(a,f_x)\,\mathrm{d}a.$$

If $\operatorname{tr}\pi(f_x)$ is nonzero, then by Frobenius reciprocity, there exists $t$ in $T$ such that $\pi$ is a constituent of $I(t)$ and a subset $W(\pi,t)$ of $W$ such that

$$\operatorname{tr}\pi(f_x) = \sum_{w\in W(\pi,t)} t(wx).$$

(3) Each constituent of $I(t)$, including $\pi$, has a nonzero vector fixed by the action of an Iwahori subgroup (see [Bo76, (4.7)] for the case of a reductive group, [FK87.2, §17] for the case of the metaplectic groups, [F11] for an extension to the tame algebra, and[MP96] for an extension to level zero representations).

(4) Regular functions play a crucial role in the study of orbital integrals of spherical functions. See [F87.2].

We shall now use the corollary, Theorem 26.14, and the results concerning spherical and regular functions, to extend the global correspondence results of [DKV84] and [F87.1] which deal with cuspidal representations of inner forms of $\mathrm{GL}(n)$. The definitions and proofs which are not given in the following discussion are detailed in these references. Put $G' = \mathrm{GL}(n)$. Let $G$ be the multiplicative group of a central simple $F$-algebra of rank $n$. The cuspidal $G'$-module $\pi' = \otimes_v \pi'_v$ and the cuspidal $G$-module $\pi = \otimes_v \pi_v$ are called *corresponding* if $\pi'_v$ and $\pi_v$ correspond for each place $v$ of $F$, where the notion of local correspondence is defined by means of the character relations; see Theorem 13.8. Fix a non-Archimedean place $u'$. Let $A$ be the set of equivalence classes of cuspidal $G'$-modules $\pi$ with a cuspidal component at $u'$, such that each component of $\pi'$ is obtained by the local correspondence. Let $A$ be the set of equivalence classes of cuspidal $G$-modules $\pi$ whose component at $u'$ corresponds to a cuspidal $G(F_{u'})$-module. Then $\pi_{u'}$ is necessarily cuspidal.

THEOREM 26.18. *The correspondence defines a bijection between the sets $A'$ and $A$. The multiplicity of each $\pi$ of $A$ in the cuspidal spectrum $L_0^2(G(F)\backslash G(\mathbb{A}))$ is one.*

REMARK 26.19. (1) This has been shown already for the subset of $\pi'$ in $A'$ with two cuspidal components and the corresponding subset of $A$.

(2) Theorem 26.18 can be extended from the context of $A'$, $A$ to the context of all cusp forms on $G'$, $G$ by known techniques, as described in Section 25 and the following chapters. It will be interesting to establish such an extension by *simple* means.

PROOF. Fix corresponding cuspidal $G'(F_{u'})$ and $G(F_{u'})$-modules $\pi'_{u'}$ and $\pi_{u'}$, and matrix coefficients $f'_{u'}$ and $f_{u'}$ thereof. These functions are matching, see Section 14, namely, have matching orbital integrals. For any functions $f'^{u'}$ on $G'(\mathbb{A}^{u'})$ and $f^{u'}$ on $G(\mathbb{A}^{u'})$, the functions $f' = f'_{u'}f'^{u'}$ and $f = f_{u'}f^{u'}$ are cuspidal; see Lemma 4.1. Fix two distinct non-Archimedean places $u$ and $u''$ of $F$, other than $u'$, with sufficiently large residual characteristics. Put $S = \{u, u', u''\}$. Objects associated to $\mathbb{A}^S$, the adèles without $S$-components, will be denoted with a superscript $S$. Let $f^S$ be any function on $G(\mathbb{A}^S)$ and $f_{u''}$ any regular function on $G(F_{u''})$. Let $f'^S$ be a matching function on $G'(\mathbb{A}^S)$ and $f'_{u''}$ a matching regular function on $G'(F_{u''})$. Put $f'^u = f'^S f_{u'} f_{u''}$ and $f^u = f^S f_{u'} f_{u''}$. Put $n_0 = \max\{n_0(f'^u), n_0(f^u)\}$. Let $f_u$ and $f'_u$ be matching spherical $n_0$-admissible functions. Since $f'_{u''}$ and $f_{u''}$ are zero on the singular set, the functions $f' = f'_u f'^u$ and $f = f_u f^u$ are zero on the $G'(\mathbb{A})$ and $G(\mathbb{A})$-orbits of any singular element $\gamma$ in $G'(F)$ and $G(F)$, respectively. Hence they are discrete. Since $f'$ and $f$ are matching, the right sides of the trace formulae (26.2.1) for $G'$ and for $G$, namely, $\sum \Phi(\gamma', f')$ and $\sum \Phi(\gamma, f)$ (see [FK87.2, §4]) are equal. By the corollary to Theorem 26.11, the left sides are equal, namely, $\sum m(\pi') \operatorname{tr} \pi'(f') = \sum m(\pi) \operatorname{tr} \pi(f)$. By virtue of the choice of $f'_{u'}$ and $f_{u'}$, the $\pi'$ and $\pi$ are cuspidal, with the cuspidal components $\pi'_{u'}$ and $\pi_{u'}$ at $u'$. Hence $m(\pi) = 1$ (by multiplicity one theorem for the cuspidal representations of $\mathrm{GL}(n)$), and each component $\pi'_v$ of $\pi'$ is relevant; see Section 19. Since $\operatorname{tr} \pi'_v(f'_v) \neq 0$ for $f'_v$ matching an $f_v$, and $\pi'_v$ is relevant, the main local correspondence theorem, Section 23, implies that $\pi'_v$ corresponds to some $\pi_v(\pi'_v)$, for each $v$. Since $f'_u$ and $f_u$ are spherical, if $\operatorname{tr} \pi'_u(f'_u)$ and $\operatorname{tr} \pi_u(f_u)$ are nonzero, then $\pi'_u$ and $\pi_u$ are unramified, and so is $\pi_u(\pi'_u)$. We write our equality in the form

$$\sum_{\pi_u} \left( \sum_{\pi^u} m(\pi) \operatorname{tr} \pi^u(f^u) - \sum_{\pi'^u} m(\pi') \operatorname{tr} \pi^u(\pi'^u)(f^u) \right) \operatorname{tr} \pi_u(f_u) = 0.$$

The sum on $\pi_u$ is over equivalence classes of unramified unitarizable $G(F_u)$-modules. The sum on $\pi^u$ is over the equivalence classes of $G(\mathbb{A}^u)$-modules such that $\pi = \pi_u \otimes \pi^u$ appears in (26.2.1). The sum over $\pi'^u$ is over those $\pi'^u = \otimes_{v \neq u} \pi_v$ such that there is a cuspidal $\pi' = \otimes \pi'_v$ with $\pi_v = \pi_v(\pi'_v)$ for all $v$. Since all the sums and products in the trace formula are absolutely convergent, and all the representations which appear there are unitarizable, Theorem 26.2.1 implies that the sums inside parentheses are equal for each $\pi_u$. We write this identity in the form

$$\sum_{\pi^{u,u''}} \left( \sum_{\pi_{u''}} m(\pi) \operatorname{tr} \pi_{u''}(f_{u''}) - \sum_{\pi_{u''}} m(\pi') \operatorname{tr}(\pi_{u''}(\pi'_{u''}))(f_{u''}) \right) \operatorname{tr} \pi^{u,u''}(f^{u,u''}) = 0.$$

The sum on $\pi^{u,u''}$ is over equivalence classes of irreducible $G(\mathbb{A}^{u,u''})$-modules. The sum on $\pi_{u''}$ is over irreducible $G(F_{u''})$-modules such that $\pi^u \pi_{u''} \pi^{u,u''}$ appears in the sum on $\pi^u$ above. The sum on $\pi_{u''}$ is over those such that the resulting $\pi^u$ occurs in the sum above. Since the function $f^S$ is arbitrary, all sums here are absolutely convergent and all representations are unitarizable, a standard argument of linear independence of characters implies the sums in the parentheses are equal for every $\pi^{u''} = \pi_u \pi^{u,u''}$.

We now use the fact that $f_{u''}$ is an arbitrary regular function. If $\operatorname{tr} \pi_{u''}(f_{u''}) \neq 0$, then $\pi_{u''}$ has a nonzero vector fixed by an Iwahori subgroup. Hence the first sum on $\pi_{u''}$ is finite by a theorem of Harish-Chandra; see [BJ79], which asserts that there are only finitely many cuspidal $G(\mathbb{A})$-modules with fixed infinitesimal character and fixed ramification at all finite places. The second sum on $\pi_{u''}$ consists of at most one term, by the rigidity theorem for cuspidal $G(\mathbb{A})$-modules.

Recall that $\operatorname{tr} \pi_{u''}(f_{u''})$ is a linear combination of characters (of the form $t \mapsto t(wx)$, where $t$ lies in $T = \{(z_i) \in \mathbb{C}^{\times n};\ \prod_i z_i = 1\}$, and $x = (x_i)$ varies over $X_* = \mathbb{Z}^n/\mathbb{Z}$, and $(z_i)(wx) = \prod_i z_{w(i)}^{x_i}$). Applying linear independence of finitely many characters, it is clear that the first sum in parentheses is empty if the second is empty and that $m(\pi) = 1$ and $\operatorname{tr} \pi_{u''}(f_{u''}) = \operatorname{tr} \pi'_{u''}(f'_{u''})$ for all matching regular $f'_{u''}$ and $f_{u''}$ otherwise. Since the Hecke algebras of $G'(F_{u''})$ and $G(F_{u''})$ with respect to an Iwahori subgroup are isomorphic, we conclude that $\pi'_{u''}$ and $\pi_{u''}$ correspond, and Theorem 26.18 follows. □

We proceed to the proof of Theorem 26.14. Fix $q \geq 1$. Let $t' = T'(q)$ be the set of $t$ in $T$ with $\bar{t} = wt^{-1}$ for some $w$ in $W$ ($w$ depends on $t$) and $q^{-1} \leq |\alpha^\vee(t)| \leq q$ for every root $\alpha$ of $A$ on $N$. The quotient $\widetilde{T} = \widetilde{T}(q)$ of $T'$ by $W$ is a compact Hausdorff space. Let $\mathbb{C}(\widetilde{T})$ be the algebra of complex-valued continuous functions on $\widetilde{T}$. Let $n_0$ be a nonnegative integer. The element $x$ of $X_*$ is called *$n_0$-admissible* if $|\nu\alpha_P(a(x))| \geq n_0$ for every maximal parabolic subgroup $P$ of $G$. This condition means that there are finitely many walls, determined by the $\alpha_P$, in the lattice $X_*$, such that $x$ is called $n_0$-admissible if it is sufficiently far (the distance depends on $n_0$) from these walls. The function $P_x(t) = \sum_{w \in W} t(wx)$ is a function on $\widetilde{T}$ which depends only on the image of $x$ in $X_*/W$. Note that $f'(x)^\vee = P_x$, and in particular $\operatorname{tr}(\pi'(t))(f'_x) = P_x(t)$. Let $C(n_0)$ be the $\mathbb{C}$-span of all $P_x(t)$ with $n_0$-admissible $x$. It is a subspace of $\mathbb{C}(\widetilde{T})$, but it is not multiplicatively closed, unless $n_0 = 0$. An element of $\mathbb{C}(\widetilde{T})$ is called *$n_0$-admissible* if it lies in $C(n_0)$.

LEMMA 26.20. *The space $C(0)$ is dense in $\mathbb{C}(\widetilde{T})$.*

PROOF. This follows from the Stone-Weierstrass theorem, since the space $\widetilde{T}$ is compact and Hausdorff and $C(0)$ is a subalgebra of $\mathbb{C}(\widetilde{T})$ which separates points and contains the scalars and the complex conjugate of each of its elements. □

Theorem 26.14 follows from the special case where $G = \mathrm{GL}(n)$ and $c_i(t) = 0$ for all $i$ in the following proposition. The general form with nonzero $c_i(t)$ is used in [F06] when $G = \mathrm{GL}(3)$ to give a short and simple proof of the trace formulae identity for the base-change lifting from $U(3)$ to $\mathrm{GL}(3, E)$ for an arbitrary test function.

PROPOSITION 26.21. *Fix $n_0 \geq 0$. Fix also a sequence $t_i$ of elements of $\widetilde{T}$, a sequence of complex numbers $c_i$, a sequence $\widetilde{T}_j$ of compact submanifolds of $\widetilde{T}$, and a sequence $c_j(t)$ of complex-valued functions on $\widetilde{T}_j$ which are measurable with respect to a bounded measure $dt$ on $\widetilde{T}_j$. Suppose that*

$$\beta = \sum_i |c_i| + \sum_j \sup_{t\in\widetilde{T}_j} |c_j(t)| + \sum_J \int_{\widetilde{T}_j} |c_j(t)|\,|dt|$$

*is finite and that for any $n_0$ admissible $x$ in $X_*$ we have*

$$\sum_i c_i P_x(t_i) = \sum_j \int_{\widetilde{T}_j} c_j(t) P_x(t)\,|dt|. \qquad (26.21.1)$$

*Then $c_i = 0$ for all $i$.*

PROOF. We begin with a definition. Let $\epsilon$ be a positive number. The points $t$ and $t'$ in $\widetilde{T}$ are called *$\epsilon$-close* if there are representatives $\tau$ and $\tau'$ of $t$ and $t'$ in $T'$ such that $|\alpha^\vee(\tau) - \alpha^\vee(\tau')| < \epsilon$ for every root $\alpha^\vee$ on $T$ (=coroot on $X_*$). Denote by $\widetilde{T}_\epsilon(t)$ the $\epsilon$-neighborhood of $t$ in $\widetilde{T}$. The quotient by $\epsilon$ of the volume of $\widetilde{T}_\epsilon(t)$ is bounded uniformly in $\epsilon$.

Suppose that $c_0 \neq 0$. Multiplying by a scalar we assume that $c_0 = 1$. The lemma implies that for every $\epsilon > 0$, there is $P = P_\epsilon$ in $C(0)$ with $P(t_0) = 1$, $|P(t)| \leq 2$ for all $t$ in $\widetilde{T}$, and $|P(t)| < \epsilon$ unless $t$ is $\epsilon^2$-close to $t_0$. Such a polynomial $P$ is called below an $\epsilon$-approximation of the delta function at $t_0$, or simply a *"delta function"* at $t_0$. Since $\beta$ is finite, for every $\epsilon > 0$, there exists $N > 0$ such that

$$\sum_{i>N} |c_i| + \sum_{j>N} \int_{\widetilde{T}_j} |c_j(t)|\,|dt| < \epsilon.$$

Take $\epsilon = 1/4(1+\beta)$. Substituting $P$ for $P_x$ in (26.21.1), if $n_0 = 0$, then we obtain a contradiction to the assumption that $c_0 = 1$. Hence the proposition is proven in the case of $n_0 = 0$. It remains to deal with a general $n_0$.

Let $x$ be an $n_0$-admissible element of $X_*$. Put $k' = 2\max_P |\nu\alpha_P(a(x))|$. For any $x'$ in $X_*$, the element $x+k'x'$ is $n_0$-admissible. Since $P_x(t)P_{x'}(t^{k'}) = \sum_{w\in W} P_{x+k'wx'}(t)$, we have that (26.21.1) applies with $P_x(t)$ replaced by $P_x(t)P_{x'}(t^{k'})$. For a fixed $x$ (and $k'$), $x'$ is arbitrary. Replacing $q$ by $q^{k'}$ in the definition of $\widetilde{T}$, we argue as in the previous paragraph and conclude that for every $r \geq 0$ we have

$$\sum_i c_i P_x(t_i) = 0. \qquad (26.21.2)$$

Here the sum ranges over all $i$ with $t_i^{k'} = t_r^{k'}$ (equality in $\widetilde{T}$). Take $r = 0$. We conclude that the equality (26.21.2) holds also for any $n_0$-admissible $x$, provided that the sum ranges over the set $I$ of all $i$ for which there is $k = k(i)$ with $t_i^k = t_0^k$. It remains to prove the following

LEMMA 26.22. *Suppose that $c_i$ is a sequence of complex numbers such that $\beta = \sum_i |c_i|$ is finite, and $t_i$ are elements of $T'$ whose images in $\widetilde{T} = T'/W$ are distinct, such that for each $i$ there is $k = k(i)$ with $t_i^k = t_0^k$. If $\sum_i c_i P(t_i) = 0$ for every $n_0$-admissible $P$, then $c_i = 0$ for all $i$.*

PROOF. We may and do assume that $c_0 = 1$ in order to derive a contradiction. If $\eta = 1/4(1+\beta)$ there is $N > 0$ such that $\sum_{i>N} |c_i| < \eta$, and a $W$-invariant polynomial $P(t) = \sum_x b(x)P_x(t)$ with $P(t_0) = 1$, $|P(t)| \le 2$ on $T'$ and $|P(t_i)| < \eta$ for $i$ ($1 \le i \le N$). This $P$ is a "delta function", and if $n_0 = 0$, then we are done. If $n_0 \neq 0$, then the "delta function" $P$ is not necessarily $n_0$-admissible. Our aim is to replace $P$ by an $n_0$-admissible "delta function" on multiplying $P$ with a suitable admissible polynomial $Q$ which (depends on $P$ and) attains the value one at $t_0$, while remaining uniformly bounded (by $2[W]$) at each $t_i$ ($i \ge 1$). For this purpose note that our assumption (that for each $i$, there is $k$ with $t_i^k = t_0^k$) implies that $t_i/t_0$ lies in the maximal compact subgroup of $T$ for all $i$. Hence for every $x$ in $X_*$, the absolute value $|t_i(x)|$ of the complex number $t_i(x)$ is independent of $i$. Take any one-admissible $\mu$ in $X_*$, such that $|t_i(\mu)| \ge |t_i(w\mu)|$ for all $w$ in $W$. Then $|P_\mu(t_i^s)| \le [W]|t_0(m)|^s$ for every positive integer $s$, and for all $i$. Put $u_w = t_0(w\mu)/|t_0(w\mu)|$ ($w \in W$), and

$$s_0 = 2n_0 + 2\max\{|\nu\alpha_P(a(x))|\,;\text{ all } P \supset A, \text{all } x \text{ with } b(x) \neq 0\}.$$

For every $\epsilon > 0$ there is $s > s_0$ such that $|u_w^s - 1| < \epsilon$ for all $w$ in $W$, and the choice of a sufficiently small $\epsilon$ guarantees that $P_\mu(t_0^s)| \ge 1/2|t_0(\mu)|^s$. Hence the $W$-invariant polynomial $Q_s(t) = P_\mu(t^s)/P_\mu(t_0^s)$ on $T'$ satisfies $Q_s(t_0) = 1$ and $|Q_s(t_i)| \le 2[W]$ for all $i$. The polynomial $Q(t) = P(t)Q_s(t)$ lies in $C(n_0)$; hence it satisfies the relation $\sum_i c_i Q(t_i) = 0$. Since $Q$ is a delta function at $t_0$, we obtain a contradiction to the assumption that $c_0 \neq 0$. This proves the lemma and completes the proof of Theorem 26.14. □

□

# Chapter 3
# Arthur's Noninvariant Trace Formula

## 1 Preliminary Definitions

### 1.1 The Group

Let $G$ be a reductive connected algebraic group defined over a number field $F$. Fix a minimal parabolic subgroup $P_0$ and a Levi component $M_{P_0}$ of $P_0$, both defined over $F$. In this chapter we work only with standard parabolic subgroups of $G$, that is, parabolic subgroups $P$, defined over $F$, which contain $P_0$. We shall refer from now on to such groups simply as "parabolic subgroups." Fix $P$. Let $N_P$ be the unipotent radical of $P$. Let $M_P$ be the unique Levi component of $P$ which contains $M_{P_0}$. Denote the split component of the center of $M_P$ by $A_P$. The groups $N_P$ and $A_P$ are defined over $F$.

### 1.2 Roots

Let $X(M_P)$ be the group of characters of $M_P$ defined over $F$. Then

$$\mathfrak{a}_P = \mathrm{Hom}(X(M_P), \mathbb{R})$$

is a real vector space. Its dimension equals that of $A_P$. Its dual space is

$$\mathfrak{a}_P^* = X(M_P) \otimes \mathbb{R}.$$

Denote the set of simple roots of $(P, A)$ by $\Delta_P$. They are elements in $X(A_P)$. They are canonically embedded in $\mathfrak{a}_P^*$. The set $\Delta_0 = \Delta_{P_0}$ is a base for a root system. In particular, we have the coroot $\alpha^\vee$ in $\mathfrak{a}_{P_0}$ for every root $\alpha \in \Delta_P$.



Y.Z. Flicker, *Arthur's Invariant Trace Formula and Comparison of Inner Forms*,
DOI 10.1007/978-3-319-31593-5_3

## 1.3 Relative Roots

Suppose that $P_1$ and $P_2$ are parabolic subgroups with $P_1 \subset P_2$. Then $\mathfrak{a}^*_{P_2}$ comes with an embedding into $\mathfrak{a}^*_{P_1}$, while $\mathfrak{a}_{P_2}$ is a natural quotient vector space of $\mathfrak{a}_{P_1}$. The group $M_{P_2} \cap P_1$ is a parabolic subgroup of $M_{P_2}$ with unipotent radical

$$N_{P_1}^{P_2} = N_{P_1} \cap M_{P_2}.$$

The set $\Delta_{P_1}^{P_2}$ of simple roots of $(M_{P_2} \cap P_1, A_{P_1})$ is a subset of $\Delta_{P_1}$. As is well known, $P_2 \longmapsto \Delta_{P_1}^{P_2}$ is a bijection between the set of parabolic subgroups $P_2$ that contain $P_1$ and the collection of subsets of $\Delta_{P_1}$. Identify $\mathfrak{a}_{P_2}$ with the subspace

$$\{H \in \mathfrak{a}_{P_1}; \alpha(H) = 0, \alpha \in \Delta_{P_1}^{P_2}\}.$$

If $\mathfrak{a}_{P_1}^{P_2}$ is defined to be the subspace of $\mathfrak{a}_{P_1}$ annihilated by $\mathfrak{a}^*_{P_2}$, then

$$\mathfrak{a}_{P_1} = \mathfrak{a}_{P_1}^{P_2} \oplus \mathfrak{a}_{P_2}.$$

The subspace $(\mathfrak{a}_{P_1}^{P_2})^*$ of $\mathfrak{a}^*_{P_1}$ spanned by $\Delta_{P_1}^{P_2}$ is in natural duality with $\mathfrak{a}_{P_1}^{P_2}$. We have

$$\mathfrak{a}^*_{P_1} = (\mathfrak{a}_{P_1}^{P_2})^* \oplus \mathfrak{a}^*_{P_2}.$$

The space $\mathfrak{a}_{P_1}^{P_2}$ also embeds into $\mathfrak{a}_{P_0}^{P_2}$, and

$$\mathfrak{a}_{P_0}^{P_2} = \mathfrak{a}_{P_0}^{P_1} \oplus \mathfrak{a}_{P_1}^{P_2}.$$

Any root $\alpha \in \Delta_{P_1}^{P_2}$ is the restriction to $\mathfrak{a}_{P_1}^{P_2}$ of unique root $\beta \in \Delta_{P_0}^{P_2}$. Define $\alpha^\vee$ to be the projection onto $\mathfrak{a}_{P_1}^{P_2}$ of the vector $\beta^\vee$ in $\mathfrak{a}_{P_0}^{P_2}$. Then

$$\{\alpha^\vee; \alpha \in \Delta_{P_1}^{P_2}\}$$

is a basis of $\mathfrak{a}_{P_1}^{P_2}$ . Let

$$\widehat{\Delta}_{P_1}^{P_2} = \{\varpi_\alpha; \alpha \in \Delta_{P_2}^{P_2}\}$$

be the corresponding dual basis for $(\mathfrak{a}_{P_1}^{P_2})^*$. If $\varpi \in \widehat{\Delta}_{P_1}^{P_2}$, define $\varpi^\vee \in \mathfrak{a}_{P_1}^{P_2}$ by

$$\alpha(\varpi^\vee) = \varpi(\alpha^\vee), \qquad \alpha \in \Delta_{P_1}^{P_2}.$$

Then $\{\varpi_\alpha^\vee\}$ and $\{\alpha\}$ is another pair of dual bases for $\mathfrak{a}_{P_1}^{P_2}$ and $(\mathfrak{a}_{P_1}^{P_2})^*$.

## 1.4 Lie Algebras

If $P_i$ is a parabolic subgroup, and $P_i$ appears in our notation as a subscript or a superscript, we shall often use only $i$, instead of $P_i$, for the subscript or superscript. For example,

$$M_0 = M_{P_0}, \quad \mathfrak{a}_1^2 = \mathfrak{a}_{P_1}^{P_2}, \quad \text{and} \quad N_1^2 = N_{P_1}^{P_2}.$$

If the letter $P$ alone is used, we shall often omit it altogether as a subscript. Thus $P = NM$, $\Delta$ is the set of simple roots of $(P, A)$, and so on. The group $G$ itself is a parabolic subgroup. We shall often write $Z$ for $A_G$ and $\mathfrak{z}$ for $\mathfrak{a}_G$.

The following proposition is simple to prove. However, it will be the ultimate justification for much of this work so we have best draw attention to it.

PROPOSITION 1.5. *Suppose that $P_1$ and $P_2$ are parabolic subgroups, with $P_1 \subset P_2$. Then*

$$\sum_{\{P;\, P_1 \subset P \subset P_2\}} (-1)^{\dim(A/A_2)} = \begin{cases} 1 \text{ if } P_1 = P_2, \\ 0 \text{ otherwise.} \end{cases}$$

*The sum is over the parabolic subgroups $P$; $A = A_P$ is uniquely determined by $P$.*

PROOF. The sum can be regarded as the sum over all subsets of $\Delta_1^2$. The result follows from the binomial theorem. □

## 1.6 Adele Groups

Let $\mathbb{A}$ (resp. $\mathbb{A}_f$) be the ring of adèles (resp. finite adèles) of the number field $F$. Denote by $F_\infty$ the direct product of $F_v$ over all Archimedean valuations $v$; we also write $v|\infty$ for these valuations. Then

$$G(\mathbb{A}) = G(F_\infty) \cdot G(\mathbb{A}_f)$$

is the restricted direct product over all valuations $v$, of the groups $G(F_v)$. Embed $\mathbb{R}$ in $F_\infty$ diagonally.

Fix a maximal compact subgroup

$$K = \prod_v K_v$$

of $G(\mathbb{A})$ so that the following properties hold:

(i) For any embedding of $G$ into $\mathrm{GL}(n)$, defined over $F$, the group $K_v$ equals $\mathrm{GL}(n, \mathcal{O}_v) \cap G(F_v)$ for almost all finite places $v$, where $\mathcal{O}_v$ is the ring of integers of the local field $F_v$.

(ii) For every finite $v$, the group $K_v$ is a special maximal compact subgroup.
(iii) The Lie algebras of $K_{F_\infty}$ and $A_0(F_\infty)$ are orthogonal with respect to the Killing form.

## 1.7 Harish-Chandra Map

Suppose that $P$ is a parabolic subgroup. If $m = (m_v)_v$ lies in $M(\mathbb{A})$, define a vector $H_M(m)$ in $\mathfrak{a}_P$ by

$$e^{\langle H_M(m), \chi\rangle} = |\chi(m)| = \prod_v |\chi(m_v)|_v, \qquad \chi \in X(M).$$

The function $H_M$ is a homomorphism of $M(\mathbb{A})$ to the additive group $\mathfrak{a}_P$. Let $M(\mathbb{A})^1$ be its kernel.

If $M = \mathbb{G}_m = \mathrm{GL}(1)$, then $\mathbb{G}_m(\mathbb{A}) = \mathbb{A}^\times$ is the group of idèles, and $\mathbb{G}_m(\mathbb{A})^1 = \mathbb{A}^1$ is the group of idèles $a$ of volume $||a|| = \prod_v |a_v|_v$ equal to 1 (the absolute values $|\cdot|_v$ on $F_v$ are normalized as usual, in particular they satisfy the product formula $||\alpha|| = 1$ for all $\alpha \in F^\times$). We embed the multiplicative group $\mathbb{R}^\times_{>0}$ of the positive real numbers, which is the connected component of 1 in $\mathbb{R}^\times$, diagonally in $F^\times_\infty$, denoting the image by $\mathbb{R}^0 \subset F^\times_\infty$. Then $\mathbb{A}^\times$ is the direct product of $\mathbb{A}^1$ and $\mathbb{R}^0$. In fact $F^\times$ is a discrete subgroup in $\mathbb{A}^\times$, and $\mathbb{A}^\times/F^\times\mathbb{R}^0$ is finite, isomorphic to the class group of $F$.

The group $M(\mathbb{A})$ is the direct product of $M(\mathbb{A})^1$ and $A(\mathbb{R}^0)$, the connected component of 1 in $A(\mathbb{R})$. By our conditions on $K$, we have the Iwasawa decomposition $G(\mathbb{A}) = P(\mathbb{A})K$. Any $x \in G(\mathbb{A})$ can be written as

$$nmak, \ n \in N(\mathbb{A}), \ m \in M(\mathbb{A})^1, \ a \in A(\mathbb{R}^0), \ k \in K.$$

We define $H_P(x) = H(x)$ to be a vector $H_M(ma) = H_M(a)$ in $\mathfrak{a}_P$. Notice that if $P_1 \subset P_2$, the space $\mathfrak{a}_{P_1}^{P_2}$ is the image of $M_{P_2}(\mathbb{A})^1$ under $H_{P_1}$.

## 1.8 Weyl Group

We shall denote the restricted Weyl group of $(G, A_0)$ by $\Omega$. The Weyl group $\Omega$ acts on $\mathfrak{a}_0$ and $\mathfrak{a}_0^*$ in the usual way. For every $s \in \Omega$ we shall fix a representative $w_s$ in the intersection of $G(F)$ with the normalizer of $A_0$. The element $w_s$ is determined modulo $M_0(F)$. If $P_1$ and $P_2$ are parabolic subgroups, let $\Omega(\mathfrak{a}_1, \mathfrak{a}_2)$ denote the set of distinct isomorphisms from $\mathfrak{a}_1$ onto $\mathfrak{a}_2$ obtained by restricting elements in $\Omega$ to $\mathfrak{a}_1$. The (standard) parabolic subgroups $P_1$ and $P_2$ are said to be *associated* if $\Omega(\mathfrak{a}_1, \mathfrak{a}_2)$ is not empty, thus $M_1$ and $M_2$ are conjugate. If $s_1$ belongs to $\Omega(\mathfrak{a}_1, \mathfrak{a}_2)$, there is a unique element $s$ in $\Omega$ whose restriction to $\mathfrak{a}_1$ is $s_1$ and such that $s^{-1}\alpha$ is a positive root (that is to say, a root of $(P_0, A_0)$) for every $\alpha \in \Delta_0^{P_1}$. Thus $\Omega(\mathfrak{a}_1, \mathfrak{a}_2)$ can be regarded as a subset of $\Omega$; in particular, $w_{s_1}$ is an element in $G(F)$ for every $s_1 \in \Omega(\mathfrak{a}_1, \mathfrak{a}_2)$.

## 1.9 Haar Measures

We shall need to adopt some conventions for choices of Haar measures. For any connected subgroup, $V$, of $N_0$, defined over $F$, we take the Haar measure on $V(\mathbb{A})$ which assigns $V(F)\backslash V(\mathbb{A})$ the volume one. Similarly, we take the Haar measure of $K$ to be one. Fix Haar measures on each of the vector spaces $\mathfrak{a}_P$. On the spaces $\mathfrak{a}_P^*$, we take the dual Haar measures. We then use the isomorphisms

$$H_P : A_P(\mathbb{R}^0) \to \mathfrak{a}_P$$

to define Haar measures on the groups $A_P(\mathbb{R}^0)$. Finally, fix a Haar measure on $G(\mathbb{A})$. For any $P$, let

$$\mathfrak{a}_P^+ = \{H \in \mathfrak{a}_P; \alpha(H) > 0,\ \alpha \in \Delta_P\},$$

and

$$(\mathfrak{a}_P^*)^+ = \{\Lambda \in \mathfrak{a}_P^*; \Lambda(\alpha^\vee) > 0,\ \alpha \in \Delta_P\}.$$

There is a vector $\rho_P$ in $\mathfrak{a}_P^+$ such that the modular function

$$\delta_P(p) = |\det(\mathrm{Ad}(p))_{\mathfrak{n}_P}|, \quad p \in P(\mathbb{A}),$$

on $P(\mathbb{A})$ equals $e^{2\rho_P(H_P(p))}$. Here $\mathfrak{n}_P$ stands for the Lie algebra of $N_P$. There are unique Haar measures on $M(\mathbb{A})$ and $M(\mathbb{A})^1$ such that for any function $h$ in $C_c(G(\mathbb{A}))$, we have

$$\begin{aligned}\int_{G(\mathbb{A})} h(x)dx &= \int_{N(\mathbb{A})}\int_{M(\mathbb{A})}\int_K h(nmk)e^{-2\rho_P(H_P(m))}\,\mathrm{d}n\,\mathrm{d}m\,\mathrm{d}k \\ &= \int_{N(\mathbb{A})}\int_{M(\mathbb{A})^1}\int_{A(\mathbb{R}^0)}\int_K h(nmak)e^{-2\rho_P(H_P(a))}\,\mathrm{d}n\,\mathrm{d}a\,\mathrm{d}m\,\mathrm{d}k.\end{aligned}$$

## 1.10 Height Function

We now recall some properties of height functions associated with rational representations of $G$. Let $V$ be a vector space defined over $F$. Suppose that $\{v_1, \cdots, v_n\}$ is a basis of $V(F)$. If $\xi_v \in V(F_v)$, and

$$\xi_v = \sum_i \xi_{i,v} v_i, \quad \xi_{i,v} \in F_v,$$

define $\|\xi_v\|_v = \max_i |\xi_{i,v}|_v$ if $v$ is finite, and if $v = \mathbb{R}$ or $= \mathbb{C}$, put

$$\|\xi_v\|_v = \left(\sum_i |\xi_{i,v}|^2\right)^{\frac{1}{2}}.$$

An element $\xi = \prod_v \xi_v$ in $V(\mathbb{A})$ is said to be *primitive* if $\|\xi_v\|_v = 1$ for almost all $v$. In this case we set $\|\xi\| = \prod_v \|\xi_v\|_v$. The function $\|\cdot\|$ is called the *height function* associated with the basis $\{v_1, \cdots, v_n\}$. For $g = (g_{ij})$ in $\mathrm{GL}(n, F_v)$, put $||g||_v = \max(|g_{ij}|_v; i, j)$. For $g = (g_{ij})$ in $\mathrm{GL}(n, \mathbb{R})$ or $\mathrm{GL}(n, \mathbb{C})$, put $||g||_{\mathbb{R}}$ or $||g||_{\mathbb{C}} = (\sum_{i,j} |g_{ij}|^2)^{1/2}$. For a primitive $g = (g_{ij})$ in $\mathrm{GL}(n, \mathbb{A})$, thus $||g_v||_v = 1$ for almost all $v$, put $||g|| = \prod_v ||g_v||_v$.

Suppose that $\Lambda : G \to \mathrm{GL}(V)$ is a homomorphism defined over $F$. Let $K_\Lambda$ be the group of elements $k \in K$ such that $\|\Lambda(k)v\| = \|v\|$ for any primitive $v \in V(\mathbb{A})$. It is possible to choose the basis $\{v_1, \cdots, v_n\}$ such that $K_\Lambda$ is of finite index in $K$, and also so that for each $a \in A_0$, the operator $\Lambda(a)$ is diagonal. We shall always assume that for a given $\Lambda$, the basis has been chosen to satisfy these two conditions. From the basis of $V(F)$, we obtain a basis for the vector space of endomorphisms of $V(F)$. Every element in $G(\mathbb{A})$ is primitive with respect to the corresponding height function, and for every primitive $v \in V(\mathbb{A})$, and every $x \in G(\mathbb{A})$, we have

$$\|\Lambda(x)v\| \leq \|\Lambda(x)\| \cdot \|v\|.$$

If $t > 0$, define

$$G_t = \{x \in G(\mathbb{A}); \|\Lambda(x)\| \leq t\}.$$

Suppose that $\Lambda$ has the further property that $G_t$ is compact for every $t$. It is known that there are positive numbers $C$ and $N$ such that for any $t$, the volume of $G_t$ (with respect to our Haar measure) is bounded by $Ct^N$. For the rest of this work, we shall simply assume that some representation $\Lambda$, satisfying this additional property, has been fixed, and we shall write $\|x\|$ for $\|\Lambda(x)\|$. This "norm" function on $G(\mathbb{A})$ satisfies, for some positive numbers $C$ and $N$, elements $x, y \in G(\mathbb{A})$, and $k_1, k_2$ belonging to a subgroup of finite index in $K$, the relations

$$\|x\| \geq 1, \ \|k_1 x k_2\| = \|x\|, \ \|xy\| \leq \|x\| \cdot \|y\|, \ \text{and} \ \|x^{-1}\| \leq C\|x\|^N.$$

Once $\|\cdot\|$ has been fixed, we shall want to consider different rational representations $\Lambda$ of $G$. In particular, suppose that the highest weight of $\Lambda$ is $\lambda$, for some element $\lambda$ in $\mathfrak{a}_0^*$. Then there are positive numbers $c_1$ and $c_2$ such that

$$c_1 e^{-\lambda(H_0(x))} \leq \|\Lambda(x)^{-1}v\| \leq c_2 e^{-\lambda(H_0(x))},$$

for all $x \in G(\mathbb{A})$. By varying the linear functional $\lambda$, one can show that for any Euclidean norm $\|\cdot\|$ on $\mathfrak{a}_0$, there is a positive number $c$ so that

$$\|H_0(x)\| \leq c(1 + \log \|x\|), \qquad x \in G(\mathbb{A}).$$

### 1.11 Chambers

Suppose that $P$ is a parabolic subgroup. Recall that there is a finite number of disjoint open subsets of $\mathfrak{a}$, called the *chambers* of $\mathfrak{a}$. Their union is the complement in $\mathfrak{a}$ of the set of hyperplanes which are orthogonal to the roots of $(P, A)$. The set $\mathfrak{a}^+$ of 1.9 is one chamber. According to Lemma 2.13 of [La76], the set of chambers is precisely the collection, indexed by all $P'$ and $s \in \Omega(\mathfrak{a}, \mathfrak{a}')$, of the open subsets $s^{-1}(\mathfrak{a}')^+$. We shall write $n(A)$ for the number of chambers. More generally, if $P_1 \subset P$, let $n_P(A_1)$ be the number of connected components in the orthogonal complement in $\mathfrak{a}_1$ of the set of hyperplanes which are orthogonal to the roots of $(P_1 \cap M, A_1)$.

## 2 The Kernel $K_P(x, y)$

### 2.1 Convolution Operator

Fix a Haar measure $dx$ on $G(\mathbb{A})^1$. Let $r$ be the regular representation of $G(\mathbb{A})^1$ on $L^2(G(F)\backslash G(\mathbb{A})^1)$. It acts by right shifts: $(r(x)\phi)(g) = \phi(gx)$. The map

$$f \mapsto r(f) = \int_{G(\mathbb{A})^1} f(x)r(x)dx$$

gives us a representation of any reasonable convolution algebra of functions $f$ on $G(\mathbb{A})^1$. For example, we could take $C_c(G(\mathbb{A})^1)$, which is defined as the topological direct limit over all compact subsets $\Gamma$ of $G(\mathbb{A})^1$, of the spaces of continuous functions on $G(\mathbb{A})^1$ supported on $\Gamma$.

### 2.2 Seminorms

We shall be more interested in the smooth functions on $G(\mathbb{A})^1$. For any place $v$, let $G(F_v)^1$ denote the intersection of $G(F_v)$ with $G(\mathbb{A})^1$. Note that $G(F_\infty)^1 = G(\mathbb{A})^1 \cap G(F_\infty)$ contains the connected component of 1 in $G(\mathbb{A})^1$. Notice that for any $v$, the compact group $K_v$ is contained in $G(F_v)^1$. Suppose that $K_0$ is an open compact subgroup of $G(\mathbb{A}_f)^1$. Then the double coset space $K_0\backslash G(\mathbb{A})^1/K_0$ is a discrete union of countably many copies of $G(F_\infty)^1$. In particular it is a differentiable manifold.

Suppose that $\Gamma$ is a compact subset of $G(\mathbb{A})^1$ such that $\Gamma = K_0\Gamma K_0$. Let $C^\infty(G(\mathbb{A})^1, \Gamma, K_0)$ be the algebra of smooth functions on $K_0\backslash G(\mathbb{A})^1/K_0$ which are supported on $\Gamma$. The universal enveloping algebra $\mathcal{U}(\mathfrak{g}(F_\infty)^1 \otimes \mathbb{C})$ of the complexification of the Lie algebra of $G(F_\infty)^1$ acts on this space on both the left and the right. We shall denote these actions by the convolution symbol. The seminorms

$$\|f\|_{X,Y} = \sup_{x \in G(\mathbb{A})^1} |(X * f * Y)(x)|, \quad f \in C^\infty(G(\mathbb{A})^1, \Gamma, K_0),$$

indexed by elements $X, Y \in \mathcal{U}(\mathfrak{g}(F_\infty)^1 \otimes \mathbb{C})$ define a topology on $C^\infty(G(\mathbb{A})^1, \Gamma, K_0)$.

Let $C_c^\infty(G(\mathbb{A})^1)$ be the topological direct limit of the spaces $C^\infty(G(\mathbb{A})^1, \Gamma, K_0)$ over all pairs $(\Gamma, K_0)$. If $m$ is any positive integer, we can define $C_c^m(G(\mathbb{A})^1)$ in the same way, except that we take only those seminorms for which the sum of the degrees of $X$ and $Y$ is no greater than $m$.

Finally, any subgroup $L$ of $G(\mathbb{A})^1$ acts on $C_c^m(G(\mathbb{A})^1)$ by

$$f^h(x) = f(hxh^{-1}), \quad x \in G(\mathbb{A})^1, \quad h \in L.$$

We shall write $C_c^m(G(\mathbb{A})^1)^L$ (and in general $X^L$ if $X$ is any set on which $L$ acts) for the set of $L$-invariant elements.

## 2.3 *Equivalent Elements*

Suppose that $f \in C_c^m(G(\mathbb{A})^1)$. Then $r(f)$ is an operator on $G(F)\backslash G(\mathbb{A})^1$ which is integral with kernel

$$K(x,y) = K_f(x,y) = \sum_{\gamma \in G(F)} f(x^{-1}\gamma y).$$

If we had the Selberg trace formula for compact quotient in mind, we would be inclined to decompose the formula for $K(x,x)$ into terms corresponding to conjugacy classes in $G(F)$. It turns out, however, that an equivalence relation in $G(F)$, weaker than conjugacy, is more appropriate to the noncompact setting.

Any $\gamma \in G(F)$ can be uniquely written as $\gamma_s\gamma_u$, where $\gamma_s$ is semisimple, $\gamma_u$ is unipotent, and the two elements commute. We shall say that elements $\gamma$ and $\gamma'$ in $G(F)$ are *equivalent* if $\gamma_s$ and $\gamma'_s$ are $G(F)$-conjugate. Let $\mathcal{O}$ be the set of equivalence classes. Every class in $\mathcal{O}$ contains one and only one conjugacy class of semisimple elements in $G(F)$.

## 2.4 *Equivalence Classes of Pairs*

If $\gamma$ is in $G(F)$, and $H$ is a subgroup of $G$, defined over $F$, let $H(\gamma)$ be the centralizer of $\gamma$ in $H$. Both $H(\gamma)$ and its identity component, $H(\gamma)^0$, are defined over $F$. If $R$ is any ring containing $F$, let $H(R,\gamma)$ be the centralizer of $\gamma$ in $H(R)$. Consider the set of pairs $(M, c)$ , where $M$ is a Levi subgroup of a parabolic group $P = MN$, and $c$ is a conjugacy class in $M(F)$ of elliptic elements. An element $\gamma$ of $M(F)$ is called (regular) *elliptic* if it does not lie in $P_M(F)$ for any proper parabolic subgroup $P_M$

of $M$. Two pairs $(M, c)$ and $(M', c')$ being defined to be equivalent if $c' = w_s c w_s^{-1}$ for some $s$ in $\Omega(\mathfrak{a}, \mathfrak{a}')$.

PROPOSITION. *There is a bijection from the set $\mathcal{O}$ of equivalence classes of rational elements onto the set of equivalence classes of pairs $(M, c)$.*

PROOF. Suppose that $\mathfrak{o}$ is a class in $\mathcal{O}$. It is clearly possible to choose a parabolic subgroup $P$ and a semisimple element $\gamma$ in $\mathfrak{o}$ such that $\gamma$ belongs to $M(F)$, but such that no $M(F)$-conjugate of $\gamma$ lies in $P_1(F)$, for $P_1$ a parabolic subgroup of $G$, $P_1 \subsetneqq P$. In other words, $\gamma$ does not lie in the group of $F$-points of any proper parabolic subgroup of $M$, defined over $F$. Elements in $M(F)$ with this property are the (regular) *elliptic* elements.

The group $M(\gamma)^0$ is a reductive group, defined over $F$, which is anisotropic modulo $A$. The group $P$ is not uniquely determined by $\mathfrak{o}$. However if $(M', \gamma')$ is another pair, associated with $\mathfrak{o}$ as above, then $\gamma' = w\gamma w^{-1}$ for some element $w \in G(F)$. Since $A'$ is the split component of the center of $G(\gamma')^0$, it equals $wAw^{-1}$. Therefore $w = w_s\eta$, for $s \in \Omega(\mathfrak{a}, \mathfrak{a}')$, and $\eta \in M(F)$. The proposition follows. □

## 2.5 *Root Spaces*

Suppose that $\mathfrak{o}$ is a class in $\mathcal{O}$ and that $P$ and $\gamma$ are as in 2.4. Let $\Sigma$ denote the set of roots of $(P, A)$. The set $\Sigma$ determines a decomposition

$$\mathfrak{n} = \bigoplus_{\alpha \in \Sigma} \mathfrak{n}_\alpha$$

of the Lie algebra of $N$. Let $\Sigma(\gamma)$ be the set of roots $\alpha$ in $\Sigma$ such that the centralizer of $\gamma$ in $\mathfrak{n}_\alpha$ is not zero. The elements in $\Sigma(\gamma)$ are, of course, characters on $A$. Let $A'$ be the intersection of their kernels. We can choose parabolic subgroups $P_1 \subset P_2$, and an element $s \in \Omega(\mathfrak{a}, \mathfrak{a}_1)$, such that $A_2 = w_s A' w_s^{-1}$. Set $\gamma_1 = w_s \gamma w_s^{-1}$. It is an elliptic element in $M_1(F)$. The split torus $A_2$ is the split component of the center of $G(\gamma_1)^0$. The group $P_1(\gamma_1)^0$ is a minimal parabolic subgroup of $G(\gamma_1)^0$. Notice that $P_1$ and $P_2$ are equal if and only if every element in the class $\mathfrak{o}$ is semisimple. In general, any element in $\mathfrak{o}$ is conjugate to $\gamma_1\nu_1$, where $\nu_1$ is a unipotent element in $P_1(\gamma_1)^0$. The unipotent element $\nu_1$ must lie in the unipotent radical of $P_1(\gamma_1)^0$, so in particular it belongs to $N_1(F)$.

PROPOSITION 2.6. *Suppose that $P = MN$ is a parabolic subgroup defined over $F$. Suppose that $\gamma$ is in $M(F)$. For all $\phi \in C_c(N(\mathbb{A}))$, we have*

$$\sum_{\delta \in N(F,\gamma_s)\backslash N(F)} \sum_{\eta \in N(F,\gamma_s)} \phi(\gamma^{-1}\delta^{-1}\gamma\eta\delta) = \sum_{\eta \in N(F)} \phi(\eta).$$

PROOF. Neither side of the putative formula changes if $\gamma$ is replaced by an $M(F)$-conjugate of itself. After noting that the previous discussion can be applied to classes in $M(F)$ as well as to $G(F)$, we can assume that there is a parabolic subgroup $P_1$, with $P_1 \subset P$ , such that $\gamma_s \in M_1(F)$, and $\gamma_u \in M(F, \gamma_s) \cap N_1(F)$.

The Lie algebra of $N$ can be decomposed into eigenspaces under the action of $A_1$. It follows that there exists a sequence

$$N = N_0 \supset N_1 \supset \cdots \supset N_r = \{1\}$$

of normal $\gamma_s$-stable subgroup of $N$ which are defined over $F$ and satisfy the following properties:

(i) $N_{k+1}\backslash N_k$ is abelian for each $k$.
(ii) If $\delta \in N_k$, and $\eta$ either belongs to $N$ or equals $\gamma_u$, then $\eta^{-1}\delta^{-1}\eta\delta$ belongs to $N_{k+1}$.

We shall show that for all $k$, $0 \leq k \leq r$,

$$\sum_{\delta \in N(F,\gamma_s)N_k(F)\backslash N(F)} \sum_{\eta \in N(F,\gamma_s)N_k(F)} \phi(\gamma^{-1}\delta^{-1}\gamma\eta\delta) \tag{2.6.1}$$

equals

$$\sum_{\delta \in N(F,\gamma_s)\backslash N(F)} \sum_{\eta \in N(F,\gamma_s)} \phi(\gamma^{-1}\delta^{-1}\gamma\eta\delta). \tag{2.6.2}$$

The assertion of the proposition is the case of $k = 0$.

The equality is immediate if $k = r$. By decreasing induction on $k$, we assume that (2.6.2) equals

$$\sum_{\delta \in N(F,\gamma_s)N_{k+1}(F)\backslash N(F)} \sum_{\eta \in N(F,\gamma_s)N_k(F)} \phi(\gamma^{-1}\delta^{-1}\gamma\eta\delta).$$

This is the sum over $\delta_1 \in N(F, \gamma_s)N_k(F)\backslash N(F)$ of

$$\begin{aligned}&\textstyle\sum_{\delta_2 \in N(F,\gamma_s)N_{k+1}(F)\backslash N(F,\gamma_s)N_k(F)} \sum_{\eta \in N(F,\gamma_s)N_{k+1}(F)} \phi(\gamma^{-1}\delta_1^{-1}\delta_2^{-1}\gamma\eta\delta_2\delta_1) \\ &\quad = \textstyle\sum_{\delta_2 \in N_k(F,\gamma_s)N_{k+1}(F)\backslash N_k(F)} \sum_\eta \phi(\gamma^{-1}\delta_1^{-1}\delta_2^{-1}\gamma\eta\delta_2\delta_1).\end{aligned}$$

For a fixed $\delta_2 \in N_k(F)$, we change variables in the sum over $\eta$. We find that

$$\begin{aligned}&\textstyle\sum_{\eta \in N(F,\gamma_s)N_{k+1}(F)} \phi(\gamma^{-1}\delta_1^{-1}\delta_2^{-1}\gamma\eta\delta_2\delta_1) = \sum_\eta \phi(\gamma^{-1}\delta_1^{-1}\gamma \cdot \gamma^{-1}\delta_2^{-1}\gamma\eta\delta_2 \cdot \delta_1) \\ &\quad = \textstyle\sum_\eta \phi(\gamma^{-1}\delta_1^{-1}\gamma \cdot \gamma_s^{-1}\delta_2^{-1}\gamma_s\eta\delta_2 \cdot \delta_1) = \sum_\eta \psi(\gamma_s^{-1}\delta_2^{-1}\gamma_s\delta_2),\end{aligned}$$

where

$$\psi(x) = \sum_{\eta \in N(F,\gamma_s)N_{k+1}(F)} \phi(\gamma^{-1}\delta_1^{-1}\gamma\eta \cdot x \cdot \delta_1)$$

is compactly supported function on the discrete set

$$N_k(F, \gamma_s)N_{k+1}(F)\backslash N_k(F).$$

The map

$$y \longmapsto N_k(\gamma_s)N_{k+1} \cdot \gamma_s^{-1}y^{-1}\gamma_s y, \quad y \in N_k(\gamma_s)N_{k+1}\backslash N_k,$$

is an isomorphism from $N_k(\gamma_s)N_{k+1}\backslash N_k$ onto itself which is defined over $F$. Therefore

$$\sum_{\delta_2 \in N_k(F,\gamma_s)N_{k+1}(F)\backslash N_k(F)} \psi(\gamma_s^{-1}\delta_2^{-1}\gamma_s\delta_2)$$

equals

$$\sum_{\eta \in N(F,\gamma_s)N_k(F)} \phi(\gamma^{-1}\delta_1^{-1}\gamma\eta\delta_1).$$

It follows that (2.6.2) is equal to (2.6.1). □

It follows from Proposition 2.6 that if $\mathfrak{o} \in \mathcal{O}$, and if $\gamma \in \mathfrak{o} \cap M(F)$, then $\gamma\eta$ belongs to $\mathfrak{o}$ for each $\eta$ in $N(F)$. In other words

$$\mathfrak{o} \cap P(F) = (\mathfrak{o} \cap M(F)) \cdot N(F).$$

A similar remark holds for the intersection of $\mathfrak{o}$ with any parabolic subgroup of $M$.

PROPOSITION 2.7. *Under the hypotheses of Proposition* 2.6, *we have*

$$\int_{N(\mathbb{A},\gamma_s)\backslash N(\mathbb{A})} \int_{N(\mathbb{A},\gamma_s)} \phi(\gamma^{-1}n_1^{-1}\gamma n_2 n_1)dn_2 dn_1 = \int_{N(\mathbb{A})} \phi(n)dn.$$

PROOF. The proof can be transcribed from the proof of Proposition 2.6 by replacing each sum over a set of rational points by the integral over the corresponding set of $\mathbb{A}$-valued points. □

### 2.8 Decomposition of the Kernel

If $\mathfrak{o} \in \mathcal{O}$, define

$$K_{\mathfrak{o}}(x,y) = \sum_{\gamma \in \mathfrak{o}} f(x^{-1}\gamma y).$$

Then

$$K(x,y) = \sum_{\mathfrak{o}} K_{\mathfrak{o}}(x,y).$$

More generally, if $P$ is a parabolic subgroup, define

$$K_{P,\mathfrak{o}}(x,y) = \sum_{\gamma \in M(F) \cap \mathfrak{o}} \int_{N(\mathbb{A})} f(x^{-1}\gamma n y)dn.$$

Then

$$K_P(x,y) = \sum_{\mathfrak{o} \in \mathcal{O}} K_{P,\mathfrak{o}}(x,y)$$

equals

$$\sum_{\gamma \in M(F)} \int_{N(\mathbb{A})} f(x^{-1}\gamma n y)dn.$$

This is just the kernel of $r_P(f)$ , where $r_P$ is the regular representation of $G(\mathbb{A})^1$ on

$$L^2(N(\mathbb{A})M(F)\backslash G(\mathbb{A})^1).$$

## 3 A Review of Eisenstein Series

### 3.1 References

In this section we shall recall those results on Eisenstein series which are needed for the trace formula. They are due to Langlands; the main ideas are in the article [La66] while details appear in [La76]. An attempt to explain the proof is at [MW95].

DEFINITION 3.2. Suppose $P$ is a parabolic subgroup. Recall: by a parabolic subgroup we mean a standard one. (1) The space of cusp forms $L^2_{\text{cusp}}(M(F)\backslash M(\mathbb{A})^1)$ on $M(\mathbb{A})^1$ is the space of functions $\phi$ in $L^2(M(F)\backslash M(\mathbb{A})^1)$ such that for any parabolic subgroup $P_1 \subsetneq P$, the integral

$$\int_{N_1(F)\cap M(F)\backslash N_1(\mathbb{A})\cap M(\mathbb{A})} \phi(nm)dn$$

is 0 for almost all $m \in M(\mathbb{A})^1$.

(2) It is known that

$$L^2_{\text{cusp}}(M(F)\backslash M(\mathbb{A})^1) = \oplus_\rho V_\rho,$$

where $\rho$ ranges over all irreducible unitary representations of $M(\mathbb{A})^1$, and each $V_\rho$ is an $M(\mathbb{A})^1$-invariant subspace of $L^2_{\text{cusp}}(M(F)\backslash M(\mathbb{A})^1)$, isomorphic under the action of $M(\mathbb{A})^1$ to a finite number of copies of $\rho$. An irreducible unitary representation $\rho$ of $M(\mathbb{A})^1$ is said to be *cuspidal* if $V_\rho \neq 0$.

(3) Suppose that $P'$ is another parabolic subgroup, and $\rho'$ is an irreducible unitary cuspidal representation of $M'(\mathbb{A})^1$. We say that the pairs $(M, \rho)$ and $(M', \rho')$ are *equivalent* if there is an $s \in \Omega(\mathfrak{a}, \mathfrak{a}')$ such that the representation

$$(s\rho)(m') = \rho(w_s^{-1} m' w_s), \quad m' \in M'(\mathbb{A})^1,$$

is unitarily equivalent to $\rho'$.

(4) Let $\mathfrak{X}$ be the set of equivalence classes of pairs. Then corresponding to any $\chi \in \mathfrak{X}$ we have a class, $\mathcal{P}_\chi$ , of associated (standard) parabolic subgroups. If $P$ is any parabolic subgroup and $\chi \in \mathfrak{X}$, set

$$L^2_{\text{cusp}}(M(F)\backslash M(\mathbb{A})^1)_\chi = \bigoplus_{\{\rho;(M,\rho)\in\chi\}} V_\rho.$$

This is a closed $M(\mathbb{A})^1$-invariant subspace of $L^2_{\text{cusp}}(M(F)\backslash M(\mathbb{A})^1)$, which is empty if $P$ does not belong to $\mathcal{P}_\chi$. We have

$$L^2_{\text{cusp}}(M(F)\backslash M(\mathbb{A})^1) = \bigoplus_{\chi\in\mathfrak{X}} L^2_{\text{cusp}}(M(F)\backslash M(\mathbb{A})^1)_\chi.$$

## 3.3 Generation

An early result in the theory of Eisenstein series, see [La66], Lemma 1, is

PROPOSITION. *Suppose that $P$ is a fixed parabolic subgroup and that $\chi \in \mathfrak{X}$. Suppose that there is a group $P_1$ in $\mathcal{P}_\chi$ which is contained in $P$. Let $\psi$ be a smooth function on $N_1(\mathbb{A})M_1(F)\backslash G(\mathbb{A})$ such that for $k \in K$, $m \in M_1(F)\backslash M_1(\mathbb{A})^1$, $a \in A_1(F)\backslash A_1(\mathbb{A})$, the function*

$$\Psi_a(m, k) = \psi(amk)$$

*vanishes for all a outside a compact subset of $A_1(F)\backslash A_1(\mathbb{A})$, transforms under $K_{F_\infty}$ according to an irreducible representation W, and as a function of m, it belongs to $L^2_{\text{cusp}}(M(F)\backslash M(\mathbb{A})^1)$. Then the function*

$$\hat{\psi}^M(m) = \sum_{\delta\in P_1(F)\cap M(F)\backslash M(F)} \psi(\delta m), \quad m \in M(F)\backslash M(\mathbb{A})^1,$$

*is square-integrable on $M(F)\backslash M(\mathbb{A})^1$.*

DEFINITION 3.4. Let $L^2(M(F)\backslash M(\mathbb{A})^1)_\chi$ be the closed span of all functions of the form $\hat{\psi}^M$, where $P_1$ runs through those groups in $\mathcal{P}_\chi$ that are contained in $P$, and $W$ varies over all irreducible representations of $K_{F_\infty}$. If there does not exist a group $P_1 \in \mathcal{P}_\chi$ which is contained in $P$, define $L^2(M(F)\backslash M(\mathbb{A})^1)_\chi$ to be $\{0\}$.

From [La66], Lemma 2, one has

PROPOSITION 3.5. *The space $L^2(M(F)\backslash M(\mathbb{A})^1)$ is orthogonal direct sum over all $\chi \in \mathfrak{X}$ of the spaces $L^2(M(F)\backslash M(\mathbb{A})^1)_\chi$.*

DEFINITION 3.6. (1) For any $P$, let $\Pi(M)$ denote the set of equivalence classes of irreducible unitary representations of $M(\mathbb{A})$.

(2) If $\zeta \in \mathfrak{a}^*_{\mathbb{C}}$ and $\pi \in \Pi(M)$, let $\pi_\zeta$ be the product of $\pi$ with the quasi-character

$$x \longmapsto e^{\zeta(H_P(x))}, \quad x \in G(\mathbb{A}).$$

(3) If $\zeta$ belongs to $i\mathfrak{a}^*$, $\pi_\zeta$ is unitary, so we obtain a free action of the group $i\mathfrak{a}^*$ on $\Pi(M)$. The set $\Pi(M)$ becomes a differentiable manifold whose connected components are the orbits of $i\mathfrak{a}^*$. We can also transfer our Haar measure on $i\mathfrak{a}^*$ to each of the orbits in $\Pi(M)$; this allow us to define a measure $d\pi$ on $\Pi(M)$.

(4) If $P_2 \supset P$, let $\Pi^{P_2}(M)$ be the space of orbits of $i\mathfrak{a}^*_2$ on $\Pi(M)$. The space $\Pi^{P_2}(M)$ inherits a measure from our measures on $\Pi(M)$ and $i\mathfrak{a}^*_2$.

(5) For $\tau \in \Pi(M)$, let $\mathcal{H}^0_P(\tau)$ be the space of smooth functions

$$\phi : N(\mathbb{A})M(F)\backslash G(\mathbb{A}) \to \mathbb{C}$$

that satisfy the following conditions:

(i) $\phi$ is right $K$-finite.

(ii) For every $x \in G(\mathbb{A})$, the function

$$m \longmapsto \phi(mx), \quad m \in M(\mathbb{A}),$$

is a matrix coefficient of $\tau$.

(iii) $\|\phi\|^2 = \int_K \int_{M(F)\backslash M(\mathbb{A})^1} |\phi(mk)|^2 dm dk$ is finite.

(6) Let $\mathcal{H}_P(\tau)$ be the completion of $\mathcal{H}^0_P(\tau)$. It is a Hilbert space. If $\phi \in \mathcal{H}_P(\tau)$, and $\zeta \in \mathfrak{a}^*_{\mathbb{C}}$, define

$$\phi_\zeta(x) = \phi(x)e^{\zeta(H_P(x))}, \quad x \in G(\mathbb{A}),$$

and

$$(I_P(\tau_\zeta, y)\phi_\zeta)(x) = \phi_\zeta(xy) \cdot \delta_P(xy)^{\frac{1}{2}} \cdot \delta_P(x)^{-\frac{1}{2}}, \quad x, y \in G(\mathbb{A}).$$

Then $I_P(\tau_\zeta)$ is a representation of $G(\mathbb{A})$ that is unitary if $\zeta \in i\mathfrak{a}^*$. Also $\mathcal{H}_P(\tau) = \{0\}$ unless there is a subrepresentation of the regular representation of $M(\mathbb{A})^1$ on $L^2(M(F)\backslash M(\mathbb{A})^1)$ that is equivalent to the restriction of $\tau$ to $M(\mathbb{A})^1$.

(7) Given $\chi \in \mathfrak{X}$, let $\mathcal{H}_P(\tau)_\chi$ be the closed subspace of $\mathcal{H}_P(\tau)$ consisting of those $\phi$ such that for all $x \in G(\mathbb{A})$, the function

$$x \longmapsto \phi(mx), \quad m \in M(F)\backslash M(\mathbb{A})^1,$$

belongs to $L^2(M(F)\backslash M(\mathbb{A})^1)_\chi$. Then

$$\mathcal{H}_P(\tau) = \bigoplus_{\chi \in \mathfrak{X}} \mathcal{H}_P(\tau)_\chi.$$

(We do know whether $\mathcal{H}_P(\tau)_\chi$ can be nonzero for more than one $\chi$.)

(8) Suppose that $K_0$ is an open compact subgroup of $G(\mathbb{A}_f)$ and that $W$ is an equivalence class of irreducible representations of $K_{F_\infty}$. Let $\mathcal{H}_P(\tau)_{\chi,K_0}$ be the subspace of functions in $\mathcal{H}_P(\tau)_\chi$ which are invariant under $K_0 \cap K$. Let $\mathcal{H}_P(\tau)_{\chi,K_0,W}$ be the space of functions in $\mathcal{H}_P(\tau)_{\chi,K_0}$ which transform under $K_{F_\infty}$ according to $W$. Fix an orthonormal basis $\mathfrak{B}_P(\tau)_\chi$ of the space $\mathcal{H}_P(\tau)_\chi$ for each $\tau$ and $\chi$, such that

$$\mathfrak{B}_P(\tau_\zeta)_\chi = \{\phi_\zeta ; \phi \in \mathfrak{B}_P(\tau)_\chi\}, \quad \zeta \in i\mathfrak{a}^*,$$

and such that every $\phi \in \mathfrak{B}_P(\tau)_\chi$ belongs to one of the spaces $\mathcal{H}_P(\tau)_{\chi,K_0,W}$.

We shall need these bases in the next section, when we give a second formula for the kernel $K(x, y)$ in terms of Eisenstein series.

It is a consequence of the decomposition of the spaces $L^2(M(F)\backslash M(\mathbb{A})^1)_\chi$, established by Langlands in Section 7 of [La76], that

PROPOSITION 3.7. *Each of the spaces $\mathcal{H}_P(\tau)_{\chi,K_0,W}$ is finite dimensional.*

DEFINITION. Suppose that $\tau \in \Pi(M)$, $\phi \in \mathcal{H}_P^0(\tau)$, $\zeta \in \mathfrak{a}_{P,\mathbb{C}}^*$, and $s \in \Omega(\mathfrak{a}, \mathfrak{a}')$. If $\operatorname{Re}\zeta$ belongs to $\rho_P + (\mathfrak{a}_P^*)^+$, the Eisenstein series and global intertwining operators are defined by

$$E(x, \phi_\zeta) = \sum_{\delta \in P(F)\backslash G(F)} \phi_\zeta(\delta x) \cdot \delta_P(\delta x)^{\frac{1}{2}},$$

and

$$(M(s, \tau_\zeta)\phi_\zeta)(x) = \int_{N'(\mathbb{A}) \cap w_s N(\mathbb{A}) w_s^{-1} \backslash N'(\mathbb{A})} \phi_\zeta(w_s^{-1} n x) \cdot \delta_P(w_s^{-1} n x)^{\frac{1}{2}} \cdot \delta_P(x)^{-\frac{1}{2}} \, dn.$$

The properties that we will need are all contained, at least implicitly, in [La76] (see especially Appendix II), and have been summarized in [Ar79]. See also [MW95]. We recall them again here.

THEOREM 3.8. (1) *The functions $E(x, \phi_\zeta)$ and $M(s, \tau_\zeta)\phi_\zeta$ can be continued as meromorphic functions in $\zeta$ to $\mathfrak{a}^*_{\mathbb{C}}$. If $\zeta \in i\mathfrak{a}^*$, the function $x \longmapsto E(x, \phi_\zeta)$ is smooth, and $M(s, \tau_\zeta)$ is a unitary operator from $\mathcal{H}_P(\tau_\zeta)$ to $\mathcal{H}_{P'}(s\tau_\zeta)$. There is an integer $N \geq 0$ such that for any $\phi$,*

$$\sup_{x \in G(\mathbb{A})}(\|x\|^{-N}|E(x, \phi_\zeta)|)$$

*is a locally bounded function on the set of $\zeta \in \mathfrak{a}^*_{\mathbb{C}}$ at which $E(x, \phi_\zeta)$ is regular.*

(2) *If $h \in C_c(G(\mathbb{A}))^K$ and $t \in \Omega(\mathfrak{a}', \mathfrak{a}'')$, the following functional equations hold.*

(i) $E(x, I_P(\tau, h)\phi) = \int_{G(\mathbb{A})} h(y)E(xy, \phi)dy.$
(ii) $E(x, M(s, \tau)\phi) = E(x, \phi).$
(iii) $M(ts, \tau) = M(t, s\tau)M(s, \tau).$

(3) *Let $\mathcal{P}$ be a class of associated parabolic subgroups. Let $\widehat{L}_{\mathcal{P}}$ be the set of collections $F = \{F_P; P \in \mathcal{P}\}$ of functions*

$$F_P : \tau \longmapsto \mathcal{H}_P(\tau), \quad \tau \in \Pi(M),$$

*such that*

(i) $F_{P'}(s\tau) = M(s, \tau)F_P(\tau), \;\; s \in \Omega(\mathfrak{a}, \mathfrak{a}'),$
(ii) $\|F\|^2 = \sum_{P \in \mathcal{P}} n(A)^{-1} \int_{\Pi(M)} \|F_P(\tau)\|^2 d\tau$ *is finite.*

*Then the map which sends $F$ to the function*

$$x \longmapsto \sum_{P \in \mathcal{P}} n(A)^{-1} \int_{\Pi(M)} E(x, F_P(\tau))d\tau,$$

*defined for $F$ in a dense subspace of $\widehat{L}_{\mathcal{P}}$, extends to a unitary map from $\widehat{L}_{\mathcal{P}}$ onto a closed, $G(\mathbb{A})$-invariant subspace $L^2_{\mathcal{P}}(G(F)\backslash G(\mathbb{A}))$ of $L^2(G(F)\backslash G(\mathbb{A}))$. Moreover, there is an orthogonal decomposition*

$$L^2(G(F)\backslash G(\mathbb{A})) = \oplus_{\mathcal{P}} \widehat{L}_{\mathcal{P}}(G(F)\backslash G(\mathbb{A})).$$

REMARK 3.9. We could equally well have defined subspaces

$$L^2_{\mathcal{P}}(G(F)\backslash G(\mathbb{A})^1)$$

of $L^2(G(F)\backslash G(\mathbb{A})^1)$. The only change in the formulation would be to integrate over $\Pi^G(M)$ instead of $\Pi(M)$. This would allow us to decompose the representation $r$ into a direct integral of the representations $I_P(\tau)$. (We denote the restriction of the representation $I_P(\tau)$ to $G(\mathbb{A})^1$ by $I_P(\tau)$ as well.)

If $\chi \in \mathfrak{X}$, we could replace $\widehat{L}_{\mathcal{P}}$ by a space of collection $F = \{F_P; P \in \mathcal{P}\}$ such that $F_P(\tau)$ belongs to $\mathcal{H}_P(\tau)_\chi$ for each $\tau$. We would obtain a decomposition of the space $L^2(G(F)\backslash G(\mathbb{A})^1)_\chi$.

More generally, if $P \subset P_2$, and if $\zeta$ is a suitable point in $\mathfrak{a}^*_{\mathbb{C}}$, define

$$E_{P_2}(x, \phi_\zeta) = \sum_{\delta \in P(F)\backslash P_2(F)} \phi_\zeta(\delta x) \cdot \delta_P(\delta x)^{\frac{1}{2}}.$$

The discussion above holds if the functions $E(x)$ are replaced by $E_{P_2}(x)$. It then amounts to a description of the decomposition of the representation $R_{P_2}$ into a direct integral of $\{I_P(\tau); P \subset P_2\}$.

## 3.10 Universal Enveloping Algebra

We shall end this section with some simple remarks on representations of the universal enveloping algebra $\mathcal{U}(\mathfrak{g}(F_\infty)^1 \otimes \mathbb{C})$ , which acts, through the representation $I_P(\tau)$, on the vector space $\mathcal{H}_P^0(\tau)$.

Consider the anti-involution $X \mapsto X^*$ of $\mathcal{U}(\mathfrak{g}(F_\infty)^1 \otimes \mathbb{C})$ given by the adjoint map. For any representation $\tau \in \Pi(M)$, elements $X, Y \in \mathcal{U}(\mathfrak{g}(F_\infty)^1 \otimes \mathbb{C})$, and test functions $h \in C_c^\infty(G(\mathbb{A})^1)$, we have

$$I_P(\tau, X)\, I_P(\tau, h)\, I_P(\tau, Y) = I_P(\tau, X * h * Y),$$

and

$$i_P(\tau, X)^* = i_P(\tau, X^*).$$

We also have $i_P(\tau, h)^* = i_P(\tau, h^*)$, where $h^*(x) = \overline{h(x^{-1})}$.

We shall denote the left invariant differential operator associated with $X$ by $r(X)$. If we wish to emphasize the fact that we are differentiating with respect to a variable $x$, we shall write $r_x(X)$. Then

$$r_x(X)E(x, \phi) = E(x, I_P(\tau, X)\phi), \quad \phi \in \mathcal{H}_P^0(\tau).$$

It follows that for every $\phi \in \mathcal{H}_P^0(\tau)$ and $X \in \mathcal{U}(\mathfrak{g}(F_\infty)^1 \otimes \mathbb{C})$, there exists a locally bounded function $c(\zeta)$, defined on the set of points $\zeta \in \mathfrak{a}^*_{\mathbb{C}}$ at which $E(x, \phi_\zeta)$ is regular, such that

$$|r_x(X)E(x, \phi_\zeta)| \leq c(\zeta)\|x\|^N, \quad x \in G(\mathbb{A})^1. \tag{3.10.1}$$

# 4 The Second Formula for the Kernel

The theory of Eisenstein series yields another formula for $K(x,y)$. The equality of this formula with that of § 2 is what will eventually lead to the trace formula.

PROPOSITION 4.1. *For any integer* $m \geq 1$, *there are functions* $g^1_\infty \in C^m_c(G(\mathbb{A})^1)^{K_{F_\infty}}$, $g^2_\infty \in C^\infty_c(G(\mathbb{R}))^{K_{F_\infty}}$, *and an element* $Z \in \mathcal{U}(\mathfrak{g}(F_\infty)^1 \otimes \mathbb{C})^{K_{F_\infty}}$ *such that* $Z * g^1_\infty + g^2_\infty$ *is the Dirac distribution at the identity in* $G(F_\infty)^1$.

PROOF. Let $\Delta$ be any elliptic element in $\mathcal{U}(\mathfrak{g}(F_\infty)^1 \otimes \mathbb{C})^{K_{F_\infty}}$. For example, $\Delta$ could be obtained as a linear combination of the Casimir operators of $G(F_\infty)^1$ and of $K_{F_\infty}$. We can assume that $\Delta$ is $K_{F_\infty}$-invariant. Let $\Omega$ be a small open $K_{F_\infty}$-invariant neighborhood of 1 in $G(F_\infty)^1$. By the existence of a fundamental solution [Ho63], p. 174, we can find, for any $n > 0$, a function $h$ on $\Omega$ such that $\Delta^n * h$ is the Dirac distribution. By the elliptic regularity theorem, we can choose $n$ so large that $h$ belongs to $C^m(\Omega)$. Since $\Delta$ is $K_{F_\infty}$-invariant, we can assume that $h \in C^m(\Omega)^{K_{F_\infty}}$. Moreover, $h$ is infinitely differentiable away from the identity. Let $\phi$ be any function in $C^\infty_c(\Omega)^{K_{F_\infty}}$ which equals 1 in a neighborhood of the identity. Then $\Delta * h\phi$ equals the Dirac distribution in a neighborhood of the identity and is smooth away from the identity. Thus $g^1_\infty = h\phi$ is in $C^m_c(\Omega)^{K_{F_\infty}}$, and $g^2_\infty$, the difference between $\Delta^n * g^1_\infty$ and the Dirac distribution, is actually in $C^\infty_c(\Omega)^{K_{F_\infty}}$. The proposition follows with $Z = \Delta^n$. □

COROLLARY 4.2. *Suppose* $m_0$ *is an integer* $> \deg Z$. *Then any* $h \in C^{m_0}_c(G(\mathbb{A})^1)$ *equals*

$$\sum_{i=1}^{2} h_i * g_i\,,$$

*where in the notation of the proposition,* $h_1 = h * Z$, $h_2 = h$, *and* $g_i$ *is the product of* $g^i_\infty$ *with a multiple of the characteristic function of an open compact subgroup of* $G(\mathbb{A}_f)$.

PROOF. Let $K_0$ be any open compact subgroup $G(\mathbb{A}_f)$ under which $h$ is bi-invariant. Define $g_i(x_\infty \cdot x_f)$, where $x_\infty \in G(F_\infty)^1$, $x_f \in G(\mathbb{A}_f)^1$, $i = 1,2$, to be $g^i_\infty(x_\mathbb{R})$ divided by the volume of $K_0$ if $x_f \in K_0$, and 0 otherwise. The corollary then follows from the proposition. □

PROPOSITION 4.3. *There is a positive integer* $N$ *and a continuous seminorm* $\|\cdot\|_0$ *on* $C_c(G(\mathbb{A})^1)$ *such that for any* $f \in C_c(G(\mathbb{A})^1)$ *and* $x, y \in G(\mathbb{A})^1$, *we have*

$$\left|\sum_{\gamma \in G(F)} f(x^{-1}\gamma y)\right| \leq \|f\|_0 \|x\|^{2N}.$$

PROOF. This claim is well known, at least in the $\Gamma\backslash G(\mathbb{R})^1$ setting, see [HC68], Lemma 9. The extension to adèle groups is easy. □

DEFINITION. A function $f$ on $G(\mathbb{A})^1$ is said to be *K-finite* if the space spanned by the left and right $K$-translates of $f$ is finite dimensional.

PROPOSITION 4.4. *There is an integer $m_0$ and a continuous seminorm $\|\cdot\|_{m_0}$ on $C_c^{m_0}(G(\mathbb{A})^1)$ such that if $X, Y \in \mathcal{U}(\mathfrak{g}(F_\infty)^1 \otimes \mathbb{C})$, $m \geq m_0 + \deg X + \deg Y$, and $f$ is a K-finite function in $C_c^m(G(\mathbb{A})^1)$, then*

$$\sum_\chi \sum_P n(A)^{-1} \int_{\Pi^G(M)} \left| \sum_{\phi \in \mathfrak{B}_P(\tau)_\chi} r(X)E(x, i_P(\tau,f)\phi) \cdot \overline{r(Y^*)E(y,\phi)} \right| d\tau \tag{4.4.1}$$

*is bounded by*

$$\|X * f * Y\|_{r_0} \cdot \|x\|^N \cdot \|y\|^N.$$

*Here N is as in Proposition* 4.3.

PROOF. Since $f$ is $K$-finite, the sum over $\phi$ is finite. Let $E(x, \tau) = E(x, \tau)_\chi$ be the vector in the algebraic direct product

$$\prod_{\phi \in \mathfrak{B}_P(\tau)_\chi} \mathbb{C}\phi$$

such that for any $K$-finite vector $\psi$ in $\mathcal{H}_P(\tau)_\chi$, we have $(\psi, E(x, \tau)) = E(x, \psi)$. Then

$$\sum_{\phi \in \mathfrak{B}_P(\tau)_\chi} r(X)E(x, i_P(\tau,f)\phi) \cdot \overline{r(Y^*)E(y,\phi)} \tag{4.4.2}$$

equals

$$\begin{aligned} &\sum_\phi E(x, i_P(\tau, X)i_P(\tau,f)\phi) \cdot \overline{E(y, i_P(\tau, Y^*)\phi)} \\ &= \sum_\phi (\phi, (i_P(\tau, X)i_P(\tau,f))^* E(x, \tau))(i_P(\tau, Y^*)^* E(y, \tau), \phi) \\ &= (i_P(\tau,X)i_P(\tau,f)i_P(\tau, Y)E(y, \tau), E(x, \tau)) \\ &= (i_P(\tau, X * f * Y)E(y, \tau), E(x, \tau)). \end{aligned}$$

Apply Corollary 4.2 with $h = X * f * Y$. (The integer $m$, upon which $Z$ depends, can be arbitrary.) Then $h = \sum_{i=1}^2 h_i * g_i$. We can choose $g_i$ to be in $C_c^m(G(\mathbb{A})^1)^K$. Since each function $h_i$ is $K$-finite, we can assume that $g_i$ is $K$-finite as well. The absolute value of (4.2) equals

$$\left| \sum_i (i_P(\tau, g_i)E(y, \tau), i_P(\tau, h_i)^* E(x, \tau)) \right|.$$

This is bounded by

$$\sum_i \|i_P(\tau, g_i)E(y,\tau)\| \cdot \|i_P(\tau, h_i)^* E(x,\tau)\|$$
$$= \sum_i (i_P(\tau, g_i^* * g_i)E(y,\tau), E(y,\tau))^{\frac{1}{2}} \cdot (i_P(h_i * h_i^*)E(x,\tau), E(x,\tau))^{\frac{1}{2}}.$$

Consider the set

$$\mathcal{S} = \{(\chi, P, \tau); x \in \mathfrak{X},\ \tau \in \Pi^G(M)\}.$$

Regarded as a disjoint union of copies of $\Pi^G(M)$, it comes equipped with a topology. The integral (4.4.1) defines a measure on $\mathcal{S}$. Suppose that $S$ is a compact subset of $\mathcal{S}$ and that $k$ is a $K$-finite function in $C_c(G(\mathbb{A})^1)$. It follows from the results on Eisenstein series summarized in Section 3 that

$$\int_{(\chi,P,\tau)\in S} (i_P(\tau, k * k^*)E(x,\tau)_\chi, E(y,\tau)_\chi)$$

is the kernel of the restriction of the operator $r(k * k^*)$ to an invariant subspace. Since the operator is positive semidefinite, the value at $x = y$ of this expression is bounded by

$$\sum_{\gamma \in G(F)} (k * k^*)(x^{-1}\gamma x).$$

By Proposition 4.3 this is in turn bounded by

$$\|k * k^*\|_0 \cdot \|x\|^{2N}.$$

Since this last expression is independent of $S$, it remains a bound if the original integral is taken over all of $\mathcal{S}$. It follows from Schwartz' inequality that (4.4.1) is bounded by

$$\|x\|^N \|y\|^N \cdot \sum_i \|g_i^* * g_i\|_0^{\frac{1}{2}} \cdot \|h_i * h_i^*\|_0^{\frac{1}{2}}.$$

Since $h_1 = h * Z$, $h_2 = h$, and $m_0 > \deg Z$, the map which sends $h$ to

$$\|h\|_{m_0} = \sum_i \|g_i^* * g_i\|_0^{\frac{1}{2}} \cdot \|h_i * h_i^*\|_0^{\frac{1}{2}}$$

is a continuous seminorm on $C_c^{m_0}(G(\mathbb{A})^1)$. The proposition follows. □

## 4.5 Test Functions

The $K$-finite functions are not dense in $C_c^m(G(\mathbb{A})^1)$. However, there is a positive integer $l_0$, depending only on $G$, such that for any $m > l_0$, the space $C_c^m(G(\mathbb{A})^1)$ is contained in the closure in $C_c^{m-l_0}(G(\mathbb{A})^1)$ of the $K$-finite functions in $C_c^{m-l_0}(G(\mathbb{A})^1)$.

DEFINITION. If $\omega = (W_1, W_2)$ is a pair of equivalence classes of irreducible representations of $K_{F_\infty}$, and $\chi_{W_i}$ is the character of $W_i$, define

$$f_\omega(x) = \deg W_1 \cdot \deg W_2 \cdot \int_{K_{F_\infty} \times K_{F_\infty}} \chi_{W_1}(k_1) f(k_1^{-1} x k_2^{-1}) \chi_{W_2}(k_2) dk_1 dk_2.$$

It follows easily from representation theory of a compact Lie group that $l_0$ may be chosen so that if $m > l_0$ and $\| \cdot \|$ is any continuous seminorm on $C_c^{m-l_0}(G(\mathbb{A})^1)$, then

$$f \longmapsto \sum_\omega \|f_\omega\|, \qquad f \in C_c^m(G(\mathbb{A})^1),$$

is a continuous seminorm on $C_c^m(G(\mathbb{A})^1)$, and $\sum_\omega f_\omega$ converges absolutely to $f$ in $C_c^{m-l_0}(G(\mathbb{A})^1)$.

DEFINITION 4.6. Suppose that the measure space $\mathcal{S}$ is defined as in the proof of Proposition 4.4. If $S$ is a measurable subset of $\mathcal{S}$ and $f$ is a $K$-finite function in $C_c^{m_0}(G(\mathbb{A})^1)$, define $I(S, f, x, y)$ to be the function

$$\int_{(\chi, P, \tau) \in S} \sum_{\phi \in \mathfrak{B}_P(\tau)_\chi} E(x, i_P(\tau, f)\phi) \overline{E(y, \phi)} d\tau.$$

For fixed $x$ and $y$, this defines a continuous linear functional on a subspace of $C_c^{m_0}(G(\mathbb{A})^1)$, the closure of which contains $C_c^{m_0+l_0}(G(\mathbb{A})^1)$. We can therefore define $I(S, f, x, y)$ for any $f \in C_c^{m_0+l_0}(G(\mathbb{A})^1)$.

PROPOSITION 4.7. *If $f \in C_c^m(G(\mathbb{A})^1)$, then $I(S, f, x, y)$ is continuously differentiable in each of $x$ and $y$ of order $m - m_0 - l_0$. If $X, Y \in \mathcal{U}(\mathfrak{g}(F_\infty)^1 \otimes \mathbb{C})$ and $m \geq m_0 + l_0 + \deg X + \deg Y$, then*

$$r_x(X) r_y(\bar{Y}^*) I(S, f, x, y)$$

*equals $I(S, X * f * Y, x, y)$.*

PROOF. Suppose first of all that $f$ is $K$-finite. If $S$ is relatively compact, the result follows from (3.10.1) and the proof of Proposition 4.4. In general, $S$ can be written as a disjoint union of relatively compact sets $S_k$. For any $n$,

$$\left| \sum_{k>n} r(\bar{Y}^*) I(S_k, X * f, x, y) \right| = \left| \sum_{k>n} I(S_k, X * f * Y, x, y) \right|.$$

In the notation of the proof of Proposition 4.4, this is bounded by

$$\sum_i \left(\sum_{k>n} I(S_k, g_i^* * g_i, y, y)\right)^{\frac{1}{2}} \left(\sum_{k>n} I(S_k, h_i * h_i^*, x, x)\right)^{\frac{1}{2}}$$

$$\leq \|y\|^N \cdot \sum_i \|g_i^* * g_i\|_0^{\frac{1}{2}} \left(\sum_{k>n} I(S_k, h_i * h_i^*, x, x)\right)^{\frac{1}{2}}.$$

As $n$ approaches $\infty$, this expression approaches 0 uniformly for $y$ in compact subsets. It follows that the function

$$\sum_k I(S_k, X * f, x, y) = I(S, X * f, x, y)$$

is differentiable in $y$ and its derivative with respect to $\bar{Y}^*$ equals

$$\sum_k I(S_k, X * f * Y, x, y) = I(S, X * f * Y, x, y).$$

The differentiability in $x$ follows the same way.

Now suppose that $f$ is an arbitrary function in $C_c^m(G(\mathbb{A})^1)$. Let $\{f_n\}$ be a sequence of $K$-finite functions that converges to $f$ in $C_c^{m-l_0}(G(\mathbb{A})^1)$. If $n_1 > n_2$,

$$|r_y(\bar{Y}^*)I(S, X * f_{n_1}, x, y) - r_y(\bar{Y}^*)I(S, X * f_{n_2}, x, y)|$$

$$= |I(S, X * f_{n_1} * Y - X * f_{n_2} * Y, x, y)|$$

$$\leq \|x\|^N \|y\|^N \cdot \|X * (f_{n_1} - f_{n_2}) * Y\|_{m_0},$$

by Proposition 4.4. Therefore the sequence

$$r_y(\bar{Y}^*)I(S, X * f_n, x, y)$$

converges uniformly for $x$ and $y$ in compact sets. In particular $I(S, X * f, x, y)$ is differentiable in $y$, and

$$r_y(\bar{Y}^*)I(S, X * f, x, y) = \lim_{n\to\infty} r_y(\bar{Y}^*)I(S, X * f_n, x, y)$$

$$= \lim_{n\to\infty} I(S, X * f_n * Y, x, y) = I(S, X * f * Y, x, y),$$

since $X * f_n * Y$ approaches $X * f * Y$ in $C_c^{m_0}(G(\mathbb{A})^1)$.

The differentiability in $x$ follows the same way. □

COROLLARY 4.8. *Suppose that* $X, Y \in \mathcal{U}(\mathfrak{g}(F_\infty)^1 \otimes \mathbb{C})$ *and* $m = m_0 + l_0 + \deg X + \deg Y$. *Then there is a continuous seminorm* $\|\cdot\|$ *on* $C_c^m(G(\mathbb{A})^1)$ *such that if* $\{S_k\}$ *is any sequence of disjoint subsets of* $S_1$*, and* $f \in C_c^m(G(\mathbb{A})^1)$*, then*

$$\sum_k |r_x(X)r_y(\bar{Y}^*)I(S_k, f, x, y)| \leq \|f\| \cdot \|x\|^N \cdot \|y\|^N.$$

PROOF. By Proposition 4.7, we may assume that $X = Y = 1$. The corollary then follows from Proposition 4.4 and the remarks immediately following its proof. □

COROLLARY 4.9. *Under the assumptions of Corollary* 4.8, *we have*

$$\sum_k r_x(X)r_y(\bar{Y}^*)I(S_k,f,x,y) = r_x(X)r_y(\bar{Y}^*)I(S,f,x,y),$$

*where $S$ is the union of the sets $S_k$.*

PROOF. Again by Proposition 4.7, we need only consider the case that $X = Y = 1$. According to Proposition 4.4 and the remark following its proof,

$$I(S,f,x,y) = \sum_\omega I(S,f_\omega,x,y) = \sum_\omega \sum_k I(S_k,f_\omega,x,y).$$

But $\sum_\omega \sum_k |I(S_k,f_\omega,x,y)|$ is bounded by $\|x\|^N \cdot \|y\|^N$ times $\sum_\omega \|f_\omega\|_{m_0}$. This last expression represents a continuous seminorm on $C_c^{m_0+l_0}(G(\mathbb{A})^1)$, and is, in particular, finite. Therefore the double series above converges absolutely. It equals

$$\sum_k \sum_\omega I(S_k,f_\omega,x,y) = \sum_k I(S_k,f,x,y),$$

as required. □

It follows from the discussion of Section 3 that to any closed subset $S$ of $\mathcal{S}$, there corresponds a closed invariant subspace of $L^2(G(F)\backslash G(\mathbb{A})^1)$. Let $P_S$ be the orthogonal projection onto this subspace. Set

$$r_S(f) = P_S r(f) P_S = r(f) P_S.$$

PROPOSITION 4.10. *The function $I(S,f,x,y)$ is the integral kernel of the operator $r_S(f)$.*

PROOF. Suppose that $f$ is a $K$-finite function in $C_c^{m_0}(G(\mathbb{A})^1)$. If $S$ is compact, the proposition follows from the results on Eisenstein series summarized in Section 3. In general, $S$ can be expressed as a disjoint union of sets $S_k$ such that for each $n$, the set

$$S^n = \cup_{k=1}^n S_k$$

is compact. If $\phi$ and $\psi$ are functions in $C_c^\infty(G(F)\backslash G(\mathbb{A})^1)$, we have

$$\begin{aligned}
(r_S(f)\phi, \psi) &= \lim_{n\to\infty} (r_{S^n}(f)\phi, \psi) \\
&= \lim_{n\to\infty} \int\int \overline{\psi}(x) I(S^n,f,x,y)\phi(y)dy\,dx \\
&= \sum_{k=1}^\infty \int\int \overline{\psi}(x) I(S_k,f,x,y)\phi(y)dy\,dx.
\end{aligned}$$

It follows from Proposition 4.4 and dominated convergence that this equals

$$\int \int \overline{\psi}(x) I(S, f, x, y) \phi(y) dy\, dx.$$

Now suppose that $f$ is an arbitrary function in $C_c^{m_0+l_0}(G(\mathbb{A}))$. Let $\{f_n\}$ be a sequence of $K$-finite functions that converges in $C_c^{m_0}(G(\mathbb{A}))$ to $f$. By the dominated convergence theorem,

$$\begin{aligned} \int \int \overline{\psi}(x)(I(S, f, x, y)\phi)(y) dy\, dx &= \lim_{n\to\infty} \int \int \overline{\psi}(x)(I(S, f_n, x, y)\phi)(y) dy\, dx \\ &= \lim_{n\to\infty} (r_S(f_n)\phi, \psi) = \lim_{n\to\infty} (r(f_n)P_S\phi, \psi). \end{aligned}$$

Proposition 4.3 allows us to use, once again, the dominated convergence theorem. The above limit equals

$$(r_S(f)\phi, \psi).$$

Hence $I(S, f, x, y)$ is the kernel of $r_S(f)$. □

DEFINITION 4.11. Suppose that $\chi$ is an element in $\mathfrak{X}$ and that $S$ is the set $\{(\chi, P, \tau)\}$ obtained by taking all $P$ and $\tau$. Then $r_S(f)$ is the projection of $r(f)$ onto the subspace $L^2(G(F)\backslash G(\mathbb{A})^1)_\chi$. We shall write $r_\chi(f)$ and $K_\chi(x, y)$ for $r_S(f)$ and $I(S, f, x, y)$.

PROPOSITION. *We have*

$$K(x, y) = \sum_{\chi \in \mathfrak{X}} K_\chi(x, y)$$

*for all $x$ and $y$.*

PROOF. It follows from Proposition 4.10 and Corollary 4.9 that $\sum_\chi K_\chi(x, y)$ equals the kernel of $r(f)$. It therefore must equal $K(x, y)$ almost everywhere. However the difference between these two functions is continuous in $x$ and $y$ (separately). The proposition follows. □

Suppose that $P$ is a parabolic subgroup. If $K(x, y)$ is replaced by $K_P(x, y)$ and $E(x, \phi)$ is replaced by $E_P(x, \phi)$, we can obtain obvious analogues of the definitions of this section, as well as of Propositions 4.3 through 4.10. Then $K_{P,\chi}(x, y)$ is defined to be

$$\sum_{P_1 \subset P} n_P(A_1)^{-1} \int_{\Pi^G(M_1)} \sum_{\phi \in \mathfrak{B}_{P_1}(\tau)_\chi} E_P(x, I_{P_1}(\tau, f)\phi) \overline{E_P(y, \phi)} d\tau$$

if $f$ is $K$-finite. If $f$ is an arbitrary function in $C_c^{m_0+l_0}(G(\mathbb{A})^1)$, then

$$K_P(x,y) = \sum_{\chi \in \mathfrak{X}} K_{P,\chi}(x,y)$$

for all $x$ and $y$.

# 5 The Modified Kernel Identity

## 5.1 *Agenda*

After comparing the formulae of Section 2 and Section 4 for $K_P(x,y)$, we note that

$$\sum_{\mathfrak{o}} K_{P,\mathfrak{o}}(x,y) = \sum_{\chi} K_{P,\chi}(x,y) \tag{5.1.1}$$

for all $x$ and $y$. In this section we shall modify each of the functions $K_{\mathfrak{o}}(x,x)$ and $K_\chi(x,x)$ so that the sum over $\mathfrak{o}$ remains equal to the sum over $\chi$. We shall later find that all of the modified functions are integrable over $G(F)\backslash G(\mathbb{A})^1$. First, we shall need a

DEFINITION 5.2. If $P_1 \subset P_2$, let

$$\tau_{P_1}^{P_2} = \tau_1^2 \text{ and } \hat{\tau}_{P_1}^{P_2} = \hat{\tau}_1^2$$

be the characteristic functions on $\mathfrak{a}_0$ of

$$\{H \in \mathfrak{a}_0; \alpha(H) > 0,\ \alpha \in \Delta_1^2\}$$

and

$$\{H \in \mathfrak{a}_0; \varpi(H) > 0,\ \varpi \in \widehat{\Delta}_1^2\}.$$

We shall denote $\tau_P^G$ and $\hat{\tau}_P^G$ simply by $\tau_P$ and $\hat{\tau}_P$.

PROPOSITION 5.3. *Let $P$ be a parabolic subgroup. Let $\|\cdot\|$ be a Euclidean norm on $\mathfrak{a}_P$. Then there are positive number $c$ and integer $N$ such that for all $x \in G(\mathbb{A})^1$ and $X \in \mathfrak{a}_P$, we have*

$$\sum_{\delta \in P(F)\backslash G(F)} \hat{\tau}_P(H(\delta x) - X) \le c(\|x\| e^{\|x\|})^N.$$

*In particular, the sum is finite.*

PROOF. Suppose that $\varpi \in \widehat{\Delta}_0$. Let $\Lambda$ be a rational representation of $G$ on the vector space $V$, with highest weight $d\varpi$, $d > 0$. Choose a height function relative to a basis on $V(F)$ as in § 1. We can assume that the basis contains a highest weight vector $v$.

According to the Bruhat decomposition, any element $\delta \in G(F)$ can be written $\pi w_s \nu$, for $\pi \in P_0(F)$, $\nu \in N_0(F)$, and $s \in \Omega$. Then

$$\|\Lambda(\delta)^{-1}v\| \geq 1.$$

There are positive numbers $c_1$, $c_2$ and integer $N_1$ such that for any $x \in G(\mathbb{A})^1$, we have

$$\|\Lambda(\delta)^{-1}v\| = \|\Lambda(x)\Lambda(\delta x)^{-1}v\|$$

$$\leq c_1\|x\|^{N_1}\|\Lambda(x)^{-1}v\| \leq c_2\|x\|^{N_1}e^{-d\varpi(H_0(\delta x))}.$$

It follows that there is a positive number $c$ such that

$$\varpi(H_0(\delta x)) \leq c(1 + \log\|x\|) \tag{5.3.1}$$

for all $\varpi \in \widehat{\Delta}_0$, $x \in G(\mathbb{A})^1$ and $\delta \in G(F)$.

For each $x$, let $\Gamma(x)$ be a fixed set of representatives of $P(F)\backslash G(F)$ in $G(F)$ such that for any $\delta \in \Gamma(x)$, $\delta x$ belongs to $\omega \mathcal{S} A(\mathbb{R}^0)K$, where $\omega$ is a fixed compact subset of $N(\mathbb{A})$ and $\mathcal{S}$ is a fixed Siegel set in $M(\mathbb{A})^1$. Then there is a compact subset $\omega_0$ of $N_0(\mathbb{A})M_0(\mathbb{A})^1$ and a point $T_0$ in $\mathfrak{a}_0$ such that for any $x$, and any $\delta \in \Gamma(x)$, the element $\delta x$ belongs to $\omega_0 A_0(\mathbb{R}^0)K$, and in addition,

$$\alpha(H_0(\delta x)) \geq \alpha(T_0), \tag{5.3.2}$$

for every $\alpha \in \Delta_0^P$.

We are interested in those $\delta$ such that $\hat{\tau}_P(H_0(\delta x) - X) = 1$, that is , such that

$$\varpi(H_0(\delta x)) > \varpi(X) \tag{5.3.3}$$

for every $\varpi \in \widehat{\Delta}_P$. The set of points $H_0(\delta x)$ in $\mathfrak{a}_0^G$ which satisfy (5.3.1), (5.3.2), and (5.3.3) is compact. In fact, it follows from our discussion that if $x \in G(\mathbb{A})^1$, $\delta \in \Gamma(x)$, and $\hat{\tau}_P(H(\delta x) - X) = 1$, then $\|\delta x\|$ is bounded by a scalar multiple of a power of $\|x\| \cdot e^{\|x\|}$. Since

$$\|\delta\| \leq \|\delta x\| \cdot \|x^{-1}\| \leq c\|\delta x\| \cdot \|x\|^N,$$

for some positive $c$ and $N$, $\|\delta\|$ too is bounded by a scalar multiple of a power of $\|x\| \cdot e^{\|x\|}$. Because $G(F)$ is a discrete subgroup of $G(\mathbb{A})^1$, the proposition follows from the fact that the volume in $G(\mathbb{A})^1$ of the set

$$\{y \in G(\mathbb{A})^1; \|y\| \leq t\}$$

is bounded by a scalar multiple of a power of $t$. □

COROLLARY 5.4. *Suppose that $T \in \mathfrak{a}_0$ and $N \geq 0$. Then we can find positive numbers $c'$ and $N'$ such that for any function $\phi$ on $P(F)\backslash G(\mathbb{A})^1$, and $x, y \in G(\mathbb{A})^1$, the sum*

$$\sum_{\delta \in P(F)\backslash G(F)} |\phi(\delta x)| \hat{\tau}_P(H(\delta x) - H(y) - T) \tag{5.4.1}$$

*is bounded by*

$$c' \|x\|^{N'} \cdot \|y\|^{N'} \cdot \sup_{u \in G(\mathbb{A})^1} (|\phi(u)| \cdot \|u\|^{-N}).$$

PROOF. The expression (5.4.1) is bounded by the product of

$$\sup_{u \in G(\mathbb{A})^1} (|\phi(u)| \cdot \|u\|^{-N})$$

and

$$\sum_{\delta \in \Gamma(x)} \|\delta x\|^N \cdot \hat{\tau}_P(H(\delta x) - H(y) - T).$$

We proved in the proposition that when $\hat{\tau}_P(H(\delta x) - H(y) - T)$ was equal to 1, $\|\delta x\|$ was bounded by a scalar multiple of a power of $\|x\| e^{\|H_P(y)+T\|}$. The corollary therefore follows from the proposition itself. □

DEFINITION 5.5. Suppose that $T$ is a fixed point in $\mathfrak{a}_0^+$. We shall say that $T$ is *suitably regular* if $\alpha(T)$ is sufficiently large for all $\alpha \in \Delta_0$.

We shall assume that $T$ is sufficiently regular whenever it is convenient, often without further comment.

For the rest of this work, $f$ will be a function in $C_c^m(G(\mathbb{A})^1)$, where $m \geq m_0 + l_0$. We shall also assume that $m$ is as large as necessary at any given time, again, without further comment.

Suppose that $x \in G(\mathbb{A})^1$. For $\mathfrak{o} \in \mathcal{O}$ and $\chi \in \mathfrak{X}$, define

$$k_{\mathfrak{o}}^T(x,f) = \sum_P (-1)^{\dim(A_P/Z)} \sum_{\delta \in P(F)\backslash G(F)} K_{P,\mathfrak{o}}(\delta x, \delta x) \cdot \hat{\tau}_P(H(\delta x) - T),$$

and

$$k_{\chi}^T(x,f) = \sum_P (-1)^{\dim(A_P/Z)} \sum_{\delta \in P(F)\backslash G(F)} K_{P,\chi}(\delta x, \delta x) \cdot \hat{\tau}_P(H(\delta x) - T).$$

PROPOSITION 5.6. *We have* $\sum_{\mathfrak{o}} k_{\mathfrak{o}}^T(x,f) = \sum_{\chi} k_{\chi}^T(x,f)$.

PROOF. The left hand side equals

$$\sum_P (-1)^{\dim(A_P/Z)} \sum_{\delta \in P(F)\backslash G(F)} \sum_{\mathfrak{o}} K_{P,\mathfrak{o}}(\delta x, \delta x) \hat{\tau}_P(H(\delta x) - T),$$

since the sums over $P$ and $\delta$ are finite. By (5.1.1) this equals

$$\sum_{P}(-1)^{\dim(A_P/Z)} \sum_{\delta\in P(F)\backslash G(F)} \sum_{\chi} K_{P,\chi}(\delta x, \delta x)\hat{\tau}_P(H(\delta x) - T),$$

which is just $\sum_{\chi} k_\chi^T(x,f)$. □

## 6 Some Geometric Lemmas

We want to show that each of the terms in the identity of Proposition 5.6 is integrable over $G(F)\backslash G(\mathbb{A})^1$. In this section we shall collect some geometric results which will be needed in estimating the integrals.

DEFINITION 6.1. If $P_1 \subset P_2$, for $H \in \mathfrak{a}_0$ set

$$\sigma_1^2(H) = \sigma_{P_1}^{P_2}(H) = \sum_{\{P_3;P_3\supset P_2\}} (-1)^{\dim(A_3/A_2)}\tau_1^3(H)\cdot\hat{\tau}_3(H).$$

For motivation for this definition, as well as for the next claim, see § 7.

PROPOSITION. *If $P_2 \supset P_1$, then $\sigma_1^2$ is the characteristic function of the set of $H \in \mathfrak{a}_1$ such that*

(i) $\alpha(H) > 0$ for all $\alpha \in \Delta_1^2$,
(ii) $\alpha(H) \leq 0$ for all $\sigma \in \Delta_1\backslash\Delta_1^2$, and
(iii) $\varpi(H) > 0$ for all $\varpi \in \widehat{\Delta}_2$.

PROOF. Fix $H \in \mathfrak{a}_1$. Consider the subset of those $\varpi$ in $\widehat{\Delta}_2$ for which $\varpi(H) > 0$. This subset is of the form $\widehat{\Delta}_R$, for a unique parabolic subgroup $R \supset P_2$. Then

$$\sigma_1^2(H) = \sum_{\{P_3;P_3\supset R\}} (-1)^{\dim(A_2/A_3)}\tau_1^3(H).$$

Suppose that $\tau_1^3(H) = 1$ for a given $P_3 \supset R$. Then $\tau_1^3(H) = 1$ for all smaller $P_3$. It follows from Proposition 1.5 that the above sum vanishes unless the original $P_3$ equals $R$. Thus

$$\sum_{\{P_3;P_3\supset R\}} (-1)^{\dim(A_2/A_3)}\tau_1^3(X)$$

is the characteristic function of

$$\{X \in \mathfrak{a}_1; \alpha(X) > 0,\ \alpha \in \Phi_1^R;\ \alpha(X) \leq 0,\ \alpha \in \Delta_1\backslash\Delta_1^R\}. \tag{6.1.1}$$

If $R = P_2$, then $\sigma_1^2(H) = 1$ if and only if $H$ belongs to the set (6.1.1), as required.

We must show that if $R$ is strictly larger than $P_2$, then $\sigma_1^2(H) = 0$. Suppose not. Then $H$ belongs to the set (6.1.1). In particular, the projection of $H$ onto $\mathfrak{a}_1^R$ lies in the positive chamber, which is contained in the set of positive linear combinations of roots in $\Delta_1^R$. Thus $\varpi(H) > 0$ for all $\varpi \in \widehat{\Delta}_1^R$. By the definition of $R$, we have $\varpi(H) > 0$ for all $\varpi \in \widehat{\Delta}_R$. From this, we shall show that if

$$H = \sum_{\alpha \in \Delta_1} c_\alpha \alpha^\vee,$$

then each $c_\alpha$ is positive. Suppose that $\varpi \in \widehat{\Delta}_R \subset \widehat{\Delta}_1$, and that $\alpha_\varpi$ is the element in $\Delta_1$ which is paired with $\varpi$. Then $c_{\alpha_\varpi} = \varpi(H)$ is positive. Therefore the projection of

$$H_R = \sum_{\varpi \in \widehat{\Delta}_R} c_{\alpha_\varpi} \alpha_\varpi^\vee$$

onto $\mathfrak{a}_1^R$ is in the negative chamber, so that if $\nu \in \widehat{\Delta}_1^R$, then $\nu(H_R)$ is negative.

If $\alpha_\nu$ is the root in $\Delta_1^R$ corresponding to $\nu$, then

$$c_{\alpha_\nu} = \nu(H) - \nu(H_R)$$

is positive. Thus each $c_\alpha$ is positive. Therefore $\varpi(H)$ is positive for each $\varpi \in \widehat{\Delta}_1$, and in particular, for each $\varpi$ in $\widehat{\Delta}_2$. Therefore $R = P_2$ so we have a contradiction.

□

COROLLARY 6.2. *Fix $T \in \mathfrak{a}_0^+$. For any $H \in \mathfrak{a}_1^G$, let $H_1^2$ be the projection of $H$ onto $\mathfrak{a}_1^2$. For any Euclidean norm $\|\cdot\|$ on $\mathfrak{a}_0$, there exists a positive number $c$ such that if $\sigma_1^2(H-T) \neq 0$, then $\alpha(H_1^2)$ is positive for each $\alpha \in \Delta_1^2$, and*

$$\|H\| \leq c(1 + \|H_1^2\|).$$

PROOF. The first condition follows directly from Proposition 6.1 and the fact that $T$ belongs to $\mathfrak{a}_0^+$. To prove the second one, write

$$H = H_1^2 + H_2.$$

The value at $H_2$ of any root in $\Delta_2$ equals $\alpha(H_2)$ for some root $\alpha$ in $\Delta_1 \backslash \Delta_1^2$. But

$$\alpha(H_2) = \alpha(H - T) - \alpha(H_1^2) + \alpha(T) < -\alpha(H_1^2) + \alpha(T),$$

by Proposition 6.1. Since

$$\varpi(H_2) = \varpi(H) > 0,$$

for each $\varpi \in \widehat{\Delta}_2$, the vector $H_2$ belongs to a compact set. In fact the norm of $H_2$ is bounded by a scalar multiple of $1 + \|H_1^2\|$ as required. □

This corollary will not be used until we begin to estimate the integrals of the functions defined in § 5.

## 6.3 Combinatorial Result

A second consequence of Proposition 6.1 is the special case that $P_2 = P_1$, and $P_1 \neq G$. Since every functional in the set $\widehat{\Delta}_1$ is negative on $-\mathfrak{a}_1^+$, the function $\sigma_{P_1}^{P_2}(H)$ is the characteristic function of the empty set. In other words,

COROLLARY. *We have*

$$\sum_{\{P_3;P_3\supset P_1\}} (-1)^{\dim(A_1/A_3)} \tau_1^3(H)\hat{\tau}_3(H) = 0$$

*for all* $H \in \mathfrak{a}_0$.

DEFINITION 6.4. Suppose that $Q$ and $P$ are parabolic subgroups with $Q \subset P$. Fix $\Lambda \in \mathfrak{a}_0^*$. Let $\epsilon_Q^P(\Lambda)$ be $(-1)$ raised to a power equal to the number of roots $\alpha \in \Delta_Q^P$ such that $\Lambda(\alpha^\vee) \leq 0$. Let

$$\phi_Q^P(\Lambda, H), \quad H \in \mathfrak{a}_0,$$

be the characteristic function of the set of $H \in \mathfrak{a}_0$ such that for any $\alpha \in \Delta_Q^P$, we have $\varpi_\alpha(H) > 0$ if $\Lambda(\alpha^\vee) \leq 0$, and $\varpi_\alpha(H) \leq 0$ if $\Lambda(\alpha) > 0$.

In the special case that none of the numbers $\Lambda(\alpha^\vee)$ or $\varpi_\alpha(H)$ is zero, these definitions give functions which occur in a combinatorial lemma of Langlands, see [Ar76]. It is desirable to have this result for general $H$ and $\Lambda$, so we shall give a different proof, based only on Proposition 1.5.

PROPOSITION 6.5. *The function*

$$\sum_{\{R;Q\subset R\subset P\}} \epsilon_Q^R(\Lambda)\, \phi_Q^R(\Lambda, H)\, \tau_R^P(H)$$

*is equal to* 0 *if* $\Lambda(\alpha^\vee) \leq 0$ *for some* $\alpha \in \Delta_Q^P$ *and to* 1 *otherwise.*

PROOF. If $R \neq P$, Corollary 6.3 implies that

$$\sum_{\{R_1;R\subset P_1\subset P\}} (-1)^{\dim(A_1/A_P)} \tau_R^1(H)\hat{\tau}_1^P(H)$$

vanishes for all $H$. Therefore

$$\sum_{\{R;Q\subset R\subset P\}} \epsilon_Q^R(\Lambda)\phi_Q^R(\Lambda, H)\tau_R^P(H)$$

is the difference between

$$\epsilon_Q^P(\Lambda)\phi_Q^P(\Lambda, H) \tag{6.5.1}$$

and

$$\sum_{\{R,P_1;Q\subset R\subset P_1\subsetneq P\}} \epsilon_Q^R(\Lambda)\phi_Q^R(\Lambda, H)\tau_R^1(H)\,(-1)^{\dim(A_1/A_P)}\hat{\tau}_1^P(H). \tag{6.5.2}$$

We shall prove the proposition by induction on $\dim(A_0/A_P)$. Define $\Delta_Q^\Lambda$ to be the set of roots $\alpha \in \Delta_Q^P$ such that $\alpha(\Lambda) > 0$. Associated with $\Delta_Q^\Lambda$ we have a parabolic subgroup $P_\Lambda$, with $Q \subset P_\Lambda \subset P$. By our induction assumption, the sum over $R$ in (6.5.2) vanishes unless $P_1 \subset P_\Lambda$, in which case it equals $(-1)^{\dim(A_1/A_P)}\hat{\tau}_1^P(H)$. Thus (6.5.2) equals

$$\sum_{\{P_1;Q\subset P_1\subset P_\Lambda\}} (-1)^{\dim(A_1/A_P)}\hat{\tau}_1^P(H) \tag{6.5.3}$$

if $P_\Lambda \neq P$, and equals (6.5.3) minus 1 if $P_\Lambda = P$. We need only show that (6.5.1) equals (6.5.3). This is a consequence of Proposition 1.5 and the definitions. □

## 6.6 A Partition

Our final aim for this section is to derive a partition of $G(F)\backslash G(\mathbb{A})$ into disjoint subsets, one for each (standard) parabolic subgroup. The partition is similar to a construction from [La76], in which disjoint subsets of $G(F)\backslash G(\mathbb{A})$ are associated with maximal parabolic subgroups. More generally, we shall partition $N(\mathbb{A})M(F)\backslash G(\mathbb{A})$, where $P = NM$ is a parabolic subgroup. The result is little more than a restatement of a basic fact from reduction theory, which we would do well to recall.

Suppose that $\omega$ is a compact subset of $N_0(\mathbb{A})M_0(\mathbb{A})^1$ and that $T_0 \in -\mathfrak{a}_0^+$. For any parabolic subgroup $P_1$, let $\mathcal{S}^{P_1}(T_0, \omega)$ be the set of

$$pak,\ \ p \in \omega,\ a \in A_0(\mathbb{R}^0),\ k \in K,$$

such that $\alpha(H_0(a) - T_0)$ is positive for each $\alpha \in \Delta_0^1$. By the properties of Siegel sets we have;

Proposition. *There are $\omega$ and $T_0$ so that for any $P_1$ we have $G(\mathbb{A}) = P_1(F)\mathcal{S}^{P_1}(T_0, \omega)$.*

The result from reduction theory is contained in the proof of [La76], Lemma 2.12:

PROPOSITION 6.7. *Any suitably regular point $T$ in $\mathfrak{a}_0^+$ has the following property. Suppose that $P_1 \subset P$ are parabolic subgroups, and that $x$ and $\delta x$ belong to $\mathcal{S}^{P_1}(T_0,\omega)$ for points $x \in G(\mathbb{A})$ and $\delta \in P(F)$. If $\alpha(H_0(x)-T) > 0$ for all $\alpha$ in $\Delta_0^P \backslash \Delta_0^{P_1}$, then $\delta$ belongs to $P_1(F)$.*

DEFINITION 6.8. Suppose that $P_1$ is a parabolic subgroup. (1) Let $\mathcal{S}^{P_1}(T_0,T,\omega)$ be the set of $x$ in $\mathcal{S}^{P_1}(T_0,\omega)$ such that $\varpi(H_0(x)-T) \le 0$ for each $\varpi \in \widehat{\Delta}_0^1$. (2) Let

$$F^{P_1}(x,T) = F^1(x,T)$$

be the characteristic function of the set of $x \in G(\mathbb{A})$ such that $\delta x$ belongs to $\mathcal{S}^{P_1}(T_0,T,\omega)$ for some $\delta \in P_1(F)$. Then $F^1(x,T)$ is left $A_1(\mathbb{R}^0)N_1(\mathbb{A})M_1(F)$ invariant, and can be regarded as the characteristic function of the projection of $\mathcal{S}^{P_1}(T_0,T,\omega)$ onto $A_1(\mathbb{R}^0)N_1(\mathbb{A})M_1(F)\backslash G(\mathbb{A})$, a compact subset of

$$A_1(\mathbb{R}^0)N_1(\mathbb{A})M_1(F)\backslash G(\mathbb{A}).$$

PROPOSITION 6.9. *Fix $P$. Let $T$ be any suitably regular point in $T_0 + \mathfrak{a}_0^+$. Then*

$$\sum_{\{P_1;P_0\subset P_1\subset P\}} \sum_{\delta\in P_1(F)\backslash G(F)} F^1(\delta x,T)\tau_1^P(H_0(\delta x)-T)$$

*equals one for all $x$ in $G(\mathbb{A})$.*

PROOF. Fix $x \in G(\mathbb{A})$. Choose $\delta \in P(F)$ such that $\delta x$ belongs to $\mathcal{S}^P(T_0,\omega)$. Apply Proposition 6.5 with $Q = P_0$, $\Lambda \in (\mathfrak{a}_0^*)^+$, and $H = H_0(\delta x) - T$. Then there is a parabolic subgroup $P_1 \subset P$ such that $\varpi(H_0(\delta x)-T) \le 0$ for all $\varpi \in \widehat{\Delta}_0^1$ and $\alpha(H_0(\delta x)-T) > 0$ for $\alpha \in \Delta_1^P$. Therefore

$$F^1(\delta x,T)\tau_1^P(H_0(\delta x)-T) = 1,$$

so the given sum is at least one.

Suppose there are elements $\delta_1,\ \delta_2 \in G(F)$, and parabolic subgroups $P_1$ and $P_2$ contained in $P$ such that

$$F^1(\delta_1 x,T)\tau_1^P(H_0(\delta_1 x)-T) = F^2(\delta_2 x,T)\tau_2^P(H_0(\delta_2 x)-T) = 1.$$

After left translating $\delta_i$ by an element in $P_i(F)$ if necessary, we may assume that

$$\delta_i x \in \mathcal{S}^{P_i}(T_0,T,\omega),\quad i = 1,2.$$

The projection of $H(\delta_i x) - T$ onto $\mathfrak{a}_0^P$ can be written

$$-\sum_{\alpha\in\Delta_0^i} c_\alpha \alpha^\vee + \sum_{\varpi\in\widehat{\Delta}^{P_i}} c_\varpi \varpi^\vee,$$

where each $c_\alpha$ and $c_\varpi$ is positive. It follows that $\alpha(H_0(\delta_i x) - T) > 0$ for every $\alpha \in \Delta_0^P \backslash \Delta_0^i$. In particular, since $T$ lies in $T_0 + \mathfrak{a}_0^+$, the element $\delta_i x$ belongs to $\mathcal{S}^P(T_0, \omega)$. The reduction theoretic result just quoted now implies that $\delta_2\delta_1^{-1}$ belongs to $P_1(F)$ and $\delta_1\delta_2^{-1}$ belongs to $P_2(F)$. In other words, $\delta_2 = \xi\delta_1$, for some element $\xi$ in $P_1(F) \cap P_2(F)$. Let $Q = P_1 \cap P_2$. Then $H_0(\delta_1 x) - T$ and $H_0(\delta_2 x) - T$ project onto the same point, say $H_Q^P$, on $\mathfrak{a}_Q^P$. If $R$ equals either $P_1$ or $P_2$, we have $\varpi(H_Q^P) \leq 0$ for some $\varpi \in \widehat{\Delta}_Q^R$ and $\alpha(H_Q^P) > 0$ for $\alpha \in \Delta_R^P$. Applying Proposition 6.5, with $\Lambda \in (\mathfrak{a}_0^*)^+$, we see that there is exactly one $R$ with $Q \subset R \subset P$, for which these inequalities hold. Therefore $P_1 = P_2$, and $\delta_1$ and $\delta_2$ belong to the same $P_1(F)$ coset in $G(F)$. This proves that the given sum is at most one. □

## 7 Integrability of $k_{\mathfrak{o}}^T(x,f)$

The aim of this section is to establish the integrability over $x \in G(F)\backslash G(\mathbb{A})^1$ of each function $k_{\mathfrak{o}}^T(x,f)$ of 5.5. In fact we will obtain

THEOREM 7.1. *For all sufficiently regular $T$, the sum*

$$\sum_{\mathfrak{o}\in\mathcal{O}} \int_{G(F)\backslash G(\mathbb{A})^1} |k_{\mathfrak{o}}^T(x,f)|dx$$

*is finite.*

PROOF. For any $x$, the function $k_{\mathfrak{o}}^T(x,f)$ equals the sum over $P$ and over $\delta \in P(F)\backslash G(F)$ of the product of

$$(-1)^{\dim(A/Z)} K_{P,\mathfrak{o}}(\delta x, \delta x) \cdot \hat{\tau}_P(H_0(\delta x) - T)$$

and

$$\sum_{\{P_1; P_0 \subset P_1 \subset P\}} \sum_{\xi \in P_1(F)\backslash P(F)} F^1(\xi\delta x, T) \cdot \tau_1^P(H_0(\xi\delta x) - T).$$

This equals the sum over $\{P_1, P; P_0 \subset P_1 \subset P\}$ and $\{\delta \in P_1(F)\backslash G(F)\}$ of

$$(-1)^{\dim(A/Z)} F^1(\delta x, T)\tau_1^P(H_0(\delta x) - T)\hat{\tau}_P(H_0(\delta x) - T) K_{P,\mathfrak{o}}(\delta x, \delta x).$$

For any $H \in \mathfrak{a}_0$, we can write

$$\tau_1^P(H)\hat{\tau}_P(H) = \sum_{\{P_2, P_3; P \subset P_2 \subset P_3\}} (-1)^{\dim(A_2/A_3)} \tau_1^3(H)\hat{\tau}_3(H)$$

by Proposition 1.5. In the notation of Section 6, this equals

$$\sum_{\{P_2;P_2\supset P\}} \sigma_1^2(H).$$

We have shown that $k_{\mathfrak{o}}^T(x,f)$ is the sum over $\{P_1,P,P_2;P_1\subset P\subset P_2\}$ and over $\delta\in P_1(F)\backslash G(F)$ of

$$(-1)^{\dim(A/Z)}F^1(\delta x,T)\sigma_1^2(H_0(\delta x)-T)K_{P,\mathfrak{o}}(\delta x,\delta x).$$

Therefore

$$\sum_{\mathfrak{o}\in\mathcal{O}}\int_{G(F)\backslash G(\mathbb{A})^1}|k_{\mathfrak{o}}^T(x,f)|dx \tag{7.1.1}$$

is bounded by the sum over $\{P_1,P_2;P_1\subset P_2\}$ and over $\mathfrak{o}\in\mathcal{O}$ of the integral over $P_1(F)\backslash G(\mathbb{A})^1$ of the product of

$$F^1(x,T)\sigma_1^2(H_0(x)-T) \tag{7.1.2}$$

with the absolute value of

$$\sum_{\{P;P_1\subset P\subset P_2\}}(-1)^{\dim(A/Z)}\sum_{\gamma\in M(F)\cap\mathfrak{o}}\int_{N(\mathbb{A})}f(x^{-1}\gamma nx)dn. \tag{7.1.3}$$

The critical part of this expression is the alternating sum over $P$. In order to exploit it, we shall show that the sum over $\gamma\in M(F)\cap\mathfrak{o}$ in (7.1.3) can be taken over a smaller set. Fix $P$, and suppose that $x\in P_1(F)\backslash G(\mathbb{A})^1$. We can assume that neither (7.1.2) nor

$$\sum_{\gamma\in M(F)\cap\mathfrak{o}}\int_{N(\mathbb{A})}f(x^{-1}\gamma nx)dn$$

vanishes. We want to show that the sum over $\gamma$ can be taken over the intersection of $\mathfrak{o}$ with the parabolic subgroup $P_1\cap M$ of $M$.

We choose a representative of $x$ in $G(\mathbb{A})^1$ of the form

$$n^*n_*mak,$$

where $k\in K$, $n^*$, $n_*$, and $m$ belong to fixed compact subsets of $N_2(\mathbb{A})$, $N_0^2(\mathbb{A})$, and $M_0(\mathbb{A})^1$, respectively, and $a\in A_0(\mathbb{R}^0)\cap G(\mathbb{A})^1$ has the property that

$$\alpha(H_0(a)-T_0)>0,\ \ \alpha\in\Delta_0^1, \tag{7.1.4}$$

and

$$\varpi(H_0(a) - T) \leq 0, \quad \varpi \in \widehat{\Delta}_0^1. \tag{7.1.5}$$

Here $T_0$ is as in Subsection 6.6. By Corollary 6.2, $\alpha(H_0(a) - T)$ is positive for any $\alpha \in \Delta_1^2$. It follows that the projection of $H_0(a) - T$ onto $\mathfrak{a}_0^2$ equals

$$\sum_{\varpi \in \widehat{\Delta}_1^2} c_\varpi \varpi^\vee - \sum_{\alpha \in \Delta_0^1} c_\alpha \alpha^\vee,$$

where each $c_\alpha$ and $c_\varpi$ is positive. Consequently,

$$\alpha(H_0(a) - T) > 0, \quad \alpha \in \Delta_0^2 \backslash \Delta_0^1. \tag{7.1.6}$$

A well-known result from reduction theory asserts that for any such $a$, the element $a^{-1}n_* ma$ belongs to a fixed compact subset of $N_0^2(\mathbb{A}) \times M_0(\mathbb{A})^1$ which is independent of $T$.

Suppose that the assertion we are trying to prove is false. Then there is a $\gamma$ in the quotient $M(F) \cap P_1(F)\backslash M(F)$ such that

$$\int_{N(\mathbb{A})} f(k^{-1}a^{-1}m^{-1}n_*^{-1}n^{*\,-1} \cdot \gamma n \cdot n^* n_* mak)dn$$

is not zero. This expression equals

$$\int_{N(\mathbb{A})} f(k^{-1}(a^{-1}mn_*a)^{-1} \cdot a^{-1}\gamma na \cdot (a^{-1}mn_*a)k)dn.$$

Therefore there is a compact subset of $G(\mathbb{A})^1$ which meets $a^{-1}\gamma N(\mathbb{A})a$. Thus $a^{-1}\gamma a$ belongs to a fixed compact subset of $M(\mathbb{A})^1$. According to the Bruhat decomposition for $M(F)$, we can write $\gamma = \nu w_s \pi$, for $\nu \in N_0^P(F)$, $\pi \in M(F) \cap P_0(F)$, and $s \in \Omega^M$, the Weyl group of $(M, A_0)$. The element $s$ cannot belong to the Weyl group of $(M_1, A_0)$, so we can find an element $\varpi \in \widehat{\Delta}_1^P$ not fixed by $s$.

Let $\Lambda$ be a rational representation of $M$ with highest weight $d\varpi$, $d > 0$. Let $v$ be a highest weight vector in $V(F)$, the space on which $G(F)$ acts. Choose a height function $\| \cdot \|$ relative to a basis of $V(F)$ as in § 1. We can assume that the basis includes the vectors $v$ and $\Lambda(w_s)v$. The component of $\Lambda(a^{-1}\nu w_s \pi a)v$ in the direction of $\Lambda(w_s)v$ is $e^{d(\varpi - s\varpi)(H_0(a))}\Lambda(w_s)v$. Therefore

$$\|\Lambda(a^{-1}\nu w_s \pi a)v\| \geq e^{d(\varpi - s\varpi)(H_0(a))}.$$

The left side of this inequality is bounded by a number which is independent of $T$. On the other hand, $\varpi - s\varpi$ is a nonnegative sum of roots in $\Delta_0^P$, and at least one element in $\Delta_0^P\backslash\Delta_0^1$ has a nonzero coefficient. It follows from (7.1.4) and (7.1.6) that the right hand side of the inequality can be made arbitrarily large by letting $T$ be sufficiently regular. This is a contradiction.

We have shown that for $T$ sufficiently regular, the absolute value of (7.1.3) equals

$$\left| \sum_{\{P;P_1\subset P\subset P_2\}} (-1)^{\dim(A/Z)} \sum_{\gamma\in P_1(F)\cap M(F)\cap\mathfrak{o}} \int_{N(\mathbb{A})} f(x^{-1}\gamma nx)dn \right|.$$

According to the remark following Proposition 2.6,

$$P_1(F)\cap M(F)\cap\mathfrak{o} = (M_1(F)\cap\mathfrak{o})N_1^P(F),$$

so the absolute value of (7.1.3) is bounded by the sum over $\gamma\in M_1(F)\cap\mathfrak{o}$ of

$$\left| \sum_{\{P;P_1\subset P\subset P_2\}} (-1)^{\dim(A/Z)} \sum_{\nu\in N_1^P(F)} \int_{N(\mathbb{A})} f(x^{-1}\gamma\nu nx)dn \right|.$$

We would like to replace the sum over $\nu$ by a sum over the rational points of $\mathfrak{n}_1^P$, the Lie algebra of $N_1^P$. Let

$$e:\mathfrak{n}_0\to N_0$$

be an isomorphism, defined over $F$, from the Lie algebra of $N_0$ onto $N_0$ which intertwines the action of $A_0$. The last expression equals

$$\left| \sum_{\{P;P_1\subset P\subset P_2\}} (-1)^{\dim(A/Z)} \sum_{\zeta\in\mathfrak{n}(F)} \int_{\mathfrak{n}(\mathbb{A})} f(x^{-1}\gamma e(\zeta+X)x)dX \right|.$$

Apply the Poisson summation formula to the sum over $\zeta$. Let $\langle\cdot,\cdot\rangle$ be a positive definite bilinear form on $\mathfrak{n}_0(F)\times\mathfrak{n}_0(F)$ for which the action of $A_0(F)$ is self-adjoint. Let $\psi$ be a nontrivial character on $\mathbb{A}/F$. We obtain

$$\left| \sum_{\{P;P_1\subset P\subset P_2\}} (-1)^{\dim(A/Z)} \sum_{\zeta\in\mathfrak{n}_1^P(F)} \int_{\mathfrak{n}_1(\mathbb{A})} f(x^{-1}\gamma e(X)x)\psi(\langle X,\zeta\rangle)dX \right|.$$

If $\mathfrak{n}_1^2(F)'$ is the set of elements in $\mathfrak{n}_1^2(F)$ which do not belong to any $\mathfrak{n}_1^P(F)$ with $P_1\subset P\subsetneqq P_2$, this equals

$$\left| \sum_{\zeta\in\mathfrak{n}_1^2(F)'} \int_{\mathfrak{n}_1(\mathbb{A})} f(x^{-1}\gamma e(X)x)\psi(\langle X,\zeta\rangle)dX \right|,$$

by Proposition 1.5. We have shown that (7.1.1) is bounded by the sum over $\{P_1, P_2; P_1 \subset P_2\}$ of the integral over $x$ in $P_1(F)\backslash G(\mathbb{A})^1$ of

$$F^1(x,T)\sigma_1^2(H(x)-T) \sum_{\gamma \in M_1(F)} \sum_{\zeta \in \mathfrak{n}_1^2(F)'} \left| \int_{\mathfrak{n}_1(\mathbb{A})} f(x^{-1}\gamma e(X)x)\psi(\langle X, \zeta \rangle)dX \right|.$$

Set

$$x = n^* n_* mak,$$

where $k \in K$, $a \in A_1(\mathbb{R}^0) \cap G(\mathbb{A})^1$, and $n^*$, $n_*$, and $m$ lie in fixed fundamental domains in $N_2(\mathbb{A})$, $N_1^2(\mathbb{A})$, and $M_1(\mathbb{A})^1$, respectively. Of course, this change of variables will add a factor of $e^{-2\rho_1(H_0(a))}$ to the integrand. The element $n^*$ is absorbed in the integral over $X$. We need only consider points for which the integrand does not vanish. Therefore $m$ and $a^{-1}n_*a$ both remain in fixed compact sets. Thus (7.1.1) is bounded by a scalar multiple of the quantity obtained by taking the sum over $P_1 \subset P_2$ and $\gamma \in M_1(F)$, the supremum as $y$ ranges over a fixed compact subset of $G(\mathbb{A})^1$, and the integral over $a$ in $A_1(\mathbb{R}^0) \cap G(\mathbb{A})^1$ of the expression

$$e^{-2\rho_1(H_0(a))}\sigma_1^2(H_0(a)-T) \cdot \sum_{\zeta \in \mathfrak{n}_1^2(F)'} \left| \int_{\mathfrak{n}_1(\mathbb{A})} f(y^{-1}a^{-1}\gamma e(X)ay)\psi(\langle X, \zeta \rangle)dX \right|.$$

This expression is the same as

$$\sigma_1^2(H_0(a)-T) \sum_{\zeta \in \mathfrak{n}_1^2(F)'} \left| \int_{\mathfrak{n}_1(\mathbb{A})} f(y^{-1}\gamma e(X)y)\psi(\langle X, \mathrm{Ad}(a)\zeta \rangle)dX \right|. \qquad (7.1.7)$$

The sum over $\gamma$ is finite. Our only remaining worry is the integral over $a$.

Let

$$\mathfrak{n}_1^2 = \oplus_\lambda \mathfrak{n}_\lambda$$

be the decomposition of $\mathfrak{n}_1^2$ into eigenspaces under the action of $A_1^2$. Each $\lambda$ stands for a linear function on $\mathfrak{a}_1^2$. Choose a basis of $\mathfrak{n}_1^2(F)$ such that each basis element lies in some $\mathfrak{n}_\lambda(F)$. The basis gives us a Euclidean norm on $\mathfrak{n}_1^2(F_\infty)$. It also allows us to speak of $\mathfrak{n}_1^2(\mathcal{O})$ and $\mathfrak{n}_\lambda(\mathcal{O})$. Note that

$$\zeta \longmapsto \int_{\mathfrak{n}_1(\mathbb{A})} f(y^{-1}\gamma e(X)y)\psi(\langle X, \mathrm{Ad}(a)\zeta \rangle)dX, \quad y \in \mathfrak{n}_Q^1(\mathbb{A}),$$

is the Fourier transform of a Schwartz-Bruhat function on $\mathfrak{n}_1^2(\mathbb{A})$ which varies smoothly with $y$. Therefore we can reduce the integral over $X$ in (7.1.7) to a finite sum of integrals over a real vector space. It follows without difficulty that for every

$n$ we can choose $N$ such that the supremum over $y$ of (7.1.7) is bounded by a scalar multiple of

$$\sigma_1^2(H_0(a) - T) \cdot \sum_{\zeta \in \mathfrak{n}_1^2(\frac{1}{N}\mathcal{O})'} \| \mathrm{Ad}(a)\zeta \|^{-n}.$$

Every $\lambda$ is a unique integral linear combination of elements in $\Delta_1^2$. Suppose that $S$ is a subset of roots $\lambda$ with the property that for any $\alpha \in \Delta_1^2$, there is a $\lambda$ in $S$ whose $\alpha$ coordinate is positive. Let $\mathfrak{n}_S(F)'$ be the set of elements in $\mathfrak{n}_1^2(F)$ whose projections onto $\mathfrak{n}_\lambda$ are nonzero if $\lambda$ belongs to $S$ and are zero otherwise. Then the above sum over $\mathfrak{n}_1^2(\frac{1}{N}\mathcal{O})'$ can be replaced by the double sum over all such $S$ and over $\zeta$ in $\mathfrak{n}_S(\frac{1}{N}\mathcal{O})'$. Clearly

$$\sigma_1^2(H_0(a) - T) \sum_{\zeta \in \mathfrak{n}_S(\frac{1}{N}\mathcal{O})'} \| \mathrm{Ad}(a)\zeta \|^{-n}$$

is bounded by

$$\sigma_1^2(H_0(a) - T) \cdot \prod_{\lambda \in S} \sum_{\zeta \in \mathfrak{n}_\lambda(\frac{1}{N}\mathcal{O})'} \| \mathrm{Ad}(a)\zeta \|^{-n_S},$$

where $\mathfrak{n}_\lambda(\frac{1}{N}\mathcal{O})'$ is the set of nonzero elements in $\mathfrak{n}_\lambda(\frac{1}{N}\mathcal{O})$ and $n_S$ is the quotient of $n$ by the number of roots in $S$. This last expression equals the product of

$$\prod_{\lambda \in S} \sum_{\zeta \in \mathfrak{n}_\lambda(\frac{1}{N}\mathcal{O})'} \|\zeta\|^{-n_S}$$

and

$$\sigma_1^2(H_0(a) - T) \cdot \prod_{\lambda \in S} e^{-n_S \lambda(H_0(a))}.$$

The first factor is finite for large enough $n$. The second factor equals

$$\sigma_1^2(H_0(a) - T) \prod_{\alpha \in \Delta_1^2} e^{-k_\alpha \alpha(H_0(a))}, \tag{7.1.8}$$

where each $k_\alpha$ is a positive real number.

The projection of $H_0(a) - T$ onto $\mathfrak{a}_1^G$ can be written

$$\sum_{\alpha \in \Delta_1^2} t_\alpha \varpi_\alpha^\vee + H^*,$$

where $H^* \in \mathfrak{a}_2^G$, and for each $\alpha \in \Delta_1^2$, the coefficient $t_\alpha$ is a positive real number. If $\sigma_1^2(H_0(a) - T) \neq 0$, it follows from Corollary 6.2 that $H^*$ belongs to a compact set whose volume is bounded by a polynomial in the numbers $\{t_\alpha\}$. Therefore there is an $N$ such that the integral of (7.1.8) is bounded by a multiple of

$$\prod_{\alpha \in \Delta_1^2} \int_0^\infty (1 + |t_\alpha|)^N e^{-k_\alpha t_\alpha} \, dt_\alpha.$$

This last expression is finite. The proof of Theorem 7.1 is complete. □

# 8 Weighted Orbital Integrals

DEFINITION 8.1. (1) For any $\mathfrak{o} \in \mathcal{O}$, set

$$J_{\mathfrak{o}}^T(f) = \int_{G(F)\backslash G(\mathbb{A})^1} k_{\mathfrak{o}}^T(x,f) dx.$$

In (3) below we define another function $j_{\mathfrak{o}}^T(x,f)$. We shall show in Theorem 8.2(2) that its integral over $x$ is equal to $J_{\mathfrak{o}}^T(f)$. Then we will, in some cases, reduce the integral of $j^T(x,f)$ to a weighted orbital integral of $f$.

(2) Given $\mathfrak{o}$ and $P$, define the function

$$J_{P,\mathfrak{o}}(x,y) = \sum_{\gamma \in M(F) \cap \mathfrak{o}} \; \sum_{\eta \in N(F,\gamma_s)\backslash N(F)} \int_{N(\mathbb{A},\gamma_s)} f(x^{-1}\eta^{-1}\gamma n \eta x) dn.$$

It is obtained from $K_{P,\mathfrak{o}}(x,y)$ by replacing part of the integral over $N(\mathbb{A})$ by the corresponding sum over $F$-rational points.

(3) Define

$$j_{\mathfrak{o}}^T(x,f) = \sum_P (-1)^{\dim(A/Z)} \sum_{\delta \in P(F)\backslash G(F)} J_{P,\mathfrak{o}}(\delta x, \delta x) \cdot \hat{\tau}_P(H(\delta x) - T).$$

THEOREM 8.2. (1) *For all sufficiently regular $T$, we have that*

$$\sum_{\mathfrak{o} \in \mathcal{O}} \int_{G(F)\backslash G(\mathbb{A})^1} |j_{\mathfrak{o}}^T(x,f)| dx$$

*is finite.*

(2) *Moreover, for any $\mathfrak{o}$,*

$$\int_{G(F)\backslash G(\mathbb{A})^1} j_{\mathfrak{o}}^T(x,f) dx = J_{\mathfrak{o}}^T(f).$$

PROOF. (1) The argument used to prove the first statement is essentially the same as the proof of Theorem 7.1. The integral

$$\sum_{\mathfrak{o}\in\mathcal{O}} \int_{G(F)\backslash G(\mathbb{A})^1} |j_{\mathfrak{o}}^T(x,f)|dx \tag{8.2.1}$$

is bounded by the sum over $\{P_1, P_2; P_1 \subset P_2\}$ and $\mathfrak{o} \in \mathcal{O}$ of the integral over $P_1(F)\backslash G(\mathbb{A})^1$ of the product of

$$F^1(x,T)\sigma_1^2(H_0(x)-T) \tag{8.2.2}$$

with the absolute value of

$$\sum_{\{P;P_1\subset P\subset P_2\}} (-1)^{\dim(A/Z)} \sum_{\gamma\in M(F)\cap\mathfrak{o}} \sum_{\eta\in N(F,\gamma_s)\backslash N(F)} \int_{N(\mathbb{A},\gamma_s)} f(x^{-1}\eta^{-1}\gamma n\eta x)dn. \tag{8.2.3}$$

As in the proof of Theorem 7.1, we observe that for $T$ sufficiently regular, the sum over $\gamma$ may be taken over $P_1(F) \cap M(F) \cap \mathfrak{o}$.

It follows from Proposition 2.6 and Poisson summation that the expression in the brackets in (8.2.3) equals the sum over $\gamma$ in $M_1(F) \cap \mathfrak{o}$ of

$$\sum_{\nu\in N_1^P(F)} \sum_{\eta\in N(F,(\gamma\nu)_s)\backslash N(F)} \int_{N(\mathbb{A},(\gamma\nu)_s)} f(x^{-1}\eta^{-1}\zeta\nu n\eta x)dn$$

$$= \sum_{\delta\in N_1^P(F,\gamma_s)\backslash N_1^P(F)} \sum_{\nu\in N_1^P(F,\gamma_s)} \sum_{\eta\in N(F,\delta^{-1}\gamma_s\delta)\backslash N(F)}$$

$$\int_{N(\mathbb{A},\delta^{-1}\gamma_s\delta)} f(x^{-1}\eta^{-1}\delta^{-1}\gamma\nu\delta n\eta x)dn$$

$$= \sum_{\eta\in N_1(F,\gamma_s)\backslash N_1(F)} \sum_{\nu\in N_1^P(F,\zeta_s)} \int_{N(\mathbb{A},\zeta_s)} f(x^{-1}\eta^{-1}\gamma n\eta x)dn$$

$$= \sum_{\eta\in N_1(F,\gamma_s)\backslash N_1(F)} \sum_{\zeta\in\mathfrak{n}_1^P(F,\gamma_s)} \int_{\mathfrak{n}_1(\mathbb{A},\gamma_s)} f(x^{-1}\eta^{-1}\cdot\gamma e(X)\cdot\eta x)\psi(\langle X,\zeta\rangle)dX.$$

Let $\mathfrak{n}_1^2(F,\gamma_s)'$ be the set, possibly empty, of elements in $\mathfrak{n}_1^2(F,\gamma_s)$ which do not belong to any $\mathfrak{n}_1^P(F,\gamma_s)$, with $P_1 \subset P \subsetneqq P_2$. Then (8.2.1) is bounded by the sum over $\{P_1, P_2; P_1 \subset P_2\}$ of the integral over $x$ in $M_1(F)N_1(\mathbb{A})\backslash G(\mathbb{A})^1$ of

$$F^1(x,T)\sigma_1^2(H_0(x)-T)\cdot \sum_{\gamma\in M_1(F)} \int_{N_1(F,\gamma_s)\backslash N_1(\mathbb{A})} dn$$

$$\cdot \sum_{\zeta\in\mathfrak{n}_1^2(F,\gamma_s)'} \left| \int_{\mathfrak{n}_1(\mathbb{A},\gamma_s)} f(x^{-1}n^{-1}\gamma e(X)nx)\psi(\langle X,\zeta\rangle)dX \right|.$$

The integrand, as a function of $n$, is left $N_2(\mathbb{A}, \gamma_s)$ invariant. We can therefore write $n = n_2 n_1$ where $n_2$ ranges over a relatively compact fundamental set of $N_1^2(F, \gamma_s)$ in $N_1^2(\mathbb{A}, \gamma_s)$ and $n_1$ belongs to $N_1(\mathbb{A}, \gamma_s)\backslash N_1(\mathbb{A})$. For any $\gamma$ the integrand will vanish for $n_1$ outside a compact set.

Next, set $x = mak$, where $k \in K$, $a \in A_1(\mathbb{R}^0) \cap G(\mathbb{A})^1$, and $m$ lies in a fixed fundamental domain in $M_1(\mathbb{A})^1$. Note that $a$ normalizes $N_1(\mathbb{A}, \gamma_s)\backslash N_1(\mathbb{A})$. If the integrand does not vanish, $m$ and $a^{-1}n_2 a$ will both remain in fixed compact sets. Moreover, the sum over $\gamma$ will be finite. It follows that (8.2.1) is bounded by a scalar multiple of the quantity obtained by taking the sum over $P_1$, $P_2$, and $\gamma$, the supremum as $y$ ranges over a fixed compact subset of $G(\mathbb{A})^1$ and the integral over $a \in A_1(\mathbb{R}^0) \cap G(\mathbb{A})^1$ of the expression

$$\sigma_1^2(H_0(x) - T) \cdot \sum_{\zeta \in \mathfrak{n}_1^2(F,\gamma_s)'} \left| \int_{\mathfrak{n}_1(\mathbb{A},\gamma_s)} f(y^{-1}\gamma e(X) y)\psi(\langle X, \mathrm{Ad}(a)\zeta\rangle) dX \right| .$$

The finitude of (8.2.1) now follows from the arguments of Section 7.

(2) Fix $\mathfrak{o}$. The integral of $j_{\mathfrak{o}}^T(x,f)$ is the sum over $P_1$ and $P_2$ of the integral over $P_1(F)\backslash G(\mathbb{A})^1$ of the product of (8.2.2) and (8.2.3). Decompose the integral over $P_1(F)\backslash G(\mathbb{A})^1$ into a double integral over $M_1(F)N_1(\mathbb{A})\backslash G(\mathbb{A})^1$ and $N_1(F)\backslash N_1(\mathbb{A})$. Then take the integral over $N_1(F)\backslash N_1(\mathbb{A})$ inside the sum over $P$ and $\gamma$ in (8.2.3). The summand is then

$$\begin{aligned}
&\int_{N_1(F)\backslash N_1(\mathbb{A})} dn_1 \cdot \sum_{\eta \in N(F,\gamma_s)\backslash N(F)} \int_{N(\mathbb{A},\gamma_s)} dn \cdot f(x^{-1}n_1^{-1}\eta^{-1} \cdot \gamma n \cdot \eta n_1 x) \\
&= \int_{N_1(F)\backslash N_1(\mathbb{A})} dn_1 \cdot \int_{N(F)\backslash N(\mathbb{A})} dn_2 \\
&\quad \cdot \sum_{\eta \in N(F,\gamma_s)\backslash N(F)N(\mathbb{A},\gamma_s)} \int_{N(\mathbb{A},\gamma_s)} dn \cdot f(x^{-1}n_1^{-1}n_2^{-1}\eta^{-1} \cdot \gamma n \cdot \eta n_2 n_1 x) \\
&= \int_{N_1(F)\backslash N_1(\mathbb{A})} dn_1 \cdot \int_{N(\mathbb{A},\gamma_s)\backslash N(\mathbb{A})} dn_2 \cdot \int_{N(\mathbb{A},\gamma_s)} dn \cdot f(x^{-1}n_1^{-1} \cdot n_2^{-1}\gamma n n_2 \cdot n_1 x) \\
&= \int_{N_1(F)\backslash N_1(\mathbb{A})} dn_1 \cdot \int_{N(\mathbb{A})} dn \cdot f(x^{-1}n_1^{-1} \cdot \gamma n \cdot n_1 x),
\end{aligned}$$

by Proposition 2.7. The integral over $N_1(F)\backslash N_1(\mathbb{A})$ can now be taken back outside the sum over $\gamma$ and $P$ and recombined with the integral over $M_1(F)N_1(\mathbb{A})\backslash G(\mathbb{A})^1$. We must of course remember that (8.2.2) is a left $N_1(\mathbb{A})$-invariant function of $x$. We end up with the sum over $P_1$ and $P_2$ of the integral over $P_1(F)\backslash G(\mathbb{A})^1$ of the product of (7.1.2) and (7.1.3). This is just the integral of $k_{\mathfrak{o}}^T(x,f)$. The theorem is proved.

□

DEFINITION 8.3. Let $\mathfrak{o}$ be a fixed equivalence class. Choose a semisimple element $\gamma_1 \in \mathfrak{o}$ and groups $P_1 \subset P_2$ as in § 2. Assume that $P_1 = P_2$, so that $\mathfrak{o}$ consists entirely of semisimple elements. Any element in $G(\gamma_1)$ normalizes $A_1$, since it is the split component of the center of $G(\gamma_1)^0$. We obtain a map from $G(\gamma_1)$ to $\Omega(\mathfrak{a}_1, \mathfrak{a}_1)$, whose kernel is $G(\gamma_1) \cap M_1$. We shall say that the class $\mathfrak{o}$ is *unramified* if the map is trivial; in other words if $G(\gamma_1)$ is contained in $M_1$. It is clear that $\mathfrak{o}$ is unramified if and only if the only $s$ in $\Omega(\mathfrak{a}_1, \mathfrak{a}_1)$ for which $w_s \gamma_1 w_s^{-1}$ is $M_1(F)$-conjugate to $\gamma_1$ is the identity.

PROPOSITION 8.4. *When $\mathfrak{o}$ is unramified, the integral $J_{\mathfrak{o}}^T(f)$ is a weighted orbital integral of $f$ given by*

$$\mathrm{vol}(A_1(\mathbb{R}^0)G(F,\gamma_1)\backslash G(\mathbb{A},\gamma_1)) \int_{G(\mathbb{A},\gamma_1)\backslash G(\mathbb{A})} f(x^{-1}\gamma_1 x) v(x,T) dx, \tag{8.4.1}$$

*where*

$$v(x,T) = \int_{Z(\mathbb{R}^0)\backslash A_1(\mathbb{R}^0)} \left( \sum_{P_2} \sum_{s \in \Omega(\mathfrak{a}_1,\mathfrak{a}_2)} \epsilon_2(s\lambda)\phi_2(s\lambda, H_0(w_s a x) - T) \right) da.$$

*The weight factor $v(x,T)$ equals the volume of the convex hull of the projection onto $\mathfrak{a}_1/\mathfrak{z}$ of*

$$\{s^{-1}T - s^{-1}H_0(w_s x);\ s \in \bigcup_{P_2} \Omega(\mathfrak{a}_1, \mathfrak{a}_2).\}$$

The proof of this proposition ends at 8.8. For the last claim of the proposition, see [Ar76], Corollary 3.5.

## 8.5 A Weyl Group

Let $\mathfrak{o}$ be an unramified class in $\mathcal{O}$. Let $\gamma_1$ and $P_1$ be as in 8.3. Since it consists entirely of semisimple elements, $\mathfrak{o}$ is an actual conjugacy class in $G(F)$. Suppose that $P$ is a parabolic subgroup and that $\gamma$ belongs to $M(F) \cap \mathfrak{o}$. Then there is a parabolic subgroup $P_2 \subset P$, and an element $\gamma_2 \in M_2(F)$, which is $M(F)$-conjugate to $\gamma$, such that the split component of the center of $G(\gamma_2)^0$ is $A_2$. Any element in $G(F)$ which conjugates $\gamma_1$ to $\gamma_2$ will conjugate $A_1$ to $A_2$. It follows that for some $s \in \Omega(\mathfrak{a}_1, \mathfrak{a}_2)$, and $\eta \in M(F)$,

$$\gamma = \eta w_s \gamma_1 w_s^{-1} \eta^{-1}. \tag{8.5.1}$$

Suppose that for a parabolic subgroup $P_3 \subset P$, and elements $s' \in \Omega(\mathfrak{a}_1, \mathfrak{a}_3)$ and $\eta' \in M(F)$, we have

$$\gamma = \eta' w_{s'} \gamma_1 w_{s'}^{-1} (\eta')^{-1}.$$

Then there is an element $\zeta \in G(F, \gamma)$ such that

$$\eta' w_{s'} = \zeta \eta w_s.$$

Since $G(\gamma) \subset M$, we have $w_{s'} = \xi w_s$ for some element $\xi \in M(F)$.

DEFINITION. Let $\Omega(\mathfrak{a}_1; P)$ be the set of elements $s$ in the union over $\mathfrak{a}_2$ of the sets $\Omega(\mathfrak{a}_1, \mathfrak{a}_2)$ such that $s\mathfrak{a}_1 = \mathfrak{a}_2$ contains $\mathfrak{a}$ and $s^{-1}\alpha$ is positive for each $\alpha \in \Delta_2^P$.

Given $\gamma$ and $P$, there is a unique $s$ in $\Omega(\mathfrak{a}_1; P)$ such that (8.5.1) holds for some $\eta \in M(F)$.

## 8.6 Rewriting a Sum

It follows from 8.5 that $J_{P,\mathfrak{o}}(y, y)$ equals

$$\sum_{s\in\Omega(\mathfrak{a}_1;P)} \sum_{\eta\in M(F,w_s\gamma_1 w_s^{-1})\backslash M(F)} \sum_{\zeta\in N(F)} f(y^{-1}\zeta^{-1}\eta^{-1}w_s\gamma_1 w_s^{-1}\eta\zeta y)$$

$$= \sum_{s} \sum_{\zeta\in M(F,w_s\gamma_1 w_s^{-1})\backslash P(F)} f(y^{-1}\zeta^{-1}w_s\gamma_1 w_s^{-1}\zeta y).$$

Therefore $j_{\mathfrak{o}}^T(x,f)$ equals the sum over $P$ of $(-1)^{\dim(A/Z)}$ times

$$\sum_{s\in\Omega(\mathfrak{a}_1;P)} \sum_{\delta\in M(F,w_s\gamma_1 w_s^{-1})\backslash G(F)} f(x^{-1}\delta^{-1}w_s\gamma_1 w_s^{-1}\delta x)\hat{\tau}_P(H_0(\delta x) - T).$$

Since the centralizer of $w_s\gamma_1 w_s^{-1}$ in $G$ is contained in $M$, this equals the sum over $\delta \in G(F, \gamma_1)\backslash G(F)$ of the product of

$$f(x^{-1}\delta^{-1}\gamma_1\delta x)$$

and

$$\sum_{P}(-1)^{\dim(A/Z)} \sum_{s\in\Omega(\mathfrak{a}_1;P)} \hat{\tau}_P(H_0(w_s\delta x) - T). \tag{8.6.1}$$

PROPOSITION 8.7. *Suppose that $\lambda$ is a point in $\mathfrak{a}_0^*$ such that $\lambda(\alpha^\vee) > 0$ for each $\alpha \in \Delta_0$. Then (8.6.1) equals*

$$\sum_{P_2} \sum_{s\in\Omega(\mathfrak{a}_1,\mathfrak{a}_2)} \epsilon_2(s\lambda)\phi_2(s\lambda, H_0(w_s\delta x) - T), \tag{8.7.1}$$

*where we have written $\epsilon_2$ and $\phi_2$ for the functions denoted by $\epsilon_{P_2}^G$ and $\phi_{P_2}^G$ in Section 6.*

PROOF. To see this, write (8.6.1) as the sum over $P_2$, over $s \in \Omega(\mathfrak{a}_1, \mathfrak{a}_2)$ and over those $P \supset P_2$ such that $s^{-1}\alpha > 0$ for all $\alpha \in \Delta_2^P$, of

$$(-1)^{\dim(A/Z)}\hat{\tau}_P(H_0(w_s\delta x) - T).$$

For a given $s$, define $P_s \supset P$ by

$$\Delta_{P_s}^{P_1} = \{\alpha \in \Delta_{P_2};\ (s\lambda)(\alpha^\vee) > 0\}.$$

Then the sum over $P$ is just the sum over $\{P;\ P_2 \subset P \subset P_s\}$. Since it is an alternating sum of characteristic functions, we can apply Proposition 1.5. The sum over $P$ will vanish, for a given $s$, unless precisely one summand is nonzero. We have shown that for all $H \in \mathfrak{a}_0$, the sum

$$\sum_{\{P \supset P_2; s^{-1}\alpha > 0, \alpha \in \Delta_2^P\}} (-1)^{\dim(A/Z)}\hat{\tau}_P(H)$$

equals the product of $(-1)^{\dim(A_s/Z)}$ with the characteristic function of

$$\{H \in \mathfrak{a}_2;\ \varpi(H) > 0, \varpi \in \widehat{\Delta}_s;\ \varpi(H) \leq 0, \varpi \in \widehat{\Delta}_2 \backslash \widehat{\Delta}_s\}.$$

This is just the function

$$\epsilon_2(s\lambda)\phi_2(s\lambda, H).$$

We have shown that (8.6.1) equals (8.7.1). □

## 8.8 *Proof of Proposition* 8.4

Substitute (8.7.1) for (8.6.1) in the expression for $j_{\mathfrak{o}}^T(x,f)$. To get $J_{\mathfrak{o}}^T(f)$, we must integrate over $G(F)\backslash G(\mathbb{A})^1$. Since the integrand is left $Z(\mathbb{R}^0)$-invariant, we can integrate over $Z(\mathbb{R}^0)G(F)\backslash G(\mathbb{A})$ instead. We could then combine the integral over $x$ and the sum over $\delta$ if we were able to prove that the resulting integral was absolutely convergent. But the resulting integral can be written as (8.4.1). The integral over $x$ can be taken over a compact set. By [Ar76], Corollary 3.3, the integral over $a$ can also be taken over $a$ compact set. It follows that $J_{\mathfrak{o}}^T(f)$ equals (8.4.1). □

# 9 A Truncation Operator

## 9.1 Definition of a Truncation Operator

Let $G$ be a reductive connected algebraic group defined over a number field $F$. Let $K$ be a maximal compact subgroup of $G(\mathbb{A})$. Let $P_0$ be a fixed minimal parabolic subgroup of $G$ defined over $F$. We use the term "parabolic subgroup" for a standard one, namely, a parabolic subgroup $P$ of $G$, defined over $F$, which contains $P_0$. We would like to prove that the terms on the right hand side of the identity given in Proposition 5.6 are integrable functions of $x$. To this end, we shall introduce a truncation operator for functions on $G(F)\backslash G(\mathbb{A})^1$.

DEFINITION. Let $T$ be a fixed, suitably regular point in $\mathfrak{a}_0^+$. If $\phi$ is a continuous function on $G(F)\backslash G(\mathbb{A})^1$, define $(\Lambda^T\phi)(x)$ to be the function

$$\sum_P (-1)^{\dim(A/Z)} \sum_{\delta \in P(F)\backslash G(F)} \int_{N(F)\backslash N(\mathbb{A})} \phi(n\delta x)dn \cdot \hat{\tau}_P(H(\delta x) - T).$$

The sum over $P$ is over all standard parabolic subgroups.

Note the similarity with the definitions of the functions $k_{\mathfrak{o}}^T(x,f)$ and $k_\chi^T(x,f)$ in 5.5. If $\phi$ is a cusp form, $\Lambda^T\phi = \phi$. It is a consequence of Corollary 5.4 that if $\phi(x)$ is *slowly increasing*, in the sense that

$$|\phi(x)| \leq C\|x\|^N,$$

for some $C$ and $N$, then so is $\Lambda^T\phi(x)$.

PROPOSITION 9.2. *Fix a parabolic subgroup $P_1$. Then for $\phi \in C(G(F)\backslash G(\mathbb{A})^1)$, we have*

$$\int_{N_1(F)\backslash N_1(\mathbb{A})} \Lambda^T\phi(n_1x)dn_1 = 0$$

*unless $\varpi(H_0(x) - T) < 0$ for each $\varpi \in \widehat{\Delta}_1$.*

DEFINITION. For any $P$, let $\Omega(\mathfrak{a}_0; P)$ be the set of $s \in \Omega$ such that $s^{-1}\alpha > 0$ for each $\alpha \in \Delta_0^P$.

To prove the proposition, we first show

**Lemma 9.2.1.** *The integral $\int_{N_1(F)\backslash N_1(\mathbb{A})} \phi(n_1x)dn_1$ equals*

$$\sum_P (-1)^{\dim(A/Z)} \sum_{s\in\Omega(\mathfrak{a}_0;P)} \int_{w_s^{-1}N_0(\mathbb{A})w_s \cap N_0^1(F)N_1(\mathbb{A})\backslash N_0^1(F)N_1(\mathbb{A})} X dn^*$$

*where*

$$X = \int_{N_s(F)\backslash N_s(\mathbb{A})} \phi(nw_s n^* x)\hat{\tau}_P(H(w_s n^* x) - T)dn,$$

*and*

$$N_s = (N_0 \cap w_s N_1 w_s^{-1} \cap M)N$$

*is the unipotent radical of a uniquely determined parabolic subgroup $P_s$ of $G$ that is contained in $P$.*

PROOF. Applying the Bruhat decomposition to $P(F)\backslash G(F)$, we find that

$$\int_{N_1(F)\backslash N_1(\mathbb{A})} \Lambda^T \phi(n_1 x)dn_1$$

equals the sum over $P$ and $s \in \Omega(\mathfrak{a}_0; P)$ of the integral over $n$ in $N(F)\backslash N(\mathbb{A})$ of the product of $(-1)^{\dim(A/Z)}$ with

$$\int_{N_1(F)\backslash N_1(\mathbb{A})} \sum_{\nu \in w_s^{-1}N_0(F)w_s \cap N_0(F)\backslash N_0(F)} \phi(nw_s \nu n_1 x) \cdot \hat{\tau}_P(H(w_s \nu n_1 x) - T)dn_1.$$

Since $N_1(F)\backslash N_1(\mathbb{A}) = N_0(F)\backslash N_0^1(F)N_1(\mathbb{A})$, this last expression equals

$$\int_{w_s^{-1}N_0(F)w_s \cap N_0(F)\backslash N_0^1(F)N_1(\mathbb{A})} \phi(nw_s n_1 x)\hat{\tau}_P(H(w_s n_1 x) - T)dn_1.$$

Decompose $w_s^{-1}N_0(F)w_s \cap N_0(F)\backslash N_0^1(F)N_1(\mathbb{A})$ as

$$\begin{aligned}
&(w_s^{-1}N_0(F)w_s \cap N_0(F)\backslash w_s^{-1}N_0(\mathbb{A})w_s \cap N_0^1(F)N_1(\mathbb{A})) \\
&\times(w_s^{-1}N_0(\mathbb{A})w_s \cap N_0^1(F)N_1(\mathbb{A})\backslash N_0^1(F)N_1(\mathbb{A})) \\
= &(w_s^{-1}N_0(F)w_s \cap N_1(F)\backslash w_s^{-1}N_0(\mathbb{A})w_s \cap N_1(\mathbb{A})) \\
&\times(w_s^{-1}N_0(\mathbb{A})w_s \cap N_0^1(F)N_1(\mathbb{A})\backslash N_0^1(F)N_1(\mathbb{A})).
\end{aligned}$$

This induces a decomposition of the measure $dn_1$ as $dn_* dn^*$. Then write

$$w_s n_* n^* = w_s n_* w_s^{-1} w_s n^* = \widetilde{n}_* w_s n^*.$$

Finally, combine the integral over $\tilde{n}_*$ with the integral over $n$ in $N(F)\backslash N(\mathbb{A})$. Because $s$ lies in $\Omega(\mathfrak{a}_0; P)$, the group $N_0 \cap w_s N_1 w_s^{-1} \cap M$ is the unipotent radical of a standard parabolic subgroup of $M$. It follows that $N_s$ is as claimed and lemma follows. □

We shall change the order of summation in the integral of Lemma 9.2.1 and consider the set of $P$ which give rise to a fixed $P_s$. Fix $s \in \Omega$. Define $S^1$ (resp. $S_1$) to be the set of $\alpha \in \Delta_0$ such that $s^{-1}\alpha$ is a positive root which is orthogonal (resp. not orthogonal) to $\mathfrak{a}_1$. If $P_s$ is one of the groups that appear in the formula of Lemma 9.2.1, $\Delta_0^s$ will be a subset of $S^1$. Those $P$ which give rise to a fixed $P_s$ are exactly the groups for which $\Delta_0^P$ is the union of $\Delta_0^s$ and a subset $S$ of $S_1$. Thus for fixed $s$ with $\Delta_0^s \subset S^1$, we will obtain an alternating sum over $S \subset S_1$ of the corresponding functions $\hat{\tau}_P$.

DEFINITION. Let $\varpi_\alpha$ be the element in $\widehat{\Delta}_0$ corresponding to $\alpha$. Let $\chi_s$ be the characteristic function of the set of $H \in \mathfrak{a}_0$ such that for $\alpha \in \Delta_0 - \Delta_0^s \cup S_1$ we have $\varpi_\alpha(H) > 0$, while $\varpi_\alpha(H) \leq 0$ for $\alpha$ in $S_1$.

**Lemma 9.2.2.** *The integral $\int_{N_1(F)\backslash N_1(\mathbb{A})} \phi(n_1 x)dn_1$ is the sum over $s \in \Omega$ and over all subsets $\Delta_0^s$ of $S^1$, of the integral over $n^*$ in*

$$w_s^{-1}N_0(\mathbb{A})w_s \cap N_0^1(F)N_1(\mathbb{A})\backslash N_0^1(F)N_1(\mathbb{A})$$

*and $n$ in $N_s(F)\backslash N_s(\mathbb{A})$ of the product of*

$$\phi(nw_sn^*x)\chi_s(H_0(w_sn^*x) - T)$$

*with $-1$ raised to a power equal to the number of roots in $\Delta_0 - S^1 \cup S_1$.*

PROOF. Apply Proposition 1.5. □

PROOF OF PROPOSITION 9.2. Suppose that for some $s$, the number $\chi_s(H_0(w_sn^*x) - T)$ is not zero. Thus if

$$H_0(w_sn^*x) - T = \sum_{\alpha \in \Delta_0} t_\alpha^\vee, \quad t_\alpha \in \mathbb{R},$$

then $t_\alpha$ is positive for $\alpha$ in $\Delta_0 - \Delta_0^s \cup S_1$ and is not positive for $\alpha \in S_1$.

If $\varpi \in \widehat{\Delta}_1$, then

$$\varpi(s^{-1}(H_0(w_sn^*x) - T)) = \sum_{\alpha \in \Delta_0} t_\alpha \varpi(s^{-1}\alpha^\vee) = \sum_{\alpha \in \Delta_0 \backslash S^1} t_\alpha \varpi(s^{-1}\alpha^\vee),$$

where $s^{-1}\alpha$ is orthogonal to $\mathfrak{a}^1$ if $\alpha \in S^1$. This last number is clearly less than or equal to 0. Now

$$s^{-1}(H_0(w_sn^*x) - T) = H_0(x) - T + s^{-1}H_0(w_svw_s^{-1}) + (T - s^{-1}T),$$

for some element $v \in N_0(\mathbb{A})$. If $\varpi \in \widehat{\Delta}_0$, it is well known that $\varpi(s^{-1}H_0(w_svw_s^{-1}))$ is nonnegative and $\varpi(T - s^{-1}T)$ is strictly positive. Therefore $\varpi(H_0(x) - T)$ is negative for any $\varpi \in \widehat{\Delta}_1$. □

From the definition of $\Lambda^T$, we obtain

COROLLARY 9.3. *We have* $\Lambda^T \Lambda^T = \Lambda^T$. □

PROPOSITION 9.4. *Suppose* $\phi_1$ *and* $\phi_2$ *are continuous functions on* $G(F)\backslash G(\mathbb{A})^1$. *Assume that* $\phi_1$ *is slowly increasing (see 9.1), and that* $\phi_2$ *is rapidly decreasing, namely: for any positive integer N, the function* $\|x\|^N \cdot |\phi_2(x)|$ *is bounded on any Siegel set. Then in* $L^2(G(F)\backslash G(\mathbb{A})^1)$, *we have*

$$(\Lambda^T \phi_1, \phi_2) = (\phi_1, \Lambda^T \phi_2).$$

PROOF. The inner product $(\Lambda^T \phi_1, \phi_2)$ is defined by an absolutely convergent integral. It equals

$$\int_{G(F)\backslash G(\mathbb{A})^1} \sum_P (-1)^{\dim(A/Z)} \sum_{\delta \in P(F)\backslash G(F)}$$

$$\cdot \int_{N(F)\backslash N(\mathbb{A})} \phi_1(n\delta x) \hat{\tau}_P(H(\delta x) - T) \overline{\phi_2(x)} \, dn \, dx$$

$$= \sum_P (-1)^{\dim(A/Z)} \int_{N(F)\backslash N(\mathbb{A})} \int_{P(F)\backslash G(\mathbb{A})^1} \phi_1(nx) \overline{\phi_2(x)} \hat{\tau}_P(H(x) - T) dx \, dn$$

$$= \sum_P (-1)^{\dim(A/Z)} \int_{N(F)\backslash N(\mathbb{A})} \int_{P(F)\backslash G(\mathbb{A})^1} \phi_1(x) \overline{\phi_2(nx)} \hat{\tau}_P(H(x) - T) dx \, dn.$$

The last expression reduces to $(\phi_1, \Lambda^T \phi_2)$. □

REMARK. The operator $\Lambda^T$ extends to an orthogonal projection on the space $L^2(G(F)\backslash G(\mathbb{A})^1)$.

## 9.5 Integrability Theorem

We would like to show that under suitable conditions, $\Lambda^T \phi(x)$ is rapidly decreasing at infinity. More precisely, we proceed to show

THEOREM. *Let* $\mathcal{S}$ *be a Siegel set in* $G(\mathbb{A})^1$. *For any pair of positive integers* $N'$ *and N, and any open compact subgroup* $K_0$ *of* $G(\mathbb{A}_f)$, *we can choose a finite subset* $\{X_i\}$ *of* $\mathcal{U}(\mathfrak{g}(F_\infty)^1 \otimes \mathbb{C})$ *and a positive integer r which satisfy the following property. Suppose that* $(S, d\sigma)$ *is a measure space and that* $\phi(\sigma, x)$ *is a measurable function from S to* $C^m(G(F)\backslash G(\mathbb{A})^1/K_0)$. *Then for any* $x \in \mathcal{S}$,

$$\int_S |\Lambda^T \phi(\sigma, x)| d\sigma$$

*is bounded by*

$$\sum_i \sup_{y \in G(\mathbb{A})^1} \left( \int_S |r(X_i)\phi(\sigma, y)| d\sigma \cdot \|y\|^{-N} \right) \cdot \|x\|^{-N}.$$

The proof begins the same way as the proofs of Theorem 7.1 and 8.2. Suppose $\phi$ is a continuous function on $G(F)\backslash G(\mathbb{A})^1$. Apply Proposition 6.9 as in the beginning of the proof of Theorem 7.1. We find that $\Lambda^T\phi(x)$ is the sum over $\{P_1, P_2; P_0 \subset P_1 \subset P_2\}$ and $\delta \in P_1(F)\backslash G(F)$, of

$$F^1(\delta x, T)\sigma_1^2(H_0(\delta x) - T)\phi_{P_1,P_2}(\delta x),$$

where

$$\phi_{P_1,P_2}(y) = \sum_{\{P; P_1 \subset P \subset P_2\}} (-1)^{\dim(A/Z)} \int_{N(F)\backslash N(\mathbb{A})} \phi(ny)dn.$$

Our first aim is to show

**Lemma 9.5.1.** *The absolute value $|\Lambda^T\phi(x)|$ is bounded by the sum over all $P_1$, $P_2$, and $\delta \in P_1(F)\backslash G(F)$ of the product of*

$$F^1(\delta x, T)\sigma_1^2(H_0(\delta x) - T)$$

*with*

$$\sum_I \sum_i \int_{N_I(F)\backslash N_I(\mathbb{A})} |r(X_i)\phi(u\delta x)| du \cdot e^{-n\beta_I(H_0(\delta x))}. \tag{9.5.2}$$

PROOF. Fix $\delta$ and $x$. We regard $\delta$ as an element in $G(F)$ which we are free to left multiply by an element in $P_1(F)$. We can therefore assume, as in Section 7, that $\delta x = n^* n_* mak$, where $k \in K$, $n^*$, $n_*$, and $m$ belong to fixed compact subsets of $N_2(\mathbb{A})$, $N^2(\mathbb{A})$, and $M_1(\mathbb{A})^1$, respectively, and $a$ is an element in $A_1(\mathbb{R}^0)$ with $\sigma_1^2(H_0(a) - T) \neq 0$. Therefore

$$\begin{aligned} \phi_{P_1,P_2}(\delta x) &= \phi_{P_1,P_2}(n^* n_* mak) = \phi_{P_1,P_2}(n_* mak) \\ &= \phi_{P_1,P_2}(aa^{-1}n_* mak) = \phi_{P_1,P_2}(ac), \end{aligned}$$

where $c$ belongs to a fixed compact subset of $G(\mathbb{A})^1$ which depends only on $G$.

The function $\phi_{P_1,P_2}$ resembles the function estimated in the corollary of [HC68], Lemma 10. We want a slightly different statement of the estimate, however, so we had best re-examine the proof.

If $\alpha \in \Delta_1^2$, let $P_\alpha$, $P_1 \subset P_\alpha \subset P_2$, be the parabolic subgroup such that $\Delta_1^\alpha = \Delta_{P_1}^{P_\alpha}$ is the complement of $\alpha$ in $\Delta_1^2$. For each $\alpha$, let $\{Y_{\alpha,1}, \ldots, Y_{\alpha,n_\alpha}\}$ be a basis of $\mathfrak{n}_\alpha^2(F)$, the Lie algebra of $N_\alpha^2(F)$. We shall assume that the basis is compatible with the action of $A_1$, so that each $Y_{\alpha,i}$ is a root vector corresponding to the root $\beta_{\alpha,i}$ of $(M_2 \cap P_1, A_1)$. We shall also assume that if $i \leq j$, the height of $\beta_{\alpha,i}$ is not less than the height of $\beta_{\alpha,j}$.

Define $\mathfrak{n}_{\alpha,j}$, $0 \leq j \leq n_\alpha$, to be the direct sum of the span of $\{Y_{\alpha,1}, \ldots, Y_{\alpha,j}\}$ with the Lie algebra of $N_2$. Let $N_{\alpha,j} = \exp \mathfrak{n}_{\alpha,j}$. Then $N_{\alpha,j}$ is a normal subgroup of $N_1$ which is defined over $F$.

If $V$ is any subgroup of $N_1$, defined over $F$, let $\pi(V)$ be the operator which sends $\phi$ to

$$\int_{V(F)\backslash V(\mathbb{A})} \phi(ny)dn, \qquad y \in G(\mathbb{A}).$$

Then $\phi_{P_1,P_2}$ is the transform of $\phi$ by the product over $\alpha \in \Delta_1^2$ of the operators

$$\pi(N_2) - \pi(N_\alpha) = \sum_{i=1}^{n_\alpha} (\pi(N_{\alpha,i-1}) - \pi(N_{\alpha,i})).$$

Let $K_0$ be an open compact subgroup of $G(\mathbb{A}_f)$. Then $G(F)\backslash G(\mathbb{A})^1/K_0$ is a differentiable manifold. We assume from now on that $\phi$ is a function on this space which is differentiable of sufficiently high order.

Suppose that $I$ is a collection of indices $\{i_\alpha; \alpha \in \Delta_1^2, 1 \leq i_\alpha \leq n_\alpha\}$. Then $N_I^- = \prod_\alpha N_{\alpha,i_\alpha-1}$ and $N_I = \prod_\alpha N_{\alpha,i_\alpha}$ are normal subgroup of $N_1$.

Let $\mathfrak{n}^I$ be the span of $\{Y_{\alpha,i_\alpha}\}$. Let $\mathfrak{n}^I(F)'$ be the set of elements $\xi = \sum_\alpha r_\alpha Y_{\alpha,i_\alpha}$, $r_\alpha \in F^\times$.

If $n$ is any positive integer, define $\xi^n = \prod_\alpha (r_\alpha^n) \in F^\times$.

By the Fourier inversion formula for the group $\mathbb{A}/F$, the function $\phi_{P_1,P_2}(y)$ is the sum over all $I$ of

$$\sum_{\xi \in \mathfrak{n}^I(F)'} \int_{\mathfrak{n}^I(F)\backslash \mathfrak{n}^I(\mathbb{A})} dX \cdot \int_{N_I^-(F)\backslash N_I^-(\mathbb{A})} du \cdot \phi(ue(X)y)\psi(\langle X, \xi\rangle).$$

Here $e$ and $\psi$ are as in §7 and $\langle , \rangle$ is the inner product defined by our basis on $\mathfrak{n}^I$.

If $n$ is a positive integer, the element $Y_I^n = \prod_\alpha (-\sqrt{-1}Y_{\alpha,i_\alpha})^n$ can be regarded as a member of $\mathcal{U}(\mathfrak{g}(F_\infty)^1 \otimes \mathbb{C})$. Then $\phi_{P_1,P_2}(y)$ equals the sum over $I$ and over $\xi \in \mathfrak{n}^I(F)'$ of

$$(\xi^n)^{-1} \int_{\mathfrak{n}^I(F)\backslash \mathfrak{n}^I(\mathbb{A})} dX \cdot \int_{N_I^-(F)\backslash N_I^-(\mathbb{A})} du \tag{9.5.3}$$

$$\cdot r_y(\mathrm{Ad}(y^{-1})Y_I^n)\phi(ue(X)y)\psi(\langle X, \xi\rangle).$$

Now, we set $y = \delta x = ac$, as above. Since $\sigma_1^2(H_0(a) - T) \neq 0$, $a$ belongs to a fixed Siegel set in $M_2(\mathbb{A})$. It follows that the integrand in (9.5.3), as a function of $X$, is invariant by an open compact subgroup of $\mathfrak{n}^I(\mathbb{A}_f)$ which is independent of $a$ and $c$. Consequently, (9.5.3) vanishes unless $\xi$ belongs to a fixed lattice, $L^I(K_0)$, in $\mathfrak{n}^I(F_\infty)$. But for an integer $n$ sufficiently large, the sum

$$\sum_{\xi \in \mathfrak{n}^I(F)' \cap L^I(K_0)} |\xi^n|^{-1}$$

is finite for all $I$. Let $c_n(K_0)$ be the supremum over all $I$ of these numbers. Then $|\phi_{P_1,P_2}(ac)|$ is bounded by

$$c_n(K_0) \sum_I \int_{N_I(F)\backslash N_I(\mathbb{A})} |(r(\mathrm{Ad}(c)^{-1}\,\mathrm{Ad}(a)^{-1} Y_I^n)\phi)(uac)|du.$$

Let $\beta_I = \sum_\alpha \beta_{\alpha, i_\alpha}$. Then $\beta_I$ is a positive sum of roots in $\Delta_1^2$. For any $n$,

$$\mathrm{Ad}(a)^{-1} Y_I^n = e^{-n\beta_I(H_0(a))} Y_I^n = e^{-n\beta_I(H_0(\delta x))} Y_I^n.$$

We can choose a finite set of elements $\{X_i\}$ in $\mathcal{U}(\mathfrak{g}(F_\infty)^1 \otimes \mathbb{C})$, depending only on $n$ and $K_0$, such that for any $P_1, P_2, I$, and $c$,

$$c_n(K_0)\,\mathrm{Ad}(c)^{-1} Y_I^n$$

is a linear combination of $\{X_i\}$. Since $c$ lies in a compact set, we may assume that each of the coefficients has absolute value less than 1. The lemma follows. □

## 9.6 *Proof of Theorem*

Substitute $\phi(\sigma)$ for $\phi$ in (9.5.2) and integrate over $\sigma$. The result is

$$\sum_I \sum_i \int_{N_I(F)\backslash N_I(\mathbb{A})} \int_S |r(X_i)\phi(\sigma, u\delta x)| d\sigma\, du \cdot e^{-n\beta_I(H_0(\delta x))}. \tag{9.6.1}$$

If $\delta x = ac$, with $a$ and $c$ as above, then

$$\|\delta x\| \leq \|a\| \cdot \|c\|.$$

We are assuming that $\sigma_1^2(H_0(a) - T) \neq 0$. Since $\beta_I$ is a positive sum of roots in $\Delta_1^2$, we conclude from Corollary 6.2 that $\|a\|$ is bounded by a fixed power of

$$e^{\beta_I(H_0(a))} = e^{\beta_I(H_0(\delta x))}.$$

It follows that for any positive integers $N$ and $N_1$, we may choose $n$ so that (9.6.1) is bounded by a scalar multiple of

$$\sum_i \sup_{y \in G(\mathbb{A})^1} \left( \int_S |r(X_i)\phi(\sigma, y)| d\sigma \cdot \|x\|^{-N'} \right) \|\delta x\|^{-N_1}.$$

It is well known (see [Bo62, PR93]) that there is a positive number $c_1$ such that for any $\gamma \in G(F)$ and $x \in \mathcal{S}$, we have

$$\|\gamma x\|^{-N_1} \leq c_1 \|x\|^{-N_1}.$$

The only thing left to estimate is

$$\sum_{\delta \in P_1(F) \backslash G(F)} F^1(\delta x, T) \sigma_1^2 (H_0(\delta x) - T).$$

The summand is the characteristic function, evaluated at $\delta x$, of a certain subset of

$$\{y \in G(\mathbb{A})^1; \varpi(H_0(y) - T) > 0, \varpi \in \widehat{\Delta}_1\}.$$

The sum is bounded by

$$\sum_{\delta \in P_1(F) \backslash G(F)} \hat{\tau}_1 (H_0(\delta x) - T).$$

It follows from Proposition 5.3 that we can find numbers $C_2 > 0$ and $N_2 > 0$ such that for all $P_1$ the last sum is bounded by $C_2 \|x\|^{N_2}$. Set $N_1 = N' + N_2$. The number $N_1$ dictates our choice of $n$, from which we obtain the differential operators $\{X_i\}$. The theorem follows with any $r$ greater than all the degrees of the operators $X_i$. □

## 9.7 *Parabolic Truncation*

In the next section, we will need to have analogues of the operators $\Lambda^T$ for different parabolic subgroup of $G$.

DEFINITION. If $P_1$ is a parabolic subgroup, and $\phi$ is a continuous function on $P_1(F) \backslash G(\mathbb{A})^1$, define

$$(\Lambda^{T,P_1} \phi)(x) = \sum_{\{R; P_0 \subset R \subset P_1\}} (-1)^{\dim(A_R/A_1)} \sum_{\delta \in R(F) \backslash P_1(F)} \phi_{N_R}(\delta x) \cdot \hat{\tau}_R^1 (H_0(\delta x) - T)$$

where

$$\phi_{N_R}(x) = \int_{N_R(F)\backslash N_R(\mathbb{A})} \phi(nx)dn.$$

PROPOSITION 9.8. *Let $P$ be a parabolic subgroup, $\phi$ a continuous function on $P(F)\backslash G(\mathbb{A})^1$. Then*

$$\sum_{\{P_1;P_0\subset P_1\subset P\}} \sum_{\delta\in P_1(F)\backslash P(F)} \Lambda^{T,P_1}\phi(\delta x)\tau_1^P(H_0(\delta x)-T)$$

*equals*

$$\phi_N(x) = \int_{N(F)\backslash N(\mathbb{A})} \phi(nx)dx. \tag{9.8.1}$$

PROOF. We need to prove that (9.8.1) is the sum over $\{R; P_0 \subset R \subset P\}$ and $\delta \in R(F)\backslash P(F)$ of the product of

$$\phi_{N_R}(\delta x) = \int_{N_R(F)\backslash N_R(\mathbb{A})} \phi(n\delta x)dn$$

with

$$\sum_{\{P_1;R\subset P_1\subset P\}} (-1)^{\dim(A_R/A_1)}\hat{\tau}_R^1(H_0(\delta x)-T)\cdot\tau_1^P(H_0(\delta x)-T). \tag{9.8.2}$$

Consider Proposition 6.5 with $\Lambda$ a point in $-(\mathfrak{a}_0^*)^+$. The sum given in that proposition then reduces to (9.8.2). It follows from Proposition 1.5 that (9.8.2) vanishes if $R \neq P$ and equals 1 if $R = P$. This establishes our proposition. □

## 10 Integrability of $k_\chi^T(x,f)$

Take $m$ to be a sufficiently large integer. Let $T$ be a suitably regular point in $\mathfrak{a}_0^+$. In 5.5 we associated with every function $f \in C_c^m(G(\mathbb{A})^1)$ a function, $k_\chi^T(x,f)$, on $G(F)\backslash G(\mathbb{A})^1$. In this section we prove

THEOREM 10.1. *For any sufficiently regular vector $T$ in $\mathfrak{a}_0^+$, the sum*

$$\sum_{\chi\in\mathfrak{X}} \int_{G(F)\backslash G(\mathbb{A})^1} |k_\chi^T(x,f)|dx$$

*is finite.*

We will not prove the theorem directly. Rather, we shall relate $k_\chi^T(x,f)$ to the truncation operators whose asymptotic properties we have just studied. We shall operate on $K_{P,\chi}(x,y)$, which of course is a function of two variables. If $P_1 \subset P_2$, we shall write $\Lambda_1^{T,P_1}$ (resp. $\Lambda_2^{T,P_1}$) for the operator $\Lambda^{T,P_1}$, acting on the first (resp. second) variable.

PROPOSITION 10.2. *For any equivalence class* $\chi \in \mathfrak{X}$, *the function* $k_\chi^T(x,f)$ *equals*

$$\sum_{\{P_1,P_2;P_0\subseteq P_1\subset P_2\}} \sum_{\delta\in P_1(F)\backslash G(F)} \sigma_1^2(H_0(\delta x)-T)\cdot \sum_{\{P;P_1\subset P\subset P_2\}} (-1)^{\dim(A/Z)}\Lambda_2^{T,P_1}K_{P,\chi}(\delta x,\delta x).$$

PROOF. The given expression is the sum over all chains $P_1 \subset P \subset P_2 \subset P_3$ and over $\delta \in P_1(F)\backslash G(F)$, of $(-1)^{\dim(A_3/A_2)}\hat{\tau}_3(H_0(\delta x)-T)$ times

$$\tau_1^3(H_0(\delta x)-T)(-1)^{\dim(A/Z)}\Lambda_2^{T,P_1}K_{P,\chi}(\delta x,\delta x).$$

As we have done many times, we appeal to Proposition 1.5. We see that the sum over $P_2$ equals 0 unless $P = P_3$. Therefore the given expression equals

$$\sum_{\{P_1,P;P_0\subset P_1\subset P\}} (-1)^{\dim(A/Z)} \sum_{\delta\in P_1(F)\backslash G(F)} \hat{\tau}_P(H_0(\delta x)-T)\cdot \tau_{P_1}^P(H_0(\delta x)-T)\cdot \Lambda_2^{T,P_1}K_{P,\chi}(\delta x,\delta x).$$

Apply Proposition 9.8 to the sum over $P_1$. We obtain

$$\sum_{\{P;P_0\subset P\}} (-1)^{\dim(A/Z)} \sum_{\delta\in P(F)\backslash G(F)} \hat{\tau}_P(H_0(\delta x)-T)\cdot \int_{N(F)\backslash N(\mathbb{A})} K_{P,\chi}(\delta x,n\delta x)dn.$$

Since for $n \in N(\mathbb{A})$, we have $K_{P,\chi}(\delta x,n\delta x) = K_{P,\chi}(\delta x,\delta x)$, the last expression equals $k_\chi^T(x,f)$, as required. □

## 10.3 A Set of Cosets

Fix $P_1 \subset P_2$. Motivated by the last proposition, we examine the sum

$$\sum_{\{P;P_1\subset P\subset P_2\}} (-1)^{\dim(A/A_2)} \int_{N_1(F)\backslash N_1(\mathbb{A})} K_P(x,ny)dn.$$

It equals

$$\sum_{\{P;P_1\subset P\subset P_2\}} (-1)^{\dim(A/A_2)} \sum_{\gamma\in P_1(F)\backslash P(F)} \int_{N_1(\mathbb{A})} \sum_{\eta\in M_1(F)} f(x^{-1}\gamma^{-1}n\eta y)dn$$

$$= \sum_P (-1)^{\dim(A/A_2)} \sum_{\gamma\in P_1(F)\backslash P(F)} K_{P_1}(\gamma x, y).$$

DEFINITION. Let $F(P_1, P_2)$ be the set of elements in $P_1(F)\backslash P_2(F)$ which do not belong to $P_1(F)\backslash P(F)$ for any $P$ with $P_1 \subset P \subsetneqq P_2$.

By Proposition 1.5 the last expression equals

$$\sum_{\gamma\in F(P_1,P_2)} K_{P_1}(\gamma x, y).$$

In this last formula, we have affected the cancelation implicit in the alternating sum over $P$. In order to exploit the equation we have just derived, we interrupt with the following

PROPOSITION 10.4. *Suppose that for each i, $1 \le i \le n$, we are given a parabolic subgroup $Q_i \supset P_1$, points $x_i$, $y_i \in G(\mathbb{A})$, and a number $c_i$ such that*

$$\sum_{i=1}^n c_i \int_{N_1(F)\backslash N_1(\mathbb{A})} K_{Q_i}(x_i, nmy_i)dn$$

*vanishes for all $m \in M_1(F)\backslash M_1(\mathbb{A})^1$. Then for any $\chi \in \mathfrak{X}$, we have that*

$$h_\chi(m) = \sum_{i=1}^n c_i \int_{N_1(F)\backslash N_1(\mathbb{A})} K_{Q_i,\chi}(x_i, nmy_i)dn$$

*also vanishes for all $m \in M_1(F)\backslash M_1(\mathbb{A})^1$.*

PROOF. Suppose that for a given $\chi' \in \mathfrak{X}$ there is a group $R$ in $P_{\chi'}$ which is contained in $P_1$. We would like to prove that for any function $\phi_{\chi'} \in L^2(M_R(F)\backslash M_R(\mathbb{A})^1)_{\chi'}$, the integral

$$\int_{M_R(F)\backslash M_R(\mathbb{A})^1} \int_{N_R^1(F)\backslash N} h_\chi(nm)\phi_{\chi'}(m)\, dn\, dm \tag{10.4.1}$$

vanishes for $\chi \ne \chi'$. Suppose that $\chi \ne \chi'$, and that $\phi \in \mathcal{H}_Q(\tau)_\chi$ for some $Q \subset Q_i$, and some $\tau \in \Pi(M_Q)$. The construction of Eisenstein series is such that if the function of $m \in M_1(F)\backslash M_1(\mathbb{A})^1$

$$m \longmapsto \int_{N_1(F)\backslash N_1(\mathbb{A})} E_{Q_i}(nmy, \phi)dn,$$

is substituted for $h_\chi$ in (10.4.1), the result is 0. It follows from the estimates of Section 4 that (10.4.1) itself is 0. The same estimates yield numbers $c > 0$ and $N > 0$ such that for any $m \in M_1(F)\backslash M_1(\mathbb{A})^1$, we have

$$\sum_{\chi\in\mathfrak{X}} |h_\chi(m)| \leq c\|m\|^N.$$

By assumption, $\sum_\chi h_\chi(m)$ equals 0. Consequently (10.4.1) is zero even when $\chi = \chi'$. The function $h_\chi$ is continuous. Because (10.4.1) vanishes for all $\phi_{\chi'}$, the function $h_\chi$ satisfies the hypotheses of [La76], Lemma 3.7. The function $h_\chi$ is therefore zero. □

To return to the proof of the theorem, we first prove

PROPOSITION 10.5. *For any $\chi \in \mathfrak{X}$, the sum*

$$\sum_{\{P;P_1\subset P\subset P_2\}} (-1)^{\dim(A/Z)}(\Lambda_2^{T,P_1}K_{P,\chi})(x,y)$$

*equals*

$$(-1)^{\dim(A_2/Z)} \sum_{\gamma\in F(P_1,P_2)} (\Lambda_2^{T,P_1}K_{P_1,\chi})(\gamma x,y).$$

PROOF. We look for conditions imposed on $x$, $y$ and $\gamma$ by the nonvanishing of

$$K_{P_1}(\gamma x, my), \quad m \in M_1(F)\backslash M_1(\mathbb{A})^1. \tag{10.5.1}$$

Set

$$y = y_1 k, \quad \text{where} \quad y_1 \in P_1(\mathbb{A}) \cap G(\mathbb{A})^1, \;\; k \in K.$$

There is a compact subset of $G(\mathbb{A})^1$, depending only on the support of $f$, which contains some point

$$x^{-1}\gamma^{-1}n\eta m y_1, \quad n \in N_1(\mathbb{A}), \quad \eta \in M_1(F),$$

whenever (10.5.1) does not vanish. Fix $\varpi \in \widehat{\Delta}_1$. Let $\Lambda$ be a rational representation of $G$ with highest weight $d\varpi$, $d > 0$. Choose a height function $\|\cdot\|$ as in Section 1. If $v$ is a highest weight vector, we can choose a number $c_1$ such that

$$\|\Lambda(x^{-1}\gamma n\eta m y_1)v\| \leq c_1$$

whenever $x^{-1}\gamma n\eta m y_1$ lies in the given compact subset of $G(\mathbb{A})^1$.

The left side of this inequality equals

$$e^{d\varpi(H_0(y_1))}\|\Lambda(x^{-1}\gamma^{-1})v\| = e^{d\varpi(H_0(y))}\|\Lambda(x^{-1}\gamma^{-1})v\|,$$

which is no less than a scalar multiple of

$$e^{d\varpi(H_0(y))}e^{-d\varpi(H_0(\gamma x))}.$$

In other words, $\varpi(H_0(\gamma x)-H_0(y))$ is no less than a fixed number.

It follows from this observation that we may choose a point $T_0\in\mathfrak{a}_0$, depending only on the support of $f$, such that

$$\hat{\tau}_1(H_0(\gamma x)-H_0(y)-T_0)=1 \tag{10.5.2}$$

whenever (10.5.1) does not vanish identically in $m$. We conclude from Proposition 10.4 that if (10.5.2) fails to hold for a given $x$, $y$ and $\gamma$, then

$$K_{P_1,\chi}(\gamma x, my),\quad m\in M_1(F)\backslash M_1(\mathbb{A})^1, \tag{10.5.3}$$

vanishes for all $\chi$ and $m$.

Combining Proposition 5.3 with what we have just shown, we conclude that for fixed $x$ and $y$, the function

$$K_{P_1,\chi}(\gamma x, y),\quad \gamma\in F(P_1,P_2),$$

vanishes unless $\gamma$ belongs to a finite subset of $F(P_1,P_2)$, independent of $\chi$. Therefore the sums in

$$\sum_{\{P;P_1\subset P\subset P_2\}}(-1)^{\dim(A/A_2)}\int_{N_1(F)\backslash N_1(\mathbb{A})}K_P(x,nmy)dn-\sum_{\gamma\in F(P_1,P_2)}K_{P_1}(\gamma x,my)$$

are finite.

Since the expression vanishes for all $m$ in $M_1(F)\backslash M_1(\mathbb{A})^1$, we can apply Proposition 10.4. We obtain an equality of functions of $y$ for each $\chi$. We are certainly at liberty to apply the truncation operator to those functions. The proposition follows. □

We have thus far shown that

Corollary 10.6. *For any equivalence class $\chi\in\mathfrak{X}$, the sum*

$$\sum_\chi\int_{G(F)\backslash G(\mathbb{A})^1}|k_\chi^T(x,f)|dx$$

*is bounded by the sum over $\{P_1,P_2;P_0\subset P_1\subset P_2\}$ of*

$$\int_{P_1(F)\backslash G(\mathbb{A})^1}\sum_\chi\sum_{\gamma\in F(P_1,P_2)}\sigma_1^2(H_0(x)-T)\left|\Lambda_2^{T,P_1}K_{P_1,\chi}(\gamma x,x)\right|dx.$$

□

## 10.7 On a Siegel Set

Let $\mathcal{S}$ be a fixed Siegel set in $M_1(\mathbb{A})^1$ with $M_1(F)\mathcal{S} = M_1(\mathbb{A})^1$. Let $\Gamma$ be a compact subset of $N_1(\mathbb{A})$ with $N_1(F)\Gamma = N_1(\mathbb{A})$. Then the last integral in 10.6 is bounded by the integral over $n \in \Gamma$, $m \in \mathcal{S} \cap P_0(\mathbb{A})$, $a \in A_1(\mathbb{R}^0) \cap G(\mathbb{A})^1$, and $k \in K$, of

$$e^{-2\rho_{P_1}(H_0(a))}\sigma_1^2(H_0(a) - T)\sum_{\chi}\sum_{\gamma}\left|\Lambda_2^{T,P_1}K_{P_1,\chi}(\gamma nmak, mak)\right|.$$

PROPOSITION. *If $a \in A_1(\mathbb{R}^0) \cap G(\mathbb{A})^1$ is such that for some $\chi$, $\gamma$, $n$, $m$, $\tilde{m}$, and $k$,*

$$\sigma_1^2(H_0(a) - T)K_{P_1,\chi}(\gamma nma, \tilde{m}ak) \tag{10.7.1}$$

*does not vanish, then for any Euclidean norm $\|\cdot\|$ on $\mathfrak{a}_0$, there is a positive number $c$ such that*

$$\|H_0(a)\| \le c(1 + \log\|m\|). \tag{10.7.2}$$

PROOF. Suppose that for $n$, $m$, $a$, and $k$ as above, and for some $\tilde{m} \in M_1(\mathbb{A})^1$, $\gamma \in F(P_1, P_2)$, and $\chi \in \mathfrak{X}$, we have

$$K_{P_1\chi}(\gamma nmak, \tilde{m}ak) \neq 0.$$

Write $\gamma = \nu w_s \pi$, for $\nu \in N_0^2(F)$, $\pi \in P_0(F)$, and $s \in \Omega^{M_2}$, the Weyl group of $(M_2, A_0)$. It follows from Proposition 10.4 that there is a fixed compact subset of $G(\mathbb{A})^1$ which contains

$$a^{-1}m^{-1}n_1 w_s p_1 a,$$

for points $n_1 \in N_0(\mathbb{A})$ and $p_1 \in M_1(\mathbb{A})^1 N_1(\mathbb{A})$. Fix $\varpi \in \widehat{\Delta}$. Let $\Lambda$ and $v$ be as above. Then $\Lambda(w_s)v$ is a weight vector, with weight $s\varpi$.

The vector

$$\Lambda(a^{-1}m^{-1}n_1 w_s p_1 a)v - e^{d(\varpi - s\varpi)(H_0(a))}e^{-ds\varpi(H_0(m))}v$$

can be written as a sum of weight vectors, with weights higher than $s\varpi$. By the construction of our height function, we have

$$e^{d(\varpi - s\varpi)(H_0(a))}e^{-ds\varpi(H_0(m))}\|v\| \le \|\Lambda(a^{-1}m^{-1}n_1 w_s p_1 a)v\|.$$

It follows that there are positive numbers $c'$ and $c$, depending only on the support of $f$, such that

$$|(\varpi - s\varpi)(H_0(a))| \le c'|s\varpi(H_0(m))| \le c(1 + \log\|m\|).$$

Since $s$ fixes $\mathfrak{a}_2$ pointwise, the inequality

$$|(\varpi - s\varpi)(H_0(a))| \leq c(1 + \log \|m\|)$$

holds for the projection of $\varpi$ onto $\mathfrak{a}_1^2$. In other words, we may take $\varpi$ to be an element in $\widehat{\Delta}_1^2$. For each such $\varpi$, the difference $\varpi - s\varpi$ is a nonnegative integral sum of roots in $\Delta_1^2$. We claim that

LEMMA. *The coefficient of the element $\alpha$ in $\Delta_1^2$ such that $\varpi = \varpi_\alpha$ is not zero.*

PROOF. Otherwise, we would have $(\varpi - s\varpi)(\varpi^\vee) = 0$, or equivalently, $s\varpi = \varpi$. This would force $s$ to belong to $\Omega^M$, for some parabolic subgroup $P$, $P_1 \subset P \subsetneqq P_2$. This contradicts the assumption that $\gamma = \nu w_s \pi$ belongs to $F(P_1, P_2)$, so the coefficient of $\alpha$ is indeed positive. □

If $a$ has the additional property that $\sigma_1^2(H_0(a) - T) \neq 0$, the proposition follows from Corollary 6.2. □

## *10.8 Proof of Theorem* 10.1

PROOF OF THEOREM 10.1. Suppose that $f$ is right invariant under an open compact subgroup $K_0$ of $G(\mathbb{A}_f)$. If $I_{P_1}(\tau,f)\phi \neq 0$ for some $\tau$ and $\phi \in \mathfrak{B}_{P_1}(\tau)_\chi$, then the function $E(y,\phi)$ is right $K_0$-invariant in $y$. Therefore for any $x$, $\gamma$, and $\chi$, the function $y \longmapsto K_{P_1,\chi}(\gamma x, y)$ is right $K_0$-invariant. It follows that (10.7.1) is right invariant in $\tilde{m}$ under the open compact subgroup

$$\bigcap_{k_1 \in K} (k_1 K_0 k_1^{-1}) \cap M_1(\mathbb{A}_f)^1$$

of $M_1(\mathbb{A}_f)^1$. We apply Theorem 9.5 with the group $G$ replaced by $M_1$.

For any positive integers $N_1$, $N_1'$, we can choose a finite set $\{X_i\}$ of elements in $\mathcal{U}(\mathfrak{m}_1(F_\infty)^1 \otimes \mathbb{C})$, the universal enveloping algebra of the complexification of the Lie algebra of $M_1(F_\infty)^1$, such that for all $n \in \Gamma$, $m \in \mathcal{S} \cap P_0(\mathbb{A})$, $\tilde{m} \in \mathcal{S}$, $a \in A_1(\mathbb{R}^0) \cap G(\mathbb{A})^1$, and $k \in K$,

$$\sum_{\gamma \in F(P_1,P_2)} \sum_{\chi} |\Lambda_2^{T,P_1} K_{P_1,\chi}(\gamma nmak, \tilde{m}ak)| \tag{10.8.1}$$

is bounded by

$$\sum_i \sup\nolimits_{u \in M_1(\mathbb{A})^1} \Big(\sum_\gamma \sum_\chi |r_u(X_i) K_{P_1,\chi}(\gamma nmak, uak)| \cdot \|u\|^{-N_1}\Big) \cdot \|\tilde{m}\|^{-N_1'}. \tag{10.8.2}$$

We can choose elements $\{Y_i\}$ in $\mathcal{U}(\mathfrak{g}(F_\infty)^1 \otimes \mathbb{C})$ such that

$$\mathrm{Ad}(ak)^{-1}X_i = \mathrm{Ad}(k)^{-1}X_i = \sum_j c_{ij}(k)Y_j,$$

where $c_{ij}(k)$ are continuous functions on $K$. Recall that $K_{P_1,\chi}(x,y)$ is ultimately defined in terms of $f$. The function $r_y(Y_j)K_{P_1,\chi}(x,y)$ is defined the same way but with $f$ replaced by $f * \bar{Y}_j^*$. The support of $f * \bar{Y}_j^*$ is contained in the support of $f$, so we can assume that (10.5.2) is valid whenever $r_y(Y_i)K_{P_1,\chi}(\gamma x, y)$ does not vanish. By Corollary 4.8, the sum

$$\sum_\chi |r_y(Y_j)K_{P_1,\chi}(x,y)|$$

is bounded by a scalar multiple of a power of $\|x\| \cdot \|y\|$. It follows from Corollary 5.4 that the expression

$$\sum_{\gamma \in F(P_1,P_2)} \sum_\chi |r_y(Y_j)K_{P_1,\chi}(\gamma x, y)|$$

$$= \sum_{\gamma \in F(P_1,P_2)} \sum_\chi |r_y(Y_j)K_{P_1,\chi}(\gamma x, y)| \cdot \hat{\tau}_1(H_0(\gamma x) - H_0(y) - T_0)$$

is also bounded by a scalar multiple of a power of $\|x\| \cdot \|y\|$.

By taking $N_1$ to be large enough, we obtain positive numbers $C_2$ and $N_2$ such (10.8.2), and therefore (10.8.1), is bounded by

$$C_2\|m\|^{N_2}\|a\|^{N_2}\|\tilde{m}\|^{-N_1'}.$$

Set $\tilde{m} = m$ in (10.8.1). Integrate the resulting expression over $n \in \Gamma$, $m \in \mathcal{S} \cap P_0(\mathbb{A})$, $k \in K$, and $a$ in the subset of elements in $A_1(\mathbb{R}^0) \cap G(\mathbb{A})^1$ which satisfy (10.7.2). There are number $C_3$ and $N_3$ such that the result is bounded by

$$C_3 \int_{\mathcal{S}} \|m\|^{N_3 - N_1'}\, dm.$$

If we set $N_1' = N_3$, this is finite. The proof of Theorem 10.1 is complete. □

PROPOSITION 10.9. *For $T$ sufficiently regular, and $m$ sufficiently large,*

$$\int_{G(F)\backslash G(\mathbb{A})^1} k_\chi^T(x,f)dx = \int_{G(F)\backslash G(\mathbb{A})^1} \Lambda_2^T K_\chi(x,x)dx,$$

*for all $f \in C_c^m(G(\mathbb{A})^1)$ and $\chi \in \mathfrak{X}$.*

PROOF. It follows from proof of Theorem 10.1 that the integral of $k_\chi^T(x,f)$ is the sum over all $P_1 \subset P_2$ of the product of $(-1)^{\dim(A_2/Z)}$ with

$$\int_{P_1(F)\backslash G(\mathbb{A})^1} \sum_{\gamma\in F(P_1,P_2)} \sigma_1^2(H_0(x)-T)\cdot \Lambda_2^{T,P_1} K_{P_1,\chi}(\gamma x,x)dx.$$

As a double integral over $x$ and $\gamma$, this converges absolutely. If $P_1 = P_2 \neq G$, the integrand is zero. If $P_1 = P_2 = G$, the result is the integral of $\Lambda_2^T K_\chi(x,x)$. We have only to show that if $P_1 \subsetneq P_2$, the result is zero. Let $\Omega(P_1,P_2)$ be the set of elements $s$ in $\Omega^{M_2}$ such that $s\alpha$ and $s^{-1}\alpha$ are positive roots for each $\alpha \in \Delta_0^1$ and such that $s$ does not belong to any $\Omega^M$, with $P_1 \subset P \subsetneq P_2$. Then the above integral equals the sum over all $s \in \Omega(P_1,P_2)$ of

$$\int_{P_1(F)\backslash G(\mathbb{A})^1} \sum_{\gamma\in P_1(F)\cap w_s^{-1}P_1(F)w_s^{-1}\backslash P_1(F)} \sigma_1^2(H_0(x)-T)\cdot \Lambda_2^{T,P_1} K_{P_1,\chi}(w_s\gamma x,x)dx.$$

Since

$$\begin{aligned}&\sigma_1^2(H_0(x)-T)\cdot \Lambda_2^{T,P_1} K_{P_1,\chi}(w_s\gamma x,x)\\ &= \sigma_1^2(H_0(\gamma x)-T)\cdot \Lambda_2^{T,P_1} K_{P_1,\chi}(w_s\gamma x,x)\end{aligned}$$

for any $\gamma \in P_1(F)$, this equals

$$\int_{(P_1(F)\cap w_s^{-1}P_1(F)w_s)\backslash G(\mathbb{A})^1} \sigma_1^2(H_0(x)-T)\cdot \Lambda_2^{T,P_1} K_{P_1,\chi}(w_s x,x)dx. \tag{10.9.1}$$

If $s \in \Omega(P_1,P_2)$, the group $w_s^{-1}P_0w_s \cap M_1$ is the standard minimal parabolic subgroup of $M_1$, since $s^{-1}\alpha > 0$ for $\alpha \in \Delta_0^1$. Therefore $M_1 \cap w_s^{-1}P_1w_s$ equals $M_1 \cap P_s$, for a unique parabolic subgroup $P_s$ of $G$, with $P_0 \subset P_s \subset P_1$. Write the integral in (10.9.1) as a double integral over

$$M_s(F)N_s(\mathbb{A})\backslash G(\mathbb{A})^1 \times (P_1(F)\cap w_s^{-1}P_1(F)w_s)\backslash M_s(F)N_s(\mathbb{A}).$$

The group $P_1 \cap w_s^{-1}P_1w_s$ is the semidirect product of the groups $M_1 \cap w_s^{-1}P_1w_s$ and $N_1 \cap w_s^{-1}P_1w_s$. The group $M_1 \cap w_s^{-1}P_1w_s$ decomposes further as the semidirect product of $M_s(F)$ and $N_s^1(F)$. Therefore (10.9.1) equals the integral over $x$ in the space $M_s(F)N_s(\mathbb{A})\backslash G(\mathbb{A})^1$ of the product of $\sigma_1^2(H_0(x)-T)$ and

$$\int_{N_s^1(F)\backslash N_s^1(\mathbb{A})} dn \int_{N_1(\mathbb{A})\cap w_s^{-1}P_1(\mathbb{A})w_s\backslash N_1(\mathbb{A})} dn_1 \cdot \Lambda_2^{T,P_1} K_{P_1,\chi}(w_sn_1nx,n_1nx). \tag{10.9.2}$$

This last expression equals

$$\int\int \Lambda_2^{T,P_1} K_{P_1,\chi}(w_s n_1 nx, nx) dn_1 dn$$
$$= \int\int \Lambda_2^{T,P_1} K_{P_1,\chi}(w_s n_1 x, nx) dn_1 dn.$$

We apply Proposition 9.2 to the parabolic subgroup $M_1 \cap P_s$ of $M_1$. Then this expression vanishes unless $\varpi(H_0(x) - T)$ is negative for each $\varpi \in \widehat{\Delta}_s^1$. On the other hand, we can assume that (10.5.2) holds, with $\gamma$, $x$, and $y$ replaced by $w_s$, $n_1x$, and $nx$, respectively. In other words,

$$\varpi(H_0(w_s n_1 x)) \geq \varpi(H_0(x)) + \varpi(T_0)$$

for each $\varpi \in \widehat{\Delta}_1$. But it is well known that

$$\varpi(H_0(w_s n_1 x)) \leq \varpi(sH_0(x)),$$

so there is a number $C$, depending only on the support of $f$, such that

$$\varpi(H_0(x) - sH_0(x)) \leq C$$

for every $\varpi$ in $\widehat{\Delta}_1$. These two conditions on $H_0(x)$, we repeat, are based on the assumption that (10.9.2) does not vanish.

We obtain a third condition by demanding that $\sigma_1^2(H_0(x) - T)$ does not vanish.

We claim that these three conditions are incompatible if $T$ is sufficiently regular.

Write the projection of $H_0(x) - T$ on $\mathfrak{a}_s^2$ as

$$-\sum_{\alpha\in\Delta_s^1} c_a \alpha^\vee + \sum_{\varpi\in\widehat{\Delta}_1^2} c_\varpi \varpi^\vee.$$

The first and third conditions on $H_0(x)$ translate to the positivity of each $c_\alpha$ and $c_\varpi$. Now the Levi component of $P_s$ equals $M_1 \cap w_s^{-1} M_1 w_s$. Therefore $s\mathfrak{a}_0^s$ is orthogonal to $\mathfrak{a}_1$. Then for $\varpi \in \widehat{\Delta}_1$, the number

$$\varpi_0(H_0(x) - sH_0(x))$$

equals

$$\varpi_0(T - sT) + \sum_{\alpha\in\Delta_s^1} c_\alpha \varpi_0(s\alpha^\vee) + \sum_{\varpi\in\widehat{\Delta}_1^2} c_\varpi \varpi_0(\varpi^\vee - s\varpi^\vee).$$

Now $\varpi^\vee - s\varpi^\vee$ is a nonnegative sum of coroots, so the sum over $\varpi$ is nonnegative. Moreover we can replace each $\alpha$ in the sum over $\Delta_s^1$ by the corresponding root in $\Delta_0^1 \backslash \Delta_0^s$. Since $s$ maps the roots in this latter set to positive roots, the sum over $\alpha$ is also nonnegative. Finally, for any $\varpi_0$, the number $\varpi_0(T - sT)$ can be made arbitrarily large for $T$ sufficiently regular. We thus contradict the second condition on $H_0(x)$. Therefore (10.9.2) is always zero so the integral of $k_\chi^T(x,f)$ equals that of $\Lambda_2^T K_\chi(x,x)$. □

# 11 The Operator $M_P^T(\tau)_\chi$

## 11.1 Truncated Eisenstein Series

For any $\chi \in \mathfrak{X}$, set

$$J_\chi^T(f) = \int_{G(F)\backslash G(\mathbb{A})^1} k_\chi^T(x,f)dx.$$

In this section we shall give another formula, which reveals a different set of properties of the distributions $J_\chi^T$. We shall build on Proposition 10.9, which is a partial step in this direction.

Fix $P$, $\tau \in \Pi(M)$, and $\chi \in \mathfrak{X}$. Suppose that $A$ is a linear operator on $\mathcal{H}_P(\tau)$ under which one of the spaces $\mathcal{H}_P(\tau)_\chi$, $\mathcal{H}_P(\tau)_{\chi,K_0}$, or $\mathcal{H}_P(\tau)_{\chi,K_0,W}$ is invariant. Here $K_0$ is an open compact subgroup of $G(\mathbb{A}_f)$ and $W$ is an equivalence class of irreducible representations of $K_{F_\infty}$. We shall write $A_\chi$, $A_{\chi,K_0}$, or $A_{\chi,K_0,W}$ for the restriction of $A$ to the subspace in question.

Suppose that $\mathcal{S}$ is Siegel set in $G(\mathbb{A})^1$. It is a consequence of Theorem 9.5 and (3.10.1) that given any integer $N' > 0$ and a vector $\phi \in \mathcal{H}_P^0(\tau)_\chi$, we can choose a locally bounded function $c(\zeta)$ on the set of $\zeta \in \mathfrak{a}_{P,\mathbb{C}}^*$ at which $E(x,\phi_\zeta)$ is regular, such that

$$|\Lambda^T E(x,\phi_\zeta)| \leq c(\zeta) \cdot \|x\|^{-N'},$$

for all $x \in \mathcal{S}$. It follows that for $\phi, \psi \in \mathcal{H}_P^0(\tau)_\chi$, the integrals

$$\int_{G(F)\backslash G(\mathbb{A})^1} \Lambda^T E(x,\phi_\zeta) \cdot \overline{\Lambda^T E(x,\psi_\eta)}dx$$

and

$$\int_{G(F)\backslash G(\mathbb{A})^1} E(x,\phi_\zeta) \cdot \overline{\Lambda^T E(x,\psi_\eta)}dx$$

converge absolutely and define meromorphic functions in $(\zeta, \bar{\eta})$ which are regular whenever the integrands are. By Corollary 9.3 and Proposition 9.4, these meromorphic functions are equal. Thus we obtain a

DEFINITION 11.2. Define a linear operator $\mathcal{M}_P^T(\tau)_\chi$ on $\mathcal{H}_P^0(\tau)_\chi$ by

$$(\mathcal{M}_P^T(\tau)_\chi\phi_1, \phi_2) = \int_{G(F)\backslash G(\mathbb{A})^1} \Lambda^T E(x, \phi_1) \cdot \overline{\Lambda^T E(x, \phi_2)} dx$$
$$= \int_{G(F)\backslash G(\mathbb{A})^1} E(x, \phi_1) \cdot \overline{\Lambda^T E(x, \phi_2)} dx,$$

for every pair $\phi_1$ and $\phi_2$ in $\mathcal{H}_P^0(\tau)_\chi$.

## 11.3 Universal Enveloping Algebra

The operator $\mathcal{M}_P^T(\tau)$ depends only on the orbit of $\tau$ in $\Pi^G(M)$. It is clear that $\mathcal{M}_P^T(\tau)_\chi$ is self-adjoint and positive definite. Notice also that

$$i_P(\tau, k) \cdot \mathcal{M}_P^T(\tau)_\chi = \mathcal{M}_P^T(\tau)_\chi \cdot i_P(\tau, k)$$

for all $k \in K$. It follows that for any $K_0$ and $W$, the operator $\mathcal{M}_P^T(\tau)_\chi$ leaves the finite dimensional space $\mathcal{H}_P(\tau)_{\chi,K_0,W}$ invariant.

Recall that in the proof of Proposition 4.1, we fixed an elliptic element $\Delta$ in $\mathcal{U}(\mathfrak{g}(F_\infty)^1 \otimes \mathbb{C})^{K_{F_\infty}}$. For any $K_0$ and $W$, the space $\mathcal{H}_P(\tau)_{\chi,K_0,W}$ is an invariant subspace for the operator $i_P(\tau, \Delta)$. Choose $\Delta$ so that for any $\chi$, $\tau$, $W$, and $K_0$ such that $\mathcal{H}_P(\tau)_{\chi,K_0,W} \neq \{0\}$, the operator $i_P(\tau, \Delta)_{\chi,K_0,W}$ is the product of the identity operator with a real number which is larger than 1. For example, we could take $\Delta$ to equal $1 + \Delta_1^*\Delta_1$, where $\Delta_1$ is a suitable linear combination of the Casimir elements for $G(F_\infty)^1$ and $K_{F_\infty}$.

If $A$ is any operator on a Hilbert space, $\|A\|_1$ denotes the trace class norm of $A$.

THEOREM 11.4. *There is a positive integer $n$ such that for any open compact subgroup $K_0$ of $G(\mathbb{A}_f)$, the sum*

$$\sum_\chi \sum_P n(A)^{-1} \int_{\Pi^G(M)} \|\mathcal{M}_P^T(\tau)_{\chi,K_0} \cdot i_P(\tau, \Delta^n)_{\chi,K_0}^{-1}\|_1 d\tau$$

*is finite.*

Assume the proof of the theorem for the moment and take $m_1 = \deg \Delta^n$. Suppose that $f$ is a function in $C_c^{m_1}(G(\mathbb{A})^1)$, which is bi-invariant under $K_0$. Then

$$\|\mathcal{M}_P^T(\tau) \cdot i_P(\tau, f)_\chi\|_1 = \|\mathcal{M}_P^T(\tau)_{\chi,K_0} \cdot i_P(\tau, f)\|_1$$
$$= \|\mathcal{M}_P^T(\tau)_{\chi,K_0} \cdot i_P(\tau, \Delta^n)^{-1} i_P(\tau, \Delta^n * f)\|_1$$
$$\leq \|\mathcal{M}_P^T(\tau)_{\chi,K_0} \cdot i_P(\tau, \Delta^n)_{\chi,K_0}^{-1}\|_1 \cdot \|i_P(\tau, \Delta^n * f)\|.$$

For any $\tau$ the norm of the operator $i_P(\tau, \Delta^n * f)$ is bounded by

$$\int_{G(\mathbb{A})^1} |(\Delta^n * f)(x)|dx.$$

Thus Theorem 11.4 implies

COROLLARY 11.5. *For every $f \in C_c^{m_1}(G(\mathbb{A})^1)$, the sum*

$$\sum_\chi \sum_P n(A)^{-1} \int_{\Pi^G(M)} \|\mathcal{M}_P^T(\tau)_\chi \cdot i_P(\tau,f)_\chi\|_1 d\tau \tag{11.5.1}$$

*is finite and in fact defines a continuous seminorm on $C_c^{m_1}(G(\mathbb{A})^1)$. In particular, the operator $\mathcal{M}_P^T(\tau)_\chi \cdot i_P(\tau,f)_\chi$ is of trace class for almost all $\tau$.* □

THEOREM 11.6. *There is an integer $m \geq m_1$ such that for any $\chi$ and any $f \in C_c^m(G(\mathbb{A})^1)$, we have*

$$J_\chi^T(f) = \sum_P n(A)^{-1} \int_{\Pi^G(M)} \mathrm{tr}(\mathcal{M}_P^T(\tau)_\chi \cdot i_P(\tau,f)_\chi) d\tau.$$

## 11.7 Preparation for Proof

We shall prove the two theorems together. Let $N$ and $m_0$ be the positive integers of Proposition 4.4. Choose an open compact subgroup, $K_0$, of $G(\mathbb{A}_f)$ and a Siegel set $\mathcal{S}$ in $G(\mathbb{A})^1$. According to Theorem 9.5 and Proposition 4.4, we may choose a finite set $\{Y_i\}$ of elements in $\mathcal{U}(\mathfrak{g}(F_\infty)^1 \otimes \mathbb{C})$ such that for $x \in G(\mathbb{A})^1$, $y \in \mathcal{S}$, $m \geq m_0 + \sum_i \deg Y_i$, and a $K$-finite function $f$ in $C_c^m(G(\mathbb{A})^1/K_0)$, we have that

$$\sum_P n(A)^{-1} \int_{\Pi^G(M)} \left| \sum_{\phi \in \mathfrak{B}_P(\tau)_\chi} E(x, i_P(\tau, \phi)f) \cdot \overline{\Lambda^T E(y, \phi)} \right| d\tau$$

is bounded by

$$\sum_i \|f * Y_i\|_{m_0} \cdot \|x\|^N \cdot \|y\|^{-N}.$$

When we set $x = y$ and integrate the above expression over $G(F)\backslash G(\mathbb{A})^1$, the result is bounded by

$$\mathrm{vol}(G(F)\backslash G(\mathbb{A})^1) \cdot \sum_i \|f * Y_i\|_{m_0}.$$

Suppose $\omega = (W_1, W_2)$ is a pair of equivalence classes of irreducible representations of $K_{F_\infty}$. We defined the function

$$f_\omega(x) = \deg W_1 \cdot \deg W_2 \cdot \int_{K_{F_\infty} \times K_{F_\infty}} \mathrm{ch}_{W_1}(k_1) f(k_1^{-1} x k_2^{-1}) \, \mathrm{ch}_{W_2}(k_2) dk_1 \, dk_2$$

and the positive integer $l_0$ in § 4.5. Let

$$m_2 = m_0 + l_0 + \sum_i \deg Y_i.$$

As we saw in § 4, the linear form in $f \in C_c^{m_2}(G(\mathbb{A})^1)$ given by

$$\|f\|_{m_2} = \mathrm{vol}(G(F)\backslash G(\mathbb{A})^1) \sum_\omega \sum_i \|f_\omega * Y_i\|_{m_0},$$

is a continuous seminorm on $C_c^{m_2}(G(\mathbb{A})^1)$. We have shown

PROPOSITION 11.8. *The sum*

$$\sum_\omega \int_{G(F)\backslash G(\mathbb{A})^1} \sum_\chi \sum_P n(A)^{-1} \tag{11.8.1}$$

$$\int_{\Pi^G(M)} \left| \sum_{\phi \in \mathfrak{B}_P(\tau)} E(x, i_P(\tau, f_\omega)\phi) \cdot \overline{\Lambda^T E(x, \phi)} \right| d\tau \, dx$$

*is bounded by* $\|f\|_{m_2}$, *for every* $f \in C_c^{m_2}(G(\mathbb{A})^1)$. □

## 11.9 A Reduction

Let $m$ be any integer larger than $m_2$ for which Proposition 10.9 is valid. If $f \in C_c^m(G(\mathbb{A})^1)$, and $\chi$ is fixed, we have

$$\sum_P n(A)^{-1} \int_{\Pi^G(M)} \sum_\omega \mathrm{tr}(\mathcal{M}_P^T(\tau)_\chi \cdot i_P(\tau, f_\omega)_\chi) d\tau$$

$$= \sum_P n(A)^{-1} \int_{\Pi^G(M)} \sum_\omega \int_{G(F)\backslash G(\mathbb{A})^1} \left( \sum_{\phi \in \mathfrak{B}_P(\tau)_\chi} E(x, i_P(\tau, f_\omega)\phi) \cdot \overline{\Lambda^T E(x, \phi)} \right) dx \cdot d\tau$$

$$= \int_{G(F)\backslash G(\mathbb{A})^1} \sum_\omega \sum_P n(A)^{-1} \int_{\Pi^G(M)} \left( \sum_\phi E(x, i_P(\tau, f_\omega)\phi) \cdot \overline{\Lambda^T E(x, \phi)} \right) d\tau \cdot dx$$

by Tonelli's theorem. The operator $\Lambda^T$ is defined in terms of sums and integrals over compact sets. If we combine Tonelli's theorem with the estimates of § 4, we find that we can take $\Lambda^T$ outside the sum over $\phi$, $P$ and $\omega$, and the integral over $\tau$. The result is

$$\int_{G(F)\backslash G(\mathbb{A})^1} \Lambda_2^T K_\chi(x,x)dx,$$

which by Proposition 10.9 equals $J_\chi^T(f)$. The proof of Theorem 11.6 will now follow from Theorem 11.4 if we take $m$ to be larger than $m_1$.

## *11.10 Proof of Theorem* 11.4

It remains to prove Theorem 11.4. We shall use Proposition 4.1. We can choose $n$ and functions $g_\infty^1 \in C_c^{m_2}(G(\mathbb{A})^1)^{K_{F_\infty}}$ and $g_\infty^2 \in C_c^\infty(G(\mathbb{A})^1)^{K_{F_\infty}}$ such that $\Delta^n * g_\infty^1 + g_\infty^2$ is the Dirac distribution at $e$ in $G(F_\infty)^1$. If $i = 1, 2$, for $x_\infty \in G(F_\infty)^1$, $x_f \in G(\mathbb{A})^1$, set

$$g_i(x_\mathbb{R} \cdot x_f) = \mathrm{vol}(K_0)^{-1} \cdot g_\infty^i(x_\mathbb{R}) \cdot \mathrm{ch}_{K_0}(x_f),$$

where $\mathrm{ch}_{K_0}$ is the characteristic function of $K_0$. Then

$$i_P(\tau, \Delta^n)_{\chi,K_0}^{-1} = i_P(\tau, g_1)_\chi + i_P(\tau, \Delta^n)^{-1} i_P(\tau, g_2)_\chi.$$

Suppose that $W$ is an irreducible $K_{F_\infty}$-type and that $\omega = (W, W)$. Then the trace of the restriction of $\mathcal{M}_P^T(\tau)_\chi \cdot i_P(\tau, \Delta^n)^{-1}$ to $\mathcal{H}_P(\tau)_{\chi,K_0,W}$ is

$$\mathrm{tr}(\mathcal{M}_P^T(\tau)_\chi \cdot i_P(\tau, g_{1,\omega})_\chi + \mathcal{M}_P^T(\tau)_\chi \cdot i_P(\tau, \Delta^n)^{-1} i_P(\tau, g_{2,\omega})_\chi).$$

Since the eigenvalues of $I_P(\tau, \Delta^n)$ are all larger than 1, this last expression is bounded by

$$\sum_{i=1}^{2} |\,\mathrm{tr}(\mathcal{M}_P^T(\tau)_\chi \cdot I_P(\tau, g_{i,\omega})_\chi)|.$$

Now the trace class norm of the operator

$$\mathcal{M}_P^T(\tau)_{\chi,K_0} \cdot I_P(\tau, \Delta^n)_{\chi,K_0}^{-1}$$

is the sum of the traces of its restriction to each of the subspaces $\mathcal{H}_P(\tau)_{\chi,K_0,W}$. Therefore

$$\sum_\chi \sum_P n(A)^{-1} \int_{\Pi^G(M)} \|\mathcal{M}_P^T(\tau)_{\chi,K_0} \cdot I_P(\tau, \Delta^n)_{\chi,K_0}^{-1}\|_1 d\tau$$

is bounded by the sum over $i = 1, 2$ of $\sum_\chi \sum_P n(A)^{-1}$ of

$$\int_{\Pi^G(M)} \sum_\omega \left| \sum_{\phi \in \mathfrak{B}_P(\tau)_\chi} \int_{G(F)\backslash G(\mathbb{A})^1} E(x, I_P(\tau, g_{i,\omega})\phi) \cdot \overline{\Lambda^T E(x,\phi)} dx \right| d\tau.$$

This in turn is bounded by $\sum_\omega \int_{G(F)\backslash G(\mathbb{A})^1} \sum_\chi \sum_P n(A)^{-1} X dx$ where

$$X = \int_{\Pi^G(M)} \left| \sum_{\phi \in \mathfrak{B}_P(\tau)_\chi} E(x, I_P(\tau, g_{i,\omega})\phi) \cdot \overline{\Lambda^T E(x,\phi)} \right| d\tau,$$

which is just (11.8.1) with $f$ replaced by $g_i$. Theorem 11.4, as well as Theorem 11.6, is now proved. □

## 12 Evaluation in a Special Case

### 12.1 Unramified Classes

In this section we shall give an explicit formula for $J_\chi^T(f)$ for a particular kind of class $\chi \in \mathfrak{X}$. These special $\chi$ we will call *unramified*; they are analogues of the unramified classes $\mathfrak{o} \in \mathcal{O}$ for which we calculate $J_\mathfrak{o}^T(f)$ in Section 8. The formula for $J_\chi^T(f)$ is a consequence of an inner product formula of Langlands which was announced in [La66], Section 9. Most of this section will be taken up with the proof, essentially that of Langlands, for the formula. First, however, we must demonstrate a connection between the truncation operator $\Lambda^T$ and the modified Eisenstein series defined by Langlands in [La66].

Fix a parabolic subgroup $P_1$ and a representation $\tau \in \Pi(M_1)$. If $\phi \in \mathcal{H}_{P_1}^0(\tau)$ and $\zeta \in \mathfrak{a}_{1,\mathbb{C}}^*$, write the Eisenstein series for $P \supset P_1$ as

$$E_P(x, \phi, \zeta) = E_P(x, \phi_\zeta).$$

If $s \in \Omega(\mathfrak{a}_1, \mathfrak{a}_2)$, define the operator $\mathcal{M}(s, \tau, \zeta) = \mathcal{M}(s, \zeta)$ by

$$\mathcal{M}(s,\zeta)\phi = (\mathcal{M}(s, \tau_\zeta)\phi_\zeta)_{-s\zeta}.$$

The operator $\mathcal{M}(s, \zeta)$ maps $\mathcal{H}_{P_1}^0(\tau)$ to $\mathcal{H}_{P_1}^0(s\tau)$.

Suppose that $\chi \in \mathfrak{X}$ is such that $P_1 \in P_\chi$. Then for all $x \in G(\mathbb{A})^1$, the function $m \longmapsto \phi(mx)$ is a cusp form in $m \in M_1(F)\backslash M_1(\mathbb{A})^1$. If $P_2$ is a second group in $\mathcal{P}_\chi$, we have the following basic formula from the theory of Eisenstein series.

PROPOSITION 12.2. *We have*

$$\int_{N_2(F)\backslash N_2(\mathbb{A})} E(nx,\phi,\zeta)dn = \sum_{s\in\Omega(\mathfrak{a}_1,\mathfrak{a}_2)} (\mathcal{M}(s,\zeta)\phi)(x)\cdot e^{(s\zeta+\rho_{P_2})(H(x))}.$$

PROOF. See [La76]. □

## 12.3 Constant Term of Eisenstein Series

A formula like this exists if $P_2$ is replaced by an arbitrary (standard) parabolic subgroup, $P$. Recall;

DEFINITION. The group $\Omega(\mathfrak{a}_1;P)$ is the union over all $\mathfrak{a}_2$ of those elements $s\in\Omega(\mathfrak{a}_1,\mathfrak{a}_2)$ such that $s\mathfrak{a}_1=\mathfrak{a}_2$ contains $\mathfrak{a}$, and $s^{-1}\alpha$ is positive for each $\alpha\in\Delta_2^P$.

PROPOSITION. *We have*

$$\int_{N(F)\backslash N(\mathbb{A})} E(nx,\phi,\zeta)dn = \sum_{s\in\Omega(\mathfrak{a}_1;P)} E_P(x,\mathcal{M}(s,\zeta)\phi,s\zeta). \tag{12.3.1}$$

The verification of this formula is left to the reader. It can be proved directly from the series definition of $E(x,\phi,\zeta)$. Alternatively, one can prove it by induction on $\dim A$, applying [La76], Lemma 3.7, to the group $M$.

Our main aim in this section is to prove

THEOREM 12.4 (LANGLANDS). *Suppose that $P_1$, $P_1'\in P_\chi$, that $\phi\in\mathcal{H}_{P_1}^0(\tau)_\chi$, $\phi'\in\mathcal{H}_{P_1'}^0(\tau')_\chi$, and that $\zeta$ and $\zeta'$ are vectors in $\mathfrak{a}_{P_1,\mathbb{C}}^*$ and $\mathfrak{a}_{P_1',\mathbb{C}}^*$ whose real parts are suitably regular points in $(\mathfrak{a}_{P_1}^*)^+$ and $(\mathfrak{a}_{P_1'}^*)^+$, respectively. Then*

$$\int_{G(F)\backslash G(\mathbb{A})^1} \Lambda^T E(x,\phi,\zeta)\overline{\Lambda^T E(x,\phi',\zeta')}dx$$

*equals the sum over $P_2\in\mathcal{P}_\chi$, $s\in\Omega(\mathfrak{a}_1,\mathfrak{a}_2)$, and $s'\in\Omega(\mathfrak{a}_1',\mathfrak{a}_2)$ of*

$$\mathrm{vol}(\mathfrak{a}_2^G/L_2)\cdot\frac{e^{(s\zeta+s'\bar{\zeta}')(T)}}{\prod_{\alpha\in\Delta_2}(s\zeta+s'\bar{\zeta}')(\alpha^\vee)}(\mathcal{M}(s,\zeta)\phi,\mathcal{M}(s',\zeta')\phi'). \tag{12.4.1}$$

□

Both sides of the identity of the theorem are meromorphic functions in $(\zeta,\bar{\zeta}')$. Therefore the identity is valid for all regular points $\zeta$ and $\zeta'$.

PROPOSITION 12.5. *Suppose that* $P_1 \in \mathcal{P}_\chi$ *as above, that* $\phi \in \mathcal{H}^0_{P_1}(\tau)_\chi$ *and that* $\zeta$ *is a point in* $\mathfrak{a}^*_{1,\mathbb{C}}$ *whose real part* $\zeta_\mathbb{R}$ *lies in* $\rho_1 + (\mathfrak{a}_1^*)^+$. *Then* $(\Lambda^T E)(x, \phi, \zeta)$ *equals*

$$\sum_{P_2} \sum_{\delta \in P_2(F)\backslash G(F)} \sum_{s \in \Omega(\mathfrak{a}_1, \mathfrak{a}_2)} \epsilon_2(s\zeta_\mathbb{R})\phi_2(s\zeta_\mathbb{R}, H_0(\delta x) - T) \tag{12.5.1}$$

$$\cdot e^{(s\zeta + \rho_2)(H_0(\delta x))}(\mathcal{M}(s, \zeta)\phi)(\delta x),$$

*with the sum over* $\delta$ *converging absolutely. The functions* $\epsilon_2$ *and* $\phi_2$ *are as in Section* 8.

PROOF. Suppose that $P_2$ and $s \in \Omega(\mathfrak{a}_1, \mathfrak{a}_2)$ are given. In the process of verifying the equality of (8.6.1) and (8.7.1), we ended up proving that for all $H \in \mathfrak{a}_0$, the expression

$$\epsilon_2(s\zeta_\mathbb{R})\phi_2(s\zeta_\mathbb{R}, H)$$

was equal to

$$\sum_{\{P; P \supset P_2, s \in \Omega(\mathfrak{a}_1; P)\}} (-1)^{\dim(A_P/Z)} \hat{\tau}_P(H).$$

Apply this to (12.5.1). Then decompose the sum over $P_2(F)\backslash G(F)$ into a sum over $P_2(F)\backslash P(F)$ and $P(F)\backslash G(F)$. The sum over $P(F)\backslash G(F)$ will be finite by Proposition 5.3. If $\alpha \in \Delta_2^P$, then $s^{-1}\alpha^\vee$ is a nonnegative sum of elements of the form $\beta^\vee$, for $\beta \in \Delta_1$. It follows that

$$(s\zeta_\mathbb{R} - \rho_2)(\alpha^\vee) = (\zeta_\mathbb{R} - \rho_1)(s^{-1}\alpha^\vee) + \rho_1(s^{-1}\alpha^\vee) - \rho_2(\alpha^\vee)$$

is positive. Therefore the sum

$$\sum_{\xi \in P_2(F)\backslash P(F)} e^{(s\zeta + \rho_2)(H_0(\xi\delta x))} \cdot (\mathcal{M}(s, \zeta)\phi)(\xi\delta x)$$

is absolutely convergent and in fact equal to $E_P(\delta x, \mathcal{M}(s, \zeta)\phi, s\zeta)$.

In particular, the original sum over $\delta$ in (12.4.1) is absolutely convergent.

We find that (12.5.1) equals

$$\sum_P (-1)^{\dim(A/Z)} \sum_{\delta \in P(F)\backslash G(F)} \{ \sum_{s \in \Omega(\mathfrak{a}_1; P)} E_P(\delta x, \mathcal{M}(s, \zeta)\phi, s\zeta)\} \hat{\tau}_P(H(\delta x) - T).$$

If the left hand side of (12.3.1) is substituted into the brackets, the result is $\Lambda^T E(x, \phi, \zeta)$. □

## 12.6 A Fourier Transform

To simplify the notation, we shall assume that $\tau(a)$ is the identity operator for all $a \in A_1(\mathbb{R}^0)$. This entails no loss of generality, since any $\tau_1 \in \Pi(M_1)$ equals $\tau_\eta$, for some such $\tau$ and some $\eta \in i\mathfrak{a}_1^*$.

DEFINITION. (1) Given $P_2$, put

$$\psi_2(x) = \sum_{s\in\Omega(\mathfrak{a}_1,\mathfrak{a}_2)} \epsilon_2(s\zeta_{\mathbb{R}})\phi_2(s\zeta_{\mathbb{R}}, H_0(x) - T) \cdot e^{(s\zeta+\rho_2)(H_0(x))}(\mathcal{M}(s,\zeta)\phi)(x).$$

(2) If $\Lambda \in i\mathfrak{a}_2^*$, and $x \in G(\mathbb{A})^1$, put

$$\Psi_2(\Lambda, x) = \int_{A_2(\mathbb{R}^0)\cap G(\mathbb{A})^1} e^{-(\Lambda+\rho_2)(H_0(ax))}\psi_2(ax)\, da.$$

This function is not hard to compute.

PROPOSITION. *The function* $\Psi_2(\Lambda, x)$ *equals*

$$\mathrm{vol}(\mathfrak{a}_2^G/L_2) \cdot \sum_{s\in\Omega(\mathfrak{a}_1,\mathfrak{a}_2)} \frac{e^{(s\zeta-\Lambda)(T)}}{\prod_{\alpha\in\Delta_2}(s\zeta-\Lambda)(\alpha^\vee)} \cdot (\mathcal{M}(s,\zeta)\phi)(x).$$

PROOF. We have to evaluate

$$\int_{A_2(\mathbb{R}^0)\cap G(\mathbb{A})^1} e^{(s\zeta-\Lambda)(H_0(ax))}\epsilon_2(s\zeta_{\mathbb{R}})\phi_2(s\zeta_{\mathbb{R}}, H_0(ax) - T)\, da.$$

Since $a \longmapsto H_2(ax)$ is a measure preserving diffeomorphism from $A_2(\mathbb{R}^0) \cap G(\mathbb{A})^1$ onto $\mathfrak{a}_2^G$, this last expression equals

$$\int_{\mathfrak{a}_2^G} e^{(s\zeta-\Lambda)(H)}\epsilon_2(s\zeta_{\mathbb{R}})\phi_2(s\zeta_{\mathbb{R}}, H - T)\, dH.$$

Make a further change of variables

$$H = \sum_{\alpha\in\Delta_2} t_\alpha \alpha^\vee, \quad t_\alpha \in \mathbb{R}.$$

Of course, we will have to multiply by the Jacobian of this change of measure. It is volume of $\mathfrak{a}_2^G$ modulo the lattice, $L_2$, spanned by $\{\alpha^\vee; \alpha \in \Delta_2\}$. The integral becomes a product of integrals of decreasing functions over half lines; it is easy to evaluate (see [Ar79], Lemma 3.4). The proposition follows. □

## 12.7 Analytic Continuation

We have been assuming that $\zeta_{\mathbb{R}}$ is a point in $\rho_1 + (\mathfrak{a}_1^*)^+$. Let us suppose from now on that it is suitably regular. Then $\Psi_2(\Lambda, x)$ can be analytically continued as a holomorphic function, for $\Lambda$ in a tube in $\mathfrak{a}_{2,\mathbb{C}}^*$ over a ball $B_{P_2}$ in $\mathfrak{a}_2^*$, centered at the origin, of arbitrarily large radius. The functions

$$\Psi_2(\Lambda) : x \longmapsto \Psi_2(\Lambda, x),$$

indexed by $\Lambda$, span a finite dimensional subspace of $L^2(M_2(F)\backslash M_2(\mathbb{A})^1 \times K)$. For fixed $\Lambda_0$ in $B_{P_2}$, $\Psi_2(\Lambda)$ is a square-integrable function from $\Lambda_0 + i(\mathfrak{a}_2^G)^*$ to this finite dimensional space.

Suppose that $P_1'$ is another group in $\mathcal{P}_\chi$. Choose a class $\tau' \in \Pi(M')$, a vector $\phi' \in \mathcal{H}_{P_1'}^0(\tau')$, and a point $\zeta' \in \mathfrak{a}_{P_1',\mathbb{C}}^*$ to satisfy the same condition as above. Define the functions $\psi_2'$ and $\Psi_2'$ associated with any other group $P_2'$ in $\mathcal{P}_\chi$. Then

$$\int_{G(F)\backslash G(\mathbb{A})^1} \Lambda^T E(x, \phi, \zeta) \cdot \overline{\Lambda^T E(x, \phi', \zeta')}\, dx \tag{12.7.1}$$

is the sum over $P_2$ and $P_2'$ in $\mathcal{P}_\chi$ of

$$\int_{G(F)\backslash G(\mathbb{A})^1} \left( \sum_{\delta \in P_2(F)\backslash G(F)} \psi_2(\delta x) \right) \overline{\left( \sum_{\delta \in P_2'(F)\backslash G(F)} \psi_2'(\delta x) \right)}\, dx.$$

This last inner product is given by a basic formula in the theory of Eisenstein series ([La66], Lemma 4.6). It equals

$$\int_{\Lambda_0 + i(\mathfrak{a}_2^G)^*} \sum_{t \in \Omega(\mathfrak{a}_2, \mathfrak{a}_2')} (\mathcal{M}(t, \Lambda)\Psi_2(\Lambda), \Psi_2'(-t\bar{\Lambda}))d\Lambda,$$

where $\Lambda_0$ is any point in $B_{P_2} \cap (\rho_2 + (\mathfrak{a}_2^*)^+)$, and $d\Lambda$ is the Haar measure on $i(\mathfrak{a}_2^G)^*$ which is dual to our Haar measure on $\mathfrak{a}_2^G$. Therefore (12.7.1) equals the sum over $P_2$ and $s \in \Omega(\mathfrak{a}_1, \mathfrak{a}_2)$, of the integral over $\Lambda$, of the product of

$$\mathrm{vol}(\mathfrak{a}_2^G/L_2)^2 \frac{e^{(s\zeta - \Lambda)(T)}}{\prod_{\alpha \in \Delta_2} (s\zeta - \Lambda)(\alpha^\vee)} \tag{12.7.2}$$

and

$$\sum_{P_2'} \sum_{t \in \Omega(\mathfrak{a}_2, \mathfrak{a}_2')} \sum_{s' \in \Omega(\mathfrak{a}_1', \mathfrak{a}_2')} \frac{e^{(s'\bar{\zeta}' + t\Lambda)(T)}}{\prod_{\alpha \in \Delta_2'} (s'\bar{\zeta}' + t\Lambda)(\alpha^\vee)} \cdot (\mathcal{M}(t, \Lambda)\mathcal{M}(s, \zeta)\phi, \mathcal{M}(s'\zeta')\phi').$$

PROPOSITION 12.8. *The function (12.7.3) is a regular function of $\Lambda$ on the tube over $\rho_2 + (\mathfrak{a}_2^*)^+$.*

PROOF. The functions $\mathcal{M}(t, \Lambda)$ are regular on this tube, so the only singularities are along hyperplanes

$$\{\Lambda; (s'\bar{\zeta}' + t\Lambda)(\alpha^\vee) = 0\},$$

for fixed $s'$, $t$, $\zeta'$, and $\alpha \in \Delta_2'$. Let $s_\alpha \in \Omega(\mathfrak{a}_2', \mathfrak{a}_2'')$ be the simple reflection belonging to $\alpha$ (see [La76], p. 35). Then $\beta = -s_\alpha \alpha$ is a root in $\Delta_{P_2''}$, and

$$\{\Lambda; (s_\alpha s'\bar{\zeta}' + st\Lambda)(\beta^\vee) = 0\}$$

is the same hyperplane. Thus the summands in (12.7.3) which are singular along a given hyperplane occur naturally in pairs. We shall show that the two residues around the hyperplane add up to 0.

Assume that $(s'\bar{\zeta}' + t\Lambda)(\alpha^\vee) = 0$. Then $(s_\alpha s'\bar{\zeta}' + s_\alpha t\Lambda)(\beta^\vee) = 0$. The inner product from the summand of (12.7.3) corresponding to $P_2''$, $s_\alpha s'$, $s_\alpha t$ equals

$$\begin{aligned} &(\mathcal{M}(s_\alpha t, \Lambda)\mathcal{M}(s, \zeta)\phi, \mathcal{M}(s_\alpha s', \zeta')\phi') \qquad (12.8.1)\\ &= (\mathcal{M}(s_\alpha, s'\zeta')^*\mathcal{M}(s_\alpha, t\Lambda) \cdot \mathcal{M}(t, \Lambda)\mathcal{M}(s, \zeta)\phi, \mathcal{M}(s', \zeta')\phi') \end{aligned}$$

by the functional equations. But

$$\mathcal{M}(s_\alpha, s'\zeta')^* = \mathcal{M}(s_\alpha, -s'\bar{\zeta}')^{-1} = \mathcal{M}(s_\alpha, t\Lambda)^{-1},$$

since $\mathcal{M}(s_\alpha, t\Lambda)$ depends only on the projection of $t\Lambda$ onto $\alpha$. Therefore (12.8.1) equals

$$(\mathcal{M}(t, \Lambda)\mathcal{M}(s, \zeta)\phi, \mathcal{M}(s', \zeta')\phi'),$$

which is the inner product from the summand of (12.7.3) corresponding to $P_2'$, $s'$, $t$. It follows that the residues of the two summands do add up to zero. Therefore (12.7.3) is regular at the hyperplane under consideration and so is regular on the tube over $\rho_2 + (\mathfrak{a}_2^*)^+$. □

Next we shall show that

PROPOSITION 12.9. *If $s \neq 1$, the integral in $\Lambda$ of the product of (12.7.2) and (12.7.3) equals zero.*

PROOF. Given such an $s$, choose a root $\alpha \in \Delta_2$ such that $(s\zeta_{\mathbb{R}})(\alpha^\vee) < 0$. Change the path of integration from $\operatorname{Re}\Lambda = \Lambda_0$ to $\operatorname{Re}\Lambda = \Lambda + N\varpi_\alpha$, where $N$ is a positive integer which we let approach $\infty$. We can do this by virtue of the regularity of (12.7.3) and the fact the numbers

$$\{\|\mathcal{M}(t, \Lambda)\|; \operatorname{Re}\Lambda = \Lambda + N\varpi_\alpha\}$$

are bounded independently of $N$. Notice that

$$|e^{-\Lambda(T)}e^{(t\Lambda)(T)}| = e^{(t\Lambda_0-\Lambda_0+N(t\varpi_\alpha-\varpi_\alpha))(T)}$$

is not greater than 1. Therefore the integral over $\mathrm{Re}\,\Lambda = \Lambda_0 + N\varpi_\alpha$ approaches 0 as $N$ approaches $\infty$. It follows that the original integral equals zero. □

PROOF OF THEOREM 12.4. We have only to set $s = 1$ in (12.7.2), multiply the result by (12.7.3), and then integrate over $\Lambda$. Make a change of variables in the integral over $\Lambda$, setting

$$\Lambda = \sum_{\alpha\in\Delta_2} z_\alpha \varpi_\alpha, \quad z_\alpha \in \mathbb{C}.$$

With this change of measures, we must multiply the result by the volume of $i(\mathfrak{a}_2^G)^*$ modulo the lattice spanned by $\{\varpi_\alpha; \alpha \in \Delta_2\}$. Since $d\Lambda$ represents the measure on $i(\mathfrak{a}_2^G)^*$ dual to that $\mathfrak{a}_2^G$, and since $\{\varpi_\alpha\}$ and $\{\alpha^\vee\}$ are duals bases, this factor equals

$$(2\pi i)^{-\dim(A_2/Z)}\,\mathrm{vol}(\mathfrak{a}_2^G/L_2)^{-1}.$$

The product of this factor with (12.7.2) then equals

$$\left(\frac{1}{2\pi i}\right)^{\dim(A_2/Z)} \mathrm{vol}(\mathfrak{a}_2^G/L_2)e^{\zeta(T)} \prod_{\alpha\in\Delta_2} \frac{e^{z_\alpha\cdot\varpi_\alpha(T)}}{\zeta(\alpha^\vee)-z_\alpha}.$$

Each $z_\alpha$ is to be integrated over the line $\Lambda_0(\alpha^\vee) + i\mathbb{R}$. We replace this contour with the line $\Lambda_0(\alpha^\vee) + N + i\mathbb{R}$, and let $N$ approach $\infty$. According to our assumptions on $\zeta$, we have $\zeta_{\mathbb{R}}(\alpha^\vee) > \Lambda_0(\alpha^\vee)$, so we pick up a residue at $z_\alpha = \zeta(\alpha^\vee)$. By the arguments of the previous section, the integral of $z_\alpha$ over the line $\Lambda_0(\alpha^\vee) + N + i\mathbb{R}$ approaches 0 as $N$ approaches $\infty$. Therefore the integral of $z_\alpha$ over $\Lambda_0(\alpha^\vee) + i\mathbb{R}$ equals the residue of the integrand at $z_\alpha = \zeta(\alpha^\vee)$. It follows that (12.7.1) is the product of $\mathrm{vol}(\mathfrak{a}_2^G/L_2)$ with the value of (12.7.3) at $s = 1$ and $\Lambda = \zeta$. Theorem 12.4 follows. □

DEFINITION 12.10. Recall that the elements of $\mathfrak{X}$ are equivalence classes of pairs $(M_1, \rho_1)$. We shall say that $\chi$ is *unramified* if for any pair $(M_1, \rho)$ in $\chi$, the only element $s \in \Omega(\mathfrak{a}_1, \mathfrak{a}_2)$ for which $s\rho = \rho$ is the identity.

For the remainder of this section, assume that $\chi$ is unramified.

Suppose that $P_1 = P_1' = P$ and that $\tau = \tau'$. If $\phi$, $\phi'$, $s$, and $s'$ are as in Theorem 12.4, then

$$(\mathcal{M}(s,\zeta)\phi, \mathcal{M}(s',\zeta')\phi') = 0$$

unless $s = s'$. It follows that for $\eta \in i\mathfrak{a}^*$, the inner product $(\mathcal{M}_P^T(\tau_\eta)_\chi \phi_\eta, \phi'_\eta)$ equals

$$\lim_{\zeta \to 0} \sum_{P_2 \in \mathcal{P}_\chi} \sum_{s \in \Omega(\mathfrak{a},\mathfrak{a}_2)} \operatorname{vol}(\mathfrak{a}_2^G/L_2) \cdot \frac{e^{(s\zeta)(T)}}{\prod_{\alpha \in \Delta_2} (s\zeta)(\alpha^\vee)} (\mathcal{M}(s, \eta + \zeta)\phi, \mathcal{M}(s', \eta)\phi').$$

We can now take $\tau$ to be any class in $\Pi(M)$. We have shown:

PROPOSITION 12.11. *Suppose that $\chi$ is unramified.*

(1) *For $P \in P_\chi$ and $\tau \in \Pi(M)$, we have*

$$\mathcal{M}_P^T(\tau)_\chi = \operatorname{vol}(\mathfrak{a}_P^G/L_P) \cdot \lim_{\zeta \to 0} \sum_{P_2 \in \mathcal{P}_\chi} \sum_{s \in \Omega(\mathfrak{a},\mathfrak{a}_2)} \frac{e^{(s\zeta)(T)} \mathcal{M}(s,\tau)^{-1} \mathcal{M}(s,\tau,\zeta)}{\prod_{\alpha \in \Delta_2} (s\zeta)(\alpha^\vee)}.$$

(2) *If $P$ does not belong to $\mathcal{P}_\chi$ and $\tau \in \Pi(M)$, then $\mathcal{H}_P(\tau)_\chi = \{0\}$.*

PROOF. (2) can be extracted from the results of [La76], Section 7. □

We can therefore write

$$J_\chi^T(f) = \sum_{P \in \mathcal{P}_\chi} n(A)^{-1} \int_{\Pi^G(M)} \operatorname{tr}(\mathcal{M}_P^T(\tau)_\chi \cdot I_P(\tau,f)_\chi) d\tau,$$

with $\mathcal{M}_P^T(\tau)_\chi$ given explicitly in 12.11(1) in terms of the global intertwining operators. If we wanted to pursue the analogy with Section 8, we might regard this formula as a linear combination of "weighted characters" of $f$.

# 13 Conclusion

The results of this chapter can be summarized as an identity for the reductive group $G$. Namely, there is an integer $m > 0$ such that for any $f \in C_c^m(G(\mathbb{A})^1)$ and any suitably regular point $T \in \mathfrak{a}_0^+$, we have

$$\sum_{\mathfrak{o} \in \mathcal{O}} J_\mathfrak{o}^T(f) = \sum_{\chi \in \mathfrak{X}} J_\chi^T(f),$$

where

$$J_\mathfrak{o}^T(f) = \int_{G(F)\backslash G(\mathbb{A})^1} k_\mathfrak{o}^T(x,f)dx = \int_{G(F)\backslash G(\mathbb{A})^1} j_\mathfrak{o}^T(x,f)dx,$$

and

$$J_\chi^T(f) = \int_{G(F)\backslash G(\mathbb{A})^1} k_\chi^T(x,f)dx$$

$$= \sum_P n(A)^{-1} \int_{\Pi^G(M)} \operatorname{tr} \mathcal{M}_P^T(\tau)_\chi \cdot I_P(\tau,f)_\chi \, d\tau.$$

Let $r_{\text{cusp}}$ be the restriction to the subspace $L^2_{\text{cusp}}(G(F)\backslash G(\mathbb{A})^1)$ of the right regular representation $r$. Let $\mathfrak{X}(G)$ be the set of classes $\chi \in \mathfrak{X}$ such that $\mathcal{P}_\chi = \{G\}$. Then $r_{\text{cusp}}$ is the direct sum over all $\chi$ in $\mathfrak{X}(G)$ of the representations $r_\chi$. If $\chi \in \mathfrak{X}(G)$ and $\pi \in \Pi(G)$, the operator $\mathcal{M}_G^T(\pi)_\chi$ is the identity operator. It follows from the finiteness of (11.5.1) that if $f$ is in $C_c^{m_1}(G(\mathbb{A})^1)$, then the operator $r_{\text{cusp}}(f)$ is of trace class. (This fact also follows from [HC68], p. 14 and Corollary 4.2.) Moreover if $f \in C_c^m(G(\mathbb{A})^1)$, for $m$ as in Theorem 11.6, then

$$\operatorname{tr}\ r_{\text{cusp}}(f) = \sum_{\chi \in \mathfrak{X}(G)} \operatorname{tr}\ r_\chi(f)$$

$$= \sum_{\chi \in \mathfrak{X}(G)} \int_{\Pi^G(G)} \operatorname{tr} I_G(\tau,f)d\tau = \sum_{\chi \in \mathfrak{X}(G)} J_\chi^T(f).$$

Thus

$$\operatorname{tr}\ r_{\text{cusp}}(f) = \sum_{\mathfrak{o} \in \mathcal{O}} J_{\mathfrak{o}}^T(f) - \sum_{\chi \in \mathfrak{X} \backslash \mathfrak{X}(G)} J_\chi^T(f).$$

# Chapter 4
# Study of Noninvariance

## 1 Notation

### 1.1 Levi Subgroups

Let $G$ be a reductive connected linear algebraic group over a field $F$. Let $M$ be a fixed Levi subgroup of $G$ over $F$. Let $\mathcal{F}(M)$ denote the set of parabolic subgroups of $G$ containing $M$. Let $\mathcal{L}(M)$ be the collection of Levi subgroups of $G$ containing $M$. Any $P \in \mathcal{F}(M)$ has a unique Levi component $M_P \in \mathcal{L}(M)$. We use this component to write $P = M_P N_P$, where $N_P$ is the unipotent radical of $P$. We let $\mathcal{P}(M)$ denote the collection of $P \in \mathcal{F}(M)$ such that $M_P = M$. Suppose $L \in \mathcal{L}(M)$. Then $M$ is a Levi subgroup of $L$. We write $\mathcal{F}^L(M)$, $\mathcal{L}^L(M)$ and $\mathcal{P}^L(M)$ for the analogous sets with $G$ replaced by $L$. Consider a pair of elements $Q \in \mathcal{P}(L)$ and $R \in \mathcal{P}^L(M)$. There is a unique element $P \in \mathcal{P}(M)$ such that $P \subset Q$ and $P \cap L = R$. We will write $Q(R)$ for this particular $P$ when we want to stress its dependence on $Q$ and $R$.

### 1.2 Lie Algebras

Let $M$ be a Levi subgroup. Denote by $A_M$ the split component of the center of $M$. Write $X(M)_F$ for the group of characters of $M$ which are defined over $F$. Set

$$\mathfrak{a}_M = \mathrm{Hom}(X(M)_F, \mathbb{R}).$$

Suppose that $M \subset L$ are Levi subgroups. There is a natural map from $\mathfrak{a}_M$ onto $\mathfrak{a}_L$. The kernel of this map is denoted by $\mathfrak{a}_M^L$.

Suppose that $P \in \mathcal{F}(M)$. We shall frequently write $A_P = A_{M_P}$ and $\mathfrak{a}_P = \mathfrak{a}_{M_P}$. The roots of $(P, A_P)$ are defined by taking the adjoint action of $A_P$ on the Lie algebra of $N_P$. They are characters on $A_P$, but we will normally regard them as elements in



Y.Z. Flicker, *Arthur's Invariant Trace Formula and Comparison of Inner Forms*,
DOI 10.1007/978-3-319-31593-5_4

the dual space, $\mathfrak{a}_P^* = X(M_P) \otimes \mathbb{R}$, of $\mathfrak{a}_P$. Denote the set of simple roots of $(P, A_P)$ by $\Delta_P$. Denote the associated set of coroots by

$$\Delta_P^\vee = \{\alpha^\vee ; \alpha \in \Delta_P\}.$$

This is a subset of $\mathfrak{a}_P$. The roots of $(P, A_P)$ divide $\mathfrak{a}_P$ into chambers. These are the connected components of the complement of the hyperplanes defined by the vanishing of the roots on $\mathfrak{a}_P$. We shall write $\mathfrak{a}_P^+$ for the chamber on which the roots in $\Delta_P$ are positive.

From now on, we take $F$ to be a local or global field of characteristic 0.

## 1.3 Harish-Chandra Homomorphism

Let $F$ be a global field. Fix a finite set $S$ of inequivalent valuations on $F$. Set

$$F_S = \prod_{v \in S} F_v.$$

It is a locally compact ring. We regard $G$ as a scheme over $F$. Since $F$ embeds diagonally in $F_S$, we can take the corresponding set, $G(F_S) = \prod_{v \in S} G(F_v)$, of $F_S$-valued points. It is a locally compact group.

Define the Harish-Chandra homomorphism

$$H_G : G(F_S) \to \mathfrak{a}_G$$

by

$$\langle H_G(x), \chi \rangle = \log |\chi(x)| = \log \left( \prod_{v \in S} |\chi(x_v)|_v \right)$$

for all $x \,(= x_v, v \in S)$ and $\chi$ in $X(G)_F$ or $\mathfrak{a}_G^*$. Write

$$\mathfrak{a}_{G,S} = H_G(G(F_S))$$

for the image of $H_G$.

## 1.4 Closure Property

Let $S_\infty$ be the set of Archimedean valuations of $F$. If $S \cap S_\infty$ is not empty, then $\mathfrak{a}_{G,S} = \mathfrak{a}_G$. If $S \cap S_\infty$ is empty, the set $\mathfrak{a}_{G,S}$ can become complicated. We are lead to the following

DEFINITION. The set $S$ is said to have the *closure property* if one of the following holds:

(1) $S$ contains an Archimedean valuation.
(2) $S$ does not contain an Archimedean valuation and all of the valuations in $S$ divide a fixed rational prime $p$.

Assume that $S$ has the closure property for the rest of this chapter. Set

$$\mathfrak{a}_{G,S}^{\vee} = \mathrm{Hom}(\mathfrak{a}_{G,S}, 2\pi\mathbb{Z}).$$

It is $\{0\}$ if $S \cap S_\infty \neq \varnothing$. It is a lattice in $\mathfrak{a}_G^*$ otherwise. Then

$$\mathfrak{a}_{G,S}^* = \mathfrak{a}_G^*/\mathfrak{a}_{G,S}^{\vee}$$

is the additive character group of $\mathfrak{a}_{G,S}$. It is a compact quotient of $\mathfrak{a}_G^*$ if $S \cap S_\infty$ is empty. It equals $\mathfrak{a}_G^*$ otherwise.

## 1.5 Measures

All integrals on groups and homogeneous spaces will be taken with respect to the invariant measures. We postpone discussing normalizations. However, we assume the measures satisfy any compatibility requirements in a given context. Later, we will discuss in some detail the measure on $N_P(\mathbb{R})$.

We will now consider the measure on $\mathfrak{a}_M$. Fix a Euclidean metric on a maximal such space, $\mathfrak{a}_{M_0}$, of a fixed minimal $F$-Levi subgroup $M_0$ in $G$, invariant under the Weyl group. This fixes Euclidean metrics on all such spaces, $\mathfrak{a}_M$, in a compatible way. Take the Haar measure on each $\mathfrak{a}_M$ associated to the Euclidean metric. Take the corresponding dual measure on $i\mathfrak{a}_M^*$. On the quotient space, $i\mathfrak{a}_{M,S}^* = i(\mathfrak{a}_M^*/\mathfrak{a}_{M,S}^{\vee})$, take the associated quotient measure. If $S \cap S_\infty = \varnothing$, we can assume that this quotient measure is dual to the discrete measure on $\mathfrak{a}_{M,S}$.

## 1.6 Maximal Compact Subgroup

Suppose that $F$ is a local field. Let $M$ be a Levi subgroup of $G$. A maximal compact subgroup $K$ of $G(F_v)$ is called *well-positioned relative to $M$* if the following holds:

(1) If $F$ is Archimedean, the Lie algebras of $A_M$ and $K$ are orthogonal with respect to the Killing form of $G$.
(2) If $F$ is non-Archimedean, the vertex of $K$ in the Bruhat-Tits building of $G$ is special, and hyperspecial if $G$ is unramified, namely, splits over an unramified field extension of $F$ and belongs to the apartment associated with a maximal split torus of $M$.

Any $K$ which is well-positioned relative to $M$ has the following properties:

(1) For any $P \in \mathcal{P}(M)$, we have $G(F) = P(F)K$.
(2) Any coset in $G(F)/M(F)$ which normalizes $M(F)$ has a representative in $K$.
(3) For any $P \in \mathcal{F}(M)$, we have $K \cap P(F) = (N_P(F) \cap K) \cdot (M_P(F) \cap K)$. If $L \in \mathcal{L}(M)$, then $K \cap L(F)$ is a maximal compact subgroup of $L(F)$ which is well-positioned relative to $M$.

See [Ti79] for further discussion of these properties.

Suppose $F$ is a global field. Let $G$ be a reductive connected linear algebraic group over $F$. Fix a minimal Levi $F$-subgroup $M_0$ of $G$. The split component $A_0 = A_{M_0}$ of the center of $M_0$ is a maximal $F$-split torus of $G$. We shall extend our definitions to a global maximal compact subgroup. A maximal compact subgroup

$$K = \prod_v K_v$$

of $G(\mathbb{A})$ is *well-positioned* relative to $M_0$ if the following conditions hold:

(1) For each valuation $v$ on $F$, the group $K_v$ is a maximal compact subgroup of $G(F_v)$, well-positioned relative to $M_0$.
(2) For every embedding of $G$ into $\mathrm{GL}(n)$, defined over $F$,

$$K_v = \mathrm{GL}(n, \mathcal{O}_v) \cap G(F_v)$$

for almost all $v$. Fix $K$. It satisfies the conditions Section 3.1.6.

Let $S$ be a finite set of places of $F$. For each $v \in S$ let $K_v$ be a fixed maximal compact subgroup of $G(F_v)$ with the properties above. Then

$$K = \prod_{v \in S} K_v$$

is a maximal compact subgroup of $G(F_S)$. Each $K_v$ is well-positioned relative to every $M \in \mathcal{L}(M_0)$. In particular, if $K_M = K \cap M(F_S)$, then $(M, K_M)$ satisfies the same conditions as $(G, K)$.

## 1.7 Induced Representations

Suppose $(\tau, V_\tau)$ is an admissible representation of $M(F_S)$. For $\lambda \in \mathfrak{a}^*_{M,\mathbb{C}}$, the twisted representation of $M(F_S)$ given by

$$\tau_\lambda(m) = \tau(m) e^{\lambda(H_M(m))}$$

is also admissible. For each $P \in \mathcal{P}(M)$, let $\tau_\lambda^G = i_P^G(\tau_\lambda)$ denote the normalized (by the factor $\delta_P^{1/2}$) induced representation. We will usually regard it as a representation of the convolution algebra of smooth, compactly supported, $K$-finite functions on $G(F_S)$. It acts on the space $V_P(\tau)$ of $K$-finite functions $\psi : K \to V_\tau$ that satisfy, for $n \in N_P(F_S) \cap K$, $m \in K_M$, and $k \in K$,

$$\psi(nmk) = \tau(m)\psi(k).$$

Such $\psi$ extends to $G(F_S)$, where it satisfies

$$\psi(nmk) = \delta_P^{1/2}(m)\tau_\lambda(m)\psi(k), \quad n \in N_P(F_S),\ m \in M_P(F_S),\ k \in K.$$

The group $G(F_S)$ acts by right translation. Considering only the restriction of the $\psi$ to $K$ trivializes the vector bundle of induced representations $\tau_\lambda^G$. They all act on the same space.

To write the action of $G(F_S)$ in the model of the $\psi$ on $K$, we write

$$x = n_P(x)m_P(x)k_P(x),$$

a decomposition into elements in $N_P(F_S)$, $M(F_S)$, and $K$. We let

$$H_P(x) = H_M(m_P(x)),$$

for $x \in G(F_S)$ and define the operator $i_P^G(\tau_\lambda, f)$ by

$$(i_P^G(\tau_\lambda, f)\psi)(k) = \int_{G(F_S)} f(y)\tau(m_P(ky))\psi(k_P(ky))e^{(\lambda+\rho_P)(H_P(ky))}\,\mathrm{d}y.$$

As usual, $\rho_F$ denotes the vector in $\mathfrak{a}_P^*$ associated with the square root $\delta_P^{1/2}$ of the modular function of $P(F_S)$. Indeed,

$$(\tau_\lambda^G(f)\psi)(g) = \int_{G(F_S)} f(y)\psi(gy)\,\mathrm{d}y$$

and

$$\psi(ky) = e^{(\lambda+\rho_P)(H_P(ky))}\tau(m_P(ky))\psi(k_P(ky))$$

when $\psi$ is viewed as a function on $G(F_S)$.

# 2 A Review of the Trace Formula

## 2.1 Minimal Levi

We wish to recall the statement of the trace formula of Chapter 3 and study how it fails to be invariant, in order to know how to modify the terms to make them invariant. Let $F$ be a global field. Fix a minimal Levi $F$-subgroup $M_0$ of our reductive connected linear $F$-group $G$.

## 2.2 Elements in a Weyl group

As in the local case, for $P \in \mathcal{F}(M_0)$ we can define the function

$$H_P(nmk) = H_{M_P}(m),$$

for $n \in N_P(\mathbb{A})$, $m \in M_P(\mathbb{A})$, and $k \in K$, from $G(\mathbb{A})$ to $\mathfrak{a}_P$. Let $W$ be the Weyl group of $(G, A_0)$. For any $s \in W$, let $w_s$ be a fixed representative of $s$ in $G(F)$. This $w_s$ is determined only modulo $M_0(F)$. But, for any $P \in \mathcal{F}(M_0)$, the vector $H_P(w_s^{-1})$ is uniquely determined. In contrast to the case of the standard maximal compact subgroup of $\mathrm{GL}(n, \mathbb{A})$, the element $w_s$ cannot always be chosen in $K$ for all $G$. However, there is an element $\tilde{w}_s \in K$ such that, for all $a \in A_0(\mathbb{A})$, we have

$$\tilde{w}_s a \tilde{w}_s^{-1} = w_s a w_s^{-1}.$$

It follows that $w_s \in KM_0(\mathbb{A})$, for every $s \in W$.

PROPOSITION 2.3. *There is a vector $T_0 \in \mathfrak{a}_0$, uniquely determined modulo $\mathfrak{a}_G$, such that, for any $P_0 \in \mathcal{P}(M_0)$ and $s \in W$, we have*

$$H_{P_0}(w_s^{-1}) = T_0 - s^{-1}T_0.$$

PROOF. The uniqueness follows from the fact that $\mathfrak{a}_G$ is the set of fixed points of the group $W$ acting on $\mathfrak{a}_0$. Since $w_s \in KM_0(\mathbb{A})$, the vector $H_{P_0}(w_s^{-1})$ is independent of $P_0 \in \mathcal{P}(M_0)$. As above, we have the set $\Delta_{P_0}$ of simple roots of $(P_0, A_0)$ in $\mathfrak{a}_0^*$, the set $\{\alpha^\vee ; \alpha \in \Delta_{P_0}\}$ of coroots in $\mathfrak{a}_0$, and the basis $\{\varpi_\alpha ; \alpha \in \Delta_{P_0}\}$ of $(\mathfrak{a}_{M_0}^G)^*$ which is dual to $\Delta_{P_0}$. Let $s_\alpha$ denote the simple reflection associated with $\alpha \in \Delta_{P_0}$. For each such $s_\alpha$, there is a real number $h_\alpha$ such that

$$H_{P_0}(w_{s_\alpha}^{-1}) = h_\alpha \alpha^\vee.$$

Define

$$T_0 = \sum_{\alpha \in \Delta_{P_0}} h_\alpha \varpi_\alpha.$$

We will use induction on the length of $s$. Let $\alpha \in \Delta_{P_0}$ be such that

$$s_1 = s_\alpha s,$$

with $l(s_1) = 1 + l(s)$. For some $m_s \in M_0(\mathbb{A})$, write $w_s^{-1} = m_s \tilde{w}_s^{-1}$. Then

$$\begin{aligned} H_{P_0}(w_{s_1}^{-1}) &= H_{P_0}(m_s) + s^{-1} H_{P_0}(w_{s_\alpha}^{-1}) \\ &= H_{P_0}(w_s^{-1}) + s^{-1} H_{P_0}(w_{s_\alpha}^{-1}). \end{aligned}$$

It follows by induction that

$$\begin{aligned} H_{P_0}(w_{s_1}^{-1}) - (T_0 - s_1^{-1} T_0) &= H_{P_0}(w_{s_1}^{-1}) - H_{P_0}(w_s^{-1}) - (T_0 - s_1^{-1} T_0) + (T_0 - s^{-1} T_0) \\ &= s^{-1} H_{P_0}(w_{s_\alpha}^{-1}) - s^{-1}(T_0 - s_\alpha T_0). \end{aligned}$$

By definition of $T_0$, this equals 0. The proposition follows. □

## 2.4 Decomposition of Haar measures

We will eventually end up with objects which are independent of any choice of Haar measures, as well as the choices of $K$ and $M_0$. For now, we will fix some measures. Suppose that $v$ is a valuation on $F$. Let $M_v$ be a Levi subgroup of $G$, defined over $F_v$, for which $K_v$ is well-positioned. We normalize the Haar measure on $K_v \cap M_v(F_v)$ so that its total volume is one. Suppose that $L_v \in \mathcal{L}(M_v)$ and $P = P_v \in \mathcal{F}^{L_v}(M_v)$. Assume that $v$ is discrete. In this case, we take the Haar measure on $N_P(F_v)$ such that $N_P(F_v) \cap K_v$ has volume one. We do the same for $M_P(F_v)$. Let $\delta_P$ denote the modular function of $P(F_v)$. For $f \in C_c(L(F_v))$, our choices give

$$\int_{L_v(F_v)} f(x)\, dx = \int_{N_P(F_v)} \int_{M_P(F_v)} \int_{K_v \cap L_v(F_v)} f(nmk) \delta_P(m)^{-1}\, dk\, dm\, dn. \tag{2.4.1}$$

Assume now that $v$ is Archimedean. We fix Haar measures on all groups $\{N_P(F_v), M_P(F_v)\}$ as above, so that (2.4.1) holds, and so that groups which are conjugate under $K_v$ have compatible Haar measures.

Suppose that $S$ is a (possibly infinite) set of valuations on $F$. Suppose that $L \in \mathcal{L}(M_0)$ and $P \in \mathcal{F}^L(M_0)$. We take the restricted product measures on the groups $\prod_{v \in S} M_P(F_v)$ and $\prod_{v \in S} N_P(F_v)$. Then the analogue of (2.4.1) holds for functions on $\prod_{v \in S} L(F_v)$. In this way, we obtain Haar measures on the groups $N_P(\mathbb{A})$ and $M_P(\mathbb{A})$. Assume $v$ is Archimedean. By further restricting our choice of measures on $N_P(F_v)$, we can assume that, for each $P$, the volume of $N_P(F) \backslash N_P(\mathbb{A})$ is one. Then our measures on adèle groups satisfy the conditions of Subsection 3.1.9.

We take the Haar measure on $\mathfrak{a}_0 = \mathfrak{a}_{M_0}$ associated with a Euclidean metric which is invariant under the Weyl group $W$ of $M_0$ in $G$. The metric gives a measure on any subspace of $\mathfrak{a}_0$. If $P \in \mathcal{P}(M_0)$, the space $i\mathfrak{a}_P^*$ is isomorphic to the group of unitary characters on $\mathfrak{a}_P$. We take the Haar measure on $i\mathfrak{a}_P^*$ which is dual to that on $\mathfrak{a}_P$. The measures on $M(\mathbb{A})$ and $\mathfrak{a}_M$ yield a measure on the kernel of the map

$$H_M : M(\mathbb{A}) \to \mathfrak{a}_M,$$

which we denote by $M(\mathbb{A})^1$. Since $M(\mathbb{A}) = M(\mathbb{A})^1 \times A(\mathbb{R}^0)$, we obtain a Haar measure on $A(\mathbb{R}^0)$, where $\mathbb{R}^0 = \mathbb{R}^\times_{>0}$ is embedded diagonally in $\prod_{v\in S_\infty} F_v^\times$. See also Subsection 3.1.7.

## 2.5 Roots, coroots, and characteristic functions

We will recall the noninvariant trace formula presented in Chapter 3. We will continue to use the notation introduced in Section 1 there. More specifically, if $P \subset Q$ are members of $\mathcal{F}(M_0)$, then $\mathfrak{a}_P^Q$ is the subspace $\mathfrak{a}_{M_P}^{M_Q} \subset \mathfrak{a}_P$. Let $\Delta_P^Q$ be the set of simple roots of $(P \cap M_Q, A_P)$. There is an associated basis $\{\alpha^\vee\,;\, \alpha \in \Delta_P^Q\}$ of $\mathfrak{a}_P^Q$. The dual basis of $(\mathfrak{a}_P^Q)^*$ is denoted by $\widehat{\Delta}_P^Q$. The sets $\Delta_P^Q$ and $\widehat{\Delta}_P^Q$ are naturally embedded in $\mathfrak{a}_0^*$. In addition, there is the characteristic function of

$$\{H \in \mathfrak{a}_0\,;\, \alpha(H) > 0, \alpha \in \Delta_P^Q\},$$

denoted by $\tau_P^Q$. Similarly, $\hat{\tau}_P^Q$ is the characteristic function of

$$\{H \in \mathfrak{a}_0\,;\, \varpi(H) > 0, \varpi \in \widehat{\Delta}_P^Q\}.$$

We use similar notation for parabolic subgroups of a Levi subgroup of $G$. For example, if $R \supset Q$, let $Q_1 = Q \cap M_R$ and $P_1 = P \cap M_R$. Then $\Delta_{P_1}^{Q_1} = \Delta_P^Q$, $\tau_{P_1}^{Q_1} = \tau_P^Q$, $\mathfrak{a}_{Q_1} = \mathfrak{a}_Q$, and so on.

## 2.6 Basic trace formula

In Chapter 3, the trace formula depends on a fixed minimal parabolic subgroup $P_0 \in \mathcal{P}(M_0)$. Until we remove this dependence, we fix $P_0$ and use $P$, $Q$, and $R$ for groups in $\mathcal{P}(M_0)$ containing $P_0$. The terms in the formula are indexed by sets $\mathcal{O}$ and $\mathfrak{X}$. (See Sections 3.2.3 and 3.3.1.) The set $\mathcal{O}$ can be defined to be the set of semisimple conjugacy classes in $G(F)$. The elements in $\mathfrak{X}$ are Weyl orbits of irreducible cuspidal automorphic representations of Levi subgroups of $G$. More specifically, the Weyl group $W$ acts on the set of pairs $\{(M, \tau^1)\}$, where $M \in \mathcal{L}(M_0)$ and $\tau^1$ is a cuspidal representation of $M(\mathbb{A})^1$. For $f \in C_c^\infty(G(\mathbb{A})^1)$, the trace formula of the previous chapter is the identity

$$\sum_{\mathfrak{o} \in \mathcal{O}} J_{\mathfrak{o}}^T(f) = \sum_{\chi \in \mathfrak{X}} J_\chi^T(f).$$

Here $T$ is a suitably regular point, depending on the support of $f$, in $\mathfrak{a}_0^+$, the positive chamber in $\mathfrak{a}_0$ defined by $P_0$. We will recall the definition of the distributions $J_{\mathfrak{o}}^T$ and $J_\chi^T$ in the next section.

We will use inductive arguments, so we need to keep track of distributions on Levi components $L = M_P$ of parabolic subgroups $P$ such that $P_0 \subset P$. We can also define the sets of equivalence classes $\mathcal{O}^L$ and $\mathfrak{X}^L$ associated with $L$. If $\mathfrak{o} \in \mathcal{O}$, then $\mathfrak{o} \cap L(F)$ is a (possibly empty) union of classes $\mathfrak{o}_1, \dots, \mathfrak{o}_n$ in $\mathcal{O}^L$. The intersection $P_0 \cap L$ is a fixed minimal parabolic subgroup of $L$. The point $T$ remains in the associated positive chamber of $\mathfrak{a}_0$. We therefore have the distribution $J_{\mathfrak{o}_i}^{L,T}$, defined on $C_c^\infty(L(\mathbb{A})^1)$. Define

$$J_{\mathfrak{o}}^{L,T} = \sum_{i=1}^n J_{\mathfrak{o}_i}^{L,T}.$$

Similarly, assume $\chi \in \mathfrak{X}$. Then $\chi$ is a $G$-Weyl orbit of irreducible cuspidal automorphic representations on Levi subgroups. Again, this decomposes into a (possibly empty) finite union of $L$-Weyl orbits $\chi_1, \dots, \chi_n$ in $\mathfrak{X}^L$. Define

$$J_\chi^{L,T} = \sum_{i=1}^n J_{\chi_i}^{L,T}.$$

The trace formula for $L$ implies that, for $f \in C_c^\infty(L(\mathbb{A})^1)$, we have

$$\sum_{\mathfrak{o} \in \mathcal{O}} J_{\mathfrak{o}}^{L,T}(f) = \sum_{\chi \in \mathfrak{X}} J_\chi^{L,T}(f). \tag{2.6.1}$$

# 3 The Distributions $J_{\mathfrak{o}}$ and $J_\chi$

## 3.1 *Dependence on truncating parameter*

In this section we shall recall the definitions of the distributions $J_{\mathfrak{o}}^T(f)$ and $J_\chi^T(f)$, originally defined in Sections 8 and 11 of Chapter 3, respectively. In addition, we will show that they are polynomials in $T$, that is, as functions of $T$, they belong to the symmetric algebra on $\mathfrak{a}_{0,\mathbb{C}}^*$. A key tool is a function of $X, H \in \mathfrak{a}_0$, denoted $\Gamma_Q'(H,X)$, used also to study invariance in the next section.

DEFINITION 3.2. *Fix $f \in C_c^\infty(G(\mathbb{A})^1)$ and $\mathfrak{o} \in \mathcal{O}$. Let $Z = A_G$. Put*

$$K_{P,\mathfrak{o}}(x,y) = \sum_{\gamma \in M_P(F)\cap\mathfrak{o}} \int_{N_P(\mathbb{A})} f(x^{-1}\gamma n y)\, dn.$$

*Then set*

$$J_{\mathfrak{o}}^T(f) = \int_{x\in G(F)\backslash G(\mathbb{A})^1} \sum_{P\supset P_0} (-1)^{\dim(A_P/Z)} \sum_{\delta\in P(F)\backslash G(F)} K_{P,\mathfrak{o}}(\delta x, \delta x)\hat{\tau}_P(H_P(\delta x) - T).$$

Fix a suitably regular point $T_1 \in \mathfrak{a}_0^+$. We will let $T$ vary in $T_1 + \mathfrak{a}_0^+$ and try to relate $J_{\mathfrak{o}}^T$ with the distributions $J_{\mathfrak{o}}^{M,T_1}$. The main issue will be expressing $\hat{\tau}_P(H_P(\delta x) - T)$ in terms of the functions $\hat{\tau}_P^Q(H_P(\delta x) - T_1)$, as $Q$ ranges over parabolic subgroups containing $P$. This suggests the following inductive definition.

DEFINITION 3.3. *Let $X \in \mathfrak{a}_0$. For each $Q \supset P_0$ define inductively a function*

$$\Gamma'_Q(H, X)$$

*of $H \in \mathfrak{a}_0$ by demanding that*

$$\hat{\tau}_Q(H - X) = \sum_{R\supset Q} (-1)^{\dim(A_R/Z)} \hat{\tau}_Q^R(H)\Gamma'_R(H, X).$$

*The definition is indeed inductive. If $\Gamma'_R(H,X)$ has been defined for all $R \supsetneq Q$, then $\Gamma'_Q(H,X)$ is specified uniquely by the formula. The function $\Gamma'_Q(H,X)$ depends only on the projections of $H$ and $X$ onto $\mathfrak{a}_Q$. Furthermore, it is invariant under $\mathfrak{a}_G$.*

## 3.4 Rewriting a function

We would like a different expression for $\Gamma'_Q(H, X)$.

PROPOSITION. *The function $\Gamma'_Q(H,X)$ is equal to*

$$\sum_{R\supset Q} (-1)^{\dim(A_R/Z)} \tau_Q^R(H)\hat{\tau}_R(H - X). \tag{3.4.1}$$

PROOF. The sum equals

$$\sum_{R_1\supset Q} \Gamma'_{R_1}(H, X) \sum_{\{R;\, Q\subset R\subset R_1\}} (-1)^{\dim(A_R/A_{R_1})} \tau_Q^R(H)\hat{\tau}_R^{R_1}(H).$$

If $Q \subsetneq R_1$, the sum over $R$ vanishes (see Corollary 3.6.3). Hence $\Gamma'_Q(H,X)$ equals (3.4.1). □

EXAMPLE 3.5. Let $G = \mathrm{GL}(3)$. Then $\mathfrak{a}_{P_0}/\mathfrak{a}_G$ is two dimensional. If $X \in \mathfrak{a}_0^+$, then $\Gamma_{P_0}(\cdot, X)$ is the characteristic function of the shaded region.

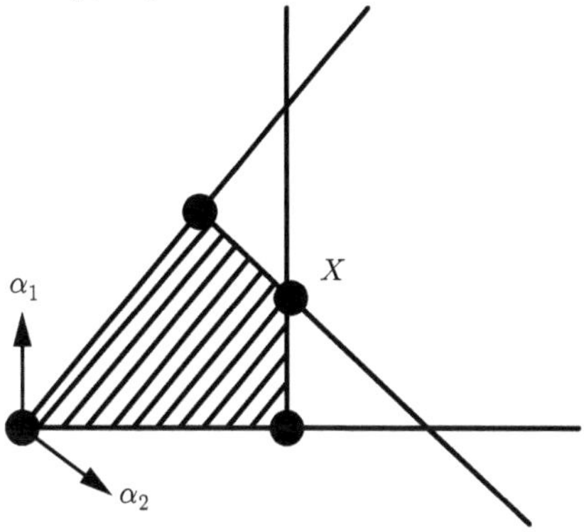


It is the algebraic signed sum of the characteristic functions of the chambers at each of the four vertices.

In general, we have the following

PROPOSITION 3.6. *For each X in a fixed compact subset of* $\mathfrak{a}_P/\mathfrak{a}_G$, *the support of the function*

$$H \mapsto \Gamma'_P(H,X), \quad H \in \mathfrak{a}_P/\mathfrak{a}_G,$$

*is contained in a fixed compact set, which is independent of X.*

PROOF. If $Q \supset P$, set $\hat{\tau}^{Q/P}(H)$ equal to the characteristic function of

$$\{H;\, \varpi(H) > 0,\, \varpi \in \widehat{\Delta}_P \smallsetminus \widehat{\Delta}_Q\}.$$

As above, we can inductively define functions $\widetilde{\Gamma}_Q(H,X)$ by demanding that, for all $Q \supset P_0$,

$$\hat{\tau}_Q(H-X) = \sum_{R \supset Q} (-1)^{\dim(A_R/Z)} \hat{\tau}^{R/Q}(H) \widetilde{\Gamma}_R(H,X).$$

Then

$$\widetilde{\Gamma}_Q(H,X) = \sum_{R \supset Q} (-1)^{\dim(A_R/Z)} \hat{\tau}^{R/Q}(H) \hat{\tau}_R(H-X).$$

By inspection, $\widetilde{\Gamma}_Q(\cdot, X)$ is the characteristic function in $\mathfrak{a}_Q/\mathfrak{a}_G$ of a parallelepiped with opposite vertices 0 and $X$, at least modulo sign. In particular, $\widetilde{\Gamma}_Q(H,X)$ is compactly supported as a function of $H \in \mathfrak{a}_Q/\mathfrak{a}_G$.

We prove the proposition by induction on $\dim G$. We have

$$\Gamma'_P(H,X) = \sum_{Q\supset P}(-1)^{\dim(A_Q,Z)}\tau_P^Q(H)\hat{\tau}_Q(H-X)$$
$$= \sum_{R\supset P}\widetilde{\Gamma}_R(H,X)\sum_{\{Q;P\subset Q\subset R\}}(-1)^{\dim(A_Q/A_R)}\tau_P^Q(H)\hat{\tau}^{R/Q}(H).$$

We noted that, if $P \neq G$, then

$$\sum_{\{Q;P\subset Q\subset G\}}(-1)^{\dim(A_Q/Z)}\tau_P^Q(H)\hat{\tau}_Q(H) = 0.$$

Therefore the outer sum may be taken over only those $R \neq G$. For $R \neq G$ and $H \in \mathfrak{a}_P^G$, decompose

$$H = H^* + H_*, \quad H^* \in \mathfrak{a}_P^R,\ H_* \in \mathfrak{a}_R^G.$$

Then $\widetilde{\Gamma}_R(H,X) = \widetilde{\Gamma}_R(H_*,X)$. Moreover,

$$\hat{\tau}^{R/Q}(H) = \hat{\tau}_Q^R(H^* - L(H_*)),$$

where

$$H_* \mapsto L(H_*)$$

is a linear map from $\mathfrak{a}_R^G$ to $\mathfrak{a}_P^R$, which is independent of $Q$. If the summand corresponding to $R$ does not vanish, $H_*$ will lie in a fixed compact set. So too will $L(H_*)$. Applying the induction assumption to the group $M_R$, we see that $H^*$ must lie in a fixed compact subset of $\mathfrak{a}_P^R$. It follows that $H$ is contained in a fixed compact subset of $\mathfrak{a}_P^G$.

DEFINITION 3.7. (1) *We define $L_P^Q$ to be the lattice generated by* $\{\alpha^\vee; \alpha \in \Delta_P^Q\}$. *Similarly, we define $\widehat{L}_P^Q$ to be the lattice generated by* $\{\varpi^\vee; \varpi \in \widehat{\Delta}_P^Q\}$.
(2) *Put*

$$\theta_Q^R(\Lambda) = \mathrm{vol}(\mathfrak{a}_Q^R/L_Q^R)^{-1}\prod_{\alpha\in\Delta_Q^R}\lambda(\alpha^\vee).$$

*Similarly, define*

$$\hat{\theta}_Q^R(\Lambda) = \mathrm{vol}(\mathfrak{a}_Q^R/\widehat{L}_Q^R)^{-1}\prod_{\varpi\in\widehat{\Delta}_Q^R}\lambda(\varpi^\vee).$$

The Fourier transform of $\Gamma'_P(\cdot, X)$ will be an entire function on $\mathfrak{a}^*_{P,\mathbb{C}}$. More precisely, we have the following proposition.

PROPOSITION 3.8. *The Fourier transform of the function of $H \in \mathfrak{a}_P/\mathfrak{a}_G$, given by*

$$H \mapsto \Gamma'_P(H, X),$$

*is*

$$\sum_{Q \supset P} (-1)^{\dim(A_P/A_Q)} e^{\lambda_Q(X)} \hat{\theta}^Q_P(\lambda)^{-1} \theta_Q(\lambda)^{-1}, \quad \lambda \in \mathfrak{a}^*_{P,\mathbb{C}}.$$

*In particular, it is a regular function of $\lambda$.*

PROOF. Let $\lambda \in \mathfrak{a}^*_{P,\mathbb{C}}$ be a point whose real part belongs to $-(\mathfrak{a}^*_P)^+$. Then

$$\int_{\mathfrak{a}^G_P} \Gamma'_P(H, X) e^{\lambda(H)}\, \mathrm{d}H = \sum_{Q \supset P} (-1)^{\dim(A_Q/Z)} \int_{\mathfrak{a}^G_P} \tau^Q_P(H) \hat{\tau}_Q(H - X) e^{\lambda(H)}\, \mathrm{d}H.$$

Make the change of variables

$$H = \sum_{\varpi \in \widehat{\Delta}^Q_P} t_\varpi \varpi^\vee + \sum_{\alpha \in \Delta_Q} t_\alpha \alpha^\vee.$$

After the change of variables, we must multiply by the volume of $\mathfrak{a}^G_P$ modulo $\widehat{L}^Q_P \times L_Q$. The result is

$$\sum_{Q \supset P} (-1)^{\dim(A_Q/Z)} (-1)^{\dim(A_P/Z)} \operatorname{vol}(\mathfrak{a}^G_P / \widehat{L}^Q_P \times L_Q) \prod_{\varpi \in \widehat{\Delta}^Q_P} \lambda(\varpi^\vee)^{-1} \cdot \prod_{\alpha \in \Delta_Q} e^{\lambda(\alpha^\vee)\varpi_\alpha(X)} \lambda(\alpha^\vee)^{-1}.$$

Here $\varpi_\alpha$ is the element in $\widehat{\Delta}_Q$ dual to $\alpha^\vee$. Let $\lambda_Q$ denote the projection of $\lambda$ onto $\mathfrak{a}^*_{Q,\mathbb{C}}$. Then

$$\prod_{\alpha \in \Delta_Q} e^{\lambda(\alpha^\vee)\varpi_\alpha(X)} = e^{\lambda_Q(X)}.$$

Using the previous definition, the proposition follows. □

PROPOSITION 3.9. *We have*

$$\int_{\mathfrak{a}_P/\mathfrak{a}_G} \Gamma'_P(H, X)\, dH = \frac{1}{q!} \sum_{Q \supset P} (-1)^{\dim(A_P/A_Q)} \lambda_Q(X)^q / (\hat{\theta}^Q_P(\lambda) \theta_Q(\lambda)).$$

*It is a polynomial in $X$ which is homogeneous of degree $q$.*

PROOF. Replace $\lambda$ by $t\lambda$, $t > 0$, in the formula and let $t$ approach 0. The limit exists and is independent of $\lambda$. Since $\hat{\theta}_P^Q(\lambda)^{-1}\theta_Q(\lambda)^{-1}$ is homogeneous of degree $q = \dim(A_P/Z)$, the proposition follows. □

We now return to the discussion of $J_{\mathfrak{o}}^T(f)$.

PROPOSITION 3.10. *We have*

$$J_{\mathfrak{o}}^T(f) = \sum_{Q \supset P_0} J_{\mathfrak{o}}^{M_Q,T_1}(f_Q) \cdot \int_{\mathfrak{a}_Q/\mathfrak{a}_G} \Gamma'_Q(H, T - T_1)\, dH. \tag{3.10.1}$$

*In particular, $J_{\mathfrak{o}}^T(f)$ is a polynomial in $T$.*

PROOF. In the expression for $J_{\mathfrak{o}}^T(f)$, make the substitution

$$\hat{\tau}_P(H_P(\delta x) - T) =$$

$$\sum_{Q \supset P} (-1)^{\dim(A_Q/Z)} \hat{\tau}_P^Q(H_P(\delta x) - T_1)\Gamma'_Q(H_Q(\delta x) - T_1, T - T_1).$$

Take the sum over $Q$ outside the sum over $P$. Write the integral over $(x, \delta)$ in $(G(F)\backslash G(\mathbb{A})^1) \times (P(F)\backslash G(F))$ as an integral over $(Q(F)\backslash G(\mathbb{A})^1) \times (P(F)\backslash Q(F))$. Then

$$J_{\mathfrak{o}}^T(f) = \sum_{Q \supset P_0} \int_{Q(F)\backslash G(\mathbb{A})^1} \sum_{\{P;\, P_0 \subset P \subset Q\}} (-1)^{\dim(A_P/A_Q)} \sum_{\delta \in P(F)\backslash Q(F)} K_{P,\mathfrak{o}}(\delta x, \delta x) \cdot X\, \mathrm{d}x$$

where $X$ is

$$\hat{\tau}_P^Q(H_P(\delta x) - T_1)\Gamma'_Q(H_Q(x) - T_1, T - T_1).$$

Decompose the integral over $x$ into an integral over $n \in N_Q(F)\backslash N_Q(\mathbb{A})$, $m \in M_Q(F)\backslash M_Q(\mathbb{A})^1$, $a \in A_Q(\mathbb{R}^0) \cap G(\mathbb{A})^1$, and $k \in K$. For $\delta \in M_Q(F)$, we have

$$K_{P,\mathfrak{o}}(\delta namk, \delta namk) = \delta_Q(a) K_{P,\mathfrak{o}}(\delta mk, \delta mk).$$

Hence, $J_{\mathfrak{o}}^T(f)$ equals the sum over $Q \supset P_0$ of the product of

$$\int_{\mathfrak{a}_Q/\mathfrak{a}_G} \Gamma'_Q(H, T - T_1)\, \mathrm{d}H$$

with

$$\int_K \int_{M_Q(F)\backslash M_Q(\mathbb{A})^1} \sum_{\{P;\, P_0 \subset P \subset Q\}} (-1)^{\dim(A_P/A_Q)} \tag{3.10.2}$$

$$\cdot \sum_{\delta \in P(F) \cap M_Q(F)\backslash M_Q(F)} K_{P,\mathfrak{o}}(\delta mk, \delta mk) \cdot \hat{\tau}_P^Q(H_P(\delta mk) - T_1).$$

Set

$$f_Q(m) = \delta_Q(m)^{1/2} \int_K \int_{N_Q(\mathbb{A})} f(k^{-1}mnk)\,\mathrm{d}n\,\mathrm{d}k.$$

It is a smooth compactly supported function on $M_Q(\mathbb{A})^1$. For $u_1, u_2 \in M_Q(\mathbb{A})^1$, we have

$$\int_K K_{P,\mathfrak{o}}(u_1 k, u_2 k)\,\mathrm{d}k = \sum_{\gamma \in M_P(F) \cap \mathfrak{o}} \int_{N_P(\mathbb{A}) \cap M_Q(\mathbb{A})} f_Q(u_1^{-1}\gamma n u_2)\,\mathrm{d}n.$$

The sum over $P$ can be regarded as a sum over standard parabolic subgroups of $M_Q$. It follows that (3.10.2) equals $J_{\mathfrak{o}}^{M_Q,T_1}(f_Q)$. □

Definition 3.11. *Let $\chi \in \mathfrak{X}$. The function $K_{P,\chi}(x,y)$ is the kernel of the restriction of the operator $r_P(f)$ to the invariant subspace $L^2(N_P(\mathbb{A})M_P(F)\backslash G(\mathbb{A})^1)_\chi$ of $L^2(N_P(\mathbb{A})M_P(F)\backslash G(\mathbb{A})^1)$. (See Subsection 3.3.3 for definitions.) It can be obtained by projecting*

$$K_P(x,y) = \sum_{\gamma \in M_P(F)} \int_{N_P(\mathbb{A})} f(x^{-1}\gamma n y)\,dn,$$

*regarded either as a function of $x$ or $y$, onto $L^2(N_P(\mathbb{A})M_P(F)\backslash G(\mathbb{A})^1)_\chi$. The kernel $K_{P,\chi}(x,y)$ is smooth as a function of $x$ or $y$. We define*

$$J_\chi^T(f) = \int_{x \in G(F)\backslash G(\mathbb{A})^1} \sum_{P \supset P_0} (-1)^{\dim(A_P/Z)} \sum_{\delta \in P(F)\backslash G(F)} K_{P,\chi}(\delta x, \delta x)\hat{\tau}_P(H_P(\delta x) - T).$$

There is an analogue of (3.10.1) for $J_\chi^T(f)$.

Proposition 3.12. *We have*

$$J_\chi^T(f) = \sum_{Q \supset P_0} J_\chi^{M_Q,T_1}(f_Q) \int_{\mathfrak{a}_Q/\mathfrak{a}_G} \Gamma_Q'(H, T - T_1)\,dH. \tag{3.12.1}$$

Proof. The argument follows that for $J_{\mathfrak{o}}^T(f)$ until we come to the integral

$$\int_K K_{P,\chi}(u_1 k, u_2 k)\,\mathrm{d}k.$$

If we allow $u_1, u_2$ to belong to $G(\mathbb{A})^1$, this integral is the kernel of the restriction of the operator

$$\int_K r_P(k) r_P(f) r_P(k)^{-1}\,\mathrm{d}k$$

to the invariant subspace $L^2(N_P(\mathbb{A})M_P(F)\backslash G(\mathbb{A}))_\chi$. Suppose $Q \supset P$. There is a representation $r^{M_Q}_{P\cap M_Q}$ of $M_Q(\mathbb{A})^1$ on the Hilbert space $L^2(N_P(\mathbb{A})M_P(F) \cap M_Q(\mathbb{A})^1\backslash M_Q(\mathbb{A})^1)$. Associated with $\chi$, we have the subspace $L^2(N_P(\mathbb{A})M_P(F) \cap M_Q(\mathbb{A})^1\backslash M_Q(\mathbb{A})^1)_\chi$, which is invariant under the operator $r^{M_Q}_{P\cap M_Q}(f_Q)$. If we take $u_1$ and $u_2$ to be elements in $M_Q(\mathbb{A})^1$, we obtain the kernel of the restriction of $r^{M_Q}_{P\cap M_Q}(f_Q)$ to this subspace. □

Analogues of (3.10.1) and (3.12.1) hold for the distributions $J^{L,T}_{\mathfrak{o}}$ and $J^{L,T}_{\chi}$. In particular, we have the following

THEOREM 3.13. *Suppose $L = L_Q$ for $Q \supset P_0$, $f \in C^\infty_c(L(\mathbb{A})^1)$, $\mathfrak{o} \in \mathcal{O}$, and $\chi \in \mathfrak{X}$. Then $J^{L,T}_{\mathfrak{o}}(f)$ and $J^{L,T}_{\chi}(f)$ are polynomial functions of $T$.*

DEFINITION 3.14. *The polynomials $J^{L,T}_{\mathfrak{o}}(f)$ and $J^{L,T}_{\chi}(f)$ can be defined for all $T$. We shall denote the values assumed at $T_0$, the vector defined by* 2.3, *by $J^L_{\mathfrak{o}}(f)$ and $J^L_{\chi}(f)$, respectively.*

*It follows from Theorem* 3.7.1 *and Theorem* 3.10.1 *that the series $\sum_{\mathfrak{o}} |J^L_{\mathfrak{o}}(f)|$ and $\sum_{\chi} |J^L_{\chi}(f)|$ converge. We obtain the identity*

$$\sum_{\mathfrak{o}\in\mathcal{O}} J^L_{\mathfrak{o}}(f) = \sum_{\chi\in\mathfrak{X}} J^L_{\chi}(f) \tag{3.14.1}$$

*from* (2.6.1).

THEOREM 3.15. *Each of the distributions $J^L_{\mathfrak{o}}$ and $J^L_{\chi}$ depend on $M_0$ and $K$, but not on a minimal parabolic subgroup of $L$. They are defined for any $L \in \mathcal{L}(M_0)$.*

PROOF. Suppose that $P'_0$ is another minimal parabolic subgroup in $\mathcal{P}(M_0)$. There is a unique element $s \in W$ such that $P'_0 = w_s^{-1}P_0w_s$, where $w_s$ is as in Subsection 2.2. If $P \supset P_0$, then $P' = w_s^{-1}Pw_s$ contains $P'_0$. Decompose an arbitrary element $y \in G(\mathbb{A})$ as

$$y = nmk, \quad n \in N_P(\mathbb{A}),\ m \in M_P(\mathbb{A}),\ k \in K.$$

Writing

$$w_s^{-1}y = w_s^{-1}nw_s \cdot w_s^{-1}mw_s \cdot w_s^{-1}k,$$

we see that

$$\begin{aligned} H_{P'}(w_s^{-1}y) &= H_{P'}(w_s^{-1}mw_s) + H_{P'}(w_s^{-1}) \\ &= s^{-1}H_P(y) + H_P(w_s^{-1}). \end{aligned}$$

Therefore

$$\begin{aligned} \hat{\tau}(H_P(y) - T) &= \hat{\tau}_P(sH_{P'}(w_s^{-1}y) - sH_P(w_s^{-1}) - T) \\ &= \hat{\tau}_{P'}(H_{P'}(w_s^{-1}y) - H_{P_0}(w_s^{-1}) - s^{-1}T). \end{aligned}$$

It follows that $J_{\mathfrak{o}}^T(f)$ equals the integral over $G(F)\backslash G(\mathbb{A})^1$ of

$$\sum_{P\supset P_0}(-1)^{\dim(A_P/Z)}\sum_{\delta\in P(F)\backslash G(F)} K_{P,\mathfrak{o}}(\delta x,\delta x)\hat{\tau}_{P'}(H_{P'}(w_s^{-1}\delta x)-H_{P_0}(w_s^{-1})-s^{-1}T)$$

$$=\sum_{P'\supset P_0'}(-1)^{\dim(A_{P'}/Z)}\sum_{\delta\in P'(F)\backslash G(F)} K_{P,\mathfrak{o}}(w_s\delta x,w_s\delta x)\hat{\tau}_{P'}(H_{P'}(\delta x)-H_{P_0}(w_s^{-1})-s^{-1}T).$$

Now,

$$\begin{aligned}K_{P,\mathfrak{o}}(w_s y,w_s y) &= \sum_{\gamma\in M_P(F)\cap\mathfrak{o}}\int_{N_P(\mathbb{A})} f(y^{-1}\cdot w_s^{-1}\gamma w_s\cdot w_s^{-1}n w_s\cdot y)\,dn\\ &= \sum_{\gamma\in M_{P'}(F)\cap\mathfrak{o}}\int_{N_{P'}(\mathbb{A})} f(y^{-1}\gamma n y)\,dn = K_{P',\mathfrak{o}}(y,y).\end{aligned}$$

We have shown that, if $P_0$ and $T$ are replaced by $P_0'$ and $H_{P_0}(w_s^{-1})+s^{-1}T$ in the definition of $J_{\mathfrak{o}}^T(f)$, the result is the same. By Proposition 2.3, we have

$$H_{P_0}(w_s^{-1})+s^{-1}T_0=T_0.$$

It follows that $J_{\mathfrak{o}}^{T_0}(f)=J_{\mathfrak{o}}(f)$ is independent of $P_0$. The same argument applies to $J_\chi^T(f)$ and to the corresponding distributions on Levi subgroups. The theorem follows. □

The argument above gives the following

Corollary 3.16. *Suppose that $L',L\in\mathcal{L}(M_0)$ and that $L'=w_s^{-1}Lw_s$ for $s\in W$. Let $f\in C_c^\infty(L(\mathbb{A})^1)$. Put*

$$f'(m')=f(\tilde{w}_s m'\tilde{w}_s^{-1})$$

*for $m'\in L'(\mathbb{A})^1$. (See 2.2 for the definition of $\tilde{w}_s$.) Then for all $\mathfrak{o}$ and $\chi$, we have*

$$J_{\mathfrak{o}}^{L'}(f')=J_{\mathfrak{o}}^L(f)\qquad\text{and}\qquad J_\chi^{L'}(f')=J_\chi^L(f).$$

## 4 Noninvariance

Definition 4.1. *For a function $f$ on $G(\mathbb{A})^1$ and $y\in G(\mathbb{A})^1$, put*

$$f^y(x)=f(yxy^{-1}).$$

*A distribution $J$ on $G(\mathbb{A})^1$ is said to be* invariant *if $J(f^y)=J(f)$ for all $f$ and $y$.*

DEFINITION 4.2. *Let*

$$v'_Q(k,y) = \int_{\mathfrak{a}_Q/\mathfrak{a}_L} \sum_{\{R\in\mathcal{F}^L(M_0)\,;\,Q\subset R\}} \tau_Q^R(H)\hat{\tau}_R(H+H_Q(ky))\,dH$$

*For* $m \in M_Q(\mathbb{A})^1$, *define*

$$f_{Q,y}(m) = \delta_Q^L(m)^{1/2} \int_{K\cap L(\mathbb{A})} \int_{N_Q(\mathbb{A})} f(k^{-1}mnk)v'_Q(k,y)\,dn\,dk. \tag{4.2.1}$$

*Consider now the special case when* $L = G$. *Then*

$$\begin{aligned} v'_Q(x,y) &= \int_{A_Q(\mathbb{R}^0)\cap G(\mathbb{A})^1} \Gamma'_Q(H_Q(ax)-T,-H_Q(k_Q(ax)y))\,da \\ &= \int_{\mathfrak{a}_Q/\mathfrak{a}_G} \Gamma'_Q(H,-H_Q(k_Q(x)y))\,dH. \end{aligned}$$

*Note that* $v'_Q(k,y)$ *is a smooth function of* $k \in K$ *and* $f_{Q,y}$ *is a smooth compactly supported function on* $M_Q(\mathbb{A})^1$.

## 4.3 Key theorem

We summarize the results we will establish in this section as a theorem. To leave room for future induction arguments, we will state the results for distributions on $L(\mathbb{A})^1$, rather than on $G(\mathbb{A})^1$.

THEOREM. *Suppose that* $L \in \mathcal{L}(M_0)$ *and* $f \in C_c^\infty(L(\mathbb{A})^1)$. *Let* $y \in L(\mathbb{A})^1$. *Then for all* $\mathfrak{o} \in \mathcal{O}$ *and* $\chi \in \mathfrak{X}$, *we have*

$$J_{\mathfrak{o}}^L(f^y) = \sum_{Q\in\mathcal{F}^L(M_0)} |W^{M_Q}||W^L|^{-1} J_{\mathfrak{o}}^{M_Q}(f_{Q,y})$$

*and*

$$J_\chi^L(f^y) = \sum_{Q\in\mathcal{F}^L(M_0)} |W^{M_Q}||W^L|^{-1} J_\chi^{M_Q}(f_{Q,y}).$$

This theorem is the basis of what follows. We will explain its role in the next section.

PROPOSITION 4.4. *We have*

$$J_{\mathfrak{o}}^T(f^y) = \sum_{Q\supset P_0} J_{\mathfrak{o}}^{M_Q,T}(f_{Q,y}). \tag{4.4.1}$$

PROOF. As before, $P_0 \in \mathcal{P}(M_0)$ is a minimal parabolic subgroup. Given $f$ and $y$, fix a suitably regular point $T \in \mathfrak{a}_0^+$. In the formula for $J_{\mathfrak{o}}^T(f^y)$, the kernel $K_{P,\mathfrak{o}}^f(\delta x, \delta x)$ will be replaced by

$$K_{P,\mathfrak{o}}^f(\delta xy^{-1}, \delta xy^{-1}) = \sum_{\gamma \in M_P(F) \cap \mathfrak{o}} \int_{N_P(\mathbb{A})} f^y(x^{-1}\delta^{-1}\gamma n \delta x)\, dn.$$

Thus, $J_{\mathfrak{o}}^T(f^y)$ equals

$$\int_{G(F)\backslash G(\mathbb{A})^1} \sum_{P \supset P_0} (-1)^{\dim(A_P/Z)} \sum_{\delta \in P(F)\backslash G(F)} K_{P,\mathfrak{o}}(\delta xy^{-1}, \delta xy^{-1}) \hat{\tau}_P(H_P(\delta x) - T)\, dx.$$

After changing variables, this is

$$\int_{G(F)\backslash G(\mathbb{A})^1} \sum_{P \supset P_0} (-1)^{\dim(A_P/Z)} \sum_{\delta \in P(F)\backslash G(F)} K_{P,\mathfrak{o}}(\delta x, \delta x) \hat{\tau}_P(H_P(\delta xy) - T)\, dx.$$

Let $k_P(\delta x)$ be any element in $K$ such that $\delta x \cdot k_P(\delta x)^{-1} \in P(\mathbb{A})$. Then

$$\begin{aligned} \hat{\tau}_P(H_P(\delta xy) - T) &= \hat{\tau}_P(H_P(\delta x) - T + H_P(k_P(\delta x) y)) \\ &= \sum_{Q \supset P} (-1)^{\dim(A_Q/Z)} \hat{\tau}_P^Q(H_P(\delta x) - T) \Gamma'_Q(H_P(\delta x) - T, -H_P(k_P(\delta x) y)). \end{aligned}$$

We proceed as in Section 3. As a function of $x$, if $Q \supset P$, then

$$\Gamma'_Q(H_P(x) - T, -H_P(k_P(x)y)) = \Gamma'_Q(H_Q(x) - T, -H_Q(k_Q(x)y))$$

is left $Q(\mathbb{A})$-invariant. The proposition now follows from the definitions of $v'_Q(x, y)$ and $f_{Q,y}(m)$.

The analogous formula holds for $J_\chi^T(f^y)$.

PROPOSITION 4.5. *Let $\chi \in \mathfrak{X}$. We have*

$$J_\chi^T(f^y) = \sum_{Q \supset P_0} J_\chi^{M_Q,T}(f_{Q,y}). \qquad (4.5.1)$$

PROOF. The proof is the same, except for the additional point that

$$r_P(f^y) = r_P(y)^{-1} r_P(f) r_P(y).$$

Therefore the kernel of the restriction of $r_P(f^y)$ to $L^2(N_P(\mathbb{A})M_P(F)\backslash G(\mathbb{A})^1)_\chi$ is

$$K_{P,\chi}(u_1 y^{-1}, u_2 y^{-1}), \quad u_1, u_2 \in G(\mathbb{A})^1.$$

Modifying the discussion above, as we obtained (3.12.1), gives (4.5.1). □

## 4.6 Proof of Theorem 4.3

We now return to the proof of theorem 4.3.

PROOF. Both sides of (4.4.1) and (4.5.1) are polynomials in $T$. If we take the values at $T = T_0$ on each side, the resulting distributions are all independent of $P_0$. The sums are still over those $Q$ containing $P_0$. Let $Q' \in \mathcal{F}(M_0)$. There is a unique $Q$ containing $P_0$ such that $Q' = w_s^{-1}Qw_s$, for some $s \in W$. Then

$$\begin{aligned} v'_{Q'}(\tilde{w}_s^{-1}k, y) &= \int_{\mathfrak{a}_{Q'}/\mathfrak{a}_G} \sum_{R' \supset Q'} (-1)^{\dim(A_{R'}/Z)} \tau_{Q'}^{R'}(H)\hat{\tau}_{R'}(H + H_{Q'}(\tilde{w}_s^{-1}ky))\, dH \\ &= \int_{\mathfrak{a}_{Q'}/\mathfrak{a}_G} \sum_{R' \supset Q'} (-1)^{\dim(A_{R'}/Z)} \tau_{Q'}^{R'}(H)\hat{\tau}_{R'}(H + s^{-1}H_Q(ky))\, dH \\ &= \int_{\mathfrak{a}_Q/\mathfrak{a}_G} \sum_{R \supset Q} (-1)^{\dim(A_R/Z)} \tau_Q^R(H)\hat{\tau}_R(H + H_Q(ky))\, dH \\ &= v'_Q(k, y). \end{aligned}$$

If $m \in M_Q(\mathbb{A})^1$, then

$$\begin{aligned} f_{Q',y}(\tilde{w}_s^{-1}m\tilde{w}_s) &= \delta_{Q'}(\tilde{w}_s^{-1}m\tilde{w}_s)^{1/2} \int_K \int_{N_{Q'}(\mathbb{A})} f(k^{-1}\tilde{w}_s^{-1}m\tilde{w}_s nk) v'_{Q'}(k, y)\, dn\, dk \\ &= \delta_Q(m)^{1/2} \int_K \int_{N_Q(\mathbb{A})} f(k^{-1}mnk) v'_{Q'}(\tilde{w}_s^{-1}k, y)\, dn\, dk \\ &= f_{Q,y}(m). \end{aligned}$$

It follows that, for all $\mathfrak{o}$ and $\chi$, $J_{\mathfrak{o}}^{M_{Q'}}(f_{Q',y}) = J_{\mathfrak{o}}^{M_Q}(f_{Q,y})$ and $J_\chi^{M_{Q'}}(f_{Q',y}) = J_\chi^{M_Q}(f_{Q,y})$. Thus the sums from (3.4.1) and (3.10.2) may be taken over all $Q \in \mathcal{F}(M_0)$. The number of $Q$ which are conjugate to a given parabolic subgroup $P_0$ equals the order of $W$ divided by the order of the Weyl group of the given Levi component. The corresponding summands must then be multiplied by $|W^{M_Q}|/|W|$. □

## 4.7 Failure to be invariant

The object of this chapter is to prepare for the next chapter, where we derive a trace formula whose terms are invariant. In particular, we will discuss the noninvariance of the distributions occurring in the trace formula of Chapter 3. The variance formulae will indicate how to modify the distributions to make them invariant.

We want an identity of the form (3.14.1) whose terms, distributions indexed by $\mathcal{O}$ and $\mathfrak{X}$, are invariant. The computation of Theorem 4.3 measures the failure of the distribution $J_{\mathfrak{o}}$ and $J_\chi$ to be invariant. We will use this information to construct the invariant distributions.

Suppose that $S$ is a fixed finite set of valuations of $F$. Write $F_S = \prod_{v \in S} F_v$. If $G'$ is any subgroup of $G$ defined over $F$, then

$$G'(F_S) = \prod_{v \in S} G'(F_v).$$

Suppose, for every $L \in \mathcal{L}(M_0)$, that $U(L)$ is a vector space of functions with common domain a subset of $L(F_S)$, complete with respect to some topology. We assume in addition that, for any $y \in L(F_S)^1 = L(F_S) \cap L(\mathbb{A})^1$, the map

$$f \mapsto f^y$$

is a continuous endomorphism of $U(L)$. Suppose further that, for any $Q \in \mathcal{F}^L(M_0)$, the map

$$f \mapsto f_{Q,y},$$

given by the formula (4.2.1), maps $U(L)$ continuously to $U(M_Q)$. By a linear form on $U(L)$, or distribution on $L$, we mean an element in $U(L)'$ , the dual topological vector space of $U(L)$. Lastly, suppose that we have a family of distributions $\{J^L \in U(L)'\}$ and a family of nonzero complex numbers $\{c(L)\}$ , both indexed by $\mathcal{L}(M_0)$, such that, for any $L$, $y \in L(F_S)$ and $f \in U(L)$, we have

$$J^L(f^y) = \sum_{Q \in \mathcal{F}^L(M_0)} c(M_Q)c(L)^{-1} J^{M_Q}(f_{Q,y}). \tag{4.7.1}$$

## 4.8 Finite set of places

The primary example of this setup comes from letting $U(L) = C_c^\infty(L(F_S)^1)$ and $c(L) = |W^L|$. If $S$ contains the Archimedean places, we can take $\{J^L\}$ to be one of the families $\{J_{\mathfrak{o}}^L\}$ or $\{J_\chi^L\}$. For, if $f \in C_c^\infty(L(F_S)^1)$, the product of $f$ with the characteristic function of

$$\prod_{v \notin S} (K_v \cap L(F_v))$$

is a function in $C_c^\infty(L(\mathbb{A})^1)$. Conversely, any function in $C_c^\infty(L(\mathbb{A})^1)$ can be obtained in this way, for a sufficiently large set $S$. It is in this sense that $J_{\mathfrak{o}}^L$ and $J_\chi^L$ are regarded as distributions on $C_c^\infty(L(F_S)^1)$.

## 4.9 A dual distribution

We wish to associate a natural family of invariant distributions to each family $\{J^L\}$. We need some additional data. For every $M \in \mathcal{L}(M_0)$, let $V(M)$ be a second complete topological vector space. For every pair $M \subset L$, assume we have a continuous map

$$\phi_M^L : U(L) \to V(M),$$

satisfying the following property. For every $y \in L(F_S)^1$, we have

$$\phi_M^L(f^y) = \sum_{Q \in \mathcal{F}^L(M)} \phi_M^{M_Q}(f_{Q,y}). \tag{4.9.1}$$

We shall sometimes write $\phi$ for $\phi_M^M$. In this case, (4.9.1) says that $\phi(f^y) = \phi(f)$, for each $y$ and $f$. For each $i \in V(M)'$, we can define the distribution $\phi'(i)$ on $U(M)$ by

$$\phi'(i) : f \mapsto i(\phi(f)).$$

By our assumptions, this distribution is invariant.

## 4.10 A surjectivity assumption

We need an additional assumption. We assume that, for every $M \in \mathcal{L}(M_0)$, the map $\phi : U(M) \to V(M)$ is onto. We further assume that the transpose, $\phi' : V(M)' \to U(M)'$, is onto.

The first statement implies that $\phi'$ is injective. The second states that any invariant distribution on $U(M)$ is of the form $\phi'(i)$. If $I$ is any invariant distribution on $U(M)$, we let $\widehat{I}$ be the unique element $i \in V(M)'$ such that $\phi'(i) = I$.

PROPOSITION 4.11. *Suppose that* $\{\phi_M^L\}$ *satisfies equation* (4.9.1) *and the assumptions of* 4.10. *Then for every family* $\{J^L\}$ *of distributions satisfying* (4.7.1), *there is a unique family* $\{I^L \in U(L)'\}$ *of invariant distributions such that, for every* $f \in U(L)$, *we have*

$$J^L(f) = \sum_{M \in \mathcal{L}^L(M_0)} c(M)c(L)^{-1}\widehat{I}^M(\phi_M^L(f)).$$

PROOF. Fix $\{J^L\}$. Assume inductively that $I^M$ has been defined for all groups $M \in \mathcal{L}(M_0)$ such that $\dim M < \dim L$. For $f \in U(L)$, define

$$I^L(f) = J^L(f) - \sum_{\substack{M \in \mathcal{L}^L(M_0) \\ M \neq L}} c(M)c(L)^{-1}\widehat{I}^M(\phi_M^L(f)).$$

We want to evaluate $I^L(f^y - f)$, for $y \in L(F_S)$. The function $f$ on $L(F_S)$ equals $f_{L,y}$. Therefore $J^L(f^y - f)$ equals the sum on the right-hand side of (4.7.1), taken over only the $Q \neq L$. The same observation gives a formula for $\phi_M^L(f^y - f)$. It follows that $I^L(f^y - f)$ is

$$\left( \sum_{\substack{Q \in \mathcal{F}^L(M_0) \\ Q \neq L}} c(M_Q)c(L)^{-1} J^{M_Q}(f_{Q,y}) \right) - \left( \sum_{M \in \mathcal{L}^L(M_0)} \sum_{\substack{Q \in \mathcal{F}^L(M) \\ Q \neq L}} c(M)c(L)^{-1} \widehat{I}^M(\phi_M^{M_Q}(f_{Q,y})) \right).$$

Note that $\mathcal{F}^L(M) \subset \mathcal{F}^L(M_0)$. A group $Q \in \mathcal{F}^L(M_0)$ belongs to $\mathcal{F}^L(M)$ if and only if $M \subset M_Q$. Therefore

$$I^L(f^y - f) = \sum_{\substack{Q \in \mathcal{F}^L(M_0) \\ Q \neq L}} c(M_Q)c(L)^{-1} \left( J^{M_Q}(f_{Q,y}) - \sum_{M \subset M_Q} c(M)c(M_Q)^{-1} \widehat{I}^M(\phi_M^{M_Q}(f_{Q,y})) \right).$$

The expression in parentheses vanishes by induction. Thus $I^L$ is an invariant distribution. □

## 4.12 Invariant trace formula

Suppose that $U(L) = C_c^\infty(L(F_S)^1)$ and that maps $\phi_M^L$, satisfying the hypotheses of the proposition, have been defined. If $S$ contains an Archimedean valuation, we can regard $J_\mathfrak{o}^L$ and $J_\chi^L$ as distributions on $C_c^\infty(L(F_S)^1)$, as noted above. By Theorem 4.3, we obtain two families, $\{I_\mathfrak{o}^L;\ \mathfrak{o} \in \mathcal{O}\}$ and $\{I_\chi^L;\ \chi \in \mathfrak{X}\}$ , of invariant distributions on $C_c^\infty(L(F_S)^1)$. The invariant trace formula is a formal consequence of the definitions.

THEOREM. *Suppose that $f \in C_c^\infty(L(F_S)^1)$. Then the series $\sum_\mathfrak{o} I_\mathfrak{o}^L(f)$ and $\sum_\chi I_\chi^L(f)$ converge absolutely, and*

$$\sum_{\mathfrak{o} \in \mathcal{O}} I_\mathfrak{o}^L(f) = \sum_{\chi \in \mathfrak{X}} I_\chi^L(f).$$

PROOF. Assume inductively that the theorem holds if $L$ is replaced by any group $M \in \mathcal{L}(M_0)$ with $\dim M < \dim L$. We have (the inner sum ranges over $M \in \mathcal{L}^L(M_0)$, $M \neq L$)

$$\sum_\mathfrak{o} |I_\mathfrak{o}^L(f)| = \sum_\mathfrak{o} \left| J_\mathfrak{o}^L(f) - \sum_M c(M)c(L)^{-1} \widehat{I}_\mathfrak{o}^M(\phi_M^L(f)) \right|.$$

It is bounded by

$$\sum_\mathfrak{o} |J_\mathfrak{o}^L(f)| + \sum_{M \subsetneq L} c(M)c(L)^{-1} \sum_{\mathfrak{o} \in \mathcal{O}} |\widehat{I}_\mathfrak{o}^M(\phi_M^L(f))|.$$

According to the remark after Proposition 3.10.1, the first term is finite. By assumption 4.10, $\phi^L_M(f)$ can be regarded as the image under $\phi^M_M$ of a function in $C_c^\infty(M(F_S)^1)$. The second term is then finite by our induction assumption. Thus $\sum_{\mathfrak{o}} |I^L_{\mathfrak{o}}(f)|$ is finite. The same argument shows that

$$\sum_{\mathfrak{o}} I^L_{\mathfrak{o}}(f) = \sum_{\mathfrak{o}} J^L_{\mathfrak{o}}(f) - \sum_{\substack{M\in\mathcal{L}^L(M_0)\\ M\neq L}} c(M)c(L)^{-1} \sum_{\mathfrak{o}} \widehat{I}^M_{\mathfrak{o}}(\phi^L_M(f)).$$

Similarly, $\sum_\chi |I^L_\chi(f)|$ is finite, and

$$\sum_{\chi} I^L_{\chi}(f) = \sum_{\chi} J^L_{\chi}(f) - \sum_{\substack{M\in\mathcal{L}^L(M_0)\\ M\neq L}} c(M)c(L)^{-1} \sum_{\chi} \widehat{I}^M_{\chi}(\phi^L_M(f)).$$

The required identity follows from (3.14.1) and the induction assumption. □

# 5 Intertwining Operators

## 5.1 *Definition of an intertwining operator*

We shall be concerned with the intertwining operators for the induced representations $\tau^G_\lambda = i^G_P(\tau_\lambda)$, acting on the space $V_P(\tau)$. Note that the space is independent of $\lambda$. See 1.7 for the definitions. For $P$ and $P' \in \mathcal{P}(M)$, consider the intertwining operator

$$\mathcal{M}_{P'|P}(\tau_\lambda) : V_P(\tau) \to V_{P'}(\tau), \qquad \psi \mapsto (k \mapsto \int \psi(n'k)\,\mathrm{d}n'),$$

$n'$ ranges over $N_{P'}(F_S) \cap N_P(F_S)\backslash N_{P'}(F_S)$. Using the decomposition $n' = n_P(n')m_P(n')k_P(n')$ of $G(F_S) = N_P(F_S)M_P(F_S)K(F_S)$, we see that it is then defined by

$$(\mathcal{M}_{P'|P}(\tau_\lambda)\psi)(k) = \int \tau(m_P(n'))\psi(k_P(n')k)\, e^{(\lambda+\rho_P)(H_P(n'))}\,\mathrm{d}n'. \tag{5.1.1}$$

The integral converges absolutely for $\lambda$ whose real part lies in a certain chamber.

We will list the elementary properties of the intertwining operators. Some of these are well known, others follow directly from the definition (5.1.1). $(\mathcal{M}_1)$ is the basic intertwining property.

PROPOSITION 5.2. $(\mathcal{M}_1)$ $\mathcal{M}_{P'|P}(\tau_\lambda) i^G_P(\tau_\lambda, f) = i^G_{P'}(\tau_\lambda, f)\mathcal{M}_{P'|P}(\tau_\lambda)$.
$(\mathcal{M}_2)$ *Let $P, P', P'' \in \mathcal{P}(M)$. Write $d(P', P)$ for the number of singular hyperplanes in $\mathfrak{a}_M$ which separate the chambers of $P'$ and $P$. If $d(P'', P) = d(P'', P') + d(P', P)$, then*

$$\mathcal{M}_{P''|P}(\tau_\lambda) = \mathcal{M}_{P''|P'}(\tau_\lambda)\mathcal{M}_{P'|P}(\tau_\lambda).$$

($\mathcal{M}_3$) *Let $L \in \mathcal{L}(M)$. Suppose that $Q \in \mathcal{P}(L)$ and $R, R' \in \mathcal{P}^L(M)$. Let $P = Q(R)$ and $P' = Q(R')$. For $\psi \in V_P(\tau)$ and $k \in K$, denote by $\psi_k$ the function of $k_1 \in K_L$ given by*

$$k_1 \mapsto \psi(k_1 k).$$

*Then*

$$(\mathcal{M}_{P'|P}(\tau_\lambda)\psi)_k = \mathcal{M}_{R'|R}(\tau_\lambda)\psi_k.$$

($\mathcal{M}_4$) *Let $(\cdot)^*$ denote the adjoint with respect to the Hermitian form on $V_P(\tau)$ given by*

$$(\psi, \psi') = \int_K (\psi(k), \psi'(k))\, dk.$$

*If $\tau$ is unitary, then*

$$\mathcal{M}_{P'|P}(\tau_\lambda)^* = \mathcal{M}_{P|P'}(\tau_{-\bar{\lambda}}).$$

($\mathcal{M}_5$) *Let $w \in K$. Define a map*

$$L(w) : V_P(\tau) \to V_{wPw^{-1}}(w\tau)$$

*by*

$$(L(w)\psi)(k) = \psi(w^{-1}k).$$

*Then*

$$L(w)\mathcal{M}_{P'|P}(\tau_\lambda)L(w)^{-1} = \mathcal{M}_{wP'w^{-1}|wPw^{-1}}((w\tau)_{w\lambda}).$$

## 5.3 Analytic continuation

It is not known if one can prove analytic continuation for general $\tau$. However, we have the following proposition.

PROPOSITION. *If $\tau$ is irreducible, then $\mathcal{M}_{P'|P}(\tau_\lambda)$ can be analytically continued as a meromorphic function to all $\lambda \in \mathfrak{a}^*_{M,\mathbb{C}}$.*

PROOF. See [HC75, KS80, Sha81]. □

The goal is to study a certain rational map constructed from the intertwining operators. However, it is necessary to use properties that hold only when the operators have been suitably normalized. In Sections 6–9, we shall discuss the normalization of the operators and related questions.

DEFINITION 5.4. (1) *Let* $\Pi(M(F_S))$ *denote the set of equivalence classes of irreducible admissible representations of* $M(F_S)$.

(2) *Let* $\Pi_{\mathrm{temp}}(M(F_S)) \subset \Pi(M(F_S))$ *be the subset of tempered representations of* $M(F_S)$.

# 6 Normalization

## 6.1 Plan for this section

In this section, $\tau$ is a representation in $\Pi(M(F_S))$. We will first state the required normalization properties. The remainder of the section will be devoted to reducing the proof of the general case to a special case. More specifically, the general case is reduced to the case of a local field $F$, $\dim(A_M/A_G) = 1$ and $\tau$ square-integrable modulo $A_M$. The special case will be discussed for real and $p$-adic groups separately in Sections 7 and 8.

THEOREM 6.2. *For* $P, P' \in \mathcal{P}(M)$ *and* $\tau \in \Pi(M(F_S))$, *there are meromorphic, scalar-valued functions*

$$m_{P'|P}(\tau_\lambda),$$

*called normalizing factors, such that the normalized intertwining operators*

$$\mathcal{R}_{P'|P}(\tau_\lambda) = m_{P'|P}(\tau_\lambda)^{-1}\mathcal{M}_{P'|P}(\tau_\lambda) : i_P^G(\tau_\lambda) \to i_{P'}^G(\tau_\lambda)$$

*have analytic continuation as meromorphic functions of* $\lambda \in \mathfrak{a}^*_{M,\mathbb{C}}$. *Furthermore, the normalized operators have the following properties:*

($\mathcal{R}_1$) $\mathcal{R}_{P'|P}(\tau_\lambda) i_P^G(\tau_\lambda, f) = i_{P'}^G(\tau_\lambda, f)\mathcal{R}_{P'|P}(\tau_\lambda)$.

($\mathcal{R}_2$) *For any* $P, P', P'' \in \mathcal{P}(M)$, *we have* $\mathcal{R}_{P''|P}(\tau_\lambda) = \mathcal{R}_{P''|P'}(\tau_\lambda)\mathcal{R}_{P'|P}(\tau_\lambda)$.

($\mathcal{R}_3$) *Let* $R, R' \in \mathcal{P}^L(M)$ *and* $Q \in \mathcal{P}(L)$. *Let* $P = Q(R)$, $P' = Q(R')$. *For* $\psi \in V_P(\tau)$ *and* $k \in K$, *we have* $(\mathcal{R}_{P'|P}(\tau_\lambda)\psi)_k = \mathcal{R}_{R'|R}(\tau_\lambda)\psi_k$.

($\mathcal{R}_4$) *If* $\tau$ *is unitary, then*

$$\mathcal{R}_{P'|P}(\tau_\lambda)^* = \mathcal{R}_{P|P'}(\tau_{-\bar{\lambda}}).$$

($\mathcal{R}_5$) *For any* $w \in K$, *we have*

$$L(w)\mathcal{R}_{P'|P}(\tau_\lambda)L(w)^{-1} = \mathcal{R}_{wP'w^{-1}|wPw^{-1}}((w\tau)_{w\lambda}).$$

($\mathcal{R}_6$) *If F is Archimedean, $\mathcal{R}_{P'|P}(\tau_\lambda)$ is a rational function of $\{\lambda(\alpha^\vee)\,;\,\alpha \in \Delta_P\}$. If F is a local field with residue field of cardinality q, then $\mathcal{R}_{P'|P}(\tau_\lambda)$ is a rational function of $\{q^{-\lambda(\alpha^\vee)}\,;\,\alpha \in \Delta_P\}$.*

($\mathcal{R}_7$) *If $\tau$ is tempered, $m_{P'|P}(\tau_\lambda)$ has neither zeroes nor poles at $\lambda$ whose real part lies in the positive chamber attached to P.*

($\mathcal{R}_8$) *Suppose that F is a non-Archimedean local field. Assume further that G and $\tau$ are unramified and that K is hyperspecial. Then if $\psi \in V_P(\tau)$ is fixed by K, the function $\mathcal{R}_{P'|P}(\tau_\lambda)\psi$ is independent of $\lambda$.*

## 6.3 Reduction

The first five properties are extensions of those of $\mathcal{M}_{P'|P}(\tau_\lambda)$. Once the normalizing factors have been defined, the analytic continuation and ($\mathcal{R}_1$) follow from the corresponding properties for $\mathcal{M}_{P'|P}(\tau_\lambda)$. Other properties, such as ($\mathcal{R}_2$), are nontrivial extensions and hold only for the normalized operators.

We next reduce the proof of the theorem to the special case of a local field mentioned in 6.1. Therefore assume in this section that the functions $m_{P'|P}(\tau_\lambda)$ have been defined and that the theorem is valid when $S = \{v\}$, $F = F_v$, $\dim(A_M/A_G) = 1$ and $\tau$ is square-integrable modulo $A_M$.

## 6.4 The rank

We will first relax the condition on the rank. Assume $S, F, G$, and $\tau$ satisfy the constraints above but that $\dim(A_M/A_G)$ is arbitrary. Given $P \in \mathcal{P}(M)$, let $\Sigma_P^r$ be the set of reduced roots of $(P, A_M)$. For $\beta \in \Sigma_P^r$, define $M_\beta \in \mathcal{L}(M)$ to be the group such that

$$\mathfrak{a}_{M_\beta} = \{H \in \mathfrak{a}_M\,;\,\beta(H) = 0\}.$$

Then $\dim(A_M/A_{M_\beta}) = 1$. Let $P_\beta$ be the unique group in $\mathcal{P}^{M_\beta}(M)$ whose simple root is $\beta$. We define the normalizing factors in terms of those of rank 1. Namely, define

$$m_{P'|P}(\tau_\lambda) = \prod_{\beta\in\Sigma_P^r\cap\Sigma_{\overline{P'}}^r} m_{\overline{P_\beta}|P_\beta}(\tau_\lambda). \tag{6.4.1}$$

The property ($\mathcal{R}_3$) follows immediately. In proving ($\mathcal{R}_2$), we may assume $d(P'', P') = 1$. If $d(P'', P) > d(P', P)$, ($\mathcal{R}_2$) holds since it holds for $\mathcal{M}_{P''|P}(\tau_\lambda)$ and $m_{P''|P}(\tau_\lambda)$ separately. On the other hand, if $d(P'', P) < d(P', P)$, we have

$$\mathcal{R}_{P'|P}(\tau_\lambda) = \mathcal{R}_{P'|P''}(\tau_\lambda)\mathcal{R}_{P''|P}(\tau_\lambda)$$

for the same reason. Reducing to the case of rank 1 by $(\mathcal{R}_3)$, we obtain

$$\mathcal{R}_{P'|P''}(\tau_\lambda) = \mathcal{R}_{P''|P'}(\tau_\lambda)^{-1},$$

so $(\mathcal{R}_2)$ follows. The analytic continuation and the remaining properties can all be reduced to the case of rank 1 by $(\mathcal{R}_2)$ and $(\mathcal{R}_3)$.

## 6.5 Tempered case

Assume that $\tau$ is tempered. It is known that $\tau$ is an irreducible constituent of an induced representation $i_R^M(\sigma)$, where $M_R$ is an admissible Levi subgroup of $M$, and $\sigma \in \Pi(M_R(F))$ is square-integrable modulo $A_R$. Then $i_P^G(\tau_\lambda)$ is canonically isomorphic to a subrepresentation of $i_{P(R)}^G(\sigma_\lambda)$. The defining integral formula shows that $\mathcal{M}_{P'|P}(\tau_\lambda)$ is identified with the restriction of $\mathcal{M}_{P'(R)|P(R)}(\sigma_\lambda)$ to the corresponding invariant subspace. If we define

$$m_{P'|P}(\tau_\lambda) = m_{P'(R)|P(R)}(\tau_\lambda), \tag{6.5.1}$$

the required properties follow from the square-integrable case.

## 6.6 General representation

Let $\tau \in \Pi(M(F))$ with no restrictions. We are still assuming that $S = \{v\}$ and $F = F_v$. The Langlands classification ([La88, BW80, Si80]) holds for $p$-adic and real groups. Therefore $\tau$ is the Langlands quotient of a representation $i_R^M(\sigma_\mu)$, where $M_R \subset M$ is an admissible Levi subgroup, $\sigma \in \Pi_{\text{temp}}(M_R(F))$, and $\mu \in \mathfrak{a}_R^*/\mathfrak{a}_M^*$ lies in the chamber attached to $R$, that is, $\tau$ is equivalent to the action of $i_R^G(\sigma_\mu)$ on the quotient of $V_R(\sigma)$ by the kernel of $\mathcal{M}_{\overline{R}|R}(\sigma_\mu)$. By $(\mathcal{R}_7)$, the function $m_{\overline{X}|X}(\sigma_\Lambda)$ has no pole or zero at $\Lambda = \mu$. Consequently, the kernel of $\mathcal{M}_{\overline{R}|R}(\sigma_\mu)$ equals the kernel of $\mathcal{R}_{\overline{R}|R}(\sigma_\mu)$. Set $\Lambda = \mu + \lambda$. Define

$$m_{P'|P}(\tau_\lambda) = m_{P'(R)|P(R)}(\sigma_\Lambda). \tag{6.6.1}$$

It follows from $(\mathcal{R}_3)$, applied to the tempered representation $\sigma$, that the induced representation $i_P^G(\tau_\lambda)$ is equivalent to the action of $i_{P(R)}^G(\sigma_\Lambda)$ on the quotient $V_{P(R)}(\sigma)/\ker \mathcal{R}_{P(\overline{R})|P(R)}(\sigma_\Lambda)$. Under this equivalence, the intertwining operator $\mathcal{R}_{P'|P}(\tau_\lambda)$ becomes $\mathcal{R}_{P'(R)|P(R)}(\sigma_\Lambda)$. The required properties of $\mathcal{R}_{P'|P}(\tau_\lambda)$, with the exception of $(\mathcal{R}_4)$, follow from the corresponding properties for $\sigma_\Lambda$.

Assume that $\tau$ is unitary. Knapp and Zuckerman ([KZ77]) observed that the unitarizability of the Langlands quotient implies that there is $w \in K_M$ such that $wRw^{-1} = \overline{R}$, $w\sigma \simeq \sigma$, $w\mu = -\mu$, and the inner product on $V_\tau$ can be obtained from

$\mathcal{R}_{\overline{R}|R}(\sigma_\mu)$ and $w$. More precisely, $\tau$ is unitarily equivalent to the action of $i_R^M(\sigma_\mu)$ on $V_R^M(\sigma)/\ker \mathcal{R}_{\overline{R}|R}(\sigma_\mu)$ under an inner product

$$\langle \Phi, \Psi \rangle = (AL(w)\mathcal{R}_{\overline{R}|R}(\sigma_\mu)\Phi, \Psi), \quad \Phi, \Psi \in V_R^M(\sigma).$$

Here $L(w)$ is as in $(\mathcal{M}_5)$ and $A$ is an intertwining operator from $w\sigma$ to $\sigma$, acting by multiplication on $V_R^M(w\sigma)$. Their observations apply to both real and $p$-adic groups. It follows that the induced representation $i_P^G(\tau)$ is unitarily equivalent to the action of $i_{P(R)}^G(\sigma_\mu)$ on $V_{P(R)}(\sigma)/\ker \mathcal{R}_{P(\overline{R})|P(R)}(\sigma_\mu)$ under the following inner product. For $\Phi, \Psi \in V_{P(R)}(\sigma)$, the new inner product is given by

$$\langle \Phi, \Psi \rangle = (AL(w)\mathcal{R}_{P(\overline{R})|P(R)}(\sigma_\mu)\Phi, \Psi).$$

Let us proceed to the adjoint condition $(\mathcal{R}_4)$. Let $\Phi \in V_{P(R)}(\sigma)$ and $\Phi' \in V_{P'(R)}(\sigma)$. Then

$$\begin{aligned}\langle \mathcal{R}_{P'(X)|P(X)}(\sigma_\mu)\Phi, \Phi' \rangle &= (AL(w)\mathcal{R}_{P'(\overline{X})|P'(X)}(\sigma_\mu)\mathcal{R}_{P'(X)|P(X)}(\sigma_\mu)\Phi, \Phi') \\ &= (AL(w)\mathcal{R}_{P'(\overline{X})|P(\overline{X})}(\sigma_\mu)\mathcal{R}_{P(\overline{X})|P(X)}(\sigma_\mu)\Phi, \Phi').\end{aligned}$$

By $(\mathcal{R}_3)$ and the definition of $A$, this equals

$$(\mathcal{R}_{P'(X)|P(X)}(\sigma_{-\mu})AL(w)\mathcal{R}_{P(\overline{X})|P(X)}(\sigma_\mu)\Phi, \Phi').$$

Applying $(\mathcal{R}_4)$ to the tempered representation $\sigma$, we see that this equals

$$(AL(w)\mathcal{R}_{P(\overline{X})|P(X)}(\sigma_\mu)\Phi, \mathcal{R}_{P(X)|P'(X)}(\sigma_\mu)\Phi') = \langle \Phi, \mathcal{R}_{P(X)|P'(X)}(\sigma_\mu)\Phi' \rangle.$$

Translating to a formula for $\tau$, we obtain

$$\mathcal{R}_{P'|P}(\tau)^* = \mathcal{R}_{P|P'}(\tau).$$

Property $(\mathcal{R}_4)$ follows for imaginary $\lambda$ by a change in the definition of $\tau$ and then for general $\lambda$ by analytic continuation.

## 6.7 *General sets of places*

We will now relax the conditions on $S$ and $F$. In other words, let these objects be arbitrary. Decompose $\tau$ as

$$\tau = \otimes_{v \in S} \tau_v, \quad \tau_v \in \Pi(M(F_v)).$$

We require that

$$m_{P'|P}(\tau_\lambda) = \prod_{v \in S} m_{P'|P}(\tau_{v,\lambda}). \tag{6.7.1}$$

Then

$$\mathcal{R}_{P'|P}(\tau_\lambda) = \otimes_{v \in S} \mathcal{R}_{P'|P}(\tau_{v,\lambda}). \tag{6.7.1$'$}$$

The theorem reduces to the case that $S$ contains one element $v$. In this case, write $M_v = M_{F_v}$ and $P_v = P_{F_v}$ for $M$ and $P$, respectively, regarded as varieties over $F_v$. Then $P \mapsto P_v$ gives an embedding of $\mathcal{P}(M)$ into $\mathcal{P}(M_v)$. We shall require that

$$m_{P'|P}(\tau_\lambda) = m_{P'_v|P_v}(\tau_v). \tag{6.7.2}$$

Then

$$\mathcal{R}_{P'|P}(\tau_\lambda) = \mathcal{R}_{P'_v|P_v}(\tau_\lambda). \tag{6.7.2$'$}$$

The definitions and properties in the theorem reduce to those for local fields, the case we dealt with in 6.6. We will discuss the base case—of local fields—in the next two sections.

# 7 Real Groups

## *7.1 Normalization*

In this section, we assume that $S = \{v\}$ and that $F = F_v$. By restricting scalars, we may assume that $F_v = \mathbb{R}$. Knapp and Stein in [KS75] and [KS80] gave a general procedure for normalizing the intertwining operators so that some of the properties of Section 6 hold. In [La76] Langlands proposed normalizing the intertwining operators in terms of $L$-functions. While the suggestion applied to any local field, it can only be done in the real case, as the $L$-functions for $p$-adic groups have not been defined in general. The aim of this section is to show that, for a natural choice of measures on the spaces $N_{P'}(\mathbb{R}) \cap N_P(\mathbb{R})$, the normalization proposed by Langlands satisfies the conditions of Theorem 6.2.

## *7.2 Choice of measure*

As we have defined them, the intertwining operators depend intrinsically on $K$. However, given $K$, there is a canonical way to choose measures on the spaces $N_{P'}(\mathbb{R}) \cap N_P(\mathbb{R})$. Let $\mathfrak{g}$ and $\mathfrak{k}$ denote the real Lie algebras of $G(\mathbb{R})$ and $K$, respectively. Let $\theta$ be the Cartan involution.

Let $B$ be a $G(\mathbb{R})$-invariant bilinear form on $\mathfrak{g}$ such that the associated quadratic form

$$X \mapsto -B(X, \theta X), \quad X \in \mathfrak{g}, \tag{7.2.1}$$

is positive definite. Let $T$ be a maximal torus of $M$ which is $\theta$-stable and defined over $\mathbb{R}$. The restriction of $B$ to the Lie algebra $\mathfrak{t}$ of $T(\mathbb{R})$ is non-degenerate. We use it to define a bilinear form, still denoted by $B$, on the dual space of $\mathfrak{t}_{\mathbb{C}}$. This form is positive definite on the real span of the roots of $(\mathfrak{g}_{\mathbb{C}}, \mathfrak{t}_{\mathbb{C}})$. For $P, P' \in \mathcal{P}(M)$, set

$$\alpha_{P'|P} = \prod_{\alpha} (B(\alpha, \alpha)/2)^{1/2},$$

where the product is taken over the roots $\alpha$ of $(\mathfrak{g}_{\mathbb{C}}, \mathfrak{t}_{\mathbb{C}})$ whose restrictions to $\mathfrak{a}_M$ are roots of both $(P', A_M)$ and $(\overline{P}, A_M)$. This number is independent of the maximal torus $T$.

The measure $\mathrm{d}n$ on $N_{P'}(\mathbb{R}) \cap N_P(\mathbb{R})$ is defined by

$$\int_{N_{P'}(\mathbb{R}) \cap N_P(\mathbb{R})} \varphi(n)\, \mathrm{d}n = \alpha_{P'|P} \int_{\mathfrak{n}_{P'} \cap \mathfrak{n}_P} \varphi(\exp X)\, \mathrm{d}X, \quad \varphi \in C_c^\infty(N_{P'}(\mathbb{R}) \cap N_P(\mathbb{R})).$$

Here $\mathrm{d}X$ is the Euclidean measure defined by the restriction of the form (7.2.1) to $\mathfrak{n}_{P'} \cap \mathfrak{n}_P$.

If $B(\cdot, \cdot)$ is replaced by $t^2 B(\cdot, \cdot)$, $t > 0$, the number $\alpha_{P'|P}$ will be replaced by $t^{-\dim(N_{P'} \cap N_P)} \alpha_{P'|P}$, while $\mathrm{d}X$ will be replaced by $t^{\dim(N_{P'} \cap N_P)} \mathrm{d}X$. Since $B(\cdot, \cdot)$ is uniquely determined up to scalar multiples on each of the simple factors of $\mathfrak{g}$, the measures $\mathrm{d}n$ are independent of $B$. We define $\mathcal{M}_{P'|P}(\tau_\lambda)$ with the associated invariant measure on $N_{P'}(\mathbb{R}) \cap N_P(\mathbb{R}) \backslash N_{P'}(\mathbb{R})$.

## 7.3 L-functions

We will now recall the definition of $L$-functions of representations of $M(\mathbb{R})$. For $\tau \in \Pi(M(\mathbb{R}))$, there is a corresponding map

$$\phi : W_{\mathbb{R}} \to {}^L M,$$

from the Weil group of $\mathbb{R}$ to the $L$-group of $M$, which is uniquely determined up to conjugation by ${}^L M$. Let $\beta$ be a finite-dimensional analytic representation of ${}^L M$. Then $\beta \circ \phi$ is a representation of $W_{\mathbb{R}}$ which has a decomposition $\oplus_\eta \eta$ into irreducible representations of $W_{\mathbb{R}}$. By definition,

$$L(s, \tau, \beta) = L(s, \beta \circ \phi) = \prod_{\eta} L(s, \eta).$$

If $\eta$ is one-dimensional, it is the pullback to $W_{\mathbb{R}}$ of a quasi-character

$$x \mapsto x^{-N} |x|^{s_1}, \quad N \in \{0, 1\}, \quad s_1 \in \mathbb{C},$$

of $\mathbb{R}^\times$. In this case,

$$L(s,\eta) = L_{\mathbb{R}}(s+s_1) \stackrel{\text{dfn}}{=} \pi^{-(s+s_1)/2}\Gamma((s+s_1)/2).$$

Otherwise, $\eta$ is the two-dimensional representation induced from a quasi-character

$$z \mapsto z^{-N}(z\bar{z})^{s_1}, \quad N \in \mathbb{Z}_{\geq 0}, \quad s_1 \in \mathbb{C},$$

or

$$z \mapsto \bar{z}^{-N}(z\bar{z})^{s_1}, \quad N \in \mathbb{Z}_{\geq 0}, \quad s_1 \in \mathbb{C},$$

of $\mathbb{C}^\times$. In this case,

$$L(s,\eta) = L_{\mathbb{C}}(s+s_1) \stackrel{\text{dfn}}{=} 2(2\pi)^{-(s+s_1)}\Gamma(s+s_1).$$

See [Ta79] for further discussion.

## 7.4 *Normalizing factors*

Let $\beta_{P'|P}$ be the adjoint representation of ${}^LM$ on the complex vector space given by ${}^L\mathfrak{n}_{P'} \cap {}^L\mathfrak{n}_P \backslash {}^L\mathfrak{n}_{P'}$. Let $\tilde{\beta}_{P'|P}$ denote the contragredient of $\beta_{P'|P}$. In the present context, the normalizing factors of Langlands can be taken to be

$$m_{P'|P}(\tau_\lambda) = L(0,\pi_\lambda,\tilde{\beta}_{P'|P})/L(1,\tau_\lambda,\tilde{\beta}_{P'|P}), \quad \lambda \in \mathfrak{a}^*_{M,\mathbb{C}}. \tag{7.4.1}$$

PROPOSITION 7.5. *The normalizing factors $m_{P|P'}(\tau_\lambda)$ satisfy the condition of Theorem* 6.2.

PROOF. It follows from the definitions that these factors satisfy the formulae (6.4.1)–(6.6.1). Therefore the reduction of the last section applies. Hence, we may assume that $\dim(A_M/A_G) = 1$ and that $\tau$ is square-integrable modulo $A_M$. We will prove Theorem 6.2 under these assumptions.

Some of the conditions of the theorem are immediate. We mentioned earlier that the analytic continuation is known. Property $(\mathcal{R}_1)$ is equivalent to $(\mathcal{M}_1)$. Condition $(\mathcal{R}_3)$ is trivial, since $\dim(A_M/A_G) = 1$. Property $(\mathcal{R}_8)$ does not apply to real groups. From definition (7.4.1), we find that, for $w \in K$, we have

$$m_{P'|P}(\tau_\lambda) = m_{wP'w^{-1}|wPw^{-1}}((w\tau)_{w\lambda}).$$

If $\tau$ is tempered, then

$$\overline{m_{P'|P}(\tau_\lambda)} = m_{P|P'}(\tau_{-\overline{\lambda}}).$$

Combining these properties with $(\mathcal{M}_4)$ and $(\mathcal{M}_5)$, they yield properties $(\mathcal{R}_4)$ and $(\mathcal{R}_5)$ of the theorem.

## 7.6 Properties

It remains to prove $(\mathcal{R}_2)$, $(\mathcal{R}_6)$ and $(\mathcal{R}_7)$. We need to examine the map $\phi$ associated with $\tau$. The Weil group $W_{\mathbb{R}}$ contains a normal subgroup $\mathbb{C}^\times$ of index 2. We fix an element $\sigma$ in the nontrivial coset such that $\sigma^2 = -1$ and $\sigma s \sigma^{-1} = \overline{z}$. The $L$-group ${}^LM = {}^LM^0 \rtimes W_{\mathbb{R}}$ comes equipped with a distinguished maximal torus ${}^LT = {}^LT^0 \rtimes W_{\mathbb{R}}$. Fix $P \in \mathcal{P}(M)$. There are embeddings

$${}^LM \subset {}^LP \subset {}^LG.$$

Following [La88] choose $\phi$ so that its image normalizes ${}^LT^0$. Then for each $z \in \mathbb{C}^\times$, $\phi(z)$ is a point in ${}^LT^0$. Recall that $L^\vee$ is the lattice of rational characters of ${}^LT^0$, and $L = \mathrm{Hom}(L^\vee, \mathbb{Z})$ is the dual lattice. For $\mu, \nu \in L \otimes \mathbb{C}$ with $\mu - \nu \in L$, we have the formula

$$\lambda^\vee(\phi(z)) = z^{\langle \mu, \lambda^\vee \rangle} \overline{z}^{\langle \nu, \lambda^\vee \rangle}, \quad \lambda^\vee \in L^\vee.$$

The expression on the right is a formal way of writing the complex number

$$z^{\langle \mu - \nu, \lambda^\vee \rangle} (z\overline{z})^{\langle \nu, \lambda^\vee \rangle} = \overline{z}^{\langle \nu - \mu, \lambda^\vee \rangle} (z\overline{z})^{\langle \mu, \lambda^\vee \rangle}.$$

For $a \in {}^LM^0$, the point

$$h = \phi(\sigma) = a \rtimes \sigma$$

normalizes ${}^LT^0$. We shall write $\overline{\sigma}$ for its adjoint action on ${}^LT^0$, $L$, and $L^\vee$. Then $\nu = \overline{\sigma}\mu$. There is a canonical injection of the space

$$\mathfrak{a}^*_{M,\mathbb{C}} = X(M) \otimes \mathbb{C}$$

into $L \otimes \mathbb{C}$. If $\tau$ is replaced by $\tau_\lambda$, $\lambda \in \mathfrak{a}^*_{M,\mathbb{C}}$, $\phi$ will be replaced by a map $\phi_\lambda$, in which $(\mu, \nu, h)$ becomes $(\mu + \lambda, \nu + \lambda, h)$.

Let $T \subset M$ be a maximal torus over $\mathbb{R}$ whose split component is $A_M$. Fix an isomorphism of $T(\mathbb{C})$ with $\mathrm{Hom}(L, \mathbb{C}^\times)$. Then $L$ and $L^\vee$ are identified with $X^*(T)$ and $X_*(T)$, respectively. In this case, $\overline{\sigma}$ is the same as the $\mathrm{Gal}(\mathbb{C}/\mathbb{R})$ action induced from $T$ (see [La88], p. 50). Let $\Sigma_P(G, T)$ be the set of roots of $(G, T)$ which restrict to roots of $(P, A_M)$. Then the eigenspaces of $\tilde{\beta}_{\overline{P}|P}(\phi_\lambda(\mathbb{C}^\times))$ are the root spaces of $\{-\alpha^\vee ; \alpha \in \Sigma_P(G,T)\}$. Consequently, the irreducible constituents of $\tilde{\beta}_{\overline{P}|P} \circ \phi_\lambda$ correspond to orbits of $\overline{\sigma}$ in $\Sigma_P(G,T)$.

Consider a two-dimensional constituent, $\eta_\lambda$, corresponding to a pair $\{\alpha, \overline{\sigma}\alpha\}$ of complex roots. Then $\eta_\lambda$ is induced from the quasi-character

$$z \mapsto z^{\langle \mu+\lambda, \alpha^\vee\rangle} \overline{z}^{\langle \nu+\lambda, \alpha^\vee\rangle} = \overline{z}^{\langle \overline{\sigma}\mu-\mu, \alpha^\vee\rangle} (z\overline{z})^{\langle \mu+\lambda, \alpha^\vee\rangle}$$

of $\mathbb{C}^\times$. Replacing $\alpha^\vee$ by $\overline{\sigma}\alpha^\vee$ if necessary, we can assume that $\langle \overline{\sigma}\mu - \mu, \alpha^\vee\rangle$ is a nonpositive integer. Consequently,

$$L(0, \eta_\lambda)/L(1, \eta_\lambda) = \Gamma_{\mathbb{C}}(\langle \mu + \lambda, \alpha^\vee\rangle)/\Gamma_{\mathbb{C}}(\langle \mu + \lambda, \alpha^\vee\rangle + 1). \qquad (7.6.1)$$

The one-dimensional constituents correspond to the real roots $\{\alpha_0\}$ in $\Sigma_P(G,T)$. There is at most one of these. If $\alpha_0$ exists, let $X_{\alpha_0^\vee}$ be a root vector for $\alpha_0^\vee$. Set

$$\mathrm{Ad}(\phi(\sigma))X_{\alpha_0^\vee} = (-1)^{N_0} X_{\alpha_0^\vee}, \quad N_0 = 0, 1.$$

For $z \in \mathbb{C}^\times$, we have

$$\phi_\lambda(z) X_{\alpha_0^\vee} = (z\overline{z})^{\langle \mu+\lambda, \alpha_0^\vee\rangle} X_{\alpha_0^\vee}.$$

Thus the one-dimensional constituent $\eta_\lambda$ comes from the quasi-character

$$x \mapsto (x/|x|)^{-N_0} |x|^{\langle \mu+\lambda, \alpha_0^\vee\rangle} = x^{-N_0} |x|^{\langle \mu+\lambda, \alpha_0^\vee\rangle + N_0}$$

of $\mathbb{R}^\times$. Consequently,

$$L(0, \eta_\lambda)/L(1, \eta_\lambda) = \Gamma_{\mathbb{R}}(\langle \mu+\lambda, \alpha_0^\vee\rangle + N_0)/\Gamma_{\mathbb{R}}(\langle \mu+\lambda, \alpha_0^\vee\rangle + N_0 + 1). \quad (7.6.2)$$

## 7.7 *Condition* $(\mathcal{R}_7)$

Condition $(\mathcal{R}_7)$ of Theorem 6.2 follows from (7.6.1) and (7.6.2). Indeed, assume $\tau$ is tempered and $\alpha$ is as in (7.6.1). Then the real part of the number $\langle \mu, \alpha^\vee\rangle$ is nonnegative. If $\lambda$ belongs to the chamber attached to $P$, the number $\langle \lambda, \alpha^\vee\rangle$ is real and positive. If $\alpha_0$ is as in (7.6.2), the real part of the number $\langle \mu+\lambda, \alpha_0^\vee\rangle$ is positive. Condition $(\mathcal{R}_7)$ follows, since the gamma function has no zeros or poles in the right half plane.

## 7.8 Condition ($\mathcal{R}_2$)

To establish ($\mathcal{R}_2$), we need to show that

$$\mathcal{R}_{P|\overline{P}}(\tau_\lambda)\mathcal{R}_{\overline{P}|P}(\tau_\lambda) = 1.$$

After incorporating $\lambda$ into the representation $\tau$ and applying analytic continuation, we may assume that $\lambda = 0$ and $\tau$ is tempered. Then ($\mathcal{R}_4$) applies and it suffices to show that

$$\mathcal{M}_{\overline{P}|P}(\tau)^*\mathcal{M}_{\overline{P}|P}(\tau) = |m_{\overline{P}|P}(\tau)|^2. \tag{7.8.1}$$

PROPOSITION 7.9. *Assume that $\tau$ is tempered. We have*

$$(\mathcal{M}_{\overline{P}|P}(\tau)^*\mathcal{M}_{\overline{P}|P}(\tau))^{-1} = (2\pi)^{-\dim N_P} \prod_{\alpha\in\Sigma_P(G,T)} |\mu(\alpha^\vee)| \cdot \left|\tanh\left(\frac{\pi\mu(\alpha_0^\vee)}{2i}\right)\right|^{\varepsilon_\tau}, \tag{7.9.1}$$

*where*

$$\varepsilon_\tau = \begin{cases} (-1)^{N_0}, & \text{if } \alpha_0 \text{ exists,} \\ 0, & \text{otherwise.} \end{cases}$$

The proof is in [Ar89, Appendix]. It relies on Harish-Chandra's explicit formula for the Plancherel density and a lemma of Langlands which interprets $\varepsilon_\tau$ as a sign occurring in Harish-Chandra's parametrization. The right side of (7.9.1) is missing certain numbers when compared to Harish-Chandra's formula. Their absence is due to the choice of measures on $N_P(\mathbb{R})$ and $N_{\overline{P}}(\mathbb{R})$.

## 7.10 Property ($\mathcal{R}_2$) follows

Given the proposition, we only have to consider the absolute values of (7.6.1) and (7.6.2). Let $s$ be an imaginary number. Then

$$|\Gamma_{\mathbb{C}}(s)/\Gamma_{\mathbb{C}}(s+1)|^{-2} = |2\pi s^{-1}|^{-2} = (2\pi)^{-2}s\bar{s}.$$

Next,

$$\begin{aligned} |\Gamma_{\mathbb{R}}(s)/\Gamma_{\mathbb{R}}(s+1)|^{-2} &= \frac{1}{\pi}\left|\Gamma\left(\frac{s}{2}\right)\right|^{-2}\left|\Gamma\left(\frac{s+1}{2}\right)\right|^2 \\ &= \frac{1}{\pi}\cdot\frac{1}{\pi}\left|\frac{s}{2i}\sinh\left(\frac{\pi s}{2i}\right)\right|\cdot\pi\left|\cosh\left(\frac{\pi s}{2i}\right)\right|^{-1} \\ &= (2\pi)^{-1}|s|\left|\tanh\left(\frac{\pi s}{2i}\right)\right|. \end{aligned}$$

Finally,

$$
\begin{aligned}
|\Gamma_{\mathbb{R}}(s+1)/\Gamma_{\mathbb{R}}(s+2)|^{-2} &= \left|\Gamma_{\mathbb{R}}(s+1)\frac{s}{2\pi}/\Gamma_{\mathbb{R}}(s)\right|^{-2} \\
&= (2\pi)^{-1}|s|\left|\coth\left(\frac{\pi s}{2i}\right)\right|.
\end{aligned}
$$

It follows from (7.6.1) and (7.6.2) that $|m_{\overline{P}|P}(\tau)|^{-2}$ equals the right side of (7.9.1). This proves (7.8.1) and therefore property $(\mathcal{R}_2)$.

## 7.11 Rationality

It remains to establish the rationality of

$$
\mathcal{R}_{\overline{P}|P}(\tau_\lambda) = m_{\overline{P}|P}(\tau_\lambda)^{-1}\mathcal{M}_{\overline{P}|P}(\tau_\lambda).
$$

Let $\beta$ be a simple root of $(P, A_M)$. The function $m_{\overline{P}|P}(\tau_\lambda)$ is a product of functions of the form (7.6.1) and (7.6.2). This implies that $m_{\overline{P}|P}(\tau_\lambda)$ and $m_{\overline{P}|P}(\tau_\lambda)^{-1}$ can be expressed as scalar multiples of products of the form

$$
\prod_{i=1}^{N}\Gamma(t_i\lambda(\beta^\vee)+\zeta_i)/\Gamma(t_i\lambda(\beta^\vee)+\eta_i), \tag{7.11.1}
$$

for $t_i$ a positive real number and $\zeta_i, \eta_i \in \mathbb{C}$.

Let $\Omega$ denote a finite set of irreducible representations in $K$. Let $\mathcal{M}_{\overline{P}|P}(\tau)_\Omega$ and $\mathcal{R}_{\overline{P}|P}(\tau)_\Omega$ for the restrictions of the respective operators to $V_P(\tau)_\Omega$, the subspace of $V_P(\tau)$ that transforms under $K$ according to $\Omega$. The space $V_P(\tau)$ is independent of $\lambda$, but the action $i_P^G(\tau_\lambda)$ depends on $\lambda$. The operator $\mathcal{M}_{\overline{P}|P}(\tau_\lambda)_\Omega$ can be expressed in terms of Harish-Chandra's $c$-function [HC76, Lemma 11.1]. A result of Wallach [Wa75, Theorem 7.2] says that the matrix coefficients of $\mathcal{M}_{\overline{P}|P}(\tau_\lambda)_\Omega$ are linear combinations of functions of the form (7.11.1). Thus the matrix coefficients of $\mathcal{R}_{\overline{P}|P}(\tau_\lambda)_\Omega$ are of this form. A result of L. Cohn [Co74, Theorem 5] says the inverse of the determinant of $\mathcal{M}_{\overline{P}|P}(\tau_\lambda)_\Omega$ is a function of the form (7.11.1). Therefore the matrix coefficients of $\mathcal{M}_{\overline{P}|P}(\tau_\lambda)_\Omega^{-1}$ are also linear combinations of functions of the form (7.11.1). The same then holds for the matrix coefficients of $\mathcal{R}_{\overline{P}|P}(\tau_\lambda)_\Omega^{-1}$.

We will need the following estimate for the gamma function. Fix $t > 0$, $\zeta, \eta \in \mathbb{C}$, and $b \in \mathbb{R}$. There are positive numbers $c$ and $n$, and a polynomial $l(z)$ such that, for $z \in \mathbb{C}$ with $\mathrm{Re} z > b$,

$$
\left|l(z)\frac{\Gamma(tz+\zeta)}{\Gamma(tz+\eta)}\right| \le c(1+|z|)^n.
$$

See, for example, [Ar83, p. 33]. It follows that we may choose $l(\cdot)$, $c$ and $n$ such that, for $\lambda \in \mathfrak{a}^*_{M,\mathbb{C}}$ with $\mathrm{Re}\lambda(\beta^\vee) \geq b$, we have

$$|l(\lambda(\beta^\vee))|(\|\mathcal{R}_{\overline{P}|P}(\tau_\lambda)_\Omega\| + \|\mathcal{R}_{\overline{P}|P}(\tau_\lambda)_\Omega^{-1}\|) \leq c(1 + |\lambda(\beta^\vee)|)^n. \qquad (7.11.2)$$

On the other hand, $(\mathcal{R}_2)$ tells us that

$$\|\mathcal{R}_{\overline{P}|P}(\tau_\lambda)_\Omega\| = \|\mathcal{R}_{P|\overline{P}}(\tau_\lambda)_\Omega^{-1}\|.$$

We may reverse the roles of $P$ and $\overline{P}$ and apply (7.11.2) to the norm on the right. Since $-\beta$ is the simple root of $(\overline{P}, A_M)$, we see, for $\lambda(\beta^\vee) \leq b$, that

$$|l(\lambda(\beta^\vee))| \cdot \|\mathcal{R}_{\overline{P}|P}(\tau_\lambda)_\Omega\| \leq c(1 + |\lambda(\beta^\vee)|)^n.$$

Combining this with (7.11.2) gives, for $\lambda \in \mathfrak{a}^*_{M,\mathbb{C}}$, the inequality

$$|l(\lambda(\beta^\vee))| \cdot \|\mathcal{R}_{\overline{P}|P}(\tau_\lambda)_\Omega\| \leq c(1 + |\lambda(\beta^\vee)|)^n.$$

Thus $\mathcal{R}_{\overline{P}|P}(\tau_\lambda)_\Gamma$, a meromorphic function of the complex variable $\lambda(\beta^\vee)$, extends to a meromorphic function on the Riemann sphere. It is therefore a rational function of $\lambda(\beta^\vee)$. This is the final property $(\mathcal{R}_6)$. Theorem 6.2 holds for real groups with the normalizing factors (7.4.1). □

REMARK 7.12. (1) Suppose $\dim(A_M/A_G) = 1$ and $G(\mathbb{R})$ has no compact Cartan subgroup. In this case, there is no real root $\alpha_0$. Since each function (7.6.1) is rational, $m_{\overline{P}|P}(\tau_\lambda)$ is rational. Consequently, $\mathcal{M}_{\overline{P}|P}(\tau_\lambda)$ is a rational function of $\lambda$.

(2) Let $E$ be an endomorphism of $V_P(\tau)$. Then

$$\mathcal{R}_{P'|P}(\tau_\lambda)^{-1} E \mathcal{R}_{P'|P}(\tau_\lambda) = \mathcal{M}_{P'|P}(\tau_\lambda)^{-1} E \mathcal{M}_{P'|P}(\tau_\lambda).$$

The rationality of $\mathcal{R}_{P'|P}(\tau_\lambda)$ therefore generalizes a result [HC76, Lemma 19.2] of Harish-Chandra.

(3) Some of the results of this section can be extracted from Shahidi's [Sha85] in which he used Whitaker functionals to investigate normalizing factors. His methods give additional information that will be useful in applications of the trace formula.

# 8 p-Adic Groups

## 8.1 Reduction

Assume now that $S = \{v\}$, $F = F_v$ and that $F_v$ is non-Archimedean with residue field of order $q$. To verify the existence of normalizing factors $m_{P'|P}(\tau_\lambda)$ such that Theorem 6.2 holds and that satisfy (6.4.1)–(6.6.1), it is enough to define them when

$\dim(A_M/A_G) = 1$ and $\tau$ is square-integrable modulo $A_M$. For $P \in \mathcal{P}(M)$ with a simple root $\beta$, Langlands observed ([CLL84]) that one can define rational functions $V_P(\tau, z)$ of one variable so that

$$m_{\overline{P}|P}(\tau_\lambda) = V_P(\tau, q^{-\lambda(\beta^\vee)})$$

satisfies the conditions of Theorem 6.2. The main requirement from $m_{\overline{P}|P}(\tau_\lambda)$ is that

$$\left(\overline{m_{\overline{P}|P}(\tau_{-\overline{\lambda}})} m_{\overline{P}|P}(\tau_\lambda)\right)^{-1} = \mu_M(\tau_\lambda), \tag{8.1.1}$$

where

$$\mu_M(\tau_\lambda) = (\mathcal{M}_{\overline{P}|P}(\tau_{-\overline{\lambda}})^* \mathcal{M}_{\overline{P}|P}(\tau_\lambda))^{-1} = (\mathcal{M}_{P|\overline{P}}(\tau_\lambda)\mathcal{M}_{\overline{P}|P}(\tau_\lambda))^{-1}$$

is Harish-Chandra's $\mu$-function, and $\tau$ is unitary and square-integrable modulo $A_M$.

## 8.2 Normalizing by L-functions

To show that related groups can be assigned the same normalizing factors, note that the general construction above will suffice, as long as one can show that the $\mu$-functions can be matched. In Chapter 6 this is carried out for the example of inner twisting of $\mathrm{GL}(n)$.

It would be useful to define the normalizing factors in terms of $L$-functions, as was done for real groups. The $L$-functions have been defined for $p$-adic fields and $G = \mathrm{GL}(n)$. Let $P$ be a standard maximal parabolic subgroup of $\mathrm{GL}(n)$. Let $\tau = \tau_1 \times \tau_2$ be an irreducible tempered representation of $M_P(F) = \mathrm{GL}(n_1, F) \times \mathrm{GL}(n_2, F)$. Assume

$$\tau_\lambda(m_1 \times m_2) = \tau(m_1 \times m_2)|\det m_1|^s|\det m_2|^{-s},$$

for $s \in \mathbb{C}$. Shahidi showed in [Sha83] that, for normalized measures on $N_P(F)$ and $N_{\overline{P}}(F)$, depending on a fixed additive character $\psi$ of $F$, the factors

$$m_{\overline{P}|P}(\tau_\lambda) = L(s, \tau_1 \times \tilde{\tau}_2)/(L(1+s, \tau_1 \times \tilde{\tau}_2) \cdot \varepsilon(s, \tau_1 \times \tilde{\tau}_2, \psi))$$

satisfy (8.1.1). Here $L(\cdot)$ and $\varepsilon(\cdot)$ are the functions defined by Jacquet, Piatetski-Shapiro, and Shalika [JPSS83]. Therefore the intertwining operators for $\mathrm{GL}(n)$ can be normalized by $L$-functions.

# 9 Standard Representations

## 9.1 Block equivalence

Assume now that $G$ and $F$ are as in Section 1. Assume that the normalizing factors $\{m_{P'|P}(\tau_\lambda)\}$ have been fixed and that they satisfy the supplementary conditions in Subsections 6.4–6.7, as well as properties $(\mathcal{R}_1)$-$(\mathcal{R}_8)$ of Theorem 6.2. This section continues the discussion of normalization. In [Vo82] Vogan introduced the notion of *block equivalence*, an equivalence relation for representations. We will compare the normalizing factors for representations related by block equivalence.

## 9.2 Finite set of places

We regard $G$ as a scheme defined over the ring $F_S$. Let $K = \prod_{v\in S} K_v$ be the fixed maximal compact subgroup of $G(F_S)$. A *well-positioned Levi subgroup over $F_S$* is a product $\mathfrak{M} = \prod_{v\in S} M_v$, where each $M_v$ is a Levi subgroup of $G$, defined over $F_v$, for which $K_v$ is well-positioned. Given $\mathfrak{M}$, we set $A_{\mathfrak{M}} = \prod_{v\in S} A_{M_v}$ and $\mathfrak{a}_{\mathfrak{M}} = \oplus_{v\in S}\mathfrak{a}_{M_v}$. A *root* of $(G, A_{\mathfrak{M}})$ is a root of $(G, A_{M_v})$ for some $v \in S$. Let $\tau = \otimes_{v\in S}\tau_v$ be an admissible representations of the group $\mathfrak{M}(F_S) = \prod_v M_v(F_v)$. Assume $\lambda = \oplus_v\lambda_v \in \mathfrak{a}^*_{\mathfrak{M},\mathbb{C}}$. We define another representation of $\mathfrak{M}(F_S)$ by setting

$$\tau_\lambda(m) = \otimes_v \tau_{v,\lambda_v}(m_v) = \otimes_v \tau_v(m_v)e^{\lambda_v(H_{M_v}(m_v))}, \quad m = \prod_v m_v.$$

Then $\tau_\lambda(m)$ is also an admissible representation of $\mathfrak{M}(F_S)$. We write

$$\tau_\lambda^G = \otimes_{v\in S}\tau_{v,\lambda_v}^{G(F_v)}$$

for the equivalence class of the representation of $G(F_S)$ induced from $\tau_\lambda = \otimes_{v\in S}\tau_{v,\lambda_v}$ on $\mathfrak{M}(F_S)$.

DEFINITION 9.3. *Let $\Sigma(G(F_S))$ denote the set of (equivalence classes of) representations of $G(F_S)$ which equal $\tau_\lambda^G$ for some $\mathfrak{M}$, with $\lambda \in \mathfrak{a}^*_{\mathfrak{M}}$ and $\tau$ a representation in*

$$\Pi_{\text{temp}}(\mathfrak{M}(F_S)) = \left\{\otimes_v\tau_v \,;\, \tau_v \in \Pi_{\text{temp}}(M_v(F_v))\right\}.$$

*The elements of $\Sigma(G(F_S))$ are called* standard representations.

Suppose that $\rho \in \Sigma(G(F_S))$ with $\lambda$ in the positive cone (see below). Then $\rho$ has a unique irreducible quotient. It is a representation in $\Pi(G(F_S))$, which we denote by $\overline{\rho}$. Moreover, $\rho \mapsto \overline{\rho}$ is a bijection from $\Sigma(G(F_S))$ onto $\Pi(G(F_S))$, by the Langlands classification, see [BW80].

## 9.4 A passage

The next proposition is a basic result. See, for example, [Vo81, Prop. 6.6.7] and the introduction to [Vo82]. We include the proof, which is based on familiar ideas.

PROPOSITION. *There are uniquely determined complex numbers*

$$\{\iota(\rho,\pi), \xi(\pi,\rho)\,;\,\pi\in\Pi(G(F_S)), \rho\in\Sigma(G(F_S))\}$$

*satisfying the following equations. For* $\rho\in\Sigma(G(F_S))$, *we have*

$$\operatorname{tr}\rho = \sum_{\pi\in\Pi(G(F_S))} \iota(\rho,\pi)\operatorname{tr}\pi. \tag{9.4.1}$$

*For* $\pi\in\Pi(G(F_S))$, *we have*

$$\operatorname{tr}\pi = \sum_{\rho\in\Sigma(G(F_S))} \xi(\pi,\rho)\operatorname{tr}\rho. \tag{9.4.2}$$

PROOF. Recall that if $\{\pi\}$ is a set of inequivalent representations in $\Pi(G(F_S))$, the functions $\{\operatorname{tr}\pi\}$ are linearly independent. The uniqueness assertion follows from this.

To prove the existence of $\{\iota(\rho,\pi)\}$ and $\{\xi(\pi,\rho)\}$, we may assume that $S=\{v\}$ and $F=F_v$. Let $\rho\in\Sigma(G(F_S))$. In the Grothendieck group, $\rho$ has a decomposition

$$\rho = \oplus_\pi m(\rho,\pi)\pi,$$

into irreducible representations of $G(F_S)$, with the $m(\rho,\pi)$ nonnegative integers. Consider this decomposition as a character identity on $G(F_S)$. Thus $\pi$ belongs to $\Pi(G(F_S))$. Consequently, we obtain (9.4.1):

$$\operatorname{tr}\rho = \sum_{\pi\in\Pi(G(F_S))} m(\rho,\pi)\operatorname{tr}\pi.$$

In other words, we may take $\iota(\rho,\pi)=m(\rho,\pi)$.

## 9.5 Inversion

The numbers $\{\xi(\pi,\rho)\}$ will be constructed by inverting (9.4.1).

LEMMA. *To invert* (9.4.1), *we need consider only the finite set of* $\pi$ *and* $\rho$ *with a given infinitesimal character.*

PROOF. Each representation in $\Sigma(G(F_S))$ or $\Pi(G(F_S))$ has an infinitesimal character

$$\chi : \mathcal{Z}(G) \to \mathbb{C}.$$

If $F$ is Archimedean, $\mathcal{Z}(G)$ is the center of the universal enveloping algebra. If $F$ is $p$-adic, we take $\mathcal{Z}(G)$ to be the Bernstein center. See [BD84, BDK86] or [F95]. The constituents of $\rho$ will have the same infinitesimal character, so $\iota(\rho, \pi)$ vanishes if $\pi$ and $\rho$ have different infinitesimal characters. Moreover, it is known that there are only finitely many representations in $\Pi(G(F_S))$ with a given infinitesimal character. For real groups, this is a result of Harish-Chandra. For $p$-adic groups, it follows from [Si81, Theorem 3.9.1]. □

## 9.6 Positive chamber

Fix a minimal parabolic subgroup $P_0$ of $G$ over $F$. Write $(\mathfrak{a}_0^*)^+$ for the positive chamber $(\mathfrak{a}_{P_0}^*)^+ \subset \mathfrak{a}_{P_0}^*$ associated with $P_0$. For $\lambda, \lambda' \in \mathfrak{a}_0^*$, write

$$\lambda' \leq \lambda$$

if $\lambda - \lambda'$ is a nonnegative, real linear combination of simple roots of $(P_0, A_{P_0})$.

Suppose that $\rho \in \Sigma(G(F_S))$. Then there are unique Levi subgroup $M$, parabolic $P \in \mathcal{P}(M)$, tempered $\tau \in \Pi_{\mathrm{temp}}(M(F_S))$, and a vector $\lambda \in (\mathfrak{a}_P^*)^+$, with $P \supset P_0$, such that $\rho = i_P^G(\tau_\lambda)$. Set $\lambda = \lambda_{\overline{\rho}}$. It is a point in the closure of $(\mathfrak{a}_0^*)^+$, uniquely determined by the representation $\overline{\rho} \in \Pi(G(F_S))$ of 9.3.

Consider the expression (9.4.1). The representation $\overline{\rho}$ occurs as a constituent of $\rho$ only as the Langlands quotient. Therefore $m(\rho, \overline{\rho}) = 1$.

LEMMA 9.7. *If $\pi$ occurs on the right side of* (9.4.1) *with positive multiplicity, then $\lambda_\pi \leq \lambda_{\overline{\rho}}$ with equality only when $\pi = \overline{\rho}$.*

PROOF. See [BW80, IV.4.13, XI.2.13]. □

This establishes that the matrix

$$(\iota(\rho', \overline{\rho})), \quad \rho, \rho' \in \Sigma(G(F_S)),$$

is unipotent. Its inverse is also a unipotent matrix. This implies that, for $\pi \in \Pi(G(F_S))$, there are integers $\xi(\pi, \rho)$ such that

$$\operatorname{tr} \pi = \sum_{\rho \in \Sigma(G(F_S))} \xi(\pi, \rho) \operatorname{tr} \rho.$$

Proposition 9.4 follows. □

DEFINITION 9.8. *Block equivalence is the equivalence relation on* $\Pi(G(F_S))$ *defined by* $\pi_1 \sim \pi_2$ *if there is* $\rho \in \Sigma(G(F_S))$ *such that* $\iota(\rho, \pi_i) \neq 0$, $i = 1, 2$. *Namely, it is generated by*

$$\{\pi \sim \overline{\rho};\ \iota(\rho, \pi) \neq 0\}.$$

Block equivalent representations have the same infinitesimal character. If $\xi(\pi, \rho) \neq 0$, then $\overline{\rho}$ and $\pi$ are block equivalent.

## 9.9 General Levi subgroup

We return to a general Levi subgroup $M$ over $F$. Suppose that $\rho \in \Sigma(M(F_S))$. Note that $\rho$ is now a representation on any Levi subgroup of $G$, not necessarily a representation of $G$ as before. For $P, P' \in \mathcal{P}(M)$ and $\lambda \in \mathfrak{a}^*_{M,\mathbb{C}}$, define

$$m_{P'|P}(\rho_\lambda) = m_{P'|P}(\overline{\rho}_\lambda)$$

and

$$\mathcal{R}_{P'|P}(\rho_\lambda) = m_{P'|P}(\rho_\lambda)^{-1}\mathcal{M}_{P'|P}(\rho_\lambda).$$

All properties but $(\mathcal{R}_4)$ of Theorem 6.2 hold for these operators. This follows by analytic continuation from the case that $\rho$ is tempered. If $\tau$ is a representation in $\Pi(M(F_S))$, set

$$\tilde{m}_{P'|P}(\tau_\lambda, \rho_\lambda) = m_{P'|P}(\tau_\lambda)^{-1} m_{P'|P}(\rho_\lambda). \tag{9.9.1}$$

PROPOSITION 9.10. *Fix* $\tau \in \Pi(M(F_S))$ *and* $\rho \in \Sigma(M(F_S))$ *with* $\tau$ *and* $\overline{\rho}$ *block equivalent. Then for any* $P$, $P'$, *and* $P''$, *we have*

$$\tilde{m}_{P''|P}(\tau_\lambda, \rho_\lambda) = \tilde{m}_{P''|P'}(\tau_\lambda, \rho_\lambda)\tilde{m}_{P'|P}(\tau_\lambda, \rho_\lambda). \tag{9.10.1}$$

*If* $F$ *is Archimedean, then* $\tilde{m}_{P'|P}(\tau_\lambda, \rho_\lambda)$ *is a rational function of* $\{\lambda(\alpha^\vee);\ \alpha \in \Delta_P\}$. *If* $F$ *is a local field of residual cardinality* $q$, *then it is a rational function of* $\{q^{-\lambda(\alpha^\vee)};\ \alpha \in \Delta_P\}$.

PROOF. Suppose that $\tau$ and $\overline{\rho}$ are block equivalent to a third representation $\tau = \overline{\sigma}$, for $\sigma \in \Sigma(M(F_S))$. Then

$$\tilde{m}_{P'|P}(\tau_\lambda, \rho_\lambda) = \tilde{m}_{P'|P}(\tau_\lambda, \sigma_\lambda)\tilde{m}_{P'|P}(\sigma_\lambda, \rho_\lambda).$$

It is therefore enough to prove the proposition when $\iota(\rho, \tau) \neq 0$. Note that $\tau$ is a constituent of $\rho$. This makes the induced space $V_P(\tau)$ into a subquotient of $V_P(\rho)$.

Let $\mathcal{M}_{P'|P}(\rho_\lambda)_\tau$ and $\mathcal{R}_{P'|P}(\rho_\lambda)_\tau$ be the operators on $V_P(\tau)$ obtained as subquotients of $\mathcal{M}_{P'|P}(\rho_\lambda)$ and $\mathcal{R}_{P'|P}(\rho_\lambda)$. The original integral formula for the intertwining operators (5.1.1) tells us that

$$\mathcal{M}_{P'|P}(\rho_\lambda)_\tau = \mathcal{M}_{P'|P}(\tau_\lambda).$$

Consequently

$$\begin{aligned}\mathcal{R}_{P'|P}(\tau_\lambda)\mathcal{R}_{P'|P}(\rho_\lambda)_\tau^{-1} &= m_{P'|P}(\tau_\lambda)^{-1}\mathcal{M}_{P'|P}(\tau_\lambda)\mathcal{M}_{P'|P}(\rho_\lambda)_\tau^{-1}m_{P'|P}(\rho_\lambda)\\ &= \tilde{m}_{P'|P}(\tau_\lambda, \rho_\lambda).\end{aligned}$$

The two assertions of the proposition follow from the properties $(\mathcal{R}_2)$ and $(\mathcal{R}_6)$, applied to $\mathcal{R}_{P'|P}(\tau_\lambda)$ and $\mathcal{R}_{P'|P}(\rho_\lambda)$. □

COROLLARY 9.11. *Let $\tau, \tau' \in \Pi(M(F_S))$ be block equivalent. Then the $\mu$-functions $\mu_M(\tau_\lambda)$ and $\mu_M(\tau'_\lambda)$ are equal.*

PROOF. By definition,

$$\mu_M(\tau_\lambda) = (\mathcal{M}_{P|\overline{P}}(\tau_\lambda)\mathcal{M}_{\overline{P}|P}(\tau_\lambda))^{-1}.$$

From $(\mathcal{R}_2)$, we obtain

$$\mu_M(\tau_\lambda) = (m_{P|\overline{P}}(\tau_\lambda)m_{\overline{P}|P}(\tau_\lambda))^{-1}.$$

Since a similar formula holds for $\tau'_\lambda$, the corollary follows from (9.10.1). □

# 10 Convex Sets and Some Related Functions

## *10.1 Smooth functions on real spaces*

Throughout this section, $M$ is a fixed Levi subgroup in $\mathcal{L}(M_0)$. We will establish some properties for smooth functions on the real vector space $i\mathfrak{a}_M^*$. Let $P \in \mathcal{P}(M)$. We saw in Proposition 3.8 that, for fixed $X \in \mathfrak{a}_P = \mathfrak{a}_M$, the function of $\lambda \in \mathfrak{a}_{P,\mathbb{C}}^*$ given by

$$\sum_{Q \supset P} (-1)^{\dim(A_P/A_Q)} e^{\lambda_Q(X)} \hat{\theta}_P^Q(\lambda)^{-1} \theta_Q(\lambda)^{-1}$$

could be extended to a smooth function on $i\mathfrak{a}_M^*$. We proved this geometrically, by exhibiting the function as the Fourier transform of a compactly supported function. We will obtain a more general statement by transcribing the proof of Proposition 3.6. In particular, we will replace $e^{\lambda(X)}$ by an arbitrary function of $\lambda$.

In the discussion of the function $\Gamma'_Q(\cdot, X)$, we used the fact that, if $P \subsetneq R$, then

$$\sum_{\{Q;P\subset Q\subset R\}} (-1)^{\dim(A_Q/A_R)} \tau_P^Q(H)\hat{\tau}_Q^R(H) = 0.$$

Suppose the real part of $\lambda$ belongs to $-(\mathfrak{a}_P^*)^+$. According to the proof of Proposition 3.8, we can integrate each summand against $e^{\lambda(H)}$. We obtain the formula

$$\sum_{\{Q;P\subset Q\subset R\}} (-1)^{\dim(A_P/A_Q)} \hat{\theta}_P^Q(\lambda)^{-1} \theta_Q^R(\Lambda)^{-1} = 0. \tag{10.1.1}$$

By analytic continuation, it is valid for all $\lambda \in \mathfrak{a}_{P,\mathbb{C}}^*$.

Definition 10.2. *Suppose that $c_P(\lambda)$ is a smooth function on $i\mathfrak{a}_M^* = i\mathfrak{a}_P^*$. Let $Q \supset P$. Recall that we defined $\lambda_Q$ to be the projection of $\lambda$ onto $i\mathfrak{a}_Q^*$. Define*

$$c_Q(\lambda) = c_P(\lambda_Q).$$

*Copying the construction of Section* 3, *we define, for each $Q \supset P$, functions*

$$c'_Q(\lambda)$$

*inductively. More specifically, we require that, for all $Q \supset P$,*

$$c_Q(\lambda)/\theta_Q(\lambda) = \sum_{R\supset Q} c'_R(\lambda)/\theta_Q^R(\lambda). \tag{10.2.1}$$

The functions $c'_Q(\lambda)$ are each defined on the complement of a finite set of hyperplanes in $i\mathfrak{a}_P^*$. It follows from (10.1.1) and (10.2.1) that, if $Q \supset P$, then

$$c'_Q(\lambda) = \sum_{R\supset Q} (-1)^{\dim(A_Q/A_R)} \hat{\theta}_Q^R(\lambda)^{-1} c_R(\lambda)\theta_R(\Lambda)^{-1}. \tag{10.2.2}$$

If $c_P(\lambda) = e^{\lambda(X)}$, then $c'_P(\lambda)$ is the smooth function mentioned above.

Proposition 10.3. *Suppose $c_P(\lambda)$ is a smooth function on $i\mathfrak{a}_P^*$. Then $c'_P(\lambda)$ extends to a smooth function on $i\mathfrak{a}_P^*$.*

Proof. For any $\lambda \in i\mathfrak{a}_P^*$, there are complex numbers $c_\varpi$ and $\lambda_Z \in \mathfrak{a}_{G,\mathbb{C}}^*$ such that

$$\lambda = \sum_{\varpi\in\widehat{\Delta}_Q} c_\varpi \varpi + \lambda_Z.$$

If $Q \supset P$, then $\widehat{\Delta}_Q$ is a subset of $\widehat{\Delta}_P$. Let $\lambda_{Q/P}$ denote the element

$$\sum_{\varpi\in\widehat{\Delta}_Q} c_\varpi \varpi + \lambda_Z$$

of $\mathfrak{a}^*_{Q,\mathbb{C}}$. If $\phi_Q(\lambda)$ is a function on $i\mathfrak{a}^*_P$ which depends only on $\lambda_{Q/P}$, set

$$\phi_{Q/P}(\lambda) = \phi_Q(\lambda_{Q/P}).$$

We then have the function $c_{Q/P}$ and, for $R \supset Q$, the function $\theta^R_{Q/P}$.

Suppose $\alpha$ is a root in $\Delta_P \smallsetminus \Delta^Q_P$. Let $\alpha^\vee_Q$ denote the projection of $\alpha^\vee$ onto $\mathfrak{a}_Q$. Then

$$\lambda_{Q/P}(\alpha^\vee_Q) = \lambda_{Q/P}(\alpha^\vee) = \lambda(\alpha^\vee).$$

Note that

$$\operatorname{vol}(\mathfrak{a}^G_Q/L^G_Q) = \operatorname{vol}(\mathfrak{a}^R_Q/L^R_Q)\operatorname{vol}(\mathfrak{a}^G_R/L^Q_R).$$

It follows that, for $R \supset Q \supset P$, we have

$$\theta_{Q/P}(\lambda) = \theta^R_{Q/P}(\lambda)\theta_{R/P}(\lambda).$$

Similarly, if $R_1 \supset R \supset Q \supset P$, then

$$\theta^{R_1}_{Q/P}(\lambda) = \theta^R_{Q/P}(\lambda)\theta^{R_1}_{R/P}(\lambda).$$

We define functions $\tilde{c}_{Q/P}(\lambda)$ on $E$ as we defined $c'_Q(\lambda)$. In particular, we require, for all $Q \supset P$, that

$$c_{Q/P}(\lambda)/\theta_{Q/P}(\lambda) = \sum_{R\supset Q} \tilde{c}_{R/P}(\lambda)/\theta^R_{Q/P}(\Lambda).$$

Then

$$\begin{aligned}
&\sum_{R\supset Q}(-1)^{\dim(A_Q/A_R)}\theta^R_{Q/P}(\lambda)^{-1}c_{R/P}(\lambda)/\theta_{R/P}(\lambda)\\
&= \sum_{\{R,R_1\,;\,R_1\supset R\supset Q\}} (-1)^{\dim(A_Q/A_R)}\theta^R_{Q/P}(\lambda)^{-1}\tilde{c}_{R_1/P}(\lambda)/\theta^{R_1}_{R/P}(\lambda)\\
&= \sum_{R_1\supset Q} \theta^{R_1}_{Q/P}(\lambda)^{-1}\tilde{c}_{R_1/P}(\lambda) \sum_{\{R;\,R_1\supset R\supset Q\}} (-1)^{\dim(A_Q/A_R)} = \tilde{c}_{Q/P}(\lambda).
\end{aligned}$$

Thus $\tilde{c}_{Q/P}(\lambda)$ equals the product of $\theta_{Q/P}(\lambda)^{-1}$ with

$$\sum_{R\supset Q}(-1)^{\dim(A_Q/A_R)}c_{R/P}(\lambda). \tag{10.3.1}$$

The function $\theta_{Q/P}(\lambda)^{-1}$ is a product of linear forms defined by the roots $\alpha \in \Delta_P \smallsetminus \Delta_P^Q$. Fix such an $\alpha$. The parabolic subgroups $R$ such that $R \supset Q$ occur in pairs, $(R, R')$. If $R$ is such that $\alpha$ does not vanish on $\mathfrak{a}_R$, we define $R'$ by setting

$$\mathfrak{a}_{R'} = \{H \in \mathfrak{a}_R\,;\,\alpha(H) = 0\}.$$

If $\lambda(\alpha^\vee) = 0$, the definitions imply that

$$c_{R/P}(\lambda) = c_{R'/P}(\lambda).$$

Since

$$\dim(A_Q/A_{R'}) = \dim(A_Q/A_R) + 1,$$

(10.3.1) vanishes whenever $\lambda(\alpha^\vee) = 0$. It follows from Taylor's theorem that (10.3.1) is divisible as a smooth function by the linear form $\lambda(\alpha^\vee)$. Therefore $\tilde{c}_{Q/P}(\lambda)$ is a smooth function of $\lambda$.

The proposition will be proved by induction on $\dim G$. Note that, if $R \supset Q \supset P$, then

$$c_Q(\lambda)/\theta_Q(\lambda) = c_Q(\lambda_Q)/\theta_Q(\lambda_Q) = c_{Q/P}(\lambda_Q)/\theta_{Q/P}(\lambda_Q)$$

and

$$\theta^R_{Q/P}(\lambda_Q) = \theta^R_Q(\lambda).$$

Therefore

$$\begin{aligned} c'_P(\lambda) &= \sum_{Q \supset P} (-1)^{\dim(A_P/A_Q)}\hat{\theta}^Q_P(\lambda)^{-1} c_Q(\lambda)/\theta_Q(\lambda) \\ &= \sum_{\{R,Q\,;\,R \supset Q \supset P\}} (-1)^{\dim(A_P/A_Q)}\hat{\theta}^Q_P(\lambda)^{-1}\tilde{c}_{R/P}(\lambda_Q)/\theta^R_Q(\lambda). \end{aligned}$$

Suppose that $R = G$. Then $\tilde{c}_{R/P}(\lambda_Q) = c_{G/P}(\lambda)$ and is independent of $Q$. In view of (10.1.1), the sum over $Q$ will vanish. It follows that $c'_P(\lambda)$ equals the sum over $\{R\,;\,P \subset R \subsetneq G\}$ of $(\tilde{c}_{R/P})'_{P\cap M_R}(\lambda)$, the function defined by (10.2.2), but with $(G, c_P(\lambda), Q)$ replaced by

$$(M_R, \tilde{c}_{R/P}(\lambda), P \cap M_R).$$

Since $\tilde{c}_{R/P}(\lambda)$ is a smooth function on $F$, the proposition follows from the induction assumption. □

EXAMPLE 10.4. To motivate the next proposition, we return to the case when $X \in \mathfrak{a}_P^+$ and

$$c_P(\lambda) = e^{\lambda(X)}.$$

Then $c'_P(\lambda)$ is the Fourier transform of the function $\Gamma'_P(\cdot, X)$. Consider the function $\Gamma'_P(\cdot, -X)$. If there is no $Q \supset P$ with $\tau_P^Q(H)\hat{\tau}_Q(H+X) = 1$, then $\Gamma'_P(H, -X)$ equals 0. Otherwise, let $R$ be the largest group with this property. Namely, there is no $Q \supset R$ such that $\tau_R^Q(H)\hat{\tau}_Q(H+X) = 1$. The group $R$ can be defined by letting $\Delta_P^R$ be the union over all such $Q$ of the sets $\Delta_P^Q$. As $X$ is in the positive chamber, $\tau_P^R(H + X)\hat{\tau}_R(H+X) = 1$. This implies $\hat{\tau}_P^R(H+X)\hat{\tau}_R(H+X) = 1$, from which one can verify that $\hat{\tau}_P(H+X) = 1$ (see [Ar76], Lemma 2.2 and the following discussion). Thus $\varpi(H+X) > 0$ for all $\varpi \in \widehat{\Delta}_P$. It follows that, for all $Q$ with $P \subset Q \subset R$, we have

$$\tau_P^Q(H)\hat{\tau}_Q(H+X) = 1.$$

Therefore

$$\Gamma'_P(H, -X) = \sum_{\{Q;\, P \subset Q \subset R\}} (-1)^{\dim(A_Q/Z)}.$$

This expression is $(-1)^{\dim(A_P/Z)}$ if $R = P$ and 0 otherwise. Thus $\Gamma'_P(\cdot, -X)$ is the product of $(-1)^{\dim(A_P/Z)}$ and the characteristic function of

$$\{H \in \mathfrak{a}_P\,;\ \alpha(H) \leq 0, \alpha \in \Delta_P\,;\ \varpi(H+X) > 0, \varpi \in \widehat{\Delta}_P\}.$$

It follows from Proposition 3.8 that $(-1)^{\dim(A_P/Z)}\Gamma'_P(-H, -X)$ and $\Gamma'_P(H, X)$ have the same Fourier transforms. Therefore modulo a set of measure 0, $\Gamma'_P(\cdot, X)$ is the characteristic function of

$$\{H \in \mathfrak{a}_P\,;\ \alpha(H) > 0, \alpha \in \Delta_P\,;\ \varpi(H-X) < 0, \varpi \in \widehat{\Delta}_P\}.$$

In other words, the figure drawn for GL(3) in Section 2 is valid in general.

DEFINITION 10.5. *A family of points $\{X_P\,;\ P \in \mathcal{P}(M)\}$ is called $A_M$-*orthogonal *if, for every pair $(P, P')$ of adjacent groups in $\mathcal{P}(M)$, the vector $X_P - X_{P'}$ is a multiple of the coroot associated with the unique root in $\Delta_P \cap (-\Delta_{P'})$.*

Such families were studied in [Ar76]. Suppose that this multiple is always positive and that each point $X_P$ lies in $(\mathfrak{a}_P^G)^+$. It follows from [Ar76, Lemma 3.2] and what we have shown that the characteristic function of the convex hull of $\{X_P\,;\ P \in \mathcal{P}(M)\}$ equals the function

$$\sum_{P \in \mathcal{P}(M)} \Gamma'_P(H, X), \quad H \in \mathfrak{a}_M^G,$$

almost everywhere.

For GL(3) the convex hull is the region

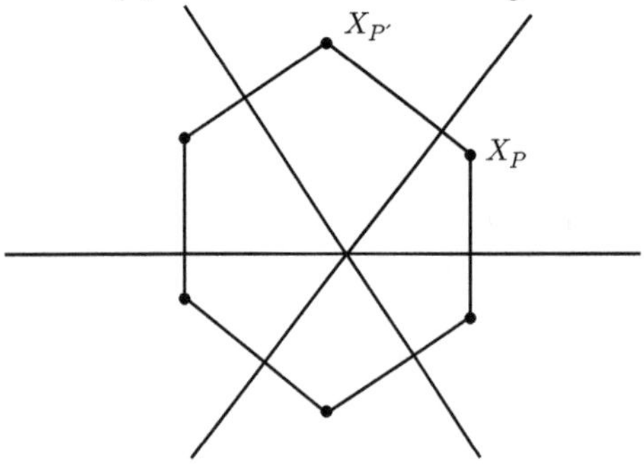


The Fourier transform of the characteristic function of the convex hull is computed in [Ar76, Lemma 3.2]. It equals

$$\sum_{P\in\mathcal{P}(M)} e^{\lambda(X_P)}/\theta_P(\lambda).$$

In particular, this function of $\lambda$ extends to a smooth function on $i\mathfrak{a}_M^*$.

DEFINITION 10.6. *Suppose that, for each $P \in \mathcal{P}(M)$, the function $c_P(\lambda)$ is smooth on $i\mathfrak{a}_M^*$. Call the collection*

$$\{c_P(\lambda)\,;\, P\in\mathcal{P}(M)\}$$

*a* $(G,M)$-family *if the following condition holds. For each pair of adjacent subgroups $P$ and $P'$ in $\mathcal{P}(M)$ and $\lambda$ belonging to the hyperplane spanned by the common wall of the chambers of $P$ and $P'$, we have*

$$c_P(\lambda) = c_{P'}(\lambda).$$

*This condition is equivalent to the following. Assume $P, P' \in \mathcal{P}(M)$ are contained in a given parabolic subgroup, $Q$, and $\lambda \in i\mathfrak{a}_Q^*$. Then the condition is $c_P(\lambda) = c_{P'}(\lambda)$. In particular, there is a well-defined function, $c_Q(\lambda)$, on $i\mathfrak{a}_Q^*$.*

The collection $\{e^{\lambda(X_P)}\}$ is a $(G,M)$-family if and only if $\{X_P\}$ is $A_M$-orthogonal.

PROPOSITION 10.7. *Let $\{c_P(\lambda)\,;\, P \in \mathcal{P}(M)\}$ be a $(G,M)$-family. Then the function*

$$c_M(\lambda) = \sum_{P\in\mathcal{P}(M)} c_P(\lambda)/\theta_P(\lambda)$$

*can be extended to a smooth function on $i\mathfrak{a}_M^*$.*

PROOF. The only possible singularities are along hyperplanes $\lambda(\alpha^\vee) = 0$, where $\alpha$ is a reduced root of $(G, A_M)$. Such a singularity occurs only in the terms corresponding to those $P$ for which either $\alpha$ or $-\alpha$ is a simple root. Such groups in $\mathcal{P}(M)$ occur in pairs $(P, P')$ where $P$ and $P'$ are adjacent and have $\alpha$ and $-\alpha$, respectively, as a simple root. If we multiply the corresponding pair of terms by $\lambda(\alpha^\vee)$, and take $\lambda$ to be a point in general position on the hyperplane $\lambda(\alpha^\vee) = 0$, the result is 0, since $c_P(\lambda) = c_{P'}(\lambda)$. Taylor's theorem implies that $c_M$ does not have a singularity on the hyperplane. □

DEFINITION 10.8. *Let $\{c_P(\lambda)\}$ be a $(G, M)$-family. We will often denote the value of $c_M(\lambda)$ at $\lambda = 0$ by $c_M$.*

PROPOSITION 10.9. *Let $p = \dim(A_M/A_G)$. For any regular $\Lambda \in \mathfrak{a}^*_{M,\mathbb{C}}$, we have*

$$c_M = \frac{1}{p!} \sum_{P \in \mathcal{P}(M)} \left( \lim_{t \to 0} \left( \frac{d}{dt} \right)^p c_P(t\Lambda) \right) / \theta_P(\Lambda). \tag{10.9.1}$$

*In particular, the expression is independent of $\Lambda$ in $\mathfrak{a}^*_{P,\mathbb{C}}$.*

PROOF. To calculate $c_M$, set $\lambda = t\Lambda$ for $t \in \mathbb{R}$ and let $t$ approach 0. □

DEFINITION 10.10. *If $Q$ contains an element of $\mathcal{P}(M)$, we will write $c'_Q$ for $c'_Q(0)$.*

From (10.2.2), Proposition 10.3 and its proof, and Proposition 10.9, we deduce

PROPOSITION 10.11. *Let $q = \dim(A_Q/A_G)$. The value of $c'_Q$ is*

$$\frac{1}{q!} \sum_{R \supset Q} (-1)^{\dim(A_Q/A_R)} \hat{\theta}^R_Q(\Lambda)^{-1} \left( \lim_{t \to 0} \left( \frac{d}{dt} \right)^q c_R(t\Lambda) \right) / \theta_R(\Lambda).$$

Fix a group $L \in \mathcal{L}(M)$. Let $Q \in \mathcal{P}(L)$, $P \in \mathcal{P}(M)$, and assume $P \subset Q$. The function of $\lambda \in i\mathfrak{a}^*_Q$ given by

$$\lambda \mapsto c_P(\lambda)$$

depends only on $Q$, and not on $P$. We have denoted it by $c_Q(\lambda)$. Then

$$\{c_Q(\lambda) \, ; \, Q \in \mathcal{P}(L)\}$$

is a $(G, L)$-family.

DEFINITION 10.12. *Fix $Q \in \mathcal{P}(L)$. If $R \in \mathcal{P}^L(M)$, let $Q(R)$ be the unique group in $\mathcal{P}(M)$ such that $Q(R) \subset Q$ and $Q(R) \cap L = R$. Define a function $c^Q_R$ on $i\mathfrak{a}^*_M$ by*

$$c^Q_R(\lambda) = c_{Q(R)}(\lambda).$$

*Then*

$$\{c^Q_R(\lambda) \, ; \, R \in \mathcal{P}^L(M)\}$$

*is an $(L,M)$-family. In particular, we have the function $c_M^Q(\lambda)$ and for $P \in \mathcal{P}^L(M)$ the function $(c_P^Q)'(\lambda)$. We have the values at $\lambda = 0$, $c_M^Q$, and $(c_P^Q)'$. In general, $c_M^Q$ depends on $Q$ and not just on $L$. If it is independent of $Q$, we will sometimes denote it by $c_M^L$. If each member of*

$$\{c_R^Q(\lambda)\,;\, R \in \mathcal{P}^L(M)\}$$

*depends only on $L$, and not on $Q$, we will denote it by $c_R^L(\lambda)$ or $c_R(\Lambda)$. This is justified, since $R$ determines $L$ uniquely.*

## 10.13 A splitting formula

Let $\{d_P(\lambda)\}$ be another $(G,M)$-family. Then

$$\{(cd)_P(\lambda) = c_P(\lambda)d_P(\lambda)\}$$

is also a $(G,M)$-family.

There is a simple formula for $(cd)_M(\lambda)$. For geometric intuition consider the case of $G = \mathrm{GL}(3)$, $M = M_0$ the diagonal subgroup, and $c_P(\lambda)d_P(\lambda) = e^{\lambda(X_P)}e^{\lambda(Y_P)}$.

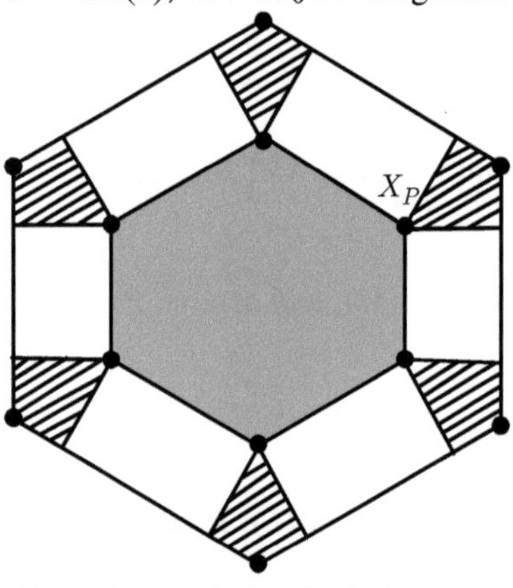


The volume of the shaded region equals $c_M$. The volumes of the six hatched regions add up to $d_M$. Each of the other six regions has volume equal to $c_M^Q d_Q'$ for a maximal parabolic subgroup $Q$.

PROPOSITION 10.14. *For all $\lambda \in i\mathfrak{a}_P^*$, we have*

$$(cd)_M(\lambda) = \sum_{Q\in\mathcal{F}(M)} c_M^Q(\lambda)d_Q'(\lambda).$$

PROOF. The value $(cd)_M(\lambda)$ equals

$$\sum_{P\in\mathcal{P}(M)} c_P(\lambda)d_P(\lambda)/\theta_P(\lambda) = \sum_{P\in\mathcal{P}(M)}\sum_{Q\supset P} c_P(\lambda)\theta_P^Q(\lambda)^{-1}d_Q'(\lambda).$$

Since $\{d_P(\lambda)\}$ is a $(G, M)$-family, each function $d'_Q(\lambda)$ is well defined, that is, it depends only on $Q$ and not on the group $P \subset Q$. Interchanging the order of summation yields the sum over $Q$ of the product of $d'_Q(\lambda)$ with

$$\sum_{\{P\in\mathcal{P}(M);\,P\subset Q\}} c_P(\lambda)/\theta_P^Q(\lambda).$$

This last expression is $c_M^Q(\lambda)$. □

COROLLARY 10.15. *We have*

$$d_M(\lambda) = \sum_{Q\in\mathcal{P}(M)} d'_Q(\lambda).$$

PROOF. For each $P \in \mathcal{P}(M)$, set

$$c_P(\lambda) = 1 = \lim_{X_P\to 0} e^{\lambda(X_P)}.$$

In view of our earlier remarks on convex hulls, $c_M^Q(\lambda)$ will vanish if $Q \in \mathcal{F}(M) \smallsetminus \mathcal{P}(M)$. If $Q \in \mathcal{P}(M)$, $c_M^Q(\lambda) = 1$. The corollary follows from Proposition 10.14. □

COROLLARY 10.16. *Suppose that, for all $L \in \mathcal{L}(M)$, the number $c_M^Q$, $Q \in \mathcal{P}(L)$, is independent of $Q$. Then*

$$(cd)_M = \sum_{L\in\mathcal{L}(M)} c_M^L d_L,$$

*where*

$$c_M^L = c_M^Q.$$

PROOF. This follows from Proposition 10.14 and Corollary 10.15. □

The results of this section hold if the functions $c_P(\lambda)$ take values in a complete topological vector space. For Proposition 10.14 we would require the space to also be an algebra.

# 11 Some Examples

## *11.1 Translations of a point*

Examples of $(G, M)$-families occur naturally in harmonic analysis. Consider the example of the Weyl group translates of a point. Let $M = M_0$. Fix $P_0 \in \mathcal{P}(M_0)$. Let $T$ be a point in $\mathfrak{a}_0$. Any $P \in \mathcal{P}(M_0)$ equals $w_s^{-1} P_0 w_s$ for a unique element $s \in W$.

Define $X_P = s^{-1}T$. Suppose that $P' = (w_{s'})^{-1}P_0 w_{s'}$ is adjacent to $P$. Then $s' = s_\alpha s$, where $s_\alpha$ is the simple reflection corresponding to $\alpha \in \Delta_{P_0}$. The point $X_P - X_{P'}$ equals $s^{-1}(T - s_\alpha^{-1}T)$. This is a multiple of $s^{-1}\alpha^\vee$. But $s^{-1}\alpha^\vee = \beta^\vee$, where $\beta$ is the unique root in $\Delta_P \cap (-\Delta_{P'})$. For $P \in \mathcal{P}(M_0)$, define

$$c_P(\lambda) = e^{\lambda(X_P)}.$$

These functions form a $(G, M)$-family.

## 11.2 Splitting

The following example will be used to study splitting. Fix $M \in \mathcal{L}(M_0)$. Let $\beta$ be a reduced root of $(G, A)$. Let $\beta^\vee$ denote the coroot. For $P \in \mathcal{P}(M)$, set $X_P^\beta = \beta^\vee$ if $\beta$ is a root of $(P, A)$, and let $X_P^\beta = 0$ otherwise. Let $\{r_\beta\}$ be a set of real numbers. Define $X_P$ to be the sum over all reduced roots $\beta$ of $r_\beta X_P^\beta$. Suppose that $P$ and $P'$ are adjacent. If $\beta$ is a root of $(P, A)$ and $(P', A)$, then $X_P^\beta = X_{P'}^\beta$. If it is a root of neither, both vectors are zero. The only reduced root of $(P, A)$ which is not a root of $(P', A)$ is the unique root $\beta$ in $\Delta_P \cap (-\Delta_{P'})$. It follows that $X_P - X_{P'}$ is always a multiple of $\beta^\vee$. Thus

$$\{c_P(\lambda) = e^{\lambda(X_P)}\,;\ P \in \mathcal{P}(M)\}$$

is a $(G, M)$-family.

Suppose that $L \in \mathcal{L}(M)$ and $Q \in \mathcal{P}(L)$. We have the $(L, M)$-family

$$\{c_R^Q(\lambda) = e^{\lambda(X_{Q(R)})}\,;\ R \in \mathcal{P}^L(M)\}.$$

Let $X_Q$ denote the sum of $r_\beta X_{Q(R)}^\beta$ over the reduced roots $\beta$ which do not vanish on $\mathfrak{a}_L$. Let $X_R$ denote the sum over the remaining $\beta$. That is, $X_R$ is the sum over the reduced roots of $(L, A_M)$. We have then the decomposition

$$X_{Q(R)} = X_R + X_Q.$$

Note that $X_R$ is independent of $Q$ and $X_Q$ is independent of $R$. Therefore

$$\begin{aligned} c_M^Q &= \lim_{\lambda \to 0} e^{\lambda(X_Q)} \sum_{R \in \mathcal{P}^L(M)} e^{\lambda(X_R)} / \theta_R^L(\lambda) \\ &= \lim_{\lambda \to 0} \sum_R e^{\lambda(X_R)} / \theta_R^L(\lambda). \end{aligned}$$

The number is independent of $Q$. If $L_1 \in \mathcal{L}(L)$ and $Q \in \mathcal{P}(L_1)$, the same is true of the $(L_1, L)$-family

$$\{c_R^Q(\lambda)\,;\, R \in \mathcal{P}^{L_1}(L)\},$$

that is, $c_L^Q$ depends only on $L_1$ and $L$, and not on $Q$. We denote it by $c_L^{L_1}$.

## 11.3 Exponentials

Fix a finite set $S$ of valuations on $F$. Fix $M \in \mathcal{L}(M_0)$. We may identify $G(F_S)$ as a subgroup of $G(\mathbb{A})$. In particular, for $x \in G(F_S)$ and $P \in \mathcal{P}(M)$, we have the vector $H_P(x)$ in $\mathfrak{a}$. Suppose that $P$ and $P'$ are adjacent and that $\alpha$ is the unique root in $\Delta_P \cap (-\Delta_{P'})$. Then $-H_P(x) + H_{P'}(x)$ is a multiple of $\alpha^\vee$. If $S$ consists of a single real valuation, this is [Ar76, Lemma 3.6]. The proof for general $S$ is the same. For $P \in \mathcal{P}(M)$, define

$$v_P(\lambda, x) = e^{-\lambda(H_P(x))}.$$

This collection of functions forms a $(G, M)$-family. Suppose that $L \in \mathcal{L}(M)$. For $\lambda \in i\mathfrak{a}_L^*$, there are two possible ways to define

$$v_Q(\lambda, x), \quad Q \in \mathcal{P}(L).$$

There is the $(G, L)$-family

$$\{e^{-\lambda(H_Q(x))}\}$$

or there is the $(G, L)$-family derived as in Section 10 from the $(G, M)$-family $\{v_P(\lambda, x)\,;\, P \in \mathcal{P}(M)\}$.

In fact, the two families are the same. For $Q \in \mathcal{P}(L)$ and $x \in L(F_S)$, we have defined, for $R \in \mathcal{P}^L(M)$, the functions

$$v_R^Q(\lambda, x) = v_{Q(R)}(\lambda, x).$$

The functions $v_R^Q(\lambda, x)$ depend only on $R$ and not on $Q$. We denote them by $v_R^L(\lambda, x)$ or $v_R(\lambda, x)$.

Sometimes, we will take $x$ to be a general point in $G(F_S)$ and use the function

$$v_M^Q(x) = \lim_{\lambda \to 0} \sum_{\{P \in \mathcal{P}(M)\,;\, P \subset Q\}} v_P(\lambda, x)/\theta_P^Q(\lambda).$$

As a function on $G(F_S)$, it is left $M(F_S)$ invariant. Indeed, since $H_P(m) = H_M(m)$ is independent of $P$, for $m \in M(F_S)$ we have

$$v_M^Q(\lambda, mx) = e^{-\lambda(H_M(m))} v_M^Q(\lambda, x).$$

Setting $\lambda = 0$ we see that $v_M^Q(mx) = v_M^Q(x)$.

## 11.4 Intertwining operators

We derive three similar examples from the intertwining operators between induced representations. Let $v$ be a valuation on $F$. Let $M_v$ be a Levi subgroup of $G$ defined over $F_v$. Unless stated otherwise, we consider only groups $M_v$ for which $K_v$ is well-positioned. Let $\tau_v \in \Pi(M_v(F_v))$. Let $\lambda_v \in \mathfrak{a}^*_{M_v,\mathbb{C}}$. If $P \in \mathcal{P}(M_v)$, we can extend the representation $\tau_{v,\lambda_v}$ to $P(F_v)$, multiply by $\delta_{P_v}^{1/2}$, and then induce to $G(F_v)$. This gives a representation $i_P(\tau_{v,\lambda_v})$ of $G(F_v)$ on a Hilbert space $V_P(\tau_v)$.

As in 1.7 we realize $V_P(\tau_v)$ as the space of square-integrable functions $\psi$ from $K_v$ to the space on which $\tau_v$ acts such that, for $k_1 \in K_v \cap P(F_v)$, we have

$$\psi(k_1 k) = \tau_v(k_1)\psi(k).$$

The space $V_P(\tau_v)$ depends only on $\tau_v$ and not on $\lambda_v$.

Suppose that $v$ is real. For $P$ and $P'$ in $\mathcal{P}(M_v)$, there are the canonically normalized intertwining operators(see Theorem 6.2)

$$\mathcal{R}_{P'|P}(\tau_v) : V_P(\tau_v) \to V_{P'}(\tau_v).$$

For all $x \in G(F_v)$, these operators satisfy

$$\mathcal{R}_{P'|P}(\tau_v) i_P(\tau_v, x) = i_{P'}(\tau_v, x)\mathcal{R}_{P'|P}(\tau_v). \tag{11.4.1}$$

These operators are unitary. In addition, for groups $P$, $P'$, and $P''$ in $\mathcal{P}(M_v)$, they satisfy

$$\mathcal{R}_{P''|P}(\tau_v) = \mathcal{R}_{P''|P'}(\tau_v)\mathcal{R}_{P'|P}(\tau_v). \tag{11.4.2}$$

Suppose that $\psi_0 \in H_P(\tau_v)$ satisfies $\psi_0(k) = \psi_0(1) \neq 0$ for all $k \in K_v$. In particular, the representation $\tau_v$ is unramified. Then for all $P'$, the normalizations have the property that

$$\mathcal{R}_{P'|P}(\tau_v)\psi_0 = \psi_0. \tag{11.4.3}$$

Suppose that $L_v \in \mathcal{L}(M_v)$ and $R \in \mathcal{P}^{L_v}(M_v)$. Consider the induced representation of $L_v(F_v)$ given by

$$\tau_v' = i_R^{L_v}(\tau_v) = i_R(\tau_v).$$

For any $Q \in \mathcal{P}(L_v)$, we have

$$V_Q(\tau_v') = V_{Q(R)}(\tau_v)$$

and

$$i_Q(\tau_v') = i_{Q(R)}(\tau_v).$$

For $Q, Q' \in \mathcal{P}(L_v)$, we have

$$\mathcal{R}_{Q'|Q}(\tau_v') = \mathcal{R}_{Q'(R)|Q(R)}(\tau_v). \tag{11.4.4}$$

If $\psi$ is a continuous function in $V_{Q(R)}(\tau_v)$ and $k \in K_v$, the function on $K_v \cap L_v(F_v)$ defined by

$$\psi_k : k_1 \mapsto \psi(k_1 k)$$

belongs to $V_R^{L_v}(\tau_v)$. If $R'$ is another group in $\mathcal{P}^{L_v}(M_v)$, then

$$(\mathcal{R}_{Q(R')|Q(R)}(\tau_v)\psi)_k = \mathcal{R}_{R'|R}(\tau_v)\psi_k. \tag{11.4.5}$$

Fix an irreducible representation $\Omega$ of $K_v$. Let $V_P(\tau_v)_\Omega$ denote the space of vectors in $V_P(\tau_v)$ which transform under $K_v$ according to $W$. It is finite dimensional. Let $\mathcal{R}_{P'|P}(\tau_v)_\Omega$ denote the restriction of $\mathcal{R}_{P'|P}(\tau_v)$ to this space. Let $D \in \mathrm{Diff}(i\mathfrak{a}_v^*)$, $\tau_v \in \Pi(M_v(F_v))$ and $\Omega \in \Pi(K_v)$. Then there are positive numbers $C$ and $N$ such that

$$\|D_{\tau_v}\mathcal{R}_{P'|P}(\tau_v)_\Omega\| \le C(1 + \|\tau_v\|)^N(1 + \|\Omega\|)^N. \tag{11.4.6}$$

These properties were established in the proof of [Ar81, Lemma 9.1].

## 11.5 A finite set of places

Let $v$ be a discrete valuation. There should be a canonical way to normalize the intertwining operators between the induced representations. However to check the required properties or even make the definitions in full generality would require a better understanding of harmonic analysis than is now available. Using Theorem 6.2, we have nonzero operators

$$\mathcal{R}_{P'|P}(\tau_v) : V_P(\tau_v) \to V_{P'}(\tau_v), \quad P, P' \in \mathcal{P}(M_v),$$

for which the properties (11.4.1)-(11.4.6) hold.

Suppose that $S$ is a finite set of valuations. Let $M \in \mathcal{L}(M_0)$. Let $\tau = \otimes_{v \in S} \tau_v$ be a representation in $\Pi(M(F_S))$. Let $P \in \mathcal{P}(M)$. Define the induced representation

$$i_P(\tau) = \otimes_{v \in S} i_P(\tau_v),$$

acting on the Hilbert space

$$V_P(\tau) = \otimes_{v \in S} V_P(\tau_v).$$

We have intertwining operators

$$\mathcal{R}_{P'|P}(\tau) = \otimes_{v \in S} \mathcal{R}_{P'|P}(\tau_v),$$

satisfying the analogues of (11.4.1)-(11.4.6).

Fix $P_0 \in \mathcal{P}(M)$. For $m \in M(F_S)$, define the representation $\tau_\lambda$ by

$$\tau_\lambda(m) = \tau(m) e^{\lambda(H_M(m))}.$$

For $P \in \mathcal{P}(M)$, we have a family of operator valued functions of $\lambda \in i\mathfrak{a}_M^*$, given by

$$\mathcal{R}_P(\lambda, \tau, P_0) = \mathcal{R}_{P|P_0}(\tau)^{-1} \mathcal{R}_{P|P_0}(\tau_\lambda).$$

PROPOSITION 11.6. *The collection*

$$\{\mathcal{R}_P(\lambda, \tau, P_0) \, ; \, P \in \mathcal{P}(M)\}$$

*is a* $(G, M)$*-family.*

PROOF. We can interpret each operator as a direct sum of operators on finite-dimensional spaces. Let $K_0$ be an open compact subgroup of $K_S$. Let $\Omega$ be an irreducible unitary representation of $K_{F_\infty}$. Denote by $V_P(\tau)_{K_0,\Omega}$ the space of vectors in $V_P(\tau)$ which are invariant under $K_0$ and which transform under $K_{F_\infty}$ according to $\Omega$. The space $V_{P_0}(\tau)_{K_0,\Omega}$ is finite dimensional. It is an invariant subspace of each operator $\mathcal{R}_P(\lambda, \tau, P_0)$. Suppose that $Q \in \mathcal{F}(M)$ and that the groups $P, P' \in \mathcal{P}(M)$ are both contained in $Q$. Then $P = Q(R)$ and $P' = Q(R')$ for uniquely determined groups $R, R' \in \mathcal{P}^{M_Q}(M)$. It follows from (11.4.5) that, if $\lambda \in i\mathfrak{a}_Q^*$, then

$$\mathcal{R}_{P'|P}(\tau_\lambda) = \mathcal{R}_{P'|P}(\tau).$$

Consequently, for any such $\lambda$, we have

$$\begin{aligned} \mathcal{R}_{P'}(\lambda, \tau, P_0) &= \mathcal{R}_{P|P_0}(\tau)^{-1} \mathcal{R}_{P'|P}(\tau)^{-1} \mathcal{R}_{P'|P}(\tau_\lambda) \mathcal{R}_{P|P_0}(\tau_\lambda) \\ &= \mathcal{R}_P(\lambda, \tau, P_0). \end{aligned}$$

The proposition follows. □

## 11.7 Normalized Intertwining Operators

More generally, let $L \in \mathcal{L}(M)$ and $P_0 \in \mathcal{P}^L(M)$. We have the $(L, M)$-family

$$\{\mathcal{R}_P(\lambda, \tau, P_0) = \mathcal{R}_P^L(\lambda, \tau, P_0)\,;\, P \in \mathcal{P}^L(M)\}$$

composed of the intertwining operators acting on $V_{P_0}^L(\tau)$. For $Q \in \mathcal{F}^L(M)$, denote the values at $\lambda = 0$ of $\mathcal{R}_M^L(\lambda, \tau, P_0)$ and $\mathcal{R}'_Q(\lambda, \tau, P_0)$ by $\mathcal{R}_M^L(\tau, P_0)$ and $\mathcal{R}'_Q(\tau, P_0)$, respectively. In general, each of these new operators is unbounded. However, it can still be regarded as a direct sum of operators on finite-dimensional spaces. It follows from the definition that, if $P_1$ is a second fixed group in $\mathcal{P}^L(M)$, then

$$\mathcal{R}_M^L(\tau, P_1) = \mathcal{R}_{P_0|P_1}(\tau)^{-1}\mathcal{R}_M^L(\tau, P_0)\mathcal{R}_{P_0|P_1}(\tau). \tag{11.7.1}$$

## 11.8 Levi subgroups

For any $L \in \mathcal{L}(M)$, we can define a scalar-valued $(L, M)$-family.

DEFINITION. Fix $f \in C_c^\infty(L(F_S))$, $\tau \in \Pi_{\text{temp}}(M(F_S))$, and $P_0 \in \mathcal{P}^L(M)$. For $P \in \mathcal{P}^L(M)$, define

$$\phi_P(\lambda, f, \tau, P_0) = \phi_P^L(\lambda, f, \tau, P_0) = \operatorname{tr}(i_{P_0}(\tau, f)\mathcal{R}_P(\lambda, \tau, P_0)).$$

This is an $(L, M)$-family. By (11.4.6) and the differentiation form of the dominated convergence theorem, each function is smooth in $\lambda$. In particular, any differentiation with respect to $\lambda$ can be interchanged with the trace operation. It follows that, for any $Q \in \mathcal{F}^L(M)$, the number $\phi'_Q(f, \tau, P_0)$ obtained a priori from the family $\{\phi_P(\lambda, f, \tau, P_0)\,;\, P \in \mathcal{P}^L(M)\}$ also equals $\operatorname{tr}(i_{P_0}(\tau, f)\mathcal{R}'_Q(\tau, P_0))$. Similarly,

$$\phi_M^L(f, \tau, P_0) = \operatorname{tr}(\mathcal{R}_M^L(\tau, P_0)i_{P_0}(\tau, f)).$$

Combining this last formula with (11.7.1), we find that $\phi_M^L(f, \tau, P_0)$ is independent of $P_0$. We denote it by $\phi_M^L(f, \tau)$. The same reasoning establishes that, for any $Q \in \mathcal{F}^L(M)$, the number $\phi_M^Q(f, \tau, P_0)$ is independent of $P_0 \in \mathcal{P}^L(M)$.

Each function $\phi_P^L(\lambda, f, \tau, P_0)$ depends only on the unitary equivalence class of $\tau$. Suppose $M_1 \in \mathcal{L}^L(M)$. Fix $Q_0 \in \mathcal{P}^L(M_1)$ and $P_0 \in \mathcal{P}^L(M)$, with $P_0 \subset Q_0$. Then $P_0 = Q_0(R_0)$ for $R_0 \in \mathcal{P}^{M_1}(M)$. Given $\tau \in \Pi_{\text{temp}}(M(F_S))$, let $\tau_1$ be the induced representation $i_{R_0}^{M_1}(\tau)$ of $M_1(F_S)$. If $Q$ is another group in $\mathcal{P}^L(M_1)$ and $\lambda \in i\mathfrak{a}_{M_1}^*$, (11.4.4) implies

$$\begin{aligned}\mathcal{R}_Q^L(\lambda,\tau_1,Q_0) &= \mathcal{R}_{Q|Q_0}(\tau_1)^{-1}\mathcal{R}_{Q|Q_0}(\tau_{1,\lambda})\\ &= \mathcal{R}_{Q(R_0)|Q_0(R_0)}(\tau)^{-1}\mathcal{R}_{Q(R_0)|Q_0(R_0)}(\tau_\lambda)\\ &= \mathcal{R}_Q^L(\lambda,\tau,P_0).\end{aligned}$$

It follows that, for $f \in C_c^\infty(L(F_S))$, we have

$$\phi_Q^L(\lambda, f, \tau, P_0) = \phi_Q^L(\lambda, f, \tau^{M_1}, Q_0). \tag{11.8.1}$$

In other words, the $(L, M_1)$-family associated with $\tau^{M_1}$ and $Q_0$ is the same as the one derived from the $(L, M)$-family $\{\phi_P^L(\lambda, f, \tau, P_0)\,;\, P \in \mathcal{P}^L(M)\}$. The $(M_1, M)$-families derived from $(L, M)$-families are related to intrinsically defined $(M_1, M)$-families.

PROPOSITION 11.9. *Suppose that $Q \in \mathcal{P}^L(M_1)$, and $P_0 \in \mathcal{P}^L(M)$, with $P_0 \subset Q$. For $f \in C(L(F_S))$, $\tau \in \Pi_{\mathrm{temp}}(M(F_S))$, and $R \in \mathcal{P}^{M_1}(M)$, we have*

$$\phi_R^Q(\lambda, f, \tau, P_0) = \phi_R^{M_1}(\lambda, f_Q, \tau, P_0 \cap M_1).$$

PROOF. Let $R_0 = P_0 \cap M_1$. Then $P_0 = Q(R_0)$. We need to evaluate the trace of an operator on $V_{P_0}^L(\tau)$. The space $V_{P_0}^L(\tau)$ can be regarded as the space of square-integrable, $K \cap Q(F_S)$-equivariant functions from $K \cap L(F_S)$ to the Hilbert space $V_{R_0}^{M_1}(\tau)$. Then $i_{P_0}^L(f)$ is an integral operator with kernel $K(k_1, k_2)$ equal to

$$\int_{M_1(F_S)} i_{R_0}^{M_1}(\tau, m)\cdot \delta_{R_0}^{M_1}(m)^{1/2}\int_{N_Q(F_S)} f(k_1^{-1}mnk_2)\,dn\,dm.$$

By (11.4.4), the operator

$$\mathcal{R}_{Q(R)}^L(\lambda,\tau,P_0), \quad R \in \mathcal{P}^{M_1}(M),$$

on $V_{P_0}^L(\tau)$ is just fiber multiplication by the operator

$$\mathcal{R}_R^{M_1}(\lambda,\tau,R_0)$$

on $V_{R_0}^{M_1}(\tau)$. Therefore the trace of

$$\mathcal{R}_{Q(R)}^L(\lambda,\tau,P_0)\,i_{P_0}^L(f)$$

equals

$$\int_{K\cap M_1(F_S)\backslash K\cap L(F_S)} \mathrm{tr}(\mathcal{R}_R^{M_1}(\lambda,\tau,R_0)K(k,k))\,dk.$$

This is just

$$\mathrm{tr}(\mathcal{R}_R^{M_1}(\lambda, \tau, R_0) i_{R_0}^{M_1}(\tau, f_Q)) = \phi_R^{M_1}(\lambda, f_Q, \tau, R_0).$$

This establishes the proposition. □

Suppose that $f \in C_c^\infty(L(F_S))$ and that $\tau \in \Pi(M(F_S))$. Fix $P_0 \in \mathcal{P}^L(M)$. For $P \in \mathcal{P}^L(M)$, consider the functions

$$\mathrm{tr}(\mathcal{R}_P(\lambda, \tau, P_0) i_{P_0}(\tau, f)).$$

They are smooth in $\lambda$ by (11.4.6). Notice that

$$\int_{i\mathfrak{a}_M^*/\mathfrak{a}_{M,S}^\vee} \mathrm{tr}(\mathcal{R}_P(\lambda, \tau_\Lambda, P_0) i_{P_0}(\tau_\Lambda, f))\, \mathrm{d}\Lambda$$

depends only on the restriction of $f$ to $L(F_S)^1$ and the orbit of $\tau$ under $i\mathfrak{a}_M^*$.

DEFINITION 11.10. *Denote the last function of $\lambda$ by*

$$\phi_P(\lambda, f, \tau, P_0),$$

*where $f$ is taken to be a function in $C_c^\infty(L(F_S)^1)$ and $\tau$ is a representation in $\Pi(M(F_S)^1)$. Then*

$$\{\phi_P(\lambda, f, \tau, P_0)\, ;\, P \in \mathcal{P}^L(M)\}$$

*is an $(L, M)$-family which satisfies properties analogous to those described above. In particular, to every $f \in C_c^\infty(L(F_S)^1)$ and $\tau \in \Pi(M(F_S)^1)$, we may associate a number*

$$\phi_M^L(f, \tau) = \lim_{\lambda \to 0} \sum_{P \in \mathcal{P}^L(M)} \phi_P(\lambda, f, \tau, P_0)/\theta_P(\lambda).$$

# 12 The Distributions $J_M(\tau_\lambda)$

## *12.1 Hecke Algebra*

Our primary objects of study are linear functionals on the Hecke algebra $\mathbb{H}(G(F_S))$. The Hecke algebra consists of the smooth, compactly supported functions on $G(F_S)$ that are $K$-finite, that is, whose left and right translates by $K$ span a finite-dimensional space. The linear functionals on $\mathbb{H}(G(F_S))$, also called distributions on $G(F_S)$, are obtained from certain rational functions constructed from the normalized intertwining operators.

DEFINITION 12.2. *Fix an irreducible representation $\tau \in \Pi(M(F_S))$. For $\nu \in i\mathfrak{a}_M^*$ and $P \in \mathcal{P}(M)$, we have the $(G, M)$-family*

$$\mathcal{R}_P(\nu, \tau_\lambda, P') = \mathcal{R}_{P|P'}(\tau_\lambda)^{-1}\mathcal{R}_{P|P'}(\tau_{\lambda+\nu}). \tag{12.2.1}$$

*The functions* (12.2.1) *are meromorphic in $\nu$ and $\lambda$ and depend on a fixed $P' \in \mathcal{P}(M)$. If $\lambda \in \mathfrak{a}_{M,\mathbb{C}}^*$ is in general position, the function* (12.2.1) *will have no poles for $\nu \in i\mathfrak{a}_M^*$. Let*

$$\mathcal{R}_M(\tau_\lambda, P') = \lim_{\nu \to 0} \sum_{P \in \mathcal{P}(M)} \mathcal{R}_P(\nu, \tau_\lambda, P')/\theta_P(\nu).$$

*The distributions $f \mapsto J_M(\tau_\lambda, f)$ on $\mathbb{H}(G(F_S))$ are defined by*

$$J_M(\tau_\lambda, f) = \operatorname{tr}(\mathcal{R}_M(\tau_\lambda, P') i_{P'}^G(\tau_\lambda, f)), \quad f \in \mathbb{H}(G(F_S)).$$

*They are independent of $P'$.*

## 12.3 Levi Subgroups

The distributions $J_M(\tau_\lambda, f)$ are meromorphic in $\lambda$. More generally, suppose that $\mathfrak{M} = \prod_{v \in S} M_v$ is a well-positioned relative to $K$ Levi subgroup of $M$ over $F_S$ and that $\sigma$ is a representation in

$$\Pi(\mathfrak{M}(F_S)) = \{\otimes_v \sigma_v \,;\, \sigma_v \in \Pi(M_v(F_v))\}\,.$$

If $\Lambda = \oplus_v \Lambda_v$ is a generic point in $\mathfrak{a}_{\mathfrak{M},\mathbb{C}}^*$, the induced representation $\sigma_\Lambda^M$ belongs to $\Pi(M(F_S))$. The associated distribution $J_M(\sigma_\Lambda^M, f)$, often denoted by $J_M(\sigma_\Lambda, f)$, extends as a meromorphic function of $\Lambda$ to $\mathfrak{a}_{\mathfrak{M},\mathbb{C}}^*$. We shall now describe its poles.

PROPOSITION 12.4. *The function $J_M(\sigma_\Lambda, f)$ is meromorphic in $\Lambda$. The poles lie along hyperplanes of the form* (12.4.2) *below.*

PROOF. Let $\Sigma_M(G, A_{\mathfrak{M}})$ denote the set of roots of $(G, A_{\mathfrak{M}})$ which do not vanish on $\mathfrak{a}_M$. Any $\beta \in \Sigma_M(G, A_{\mathfrak{M}})$ belongs to $\Sigma_M(G, A_{M_v})$, for a unique $v \in S$. Let $q_\beta(\Lambda)$ denote $\Lambda_v(\beta^\vee)$ if $F_v$ is Archimedean and $q_v^{-\Lambda_v(\beta^\vee)}$ if $F_v$ is non-Archimedean of residual cardinality $q_v$. The properties $(\mathcal{R}_2)$, $(\mathcal{R}_3)$, $(\mathcal{R}_6)$, and (6.7.1′) of Section 6 tell us that the matrix coefficients of the operators

$$\mathcal{R}_{P|P'}(\sigma_\Lambda^M), \quad P, P' \in \mathcal{P}(M),$$

are all rational functions in the variables

$$\{q_\beta(\Lambda)\,;\ \beta \in \Sigma_M(G, A_{\mathfrak{M}})\} \tag{12.4.1}$$

whose poles lie along hyperplanes of the form

$$\Lambda(\beta^\vee) = c, \quad \beta \in \Sigma_M(G, A_{\mathfrak{M}}), \quad c \in \mathbb{C}. \tag{12.4.2}$$

The same is therefore true of the matrix coefficients of $\mathcal{R}_M(\sigma_\Lambda^M, P')$. □

## 12.5 Standard Representations

It is important to relate the distributions $J_M(\tau_\lambda)$ to similar objects defined for standard representations. Suppose that $\rho \in \Sigma(M(F_S))$. Then there is an $\mathfrak{M}$ as in 12.3 such that

$$\rho = \sigma_{\Lambda_0}^M, \quad \sigma \in \Pi_{\mathrm{temp}}(\mathfrak{M}(F_S)), \quad \Lambda_0 \in \mathfrak{a}_{\mathfrak{M}}^*.$$

For $\lambda \in \mathfrak{a}_{M,\mathbb{C}}^*$, define

$$J_M(\rho_\lambda, f) = \lim_{\Lambda \to \Lambda_0 + \lambda} J_M(\sigma_\Lambda, f).$$

Here $\Lambda$ ranges over the points of $\mathfrak{a}_{\mathfrak{M},\mathbb{C}}^*$ for which $\sigma_\Lambda^M$ belongs to $\Pi(M(F_S))$.
We have the $(G, M)$-family

$$\mathcal{R}_P(\nu, \rho_\lambda, P') = \mathcal{R}_{P|P'}(\rho_\lambda)^{-1}\mathcal{R}_{P|P'}(\rho_{\lambda+\nu}), \quad \nu \in i\mathfrak{a}_M^*, P \in \mathcal{P}(M).$$

This $(G, M)$ family defines $\mathcal{R}_M(\rho_\lambda, P')$ , and thus we have

$$J_M(\rho_\lambda, f) = \mathrm{tr}(\mathcal{R}_M(\rho_\lambda, P') i_{P'}^G(\rho_\lambda, f)).$$

The remarks above imply that $J_M(\rho_\lambda, f)$ is a well-defined meromorphic function of $\lambda$ on $\mathfrak{a}_{M,\mathbb{C}}^*$.

DEFINITION 12.6. (1) *Suppose $\tau \in \Pi(M(F_S))$. The distributions $J_M(\tau_\lambda)$ and $J_M(\rho_\lambda)$ will be related by a $(G, M)$-family of scalar-valued functions. Namely, for $P \in \mathcal{P}(M)$ and $\nu \in i\mathfrak{a}_M^*$, put*

$$m_P(\nu, \tau_\lambda, \rho_\lambda, P') = \Delta(\tau, \rho)\tilde{m}_{P|P'}(\tau_\lambda, \rho_\lambda)^{-1}\tilde{m}_{P|P'}(\tau_{\lambda+\nu}, \rho_{\lambda+\nu}). \tag{12.6.1}$$

*See 9.9 for the definition of $\tilde{m}_{P|P'}$. These functions vanish unless $\tau$ and $\overline{\rho}$ are block equivalent. The compatibility condition required of a $(G, M)$-family follows from* (6.6.1) *and* (9.10.1).

(2) *Suppose* $L \in \mathcal{L}(M)$. *Replacing* $(G,M)$ *by* $(L,M)$ *in* (12.6.1), *we obtain the* $(L,M)$*-family*

$$m_R^L(\nu,\tau_\lambda,\rho_\lambda,R'),\quad R,R'\in\mathcal{P}^L(M).$$

(3) *For any* $Q\in\mathcal{P}(L)$,

$$m_R^Q(\nu,\tau_\lambda,\rho_\lambda,P') = m_{Q(R)}(\nu,\tau_\lambda,\rho_\lambda,P'),\quad R\in\mathcal{P}^L(M)$$

*is also an* $(L,M)$*-family.*

(4) *These two* $(L,M)$*-families are distinct. However,* (9.10.1) *implies the associated numbers*

$$m_M^L(\tau_\lambda,\rho_\lambda,R') = \lim_{\nu\to 0}\sum_R m_R^L(\nu,\tau_\lambda,\rho_\lambda,R')/\theta_R(\nu)$$

*and*

$$m_M^Q(\tau_\lambda,\rho_\lambda,P') = \lim_{\nu\to 0}\sum_R m_R^Q(\nu,\tau_\lambda,\rho_\lambda,P')/\theta_R(\nu)$$

*are equal. We denote their common value by* $m_M^L(\tau_\lambda,\rho_\lambda)$. *It is independent of* $R'$, $P'$, *and* $Q$.

## 12.7 A Generalization

We have the following generalization of (9.4.2)

PROPOSITION. *Let* $\tau\in\Pi(M(F_S))$ *and* $f\in\mathbb{H}(G(F_S))$. *Then*

$$J_M(\tau_\lambda,f) = \sum_{L\in\mathcal{L}(M)}\sum_{\rho\in\Sigma(M(F_S))} m_M^L(\tau_\lambda,\rho_\lambda)J_L(\rho_\lambda^L,f).$$

PROOF. By definition,

$$J_M(\tau_\lambda,f) = \lim_{\nu\to 0}\sum_{P\in\mathcal{P}(M)}\mathrm{tr}(\mathcal{R}_{P|P'}(\tau_\lambda)^{-1}\mathcal{R}_{P|P'}(\tau_{\lambda+\nu})i_{P'}^G(\tau_\lambda,f))/\theta_P(\nu).$$

We write this as

$$\lim_{\nu\to 0}\sum_{P\in\mathcal{P}(M)}(m_{P|P'}(\tau_\lambda)^{-1}m_{P|P'}(\tau_{\lambda+\nu}))^{-1}T_P(\nu,\tau_\lambda),$$

where

$$T_P(\nu,\tau_\lambda) = \mathrm{tr}(\mathcal{M}_{P|P'}(\tau_\lambda)^{-1}\mathcal{M}_{P|P'}(\tau_{\lambda+\nu})i_{P'}^G(\tau_\lambda,f))$$

for any admissible representation $\tau$ of $M(F_S)$.

The integral formula (5.1.1) implies that $T_P(\nu, \tau_\lambda)$, as a function of $\tau$, depends only on $\operatorname{tr}\tau$. In fact,

$$\operatorname{tr}\tau \mapsto T_P(\nu, \tau_\lambda)$$

extends to a linear functional on the vector space spanned by the functions $\{\operatorname{tr}\tau\}$. It follows from Proposition 9.4 that

$$\begin{aligned} T_P(\nu, \tau_\lambda) &= \sum_{\rho \in \Sigma(M(F_S))} \xi(\tau, \rho) T_P(\nu, \rho_\lambda) \\ &= \sum_{\rho} \xi(\tau, \rho) m_{P|P'}(\rho_\lambda)^{-1} m_{P|P'}(\rho_{\lambda+\nu}) \operatorname{tr}(\mathcal{R}_{P|P'}(\rho_\lambda)^{-1} \mathcal{R}_{P|P'}(\rho_{\lambda+\nu}) i^G_{P'}(\rho_\lambda, f)). \end{aligned}$$

Therefore $J_M(\tau_\lambda, f)$ equals the sum over $\rho \in \Sigma(M(F_S))$ of

$$\lim_{\nu \to 0} \sum_{P \in \mathcal{P}(M)} m_P(\nu, \tau_\lambda, \rho_\lambda, P') \operatorname{tr}(\mathcal{R}_P(\nu, \rho_\lambda, P') i^G_{P'}(\rho_\lambda, f)) / \theta_P(\nu).$$

This last expression is built out of a product of two $(G, M)$-families. Applying Corollary 10.16, we obtain

$$J_M(\tau_\lambda, f) = \sum_{L \in \mathcal{L}(M)} \sum_{\rho \in \Sigma(M(F_S))} m^L_M(\tau_\lambda, \rho_\lambda) \operatorname{tr}(\mathcal{R}_L(\rho_\lambda, P') i^G_{P'}(\rho_\lambda, f)).$$

An argument similar to the derivation of (11.8.1) establishes that

$$\operatorname{tr}(\mathcal{R}_L(\rho_\lambda, P') i^G_{P'}(\rho_\lambda, f)) = J_L(\rho^L_\lambda, f).$$

The proposition follows. □

# 13 The Distributions $J_{M,\gamma}$ and $J_{M,\pi}$

## *13.1 Local Approximations*

In this section, $S$ is a finite set of valuations on $F$, and $M \subset L$ are fixed groups in $\mathcal{L}(M_0)$. We will use the examples of Section 11 to define two families of tempered distributions on $L(F_S)$. They are to be regarded as local approximations to the distributions $J^L_{\mathfrak{o}}$ and $J^L_\chi$.

## 13.2 A Norm

If $H$ is a locally compact group, let $\Pi(H)$ denote as usual the set of equivalence classes of irreducible (continuous) unitary representations of $H$. If the notion of a *tempered* representation is defined for $H$, we will let $\Pi_{\mathrm{temp}}(H)$ stand for those classes that are tempered. Suppose that $v$ is a valuation on $F$. If $v$ is discrete, the residual characteristic is a rational prime power $q_v$. If $v$ is real, set $q_v = e$. Suppose that $M_v$ is a Levi subgroup of $G$ defined over $F_v$.

Harish-Chandra defines the map $H_{M_v}$ from $M_v(F_v)$ to $\mathfrak{a}_v = \mathrm{Hom}(X(M_v)_{F_v}, \mathbb{R})$ by setting

$$q_v^{\langle H_{M_v}(m_v),\chi\rangle} = |\chi_v(m_v)|_v, \quad \chi \in X(M_v)_{F_v}, \quad m_v \in M_v(F_v).$$

Suppose that $\tau_v \in \Pi(M_v(F_v))$. If $\lambda_v$ is a vector in $\mathfrak{a}^*_{M_v,\mathbb{C}}$, we set

$$\tau_{v,\lambda_v}(m_v) = \tau_v(m_v) \cdot q_v^{\lambda_v(H_{M_v}(m_v))}.$$

Suppose that $M'_v \subset M_v$ is another Levi subgroup of $G$ defined over $F_v$ and that $\sigma_v \in \Pi(M'_v(F_v))$. If $P_v \in \mathcal{P}^{M_v}(M'_v)$, we can extend $\sigma_v$ to the parabolic subgroup $P_v(F_v)$ and then induce up to $M_v(F_v)$. The class of the resulting representation of $M_v(F_v)$ is independent of $P_v$. We denote it by $\sigma_v^{M_v}$.

DEFINITION. It is convenient to define, in a noncanonical way, a "norm" function on $\Pi(M_v(F_v))$. If $v$ is discrete and $\tau_v \in \Pi(M_v(F_v))$, set $\|\tau_v\| = 0$.

However, if $v$ is Archimedean, let $\Delta$ be a fixed left invariant differential operator on $M_v(F_v)$ of order two. We assume that $\Delta$ is positive definite and that it commutes with right translations on $M_v(F_v)$ by $K_v \cap M_v(F_v)$. For any $\tau_v \in \Pi(M_v(F_v))$, we obtain an operator $\tau_v(\Delta)$ on the space on which $\tau_v$ acts. Let $\|\tau_v\|$ be its smallest eigenvalue. It is a positive number. We can follow the same prescription to define $\|\Omega\|$ for any $\Omega$ in $\Pi(K_{F_\infty})$.

## 13.3 Finite Set of Places

Now suppose that $S$ is a finite set of valuations on $F$. Suppose that for each $v \in S$, $M_v$ is a Levi subgroup of $G$ defined over $F_v$. The group $\mathfrak{M} = \prod_{v\in S} M_v$ will be called a *Levi S-subgroup* of G, and we write $\mathfrak{M}(F_S) = \prod_{v\in S} M_v(F_v)$. Any $\tau \in \Pi(\mathfrak{M}(F_S))$ is a unique tensor product $\otimes_{v\in S}\tau_v$ of irreducible representations of the groups $M_v(F_v)$. If $\zeta = \oplus_{v\in S}\zeta_v$ is a vector in $\oplus_v \mathfrak{a}^*_{M_v,\mathbb{C}}$, we put

$$\tau_\zeta = \otimes_{v\in S}\tau_{v,\zeta_v}.v$$

We also put

$$\|\tau\| = \sup_{v\in S} \|\tau_v\|.$$

If $\mathfrak{M}' = \prod_{v\in S} M'_v$ is contained in $\mathfrak{M}$, and $\sigma = \otimes_{v\in S}\sigma_v$ belongs to $\Pi(\mathfrak{M}(F'_S))$, set

$$\sigma^{\mathfrak{M}} = \otimes_{v\in S}\sigma_v^{M_v}.$$

Most of the time we will take each $M_v$ equal to a fixed $M \in \mathcal{L}(M_0)$. So we will write $\sigma^M$ for $\sigma^{\mathfrak{M}}$. In this situation, we shall sometimes want to embed a vector $\lambda \in \mathfrak{a}^*_{\mathbb{C}}$ diagonally into $\oplus_{v\in S}\mathfrak{a}^*_{v,\mathbb{C}}$ by

$$\lambda \mapsto \zeta = \oplus_{v\in S}(\log q_v)^{-1}\lambda.$$

We shall then write $\tau_\lambda$ for $\tau_\zeta$, so that we have

$$\tau_\lambda(m) = \tau(m)e^{\lambda(H_M(m))}, \quad m \in M(F_S).$$

## 13.4 Schwarz Space

Only the Schwarz spaces $\mathcal{C}(M(F_v))$, $v \in S$, appear in the literature. However, Harish-Chandra's definition (see [HC66, HC73]) extends easily to $M(F_S)$. Indeed, if $v \in S$ let $\Xi_v^M$ and $\sigma_v^M$ be the functions on $M(F_S)$ used in the definition of $\mathcal{C}(M(F_v))$. Given $m = \prod_{v\in S} m_v$ in $M(F_S)$, set

$$\Xi^M(m) = \prod_{v\in S} \Xi_v^M(m_v)$$

and

$$\sigma^M(m) = \prod_{v\in S} \sigma_v^M(m_v).$$

If the Archimedean valuation $v$ belongs to $S$, let $X_L$ and $X_R$ be the operators on $C_c^\infty(M(F_S))$ which act, through $C_c^\infty(M(F_v))$, as left and right invariant differential operators. If no Archimedean valuation belongs to $S$, set $X_L = X_R = 1$. For any $n \geq 0$ and $f \in C_c^\infty(M(F_S))$, put

$$\|f\|_{X_L,X_R,n} = \sup_{m\in M(F_S)} \left\{|(X_L X_R f)(m)|\,\Xi^M(m)\left(1+\sigma^M(m)\right)^{-n}\right\}.$$

DEFINITION 13.5. *For any open compact subgroup $K_0$ of*

$$K_S = \prod_{\{v\in S;\, v \text{ discrete}\}} K_v$$

*let $\mathcal{C}_{K_0}(M(F_S))$ be the space of smooth, $K_0 \cap M(F_S)$ bi-invariant functions $f$ on $M(F_S)$ such that*

$$\|f\|_{X_L,X_R,n} < \infty$$

*for all $X_L$, $X_R$, and $n$. The seminorms $\|\cdot\|_{X_L,X_R,n}$ induce a topology on $\mathcal{C}_{K_0}(M(F_S))$. The Schwarz space $\mathcal{C}(M(F_S))$ can then be defined as the topological direct limit, over all $K_0$, of the spaces $\mathcal{C}_{K_0}(M(F_S))$.*

EXAMPLE 13.6. Orbital integrals. Important examples of invariant tempered distributions on $M$ are the orbital integrals. Let $M(F_S)_{\mathrm{reg}}$ be the set of regular semisimple elements in $M(F_S)$. An element $\gamma$ belongs to $M(F_S)_{\mathrm{reg}}$ if and only if $M(F_S)_\gamma$, the centralizer of $\gamma$ in $M(F_S)$, is of the form $\mathcal{T}_S = \prod_{v\in S} T_v(F_v)$, where each $T_v$ is a maximal torus of $M$ defined over $F_v$. Let

$$D^M(m), \quad m \in M(F_S),$$

be the coefficient of degree equal to the rank of $M$ in the characteristic polynomial of $1 - \mathrm{Ad}(m)$. If $m = \prod_{v\in S} m_v$ belongs to $M(F_S)$, set

$$|D^M(m)| = \prod_{v\in S} |D^M(m_v)|_v.$$

The orbital integral of $f \in \mathcal{C}(M(F_S))$ at $\gamma \in M(F_S)_{\mathrm{reg}}$ is then defined as

$$I_\gamma^M(f) = |D^M(\gamma)|^{1/2} \int_{M(F_S)_\gamma \backslash M(F_S)} f(x^{-1}\gamma x)\, dx.$$

This $I_\gamma^M$ is an invariant tempered distribution.

EXAMPLE 13.7. Weighted orbital integrals. Our first distributions are similar to those studied in [Ar76]. They are obtained by taking orbital integrals on $L(F_S)$, weighted by the functions $v_M^L(x)$ (see [Ar76]). We need the following to guarantee that they are tempered.

PROPOSITION 13.8. *If $\gamma \in L(F_S)_{\mathrm{reg}} \cap M$ and $n$ is sufficiently large, then*

$$\int_{L(F_S)_\gamma \backslash L(F_S)} \Xi^L(x^{-1}\gamma x)(1 + \sigma^L(x^{-1}\gamma x))^{-n} v_M^L(x)\, dx$$

*is finite. Since $v_M^L(x)$ is left $M$-invariant, the integrand is $L(F_S)_\gamma$-invariant.*

PROOF. Write the variable $x$ as $\prod_{v\in S} x_v$. Let $\gamma = \prod_{v\in S} \gamma_v$. Since

$$H_P(x) = \sum_{v\in S} H_P(x_v)$$

and

$$v_M^L(x) = \frac{1}{q!} \sum_{P \in \mathcal{P}^L(M)} (-\lambda(H_P(x)))^q / \theta_P(\lambda),$$

$q = \dim(A_M/A_L)$, we can rewrite the integral as a sum of products of integrals over $L(F_v)_{\gamma_v} \backslash L(F_v)$. If $v$ is Archimedean, the convergence of the resulting integral can be proved as in [Ar76, Lemma 7.2]. If $v$ is discrete, one knows that any finite-dimensional $L(F_v)$-module over $F_v$ has a basis of eigenvectors for any given split torus in $L(F_v)$, such that the corresponding lattice is stabilized by $K_v$. This fact, together with [HC73, Lemma 13] (see also the proofs of Corollaries 4.7.3 and 4.8.4 of [Si80]), allows us to transcribe the Archimedean proof. The argument is the same, so we do not present the details. □

DEFINITION 13.9. *If $\gamma \in L(F_S)_{\mathrm{reg}} \cap M$ and $f \in \mathcal{C}(L(F_S))$, define $J_M^L(\gamma,f)$ to be*

$$J_{M,\gamma}^L(f) = |D^L(\gamma)|^{1/2} \int_{L(F_S)_\gamma \backslash L(F_S)} f(x^{-1}\gamma x) v_M^L(x)\, dx.$$

*By Proposition 13.8, the integral converges absolutely, and each $J_{M,\gamma}^L$ is a tempered distribution.*

If $Q \in \mathcal{F}^L(M)$, we have

$$J_{M,\gamma}^{M_Q}(f) = |D^L(\gamma)|^{1/2} \int_{L(F_S)_\gamma \backslash L(F_S)} f(x^{-1}\gamma x) v_M^Q(x)\, \mathrm{d}x. \tag{13.9.1}$$

This follows from the change of variables formula

$$|D^L(\gamma)|^{1/2} \int_{N_Q(F_S)} \phi\left((m^{-1}\gamma m)^{-1} n^{-1} (m^{-1}\gamma m) n\right)\, \mathrm{d}n = |D^{M_Q}(\gamma)|^{1/2} \delta_Q^L(\gamma)^{1/2} \int_{N_Q(F_S)} \phi(n)\, \mathrm{d}n,$$

for $m \in M_Q(F_S)$. If $L = M$, then $J_{M,\gamma}^L$ is just the ordinary orbital integral $I_\gamma^M$, defined in 13.6.

PROPOSITION 13.10. *For $\gamma \in L(F_S)_{\mathrm{reg}} \cap M$, $f \in \mathcal{C}(L(F_S))$, and $y \in L(F_S)$, we have*

$$J_{M,\gamma}^L(f^y) = \sum_{Q \in \mathcal{F}^L(M)} J_{M,\gamma}^{M_Q}(f_{Q,y}).$$

PROOF. With a change of variables, we see that $J_{M,\gamma}^L(f^y)$ equals

$$|D^L(\gamma)|^{1/2} \int_{L(F_S)_\gamma \backslash L(F_S)} f(x^{-1}\gamma x) v_M^L(xy)\, \mathrm{d}x.$$

If $P \in \mathcal{P}^L(M)$, then

$$v_P^L(\lambda, xy) = e^{-\lambda(H_P(xy))} = e^{-\lambda(H_P(k_P(x)y))}e^{-\lambda(H_P(x))}$$

where $k_P(x)$ is any element in $K_S = \prod_{v \in S} K_v$ such that $xk_P(x)^{-1}$ belongs to $P(F_S)$. This equals

$$u_P(\lambda, x, y)v_P^L(x),$$

where

$$u_P(\lambda, x, y) = e^{-\lambda(H_P(k_P(x)y))}, \quad P \in \mathcal{P}^L(M),$$

is an $(L, M)$-family. It follows from Proposition 10.14 that

$$v_M^L(xy) = \sum_{Q \in \mathcal{F}^L(M)} v_M^Q(x)v_Q'(x, y).$$

We see from the discussion of Section 10 that $v_Q'(x, y)$ is the same as the function defined in Section 3. As a function of $x$, the function $v_Q'(x, y)$ is left $Q(F_S)$-invariant. We write

$$|D^L(\gamma)|^{1/2} \int_{L(F_S)_\gamma \backslash L(F_S)} f(x^{-1}\gamma x)v_M^Q(x)v_Q'(x, y)\, \mathrm{d}x$$

as

$$|D^L(\gamma)|^{1/2} \int_{K \cap L(F_S)} \int_{N_Q(F_S)} \int_{M_Q(F_S)_\gamma \backslash M_Q(F_S)} f(k^{-1}n^{-1}m^{-1}\gamma mnk)v_M^Q(m)v_Q'(k, y)\, \mathrm{d}m\, \mathrm{d}n\, \mathrm{d}k.$$

This equals

$$|D^{M_Q}(\gamma)|^{1/2} \int_{M_Q(F_S)_\gamma \backslash M_Q(F_S)} f_{Q,y}(m^{-1}\gamma m)v_M^{M_Q}(m)\, \mathrm{d}m = J_{M,\gamma}^{M_Q}(f_{Q,y}).$$

The proposition follows. □

## 13.11 Intertwining Operators

Our second family of distributions has already been defined. We introduce new notation only to point out the analogy with the distributions just defined.

Definition. If $\tau \in \Pi_{\mathrm{temp}}(M(F_S))$ and $f \in \mathcal{C}(L(F_S))$, define $J_M^L(\tau, f)$ to be

$$J_{M,\tau}^L(f) = \phi_M^L(f, \tau) = \mathrm{tr}\left(\mathcal{R}_M^L(\tau, P_0)i_{P_0}^L(\tau, f)\right)$$

for any $P_0 \in \mathcal{P}^L(M)$. If $Q \in \mathcal{F}^L(M)$ we have the analogue of (13.9.1),

$$J_{M,\tau}^{M_Q}(f_Q) = \operatorname{tr}\left(\mathcal{R}_M^Q(\tau, P_0) i_{P_0}^L(\tau, f)\right), \tag{13.11.1}$$

for any $P_0 \in \mathcal{P}^L(M)$. This formula follows from Proposition 11.9 along with the fact, noted in 11.8, that the right-hand side is independent of $P_0$.

PROPOSITION 13.12. *For $\tau \in \Pi_{\text{temp}}(M(F_S))$, $f \in \mathcal{C}(L(F_S))$, and $y \in L(F_S)$, we have*

$$J_{M,\tau}^L(f^y) = \sum_{Q \in \mathcal{F}^L(M)} J_{M,\tau}^{M_Q}(f_{Q,y}).$$

PROOF. Fix $P_0 \in \mathcal{P}^L(M)$. Then $J_{M,\tau}^L(f^y) = \phi_M^L(\tau, f^y)$ equals

$$\lim_{\lambda \to 0} \sum_{P \in \mathcal{P}^L(M)} \operatorname{tr}\left(\mathcal{R}_P(\lambda, \tau, P_0) i_{P_0}(\tau, f^y)\right) / \theta_P(\lambda).$$

Substituting for

$$i_{P_0}(\tau, f^y) = i_{P_0}(\tau, y)^{-1} i_{P_0}(\tau, f) i_{P_0}(\tau, y)$$

and

$$\mathcal{R}_P(\lambda, \tau, P_0) = \mathcal{R}_{P|P_0}(\tau)^{-1} \mathcal{R}_{P|P_0}(\tau_\lambda)$$

yields the trace of

$$\lim_{\lambda \to 0} \sum_{P \in \mathcal{P}^L(M)} i_{P_0}(\tau, f) i_{P_0}(\tau, y) \mathcal{R}_{P|P_0}(\tau)^{-1} \mathcal{R}_{P|P_0}(\tau_\lambda) i_{P_0}(\tau, y)^{-1} \theta_P(\lambda)^{-1}.$$

In this expression, we can replace $i_{P_0}(\tau, y)^{-1}$ by $i_{P_0}(\tau_\lambda, y)^{-1}$ without changing the final limit. We obtain the trace of the operator

$$i_{P_0}(\tau, f) \cdot \lim_{\lambda \to 0} \sum_{P \in \mathcal{P}^L(M)} \mathcal{R}_{P|P_0}(\tau)^{-1} \cdot U_P(\lambda, \tau, y) \theta_P(\lambda)^{-1} \cdot \mathcal{R}_{P|P_0}(\tau_\lambda), \tag{13.12.1}$$

where

$$U_P(\lambda, \tau, y) = i_P(\tau, y) i_P(\tau_\lambda, y)^{-1}.$$

One sees from the definition of induced representations that for $\psi \in V_P(\tau) = V_P^L(\tau)$ and $k \in K \cap L(F_S)$, that with

$$e^{-\lambda(H_P(ky))} = u_P(\lambda, k, y),$$

we have

$$(U_P(\lambda,\tau,y)\psi)(k) = u_P(\lambda,k,y)\psi(k). \tag{13.12.2}$$

In particular, any derivative in $\lambda$ of $U_P(\lambda,\tau,y)$ is a bounded operator. By (10.2.1),

$$U_P(\lambda,\tau,y)/\theta_P(\lambda) = \sum_{\{Q\in\mathcal{F}^L(M)\,;\,Q\supset P\}} U'_Q(\lambda,\tau,y)/\theta_P^Q(\lambda).$$

Substitute this expression in (13.12.1). Take the sum over $P$ inside the sum over $Q$.
We shall show:

LEMMA 13.13. *The limit in $\lambda$ can be taken inside the sum over $Q$, that is,*

$$i_{P_0}(\tau,f) \sum_{\{P\in\mathcal{P}^L(M)\,;\,P\subset Q\}} \mathcal{R}_{P|P_0}(\tau)^{-1}U'_Q(\lambda,\tau,y)\mathcal{R}_{P|P_0}(\tau_\lambda)/\theta_P^Q(\lambda)$$

*has a limit as $\lambda$ approaches* 0.

PROOF. Suppose that $P_0 \subset Q$. If we can show that the limit exists in this case, then the limit will exist for an arbitrary $P'_0$ in $\mathcal{P}^L(M)$. In fact, it will just be the conjugate by $\mathcal{R}_{P'_0|P_0}(\tau)$ of the limit for $P_0$. When we evaluate the trace, the two limits will be equal. Therefore we may assume $P_0 \subset Q$. Now it is clear that if $m \in M_Q(\mathbb{A})$, then

$$u_Q(\lambda,mx,y) = u_Q(\lambda,x,y).$$

It follows from this fact, and from (13.12.2) and (11.4.5), that if $P \in \mathcal{P}^L(M)$, $P \subset Q$, then

$$\mathcal{R}_{P|P_0}(\tau)^{-1}U_P(\lambda_Q,\tau,y) = U_{P_0}(\lambda_Q,\tau,y)\mathcal{R}_{P|P_0}(\tau)^{-1},$$

where we recall that $\lambda_Q$ is the projection of $\lambda$ onto $i\mathfrak{a}_Q^*$. Therefore

$$\mathcal{R}_{P|P_0}(\tau)^{-1}U'_Q(\lambda,\tau,y) = U'_Q(\lambda,\tau,y)\mathcal{R}_{P|P_0}(\tau)^{-1}.$$

Our notation here is confusing. The operator $U'_Q(\lambda,\tau,y)$ on the left side has been obtained, via (10.2.2), from $U_P(\lambda_Q,\tau,y)$, while on the right side, $U'_Q(\lambda,\tau,y)$ is obtained from $U_{P_0}(\lambda_Q,\tau,y)$ and is independent of $P$. Now, the existence of the required limit follows from Proposition 10.7. □

## 13.14 *Proof of Proposition* 13.12

We now complete the proof of Proposition 13.12. We have that $\phi_M^L(\tau, f^y)$ equals the sum over $Q \in \mathcal{F}^L(M)$ of the trace of the operator

$$i_{P_0}(\tau, f) U_Q'(\tau, y) \mathcal{R}_M^Q(\tau, P_0) \tag{13.14.1}$$

where $P_0$ can be taken to be any group in $\mathcal{P}^L(M)$ with $P_0 \subset Q$.

This operator acts on $V_{P_0}(\tau) = V_{P_0}^L(\tau)$. The space $V_{P_0}(\tau)$ can be interpreted as the space of square-integrable, $(K \cap Q(F_S))$-equivariant functions from $K \cap L(F_S)$ to the Hilbert space $V_{P_0 \cap M_Q}^{M_Q}(\tau)$. Now, (11.4.5) and (13.12.2) tell us how to interpret the operators $\mathcal{R}_M^Q(\tau, P_0)$ and $U_Q'(\tau, y)$ in this picture. Then (13.14.1) becomes an integral operator with kernel $K(k_1, k_2)$ equal to

$$\int_{M_Q(F_S)} i_{P_0 \cap M_Q}^{M_Q}(\tau, m) \cdot \delta_Q(m)^{1/2} \int_{N_Q(F_S)} f(k_1^{-1} m n k_2) v_Q'(k_2, y)\, \mathrm{d}n\, \mathrm{d}m \cdot \mathcal{R}_M^{M_Q}(\tau, P_0 \cap M_Q).$$

Therefore

$$\int_{K \cap M_Q(F_S) \backslash K \cap L(F_S)} \operatorname{tr} K(k, k)\, \mathrm{d}k$$

equals

$$\operatorname{tr}\left(\mathcal{R}_M^{M_Q}(\tau, P_0 \cap M_Q) i_{P_0 \cap M_Q}^{M_Q}(\tau, f_{Q,y})\right).$$

This is just $J_{M,\tau}^{M_Q}(f_{Q,y})$. The proposition follows. □

# 14 Hecke Invariance

## 14.1 *Invariance Formula*

In Proposition 13.12 we proved a formula

$$J_M(\tau_\lambda, f^y) = \sum_{Q \in \mathcal{F}(M)} J_M^{M_Q}(\tau_\lambda, f_{Q,y}) \tag{14.1.1}$$

for the behavior under conjugation, in fact for Schwarz functions $f$ and tempered representations $\tau$. Here $J_M^{M_Q}$ stands for the distribution on $M_Q(F_S)$, while

$$f^y(x) = f(yxy^{-1}), \quad y \in G(\mathbb{A}),$$

and

$$f_{Q,y}(m) = \delta_Q(m)^{1/2} \int_K \int_{N_Q(F_S)} f(k^{-1}mnk)v'_Q(ky)\,\mathrm{d}n\,\mathrm{d}k, \qquad m \in M_Q(F_S).$$

## 14.2 Hecke Algebra

This formula must be modified for the present situation, because conjugation at the Archimedean place does not preserve the Hecke space.

Take $f \in \mathbb{H}(G(F_S))$. If $y \in G(F_S)$, the functions

$$(L_y f)(x) = f(y^{-1}x), \qquad \text{and} \qquad (R_y f)(x) = f(xy^{-1})$$

do not in general belong to $\mathbb{H}(G(F_S))$.

However, if $h$ is in the Hecke algebra of the group

$$G(F_S)^1 = \{y \in G(F_S)\,;\, H_G(y) = 0\},$$

the functions

$$h \mapsto (L_h f)(x) = \int_{G(F_S)^1} h(y)(L_y f)(x)\,\mathrm{d}y = (h * f)(x)$$

and

$$x \mapsto (R_h f)(x) = \int_{G(F_S)^1} h(y)(R_y f)(x)\,\mathrm{d}y = (f * h)(x)$$

do belong to $\mathbb{H}(G(F_S))$.

More generally, for any $Q \in \mathcal{F}(M)$, the functions

$$m \mapsto (L_{Q,h} f)(m) = \int_{G(F_S)^1} h(y)(L_y f)_{Q,y^{-1}}(m)\,\mathrm{d}y$$

and

$$m \mapsto (R_{Q,h} f)(m) = \int_{G(F_S)^1} h(y)(R_y f)_{Q,y^{-1}}(m)\,\mathrm{d}y$$

belong to $\mathbb{H}(M_Q(F_S))$. Observe that $L_{G,h} f = L_h f$ and $R_{G,h} f = R_h f$.

PROPOSITION 14.3. *Fix $f \in \mathbb{H}(G(F_S))$ and $h \in \mathbb{H}(G(F_S)^1)$. Then*

$$J_M(\tau_\lambda, L_h f) = \sum_{Q \in \mathcal{F}(M)} J_M^{M_Q}(\tau_\lambda, R_{Q,h} f)$$

*and*

$$J_M(\tau_\lambda, R_h f) = \sum_{Q\in\mathcal{F}(M)} J_M^{M_Q}(\tau_\lambda, L_{Q,h} f)$$

*for any* $\tau \in \Pi(M(F_S))$.

PROOF. Suppose first that $\tau \in \Pi_{\text{temp}}(M(F_S))$. Then the formula (14.1.1) may be applied. Since

$$L_y f = (R_y f)^{y^{-1}},$$

we have

$$J_M(\tau_\lambda, L_y f) = \sum_{Q\in\mathcal{F}(M)} J_M^{M_Q}(\tau_\lambda, (R_y f)_{Q,y^{-1}}).$$

Multiply both sides by $h(y)$ and integrate over $y \in G(F_S)^1$. We obtain

$$J_M(\tau_\lambda, L_h f) = \sum_{Q\in\mathcal{F}(M)} J_M^{M_Q}(\tau_\lambda, R_{Q,h} f),$$

which is the first of the required formulae. Observe that if

$$\tau = \sigma_\Lambda^M, \quad \sigma \in \Pi_{\text{temp}}(\mathfrak{M}(F_S)), \quad \Lambda \in i\mathfrak{a}_{\mathfrak{M}}^*,$$

for some $\mathfrak{M} = \prod_{v\in S} M_v$, with the $M_v$ $F_v$-Levi subgroups of the $F$-Levi subgroup $M$, then each side of the formula can be analytically continued to any $\Lambda \in \mathfrak{a}_{\mathfrak{M},\mathbb{C}}^*$. The formula therefore holds if $\tau$ is replaced by any standard representation in $\Sigma(M(F_S))$.

Now, suppose that $\tau \in \Pi(M(F_S))$ is arbitrary. Combining Proposition 12.7 with what we have just proved, we see that

$$\begin{aligned} J_M(\tau_\lambda, L_h f) &= \sum_{L\in\mathcal{L}(M)} \sum_{\rho\in\Sigma(M(F_S))} m_M^L(\tau_\lambda, \rho_\lambda) \sum_{Q\in\mathcal{F}(L)} J_L^{M_Q}(\rho_\lambda^L, R_{Q,h} f) \\ &= \sum_{Q\in\mathcal{F}(M)} \sum_{L\in\mathcal{L}^{M_Q}(M)} \sum_{\rho\in\Sigma(M(F_S))} m_M^L(\tau_\lambda, \rho_\lambda) J_L^{M_Q}(\rho_\lambda^L, R_{Q,h} f) \\ &= \sum_{Q\in\mathcal{F}(M)} J_M^{M_Q}(\tau_\lambda, R_{Q,h} f). \end{aligned}$$

Thus the first of the required formulae holds for general $\tau$, not only for tempered ones.

The second required formula is established analogously. □

DEFINITION 14.4. *A distribution $I$ on $\mathbb{H}(G(F_S))$ is called* invariant *if*

$$I(L_h f - R_h f) = 0$$

*for each $f \in \mathbb{H}(G(F_S))$ and $h \in \mathbb{H}(G(F_S)^1)$.*

Proposition 14.3 gives the following

COROLLARY 14.5. *We have*

$$J_M(\tau_\lambda, L_h f - R_h f) = \sum_{Q \neq G} J_M^{M_Q}(\tau_\lambda, R_{Q,h} f) = -\sum_{Q \neq G} J_M^{M_Q}(\tau_\lambda, L_{Q,h} f).$$

This gives the obstruction to $J_M(\tau_\lambda)$ being invariant.

# 15 The Distributions $J_M(\tau, X)$

## 15.1 Definitions

The distributions $J_M(\tau_\lambda)$ occur in the trace formula via their integrals over $\tau$. Recall the following definitions from Section 1.

DEFINITION. (1) The group

$$\mathfrak{a}_{M,S} = \{H_M(m)\,;\, m \in M(F_S)\}$$

is a subgroup of $\mathfrak{a}_M$. It equals $\mathfrak{a}_M$ and $\mathfrak{a}_{M,S}^\vee = \mathfrak{a}_M^\vee = \{0\}$ if $S$ contains an Archimedean place. It is a lattice in $\mathfrak{a}_M$ otherwise. The additive character group

$$\mathfrak{a}_{M,S}^* = \mathfrak{a}_M^* / \mathfrak{a}_{M,S}^\vee, \quad \mathfrak{a}_{M,S}^\vee = \mathrm{Hom}(\mathfrak{a}_{M,S}, \mathbb{Z}),$$

equals $\mathfrak{a}_M^*$ in the first case and is a compact quotient of $\mathfrak{a}_M^*$ in the second.

(2) Suppose that $\tau \in \Pi(M(F_S))$ is such that $J_M(\tau_\lambda, f)$ is regular for $\lambda \in i\mathfrak{a}_M^*$. This holds, for example, if $\tau$ is unitary (by property $(\mathcal{R}_4)$ of Theorem 6.2). If $X \in \mathfrak{a}_{M,S}$, define the Fourier transform

$$J_M(\tau, X, f) = \int_{i\mathfrak{a}_{M,S}^*} J_M(\tau_\lambda, f) e^{-\lambda(X)}\, d\lambda, \quad f \in \mathbb{H}(G(F_S)).$$

(3) For a general representation $\tau \in \Pi(M(F_S))$, we define

$$J_M(\tau, X, f) = \sum_{P \in \mathcal{P}(M)} \omega_P J_M(\tau_{\varepsilon_P}, X, f) e^{-\varepsilon_P(X)},$$

where each $\varepsilon_P$ is a small point in the chamber $(\mathfrak{a}_P^*)^+$ where $J_M(\tau_{\varepsilon_P}, X, f)$ is defined, and

$$\omega_P = \mathrm{vol}(\mathfrak{a}_P^+ \cap B) / \mathrm{vol}(B),$$

with $B$ a ball in $\mathfrak{a}_M$ centered at the origin.

We see that these two definitions are compatible by changing the contour of integration.

## 15.2 Transformation Properties

The distributions $J_M(\tau, X)$ have some simple transformation properties.

PROPOSITION. *If $\lambda$ is any element in $\mathfrak{a}_G^* + i\mathfrak{a}_{M,S}^*$, we have*

$$J_M(\tau_\lambda, X, f) = J_M(\tau, X, f)e^{\lambda(X)}. \tag{15.2.1}$$

PROOF. This follows from (6.7.2′). □

Proposition 14.3 can be used to describe the behavior under convolution.

PROPOSITION 15.3. *We have*

$$J_M(\tau, X, L_h f) = \sum_{Q \in \mathcal{F}(M)} J_M^{M_Q}(\tau, X, R_{Q,h} f), \quad h \in \mathbb{H}(G(F_S)^1), \tag{15.3.1}$$

*and the analogous formula, with L replaced by R and R by L.*

PROOF. This follows from Proposition 14.3. Multiply the two formulae of that proposition by $e^{-\lambda(X)}$ and then integrate over $\lambda$. The first formula of Proposition 14.3 gives (15.3.1). □

We shall sometimes need to define $J_M(\tau, X, f)$ when $f$ is not quite in the Hecke space.

DEFINITION 15.4. (1) *Suppose that $Z$ is a point in $\mathfrak{a}_{G,S}$. Let $f^Z$ denote the restriction of a given function $f \in \mathbb{H}(G(F_S))$ to*

$$G(F_S)^Z = \{x \in G(F_S)\, ;\, H_G(x) = Z\}.$$

(2) *The Haar measures on $G(F_S)$ and $\mathfrak{a}_{G,S}$ determine measures on the spaces $G(F_S)^Z$. For any $\tau \in \Pi(M(F_S))$, we have that*

$$i_{P'}^G(\tau, f^Z) = \int_{G(F_S)^Z} f^Z(x) i_{P'}^G(\tau, x)\, dx, \quad P' \in \mathcal{P}(M),$$

*is an operator on $V_{P'}(\tau)$.*

(3) *Define*

$$J_M(\tau_\lambda, f^Z) = \operatorname{tr} \mathcal{R}_M(\tau_\lambda, P') i_{P'}^G(\tau_\lambda, f^Z).$$

*It is clear that*

$$J_M(\tau_{\lambda + \lambda_G}, f^Z) = J_M(\tau_\lambda, f^Z)e^{\lambda_G(Z)}$$

*for any point $\lambda_G \in \mathfrak{a}_{G,\mathbb{C}}^*$.*

(4) *Take*

$$Z = h_G(X), \quad X \in \mathfrak{a}_{M,S},$$

*where $h_G : \mathfrak{a}_{M,S} \to \mathfrak{a}_G$ denotes the projection from $\mathfrak{a}_{M,S}$ onto $\mathfrak{a}_G$.*

PROPOSITION 15.5. *We have*

$$J_M(\tau, X, f) = \sum_{P \in \mathcal{P}(M)} \omega_P \int_{\varepsilon_P + i\mathfrak{a}_{M,S}^*/\mathfrak{a}_{G,S}^*} J_M(\tau_\lambda, f^Z) e^{-\lambda(X)}\, d\lambda, \tag{15.5.1}$$

*for $Z = Z(X) = h_G(X)$ and $\{\varepsilon_P\}$ as above.*

PROOF. This follows from the Fourier inversion formula on $\mathfrak{a}_{G,S}$. □

COROLLARY 15.6. *The function $(\tau, X) \mapsto J_M(\tau, X, f)$ depends only on $f^Z$. In particular, it can be defined for any function $f$ which has the same restriction to $G(F_S)^Z$ as some function in $\mathbb{H}(G(F_S))$.*

## 15.7 Expansion in Standard Representations

As with $J_M(\tau_\lambda)$, the distribution $J_M(\tau, X)$ has an expansion in terms of standard representations. If $\rho$ belongs to $\Sigma(M(F_S))$, we can define $J_M(\rho, X)$ in terms of $J_M(\rho_\lambda)$ by mimicking the discussion above. This new distribution then satisfies the obvious analogues of (15.2.1)–(15.5.1).

PROPOSITION. *Fix $\tau \in \Pi(M(F_S))$, $X \in \mathfrak{a}_{M,S}$, and $f \in \mathbb{H}(G(F_S))$. Then*

$$J_M(\tau, X, f) = \sum_{P \in \mathcal{P}(M)} \omega_P \sum_{L \in \mathcal{L}(M)} m_{M,\varepsilon_P}^L(\tau, X, J_L(f)),$$

*where for any point $\mu \in \mathfrak{a}_M^*$ in general position, the function*

$$m_{M,\mu}^L(\tau, X, J_L(f))$$

*equals*

$$\int_{\mu + i\mathfrak{a}_{M,S}^*/i\mathfrak{a}_{L,S}^*} \sum_{\rho \in \Sigma(M(F_S))} m_M^L(\tau_\lambda, \rho_\lambda) J_L(\rho_\lambda^L, h_L(X), f) e^{-\lambda(X)}\, d\lambda.$$

PROOF. Observe that

$$\begin{aligned} J_M(\tau, X, f) &= \sum_{P \in \mathcal{P}(M)} \omega_P \int_{\varepsilon_P + i\mathfrak{a}_{M,S}^*/i\mathfrak{a}_{G,S}^*} J_M(\tau_\lambda, f^{h_G(X)}) e^{-\lambda(X)}\, \mathrm{d}\lambda \\ &= \sum_P \omega_P \int \sum_{L \in \mathcal{L}(M)} \sum_{\rho \in \Sigma(M(F_S))} m_M^L(\tau_\lambda, \rho_\lambda) J_L(\rho_\lambda^L, f^{h_G(X)}) e^{-\lambda(X)}\, \mathrm{d}\lambda, \end{aligned}$$

by (15.5.1) and Proposition 12.7. Since $\varepsilon_P$ is in general position, the function

$$m_M^L(\tau_\lambda, \rho_\lambda) J_L(\rho_\lambda^L, f^{h_G(X)}) e^{-\lambda(X)} \tag{15.7.1}$$

has no singularities which meet $\varepsilon_P + i\mathfrak{a}_{M,S}^*/i\mathfrak{a}_{G,S}^*$. By a standard estimate, it is integrable over this space. We may therefore take the integral above inside the sum over $L$ and $\rho$. We then decompose the resulting integral into a double integral over

$$(\varepsilon_P + i\mathfrak{a}_{M,S}^*/i\mathfrak{a}_{L,S}^*) \times (i\mathfrak{a}_{L,S}^*/i\mathfrak{a}_{G,S}^*).$$

It becomes

$$\int_{\varepsilon_P + i\mathfrak{a}_{M,S}^*/i\mathfrak{a}_{L,S}^*} m_M^L(\tau_\lambda, \rho_\lambda) J_L(\rho_\lambda^L, h_L(X), f) e^{-\lambda(X)}\, \mathrm{d}\lambda.$$

The proposition follows. □

## 15.8 Another Point of View

We should keep in mind that $J_M(\tau, X, f)$ is a function not only of $f$ but also of $(\tau, X)$. Interpreted one way it is a family of distributions, and the other way it is a transform. We shall use different notation when we want to emphasize this second point of view.

DEFINITION 15.9. *Write $\phi_M$ for the map which takes $f \in \mathbb{H}(G(F_S))$ to the function*

$$\phi_M(f) : \tau \mapsto \phi_M(f, \tau) = J_M(\tau, f)$$

*on $\Pi_{\text{temp}}(M(F_S))$. This is a linear combination of matrix entries of $i_{P'}^G(\tau, f)$, which for*

$$\tau = \sigma_\Lambda^M, \quad \sigma \in \Pi_{\text{temp}}(\mathfrak{M}(F_S)), \ \Lambda \in i\mathfrak{a}_{\mathfrak{M}}^*,$$

*as in Section* 12, *has coefficients which are rational functions of the variables* (12.4.1). *In particular, it cannot be extended to all nontempered $\tau$.*

## 15.10 Extension

We shall show that

$$(\tau, X) \mapsto \phi_M(f, \tau, X) = \int_{i\mathfrak{a}_{M,S}^*} \phi_M(f, \tau_\lambda) e^{-\lambda(X)}\, \mathrm{d}\lambda = J_M(\tau, X, f)$$

does extend to $\Pi(M(F_S))$. We shall also compare its values at arbitrary $(\tau, X)$ with $J_M(\tau, X, f)$. Both of these questions are related to the residues of the function

$$\phi_M(f, \sigma_\Lambda^M) = J_M(\sigma_\Lambda^M, f).$$

We shall devote the next two sections to a study of these residues.

# 16 Residues

## 16.1 To Investigate

Suppose that $\mathfrak{M} = \prod_{v \in S} M_v$ is a well-positioned relative to $K$ Levi subgroup of $M$ over $F_S$. For each $v$, write $S_v = \{v\}$. Define $\mathfrak{a}_{M_v, S_v}$ to be the image of the map

$$H_{M_v} : G(F_v) \to \mathfrak{a}_{M_v}.$$

Set

$$\mathfrak{a}_{\mathfrak{M},S} = \oplus_{v \in S} \mathfrak{a}_{M_v, S_v},$$

$$\mathfrak{a}_{\mathfrak{M},S}^\vee = \mathrm{Hom}(\mathfrak{a}_{\mathfrak{M},S}, 2\pi\mathbb{Z}) = \oplus_{v \in S} \mathrm{Hom}(\mathfrak{a}_{M_v, S_v}, 2\pi\mathbb{Z}),$$

and

$$\mathfrak{a}_{\mathfrak{M},S}^* = \mathfrak{a}_{\mathfrak{M}}^* / \mathfrak{a}_{\mathfrak{M},S}^\vee.$$

For the next two sections, we keep $\mathfrak{M}$ fixed. We also fix a representation $\sigma \in \Pi(\mathfrak{M}(F_S))$ and a function $a_\Lambda = a(\Lambda)$ which is defined and analytic on a neighborhood of some point $\Lambda_0$ in

$$\mathfrak{a}_{\mathfrak{M},\mathbb{C}}^* / i\mathfrak{a}_{\mathfrak{M},S}^\vee = \mathfrak{a}_{\mathfrak{M}}^* + i\mathfrak{a}_{\mathfrak{M},S}^*.$$

We propose to investigate the residues in $\Lambda$ of the functions

$$a_\Lambda J_M(\sigma_\Lambda, f), \quad f \in \mathbb{H}(G(F_S)). \tag{16.1.1}$$

It is clear from the discussion of Section 10 that the singularities of each of these functions lie along a set of hyperplanes of the form (12.4.2) which is finite modulo $i\mathfrak{a}_{\mathfrak{M},S}^\vee$.

DEFINITION 16.2. (1) *Consider a sequence*

$$\mathfrak{M} = \mathfrak{M}_0 \subset \mathfrak{M}_1 \subset \cdots \subset \mathfrak{M}_r = \mathfrak{L} \tag{16.2.1}$$

*of embedded Levi subgroups of G over $F_S$. We assume that for each i,* $1 \leq i \leq r$,

$$\mathfrak{a}_{\mathfrak{M}_i} = \{H \in \mathfrak{a}_{\mathfrak{M}_{i-1}}\,;\ \beta_i(H) = 0\},$$

*for some root $\beta_i$ of $(G, A_{\mathfrak{M}_{i-1}})$. The roots $\{\beta_i\}$ are uniquely determined up to scalar multiples.*

(2) *For each i let $E_i$ be a fixed nonzero real multiple of $\beta_i$. Then the set*

$$\mathcal{E}_\Omega = (E_1, \dots, E_r)$$

*determines the sequence* (16.2.1)*.*

(3) *Fix a linear functional $\Lambda_\Omega \in \mathfrak{a}^*_{\mathfrak{M},\mathbb{C}}$ which vanishes on $\mathfrak{a}_{\mathfrak{L}}$. We shall call the pair*

$$\Omega = (\mathcal{E}_\Omega, \Lambda_\Omega)$$

*a* residue datum *for $\mathfrak{L}$.*

(4) *Take $\Lambda_0$ to be a fixed point in general position in the affine subspace $\Lambda_\Omega + \mathfrak{a}^*_{\mathfrak{L},\mathbb{C}}$ of $\mathfrak{a}^*_{\mathfrak{M},\mathbb{C}}$. Set*

$$\Lambda(z) = \Lambda_0 + z_1 E_1 + \cdots + z_r E_r$$

*for $z = (z_1, \dots, z_r)$ in $\mathbb{C}^r$.*

(5) *Let $\Gamma_1, \dots, \Gamma_r$ be small positively oriented circles about the origin in the complex plane such that for each i, the radius of $\Gamma_i$ is much smaller than that of $\Gamma_{i+1}$.*

(6) *Consider a meromorphic function $\psi(\Lambda)$ on a neighborhood of $\Lambda_0$ in $\mathfrak{a}^*_{\mathfrak{M},\mathbb{C}}$ whose singularities lie along hyperplanes of the form* (12.4.2)*. Then*

$$(2\pi i)^{-r} \int_{\Gamma_1} \cdots \int_{\Gamma_1} \psi(\Lambda(z))\, dz_1 \cdots dz_r$$

*is a meromorphic function of $\Lambda_0$. We denote it by*

$$\mathrm{Res}_{\Omega,\Lambda\to\Lambda_0}\, \psi(\Lambda) = \mathrm{Res}_\Omega\, \psi(\Lambda_0).$$

*We shall study it with $\psi(\Lambda)$ equal to the function* (16.1.1) *above.*

(7) *Define a sequence*

$$M = M_0 \subset M_1 \subset \cdots \subset M_r = L$$

*of elements in $\mathcal{L}(M)$ inductively by*

$$\mathfrak{a}_{M_i} = \{H \in \mathfrak{a}_{M_{i-1}}\,;\ E_i(H) = 0\}, \quad 1 \leq i \leq r.$$

PROPOSITION 16.3. *If $L = G$, then*

$$\mathrm{Res}_{\Omega,\Lambda\to\Lambda_0}(a_\Lambda J_M(\sigma_\Lambda, f)), \quad f \in \mathbb{H}(G(F_S)),$$

*is an invariant distribution.*

PROOF. We shall use the formula

$$J_M(\sigma_\Lambda, L_h f - R_h f) = \sum_{\{Q\in\mathcal{F}(M)\,;\,Q\neq G\}} J_M^{M_Q}(\sigma_\Lambda, R_{Q,h} f), \quad h \in \mathbb{H}(G(F_S)^1)$$

of Corollary 14.3. Fix $Q \in \mathcal{F}(M)$ with $Q \neq G$. Let $i$ be the smallest integer such that $M_i$ is not contained in $M_Q$. The partial residue

$$\int_{\Gamma_{i-1}} \cdots \int_{\Gamma_1} a_{\Lambda(z)} J_M^{M_Q}(\sigma_{\Lambda(z)}, R_{Q,h} f)\, \mathrm{d}z_1 \cdots \mathrm{d}_{z_{i-1}}$$

is a meromorphic function of $(z_i, \ldots, z_r)$. Its poles lie along affine hyperplanes obtained from roots of $(G, A_{\mathfrak{M}_{i-1}})$ which vanish on $\mathfrak{a}_{M_Q}$. The hyperplane $z_i = 0$ is defined by any root which is a multiple of $E_i$. Our choice of $i$ means that $E_i$ does not vanish on $\mathfrak{a}_{M_Q}$. Consequently $z_i = 0$ is not a singular hyperplane of the function. It follows that

$$\mathrm{Res}_{\Omega,\Lambda\to\Lambda_0}(a_\Lambda J_M^{M_Q}(\sigma_\Lambda, R_{Q,h} f)) = 0.$$

Therefore

$$\mathrm{Res}_{\Omega,\Lambda\to\Lambda_0}(a_\Lambda J_M(\sigma_\Lambda, L_h f - R_h f)) = 0,$$

and the proposition follows. □

## 16.4 Induction

We return to the case that $L$ is arbitrary. The goal is to provide a simple formula for

$$\mathrm{Res}_{\Omega,\Lambda\to\Lambda_0}(a_\Lambda J_M(\sigma_\Lambda, f)).$$

We shall postpone this until the next section. In the meantime, we shall make some comments of a general nature.

The distribution

$$\nabla(f) = \mathrm{Res}_{\Omega,\Lambda\to\Lambda_0}(a_\Lambda J_M^L(\sigma_\Lambda, f)), \quad f \in \mathbb{H}(L(F_S)),$$

is invariant—see Proposition 16.3 and Corollary 14.3—in the sense that

$$\nabla(f_1 * f_2) = \nabla(f_2 * f_1).$$

Let us fix a parabolic subgroup $R$ in $\mathcal{P}^L(M)$. Then for each $f \in \mathbb{H}(L(F_S))$, the number $\nabla(f)$ can be obtained from the Taylor series of $i_R^G(\sigma^M_{\Lambda(z)}, f)$ about $z = 0$. In fact there is a positive integer $N$, independent of $f$ and also of the function $a_\Lambda$ above, such that $\nabla(f)$ depends only on the Taylor coefficients of total degree no greater than $N$.

We shall let $\rho$ denote the representation of $\mathbb{H}(L(F_S))$ obtained by taking the Taylor series of $i_R^G(\sigma^M_{\Lambda(z)}, f)$ modulo terms of degree greater than $N$. It acts on the space of power series in $z$, taken modulo terms of degree greater than $N$, with values in $V_R(\sigma^M_{\Lambda_0})$, the Hilbert space on which $i_R^G(\sigma^M_\Lambda)$ acts.

We can of course also regard $i_R^G(\sigma^M_{\Lambda(z)})$ as a representation of the group $L(F_S)$, so that $\rho$ is the representation of $\mathbb{H}(L(F_S))$ associated with an admissible representation $(\rho, V_\rho)$ of $L(F_S)$. By construction,

$$\tau(f) \mapsto \langle \nabla, \rho(f) \rangle := \nabla(f)$$

is a well-defined invariant form on the algebra

$$\mathcal{A}_\rho = \{\rho(f) \, ; f \in \mathbb{H}(L(F_S))\}$$

of operators on $V_\rho$.

Let $Q$ be an element of $\mathcal{P}(L)$, and form the induced representation

$$i_Q^G(\rho_\lambda), \quad \lambda \in \mathfrak{a}^*_{L,\mathbb{C}}.$$

DEFINITION 16.5. (1) *An operator $A(\rho)$ on $V_Q(\rho)$ is called* admissible *if it is represented by a $K$-finite kernel $A(\rho; k_1, k_2)$ with values in the algebra $\mathcal{A}_\rho$ of operators on $V_\rho$.*

(2) *Define a linear form*

$$T_\nabla(A(\rho)) = \int_K \langle \nabla, A(\rho; k, k) \rangle \, dk$$

*on the space of admissible operators. Since $\nabla$ is invariant, we have*

$$T_\nabla(A(\rho)B(\rho)) = T_\nabla(B(\rho)A(\rho)) \tag{16.5.1}$$

*for every pair $A(\rho)$ and $B(\rho)$ of admissible operators.*

(3) *For any function $f \in \mathbb{H}(G(F_S))$, the operator $i_Q^G(\rho, f)$ is admissible. Its kernel is*

$$\int_{M_Q(F_S)} \int_{N_Q(F_S)} f(k_1^{-1} m n k_2) \rho(m) e^{(\lambda + \rho_Q)(H_Q(m))} \, dn \, dm.$$

*We can refer to*

$$f \mapsto T_\nabla(i_Q^G(\rho,f))$$

*as the distribution on* $G(F_S)$ induced from $\nabla$.

The linear form $T_\nabla(A(\rho))$ is of course closely related to our study of residues. Set $P = Q(R)$ and write $\tau = \sigma^M_{\Lambda_0}$. By induction in stages, we can identify $V_P(\tau)$, the Hilbert space on which $i_P^G(\sigma^M_\Lambda)$ acts, with $V_Q(V_R(\tau))$. Then the operator

$$\mathcal{R}_M^Q(\sigma^M_\Lambda, P) = \lim_{\nu\to 0} \sum_{\{P_1\in\mathcal{P}(M);\, P_1\subset Q\}} \mathcal{R}_{P_1|P}(\sigma^M_\Lambda)^{-1}\mathcal{R}_{P_1|P}(\sigma^M_{\Lambda+\nu})/\theta_{P_1\cap L}(\nu)$$

acts on $V_P(\tau)$ through the fiber. It transforms the values of a given function by the operator $\mathcal{R}_M^L(\sigma^M_\Lambda, R)$ on $V_R(\tau)$.

Now, suppose that $A(\sigma^M_\Lambda)$ is a holomorphic function with values in the space of operators on $V_P(\tau)$. By taking the Taylor series of $A(\sigma^M_{\Lambda(z)})$, modulo terms of degree greater than $N$, we obtain an operator $A(\rho)$ on $V_P(\rho)$. It is clear that $A(\sigma^M_\Lambda) \mapsto A(\rho)$ is an algebra homomorphism and that each $i_P^G(\sigma^M_\Lambda, f)$ maps to $i_Q^G(\rho,f)$.

DEFINITION 16.6. *Call* $A(\sigma^M_\Lambda)$ admissible *if the corresponding operator* $A(\rho)$ *is admissible. In this case we have*

$$\mathrm{Res}_{\Omega,\Lambda\to\Lambda_0}\, a_\Lambda\, \mathrm{tr}(A(\sigma^M_\Lambda)\mathcal{R}_M^Q(\sigma^M_\Lambda, P)) = T_\nabla(A(\rho)). \qquad (16.6.1)$$

*This provides the connection with residues.*

## 16.7 Restriction is Admissible

Let $Q'$ be another element in $\mathcal{P}(L)$, and set $P' = Q'(R)$. It is clear that the definition of admissible operator can be extended to linear transformations from $V_Q(\rho)$ to $V_{Q'}(\rho)$. Formula (16.5.1) and the correspondence $A(\sigma^M_\Lambda) \mapsto A(\rho)$ also have obvious extensions. It is easily deduced from (5.1.1) that the intertwining operator $\mathcal{M}_{P'|P}(\sigma^M_{\Lambda+\lambda})$ maps to $\mathcal{M}_{Q'|Q}(\rho_\lambda)$. In particular, $\mathcal{M}_{Q'|Q}(\rho_\lambda)$ can be analytically continued to a meromorphic function of $\lambda$ on $\mathfrak{a}^*_{L,\mathbb{C}}$. Now, $\mathcal{M}_{Q'|Q}(\rho_\lambda)$ is not admissible as it stands. However, let $\Gamma$ be a finite subset of $\Pi(K)$. Let $(\cdot)_\Gamma$ denote the subspace that transforms under $K$ according to representations in $\Gamma$. We shall show:

PROPOSITION. *The restriction of* $\mathcal{M}_{Q'|Q}(\rho_\lambda)$ *to* $V_Q(\rho)_\Gamma$ *is an admissible operator.*

PROOF. Let $\Gamma_L$ be the set of irreducible representations of $K_L$ which occur as constituents of restrictions to $K_L$ of representations in $\Gamma$. Define

$$\theta_{\Gamma_L}(k_L) = \sum_{\mu\in\Gamma_L} \mathrm{tr}\,\mu(k_L), \quad k_L \in K_L.$$

Set

$$E_{\Gamma}v = \int_{K_L} \theta_{\Gamma_L}(k_L^{-1})\tau(k_L)v \, \mathrm{d}k_L,$$

for any $v \in V_\rho$. Then $E_\rho$ is the projection of $V_\rho$ onto the finite-dimensional subspace $(V_\rho)_{\Gamma_L}$. If $\varphi$ is any vector in $V_Q(\rho)_\Gamma$, the value of $\mathcal{M}_{Q'|Q}(\rho_\lambda)\varphi$ at $k \in K$ equals

$$\int_{N_{Q'}(F_S)\cap N_Q(F_S)\backslash N_{Q'}(F_S)} E_\Gamma \rho(m_Q(n))E_\Gamma \varphi(k_Q(n)k)e^{(\lambda+\rho_Q)(H_Q(n))} \, \mathrm{d}n. \qquad (16.7.1)$$

This follows from (5.1.1) and the fact that $\mathcal{M}_{Q'|Q}(\rho_\lambda)$ maps $V_Q(\rho)_\Gamma$ to $V_{Q'}(\rho)_\Gamma$.

We claim that for each $m \in L(F_S)$, the operator

$$E_\Gamma \rho(m) E_\Gamma$$

belongs to the algebra $\mathcal{A}_\rho$. To see this, choose a sequence $\{f_i\}$ of functions in $C_c^\infty(L(F_S))$ that approach the Dirac measure at $m$. Then the matrix coefficients of the operators $\rho(f_i)$ approach those of $\rho(m)$. But the functions

$$f_{i,\Gamma}(m') = \int_{K_L}\int_{K_L} \theta_{\Gamma_L}(k_1)f_i(k_1 m' k_2)\theta_{\Gamma_L}(k_2) \, \mathrm{d}k_1 \, \mathrm{d}k_2, \quad m' \in L(F_S),$$

all belong to $\mathbb{H}(L(F_S))$, and

$$\rho(f_{i,\Gamma}) = E_\Gamma \rho(f_i) E_\Gamma.$$

In particular, $\rho(f_{i,\Gamma})$ converges to $E_\Gamma\rho(m)E_\Gamma$. This shows that $E_\Gamma\rho(m)E_\Gamma$ belongs to the closure of the subspace

$$\{E_\Gamma \rho(f) E_\Gamma \, ; f \in \mathbb{H}(L(F_S))\}$$

of $\mathcal{A}_\rho$. Since the subspace is actually finite dimensional, the claim follows.

Now, left translation on any space of $K$-finite functions on $K$ is an integral operator with $K$-finite kernel. It follows from (16.7.1) that the restriction of $\mathcal{M}_{Q'|Q}(\rho_\lambda)$ to $V_Q(\rho)_\Gamma$ is an admissible operator. □

## 16.8 Equalities of Residues

The following proposition is a consequence of this discussion. We shall use it in the next section.

PROPOSITION. *Suppose that we are given a finite sum*

$$A(\sigma_\Lambda^M) = \sum_{i=1}^{n} a_{i,\Lambda} A_i(\sigma_\Lambda^M),$$

*where for each i, the function* $a_{i,\Lambda}$ *is holomorphic on a neighborhood of* $\Lambda_0$ *and*

$$A_i(\sigma_\Lambda^M) : V_P(\tau) \to V_{P'}(\tau)$$

*is admissible. Then*

$$\mathrm{Res}_{\Omega,\Lambda\to\Lambda_0}\, \mathrm{tr}(\mathcal{R}_{P'|P}(\sigma_\Lambda^M)^{-1} A(\sigma_\Lambda^M)\mathcal{R}_M^Q(\sigma_\Lambda^M, P))$$

*equals*

$$\mathrm{Res}_{\Omega,\Lambda\to\Lambda_0}\, \mathrm{tr}(A(\sigma_\Lambda^M)\mathcal{R}_{P'|P}(\sigma_\Lambda^M)^{-1}\mathcal{R}_M^{Q'}(\sigma_\Lambda^M, P')).$$

PROOF. Since both expressions are linear in $A(\sigma_\Lambda^M)$, we can assume that $n = 1$. Write

$$\mathcal{R}_{P'|P}(\sigma_\Lambda^M)^{-1} = m_{P'|P}(\sigma_\Lambda^M)\mathcal{M}_{P'|P}(\sigma_\Lambda^M)^{-1}.$$

Each of these three functions is holomorphic in a neighborhood of $\Lambda_0$. This follows from the general position of $\Lambda_0$ and the fact that $P \cap L = P' \cap L = R$. Define $\nabla$ as above, with

$$a_\Lambda = m_{P'|P}(\sigma_\Lambda^M)a_{1,\Lambda}.$$

Then by (16.6.1), our two expressions equal

$$T_\nabla(\mathcal{M}_{Q'|Q}(\rho)^{-1}A_1(\rho))$$

and

$$T_\nabla(A_1(\rho)\mathcal{M}_{Q'|Q}(\rho)^{-1}),$$

respectively. We can certainly replace $\mathcal{M}_{Q'|Q}(\rho)^{-1}$ by its restriction to a subspace $V_{Q'}(\rho)_\Gamma$. The operator is then admissible, so the proposition follows from an obvious variant of (16.5.1). □

# 17 Proof of Proposition 17.3

## *17.1 Formula for a Residue*

We shall now establish the formula for

$$\mathrm{Res}_{\Omega,\Lambda\to\Lambda_0}(a_\Lambda J_M(\sigma_\Lambda, f)).$$

It will be given in terms of a certain operator $\Gamma_\Omega(\sigma_\Lambda^M, P_0)$, which we must first describe.

We continue with the notation of the last section.

## 17.2 Embedded Spaces

DEFINITION 17.2.1. The embedded subspaces

$$\mathfrak{a}_L = \mathfrak{a}_{M_r} \subset \mathfrak{a}_{M_{r-1}} \subset \cdots \subset \mathfrak{a}_{M_0} = \mathfrak{a}_M$$

are of successive codimension 0 or 1. If $\mathfrak{a}_{M_i}$ is of codimension 1 in $\mathfrak{a}_{M_{i-1}}$, let $\varepsilon_i$ be the *unit* vector in $\mathfrak{a}_{M_{i-1}}^*$ in the direction of the restriction of $E_i$ to $\mathfrak{a}_{M_{i-1}}$. (Recall that we have fixed Euclidean norms on $\mathfrak{a}$ and $\mathfrak{a}_M^*$.) If $\mathfrak{a}_{M_i} = \mathfrak{a}_{M_{i-1}}$, take $\varepsilon_i$ to be the zero vector. Then the nonzero vectors in $\{\varepsilon_1, \dots, \varepsilon_r\}$ form an orthonormal basis of $(\mathfrak{a}_M^L)^*$. Let $R_0$ be the unique parabolic in $\mathcal{P}^L(M)$ for which the Levi components $M_j$ are all standard and on whose chamber $\mathfrak{a}_{R_0}^+$ the functions $\varepsilon_j$ are all nonnegative. Similarly, for $1 \leq i \leq r$, let $R_i \in \mathcal{P}^L(M)$ be the parabolic for which each $M_i$ is standard and such that the functions $\{\varepsilon_1, \dots, -\varepsilon_i, \dots, \varepsilon_r\}$ are all nonnegative on $\mathfrak{a}_{R_i}^+$. Fix $Q_0 \in \mathcal{P}(L)$, and define parabolics

$$P_i = Q_0(R_i), \quad 0 \leq i \leq r,$$

in $\mathcal{P}(M)$. Set

$$\mathcal{R}'_{P_0|P_i}(\sigma_\Lambda^M, \varepsilon_i) = \lim_{t\to 0} \frac{d}{dt} \mathcal{R}_{P_0|P_i}(\sigma_{\Lambda+t\varepsilon_i}^M), \quad 1 \leq i \leq r.$$

Taking an $r$-fold product of logarithmic derivatives, we define on $V_{P_0}(\tau)$ an operator

$$\Gamma_\Omega(\sigma_\Lambda^M, P_0) = \mathcal{R}'_{P_0|P_r}(\sigma_\Lambda^M, \varepsilon_r)\mathcal{R}_{P_0|P_r}(\sigma_\Lambda^M)^{-1} \cdots \mathcal{R}'_{P_0|P_1}(\sigma_\Lambda^M, \varepsilon_1)\mathcal{R}_{P_0|P_1}(\sigma_\Lambda^M)^{-1}.$$

Observe that if any of the vectors $\{\varepsilon_1, \dots, \varepsilon_r\}$ is 0, the operator $\Gamma_\Omega(\sigma_\Lambda^M, P_0)$ vanishes.

PROPOSITION 17.3. *The distribution*

$$\operatorname{Res}_{\Omega,\Lambda\to\Lambda_0}(a_\Lambda J_M(\sigma_\Lambda, f))$$

*equals*

$$\operatorname{Res}_{\Omega,\Lambda\to\Lambda_0}(a_\Lambda \operatorname{tr}\left[i_{P_0}(\sigma_\Lambda^M, f)\mathcal{R}_L(\sigma_\Lambda^M, P_0)\Gamma_\Omega(\sigma_\Lambda^M, P_0)\right]).$$

PROOF. The proof will be by induction on the length of the residue datum $\Omega$. Assume that the proposition holds for any datum of length $r$. Let $\Omega'$ be a datum of length $(r+1)$ associated with a sequence

$$\mathfrak{M} = \mathfrak{M}_0 \subset \mathfrak{M}_1 \subset \cdots \subset \mathfrak{M}_r \subset \mathfrak{M}_{r+1} = \mathfrak{L}'.$$

It is clear that $\Omega'$ is obtained from a datum $\Omega$ of length $r$, for which we follow the notation above. The only additional information in $\Omega'$ is $E_{r+1}$, a multiple of some root $\beta_{r+1}$ of $(G, A_{\mathfrak{L}})$, and a point $\Lambda_{\Omega'}$ in $(\Lambda_\Omega + \mathbb{C}E_{r+1})$. Let $\Lambda'_0$ be a point in general position in $\Lambda_{\Omega'} + \mathfrak{a}^*_{\mathfrak{L}',\mathbb{C}}$, and set

$$\Lambda_0 = \Lambda'_0 + z_{r+1}E_{r+1},$$

with $z_{r+1}$ a variable point in $\mathbb{C}$. The operator $\mathrm{Res}_{\Omega',\Lambda\to\Lambda'_0}$ can be calculated by first applying $\mathrm{Res}_{\Omega,\Lambda\to\Lambda_0}$ and then integrating $z_{r+1}$ over a small circle about the origin. It follows from our induction hypothesis that

$$\mathrm{Res}_{\Omega',\Lambda\to\Lambda'_0}(a_\Lambda J_M(\sigma_\Lambda, f)) \tag{17.3.1}$$

equals the residue about $z_{r+1} = 0$ of

$$\mathrm{Res}_{\Omega,\Lambda\to\Lambda_0}(a_\Lambda \,\mathrm{tr}\left[i_{P_0}(\sigma^M_\Lambda, f)\mathcal{R}_L(\sigma^M_\Lambda, P_0)\Gamma_\Omega(\sigma^M_\Lambda, P_0)\right]). \tag{17.3.2}$$

We recall here that $P_0 = Q_0(R_0)$, where $Q_0 \in \mathcal{P}(L)$ is arbitrary, and $R_0 \in \mathcal{P}^L(M)$ is chosen to be compatible with the directions $\{\varepsilon_1, \ldots, \varepsilon_r\}$. The operator $\mathcal{R}_L(\sigma^M_\Lambda, P_0)$ is obtained from the $(G, L)$-family

$$\mathcal{R}_{Q(R_0)|P_0}(\sigma^M_\Lambda)^{-1}\mathcal{R}_{Q(R_0)|P_0}(\sigma^M_{\Lambda+\nu}), \quad \nu \in \mathfrak{a}^*_{L,\mathbb{C}}, \quad Q \in \mathcal{P}(L).$$

Applying (10.9.1) to this family, we see that (17.3.2) equals

$$\mathrm{Res}_{\Omega,\Lambda\to\Lambda_0} \frac{a_\Lambda}{r!} \sum_{\{P=Q(R_0)\,;\,Q\in\mathcal{P}(L)\}} \mathrm{tr}\left(i_{P_0}(\sigma^M_\Lambda, f)\mathcal{R}_{P|P_0}(\sigma^M_\Lambda)^{-1}\mathcal{R}^{(r)}_{P|P_0}(\sigma^M_\Lambda, \nu)\Gamma_\Omega(\sigma^M_\Lambda, P_0)\right)/\theta_Q(\nu),$$

where

$$\mathcal{R}^{(r)}_{P|P_0}(\sigma^M_\Lambda, \nu) = \lim_{t\to 0}\left(\frac{d}{dt}\right)^r \mathcal{R}_{P|P_0}(\sigma^M_{\Lambda+t\nu}).$$

This expression does not depend on the point $\nu \in \mathfrak{a}^*_{L,\mathbb{C}}$. The only constituent of the expression which could contribute a pole along the hyperplane $\Lambda_0(\beta^\vee_{r+1}) = 0$ is the function

$$\mathcal{R}_{P|P_0}(\sigma^M_\Lambda)^{-1}\mathcal{R}^{(r)}_{P|P_0}(\sigma^M_\Lambda, \nu) = \mathcal{R}_{Q(R_0)|Q_0(R_0)}(\sigma^M_\Lambda)^{-1}\mathcal{R}^{(r)}_{Q(R_0)|Q_0(R_0)}(\sigma^M_\Lambda, \nu).$$

But each singular hyperplane of this function is defined by a coroot whose restriction to $\mathfrak{a}_L^*$ separates the chambers $(\mathfrak{a}_Q^*)^+$ and $(\mathfrak{a}_{Q_0}^*)^+$. Consequently, the function gives no contribution to the residue (17.3.1) unless the restriction of $E_{r+1}$ to $\mathfrak{a}_L$ defines a hyperplane of this sort. In particular, (17.3.1) vanishes if the restriction of $E_{r+1}$ to $\mathfrak{a}_L$ is zero. Combining this with our induction assumption, we obtain the required assertion that (17.3.1) vanishes in case any of the vectors $\{\varepsilon_1, \ldots, \varepsilon_{r+1}\}$ is zero.

We can then assume $\varepsilon_1, \ldots, \varepsilon_{r+1}$ are all nonzero. We must fix an arbitrary parabolic $Q_0' \in \mathcal{P}(L')$. Taken together with the unit vectors $\{\varepsilon_1, \ldots, \varepsilon_{r+1}\}$, it determines unique parabolics

$$P_i' = Q_0'(R_i'), \quad 0 \leq i \leq r+1,$$

in $\mathcal{P}(M)$ by the conventions above. The parabolic $Q_0 \in \mathcal{P}(L)$, which has been arbitrary, we now take to be the unique parabolic which is contained in $Q_0'$ and for which the function $(-\varepsilon_{r+1})$ is positive on $\mathfrak{a}_{Q_0}^+$. Then

$$P_0 = Q_0(R_0) = P_{r+1}'.$$

Given $P = Q(R_0)$, with $Q \in \mathcal{P}(L)$, we note that the hyperplane in $\mathfrak{a}_L$ defined by $\beta_{r+1}$ separates the chambers $\mathfrak{a}_Q^+$ and $\mathfrak{a}_{Q_0}^+$ if and only if $d(P, P_0) > d(P, P_0')$. Writing

$$i_{P_0}(\sigma_\Lambda^M, f)\mathcal{R}_{P|P_0}(\sigma_\Lambda^M)^{-1} = \mathcal{R}_{P_0'|P_0}(\sigma_\Lambda^M)^{-1} i_{P_0'}(\sigma_\Lambda^M, f)\mathcal{R}_{P|P_0'}(\sigma_\Lambda^M)^{-1}$$

for each such $P$, we see that (17.3.1) can be obtained by summing the product of $(r!\theta_Q(\nu))^{-1}$ and

$$\begin{aligned}\operatorname{Res}_{\Omega,\Lambda\to\Lambda_0}\big(a_\Lambda \operatorname{tr}\big(&\mathcal{R}_{P_0'|P_0}(\sigma_\Lambda^M)^{-1} i_{P_0'}(\sigma_\Lambda^M, f)\mathcal{R}_{P|P_0'}(\sigma_\Lambda^M)^{-1}\\ &\times \mathcal{R}_{P|P_0}^{(r)}(\sigma_\Lambda^M, \nu)\Gamma_\Omega(\sigma_\Lambda^M, P_0)\big)\big)\end{aligned} \tag{17.3.3}$$

over $P = Q(R_0)$ in the set

$$\{P = Q(R_0)\,;\, Q \in \mathcal{P}(L), d(P, P_0) > d(P, P_0')\}, \tag{17.3.4}$$

and then taking the residue about $z_{r+1} = 0$.

The operator $\Gamma_\Omega(\sigma_\Lambda^M, P_0)$ acts on the vector space $V_{P_0}(\tau) = V_{Q_0}(V_{R_0}(\tau))$ through the fiber. It transforms a given function from $K$ to $V_{R_0}(\tau)$ by the operator $\Gamma_\Omega(\sigma_\Lambda^M, R_0)$. The other operators in (17.3.3) are products of scalar-valued functions of $\Lambda$ with admissible operators. We can therefore apply Proposition 17.3 inductively, with $G$ replaced by $L$, to the fibers of these operators. It follows that the expression (17.3.3) is left unchanged if $\Gamma_\Omega(\sigma_\Lambda^M, P_0)$ is replaced by $\mathcal{R}_M^{Q_0}(\sigma_\Lambda^M, P_0)$. Consequently, we may apply Proposition 16.8 to commute the operators $\mathcal{R}_{P_0'|P_0}(\sigma_\Lambda^M)^{-1}$ and

$$A(\sigma_\Lambda^M) = i_{P_0'}(\sigma_\Lambda^M, f)\mathcal{R}_{P|P_0'}(\sigma_\Lambda^M)^{-1}\mathcal{R}_{P|P_0}^{(r)}(\sigma_\Lambda^M, \nu)$$

in (17.3.3). As a result, (17.3.3) equals

$$\mathrm{Res}_{\Omega,\Lambda\to\Lambda_0}\, a_\Lambda \operatorname{tr}\Big(i_{P'_0}(\sigma_\Lambda^M, f)\mathcal{R}_{P|P'_0}(\sigma_\Lambda^M)^{-1}\mathcal{R}^{(r)}_{P|P_0}(\sigma_\Lambda^M,\nu)\mathcal{R}_{P'_0|P_0}(\sigma_\Lambda^M)^{-1}\Gamma_\Omega(\sigma_\Lambda^M, P'_0)\Big).$$

By Leibnitz's rule,

$$\mathcal{R}_{P|P'_0}(\sigma_\Lambda^M)^{-1}\mathcal{R}^{(r)}_{P|P_0}(\sigma_\Lambda^M,\nu)\mathcal{R}_{P'_0|P_0}(\sigma_\Lambda^M)^{-1}$$

equals

$$\sum_{j=0}^{r}\frac{r!}{(r-j)!j!}\mathcal{R}_{P|P'_0}(\sigma_\Lambda^M)^{-1}\mathcal{R}^{(r-j)}_{P|P'_0}(\sigma_\Lambda^M,\nu)\mathcal{R}^{(j)}_{P'_0|P_0}(\sigma_\Lambda^M,\nu)\mathcal{R}_{P'_0|P_0}(\sigma_\Lambda^M)^{-1}.$$

It is only the operators $\mathcal{R}_{P'_0|P_0}(\cdot)$ which can contribute a pole along $\Lambda_0(\beta^\vee_{r+1}) = 0$. If $j = 0$, these operators cancel. The corresponding term can therefore be left out of the resulting formula for (17.3.1). Recombining the residues in $z_{r+1}$ and $\Omega$, we express (17.3.1) finally as

$$\mathrm{Res}_{\Omega',\Lambda\to\Lambda'_0}\, a_\Lambda \operatorname{tr}\Big(i_{P'_0}(\sigma_\Lambda^M, f)\mathcal{R}(\Lambda)\Gamma_\Omega(\sigma_\Lambda^M, P'_0)\Big),$$

where $\mathcal{R}(\Lambda)$ is the operator

$$\sum_{P}\theta_Q(\nu)^{-1}\sum_{j=1}^{r}\frac{1}{(r-j)!j!}\mathcal{R}_{P|P'_0}(\sigma_\Lambda^M)^{-1}\mathcal{R}^{(r-j)}_{P|P'_0}(\sigma_\Lambda^M,\nu)\mathcal{R}^{(j)}_{P'_0|P_0}(\sigma_\Lambda^M,\nu)\mathcal{R}_{P'_0|P_0}(\sigma_\Lambda^M)^{-1}. \tag{17.3.5}$$

Here $P = Q(R_0)$ is summed over the set (17.3.4).

The point $\nu$ intervenes only in the expression (17.3.5). Since the final residue is independent of $\nu$, we may choose the point any way we wish. Set $\nu = \nu' + s\varepsilon_{r+1}$, with $\nu'$ a point in general position in $\mathfrak{a}^*_{L',\mathbb{C}}$, and $s$ a small complex number which approaches 0. Note that the function

$$\theta_Q(\nu)^{-1} = \mathrm{vol}(\mathfrak{a}_L^G/\mathbb{Z}(\Delta_Q^\vee))/\prod_{\alpha\in\Delta_Q}(\nu' + s\varepsilon_{r+1})(\alpha^\vee)$$

is the only term which can contribute a singularity in $s$ to (17.3.5). It has at most a pole of order 1 at $s = 0$, and this occurs precisely when some root $\alpha \in \Delta_Q$ vanishes on $\mathfrak{a}_{L'}$, that is, when the parabolic $P = Q(R_0)$ equals $Q'(R'_0)$ for some $Q' \in \mathcal{P}(L')$. On the other hand,

$$\lim_{s\to 0}\mathcal{R}^{(r-j)}_{P|P'_0}(\sigma_\Lambda^M,\nu) = \mathcal{R}^{(r-j)}_{P|P'_0}(\sigma_\Lambda^M,\nu'),$$

while

$$\mathcal{R}^{(j)}_{P'_0|P_0}(\sigma_\Lambda^M,\nu) = \mathcal{R}^{(j)}_{P'_0|P_0}(\sigma_\Lambda^M,s\varepsilon_{r+1}) = s^j\mathcal{R}^{(j)}_{P'_0|P_0}(\sigma_\Lambda^M,\varepsilon_{r+1}),$$

since $P_0' \cap L' = P_0 \cap L'$. Therefore the only summands in (17.3.5) which do not approach 0 are those with $j = 1$ and $P = Q(R_0) = Q'(R_0')$, where $Q' \in \mathcal{P}(L')$. Observe that if $P$ is of this form, and $\alpha_1$ is the unique root in $\Delta_Q$ which vanishes on $\mathfrak{a}_{L'}$, then

$$\lim_{s\to 0} \theta_Q(\nu)^{-1}s = \mathrm{vol}(\mathfrak{a}_L^G/\Delta_Q^\vee)\|\alpha_1^\vee\|^{-1}/\prod_{\alpha\in\Delta_Q\setminus\{\alpha_1\}} \nu'(\alpha^\vee)$$

$$= \mathrm{vol}(\mathfrak{a}_{L'}^G/\mathbb{Z}(\Delta_{Q'}^\vee))/\prod_{\alpha'\in\Delta_{Q'}} \nu'((\alpha')^\vee) = \theta_{Q'}(\nu')^{-1}.$$

Consequently, the value of (17.3.5) at $s = 0$ is

$$\frac{1}{(r-1)!}\sum_{\{P=Q'(R_0')\,;\,Q'\in\mathcal{P}(L')\}} \theta_{Q'}(\nu')^{-1}\mathcal{R}_{P|P_0'}(\sigma_\Lambda^M)^{-1}\mathcal{R}_{P|P_0'}^{(r-1)}(\sigma_\Lambda^M,\nu')\mathcal{R}'_{P_0'|P_0}(\sigma_\Lambda^M,\varepsilon_{r+1})\mathcal{R}_{P_0'|P_0}(\sigma_\Lambda^M)^{-1}.$$

Appealing again to (10.9.1), we see that this equals

$$\mathcal{R}_{L'}(\sigma_\Lambda^M,P_0')\mathcal{R}'_{P_0'|P_0}(\sigma_\Lambda^M,\varepsilon_{r+1})\mathcal{R}_{P_0'|P_0}(\sigma_\Lambda^M)^{-1}.$$

Consequently, (17.3.1) equals

$$\mathrm{Res}_{\Omega',\Lambda\to\Lambda_0'}\, a_\Lambda\, \mathrm{tr}(i_{P_0'}(\sigma_\Lambda^M,f)\mathcal{R}_{L'}(\sigma_\Lambda^M,P_0')\mathcal{R}'_{P_0'|P_0}(\sigma_\Lambda^M,\varepsilon_{r+1})$$
$$\times\, \mathcal{R}_{P_0'|P_0}(\sigma_\Lambda^M)^{-1}\Gamma_\Omega(\sigma_\Lambda^M,P_0')). \tag{17.3.6}$$

We are now essentially done. The parabolics $P_i$, $0 \le i \le r$, are all contained in $Q_0$. Similarly, the parabolics $P_i'$, $0 \le i \le r$, are all contained in another fixed element of $\mathcal{P}(L)$. Since $P_i \cap L = P_i' \cap L$, we have

$$\mathcal{R}'_{P_0|P_i}(\sigma_\Lambda^M,\varepsilon_i)\mathcal{R}_{P_0|P_i}(\sigma_\Lambda^M)^{-1} = \mathcal{R}'_{P_0'|P_i'}(\sigma_\Lambda^M,\varepsilon_i)\mathcal{R}_{P_0'|P_i'}(\sigma_\Lambda^M)^{-1}.$$

We have already noted that $P_0 = P'_{r+1}$. It follows that

$$\mathcal{R}'_{P_0'|P_0}(\sigma_\Lambda^M,\varepsilon_{r+1})\mathcal{R}_{P_0'|P_0}(\sigma_\Lambda^M)^{-1}\Gamma_\Omega(\sigma_\Lambda^M,P_0') = \Gamma_{\Omega'}(\sigma_\Lambda^M,P_0').$$

Substituting in (17.3.6) gives the required formula

$$\mathrm{Res}_{\Omega',\Lambda\to\Lambda_0'}\, a_\Lambda\, \mathrm{tr}\left(i_{P_0'}(\sigma_\Lambda^M,f)\mathcal{R}_{L'}(\sigma_\Lambda^M,P_0')\Gamma_{\Omega'}(\sigma_\Lambda^M,P_0')\right)$$

for (17.3.1). This completes the induction step and the proof of the proposition. □

## 17.4 Vanishing Assertion

In this work we shall use Proposition 17.3 only as a vanishing assertion. Let $h_{\mathfrak{L}}$ and $h_M$ denote the natural projections of $\mathfrak{a}_{\mathfrak{M}}$ onto $\mathfrak{a}_{\mathfrak{L}}$ and $\mathfrak{a}_M$, respectively.

COROLLARY. *The distribution*

$$\mathrm{Res}_{\Omega,\Lambda\to\Lambda_0}(a_\Lambda J_M(\sigma_\Lambda, f))$$

*vanishes unless*

$$\ker(h_{\mathfrak{L}}) \cap \ker(h_M) = \{0\}.$$

PROOF. The projection $h_{\mathfrak{L}}$ is associated with a canonical splitting

$$\mathfrak{a}_{\mathfrak{M}} = \mathfrak{a}_{\mathfrak{M}}^{\mathfrak{L}} \oplus \mathfrak{a}_{\mathfrak{L}}.$$

A similar assertion holds for $h_M$. Consider the associated dual projections $\mathfrak{a}_{\mathfrak{M}}^* \to \mathfrak{a}_{\mathfrak{L}}^*$ and $\mathfrak{a}_{\mathfrak{M}}^* \to \mathfrak{a}_M^*$. The kernel of the first one is spanned by $\{E_1, \ldots, E_r\}$. But if these vectors have images in $\mathfrak{a}_M^*$ which are linearly dependent, the operator $\Gamma_\Omega(\sigma_\Lambda^M, P_0)$ is defined to be 0. The corollary therefore follows from the proposition. □

# 18 Changes of Contour

The reason for studying residues is to be able to deform contour integrals. In this section we shall set up a scheme for keeping track of residues that arise from changes of contour. It is similar to the procedure used in the proof of the Paley-Wiener theorem [Ar83, sect. II.2] and was originally motivated by Langlands' theory of Eisenstein series [La76, sect. 7].

DEFINITION 18.1. (1) *Suppose that $\mathfrak{M}$ and $\sigma \in \Pi(\mathfrak{M}(F_S))$ are as in the last two sections. Let $\mu$ be a fixed point in general position in $\mathfrak{a}_{\mathfrak{M}}^*$. Suppose also that for each Levi subgroup $\mathfrak{L}$ over $F_S$ which contains $\mathfrak{M}$, we have fixed a point $\nu_{\mathfrak{L}}$ in general position in $\mathfrak{a}_{\mathfrak{L}}^*$.*

(2) *Let $\Gamma$ be a finite subset of $\Pi(K)$, the set of equivalence classes of irreducible representations of $K$. We write $\mathbb{H}(G(F_S))_\Gamma$ for the space of functions in $\mathbb{H}(G(F_S))$ which transform on each side under $K$ according to representations in $\Gamma$.*

(3) *The residue scheme will be determined in a canonical way from the point $\mu$, the collection*

$$\mathcal{N} = \{\nu_{\mathfrak{L}};\ \mathfrak{L} \supset \mathfrak{M}\},$$

*and the set*

$$\Psi = \{\psi(\Lambda) = e^{-\Lambda(\mathfrak{X})} J_M(\sigma_\Lambda, f)\,;\ \mathfrak{X} \in \mathfrak{a}_{\mathfrak{M},S}, f \in \mathbb{H}(G(F_S))_\Gamma\}$$

*of functions on* $\mathfrak{a}^*_{\mathfrak{M},\mathbb{C}}$. *Note that the singularities of all the functions in* $\Psi$ *form a set of hyperplanes of the form* (12.4.2) *which is finite modulo* $i\mathfrak{a}^\vee_{\mathfrak{M},S}$.

*Our assumption on the general position of* $\nu_{\mathfrak{L}}$ *implies that if* $\psi(\Lambda)$ *belongs to* $\Psi$ *and* $\Omega$ *is a residue datum for* $\mathfrak{L}$, *then the function*

$$\mathrm{Res}_\Omega\, \psi(\Lambda_\Omega + \lambda) = \mathrm{Res}_{\Omega, \Lambda \to (\Lambda_\Omega + \lambda)}\, \psi(\Lambda), \quad \lambda \in \mathfrak{a}^*_{\mathfrak{L},\mathbb{C}},$$

*is regular on* $\nu_{\mathfrak{L}} + i\mathfrak{a}^*_{\mathfrak{L}}$.

PROPOSITION 18.2. *For each* $\mathfrak{L}$ *there is a finite set*

$$R_{\mathfrak{L}} = R_{\mathfrak{L}}(\mu, \mathcal{N})$$

*of residue data for* $\mathfrak{L}$ *such that*

$$\int_{\mu + i\mathfrak{a}^*_{\mathfrak{M},S}} \psi(\Lambda)\, d\Lambda = \sum_{\mathfrak{L}} \int_{\nu_{\mathfrak{L}} + i\mathfrak{a}^*_{\mathfrak{L},S}} \left( \sum_{\Omega \in R_{\mathfrak{L}}} \mathrm{Res}_\Omega\, \psi(\Lambda_\Omega + \lambda) \right) d\Lambda \qquad (18.2.1)$$

*for any function* $\psi$ *in* $\Psi$.

PROOF. The construction is similar to that of [Ar83, pp. 45–51], so our discussion will be brief. We shall define the sets $R_{\mathfrak{L}}$ by induction on $\dim(\mathfrak{a}^*_{\mathfrak{M}}/\mathfrak{a}^*_{\mathfrak{L}})$. In the process, we shall associate to each $\Omega \in R_{\mathfrak{L}}$ a point $\mu_\Omega$ in $\mathfrak{a}^*_{\mathfrak{L}}$.

If $\mathfrak{L} = \mathfrak{M}$, take $R_{\mathfrak{L}}$ to consist only of the trivial residue datum $\Omega_0$, with $\mathcal{E}_{\Omega_0}$ empty and $\Lambda_{\Omega_0} = 0$. Set $\mu_{\Omega_0} = \mu$. Now assume inductively that we have defined the sets $R_{\mathfrak{L}}$ and also points $\{\mu_\Omega \in \mathfrak{a}^*_{\mathfrak{L}}\,;\ \Omega \in R_{\mathfrak{L}}\}$, for each $\mathfrak{L}$ with $\dim(\mathfrak{a}^*_{\mathfrak{M}}/\mathfrak{a}^*_{\mathfrak{L}}) = r$. Fix a Levi subgroup $\mathfrak{L}'$ over $F_S$ with $\dim(\mathfrak{a}^*_{\mathfrak{M}}/\mathfrak{a}^*_{\mathfrak{L}'}) = r + 1$. Then $R_{\mathfrak{L}'}$ will be defined as a union over all $\mathfrak{L} \supset \mathfrak{L}'$ with $\dim(\mathfrak{a}^*_{\mathfrak{M}}/\mathfrak{a}^*_{\mathfrak{L}}) = r$, and over all $\Omega \in R_{\mathfrak{L}}$, of certain sets. Consider such an $\mathfrak{L}$ and a residue datum

$$\Omega = (\mathcal{E}_\Omega, \Lambda_\Omega) = ((E_1, \ldots, E_r), \Lambda_\Omega),$$

in $R_{\mathfrak{L}}$. By our general position assumption, $\nu_{\mathfrak{L}} - \mu_\Omega$ does not belong to $\mathfrak{a}^*_{\mathfrak{L}'}$. Let $E_{r+1}$ be the unit vector in $\mathfrak{a}^*_{\mathfrak{L}}$ which is orthogonal to $\mathfrak{a}^*_{\mathfrak{L}'}$ and whose inner product with the vector $\nu_{\mathfrak{L}} - \mu_\Omega$ is positive. The inner product on $\mathfrak{a}^*_{\mathfrak{L}}$ is constructed in the same way as that on $\mathfrak{a}^*_M$. We shall describe the subset of $R_{\mathfrak{L}'}$ associated with $\mathfrak{L}$ and $\Omega$. It is parametrized by the orbits under $i\mathfrak{a}^\vee_{\mathfrak{M},S}$ of those singular hyperplanes of the function

$$\lambda \mapsto \mathrm{Res}_\Omega\, \psi(\Lambda_\Omega + \lambda), \quad \lambda \in \mathfrak{a}^*_{\mathfrak{L},\mathbb{C}},\ \psi \in \Psi, \qquad (18.2.2)$$

which are of the form

$$\zeta' E_{r+1} + \mathfrak{a}^*_{\mathfrak{L}',\mathbb{C}}, \quad \zeta' \in \mathbb{C},$$

and which intersect the set

$$\{t\mu_\Omega + (1-t)\nu_{\mathfrak{L}} + \lambda\,;\, 0 < t < 1, \lambda \in i\mathfrak{a}^*_{\mathfrak{L}}\}. \tag{18.2.3}$$

The residue datum $\Omega' = (\mathcal{E}_{\Omega'}, \Lambda_{\Omega'})$ attached to such a singular hyperplane is defined by

$$\mathcal{E}_{\Omega'} = (E_1, \dots, E_r, E_{r+1})$$

and

$$\Lambda_{\Omega'} = \Lambda_\Omega + \zeta' E_{r+1}.$$

We then take $\mu_{\Omega'}$ to be the unique point in $\mathfrak{a}^*_{\mathfrak{L}'}$ such that $\Lambda_{\Omega'} + \mu_{\Omega'}$ belongs to the set (18.2.3).

The inductive definition is set up to account for changes of contours of integration. Standard estimates (such as inequality (20.3.5) below) allow us to control the growth of a function (18.2.2) on the set (18.2.3), at least away from the singular hyperplanes. We can therefore deform the integral of (18.2.2) over $(\mu_\Omega + i\mathfrak{a}^*_{\mathfrak{L},S})$ to an integral over $(\nu_{\mathfrak{L}} + i\mathfrak{a}^*_{\mathfrak{L},S})$. In the process, we pick up residues at the singular hyperplanes. The general position of $\nu_{\mathfrak{L}}$ means that the singularities can be handled separately. It follows from our definition that the sum over $(\mathfrak{L}, \Omega)$, with $\Omega \in R_{\mathfrak{L}}$ and $\dim(\mathfrak{a}^*_{\mathfrak{M}}/\mathfrak{a}^*_{\mathfrak{L}}) = r$, of the expression

$$\int_{\mu_\Omega + i\mathfrak{a}^*_{\mathfrak{L},S}} \operatorname{Res}_\Omega \psi(\Lambda_\Omega + \lambda)\, d\lambda - \int_{\nu_{\mathfrak{L}} + i\mathfrak{a}^*_{\mathfrak{L},S}} \operatorname{Res}_\Omega \psi(\Lambda_\Omega + \lambda)\, d\lambda,$$

equals the sum over $(\mathfrak{L}', \Omega')$, with $\Omega' \in R_{\mathfrak{L}'}$ and $\dim(\mathfrak{a}^*_{\mathfrak{M}}/\mathfrak{a}^*_{\mathfrak{L}'}) = r+1$, of

$$\int_{\mu_{\Omega'} + i\mathfrak{a}^*_{\mathfrak{L}',S}} \operatorname{Res}_{\Omega'} \psi(\Lambda_{\Omega'} + \lambda)\, d\lambda.$$

The required identity (18.2.1) is then obtained by applying this last formula repeatedly, as $r$ increases from 0 to $\dim(\mathfrak{a}_L/\mathfrak{a}_G)$. □

REMARK. It is clear that the construction applies to any family of functions on $\mathfrak{a}^*_{\mathfrak{M},\mathbb{C}}/i\mathfrak{a}^\vee_{\mathfrak{M},S}$ whose singularities and growth properties are similar to those of $\Psi$.

# 19 The Spaces $\mathbb{H}_{\mathrm{ac}}(G(F_S))$ and $\mathbb{I}_{\mathrm{ac}}(G(F_S))$

## 19.1 Objects to Study

As an application of our discussion on residues, we will study the function

$$\phi_M(f, \tau, X), \quad \tau \in \Pi_{\mathrm{temp}}(M(F_S)), \quad X \in \mathfrak{a}_{M,S}.$$

In particular, we shall show that, as a function of the parameters on $\Pi_{\mathrm{temp}}(M(F_S))$, it can be analytically continued to an entire function. We will come to this in the next section. In the present section we shall simply describe some spaces of functions, in order to illustrate the properties of $\phi_M$. These spaces will be used in the next chapter on the invariant trace formula.

DEFINITION 19.2. (1) *We shall consider $\mathbb{H}(G(F_S))$ as a topological vector space.*

*Fix a positive function*

$$\|x\| = \prod_{v \in S} \|x_v\|_v, \quad x \in G(F_S),$$

*on $G(F_S)$, as in* [Ar82.I, Section2]. *We assume in particular that $\|\cdot\|$ satisfies* [Ar82.I, conditions (i)-(iii), p. 1253].

(2) *Suppose that $N$ is a positive number and that $\Gamma$ is a finite subset of $\Pi(K)$. We define $\mathbb{H}_N(G(F_S))_\Gamma$ to be the space of smooth functions on $G(F_S)$ which are supported on the set*

$$G(F_S, N) = \{x \in G(F_S)\,;\, \log \|x\| \le N\}$$

*and which transform on each side under $K$ according to representations in $\Gamma$. The topology on $\mathbb{H}_N(G(F_S))_\Gamma$ is that given by the seminorms*

$$\|f\|_D = \sup_{x \in G(F_S)} |Df(x)|, \quad f \in \mathbb{H}_N(G(F_S))_\Gamma,$$

*where $D$ is a differential operator on $G(F_{S \cap S_\infty})$. As usual, we are writing $S_\infty$ for the set of Archimedean valuations of $F$.*

(3) *We then define $\mathbb{H}(G(F_S))$ as a topological direct limit. Namely, set*

$$\mathbb{H}(G(F_S))_\Gamma = \varinjlim_N \mathbb{H}_N(G(F_S))_\Gamma$$

*and*

$$\mathbb{H}(G(F_S)) = \varinjlim_\Gamma \mathbb{H}(G(F_S))_\Gamma.$$

(4) *Suppose that* $f \in \mathbb{H}(G(F_S))$. *For* $\pi \in \Pi_{\text{temp}}(G(F_S))$, *we have the invariant Fourier transform*

$$f_G : \pi \mapsto f_G(\pi) = \operatorname{tr} \pi(f).$$

*However, it is convenient for us to take a slightly different point of view. For* $\pi \in \Pi_{\text{temp}}(G(F_S))$ *and* $Z \in \mathfrak{a}_{G,S}$, *define*

$$f_G(\pi, Z) = \operatorname{tr} \pi(f^Z).$$

(*See Definition* 15.4 *for* $f^Z$.) *Then*

$$f_G(\pi, Z) = \int_{i\mathfrak{a}_{G,S}^*} f_G(\pi_\lambda) e^{-\lambda(Z)}\, d\lambda, \qquad f_G(\pi_\lambda) = \operatorname{tr} \pi_\lambda(f).$$

Thus $f_G$ can be interpreted in two ways. It is either a function on $\Pi_{\text{temp}}(G(F_S))$, or, via the Fourier transform on $\mathfrak{a}_{G,S}$, a function on $\Pi_{\text{temp}}(G(F_S)) \times \mathfrak{a}_{G,S}$. Note that the situation is analogous to that of the function $\phi_M(f)$. Indeed, $f_G$ is just the special case that $M = G$. We will generally lean toward the second interpretation. The map $f \mapsto f_G$ will be regarded as sending $\mathbb{H}(G(F_S))$ to a space of functions on $\Pi_{\text{temp}}(G(F_S)) \times \mathfrak{a}_{G,S}$.

The work of Clozel-Delorme [CD84, CD85] and Bernstein-Deligne-Kazhdan [BDK86] and [F95] provides a characterization of the image.

DEFINITION 19.3. (1) *We fix Euclidean inner products and Haar measures on the various spaces associated with Levi subgroups* $\mathfrak{M} = \prod_{v \in S} M_v$. *We do this for each* $v \in S$ *separately, by following the conventions of Section* 1, *with* $F$ *replaced by* $F_v$. *We obtain Euclidean norms on the spaces* $\mathfrak{a}_{\mathfrak{M}}$ *and* $\mathfrak{a}_{\mathfrak{M}}^*$ *and Haar measures on the groups* $\mathfrak{a}_{\mathfrak{M}}$, $\mathfrak{a}_{\mathfrak{M},S}$, $i\mathfrak{a}_{\mathfrak{M}}^*$, *and* $i\mathfrak{a}_{\mathfrak{M},S}^*$.

(2) *For any positive number* $N$, *let* $C_N^\infty(\mathfrak{a}_{\mathfrak{M},S})$ *denote the topological vector space of smooth functions on* $\mathfrak{a}_{\mathfrak{M},S}$ *which are supported on*

$$\{\mathfrak{X} \in \mathfrak{a}_{\mathfrak{M},S}\,;\ \|\mathfrak{X}\| \le N\}.$$

(3) *Suppose that* $\Gamma$ *is a finite subset of* $\Pi(K)$ *and* $N$ *is a positive number. We define* $\mathbb{I}_N(G(F_S))_\Gamma$ *to be the space of functions*

$$\varphi : \Pi_{\text{temp}}(G(F_S)) \times \mathfrak{a}_{G,S} \to \mathbb{C}$$

*which satisfy the following three conditions*:

(i) *If* $\lambda$ *is any element in* $i\mathfrak{a}_{G,S}$, *then*

$$\varphi(\pi_\lambda, Z) = \varphi(\pi, Z) e^{\lambda(Z)}.$$

(ii) *Suppose that the restriction of $\pi$ to $K$ does not contain any representation in $\Gamma$. Then $\varphi(\pi, Z) = 0$.*
(iii) *Suppose $\mathfrak{M}$ is a well-positioned relative to $K$ Levi subgroup of $G$ over $F_S$ and $\sigma \in \Pi_{\mathrm{temp}}(\mathfrak{M}(F_S))$. Then the integral*

$$\varphi(\sigma, \mathfrak{X}) = \int_{i\mathfrak{a}^*_{\mathfrak{M},S}/i\mathfrak{a}^*_{G,S}} \varphi(\sigma^G_\Lambda, h_G(\mathfrak{X}))e^{-\Lambda(\mathfrak{X})}\, d\Lambda, \quad \mathfrak{X} \in \mathfrak{a}_{\mathfrak{M},S},$$

*converges to a function of $\mathfrak{X}$ which belongs to $C^\infty_N(\mathfrak{a}_{\mathfrak{M},S})$.*

(4) *We give $\mathbb{I}_N(G(F_S))_\Gamma$ the topology provided by the seminorms*

$$\varphi \mapsto \|\varphi(\sigma, \cdot)\|_{\mathfrak{M},\sigma},$$

*with $\mathfrak{M}$ and $\sigma$ as above, and $\|\cdot\|_{\mathfrak{M},\sigma}$ a continuous seminorm on $C^\infty_N(\mathfrak{a}_{\mathfrak{M},S})$. We then define $\mathbb{I}(G(F_S))$ in the same way that we defined $\mathbb{H}(G(F_S))$. Namely, we set*

$$\mathbb{I}(G(F_S))_\Gamma = \varinjlim_N \mathbb{I}_N(G(F_S))_\Gamma$$

*and*

$$\mathbb{I}(G(F_S)) = \varinjlim_\Gamma \mathbb{I}(G(F_S))_\Gamma.$$

*Note that the first condition implies that the integral*

$$\int_{\mathfrak{a}_{G,S}} \varphi(\pi, Z)\, dZ$$

*is actually a Fourier transform on $\mathfrak{a}_{G,S}$. The other two conditions are taken from* [CD84] *and* [BDK86] *or* [F95]. *For example, condition* (ii) *asserts that the function $\pi \mapsto \varphi(\pi, Z)$ is supported on finitely many components, in the sense of* [BDK86] *or* [F95]. *Condition* (iii) *requires that for every $\mathfrak{M}$ and $\sigma$, the function*

$$\int_{\mathfrak{a}_{G,S}} \varphi(\sigma^G_\Lambda, Z)\, dZ, \quad \Lambda \in i\mathfrak{a}^*_{\mathfrak{M},S},$$

*belongs to the Paley-Wiener space on $i\mathfrak{a}^*_{\mathfrak{M},S}$. In particular, if $S$ consists of one discrete valuation, the function is a finite Fourier series on the torus $i\mathfrak{a}^*_{\mathfrak{M},S}$.*

(5) *The function $\phi_M(f)$ does not in general belong to $\mathbb{I}(M(F_S))$. To accommodate it, we must extend our definitions slightly. Suppose again that $\Gamma$ is a finite subset of $\Pi(K)$. Define $\mathbb{H}_{\mathrm{ac}}(G(F_S))_\Gamma$ to be the space of functions $f$ on $G(F_S)$ such that, for any $b \in C^\infty_c(\mathfrak{a}_{G,S})$, the function*

$$f^b(x) = f(x)b(H_G(x)), \quad x \in G(F_S),$$

*belongs to* $\mathbb{H}(G(F_S))_\Gamma$*. The subscript "ac" stands for* almost compact *support. Give* $\mathbb{H}_{\mathrm{ac}}(G(F_S))_\Gamma$ *the topology defined by the seminorms*

$$f \mapsto \|f^b\|, \quad f \in \mathbb{H}_{\mathrm{ac}}(G(F_S))_\Gamma,$$

*where* $b$ *is any function as above, and* $\|\cdot\|$ *is a continuous seminorm on* $\mathbb{H}(G(F_S))_\Gamma$.

(6) *Similarly, define* $\mathbb{I}_{\mathrm{ac}}(G(F_S))_\Gamma$ *to be the space of functions* $\varphi$ *on* $\Pi_{\mathrm{temp}}(G(F_S)) \times \mathfrak{a}_{G,S}$ *such that, for any* $b$ *as above, the function*

$$\varphi^b(\pi, Z) = \varphi(\pi, Z)b(Z), \quad \pi \in \Pi_{\mathrm{temp}}(G(F_S)), \quad Z \in \mathfrak{a}_{G,S},$$

*belongs to* $\mathbb{I}(G(F_S))_\Gamma$*. We topologize* $\mathbb{I}_{\mathrm{ac}}(G(F_S))_\Gamma$ *the same way, by the seminorms*

$$\varphi \mapsto \|\varphi^b\|, \quad \varphi \in \mathbb{I}_{\mathrm{ac}}(G(F_S))_\Gamma,$$

*with* $\|\cdot\|$ *a continuous seminorm on* $\mathbb{I}(G(F_S))_\Gamma$.

(7) *We then define* $\mathbb{H}_{\mathrm{ac}}(G(F_S))$ *and* $\mathbb{I}_{\mathrm{ac}}(G(F_S))$ *as topological direct limits*

$$\mathbb{H}_{\mathrm{ac}}(G(F_S)) = \varinjlim_{\Gamma} \mathbb{H}_{\mathrm{ac}}(G(F_S))_\Gamma$$

*and*

$$\mathbb{I}_{\mathrm{ac}}(G(F_S)) = \varinjlim_{\Gamma} \mathbb{I}_{\mathrm{ac}}(G(F_S))_\Gamma.$$

(8) *While we are at it, we shall define a useful space of functions that lies between* $\mathbb{H}(G(F_S))$ *and* $\mathbb{H}_{\mathrm{ac}}(G(F_S))$*. We shall say that a function* $f \in \mathbb{H}_{\mathrm{ac}}(G(F_S))$ *is* moderate *if there are positive numbers* $c$ *and* $d$ *such that* $f$ *is supported on the set*

$$\{x \in G(F_S)\,;\, \log\|x\| \leq c(\|H_G(x)\| + 1)\},$$

*and such that*

$$\sup_{x \in G(F_S)}(|\Delta f(x)| \exp\{-d\|H_G(x)\|\}) < \infty,$$

*for any left invariant differential operator* $\Delta$ *on* $G(F_{S_\infty} \cap S)$.

(9) *We shall also say that a function* $\varphi \in \mathbb{I}_{\mathrm{ac}}(G(F_S))$ *is* moderate *if for every Levi subgroup* $\mathfrak{M}$ *over* $F_S$*, and every* $\sigma \in \Pi_{\mathrm{temp}}(\mathfrak{M}(F_S))$*, the function* $\varphi(\sigma, \cdot)$ *has similar support and growth properties. Namely, there are positive numbers* $c_1$ *and* $d_1$ *such that* $\varphi(\sigma, \cdot)$ *is supported on the set*

$$\{\mathfrak{X} \in \mathfrak{a}_{\mathfrak{M},S}\,;\, \|\mathfrak{X}\| \leq c_1(\|h_G(\mathfrak{X})\| + 1)\},$$

*and such that*

$$\sup_{\mathfrak{X}\in\mathfrak{a}_{\mathfrak{M},S}}(|\Delta_1\phi(\sigma,\mathfrak{X})|\exp\{-d_1\|h_G(\mathfrak{X})\|\}) < \infty,$$

*for any left invariant differential operator* $\Delta_1$ *on* $\mathfrak{a}_{\mathfrak{M},S\cap S_\infty}$.

*The notion of moderate function will be a crucial ingredient in a certain convergence estimate required for the comparison of trace formulae. We shall see this in the next chapter, where we shall also show that* $f \mapsto f_G$ *maps the moderate functions in* $\mathbb{H}_{\mathrm{ac}}(G(F_S))$ onto *those in* $\mathbb{I}_{\mathrm{ac}}(G(F_S))$.

# 20 The Map $\phi_M$

## 20.1 Key Theorem

We defined the function

$$\phi_M(f) : (\tau, X) \mapsto \phi_M(f, \tau, X), \quad (\tau, X) \in \Pi_{\mathrm{temp}}(M(F_S)) \times \mathfrak{a}_{M,S},$$

in Section 15. Let $Z$ be the projection of $X$ onto $\mathfrak{a}_{G,S}$. The value of $\phi_M(f, \tau, X)$ depends only on $f^Z$. Consequently, $\phi_M(f)$ is defined for any $f \in \mathbb{H}_{\mathrm{ac}}(G(F_S))$. In this section we shall establish the following theorem. It is one of the main results of this chapter.

THEOREM. *The function* $\phi_M$ *maps* $\mathbb{H}_{\mathrm{ac}}(G(F_S))$ *to* $\mathbb{I}_{\mathrm{ac}}(M(F_S))$.

It is convenient to study a slightly more general map.

DEFINITION 20.2. (1) *Suppose* $\mu \in \mathfrak{a}_M^*$. *Let* $\phi_{M,\mu}(f)$ *be the function whose value at a point* $(\tau, X)$ *in*

$$\Pi_{\mathrm{temp}}(M(F_S)) \times \mathfrak{a}_{M,S}$$

*equals*

$$\phi_{M,\mu}(f, \tau, X) = J_M(\tau_\mu, X, f)e^{-\mu(X)}.$$

*This too is defined if* $f$ *is any function in* $\mathbb{H}_{\mathrm{ac}}(G(F_S))$.

*If* $\lambda$ *belongs to* $i\mathfrak{a}_M^*$, *we have*

$$\phi_{M,\mu}(f, \tau_\lambda, X) = \phi_{M,\mu}(f, \tau, X)e^{\lambda(X)}, \tag{20.2.1}$$

*from* (15.2.1). *If* $h$ *belongs to* $\mathbb{H}(G(F_S)^1)$, (15.3.1) *tells us that*

$$\phi_{M,\mu}(L_h f) = \sum_{Q\in\mathcal{F}(M)} \phi_{M,\mu}^{M_Q}(R_{Q,h} f). \tag{20.2.2}$$

(2) *Suppose for a moment that* $f$ *belongs to* $\mathbb{H}(G(F_S))$. *Suppose also that* $b \in C_c^\infty(\mathfrak{a}_{M,S})$, *that* $\mathfrak{M} = \prod_{v \in S} M_v$ *is a well-positioned relative to K Levi subgroup of* $M$ *over* $F_S$ *and that* $\sigma \in \Pi_{\mathrm{temp}}(\mathfrak{M}(F_S))$. *We shall need to study*

$$\phi^b_{M,\mu}(f, \sigma, \mathfrak{X})$$

*as a function of* $\mathfrak{X} \in \mathfrak{a}_{\mathfrak{M},S}$. *From the definition of the last section, we have*

$$\begin{aligned}\phi^b_{M,\mu}(f, \sigma, \mathfrak{X}) &= b(h_M(\mathfrak{X})) \int_{i\mathfrak{a}^*_{\mathfrak{M},S}/i\mathfrak{a}^*_{M,S}} \phi_{M,\mu}(f, \sigma^M_\Lambda, h_M(\mathfrak{X})) e^{-\Lambda(\mathfrak{X})}\, d\Lambda \\ &= b(h_M(\mathfrak{X})) \sum_{P \in \mathcal{P}(M)} \omega_P \int_{\varepsilon_P + \mu + i\mathfrak{a}^*_{\mathfrak{M},S}} J_M(\sigma_\Lambda, f) e^{-\Lambda(\mathfrak{X})}\, d\Lambda.\end{aligned}$$

*Assume, for simplicity, that* $J_M(\sigma_\Lambda, f)$ *is analytic for* $\Lambda$ *in* $\mu + i\mathfrak{a}^*_{\mathfrak{M},S}$. *Then*

$$\phi^b_{M,\mu}(f, \sigma, \mathfrak{X}) = b(h_M(\mathfrak{X})) \int_{\mu + i\mathfrak{a}^*_{\mathfrak{M},S}} J_M(\sigma_\Lambda, f) e^{-\Lambda(\mathfrak{X})}\, d\Lambda.$$

Our main concern will be to show that this function is compactly supported in $\mathfrak{X}$. As in the proof of the classical Paley-Wiener theorem, this entails changing contours of integration. We will use Proposition 18.2.1 to account for the resulting residues.

THEOREM 20.3. *For each* $\mu \in \mathfrak{a}^*_M$, $\phi_{M,\mu}$ *maps* $\mathbb{H}_{\mathrm{ac}}(G(F_S))$ *continuously to* $\mathbb{I}_{\mathrm{ac}}(M(F_S))$.

PROOF. Fix a function $b \in C_c^\infty(\mathfrak{a}_{M,S})$. Then

$$\begin{aligned}\phi^b_{M,\mu}(f, \tau, X) &= b(X)\phi_{M,\mu}(f, \tau, X) \\ &= b(X) \sum_{P \in \mathcal{P}(M)} \omega_P \int_{\varepsilon_P + \mu + i\mathfrak{a}^*_{M,S}/i\mathfrak{a}^*_{G,S}} J_M(\tau_\lambda, f^{h_G(X)}) e^{-\lambda(X)}\, d\lambda,\end{aligned}$$

for any $f \in \mathbb{H}_{\mathrm{ac}}(G(F_S))$. Thus if $b'$ is any function in $C_c^\infty(\mathfrak{a}_{G,S})$ which equals 1 on the image in $\mathfrak{a}_{G,S}$ of the support of $b$, we have

$$\phi^b_{M,\mu}(f) = \phi^b_{M,\mu}(f^{b'}).$$

We may therefore assume that $f$ belongs to $\mathbb{H}(G(F_S))$.

More precisely, we need only establish that

$$f \mapsto \phi^b_{M,\mu}(f)$$

is a continuous map from $\mathbb{H}(G(F_S))$ to $\mathbb{H}(M(F_S))$.

Choose a positive number $N$ and a finite set $\Gamma \subset \Pi(K)$. Let $\Gamma_M$ be the set of irreducible representations of $K_M$ that are constituents of the restrictions of $\Gamma$ to $K_M$. The theorem will follow if we can show that, for some $N_M > 0$, the function $\phi^b_{M,\mu}$ maps $\mathbb{H}_N(G(F_S))_\Gamma$ continuously to $\mathbb{I}_{N_M}(M(F_S))_{\Gamma_M}$.

In order to prove that a function

$$\phi^b_{M,\mu}(f), \quad f \in \mathbb{H}_N(G(F_S))_\Gamma,$$

belongs to a space $\mathbb{I}_{N_M}(M(F_S))_{\Gamma_M}$, we must establish three conditions. The first condition is just (20.2.1), while the second follows immediately from Frobenius reciprocity and the definition of $\Gamma_M$.

The third condition, of course, is the main point. Fix $\mathfrak{M}$ and $\sigma$ as above. We shall show that the function

$$\mathfrak{X} \mapsto \phi^b_{M,\mu}(f,\sigma,\mathfrak{X}), \quad \mathfrak{X} \in \mathfrak{a}_{\mathfrak{M},S}, \tag{20.3.1}$$

belongs to the space $C^\infty_{N_M}(\mathfrak{a}_{\mathfrak{M},S})$, with $N_M$ depending only on $N$ and $\Gamma$ and that it varies continuously with $f \in \mathbb{H}_N(G(F_S))_\Gamma$. This will establish the third condition and complete the proof of the theorem.

We shall combine Proposition 18.2 with Corollary 17.4. However, we first observe that it is sufficient to prove the assertion with $\mu$ replaced by any of the points $\varepsilon_P + \mu$ in the formula $\phi^b_{M,\mu}(f,\sigma,\mathfrak{X})$ above. We may therefore assume that each function $J_M(\sigma_\Lambda, f)$ is analytic for $\Lambda \in (\mu + i\mathfrak{a}^*_{\mathfrak{M},S})$. Consequently,

$$\phi^b_{M,\mu}(f,\sigma,\mathfrak{X}) = b(h_M(\mathfrak{X})) \int_{\mu + i\mathfrak{a}^*_{\mathfrak{M},S}} J_M(\sigma_\Lambda, f) e^{-\Lambda(\mathfrak{X})}\, d\Lambda.$$

Next, we assign a chamber $c_\mathfrak{L}$ in $\mathfrak{a}_\mathfrak{L}$ to each $\mathfrak{L} \supset \mathfrak{M}$. There are only finitely many such assignments, and $\mathfrak{a}_{\mathfrak{M},S}$ is the corresponding finite union of the sets

$$\{\mathfrak{X} \in \mathfrak{a}_{\mathfrak{M},S}\, ;\, h_\mathfrak{L}(\mathfrak{X}) \in \overline{c}_\mathfrak{L}, \mathfrak{L} \supset \mathfrak{M}\}. \tag{20.3.2}$$

We may therefore assume that $\mathfrak{X}$ actually belongs to a given set (20.3.2). For each $\mathfrak{L}$, let $c^*_\mathfrak{L}$ be the associated chamber in $\mathfrak{a}^*_\mathfrak{L}$. Let $\nu_\mathfrak{L}$ be a highly regular point in general position in $c^*_\mathfrak{L}$. Applying Proposition 18.2, with $\mathcal{N} = \{\nu_\mathfrak{L}\}$, we see that $\phi^b_{M,\mu}(f,\sigma,\mathfrak{X})$ equals

$$b(h_M(\mathfrak{X})) \sum_{\mathfrak{L} \supset \mathfrak{M}} \sum_{\Omega \in R_\mathfrak{L}} \int_{\nu_\mathfrak{L} + i\mathfrak{a}^*_{\mathfrak{L},S}} \mathrm{Res}_{\Omega, \Lambda \to \Lambda_\Omega + \lambda} (e^{-\Lambda(\mathfrak{X})} J_M(\sigma_\Lambda, f))\, d\lambda.$$

Corollary 17.4 (with $a_\Lambda = e^{-\Lambda(\mathfrak{X})}$) then provides an important condition on $\mathfrak{L}$ in order that the integral not vanish. We are thus able to write $\phi^b_{M,\mu}(f,\sigma,\mathfrak{X})$ as the sum over those $\mathfrak{L} \supset \mathfrak{M}$ with

$$\ker(h_\mathfrak{L}) \cap \ker(h_M) = \{0\}, \tag{20.3.3}$$

and over $\Omega \in R_{\mathfrak{L}}$, of the product of $b(h_M(\mathfrak{X}))$ with

$$\int_{\nu_{\mathfrak{L}}+i\mathfrak{a}^*_{\mathfrak{L},S}} \mathrm{Res}_{\Omega,\Lambda\to\Lambda_\Omega+\lambda}(e^{-\Lambda(\mathfrak{X})}\,\mathrm{tr}(\mathcal{R}_M(\sigma^M_\Lambda, P_0)i_{P_0}(\sigma^M_\Lambda, f)))\,\mathrm{d}\lambda. \tag{20.3.4}$$

As always, $P_0$ is any fixed element in $\mathcal{P}(M)$.

Our next step is to deform the contour of integration in (20.3.4). Our assumption on $\nu_{\mathfrak{L}}$ ensures that none of the singularities of the integrand meets the tube over the translated chamber $(\nu_{\mathfrak{L}} + c^*_{\mathfrak{L}})$. Now, a standard argument shows that there is a positive number $A$, depending at most on $\Gamma$, such that

$$\|i_{P_0}(\sigma^M_\Lambda, f)\|_\sigma \leq c_n(f)e^{AN\|\mathrm{Re}(\Lambda)\|}(1 + \|\mathrm{Im}(\Lambda_\infty)\|)^{-n}, \tag{20.3.5}$$

for any $f \in \mathbb{H}_N(G(F_S))_\Gamma$, $n \in \mathbb{R}_{>0}$, and $\Lambda \in \mathfrak{a}^*_{\mathfrak{M},\mathbb{C}}$. (See, e.g., the first steps in the proof of [Ar83, Lemma III.3.1].) Here $\|\cdot\|_\sigma$ is any norm on the finite-dimensional space $V_{P_0}(\sigma)_\Gamma$, $c_n(\cdot)$ is a continuous seminorm on $\mathbb{H}_N(G(F_S))_\Gamma$, and $\Lambda_\infty$ is the projection of $\Lambda$ onto

$$\oplus_{v\in S\cap S_\infty}\mathfrak{a}^*_{M_v,\mathbb{C}}.$$

Since $\mathcal{R}_M(\sigma^M_\Lambda, P_0)$ is a rational expression in the variables (12.4.2), the function

$$\mathrm{tr}(\mathcal{R}_M(\sigma^M_\Lambda, P_0)i_{P_0}(\sigma^M_\Lambda, f))$$

satisfies a similar estimate for $\Lambda$ in the tube over $(\nu_{\mathfrak{L}}+c^*_{\mathfrak{L}})$. We can therefore deform the contour of integration in (20.3.4) to $t\nu_{\mathfrak{L}} + i\mathfrak{a}^*_{\mathfrak{L},S}$, where $t$ is a real number which approaches infinity.

If $\mathfrak{X}$ belongs to the support of (20.3.4), we obtain an inequality

$$|\nu_{\mathfrak{L}}(\mathfrak{X})| \leq AN\|\nu_{\mathfrak{L}}\|,$$

with $A$ depending at most on $\Gamma$. But we are already assuming that $h_{\mathfrak{L}}(\mathfrak{X})$ belongs to the closure of $c_{\mathfrak{L}}$. Since $\nu_{\mathfrak{L}}$ is strictly positive on the complement of the origin in this set, we can estimate $\|h_{\mathfrak{L}}(\mathfrak{X})\|$ in terms of $|\nu_{\mathfrak{L}}(\mathfrak{X})|$. Consequently, there is an $A_1$, depending only on $\Gamma$, such that

$$\|h_{\mathfrak{L}}(\mathfrak{X})\| \leq A_1 N, \tag{20.3.6}$$

if $\mathfrak{X}$ belongs to the support of (20.3.4).

The condition (20.3.3) implies that

$$\|\mathfrak{X}\| \leq \|h_{\mathfrak{L}}(\mathfrak{X})\| + \|h_M(\mathfrak{X})\|. \tag{20.3.7}$$

Imposing the additional requirement that $b(h_M(\mathfrak{X})) \neq 0$, we then combine (20.3.6) and (20.3.7). We obtain

$$\|\mathfrak{X}\| \leq A_1 N + A_2,$$

for a fixed positive number $A_2$. It follows from this that the original function (20.3.1) is supported on a ball whose radius depends only on $N$ and $\Gamma$.

The proof of the theorem is essentially complete. The only additional point is to establish the continuous dependence of (20.3.1) on $f$. In this regard, it is simplest to represent the value of a continuous seminorm on the function (20.3.1) in terms of the Fourier transform. The required inequality then follows easily from the estimate (20.3.5). Thus the properties of (20.3.1) are as promised, and the theorem is proved. □

COROLLARY 20.4. *For each $\mu \in \mathfrak{a}^*_{\mathfrak{M}}$, the map $\phi_{M,\mu}$ takes moderate functions in $\mathbb{H}_{ac}(G(F_S))$ to moderate functions in $\mathbb{I}_{ac}(M(F_S))$.*

PROOF. Suppose that $f$ is a moderate function in $\mathbb{H}_{ac}(G(F_S))$. In order to show that $\phi_{M,\mu}(f)$ is a moderate function in $\mathbb{I}_{ac}(M(F_S))$, we must verify two conditions. For the support condition, we must look back at the proof of the theorem. Note that the integral (20.3.4) depends only on the function

$$f^Z, \quad Z = h_G(\mathfrak{X}).$$

By assumption, $f$ is supported on a set

$$\{x \in G(F_S)\,;\, \log \|x\| \leq c(\|H_G(x)\| + 1)\},$$

so we may identify $f^Z$ with the restriction to $G(F_S)^Z$ of a function in $\mathbb{H}_N(G(F_S))$, where

$$N = c(\|Z\| + 2) = c(\|h_G(\mathfrak{X})\| + 2).$$

The inequality (20.3.6) can therefore be written

$$\|h_{\mathfrak{L}}(\mathfrak{X})\| \leq A_1 c(\|h_G(\mathfrak{X})\| + 2).$$

Combined with (20.3.7), this becomes

$$\|\mathfrak{X}\| \leq A_1 c(\|h_G(\mathfrak{X})\| + 2) + \|h_M(\mathfrak{X})\|.$$

Since

$$\|h_G(\mathfrak{X})\| \leq \|h_M(\mathfrak{X})\|,$$

the function

$$\mathfrak{X} \mapsto \phi_{M,\mu}(f, \sigma, \mathfrak{X}), \quad \mathfrak{X} \in \mathfrak{a}_{\mathfrak{M},S},$$

is supported on a set

$$\{\mathfrak{X} \in \mathfrak{a}_{\mathfrak{M},S}\,;\ \|\mathfrak{X}\| \leq c_1(\|h_M(\mathfrak{X})\| + 1)\}.$$

This is the required support condition. The growth condition on $\phi_{M,\mu}(f, \sigma, \mathfrak{X})$ is a routine matter. It follows easily from the given growth condition on $f$ and the appropriate variant of the estimate (20.3.5). Therefore $\phi_{M,\mu}(f)$ is a moderate function. □

COROLLARY 20.5. *The linear transformation $\phi_M$ maps the space $\mathbb{H}_{\mathrm{ac}}(G(F_S))$ continuously to $\mathbb{I}_{\mathrm{ac}}(M(F_S))$. The image of a moderate function in $\mathbb{H}_{\mathrm{ac}}(G(F_S))$ is a moderate function in $\mathbb{I}_{\mathrm{ac}}(M(F_S))$.*

PROOF. This is the previous theorem when $\mu = 0$. □

# Chapter 5
# The Invariant Trace Formula

The well-known Poisson summation formula applies to a lattice $\Gamma$ in $\mathbb{R}$ and a function $f \in C_c^\infty(\mathbb{R})$. It can be written

$$\sum_{\gamma \in \mathbb{R}} a^\Gamma(\gamma) f(\gamma) = \sum_{\lambda \in \mathbb{R}} \hat{a}^\Gamma(\lambda) \hat{f}(\lambda), \tag{0.5.1}$$

where $\hat{f}$ is the Fourier transform of $f$, while

$$a^\Gamma(\gamma) = \begin{cases} \mathrm{vol}(\mathbb{R}/\Gamma), & \gamma \in \Gamma, \\ 0, & \gamma \notin \Gamma, \end{cases}$$

and

$$\hat{a}^\Gamma(\lambda) = \begin{cases} 1, & \lambda\Gamma \subset \mathbb{Z}, \\ 0, & \text{otherwise.} \end{cases}$$

Note the general structure of the terms. The functions $f(\gamma)$ and $\hat{f}(\lambda)$ are independent of $\Gamma$, while the coefficients $a^\Gamma(\gamma)$ and $\hat{a}^\Gamma(\lambda)$ are independent of $f$. The Poisson summation formula has a number of applications. They all involve playing some of the terms off against the others.

The Poisson summation formula has a generalization to a discrete subgroup of a general locally compact (unimodular) group with compact quotient. It is the Selberg trace formula. For example, suppose that $G$ is a semisimple algebraic group defined and anisotropic over $F$. Then $G(F)$ is a discrete subgroup of the locally compact group $G(\mathbb{A})$. The quotient $G(F)\backslash G(\mathbb{A})$ is compact. The *Selberg trace formula* is

$$\sum_{\gamma \in (G(F))} a^G(\gamma) I_G(\gamma, f) = \sum_{\pi \in \Pi(G)} a^G(\pi) I_G(\pi, f), \quad f \in C_c^\infty(G(\mathbb{A})). \tag{0.5.2}$$



Y.Z. Flicker, *Arthur's Invariant Trace Formula and Comparison of Inner Forms*,
DOI 10.1007/978-3-319-31593-5_5

Here $(G(F))$ is the set of conjugacy classes in $G(F)$, $\Pi(G)$ is a set of equivalence classes of irreducible unitary representations of $G(\mathbb{A})$, and

$$a^G(\gamma) = \mathrm{vol}(G(F,\gamma)\backslash G(\mathbb{A},\gamma)),$$

$$a^G(\pi) = \mathrm{Multiplicity}(\pi, L^2(G(F)\backslash G(\mathbb{A})),$$

$$I_G(\gamma,f) = \int_{G(\mathbb{A},\gamma)\backslash G(\mathbb{A})} f(x^{-1}\gamma x)\,\mathrm{d}x,$$

$$I_G(\pi,f) = \mathrm{tr}\int_{G(\mathbb{A})} f(x)\pi(x)\,\mathrm{d}x.$$

Again, the terms have the same general structure. The functions $I_G(\gamma,f)$ and $I_G(\pi,f)$ are invariant distributions on $G(\mathbb{A})$ which do not depend on the discrete subgroup $G(F)$. The coefficients $a^G(\gamma)$ and $a^G(\pi)$ depend strongly on $G(F)$, but are independent of $f$. The Selberg trace formula also has many applications. Again, one obtains information about one set of terms from a knowledge of the others.

If $G$ is not anisotropic over $F$, the quotient $G(F)\backslash G(\mathbb{A})$ is no longer compact, and the situation changes rather drastically. The terms in (0.5.2) diverge, in several senses, and in general are not defined. There are natural ways to truncate the integrals that diverge, however, and one ends up with a trace formula that appears quite complicated. This is the formula obtained in Chapter 3.

In this chapter we shall show that the general structure of the trace formula is rather simple. We shall establish an identity of the general form

$$\sum_M |W_0^M||W_0^G|^{-1} \sum_{\gamma\in(M(F))} a^M(\gamma)I_M(\gamma,f) = \sum_M |W_0^M||W_0^G|^{-1} \int_{\Pi(M)} a^M(\tau)I_M(\tau,f)\,\mathrm{d}\tau, \tag{0.5.3}$$

in which $M$ ranges over the finite set of rational Levi subgroups of $G$.

The terms corresponding to $M \neq G$ represent contributions from the boundary. They are what is left of the original integrals that had to be truncated.

The functions $a^M(\gamma)$ and $a^M(\tau)$ depend only on the group $M$ and not on its embedding in $G$. They are global in nature, in that they depend on the rational structure of $M$.

The functions $I_M(\gamma,f)$ and $I_M(\tau,f)$ are invariant linear forms in $f$. They are local objects that are independent of the discrete subgroup $G(F)$ of $G(\mathbb{A})$.

The applications of the general trace formula include getting information about the discrete spectrum, which is a priori wrapped up in the definition of the function $a^G(\pi)$, from the other terms in the trace formula. A theorem of this kind is the subject of the next chapter.

We shall discuss the global theory of (0.5.3) and the proof of the formula itself in this chapter. First, we shall study the functions $I_M(\gamma,f)$ and $I_M(\tau,f)$. These are interesting objects in their own right. If $M = G$, $I_M(\gamma,f)$ is just the orbital integral over $\gamma$ and $I_M(\tau,f)$ is the character of $\tau$. For general $M$ they are more complicated, but they retain many of the essential properties of the special case.

Take $G$ to be a connected reductive group over a number field $F$. If $S$ is a finite set of valuations of $F$, one can define

$$I_M(\gamma,f), \quad \gamma \in M(F_S),$$

and

$$I_M(\tau,f), \quad \tau \in \Pi_{\mathrm{unit}}(M(F_S))$$

as invariant linear forms on the Hecke algebra of $G(F_S)$.

It is important to express them in terms of the local groups $G(F_v)$. In Section 9, we shall prove *splitting formulae* for $I_M(\gamma,f)$ and $I_M(\tau,f)$ in terms of the corresponding objects on the groups $G(F_v)$, $v \in S$.

A related question concerns the case that the data $\gamma$ and $\tau$ come from a proper Levi subgroup $M_1$ of $M$. In Section 8, we shall prove *descent formulae* for $I_M(\gamma,f)$ and $I_M(\tau,f)$ in terms of the corresponding objects for $M_1$.

Both sets of results will be proved from Proposition 7.6, which gives a general descent property for $(G,M)$-families. This in turn is closely related to a similar property for convex polytopes, for which we refer to Section 11.

It is perhaps helpful to think of the distributions $I_M(\gamma,f)$ and $I_M(\tau,f)$ themselves in terms of convex polytopes. Indeed, the chambers of the restricted Weyl group are dual to a certain convex polytope $\Pi_0$. The groups $M$ are parametrized by subspaces which intersect the faces of $\Pi_0$ orthogonally. If we project $\Pi_0$ onto such a subspace, we obtain another convex polytope $\Pi_M$. The geometry of $\Pi_M$ then governs the descent and splitting properties of the corresponding distributions.

The invariant distributions $I_M(\gamma,f)$ are obtained from the weighted orbital integrals $J_M(\gamma,f)$ studied in [Ar88.1]. In Section 2 we shall list the various properties that $I_M(\gamma,f)$ inherits from $J_M(\gamma,f)$. They all generalize well-known properties of ordinary orbital integrals. For example, the value of $I_M(\gamma,f)$ at a general point $\gamma \in M(F_S)$ can be approximated by its values at $G$-regular points in $M(F_S)$. If $S$ consists of one Archimedean valuation, $I_M(\gamma,f)$ satisfies a differential equation in $\gamma$. It also has a simple formula for the jump across the singular hyperplane of a real root. If $S$ consists of one discrete valuation, $I_M(\gamma,f)$ satisfies a germ expansion in $\gamma$.

The distributions $I_M(\tau,f)$ are the values at $X = 0$ (and $\tau$ unitary) of a more general family of invariant distributions

$$I_M(\tau,X,f), \quad \tau \in \Pi(M(F_S)),\ X \in \mathfrak{a}_{M,S},$$

which we introduce in Section 3. These are defined in terms of the weighted characters

$$J_M(\tau,X,f) = \int_{i\mathfrak{a}_{M,S}} J_M(\tau_\lambda,f)e^{-\lambda(X)}\,\mathrm{d}\lambda,$$

studied in Chapter 4. It will follow from the definition that $I_M(\tau, X, f)$ is trivial if $\tau$ is tempered, that is, it is $f_G(\pi, X)$ if $M = G$ and $\tau = \pi$, and 0 otherwise. See Proposition 3.4. However, for general $\tau$, the distribution is more interesting. It turns out to be closely related to the residues (in $\tau_\lambda$) of $J_M(\tau_\lambda, f)$. There are hints of this in Proposition 3.5 and 3.6.

The distributions $I_M(\gamma)$ and $I_M(\tau, X)$ are not independent of each other. This is fortunate because it enhances the possibility of playing them off against each other in the trace formula. If $\gamma$ is restricted to a maximal torus $T(F_S)$ in $M(F_S)$, the weighted orbital integral $J_M(\gamma, f)$ is compactly supported in $\gamma$. However, $I_M(\gamma, f)$ is not compactly supported in $\gamma$. The distributions $I_M(\tau, X, f)$ may be viewed as the obstruction to being compactly supported.

In Section 4 we shall study various objects which arise naturally when one tries to analyze the asymptotic behavior of $I_M(\gamma, f)$. We shall define new invariant distributions, ${}^cI_M(\gamma, f)$ and ${}^cI_M(\tau, X, f)$, by improving the support properties at the expense of properties of smoothness. In particular, we shall show that ${}^cI_M(\gamma, f)$ is compactly supported if $\gamma$ lies in $T(F_S)$ (Proposition 4.10).

We shall also define certain maps $\theta_M$ and ${}^c\theta_M$ that provide expansions for $I_M$ and ${}^cI_M$ in terms of each other. These maps are in fact determined by the asymptotic behavior of $I_M(\gamma, f)$. This sets the stage for Proposition 5.8. The result is an important formula for ${}^cI_M(\tau, X, f)$ as a contour integral involving ${}^c\theta_M(f)$. It follows that the distributions ${}^cI_M(\tau, X, f)$ and $I(\tau, X, f)$ may be determined, at least in principle, from the asymptotic behavior of $I_M(\gamma, f)$.

In Section 6 we shall give a simple example of how Proposition 5.8 can be applied in practice. It is not known in general that an invariant distribution annihilates functions whose orbital integrals vanish. In the non-Archimedean case, this is proven in [F16]. See also our inductive elaboration in 2.2.25 of [GK75, Proposition 1, p. 100]. In Theorem 6.2 we shall show that this property holds for $I_M(\tau, X)$ provided that it holds for $I_M(\gamma)$. We will establish the property for $I_M(\gamma)$ with Theorem 15.1, together with showing that $I_M(\gamma)$ is invariant, using the global trace formula. The analogous local result—that the local distribution $f \mapsto I_M(\gamma, f)$ is supported on characters—is proven by purely local means (the local trace formula) in [Ar93, Corollary 5.3], also in the Archimedean case.

We have already mentioned the descent and splitting formulae that are proved in Sections 7–9. To illustrate the descent formulae, we shall discuss the example of $\mathrm{GL}(n)$. We shall show that our invariant distributions often vanish on functions associated with the comparison with central simple algebras. These vanishing formulae (Proposition 10.7 and 10.8) will in fact be required for this comparison. Together with the global vanishing results of Section 18, they are the starting point for a comparison of the full trace formula of $\mathrm{GL}(n)$ with the trace formula for its inner forms, undertaken in the next chapter.

# 1 Invariant Harmonic Analysis

## 1.1 The Group

Let $G$ be a connected reductive linear algebraic group over a local or global field $F$ of characteristic 0. In this chapter, $S$ always stands for a finite set of valuations of $F$ with the closure property, see 4.1.4. This simply means that, if $S$ contains no Archimedean valuations, it consists entirely of valuations which divide a fixed rational prime $p$. We fix a maximal compact subgroup

$$K = \prod_{v \in S} K_v$$

of $G(F_S)$, such that the group $K_v$ is special for every non-Archimedean valuation $v \in S$. Having fixed $K$, we can form the Hecke space $\mathbb{H}(G(F_S))$. It consists of the smooth, compactly supported functions on $G(F_S)$ which are left and right $K$-finite.

## 1.2 Distributions

The Hecke space seems to be the correct space of test functions to use in the trace formula. We are interested in the continuous linear functionals on $\mathbb{H}(G(F_S))$, also called distributions, on $G(F_S)$ which make up the individual terms in the trace formula. The local properties of the spectral side were introduced and discussed in the previous chapter. We will shortly recall the definition and basic properties of the local distributions on the geometric side. They were originally studied in [Ar88.1]. This chapter unites the studies of the local distributions. We shall attach *invariant* distributions to each of the distributions in the two families. By studying the parallel behavior of these, we shall find that the two families are really quite closely related.

## 1.3 Levi Subgroups

We shall routinely adopt the notation of Section 4.1. In particular, the letter $M$ is always understood to be a Levi subgroup of $G$ which is well-positioned with respect to $K$. More precisely, we require that each $M(F_v)$ be well-positioned relative to $K_v$ in the sense of Subsection 4.1.6. Recall that $\mathcal{L}(M)$ denotes the collection of Levi subgroups of $G$ which contain $M$, $\mathcal{L}_0(M)$ the subset of $M_1 \in \mathcal{L}(M)$ with $M_1 \neq G$, $\mathcal{P}(M)$ is the set of parabolic subgroups with Levi component $M$, and $\mathcal{F}(M)$ denotes the set of parabolic subgroups

$$P = M_P N_P, \quad M_P \in \mathcal{L}(M),$$

which contain $M$. Recall that we also have the real vector space

$$\mathfrak{a}_M = \mathrm{Hom}(X(M)_F, \mathbb{R}),$$

which we assume has been assigned a suitable Euclidean metric. This provides a Euclidean metric by restriction on any subspace of $\mathfrak{a}_M$. We also have the dual space $\mathfrak{a}_M^* = X(M)_F \otimes \mathbb{R}$.

DEFINITION 1.4. (1) Recall that $\mathfrak{a}_{G,S}$ is the subgroup $H_G(G(F_S))$ of $\mathfrak{a}_G$. See 4.1.3 for details.

(2) In Section 4.19, we defined the *Paley-Wiener space* $\mathbb{I}(G(F_S))$ of functions on

$$\Pi_{\mathrm{temp}}(G(F_S)) \times \mathfrak{a}_{G,S}.$$

(3) For $\pi \in \Pi_{\mathrm{temp}}(G(F_S))$ and $X \in \mathfrak{a}_{G,S}$, set

$$f_G(\pi, X) = \int_{i\mathfrak{a}_{G,S}^*} \mathrm{tr}\, \pi_\lambda(f) \cdot e^{-\lambda(X)}\, \mathrm{d}\lambda .$$

The map $T : \mathbb{H}(G(F_S)) \to \mathbb{I}(G(F_S))$ given by $f \mapsto f_G$ is continuous.

(4) For any $P \in \mathcal{P}(M)$, the constant term $f_P$ is the function

$$m \mapsto \delta_P(m)^{1/2} \int_K \int_{N_P(F_S)} f(k^{-1}mnk)\, \mathrm{d}n\, \mathrm{d}k$$

in $\mathbb{H}(M(F_S))$.

(5) We let $\tau_\lambda^G = i_P^G(\tau_\lambda)$ be the representation parabolically induced from $\tau_\lambda$, normalized by $\delta_P^{1/2}$ to preserve unitarity. It is tempered if $\tau$ is.

(6) More generally, for $\tau \in \Pi_{\mathrm{temp}}(M(F_S))$ and $X \in \mathfrak{a}_{M,S}$, we consider the function

$$f_M(\tau, X) = (f_P)_M(\tau, X) = \int_{i\mathfrak{a}_{M,S}^*} \mathrm{tr}\, i_P^G(\tau_\lambda, f) \cdot e^{-\lambda(X)}\, \mathrm{d}\lambda .$$

LEMMA 1.5. *The map $f \mapsto f_M$ is a continuous linear map from $\mathbb{H}(G(F_S))$ to $\mathbb{I}(M(F_S))$.*

We now recall some spaces of functions introduced in Section 4.19.

DEFINITION 1.6. (1) Let $\Gamma$ denote a finite subset of $\Pi(K)$.

(2) Define $\mathbb{H}_{\mathrm{ac}}(G(F_S))_\Gamma$ to be the space of functions $f$ on $G(F_S)$ such that, for any $b \in C_c^\infty(\mathfrak{a}_{G,S})$, the function

$$f^b(x) = f(x)b(H_G(x))$$

belongs to $\mathbb{H}(G(F_S))_\Gamma$; the latter is defined in Definition 4.19.2(3).

(3) Similarly, $\mathbb{I}_{\mathrm{ac}}(G(F_S))_\Gamma$ is the space of functions $\varphi$ on $\Pi_{\mathrm{temp}}(G(F_S)) \times \mathfrak{a}_{G,S}$ such that, for every $b$, the function

$$\varphi^b(\pi, X) = \varphi(\pi, X)b(X)$$

belongs to $\mathbb{I}(G(F_S))_\Gamma$; the latter is defined in Definition 4.19.3(4).

(4) Put

$$\mathbb{H}_{\mathrm{ac}}(G(F_S)) = \varinjlim_{\Gamma} \mathbb{H}_{\mathrm{ac}}(G(F_S))_\Gamma.$$

(5) Put

$$\mathbb{I}_{\mathrm{ac}}(G(F_S)) = \varinjlim_{\Gamma} \mathbb{I}_{\mathrm{ac}}(G(F_S))_\Gamma.$$

There is an important map $\phi_M$ which sends $\mathbb{H}(G(F_S))$ to a space of functions on

$$\Pi_{\mathrm{temp}}(M(F_S)) \times \mathfrak{a}_{M,S}.$$

It is discussed in Sections 4.15 and 4.20. Its image is not contained in $\mathbb{I}(M(F_S))$. However, we have the following proposition.

PROPOSITION 1.7. (1) *The map $\phi_M$ can be defined on $\mathbb{H}_{\mathrm{ac}}(G(F_S))$, with image in $\mathbb{I}_{\mathrm{ac}}(M(F_S))$.*

(2) *The map $f \mapsto f_M$ extends to a continuous map from $\mathbb{H}_{\mathrm{ac}}(G(F_S))$ into $\mathbb{I}_{\mathrm{ac}}(M(F_S))$. In particular,*

$$T : f \mapsto f_G, \quad f \in \mathbb{H}_{\mathrm{ac}}(G(F_S)),$$

*maps $\mathbb{H}_{\mathrm{ac}}(G(F_S))$ continuously into $\mathbb{I}_{\mathrm{ac}}(G(F_S))$.*

PROOF. (1) This is Corollary 4.20.4.

(2) This follows directly from the definitions.

□

PROPOSITION 1.8. *The map*

$$T : f \mapsto f_G, \quad f \in \mathbb{H}_{\mathrm{ac}}(G(F_S)),$$

*is an open, surjective map from $\mathbb{H}_{\mathrm{ac}}(G(F_S))$ onto $\mathbb{I}_{\mathrm{ac}}(G(F_S))$.*

PROOF. It is enough to establish the result with the spaces $\mathbb{H}_{\mathrm{ac}}(G(F_S))$ and $\mathbb{I}_{\mathrm{ac}}(G(F_S))$ replaced by $\mathbb{H}(G(F_S))$ and $\mathbb{I}(G(F_S))$. Indeed, the topologies on the larger spaces are defined so that the openness assertion extends immediately. One extends the surjectivity to the larger spaces by a partition of unity argument on $\mathfrak{a}_{G,S}$. It is also clear that the valuations in $S$ may be treated separately. We shall therefore assume that $S = \{v\}$ consists of one valuation and that $F$ is a local field. Then $F_S = F_v = F$.

Suppose first that $F$ is non-Archimedean. The surjectivity of the map $\mathbb{H}(G(F)) \to \mathbb{I}(G(F_S))$ follows directly from the trace Paley-Wiener theorem [BDK86, F95]. The openness is clear since $\mathbb{H}(G(F))$ and $\mathbb{I}(G(F))$ are topological direct limits of finite-dimensional spaces.

If $F$ is Archimedean, the surjectivity has been proved by Clozel and Delorme [CD84, CD85]. The openness assertion can also be extracted from the work of Clozel and Delorme. For implicit in their proof of surjectivity is the construction of a continuous section

$$\mathbb{I}(G(F_S)) \to \mathbb{H}(G(F_S)).$$

(See [Ar88.3, Appendix].) □

DEFINITION 1.9. Suppose that $\theta$ is a continuous linear map from $\mathbb{H}_{\mathrm{ac}}(G(F_S))$ to another topological vector space $V$. We shall say that $\theta$ is *supported on characters* if it vanishes on the kernel of $T$. That is, if $\theta(f) = 0$ for every function $f \in \mathbb{H}_{\mathrm{ac}}(G(F_S))$ such that $f_G = 0$.

LEMMA 1.10. *If $\theta$ is supported on characters, there is a unique continuous map*

$$\hat{\theta} : \mathbb{I}_{\mathrm{ac}}(G(F_S)) \to V$$

*such that, for $f \in \mathbb{H}_{\mathrm{ac}}(G(F_S))$,*

$$\hat{\theta}(f_G) = \theta(f).$$

PROOF. This is an immediate consequence of Proposition 1.8. □

## 1.11 Supported on Characters

Consider the special case that $V = \mathbb{C}$. Then $\theta$ is supported on characters if and only if it lies in the image of the transpose map

$$T' : \mathbb{I}'_{\mathrm{ac}}(G(F_S)) \to \mathbb{H}'_{\mathrm{ac}}(G(F_S))$$

given by

$$\hat{\theta} \mapsto \theta.$$

In other words, the function $\hat{\theta}$ is the inverse image of $\theta$ under $T'$. We shall often refer to elements in the dual spaces $\mathbb{H}'_{\mathrm{ac}}(G(F_S))$ and $\mathbb{I}'_{\mathrm{ac}}(G(F_S))$ as distributions on $\mathbb{H}_{\mathrm{ac}}(G(F_S))$ and $\mathbb{I}_{\mathrm{ac}}(G(F_S))$.

Any map

$$\theta : \mathbb{H}_{\mathrm{ac}}(G(F_S)) \to V$$

which is supported on characters is also invariant. That is,

$$\theta(L_h f) = \theta(R_h f), \quad h \in \mathbb{H}(G(F_S)^1), f \in \mathbb{H}_{\mathrm{ac}}(G(F_S)).$$

See Section 4.19 for the definitions of $L_h$ and $R_h$. Conversely, it is likely that every map which is invariant is supported on characters. In the non-Archimedean case, this is proven in [F16]. See also our inductive elaboration in 2.2.25 of [GK75, Proposition 1, p. 100], together with [Ka86.1, Appendix]. However, we shall not try to prove this in the Archimedean case. We shall be content simply to show that those invariant maps and distributions which arise from the trace formula are supported on characters. The proof will be based on a long induction, completed by Corollary 15.7. We will use a global argument introduced by Kazhdan. The proof does not require that we keep track of which maps are invariant. However, we shall do so, in order to motivate our constructions. In fact, the reader might find it easier to proceed as if it were known that all invariant maps were supported on characters. In the Archimedean case [Ar93, Corollary 5.3] proves by purely local means (the local trace formula) that the local distribution $f \mapsto I_M(\gamma, f)$ is supported on characters.

## 2 The Invariant Distributions $I_M(\gamma)$

### 2.1 Weighted Orbital Integrals

We shall introduce one of the two families of distributions which occur in the trace formula. These distributions are parametrized by elements in $M(F_S)$ and are obtained from weighted orbital integrals.

Suppose that $\gamma$ is an element in $M(F_S)$. The weighted orbital integral

$$J_M(\gamma, f), \quad f \in C_c^\infty(G(F_S)),$$

is defined in [Ar88.1]. It is a distribution which depends only on the restriction of $f$ to

$$G(F_S)^Z = \{x \in G(F_S) \,;\, H_G(x) = Z\},$$

for $Z = H_G(\gamma)$. The restriction of any function in $\mathbb{H}_{ac}(G(F_S))$ to this set coincides with that of a function in $\mathbb{H}(G(F_S))$. Consequently, $J_M(\gamma)$ may be regarded as a distribution on $\mathbb{H}_{ac}(G(F_S))$.

Let us review the definition of $J_M(\gamma)$ as a linear form on $C_c^\infty(G(F_S))$.

DEFINITION 2.2. (1) Let $M_v$ be a Levi subgroup of $G$ defined over $F_v$. We have the map

$$H_{P_v} : G(F_v) \to \mathfrak{a}_{M_v}, \quad P_v \in \mathcal{P}(M_v),$$

defined by

$$H_{P_v}(n_v m_v k_v) = H_M(m_v), \quad n_v \in N_{P_v}(F_v),\ m_v \in M_v(F_v),\ k_v \in K_v.$$

Then

$$\{Y_{P_v} = -H_{P_v}(x_v)\,;\ P_v \in \mathcal{P}(M_v)\}$$

is a positive $(G, M_v)$-orthogonal set, and thus

$$\left\{v_{P_v}(\lambda, x_v) = e^{-\lambda(H_P(x_v))}\,;\ P_v \in \mathcal{P}(M_v), \lambda \in i\mathfrak{a}_{M_v}^*\right\}$$

is a $(G, M_v)$-family of functions. The associated smooth function

$$v_{M_v}(\lambda, x_v) = \sum_{P_v \in \mathcal{P}(M_v)} v_{P_v}(\lambda, x_v)/\theta_{P_v}(\lambda)$$

is the Fourier transform of the characteristic function of the convex hull in $\mathfrak{a}_{M_v}^G$ of the projection of the $(G, M_v)$-orthogonal set onto $\mathfrak{a}_{M_v}^G$. The number

$$v_{M_v}(x_v) = v_M(0, x_v) = \lim_{\lambda \to 0} \sum_{P_v \in \mathcal{P}(M_v)} v_{P_v}(\lambda, x_v)/\theta_{P_v}(\lambda)$$

is the volume of this convex hull.

(2) The same remarks apply in the global situation. Let $M \in \mathcal{L}$ be a Levi subgroup over $F$ and let $x \in G(\mathbb{A})$. Then we have the $(G, M)$-family

$$\left\{v_P(\lambda, x) = e^{-\lambda(H_P(x))}\,;\ P \in \mathcal{P}(M), \lambda \in i\mathfrak{a}_M^*\right\},$$

where

$$H_P : G(\mathbb{A}) \to \mathfrak{a}_M.$$

The number

$$v_M(x) = \lim_{\lambda \to 0} \sum_{P \in \mathcal{P}(M)} v_P(\lambda, x)/\theta_P(\lambda)$$

is the volume of the convex hull in $\mathfrak{a}_M^G$ of the points $\{-H_P(x)\,;\, P \in \mathcal{P}(M)\}$. These weight factors were introduced in a different context in Section 3.8.

(3) Let $x = (x_v) \in G(\mathbb{A})$. Then

$$H_P(x) = \sum_v H_P(x_v), \quad P \in \mathcal{P}(M).$$

For almost all $v$ the term $H_P(x_v)$ vanishes, and we obtain the finite sum

$$H_P(x) = \sum_{v \in S} H_P(x_v)$$

where $S \supset S_\infty$ is a finite set of valuations. Regarding $x$ as a point in $G(F_S)$, we have the product

$$v_P(\lambda, x) = \prod_{v \in S} v_P(\lambda, x_v), \quad \lambda \in i\mathfrak{a}_M^*,\ P \in \mathcal{P}(M).$$

We may thus use the descent and splitting formulae, see Sections 8 and 9, to express the volume $v_M(x)$ in terms of $v_M(x_v)$.

The weighted orbital integral $J_M(\gamma)$ can be defined simply only when $M_\gamma = G_\gamma$. In this case, we make the following

DEFINITION 2.3. Let $\gamma \in M(F_S)$. Suppose that $G_\gamma = M_\gamma$. The weighted orbital integral $J_M(\gamma) = J_M^G(\gamma)$ is a linear form on $C_c^\infty(G(F_S))$ defined by

$$J_M(\gamma, f) = |D_G(\gamma)|^{1/2} \int_{G_\gamma(F_S)\backslash G(F_S)} f(x^{-1}\gamma x) v_M(x)\, dx.$$

The integral is well defined, since $v_M(mx) = v_M(x)$ for $m \in M(F_S)$. Note that $M_\gamma = G_\gamma$ for $\gamma \in M(F_S)$ which are the image of an element in $M(F)$ which represents an unramified class $\mathfrak{o} \in \mathcal{O}$. We now state the main theorem on the existence of weighted orbital integrals for more general points. It is Theorem 5.2 of [Ar88.1].

THEOREM 2.4. *For each $M \in \mathcal{L}$, there are canonical functions*

$$m_M^L(\gamma, a), \quad \gamma \in M(F_S),\ L \in \mathcal{L}(M),\ a \in A_{M,\mathrm{reg}}(F_S),$$

*such that the limit*

$$J_M(\gamma, f) = \lim_{a \to e} \sum_{L \in \mathcal{L}(M)} m_M^L(\gamma, a) J_L(a\gamma, f)$$

*exists and equals the integral of $f$ with respect to a Borel measure on $\gamma^G$.*

The functions $m^L_M(\gamma, a)$ here are defined in [Ar88.1], Section 5, and especially Lemma 5.1, where they are denoted by $r^L_M(\gamma, a)$, in terms of a certain $(G, M)$-family. These functions depend only on $M$, $L$, $\gamma$, and $a$, and not on the group $G$. The limit is taken over $a$ in $A_{M,\mathrm{reg}}(F_S)$ sufficiently close to the identity, where $A_{M,\mathrm{reg}}(F_S)$ is the set of points in $A_M(F_S)$ whose centralizer in $G(F_S)$ equals $M(F_S)$.

## 2.5 Invariance

Arguing as in the proof of Proposition 4.14.3, we can transform the formula

$$J_M(\gamma, f^y) = \sum_{Q \in \mathcal{F}(M)} J_M^{M_Q}(\gamma, f_{Q,y}), \quad f \in C_c^\infty(G(F_S)), \tag{2.5.1}$$

established in Proposition 4.13.8 into

$$J_M(\gamma, L_h f) = \sum_{Q \in \mathcal{F}(M)} J_M^{M_Q}(\gamma, R_{Q,h} f), \quad h \in \mathbb{H}_{\mathrm{ac}}(G(F_S)^1).$$

A similar formula,

$$\phi_M(L_h f) = \sum_{Q \in \mathcal{F}(M)} \phi_M^{M_Q}(R_{Q,h} f),$$

holds for the map

$$\phi_M : \mathbb{H}_{\mathrm{ac}}(G(F_S)) \to \mathbb{I}_{\mathrm{ac}}(M(F_S))$$

(See (20.2.2) in Chapter 4). This suggests that we define an invariant distribution on $\mathbb{H}_{\mathrm{ac}}(G(F_S))$,

$$I_M(\gamma, f) = I_M^G(\gamma, f),$$

inductively. More specifically, we would like to set

$$J_M(\gamma, f) = \sum_{L \in \mathcal{L}(M)} \widehat{I}_M^L(\gamma, \phi_L(f)), \quad f \in \mathbb{H}_{\mathrm{ac}}(G(F_S)).$$

However, we cannot say that $I_M(\gamma)$ is supported on characters, so we do not know that $\widehat{I}_M(\gamma)$ is defined. We must proceed as follows.

## 2.6 First Induction Assumption

Recall that $\mathcal{L}_0(M)$ denotes the set of elements $L \in \mathcal{L}(M)$ with $L \neq G$. *We assume inductively that, for every* $L \in \mathcal{L}_0(M)$ *and every finite set* $S$ *of valuations with the closure property, the distributions* $I^L_M(\gamma)$ *are defined and are* supported on characters.

DEFINITION 2.7. For $f \in \mathbb{H}(G(F_S))$ put

$$I_M(\gamma,f) = J_M(\gamma,f) - \sum_{L\in\mathcal{L}_0(M)} \widehat{I}^L_M(\gamma, \phi_L(f)). \tag{2.7.1}$$

The invariance of $I_M(\gamma)$ follows from the two formulae above defining (2.5.1) by Proposition 4.4.11. We shall carry this first induction assumption throughout most of this chapter. The arguments will be completed only by Corollary 15.7, in which we shall show that $I^G_M(\gamma)$ is also supported on characters. Only then will $\widehat{I}_M(\gamma)$ be defined and we will be able to write

$$J_M(\gamma,f) = \sum_{L\in\mathcal{L}(M)} \widehat{I}^L_M(\gamma, \phi_L(f)).$$

## 2.8 Induced Classes

In the paper [Ar88.1], the local behavior of $J_M(\gamma,f)$ as a function of $\gamma$ is investigated. It is easy to see that $I_M(\gamma,f)$ has similar properties. They can all be established inductively from the corresponding properties of $J_M(\gamma,f)$.

PROPOSITION 2.9. *For all* $\gamma$ *in* $M(F_S)$ *and* $f \in \mathbb{H}(G(F_S))$, *we have*

$$I_M(\gamma,f) = \lim_{a\to e} \sum_{L\in\mathcal{L}(M)} m^L_M(\gamma, a) I_L(a\gamma,f), \tag{2.9.1}$$

*with* $a \in A_{M,\mathrm{reg}}(F_S)$.

PROOF. Assume inductively that

$$I^{M_1}_M(\gamma, g) = \lim_{a\to e} \sum_{L\in\mathcal{L}^{M_1}(M)} m^L_M(\gamma, a) I^{M_1}_L(a\gamma, g),$$

for any $M_1 \in \mathcal{L}_0(M)$ and $g \in \mathbb{I}_{\mathrm{ac}}(M_1(F_S))$. A similar formula then holds if $I^{M_1}_M$ is replaced by $\widehat{I}^{M_1}_M$. It follows from the definition (2.7.1) that $I_M(\gamma,f)$ equals

$$\lim_{a\to e} \sum_{L\in\mathcal{L}(M)} m^L_M(\gamma, a) \left( J_L(a\gamma,f) - \sum_{M_1\in\mathcal{L}_0(L)} \widehat{I}^{M_1}_L(a\gamma, \phi_{M_1}(f)) \right).$$

Applying the definition again, we obtain the proposition. □

It is convenient to introduce the notion of an induced space of orbits, as in [Ar88.1, Section 6].

DEFINITION 2.10. Let $\gamma$ be an element of $M(F_S)$. Define the *induced orbit* $\gamma^G$ of $\gamma$ to $G$ ($G$ a reductive connected $F$-group) to be the union of the $G(F_S)$-orbits $\{\gamma_i\}$ in $G(F_S)$ which for every $P \in \mathcal{P}(M)$ intersect $\gamma N_P(F_S)$ in an open set of $\gamma N_P(F_S)$. See [Ar88.1, Section 6]. There are only finitely many such $G(F_S)$-orbits. They all belong to one geometric orbit. If $\gamma$ is $G$-regular, $\gamma^G$ consists of a single orbit, of $\gamma$ itself. If $\gamma$ is unipotent, then each $\gamma_i$ is contained in the induced geometric conjugacy class of $\gamma$.

## 2.11 *Induced Orbits*

More generally, suppose that $L_1 \in \mathcal{L}(M)$. The induced space $\gamma^{L_1} \subset L_1(F_S)$ is defined in Section 2.6 as in [Ar88.1, Section 6]. It is a finite union of $L_1(F_S)$-orbits. Corollary 6.3 of [Ar88.1] asserts that

$$J_{L_1}(\gamma^{L_1},f) = \lim_{a \to e} \sum_{L \in \mathcal{L}(L_1)} m_{L_1}^{L}(\gamma, a) J_L(a\gamma,f),$$

with $a \in A_{M,\text{reg}}(F_S)$, for all $\gamma$ in $M(F_S)$ and $f \in \mathbb{H}(G(F_S))$.

Inductively from this, we get the following formula.

PROPOSITION. *For all $\gamma$ in $M(F_S)$ and $f \in \mathbb{H}(G(F_S))$, we have*

$$I_{L_1}(\gamma^{L_1},f) = \lim_{a \to e} \sum_{L \in \mathcal{L}(L_1)} m_{L_1}^{L}(\gamma, a) I_L(a\gamma,f), \tag{2.11.1}$$

*with $a \in A_{M,\text{reg}}(F_S)$. In particular, the limit on the right exists.*

## 2.12 *An Equivalence*

Suppose that $\sigma \in M(F_S)$ is a semisimple element such that $G_\sigma$ is contained in $M$. Then

$$J_M(\gamma,f) \overset{(M,\sigma)}{\sim} 0, \quad \gamma \in \sigma M_\sigma(F_S),$$

in the notation of [Ar88.1, Lemma 2.2]. This means the following: $J_M(\gamma,f)$ coincides with the orbital integral of a smooth function of compact support on $M(F_S)$ for $\gamma$ near $\sigma$ in $\sigma M_\sigma(F_S)$.

It follows inductively from (2.7.1) that the same property holds for the invariant distributions. Namely, we have the following proposition, for all $f \in \mathbb{H}(G(F_S))$.

PROPOSITION. *We have*

$$I_M(\gamma,f) \overset{(M,\sigma)}{\sim} 0, \quad \gamma \in \sigma M_\sigma(F_S). \tag{2.12.1}$$

PROPOSITION 2.13. *The distribution $I_M(\gamma)$ depends only on the $M(F_S)$-orbits of $\gamma$.*

PROOF. The same is true of $J_M(\gamma)$, see [Ar88.1, Lemma 2.2]. □

## 2.14 Invariance Property

More generally, suppose that $y$ belongs to $M(F_S)G(F)$. Then $y^{-1}My$ is another Levi subgroup of $G$. If $f$ belongs to $\mathbb{H}(G(F_S))$, the function

$$f^y(x) = f(yxy^{-1})$$

belongs to the Hecke space with respect to the maximal compact subgroup $y^{-1}Ky$.

PROPOSITION. *We have*

$$I_{y^{-1}My}(y^{-1}\gamma y,f) = I_M(\gamma,f). \tag{2.14.1}$$

PROOF. We have the formula

$$J_{y^{-1}My}(y^{-1}\gamma y,f^y) = J_M(\gamma,f).$$

See the remark following the proof of [Ar88.1, Lemma 8.1]. It follows from (2.7.1) that

$$I_{y^{-1}My}(y^{-1}\gamma y,f^y) = I_M(\gamma,f). \tag{2.14.2}$$

Suppose that $y$ belongs to $M(F_S)G(F) \cap K$. Then the proposition follows since $I_M(\gamma)$ is invariant. □

## 2.15 Non-Archimedean Place

Consider the case that $S$ consists of one non-Archimedean valuation $v$ and write $F$ for $F_v = F_S$. Let $\sigma$ be a semisimple element in $M(F)$. Proposition 9.1 of [Ar88.1] establishes a germ expansion

$$J_M(\gamma,f) \overset{(M,\sigma)}{\sim} \sum_{L\in\mathcal{L}(M)} \sum_{\delta\in(\sigma\mathcal{U}_{L_\sigma}(F))} g_M^L(\gamma,\delta)J_L(\delta,f).$$

Here $(\sigma\mathcal{U}_{L_\sigma}(F))$ are the conjugacy classes and the $g_M^L(\gamma,\delta)$ are (uniquely determined) $(M,\sigma)$-equivalence classes of functions of $\gamma$. See [Ar88.1, Section 9] for further explanation.

The following corollary follows inductively from (2.7.1).

COROLLARY. *We have that*

$$I_M(\gamma,f) \overset{(M,\sigma)}{\sim} \sum_{L\in\mathcal{L}(M)} \sum_{\delta\in(\sigma\mathcal{U}_{L_\sigma}(F))} g_M^L(\gamma,\delta) I_L(\delta,f). \tag{2.15.1}$$

## 2.16 Archimedean Place

Consider finally the case that $F = F_v = F_S$ is an Archimedean local field. Suppose that $T$ is a "maximal torus" of $G$ over $F$, in the sense of [Ar88.1, Section 1]. If $z$ belongs to the center of the associated universal enveloping algebra, we have the differential equation

$$J_M(\gamma, zf) = \sum_{L\in\mathcal{L}(M)} \partial_M^L(\gamma, z_L) J_L(\gamma,f),$$

for $\gamma$ in the open set $T_{\mathrm{reg}}(F)$ of $G$-regular elements in $T(F)$ [Ar88.1, Proposition 11.1]. Using the definition (2.7.1) inductively again, we convert this to a differential equation for the invariant distributions.

COROLLARY. *We have*

$$I_M(\gamma, zf) = \sum_{L\in\mathcal{L}(M)} \partial_M^L(\gamma, z_L) I_L(\gamma,f), \quad \gamma \in T_{\mathrm{reg}}(F), \tag{2.16.1}$$

## 2.17 Jumps

The behavior of $I_M(\gamma,f)$ as $\gamma$ approaches the singular set is also identical with that of $J_M(\gamma,f)$. In particular, the jump around a semiregular point of noncompact type can be computed for any derivative of $J_M(\gamma,f)$. It is given by the following lemma.

LEMMA. *We have*

$$\lim_{r\to 0}(\partial(u) I_M^\beta(\gamma_r,f) - \partial(u) I_M^\beta(\gamma_{-r},f)) = n_\beta \lim_{s\to 0}(\partial(u_1) I_{M_1}(\delta_s,f)). \tag{2.17.1}$$

PROOF. This is the analogue of [Ar88.1, Proposition 13.1], where the notation is explained. □

Similarly, [Ar88.1, Proposition 13.2] becomes

LEMMA 2.18. *For every $f \in \mathbb{H}(G(F_S))$, there is a real $c(f) > 0$ such that*

$$|\partial(u) I_M(\gamma, f)| \leq c(f) |D^G(\gamma)|^{-q}, \quad \gamma \in G_{\mathrm{reg}}. \tag{2.18.1}$$

PROOF. These results follow inductively from (2.7.1). □

We conclude this section with a proposition which will be needed for global applications.

PROPOSITION 2.19. *Suppose that $v$ is an unramified finite valuation and that $f$ is a function in $\mathbb{H}_{\mathrm{ac}}(G(F_v))$ which is bi-invariant under $K_v$. Then*

$$I_M(\gamma, f) = J_M(\gamma, f), \quad \gamma \in M(F_v).$$

PROOF. Suppose that $L \in \mathcal{L}_0(M)$. Then

$$\phi_L(f, \tau, X) = \int_{i\mathfrak{a}^*_{L,v}/i\mathfrak{a}^*_{G,v}} \mathrm{tr}(\mathcal{R}_L(\tau_\lambda, Q_0) i_{Q_0}(\tau_\lambda, f^Z))\, d\lambda,$$

in the notation of Section 4.15. Here $\tau$ is a representation in $\Pi_{\mathrm{temp}}(L(F_v))$, $X$ is a point in $\mathfrak{a}_{L,v}$ whose projections onto $\mathfrak{a}_{G,v}$ equals $Z$, and $Q_0$ is any element in $\mathcal{P}(L)$. Since $f$ is bi-invariant under $K_v$, the operator $i_{Q_0}(\tau_\lambda, f^Z)$ vanishes unless $\tau$ is unramified. Suppose then that $\tau$ is unramified. Let $\psi$ be a vector in the space on which $i_{Q_0}(\tau_\lambda)$ acts which is fixed by $K_v$. By condition $(\mathcal{R}_8)$ of Theorem 4.6.2, the normalized intertwining operators

$$\mathcal{R}_{Q|Q_0}(\tau_\lambda), \quad Q \in \mathcal{P}(L),$$

take values at $\psi$ which are independent of $\lambda$. Recalling Definition 4.12.2, we see that

$$\begin{aligned} \mathcal{R}_L(\tau_\lambda, Q_0)\psi &= \lim_{\nu \to 0} \sum_{Q \in \mathcal{P}(L)} (\mathcal{R}_{Q|Q_0}(\tau_\lambda)^{-1} \mathcal{R}_{Q|Q_0}(\tau_{\lambda+\nu})\psi)/\theta_Q(\nu) \\ &= \left( \lim_{\nu \to 0} \sum_{Q \in \mathcal{P}(L)} \theta_Q(\nu)^{-1} \right) \psi = 0 \end{aligned}$$

It thus follows that the function $\phi_L(f)$ vanishes. The proposition is then an immediate consequence of the definition (2.7.1). □

# 3 The Invariant Distributions $I_M(\tau, X)$

## 3.1 Weighted Characters

Next we shall define the other invariant distributions which occur in the trace formula. These distributions are parametrized by pairs

$$(\tau, X), \quad \tau \in \Pi(M(F_S)), \; X \in \mathfrak{a}_{M,S}.$$

They are related to the weighted characters

$$J_M(\tau, X, f), \quad f \in \mathbb{H}(G(F_S)),$$

studied in Chapter 4.

Corollary 4.15.6 states that $J_M(\tau, X, f)$ depends only on the restriction of $f$ to $G(F_S)^Z$ for $Z = h_G(X)$. Thus, as with the weighted orbital integrals, the weighted characters may be regarded as linear forms on $\mathbb{H}_{\text{ac}}(G(F_S))$. It follows from Proposition 4.14.3 that

$$J_M(\tau, X, L_h f) = \sum_{Q \in \mathcal{F}(M)} J_M^{M_Q}(\tau, X, R_{Q,h} f),$$

for any $f \in \mathbb{H}_{\text{ac}}(G(F_S))$ and $h \in \mathbb{H}(G(F_S)^1)$.

Since a similar formula holds for the map $\phi_M$, we make a

DEFINITION 3.2. Define the invariant distribution

$$I_M(\tau, X, f) = I_M^G(\tau, X, f), \quad f \in \mathbb{H}_{\text{ac}}(G(F_S)),$$

inductively by setting

$$I_M(\tau, X, f) = J_M(\tau, X, f) - \sum_{L \in \mathcal{L}_0(M)} \widehat{I}_M^L(\tau, X, \phi_L(f)). \tag{3.2.1}$$

Included in the definition is the following induction assumption.

## 3.3 Second Induction Assumption

*Assume that, for any $L \in \mathcal{L}_0(M)$ and any pair $(\tau, X)$, the distribution $I_M^L(\tau, X)$ is supported on characters.*

This is our second induction hypothesis. Before we are done, we will be forced to take on several more of the same kind. All but one of these will be resolved shortly.

We shall show in Section 6 that our induction hypotheses all follow from that of 2.6. But as we have already remarked, we shall carry assumption 2.6 for much of the chapter.

PROPOSITION 3.4. *Suppose that $\tau$ is tempered. Then*

$$I_M(\tau,X,f) = \begin{cases} f_G(\tau,X), & M = G, \\ 0, & M \neq G. \end{cases}$$

PROOF. If $M = G$, we have $I_G(\tau,X,f) = f_G(\tau,X)$, by definition, even if $\tau$ is not tempered. If $M \neq G$, definition 4.15.9 also implies that

$$J_M(\tau,X,f) = \phi_M(f,\tau,X) = \widehat{I}_M^M(\tau,X,\phi_M(f)),$$

as long as $\tau$ is tempered. The proposition follows inductively from (3.2.1). □

## 3.5 A Difference

At first glance, one might guess that the proposition holds for arbitrary $\tau$. However, this is decidedly not the case. If $\tau$ is not tempered and $M \neq G$, the difference

$$J_M(\tau,X,f) - \phi_M(f,\tau,X)$$

is no longer 0. For $J_M(\tau,X,f)$ is defined directly as an integral over $\{\tau_\lambda\}$, whereas $\phi_M(f,\tau,X)$ is defined by analytic continuation from such integrals taken over tempered representations. One finds that the difference depends in a complicated way on the residues discussed in Section 4.16.

On the other hand, $I_M(\tau,X,f)$ does not assume too many values. Set

$$I_{M,\mu}(\tau,X,f) = I_M(\tau_\mu,X,f)e^{-\mu(X)}, \quad \mu \in \mathfrak{a}_M^*.$$

Consider this expression as a function of $\mu$.

PROPOSITION. (1) *As a function of $\mu$, $I_{M,\mu}(\tau,X,f)$ is locally constant on the complement of a finite set of hyperplanes of the form $\mu(\alpha^\vee) = N$, for $N \in \mathbb{R}$ and $\alpha$ a root of $(G, A_M)$.*

(2) *For each $P \in \mathcal{P}(M)$, let $\varepsilon_P$ be a small point in the chamber $(\mathfrak{a}_P^*)^+$. Set*

$$\omega_P = \mathrm{vol}(\mathfrak{a}_P^+ \cap B)/\,\mathrm{vol}(B),$$

*where $B$ is a ball in $\mathfrak{a}_M$ centered at the origin. Note that the numbers $\omega_P$, $P \in \mathcal{P}(M)$, need not be equal. Then*

$$I_{M,\mu}(\tau,X,f) = \sum_{P\in\mathcal{P}(M)} \omega_P I_{M,\mu+\varepsilon_P}(\tau,X,f).$$

PROOF. The definition (3.2.1) may be rewritten

$$I_{M,\mu}(\tau,X,f) = J_{M,\mu}(\tau,X,f) - \sum_{L\in\mathcal{L}_0(M)} \widehat{I}^L_{M,\mu}(\tau,X,\phi_L(f)), \tag{3.2.1*}$$

where

$$J_{M,\mu}(\tau,X,f) = J_M(\tau_\mu,X,f)e^{-\mu(X)}.$$

The first assertion (1) of the proposition will follow inductively from this if we can establish the corresponding statement for $J_{M,\mu}(\tau,X,f)$. We may assume that $f$ belongs to $\mathbb{H}(G(F_S))$. Then if $\mu \in \mathfrak{a}^*_M$ is in general position, we have

$$J_{M,\mu}(\tau,X,f) = \int_{\mu+i\mathfrak{a}^*_{M,S}} J_M(\tau_\lambda,f)e^{-\lambda(X)}\,\mathrm{d}\lambda.$$

The required assertion then follows from the properties of the functions $J_M(\tau_\lambda,f)$, see Section 4.12. This proves (1).

Assume inductively that (2) holds if $G$ is replaced by any element $L \in \mathcal{L}_0(M)$. Then

$$\widehat{I}^L_{M,\mu}(\tau,X,\phi_L(f)) = \sum_{R\in\mathcal{P}^L(M)} \omega_R \widehat{I}^L_{M,\mu+\varepsilon_R}(\tau,X,\phi_L(f)).$$

If we apply the assertion (1) to $L$, we see that this may be written as

$$\sum_{P\in\mathcal{P}(M)} \omega_P \widehat{I}^L_{M,\mu+\varepsilon_P}(\tau,X,\phi_L(f)).$$

But it is an immediate consequence of the definition, Section 4.15, of $J_M(\tau_\mu,X,f)$ that

$$J_{M,\mu}(\tau,X,f) = \sum_{P\in\mathcal{P}(M)} \omega_P J_{M,\mu+\varepsilon_P}(\tau,X,f).$$

The second part (2) of the proposition follows from (3.2.1*). □

REMARK. The reader might want to keep a special case in mind. Suppose that $F = \mathbb{R}$, $\tau$ is tempered, and $M = A_M$ (so in particular, $G$ is a split group). Then from the reducibility properties of the representation $i_P(\tau_\mu)$, one can see that the singular hyperplanes are all of the form

$$\mu(\alpha^\vee) = n, \quad \alpha \in \Sigma(G,A_M),\ n \in \mathbb{Z}.$$

Therefore $I_{M,\mu}(\tau,X,f)$ is constant on the affine Weyl chambers of $\mathfrak{a}^*_M$.

PROPOSITION 3.6. *Suppose that* $\tau \in \Pi_{\text{unit}}(M(F_S))$. *Then the function*

$$I_{M,\mu}(\tau, X, f)$$

*of* $\mu \in \mathfrak{a}_M^*$ *is constant for* $\mu$ *in a neighborhood of the origin.*

PROOF. First, consider the function $J_{M,\mu}(\tau, X, f)$. As in the proof of the last proposition, we can assume that $f$ belongs to $\mathbb{H}(G(F_S))$, so that

$$J_{M,\mu}(\tau, X, f) = \int_{\mu + i\mathfrak{a}_{M,S}^*} J_M(\tau_\lambda, f) e^{-\lambda(X)} \, d\lambda.$$

By definition, see Section 4.12, we have

$$J_M(\tau_\lambda, f) = \mathrm{tr}(\mathcal{R}_M(\tau_\lambda, P_0) i_{P_0}(\tau_\lambda, f)),$$

where $\mathcal{R}_M(\tau_\lambda, P_0)$ is constructed from the normalized intertwining operators

$$\mathcal{R}_{P|P_0}(\tau_\lambda) : i_{P_0}(\tau_\lambda) \to i_P(\tau_\lambda), \quad P, P_0 \in \mathcal{P}(M).$$

In particular, $J_M(\tau_\lambda, f)$ is regular at any point $\lambda$ where the intertwining operators are all regular. But by Theorem 4.6.2, the operators $\mathcal{R}_{P|P_0}(\tau_\lambda)$ are unitary whenever $\tau_\lambda$ is unitary. It follows that $J_M(\tau_\lambda, f)$ is regular if the real part of $\lambda$ is near 0. By changing the contour in the integral above, we see that $J_{M,\mu}(\tau, X, f)$ is constant for $\mu$ near 0. The proposition then follows inductively from the formula (3.2.1*). □

For future reference, we state a variant of the last proposition. Its proof is similar.

PROPOSITION 3.7. *Suppose that* $\tau \in \Pi_{\text{unit}}(M(F_S))$ *and that* $L \in \mathcal{L}(M)$. *Then the function*

$$I_L(\tau_\lambda^L, h_L(X), f)$$

*of* $\lambda \in \mathfrak{a}_{M,\mathbb{C}}^*$ *is analytic if the real part of* $\lambda$ *is near* 0.

## 3.8 Standard Representations

It is sometimes useful to take a standard representation $\rho \in \Sigma(M(F_S))$ instead of the irreducible $\tau$. See Section 4.9 for the definition of standard representations. We noted in Section 4.15 that the distributions $J_M(\rho, X, f)$ could be defined in the same way as $J_M(\tau, X, f)$. We then showed, in Proposition 4.15.7, that for any $\tau \in \Pi(M(F_S))$, the integral $J_M(\tau, X, f)$ had an expansion

$$\sum_P \omega_P \sum_L \sum_\rho \int_{\varepsilon_P + i\mathfrak{a}_{M,S}^*/i\mathfrak{a}_{L,S}^*} m_M^L(\tau_\lambda, \rho_\lambda) J_L(\rho_\lambda^L, h_L(X), f) e^{-\lambda(X)} \, d\lambda,$$

with $P$, $L$, and $\rho$ summed over $\mathcal{P}(M)$, $\mathcal{L}(M)$, and $\Sigma(M(F_S))$, respectively. The notation here follows Chapter 4. In particular, $m_M^L(\tau_\lambda, \rho_\lambda)$ is a meromorphic function obtained from the ratios of the normalizing factors for $\tau_\lambda$ and $\rho_\lambda$. See Subsection 4.12.6 for the definition. Arguing as in the proof of Proposition 3.5(2), we obtain a similar expansion

$$I_M(\tau, X, f) = \sum_P \omega_P \sum_L \sum_\rho \int_{\varepsilon_P + i\mathfrak{a}_{M,S}^*/i\mathfrak{a}_{L,S}^*} m_M^L(\tau_\lambda, \rho_\lambda) I_L(\rho_\lambda^L, h_L(X), f) e^{-\lambda(X)} \, d\lambda \tag{3.8.1}$$

in terms of the invariant distributions defined by the analogue of (3.2.1).

# 4 Some Further Maps and Distributions

## 4.1 Relating Distributions

In this section we shall study some supplementary maps and distributions. These do not appear in the trace formula, but they will be needed to relate the two families of distributions we have already described.

The function $X \mapsto \phi_M(f, \tau, X)$ does not have compact support. Our first task will be to define a different map, ${}^c\phi_M$, with the property that, for any $f \in \mathbb{H}(G(F_S))$, the function

$$X \mapsto {}^c\phi_M(f, \tau, X)$$

does have compact support. However, the latter function turns out not to be smooth in $X$. In order to describe it properly, we must first introduce some larger function spaces.

DEFINITION 4.2. (1) Suppose that $\Phi$ is a finite set of hyperplanes in a Euclidean space $\mathfrak{a}$. The complement of $\Phi$ in $\mathfrak{a}$ is a union of a finite set, $\mathcal{C}$, of open connected components. Given a pair $(c, X)$, $c \in \mathcal{C}$, $X \in \mathfrak{a}$, we set

$$\omega(c, X) = \mathrm{vol}(c \cap B_X) / \mathrm{vol}(B_X).$$

Here $B_X$ is a small ball in $\mathfrak{a}$ centered at $X$. The function $X \mapsto \omega(c, X)$ vanishes for any $X$ outside the closure, $\overline{c}$, of $c$. Notice also that $\sum_{c \in \mathcal{C}} \omega(c, X) = 1$. As a function of $X$, $\omega(c, X)$ is locally constant on the strata of $\mathfrak{a}$ defined by intersections of planes in $\Phi$.

Suppose that $\Phi'$ is a subset of $\Phi$. Then any element $c'$ in the corresponding set $\mathcal{C}'$ of components is a union of elements in $\mathcal{C}$ together with a set of measure 0. It is obvious that

$$\omega(c', X) = \sum_{\{c \in \mathcal{C}\,;\, c \subseteq c'\}} \omega(c, X). \tag{4.2.1}$$

(2) Take $\mathfrak{a}$ to be $\mathfrak{a}_G$. For a given set $\Phi$ of hyperplanes, define $\mathbb{H}^{\Phi}(G(F_S))$ to be the space of functions $f$ on $G(F_S)$ such that

$$f(x) = \sum_{c \in \mathcal{C}} \omega(c, H_G(x)) f_c(x), \quad x \in G(F_S), \tag{4.2.2}$$

where each $f_c$ belongs to $\mathbb{H}(G(F_S))$.

(3) Similarly, let $\mathbb{I}^{\Phi}(G(F_S))$ be the space of functions

$$\varphi : \Pi_{\text{temp}}(G(F_S)) \times \mathfrak{a}_G \to \mathbb{C}$$

of the form

$$\varphi(\pi, X) = \sum_{c \in \mathcal{C}} \omega(c, X) \varphi_c(\pi, X),$$

with $\varphi_c \in \mathbb{I}(G(F_S))$.

(4) As in Section 4.19, we assign topologies to these spaces. Let $\mathbb{H}_N^{\Phi}(G(F_S))_{\Gamma}$ denote the space of functions $f$ such that each $f_c$ belongs to the space

$$\mathbb{H}_N(G(F_S))_{\Gamma}$$

defined in Subsection 4.19.2. Let $S_\infty$ be the set of Archimedean valuations in $S$. For any differential operator $D$ on $G(F_{S \cap S_\infty})$, there is a seminorm

$$\|f\|_D = \sup_{c \in \mathcal{C}} \sup_{\{x \in G(F_S)\, ;\, H_G(x) \in c\}} |Df_c(x)|.$$

We take the topology on $\mathbb{H}_N^{\Phi}(G(F_S))_{\Gamma}$ defined by this collection of seminorms. Then we take $\mathbb{H}^{\Phi}(G(F_S))$ to be a topological direct limit

$$\varinjlim_{\Gamma} \varinjlim_{N} \mathbb{H}_N^{\Phi}(G(F_S))_{\Gamma}.$$

The space $\mathbb{I}^{\Phi}(G(F_S))$ is given a similar topology.

(5) The collection of all $\Phi$ is a partially ordered set. Define

$$\widetilde{\mathbb{H}}(G(F_S)) = \varinjlim_{\Phi} \mathbb{H}^{\Phi}(G(F_S)),$$

and

$$\widetilde{\mathbb{I}}(G(F_S)) = \varinjlim_{\Phi} \mathbb{I}^{\Phi}(G(F_S)).$$

If $S$ contains no Archimedean valuations, $\mathfrak{a}_{G,S}$ is just a lattice in $\mathfrak{a}_G$. In this case, the spaces $\widetilde{\mathbb{H}}(G(F_S))$ and $\widetilde{\mathbb{I}}(G(F_S))$ equal $\mathbb{H}(G(F_S))$ and $\mathbb{I}(G(F_S))$, respectively. In general, however, they are proper extensions.

(6) We of course also have spaces $\widetilde{\mathbb{H}}(L(F_S))$ and $\widetilde{\mathbb{I}}(L(F_S))$ for each $L \in \mathcal{L}(M)$. In a similar fashion, we can define extensions $\widetilde{\mathbb{H}}_{\mathrm{ac}}(L(F_S))$ and $\widetilde{\mathbb{I}}_{\mathrm{ac}}(L(F_S))$ of the spaces $\mathbb{H}_{\mathrm{ac}}(L(F_S))$ and $\mathbb{I}_{\mathrm{ac}}(L(F_S))$.

## 4.3 Properties of Maps

For $L \in \mathcal{L}(M)$, suppose that $\mathbb{H}$ is one of the spaces $\mathbb{H}^{\Phi}(L(F_S))$, $\widetilde{\mathbb{H}}(L(F_S))$, or $\widetilde{\mathbb{H}}_{\mathrm{ac}}(L(F_S))$, and $\mathbb{I}$ is the corresponding space $\mathbb{I}^{\Phi}(L(F_S))$, $\widetilde{\mathbb{I}}(L(F_S))$, or $\widetilde{\mathbb{I}}_{\mathrm{ac}}(L(F_S))$. For $f \in \mathbb{H}$, set $f_L(\tau) = f_G(\tau^G)$.

PROPOSITION. *The map $f \mapsto f_L$, where $f_L(\tau) = f_G(\tau^G)$, is a continuous, open, surjective map from $\mathbb{H}$ onto $\mathbb{I}$.*

PROOF. As in Proposition 1.8, the proposition follows from its analogue for $\mathbb{H} = \mathbb{H}(G(F_S))$ and $\mathbb{I} = \mathbb{I}(G(F_S))$. □

## 4.4 Hyperplanes

In Subsection 4.20.1, for each $\mu \in \mathfrak{a}_M^*$, we defined a map

$$f \mapsto \phi_{M,\mu}(f), \quad f \in \mathbb{H}_{\mathrm{ac}}(G(F_S)).$$

Theorem 4.20.3 established that $\phi_{M,\mu}$ maps $\mathbb{H}_{\mathrm{ac}}(G(F_S))$ continuously to $\mathbb{I}_{\mathrm{ac}}(M(F_S))$. The values of the function are defined by

$$\phi_{M,\mu}(f,\tau,X) = J_{M,\mu}(\tau,X,f), \quad \tau \in \Pi_{\mathrm{temp}}(M(F_S)),\ X \in \mathfrak{a}_{M,S}.$$

The value depends only on $f^{h_G(X)}$, so it follows that $\phi_{M,\mu}$ can be defined for any $f \in \widetilde{\mathbb{H}}_{\mathrm{ac}}(G(F_S))$. The map sends $\widetilde{\mathbb{H}}_{\mathrm{ac}}(G(F_S))$ continuously to $\widetilde{\mathbb{I}}_{\mathrm{ac}}(M(F_S))$. This applies in particular to $\phi_M$, which is the case where $\mu = 0$. It follows easily that the distributions $I_M(\tau,X)$ and $I_M(\gamma)$ can be defined on $\widetilde{\mathbb{H}}_{\mathrm{ac}}(G(F_S))$.

The most familiar set of hyperplanes in $\mathfrak{a}_M$ is the collection

$$\Phi = \{\mathfrak{a}_L\,;\ L \in \mathcal{L}(M), \dim(A_M/A_L) = 1\}.$$

The associated components are just the usual chambers $\{\mathfrak{a}_P^+\,;\ P \in \mathcal{P}(M)\}$. We shall write

$$\omega_P(X) = \omega(\mathfrak{a}_P^+, X), \quad P \in \mathcal{P}(M),$$

so that $\omega_P(0)$ is the number $\omega_P$ referred to in Section 3.

DEFINITION 4.5. (1) For each $P \in \mathcal{P}(M)$, let $\nu_P$ be a point in the associated chamber $(\mathfrak{a}_P^*)^+$ in $\mathfrak{a}_M^*$ whose distance from the walls is very large. The function $\phi_{M,\nu_P}(f, \underset{\sim}{\tau}, X)$ is then independent of $\nu_P$.

(2) For $f \in \widetilde{\mathbb{H}}_{\rm ac}(G(F_S))$, $\tau \in \Pi_{\rm temp}(M(F_S))$, and $X \in \mathfrak{a}_{M,S}$, define

$$^c\phi_M(f, \tau, X) = \sum_{P \in \mathcal{P}(M)} \omega_P(X)\phi_{M,\nu_P}(f, \tau, X).$$

## 4.6 A Continuous Map

We have already agreed that $\phi_{M,\nu_P}$ maps $\widetilde{\mathbb{H}}_{\rm ac}(G(F_S))$ continuously to $\widetilde{\mathbb{I}}_{\rm ac}(M(F_S))$. It follows easily from the definitions that $f \mapsto {}^c\phi_M(f)$ is a continuous map from $\widetilde{\mathbb{H}}_{\rm ac}(G(F_S))$ to $\widetilde{\mathbb{I}}_{\rm ac}(M(F_S))$.

The reason for introducing the distribution ${}^c\phi_M$ is that it maps functions of compact support to functions of compact support.

PROPOSITION. *The function ${}^c\phi_M$ maps $\widetilde{\mathbb{H}}(G(F_S))$ continuously to $\widetilde{\mathbb{I}}(M(F_S))$.*

PROOF. We must show that there is a positive integer $N$, depending only on the support of $f \in \widetilde{\mathbb{H}}(G(F_S))$, such that ${}^c\phi_M(f, \tau, X)$ is supported in the ball in $\mathfrak{a}_{M,S}$ of radius $N$. Looking back at the definition of ${}^c\phi_M$, we see that it is sufficient to show that, for any $P \in \mathcal{P}(M)$ and for $X$ in the closure of $\mathfrak{a}_P^+ \cap \mathfrak{a}_{M,S}$, our $\phi_{M,\nu_P}(f, \tau, X)$ is supported as a function of $X$ on the ball of radius $N$.

Consider the decomposition (4.2.2) for $f$. We can of course assume that the functions on the right-hand side of this formula are each supported on a set which depends only on the support of $f$. We may therefore assume that $f$ itself belongs to $\mathbb{H}(G(F_S))$. Then

$$\phi_{M,\nu_P}(f, \tau, X) = \int_{\nu_P + i\mathfrak{a}_{M,S}^*} e^{-\lambda(X)} J_M(\tau_\lambda, f)\, d\lambda. \tag{4.6.1}$$

We need only show that as a function of $X \in \mathfrak{a}_P^+$, (4.6.1) is supported on a ball which depends only on the support of $f$. The proof of this fact is straightforward and is similar to an argument used in the derivation of Theorem 4.20.3. Indeed, we have

$$J_M(\tau_\lambda, f) = \mathrm{tr}(\mathcal{R}_M(\tau_\lambda, P_0) i_{P_0}(\tau_\lambda, f)),$$

in the notation of Chapter 4. There is a standard estimate for the function $i_{P_0}(\tau_\lambda, f)$, see Chapter 4, equation (20.3.5). Combined with the rationality properties of $\mathcal{R}_M$ and the classical Paley-Wiener theorem, it yields the required assertion. □

The map ${}^c\phi_M$ behaves the same way under conjugation as $\phi_M$.

PROPOSITION 4.7. *If $f \in \widetilde{\mathbb{H}}_{\rm ac}(G(F_S))$ and $h \in \mathbb{H}(G(F_S)^1)$, we have*

$$ {}^{c}\phi_M(L_h f) = \sum_{Q \in \mathcal{F}(M)} {}^{c}\phi_M^{M_Q}(R_{Q,h} f). $$

PROOF. According to Chapter 4, (20.2.2), we have

$$ \phi_{M,\nu_P}(L_h f) = \sum_{Q \in \mathcal{F}(M)} \phi_{M,\nu_P}^{M_Q}(R_{Q,h} f), $$

for each $P \in \mathcal{P}(M)$. Therefore

$$ {}^{c}\phi_M(L_h f, \tau, X) = \sum_{Q \in \mathcal{F}(M)} \sum_{P \in \mathcal{P}(M)} \omega_P(X) \phi_{M,\nu_P}^{M_Q}(R_{Q,h} f, \tau, X). $$

Fix $Q \in \mathcal{F}(M)$ and set $L = M_Q$. If $P \in \mathcal{P}(M)$, the point $\nu_P$ certainly belongs to the chamber $(\mathfrak{a}_{L\cap P}^*)^+$ and it is far from the walls. In particular, $\phi_{M,\nu_P}^{M_Q}$ depends only on the set $L \cap P$ in $\mathcal{P}^L(M)$. Any $R \in \mathcal{P}^L(M)$ is of this form. By (4.2.1),

$$ \omega_R(X) = \sum_{\{P \in \mathcal{P}(M)\,;\, P \cap L = R\}} \omega_P(X). $$

The proposition follows. □

## 4.8 Third and Fourth Induction Assumptions

To make the following definition 4.9, we introduce our third and fourth induction assumptions. *We assume, for any $L \in \mathcal{L}_0(M)$, that the distributions ${}^{c}I_M^L(\gamma)$ and ${}^{c}I_M^L(\tau, X)$ are supported on characters.*

DEFINITION 4.9. By Proposition 4.7, the distributions ${}^{c}\phi_M(f)$ and $\phi_M(f)$ have the same formal behavior under commutation. We can therefore copy the construction of the distributions $I_M(\gamma)$ and $I_M(\tau, X)$, but with ${}^{c}\phi_M$ playing the role of $\phi_M$. We obtain invariant distributions

$$ \{{}^{c}I_M(\gamma)\,;\, \gamma \in M(F_S)\} $$

and

$$ \{{}^{c}I_M(\tau, X)\,;\, (\tau, X) \in \Pi(M(F_S)) \times \mathfrak{a}_{M,S}\} $$

on $\widetilde{\mathbb{H}}_{\mathrm{ac}}(G(F_S))$. For any $f \in \widetilde{\mathbb{H}}_{\mathrm{ac}}(G(F_S))$, they satisfy

$$ {}^{c}I_M(\gamma, f) = J_M(\gamma, f) - \sum_{L \in \mathcal{L}_0(M)} {}^{c}\widehat{I}_M^L(\gamma, {}^{c}\phi_L(f)) \tag{4.9.1} $$

and

$$^{c}I_M(\tau, X, f) = J_M(\tau, X, f) - \sum_{L \in \mathcal{L}_0(M)} {}^{c}\widehat{I}^L_M(\tau, X, {}^{c}\phi_L(f)), \tag{4.9.2}$$

respectively.

The significance of ${}^{c}I_M(\gamma)$ is the following proposition.

PROPOSITION 4.10. *Suppose that $T$ is a "maximal torus" of $M$ over $F_S$ (in the sense of* [Ar88.1, Section 1]). *Then for any $f \in \widetilde{\mathbb{H}}(G(F_S))$, the function*

$$\gamma \mapsto {}^{c}I_M(\gamma, f), \quad \gamma \in T(F_S),$$

*has compact support.*

PROOF. It follows from Lemma 2.1 and definition (6.5) of [Ar88.1] that the function

$$\gamma \mapsto J_M(\gamma, f), \quad \gamma \in T(F_S),$$

has compact support. Assume inductively that the proposition holds if $G$ is replaced by any $L \in \mathcal{L}_0(M)$. By Proposition 4.6, the function ${}^{c}\phi_L(f)$ belongs to $\widetilde{\mathbb{I}}(L(F_S))$. Proposition 4.3 then tells us that it is the image of a function on $\widetilde{\mathbb{H}}(L(F_S))$. Applying the induction assumption, we obtain the compact support of

$$\gamma \mapsto \widehat{I}^L_M(\gamma, \phi_L(f)), \quad \gamma \in T(F_S).$$

The proposition follows from (4.9.1). □

## 4.11 Similar Properties

The distribution ${}^{c}I_M(\tau, X)$ is to be regarded as a companion of $I_M(\tau, X)$. The two have some rather similar properties. For example, if $\tau \in \Pi(M(F_S))$, $X \in \mathfrak{a}_{M,S}$, and $f \in \widetilde{\mathbb{H}}_{\mathrm{ac}}(G(F_S))$ are fixed, the function

$${}^{c}I_{M,\mu}(\tau, X, f) = {}^{c}I_M(\tau_\mu, X, f)e^{-\mu(X)}, \quad \mu \in \mathfrak{a}^*_M, \tag{4.11.1}$$

satisfies the analogue of Proposition 3.5. It is locally constant on the complement of a finite set of hyperplanes defined by roots. It satisfies the mean value property

$${}^{c}I_{M,\mu}(\tau, X, f) = \sum_{P \in \mathcal{P}(M)} \omega_P \cdot {}^{c}I_{M,\mu+\varepsilon_P}(\tau, X, f). \tag{4.11.2}$$

Moreover, when $\tau$ is tempered and $X$ is in general position, there is an open set on which (4.11.1) vanishes. However, for ${}^cI_{M,\mu}$ the open set is an infinite chamber which depends on $X$. See the next corollary.

PROPOSITION 4.12. *Suppose that $\tau$ is tempered, $f \in \widetilde{\mathbb{H}}_{ac}(G(F_S))$, and $M \neq G$. Then*

(1)

$$I_M(\tau, X, f) = 0$$

*and*

(2)

$$\sum_{P \in \mathcal{P}(M)} \omega_P(X) \cdot {}^cI_{M,\nu_P}(\tau, X, f) = 0.$$

PROOF. Assertion (1) is just Proposition 3.4. We have included it here only for the sake of comparison.

For assertion (2), we begin by observing that

$$\begin{aligned} {}^c\phi_M(f, \tau, X) &= \sum_{P \in \mathcal{P}(M)} \omega_P(X)\phi_{M,\nu_P}(f, \tau, X) \\ &= \sum_{P \in \mathcal{P}(M)} \omega_P(X)e^{-\nu_P(X)} J_M(\tau_{\nu_P}, X, f). \end{aligned}$$

Therefore the given expression,

$$\sum_{P \in \mathcal{P}(M)} \omega_P(X) \cdot {}^cI_{M,\nu_P}(\tau, X, f),$$

is equal to the difference between ${}^c\phi_M(f, \tau, X)$ and

$$\sum_{P \in \mathcal{P}(M)} \omega_P(X) \sum_{L \in \mathcal{L}_0(M)} {}^c\widehat{I}^L_{M,\nu_P}(\tau, X, {}^c\phi_L(f)).$$

Since ${}^c\widehat{I}^L_{M,\nu_P}$ depends only on the element $R = P \cap L$ in $\mathcal{P}^L(M)$, we can argue as in the proof of Proposition 4.7. The last expression becomes

$$\sum_{L \in \mathcal{L}_0(M)} \sum_{R \in \mathcal{P}^L(M)} \omega_R(X) \cdot {}^c\widehat{I}^L_{M,\nu_R}(\tau, X, {}^c\phi_L(f)).$$

We can assume inductively that the summand corresponding to any $L \neq M$ vanishes. But the summand corresponding to $L = M$ is just equal to ${}^c\phi_M(f, \tau, X)$. It follows that the original expression vanishes. □

COROLLARY 4.13. *Suppose that $\tau$, $f$, and $M$ are as in the proposition and that $X$ belongs to a chamber $\mathfrak{a}_P^+$, $P \in \mathcal{P}(M)$. Then*

$$ {}^cI_M(\tau_{\nu_P}, X, f) = 0. $$

## 4.14 Fifth and Sixth Induction Assumptions

To make the following definition, we need additional induction assumptions, our fifth and sixth. *We assume, for any $L \in \mathcal{L}_0(M)$, that the maps $\theta_M^L$ and ${}^c\theta_M^L$ are supported on characters.*

If we try to compare $\phi_M(f)$ and ${}^c\phi_M(f)$ directly, we are lead to the following definitions.

DEFINITION 4.15. We define invariant maps

$$ \theta_M, {}^c\theta_M : \widetilde{\mathbb{H}}_{\mathrm{ac}}(G(F_S)) \to \widetilde{\mathbb{I}}_{\mathrm{ac}}(M(F_S)) $$

inductively by

$$ \theta_M(f) = {}^c\phi_M(f) - \sum_{L \in \mathcal{L}_0(M)} \hat{\theta}_M^L(\phi_L(f)) \tag{4.15.1} $$

and

$$ {}^c\theta_M(f) = \phi_M(f) - \sum_{L \in \mathcal{L}_0(M)} {}^c\hat{\theta}_M^L({}^c\phi_L(f)), \tag{4.15.2} $$

for any $f \in \widetilde{\mathbb{H}}_{\mathrm{ac}}(G(F_S))$.

PROPOSITION 4.16. *Suppose that $\tau \in \Pi_{\mathrm{temp}}(M(F_S))$ and $f \in \widetilde{\mathbb{H}}_{\mathrm{ac}}(G(F_S))$. Then*

$$ \theta_M(f, \tau, X) = \sum_{P \in \mathcal{P}(M)} \omega_P(X) \cdot I_{M,\nu_P}(\tau, X, f), \tag{4.16.1} $$

*and*

$$ {}^c\theta_M(f, \tau, X) = {}^cI_M(\tau, X, f). \tag{4.16.2} $$

PROOF. According to the definition, $\theta_M(f, \tau, X)$ equals the difference between ${}^c\phi_M(f, \tau, X)$ and $\sum_{L \in \mathcal{L}_0(M)} \hat{\theta}_M^L(\phi_L(f), \tau, X)$. By induction, we can assume that

$$ \hat{\theta}_M^L(\phi_L(f), \tau, X) = \sum_{R \in \mathcal{P}^L(M)} \omega_R(X) \cdot \widehat{I}_{M,\nu_R}^L(\tau, X, \phi_L(f)), $$

for any $L \in \mathcal{L}_0(M)$. The summand on the right is independent of $\nu_R$ as long as the point remains highly regular in $(\mathfrak{a}_R^*)^+$. It follows from (4.2.1) that

$$\hat{\theta}_M^L(\phi_L(f), \tau, X) = \sum_{P \in \mathcal{P}(M)} \omega_P(X) \cdot \widehat{I}^L_{M,\nu_P}(\tau, X, \phi_L(f)).$$

We must subtract the sum over $L \in \mathcal{L}_0(M)$ of this expression from the function

$${}^c\phi_M(f, \tau, X) = \sum_{P \in \mathcal{P}(M)} \omega_P(X) \cdot \phi_{M,\nu_P}(f, \tau, X).$$

Since

$$\phi_{M,\nu_P}(f, \tau, X) = J_{M,\nu_P}(\tau, X, f),$$

the result is just

$$\sum_{P \in \mathcal{P}(M)} \omega_P(X) \cdot I_{M,\nu_P}(\tau, X, f).$$

The equality of this expression with $\theta_M(f, \tau, X)$ is the required formula (4.16.1).

The second formula (4.16.2) follows by a similar inductive argument from (4.9.2) and (4.15.2). □

PROPOSITION 4.17. *Suppose that $f \in \widetilde{\mathbb{H}}_{ac}(G(F_S))$, $\gamma \in M(F_S)$, $\tau \in \Pi(M(F_S))$, and $X \in \mathfrak{a}_{M,S}$. Then the following formulae hold:*

$$I_M(\gamma, f) = {}^cI_M(\gamma, f) + \sum_{L \in \mathcal{L}_0(M)} {}^c\widehat{I}^L_M(\gamma, \theta_L(f)). \tag{4.17.1}$$

$${}^cI_M(\gamma, f) = I_M(\gamma, f) + \sum_{L \in \mathcal{L}_0(M)} \widehat{I}^L_M(\gamma, {}^c\theta_L(f)). \tag{4.17.2}$$

$$I_M(\tau, X, f) = {}^cI_M(\tau, X, f) + \sum_{L \in \mathcal{L}_0(M)} {}^c\widehat{I}^L_M(\tau, X, \theta_L(f)). \tag{4.17.3}$$

$${}^cI_M(\tau, X, f) = I_M(\tau, X, f) + \sum_{L \in \mathcal{L}_0(M)} \widehat{I}^L_M(\tau, X, {}^c\theta_L(f)). \tag{4.17.4}$$

$${}^c\theta_M(f) + \sum_{L \in \mathcal{L}_0(M)} {}^c\hat{\theta}^L_M(\theta_L(f)) = \theta_M(f) + \sum_{L \in \mathcal{L}_0(M)} \hat{\theta}^L_M({}^c\theta_L(f)) = \begin{cases} 1, & M = G, \\ 0, & M \neq G. \end{cases} \tag{4.17.5}$$

REMARK. We should keep in mind what will eventually be proven, namely, that the distributions and maps above are all supported on characters. Once we know that, we will be able to change the right-hand side of each formula to a single sum over $L \in \mathcal{L}(M)$.

PROOF. We assume inductively that each formula holds when $G$ is replaced by a proper Levi subgroup. The formulae for $G$ are then easily established from the definitions. We shall prove only (4.17.1).

It follows from the definitions (2.7.1) and (4.9.1) that

$$I_M(\gamma,f) - {}^cI_M(\gamma,f) = \sum_{L\in\mathcal{L}_0(M)} {}^c\widehat{I}^L_M(\gamma, {}^c\phi_L(f)) - \sum_{L_1\in\mathcal{L}_0(M)} \widehat{I}^{L_1}_M(\gamma, \phi_{L_1}(f)).$$

By (4.15.1) the first of these sums equals

$$\sum_{L\in\mathcal{L}_0(M)} {}^c\widehat{I}^L_M(\gamma, \theta_L(f)) + \sum_{L_1\in\mathcal{L}_0(M)} \sum_{L\in\mathcal{L}^{L_1}(M)} {}^c\widehat{I}^L_M(\gamma, \hat{\theta}^{L_1}_L(\phi_{L_1}(f))).$$

Applying (4.17.1) inductively to each $L_1 \in \mathcal{L}_0(M)$, we obtain

$$\sum_{L\in\mathcal{L}^{L_1}(M)} {}^c\widehat{I}^L_M(\gamma, \hat{\theta}^{L_1}_L(\phi_{L_1}(f))) = \widehat{I}^{L_1}_M(\gamma, \phi_{L_1}(f)).$$

Formula (4.17.1) then follows for $G$. □

# 5 A Contour Integral

## *5.1 Meromorphic Functions*

Formulae (4.17.1) and (4.17.2) can be taken as motivation for the introduction of the maps $\theta_M$ and ${}^c\theta_M$. The two formulae describe the asymptotic behavior of $I_M(\gamma)$. Their value lies in the fact that they consist entirely of invariant distributions. Of course it is the compact support (Proposition 4.10) of ${}^cI_M(\gamma)$ that is essential here. We point out that this property has come at the expense of properties of smoothness. The original distribution $I_M(\gamma,f)$ is not smooth in $\gamma$, but its singularities are not too bad. For example, if $F = \mathbb{R}$, (2.17.1) provides a simple formula for its jumps across singular hyperplanes. The singularities of ${}^c\widehat{I}_M(\gamma,f_G)$ are more complicated. The same sort of thing is true of ${}^c\phi_M$, ${}^cI_M(\tau,M)$, and ${}^c\theta_M$. Each of these objects has better support properties than the original one but has worse properties of smoothness.

The distributions $\{I^L_M(\tau,X)\}$ and $\{{}^cI^L_M(\tau,X)\}$ and the maps $\{\theta^L_M\}$ and $\{{}^c\theta^L_M\}$ are closely related. It turns out that all of these objects can be computed from each other. By (4.17.5), either of the two sets of maps can be computed from the other one. By Proposition 4.16, the maps can in turn be computed from either of the families of distributions. The other family of distributions could then be obtained from (4.17.3) and (4.17.4). To complete the picture, we need to establish a formula for ${}^cI_M(\tau, X)$ in terms of the map ${}^c\theta_M$. In this section we shall show how to write ${}^cI_M(\tau, X)$ as a sum of contour integrals of a certain meromorphic function. This meromorphic function is derived from ${}^c\theta_M$ in the same way that the weighted character

$$\phi_M(f, \tau_\lambda) = J_M(\tau_\lambda, f)$$

can be obtained from the map $\phi_M$. We shall review this latter construction first.

## 5.2 Test Functions

Suppose that $\tau \in \Pi_{\text{temp}}(M(F_S))$. Let $\Gamma$ be a finite subset of $\Pi(K)$. For the moment, *take $f$ to be a function in $\mathbb{H}(G(F_S))_\Gamma$*. See Definition 4.19.2(3) for the definition of $\mathbb{H}(G(F_S))_\Gamma$. The original definition of $\phi_M(f)$ was given in Definition 4.15.9. Recall that

$$\phi_M(f, \tau, X) = \int_{i\mathfrak{a}^*_{M,S}} \phi_M(f, \tau_\lambda) e^{-\lambda(X)} \, d\lambda$$

where

$$\phi_M(f, \tau_\lambda) = J_M(\tau_\lambda, f) = \operatorname{tr}(\mathcal{R}_M(\tau_\lambda, P_0) i_{P_0}(\tau_\lambda, f)), \quad \lambda \in \mathfrak{a}^*_{M,\mathbb{C}}$$

in the notation of Section 4.15. The function $\lambda \mapsto \phi_M(f, \tau_\lambda)$ is meromorphic. It has finitely many poles, which lie along hypersurfaces of the form

$$q_{v,\alpha}(\lambda) - c = 0, \quad c \in \mathbb{C},$$

where $\alpha$ is a root of $(G, A_M)$ and $v$ is a valuation of $F$. (As in Chapter 4, $q_{v,\alpha}(\lambda)$ equals $\lambda(\alpha^\vee)$ if $v$ is Archimedean and equals $q_v^{-\lambda(\alpha^\vee)}$ if $v$ is a discrete valuation with residue field of order $q_v$.) In fact, there is a finite product

$$q_{\tau,\Gamma}(\lambda) = \prod_{(v,\alpha)} (q_{v,\alpha}(\lambda) - c_{v,\alpha}), \quad c_{v,\alpha} \in \mathbb{C}, \tag{5.2.1}$$

where the set over which the pairs $(v, \alpha)$ range depends only on $\tau$ and $\Gamma$, such that the function

$$\lambda \mapsto q_{\tau,\Gamma}(\lambda) \phi_M(f, \tau_\lambda)$$

belongs to the rapidly decreasing Paley-Wiener space on $\mathfrak{a}^*_M + i\mathfrak{a}^*_{M,S}$. If $\mathfrak{a}_{M,S} = \mathfrak{a}_M$, the definition of the Paley-Wiener space is standard. Otherwise $\mathfrak{a}_{M,S}$ is a lattice and $i\mathfrak{a}_{M,S}^*$ is compact. In this case, the definition is similar, except that we impose no growth condition in the imaginary direction. More generally, $\phi_M(f, \tau_\lambda)$ is meromorphic in $\tau$. In other words, if $\mathfrak{M}$ is a Levi subgroup of $M$ over $F_S$, and

$$\tau = \sigma_\Lambda^M, \quad \sigma \in \Pi_{\mathrm{temp}}(\mathfrak{M}(F_S)), \quad \Lambda \in i\mathfrak{a}_{\mathfrak{M}}^*,$$

in the notation of Section 4.12, the resulting function of $\Lambda$ extends to a meromorphic function on $\mathfrak{a}_{\mathfrak{M},S}^*$. From the Fourier inversion formula on $\mathfrak{a}_{M,S}$, we obtain

$$\phi_M(f, \tau_\lambda) = \int_{\mathfrak{a}_{M,S}} \phi_M(f, \tau, X) e^{\lambda(X)} \, \mathrm{d}X.$$

Now, suppose that $f$ *belongs to the larger space* $\widetilde{\mathbb{H}}(G(F_S))_\Gamma$. Then $f$ has compact support, and we can still define

$$i_{P'}(\tau_\lambda, f) = \int_{G(F_S)} f(x) i_{P'}(\tau_\lambda, x) \, \mathrm{d}x, \quad P' \in \mathcal{P}(M),$$

and

$$\phi_M(f, \tau_\lambda) = J_M(\tau_\lambda, f) = \mathrm{tr}(\mathcal{R}_M(\tau_\lambda, P') i_{P'}(\tau_\lambda, f)).$$

Again, $\phi_M(f, \tau_\lambda)$ is meromorphic in $\tau$. In particular, it is a meromorphic function of $\lambda$. There is a function $q_{\tau,\Gamma}(\lambda)$ of the form above such that for $f \in \widetilde{\mathbb{H}}(G(F_S))_\Gamma$ the function

$$\lambda \mapsto q_{\tau,\Gamma}(\lambda) \phi_M(f, \tau_\lambda)$$

belongs to the slowly increasing Paley-Wiener space on $\mathfrak{a}_M^* + i\mathfrak{a}_{M,S}^*$.

PROPOSITION 5.3. (1) *The function*

$$\phi_M(f, \tau, X), \quad X \in \mathfrak{a}_{M,S},$$

*is rapidly decreasing on* $\mathfrak{a}_{M,S}$.

(2) *We have*

$$\phi_M(f, \tau_\lambda) = \int_{\mathfrak{a}_{M,S}} \phi_M(f, \tau, X) e^{\lambda(X)} \, dX. \tag{5.3.1}$$

PROOF. By the definition 4.15.10, we have

$$\phi_M(f, \tau, X) = \int_{i\mathfrak{a}_{M,S}^*/i\mathfrak{a}_{G,S}^*} \mathrm{tr}(\mathcal{R}_M(\tau_\lambda, P') i_{P_0}(\tau_\lambda, f^{h_G(X)})) e^{-\lambda(X)} \, \mathrm{d}\lambda,$$

where $h_G(X) = Z$ is the projection of $X$ onto $\mathfrak{a}_G$, and

$$i_{P'}(\tau_\lambda, f^{h_G(X)}) = \int_{G(F_S)^Z} f(x) i_{P'}(\tau_\lambda, x)\,\mathrm{d}x.$$

As in (4.2.2), we can write $f(x)$ as a finite sum

$$\sum_{c \in C} \omega(c, H_G(x)) f_c(x),$$

where each function $f_c$ belongs to $\mathbb{H}(G(F_S))$. Then

$$\phi_M(f, \tau, X) = \sum_{c \in C} \omega(c, X) \phi_M(f_c, \tau, X).$$

Since each function $\phi_M(f_c, \tau, X)$ is rapidly decreasing in $X$, the same is true of $\phi_M(f, \tau, X)$.

To prove the second assertion of the proposition, note that

$$i_{P'}(\tau_\lambda, f) = \int_{\mathfrak{a}_{G,S}} i_{P'}(\tau_\lambda, f^Z)\,\mathrm{d}Z,$$

since $f$ has compact support. Consequently,

$$\phi_M(f, \tau_\lambda) = \int_{\mathfrak{a}_{G,S}} \operatorname{tr}(\mathcal{R}_M(\tau_\lambda, P') i_{P'}(\tau_\lambda, f^Z))\,\mathrm{d}Z.$$

The required formula (5.3.1) then follows from the Fourier inversion formula on $i\mathfrak{a}^*_{M,S}/i\mathfrak{a}^*_{G,S}$. □

DEFINITION 5.4. We continue to assume that $f \in \widetilde{\mathbb{H}}(G(F_S))$. Copying the formula (5.3.1), we shall define

$${}^c\theta_M(f, \tau_\lambda) = \int_{\mathfrak{a}_{M,S}} {}^c\theta_M(f, \tau, X) e^{\lambda(X)}\,\mathrm{d}X. \tag{5.4.1}$$

For the absolute convergence of the integral, we require a proposition.

PROPOSITION 5.5. *The function*

$${}^c\theta_M(f, \tau, X), \quad X \in \mathfrak{a}_{M,S},$$

*is rapidly decreasing on* $\mathfrak{a}_{M,S}$.

PROOF. Definition (4.15.2) is

$${}^c\theta_M(f, \tau, X) = \phi_M(f, \tau, X) - \sum_{L \in \mathcal{L}_0(M)} {}^c\hat{\theta}^L_M({}^c\phi_L(f), \tau, X). \tag{5.5.1}$$

According to Proposition 4.6, each function ${}^c\phi_L(f)$ belongs to $\widetilde{\mathbb{I}}(L(F_S))$. The proposition then follows inductively from Propositions 4.3 and 5.3. □

For future reference, we record a corollary. It is proved in exactly the same way.

COROLLARY 5.6. *Suppose that $S \cap S_\infty \neq \varnothing$ (so that $\mathfrak{a}_{M,S} = \mathfrak{a}_M$) and that $D$ is an invariant differential operator on $\mathfrak{a}_M$. Then $D^c\theta_M(f, \tau, X)$ is a rapidly decreasing (piecewise smooth) function of $X \in \mathfrak{a}_M$.*

## 5.7 Paley-Wiener Space

We can now take up the study of the function (5.4.1). Suppose that $f$ belongs to $\widetilde{\mathbb{H}}(G(F_S))_\Gamma$. It follows inductively from (5.5.1) and (5.3.1) that

$$ {}^c\theta_M(f, \tau_\lambda) = \phi_M(f, \tau_\lambda) - \sum_{L \in \mathcal{L}_0(M)} {}^c\hat{\theta}_M^L({}^c\phi_L(f), \tau_\lambda). $$

This formula in turn tells us that ${}^c\theta_M(f, \tau_\lambda)$ has properties which are similar to those of $\phi_M(f, \tau_\lambda)$. In particular, ${}^c\theta_M(f, \tau_\lambda)$ is analytic in $\tau$ and therefore also in $\lambda$. Moreover, there is a function $q_{\tau,\Gamma}(\lambda)$ of the form (5.2.1) such that

$$ \lambda \mapsto q_{\tau,\Gamma}(\lambda) \cdot {}^c\theta_M(f, \tau_\lambda) $$

belongs to the slowly increasing Paley-Wiener space on $\mathfrak{a}_M^* + i\mathfrak{a}_{M,S}^*$. Observe that the functions $\phi_M(f, \tau_\lambda)$ and ${}^c\theta_M(f, \tau_\lambda)$ can be analytically continued in $\tau$. They may therefore both be defined, as meromorphic functions of $\lambda$, if $\tau$ is replaced by a standard representation $\rho \in \Sigma(M(F_S))$.

Let us now take $\tau$ to be any representation in $\Pi(M(F_S))$. Recall that the functions $m_M^L(\tau_\lambda, \rho_\lambda)$ were defined in 4.12.6(4) and were shown to be rational functions of $\{q_{v,\alpha}(\lambda)\}$, see Proposition 4.9.10. Motivated by Proposition 4.12.7, we make the following

DEFINITION. We set

$$ \phi_M(f, \tau_\lambda) = \sum_{L \in \mathcal{L}(M)} \sum_{\rho \in \Sigma(M(F_S))} m_M^L(\tau_\lambda, \rho_\lambda)\phi_L(f, \rho_\lambda) \tag{5.7.1} $$

and

$$ {}^c\theta_M(f, \tau_\lambda) = \sum_{L \in \mathcal{L}(M)} \sum_{\rho \in \Sigma(M(F_S))} m_M^L(\tau_\lambda, \rho_\lambda){}^c\theta_L(f, \rho_\lambda). \tag{5.7.2} $$

Then we have

$$ {}^c\theta_M(f, \tau_\lambda) = \phi_M(f, \tau_\lambda) - \sum_{L \in \mathcal{L}_0(M)} {}^c\hat{\theta}_M^L({}^c\phi_L(f), \tau_\lambda). \tag{5.7.3} $$

Once more, $\phi_M(f, \tau_\lambda)$ and ${}^c\theta_M(f, \tau_\lambda)$ are meromorphic in $\lambda$. Again, there is a function $q_{\tau,\Gamma}(\lambda)$ of the form (5.2.1) whose product with either of them belongs to the slowly increasing Paley-Wiener space on $\mathfrak{a}_M^* + i\mathfrak{a}_{M,S}^*$.

PROPOSITION 5.8. *Suppose that* $\tau \in \Pi(M(F_S))$ *and* $f \in \widetilde{\mathbb{H}}(G(F_S))$. *Then*

$$ {}^cI_M(\tau, X, f) = \lim_{\beta \to \delta_0} \sum_{P \in \mathcal{P}(M)} \omega_P \int_{\varepsilon_P + i\mathfrak{a}_{M,S}^*} \hat{\beta}(\lambda)\, {}^c\theta_M(f, \tau_\lambda) e^{-\lambda(X)}\, d\lambda, $$

*where* $X$ *lies in the complement of a finite set of hyperplanes, and* $\beta$ *is a test function in* $C_c^\infty(\mathfrak{a}_{M,S})$ *which approaches the Dirac measure at the origin.*

REMARK. (1) The function $\hat{\beta}$ belongs to the rapidly decreasing Paley-Wiener space on $\mathfrak{a}_M^* + i\mathfrak{a}_{M,S}^*$, so the existence of the integrals over $\varepsilon_P + i\mathfrak{a}_{M,S}^*$ follows from the remarks above.

(2) If none of the poles of ${}^c\theta_M(f, \tau_\lambda)$ meet $i\mathfrak{a}_{M,S}^*$, the right-hand side of the formula simplifies to

$$ \lim_{\beta} \int_{i\mathfrak{a}_{M,S}^*} \hat{\beta}(\lambda)\, {}^c\theta_M(f, \tau_\lambda) e^{-\lambda(X)}\, d\lambda. $$

(3) Suppose that $S$ consists of one discrete valuation. Then $\mathfrak{a}_{M,S}$ is a lattice in $\mathfrak{a}_M$, and $\beta$ may be taken to be the Dirac measure. It can be removed from the formula. The formula in this case holds for all values of $X$.

PROOF. We shall actually show

$$ \int_{\mathfrak{a}_{M,S}} \beta(Y)\, {}^cI_M(\tau, X - Y, f)\, dY = \sum_{P \in \mathcal{P}(M)} \omega_P \int_{\varepsilon_P + i\mathfrak{a}_{M,S}^*} \hat{\beta}(\lambda)\, {}^c\theta_M(f, \tau_\lambda) e^{-\lambda(X)}\, d\lambda, \tag{5.8.1} $$

for any $X \in \mathfrak{a}_{M,S}$ and $\beta \in C_c^\infty(\mathfrak{a}_{M,S})$. It follows easily from (4.9.2) and Proposition 4.6 that ${}^cI_M(\tau, X, f)$ is a piecewise smooth function of $X$, whose singularities lie along a finite set of hyperplanes. The required formula of the proposition would then hold for $X$ in the complement of these hyperplanes.

We shall first derive an analogue of (5.8.1) for $J_M(\tau, X, f)$. Since $f$ has compact support, the function

$$ J_M(\tau_\lambda, f) = \mathrm{tr}(\mathcal{R}_M(\tau_\lambda, P') i_{P'}(\tau_\lambda, f)) $$

exists. It in fact equals the function $\phi_M(f, \tau_\lambda)$ introduced above. This is just the definition if $\tau$ is tempered. The general case follows from analytic continuation, Proposition 4.12.7, and formula (5.7.1). Consequently

$$ \phi_M(f, \tau_\lambda) = \int_{\mathfrak{a}_{G,S}} J_M(\tau_\lambda, f^Z)\, dZ, $$

where

$$J_M(\tau_\lambda, f^Z) = \operatorname{tr}(\mathcal{R}_M(\tau_\lambda, P')i_{P'}(\tau_\lambda, f^Z)).$$

By definition,

$$J_M(\tau, X, f) = \sum_{P\in\mathcal{P}(M)} \omega_P \int_{\varepsilon_P + i\mathfrak{a}^*_{M,S}/i\mathfrak{a}^*_{G,S}} J_M(\tau_\lambda, f^{h_G(X)})e^{-\lambda(X)}\,\mathrm{d}\lambda.$$

Combined with the Fourier inversion formula in $i\mathfrak{a}^*_{M,S}$, these facts lead to the formula

$$\int_{\mathfrak{a}_{M,S}} \beta(Y)J_M(\tau, X-Y, f)\,\mathrm{d}Y = \sum_{P\in\mathcal{P}(M)} \omega_P \int_{\varepsilon_P + i\mathfrak{a}^*_{M,S}} \hat{\beta}(\lambda)\phi_M(f, \tau_\lambda)e^{-\lambda(X)}\,\mathrm{d}\lambda. \tag{5.8.2}$$

We now prove (5.8.1). According to (4.9.2), the left-hand side of (5.8.1) equals the difference between the left-hand side of (5.8.2) and

$$\sum_{L\in\mathcal{L}_0(M)} \int_{\mathfrak{a}_{M,S}} \beta(Y)^c\widehat{I}^L_M(\tau, X-Y, {}^c\phi_L(f))\,\mathrm{d}Y.$$

Assume inductively that (5.8.1) holds for $L$. Then the last expression can be written as

$$\sum_{L\in\mathcal{L}_0(M)} \sum_{R\in\mathcal{P}^L(M)} \omega_R \int_{\varepsilon_R + i\mathfrak{a}^*_{M,S}} \hat{\beta}(\lambda)^c\hat{\theta}^L_M({}^c\phi_L(f), \tau_\lambda)e^{-\lambda(X)}\,\mathrm{d}\lambda.$$

Since ${}^c\phi_L(f)$ belongs to $\widetilde{\mathbb{I}}(L(F_S))$, the function

$$\zeta \mapsto {}^c\hat{\theta}^L_M({}^c\phi_L(f), \tau_{\lambda+\zeta}), \quad \zeta \in \mathfrak{a}^*_{L,\mathbb{C}},$$

is entire. We can therefore translate the contour integration by any vector in $\mathfrak{a}^*_L$. The expression may consequently be written as

$$\sum_{L\in\mathcal{L}_0(M)} \sum_{P\in\mathcal{P}(M)} \omega_P \int_{\varepsilon_P + i\mathfrak{a}^*_{M,S}} \hat{\beta}(\lambda)^c\hat{\theta}^L_M({}^c\phi_L(f), \tau_\lambda)e^{-\lambda(X)}\,\mathrm{d}\lambda.$$

In particular, the sum over $L$ can be taken inside the integral over $\lambda$. Thus the left-hand side of (5.8.1) equals

$$\sum_P \omega_P \int_{\varepsilon_P + i\mathfrak{a}^*_{M,S}} \hat{\beta}(\lambda)\left(\phi_M(f, \tau_\lambda) - \sum_{L\in\mathcal{L}_0(M)} {}^c\hat{\theta}^L_M({}^c\phi_L(f), \tau_\lambda)\right) e^{-\lambda(X)}\,\mathrm{d}\lambda.$$

By (5.7.3), this is just the required right-hand side of (5.8.1). The proposition is proved. □

COROLLARY 5.9. *Let $\beta_1$ be a function in $C_c^\infty(\mathfrak{a}_{M,S})$ which is symmetric about the origin. Set*

$$\beta_\varepsilon(Y) = \varepsilon^{-\dim \mathfrak{a}_M}\beta_1(\varepsilon^{-1}Y), \quad \varepsilon > 0,\ Y \in \mathfrak{a}_{M,S}.$$

*Then*

$$^cI_M(\tau,X,f) = \lim_{\varepsilon\to 0}\sum_P \omega_P \int_{\varepsilon_P + i\mathfrak{a}^*_{M,S}} \hat{\beta}_\varepsilon(\lambda)\,^c\theta_M(f,\tau_\lambda)e^{-\lambda(X)}\,d\lambda.$$

*In particular, we can determine $^cI_M(\tau,X,f)$ from $^c\theta_M$ for all values of X.*

PROOF. It is not hard to show from our definitions that, for any $X \in \mathfrak{a}_{M,S}$, we have

$$^cI_M(\tau,X,f) = \lim_{\varepsilon\to 0}\int_{\mathfrak{a}_{M,S}} \beta_\varepsilon(Y)\,^cI_M(\tau,X-Y,f)\,dY.$$

The required formula follows for any $X$ from (5.8.2). □

# 6 Reduction of Induction Hypotheses

## *6.1 Supported on Characters*

The distributions $I_M(\gamma)$ do not have compact support in $\gamma$. This circumstance is behind the existence of the distributions $I_M(\tau,X)$. This is also the reason we defined the supplementary distributions and the maps $\theta_M$ and $^c\theta_M$. The implication is that these objects could all be computed from an adequate knowledge of the asymptotic behavior of $I_M(\gamma)$. This will be the role of the integral formula in Proposition 5.8. The formula is actually more suited to comparing distributions on different groups than to evaluating them on a single group. The same is of course true of the trace formula itself. However, we can give one illustration here of how the integral formula may be applied. We shall show that $^cI_M(\gamma)$, $I_M(\tau,X)$, $^cI_M(\tau,X)$, $\theta_M$, and $^c\theta_M$ *are all supported on characters, provided that the same is true of* $I_M(\gamma)$. In other words, we shall show that the induction hypotheses of 3.3, 4.8 and 4.14 may be subsumed in those of 2.6.

THEOREM 6.2. *Fix a Levi subgroup M and a function $f \in \widetilde{\mathbb{H}}_{ac}(G(F_S))$ such that $f_G = 0$. Assume that $I_L(\delta,f) = 0$ for each $L \in \mathcal{L}(M)$ and $\delta \in L(F_S)$. Then*

(1)

$$^cI_M(\gamma,f) = 0, \quad \gamma \in M(F_S),$$

(2)

$$\theta_M(f) = {}^c\theta_M(f) = 0,$$

*and*

(3)

$$I_M(\tau, X, f) = {}^cI_M(\tau, X, f) = 0, \quad \tau \in \Pi(M(F_S)),\ X \in \mathfrak{a}_{M,S}.$$

*In particular, the induction hypotheses of Subsections* 3.3, 4.8, *and* 4.14 *are all implied by the original induction assumption of* 2.6.

PROOF. If $M = G$, the definitions imply that

$$I_M(\tau, X, f) = {}^cI_M(\tau, X, f) = f_G(\tau, X) = 0,$$
$${}^cI_M(\gamma, f) = I_M(\gamma, f) = 0,$$

and

$$\theta_M(f) = {}^c\theta_M(f) = f_G = 0.$$

We may therefore assume that $M \neq G$. We may also take $f$ to be a function in $\mathbb{H}(G(F_S))$. For if $Z$ equals either $h_G(X)$ or $H_G(\gamma)$, the restriction of any given function in $\widetilde{\mathbb{H}}_{\mathrm{ac}}(G(F_S))$ to the set $G(F_S)^Z$ coincides with that of some function in $\mathbb{H}(G(F_S))$.

Assume inductively that the theorem has been proved if $M$ is replaced by any $L \in \mathcal{L}(M)$ with $L \neq M$. By (4.17.2) we have

$${}^cI_M(\gamma, f) = I_M(\gamma, f) + \sum_{L \in \mathcal{L}_0(M)} \widehat{I}^L_M(\gamma, {}^c\theta_L(f)), \quad \gamma \in M(F_S).$$

Our latest induction assumption then implies that ${}^c\theta_L(f) = 0$ if $L \neq M$. Combining this with the hypothesis of the theorem, we obtain

$${}^cI_M(\gamma, f) = \widehat{I}^M_M(\gamma, {}^c\theta_M(f)). \tag{6.2.1}$$

Since $f$ belongs to $\mathbb{H}(G(F_S))$, Proposition 4.10 tells us that the left-hand side has bounded support as a function of $\gamma$ in the space of $M(F_S)$-orbits in $M(F_S)$. The same is therefore true of the right-hand side. For a given $X \in \mathfrak{a}_{M,S}$, the right-hand side is the orbital integral in

$$\{\gamma \in M(F_S)\,;\, H_M(\gamma) = X\}$$

of a function defined on $M(F_S)^X$. The tempered characters of this function are just

$${}^c\theta_M(f, \tau, X), \quad \tau \in \Pi_{\mathrm{temp}}(M(F_S)).$$

Therefore this last expression is compactly supported in $X \in \mathfrak{a}_{M,S}$. It follows that

$$^{c}\theta_M(f, \tau_\lambda) = \int_{\mathfrak{a}_{M,S}} {}^{c}\theta_M(f, \tau, X) e^{\lambda(X)} \, dX$$

is an entire function of $\lambda \in \mathfrak{a}^*_{M,\mathbb{C}}$.

Take a representation $\tau \in \Pi_{\text{temp}}(M(F_S))$ and a point $\mu \in \mathfrak{a}^*_{M,S}$ in general position. Apply Proposition 5.8 to the representation $\tau_\mu$. We obtain

$$\begin{aligned} ^{c}I_M(\tau_\mu, X, f) &= \lim_{\beta \to \delta_0} \int_{i\mathfrak{a}^*_{M,S}} \hat{\beta}(\lambda) \cdot {}^{c}\theta_M(f, \tau_{\mu+\lambda}) e^{-\lambda(X)} \, d\lambda \\ &= e^{\mu(X)} \lim_{\beta} \int_{\mu + i\mathfrak{a}^*_{M,S}} \hat{\beta}(\lambda - \mu) \cdot {}^{c}\theta_M(f, \tau_\lambda) e^{-\lambda(X)} \, d\lambda . \end{aligned}$$

Remember that $\beta$ is allowed to be any function which approaches the Dirac measure at the origin. But $\lambda \mapsto \hat{\beta}(\lambda - \mu)$ is the Fourier-Laplace transform of a function $X \mapsto e^{-\mu(X)}\beta(X)$ which also approaches the Dirac measure at the origin. We may therefore replace $\hat{\beta}(\lambda - \mu)$ by $\hat{\beta}(\lambda)$. We obtain

$$e^{-\mu(X)} \cdot {}^{c}I_M(\tau_\mu, X, f) = \lim_{\beta} \int_{\mu + i\mathfrak{a}^*_{M,S}} \hat{\beta}(\lambda) \cdot {}^{c}\theta_M(f, \tau_\lambda) e^{-\lambda(X)} \, d\lambda .$$

Now, the integrand on the right is entire in $\lambda$. It follows that the integral over $\mu + i\mathfrak{a}^*_{M,S}$ can be deformed to any other translate of $i\mathfrak{a}^*_{M,S}$. The outcome is that the function

$$e^{-\mu(X)} \cdot {}^{c}I_M(\tau_\mu, X, f) = {}^{c}I_{M,\mu}(\tau, X, f)$$

is independent of $\mu$. At least this is true for almost all $\mu$ and $X$. But by formula (4.11.2), the value of this function at any $\mu$ and $X$ can be expressed in terms of its values at nearby points in general position. It follows that the function is independent of $\mu$ without exception. Deforming $\mu$ to each of the points $\nu_P$, $P \in \mathcal{P}(M)$, we obtain (see 4.4 for $\omega_P(X)$)

$$^{c}I_{M,\mu}(\tau, X, f) = \sum_{P \in \mathcal{P}(M)} \omega_P(X) \cdot {}^{c}I_{M,\nu_P}(\tau, X, f).$$

It thus follows from Proposition 4.12 that

$$^{c}I_{M,\mu}(\tau, X, f) = 0.$$

Set $\mu = 0$, and combine the last formula with that of Proposition 4.16. The result is that, for any $\tau \in \Pi_{\text{temp}}(M(F_S))$ and $X \in \mathfrak{a}_{M,S}$, we have

$$^{c}\theta_M(f, \tau, X) = {}^{c}I_M(\tau, X, f) = {}^{c}I_{M,0}(\tau, X, f) = 0.$$

Therefore the function ${}^c\theta_M(f)$ vanishes. The assertions of the theorem can now be easily proved. The required formula (1) follows immediately from (6.2.1). Formula (2) follows from (4.17.5) and the fact that the function ${}^c\theta_L(f)$, $L \in \mathcal{L}(M)$, all vanish. To establish (3), fix an arbitrary representation $\tau$ in $\Pi(M(F_S))$ and consider the function ${}^c\theta_M(f, \tau_\lambda)$. The vanishing of ${}^c\theta_M(f)$ means that the function is zero if $\tau$ is tempered. By analytic continuation from the tempered case, it follows that ${}^c\theta_M(f, \rho_\lambda) = 0$ for any standard representation $\rho \in \Sigma(M(F_S))$. A similar formula is of course valid if $M$ is replaced by any element $L \in \mathcal{L}(M)$. Consequently, the expansion (5.7.2) implies that ${}^c\theta_M(f, \tau_\lambda) = 0$ in general. Apply Proposition 5.8. The formula

$${}^cI_M(\tau, X, f) = 0, \quad X \in \mathfrak{a}_{M,S},$$

follows. But with what has already been proved, the formula (4.17.4) simplifies to

$$I_M(\tau, X, f) = {}^cI_M(\tau, X, f).$$

This gives the final assertion (3). □

# 7 A Property of $(G, M)$-Families

## *7.1 Descent and Splitting*

We would like to investigate the *descent* and *splitting properties* of our distributions. We shall establish splitting formulae in Section 9. They reduce questions about the distributions to the case that $S$ contains one valuation. The descent formulae, which we shall prove in Section 8, reduce such questions further to the case that the data which parametrize the distributions are elliptic.

In this section we shall discuss a general descent formula for $(G, M)$-families. The formula, proven in Section 11, will make the behavior of our distributions appear more transparent. In particular, it will provide a simple interpretation of the coefficients that appear in the expansions of the distributions.

## *7.2 Related Functions*

Suppose that

$$c_P(\lambda), \quad P \in \mathcal{P}(M), \quad \lambda \in i\mathfrak{a}_M^*,$$

is a $(G,M)$-family (see Definition 4.10.6). Then

$$c_M(\lambda) = \sum_{P\in\mathcal{P}(M)} c_P(\lambda)/\theta_P(\lambda), \quad \lambda \in i\mathfrak{a}_M^*,$$

is a smooth function whose value at $\lambda = 0$ we generally denote $c_M$ (see Proposition 4.10.7 and Definition 4.10.8). Recall that for each $L \in \mathcal{L}(M)$, there is an associated $(G,L)$-family

$$c_Q(\lambda), \quad Q \in \mathcal{P}(L),\ \lambda \in i\mathfrak{a}_L^*,$$

and for every $Q \in \mathcal{F}(M)$ there is an associated $(M_Q, M)$-family

$$c_R(\lambda), \quad R \in \mathcal{P}^{M_Q}(M),\ \lambda \in i\mathfrak{a}_M^*.$$

For each of these, we have the corresponding functions $c_L(\lambda)$ and $c_M^Q(\lambda)$ . See definition 4.10.12 for more details. We shall find a formula for $c_L(\lambda)$ in terms of the functions $c_M^Q(\lambda)$.

We shall actually study a family of functions derived from $\{c_P(\lambda)\}$ which is larger than the collection $\{c_L(\lambda)\,;\, L \in \mathcal{L}(M)\}$. This comes from a class of subspaces of $\mathfrak{a}_M$ which was introduced in [La76, Section 2].

DEFINITION 7.3. (1) Suppose first that $\mathfrak{b}$ is any vector subspace of $\mathfrak{a}_M$ which contains $\mathfrak{a}_G$. Then

$$\mathfrak{a}_M = \mathfrak{a}_M^{\mathfrak{b}} \oplus \mathfrak{b}^G \oplus \mathfrak{a}_G, \quad \mathfrak{a}_M^{\mathfrak{b}} = \mathfrak{b}^\perp, \ \mathfrak{b}^G = \mathfrak{b} \cap \mathfrak{a}_G^\perp,$$

where $\mathfrak{a}_M^{\mathfrak{b}}$ (resp. $\mathfrak{b}^G$) is the respective orthogonal complements of $\mathfrak{b}$ (resp. $\mathfrak{a}_G$) in $\mathfrak{a}_M$ (resp $\mathfrak{b}$).

(2) By a root $\beta$ of $\mathfrak{b}$, we mean the restriction to $\mathfrak{b}$ of a root of $(G, A_M)$.

(3) For any such $\beta$, let $\Sigma(\beta)$ be the set of roots of $(G, A_M)$ whose restriction to $\mathfrak{b}$ equals $\beta$.

(4) We say that $\mathfrak{b}$ is *special* if, for every such $\beta$, the linear function $\sum_{\alpha\in\Sigma(\beta)} m_\alpha \alpha$, in which $m_\alpha$ stands for the multiplicity of $\alpha$, vanishes on $\mathfrak{a}_M^{\mathfrak{b}}$. Assume that $\mathfrak{b}$ is special.

(5) The roots partition $\mathfrak{b}$ into a finite set of chambers. Each of these chambers defines a system of positive roots, by taking the roots positive on the chamber. We shall write $\mathcal{P}(\mathfrak{b})$ for the collection of such systems of positive roots.

(6) Write

$$\mathfrak{b}_{\mathfrak{p}}^+, \quad \mathfrak{p} \in \mathcal{P}(\mathfrak{b}),$$

for the chambers in $\mathfrak{b}$ corresponding to the system of positive roots.

(7) According to [La76, Lemma 2.2], every positive system $\mathfrak{p}$ in $\mathcal{P}(\mathfrak{b})$ has a uniquely determined subset $\Delta_{\mathfrak{p}}$ which has the usual properties of simple roots.

Namely, $\Delta_{\mathfrak{p}}$ is linearly independent, and every element in $\mathfrak{p}$ can be represented as a nonnegative integral linear combination of roots in $\Delta_{\mathfrak{p}}$. Suppose that $\mathfrak{p} \in \mathcal{P}(\mathfrak{b})$. Then there is a unique element $Q \in \mathcal{F}(M)$ such that the chamber $\mathfrak{b}_{\mathfrak{p}}^{+}$ is contained in $\mathfrak{a}_Q^{+}$. The restriction to $\mathfrak{b}$ of any root of $(Q, A_{M_Q})$ belongs to $\mathfrak{p}$. It follows easily that $\Delta_{\mathfrak{p}}$ is the restriction to $\mathfrak{b}$ of a subset of the simple roots $\Delta_Q$.

(8) Many of the constructions for the space $\mathfrak{a}_M$ can be carried over to $\mathfrak{b}$. For example, if $\mathfrak{p} \in \mathcal{P}(\mathfrak{b})$, one can define "coroots" $\Delta_{\mathfrak{p}}^{\vee} = \{\alpha^{\vee} ; \alpha \in \Delta_{\mathfrak{p}}\}$, and one can then set

$$\theta_{\mathfrak{p}}(\lambda) = (\mathrm{vol}(\mathfrak{b}^G/\mathbb{Z}(\Delta_{\mathfrak{p}}^{\vee})))^{-1} \prod_{\alpha \in \Delta_{\mathfrak{p}}} \lambda(\alpha^{\vee}), \quad \lambda \in i\mathfrak{b}^*.$$

(9) One can also introduce the notion of an $(\mathfrak{a}_G, \mathfrak{b})$-*family* of functions

$$c_{\mathfrak{p}}(\lambda), \quad \mathfrak{p} \in \mathcal{P}(\mathfrak{b}), \quad \lambda \in i\mathfrak{b}^*,$$

by copying the definitions of a $(G, M)$-family. For any such family, the number

$$c_{\mathfrak{b}} = \lim_{\lambda \to 0} c_{\mathfrak{b}}(\lambda) = \lim_{\lambda \to 0} \sum_{\mathfrak{p} \in \mathcal{P}(\mathfrak{b})} c_{\mathfrak{p}}(\lambda)/\theta_{\mathfrak{p}}(\lambda)$$

is defined. Pursuing the analogy further, we let $\mathcal{L}(\mathfrak{b})$ denote the finite collection of subspaces of $\mathfrak{b}$ of the form

$$\mathfrak{b}_1 = \{H \in \mathfrak{b} ; \beta_1(H) = \cdots = \beta_l(H) = 0\},$$

for roots $\beta_1, \ldots, \beta_l$ of $\mathfrak{b}$. Any such $\mathfrak{b}_1$ is also a special subspace of $\mathfrak{a}_M$. We write $\mathcal{F}(\mathfrak{b})$ for the set of positive systems $\mathfrak{q} \in \mathcal{P}(\mathfrak{b}_1)$, where $\mathfrak{b}_1 = \mathfrak{b}_{\mathfrak{q}}$ ranges over the spaces in $\mathcal{L}(\mathfrak{b})$. For any $(\mathfrak{a}_G, \mathfrak{b})$-family, and elements $\mathfrak{b}_1 \in \mathcal{L}(\mathfrak{b})$ and $\mathfrak{q} \in \mathcal{F}(\mathfrak{b})$, there is associated an $(\mathfrak{a}_G, \mathfrak{b}_1)$-family and a $(\mathfrak{b}_{\mathfrak{q}}, \mathfrak{b})$-family.

(10) Suppose that $\{c_P(\lambda) ; P \in \mathcal{P}(M)\}$ is a $(G, M)$-family. If $\mathfrak{b}$ is a special subspace of $\mathfrak{a}_M$, we shall write $M_{\mathfrak{b}}$ for the maximal element in $\mathcal{L}(M)$ such that $\mathfrak{a}_{M_{\mathfrak{b}}}$ contains $\mathfrak{b}$. Then $M_{\mathfrak{b}}$ is the Levi subgroup defined by the roots of $(G, A_M)$ which vanish on $\mathfrak{b}$. Consider the associated $(G, M_{\mathfrak{b}})$-family

$$\{c_Q(\nu) ; Q \in \mathcal{P}(M_{\mathfrak{b}}), \nu \in i\mathfrak{a}_{M_{\mathfrak{b}}}^*\}.$$

For any $\mathfrak{p} \in \mathcal{P}(\mathfrak{b})$, there is a unique element $Q \in \mathcal{P}(M_{\mathfrak{b}})$ such that $\mathfrak{b}_{\mathfrak{p}}^{+}$ is contained in $\mathfrak{a}_Q^{+}$. Define $c_{\mathfrak{p}}(\nu) = c_Q(\nu)$, for $\nu$ restricted to the subspace $i\mathfrak{b}^*$ of $i\mathfrak{a}_{M_{\mathfrak{b}}}^*$. Then

$$c_{\mathfrak{p}}(\nu), \quad \mathfrak{p} \in \mathcal{P}(\mathfrak{b}), \quad \nu \in i\mathfrak{b}^*,$$

is an $(\mathfrak{a}_G, \mathfrak{b})$-family.

The main result stated in this section, Proposition 7.6, will be an expansion for $c_{\mathfrak{b}}$ in terms of $\{c_M^Q\,;\,Q \in \mathcal{F}(M)\}$. The coefficients will be certain numbers

$$d_M^Q(\mathfrak{b}, L), \quad L \in \mathcal{L}(M),$$

which we define as follows.

DEFINITION 7.4. (1) For a given element $L \in \mathcal{L}(M)$, consider the natural map

$$\mathfrak{a}_M^{\mathfrak{b}} \oplus \mathfrak{a}_M^L \to \mathfrak{a}_M^G.$$

If the map is not an isomorphism, $d_M^G(\mathfrak{b}, L)$ is defined to be 0. If the map is an isomorphism, we set $d_M^G(\mathfrak{b}, L)$ equal to the volume in $\mathfrak{a}_M^G$ of the parallelogram generated by the orthonormal bases of $\mathfrak{a}_M^{\mathfrak{b}}$ and $\mathfrak{a}_M^L$. Notice that in this case, the natural map from $\mathfrak{a}_M^L$ to $\mathfrak{b}^G$ is also an isomorphism.

(2) If $\Omega$ is any bounded measurable subset of $\mathfrak{a}_M^L$ and $\widetilde{\Omega}$ is its image in $\mathfrak{b}^G$, then

$$\mathrm{vol}(\widetilde{\Omega}) = d_M^G(\mathfrak{b}, L)\,\mathrm{vol}(\Omega).$$

(3) In the special case that $\mathfrak{b} = \mathfrak{a}_{M_1}$, $M_1 \in \mathcal{L}(M)$, we shall write

$$d_M^G(M_1, L) = d_M^G(\mathfrak{b}, L).$$

## 7.5 *Dimensions*

Fix a small point $\xi$ in $\mathfrak{a}_M^{\mathfrak{b}}$. Consider an element $L \in \mathcal{L}(M)$ with

$$\dim(\mathfrak{a}_M^{\mathfrak{b}}) + \dim(\mathfrak{a}_M^L) \geq \dim(\mathfrak{a}_M^G).$$

Assume that $\xi$ is in general position in $\mathfrak{a}_M^{\mathfrak{b}}$. Then the affine space $\xi + \mathfrak{b}^G$ does not intersect $\mathfrak{a}_L^G$ unless $\mathfrak{a}_M^G = \mathfrak{a}_M^{\mathfrak{b}} \oplus \mathfrak{a}_M^L$, or equivalently, unless $d_M^G(\mathfrak{b}, L) \neq 0$. In this case, the spaces $\xi + \mathfrak{b}^G$ and $\mathfrak{a}_L^G$ intersect at one point. The point is nonsingular, and so belongs to a chamber $\mathfrak{a}_Q^+$, for a unique element $Q = Q_L$ in $\mathcal{P}(L)$. Thus $\xi$ determines a section

$$L \mapsto Q_L$$

from the set

$$\{L \in \mathcal{L}(M)\,;\, d_M^G(\mathfrak{b}, L) \neq 0\}$$

into the fibers $\mathcal{P}(L)$.

PROPOSITION 7.6. *Suppose that*

$$c_P(\lambda), \quad P \in \mathcal{P}(M), \quad \lambda \in i\mathfrak{a}_M^*,$$

*is a $(G, M)$-family. Then for any $\nu \in i\mathfrak{b}^*$, we have*

$$c_{\mathfrak{b}}(\nu) = \sum_{L \in \mathcal{L}(M)} d_M^G(\mathfrak{b}, L) c_M^{Q_L}(\nu).$$

The proof of this proposition requires a study of convex polytopes. It can be found in Section 11. In the rest of this section, we shall derive some simple consequences of the proposition.

Most of the applications of the proposition concern only the case $\nu = 0$, so we state this separately.

COROLLARY 7.7. *We have*

$$c_{\mathfrak{b}} = \sum_{L \in \mathcal{L}(M)} d_M^G(\mathfrak{b}, L) c_M^{Q_L}.$$

For certain natural $(G, M)$-families, Corollary 7.7 provides a formula which is independent of the section $L \mapsto Q_L$.

COROLLARY 7.8. *Suppose that for any $L \in \mathcal{L}(M)$, the number*

$$c_M^L = c_M^Q, \quad Q \in \mathcal{P}(L),$$

*is independent of $Q$. Then*

$$c_{\mathfrak{b}} = \sum_{L \in \mathcal{L}(M)} d_M^G(\mathfrak{b}, L) c_M^L.$$

Another special case of Corollary 7.7 applies to products of $(G, M)$-families.

COROLLARY 7.9. *We have*

$$(cd)_M = \sum_{L_1, L_2 \in \mathcal{L}(M)} d_M^G(L_1, L_2) c_M^{Q_1} d_M^{Q_2},$$

*where $(Q_1, Q_2)$ stands for the value of the section at $(L_1, L_2)$.*

PROOF. Instead of $(G, M)$, we take the pair

$$(\mathcal{G}, \mathcal{M}) = (G \times G, M \times M).$$

Then $\mathfrak{a}_{\mathcal{M}} = \mathfrak{a}_M \oplus \mathfrak{a}_M$, and $\mathcal{L}(\mathcal{M})$ consists of the set of pairs

$$\mathcal{L} = (L_1, L_2), \quad L_i \in \mathcal{L}(M).$$

Take $\mathfrak{b}$ to be the direct sum of $\mathfrak{a}_{\mathcal{G}}$ with the space $\mathfrak{a}_M$, embedded diagonally in $\mathfrak{a}_{\mathcal{M}}$. It is a special subspace. In order to apply the proposition, we must fix a small point

$$\Xi = (H, -H), \quad H \in \mathfrak{a}_M,$$

in general position in the orthogonal complement of $\mathfrak{b}$. For any pair $\mathcal{L} = (L_1, L_2)$ in $\mathcal{L}(M)$, one checks that

$$d^{\mathcal{G}}_{\mathcal{M}}(\mathfrak{b}, L) = 2^{(1/2)\dim(\mathfrak{a}^G_M)} d^G_M(L_1, L_2).$$

If this number is nonzero, we have

$$\mathfrak{a}^G_M = \mathfrak{a}^{L_1}_M \oplus \mathfrak{a}^{L_2}_M = \mathfrak{a}^G_{L_1} \oplus \mathfrak{a}^G_{L_2},$$

and we can write

$$H = \frac{1}{2}H_1 - \frac{1}{2}H_2, \quad H_1 \in \mathfrak{a}^G_{L_1},\ H_2 \in \mathfrak{a}^G_{L_2}.$$

For each $i = 1, 2$, $H_i$ is a point in general position in $\mathfrak{a}^G_{L_i}$, and belongs to a chamber $\mathfrak{a}^+_{Q_i}$, for a unique element $Q_i \in \mathcal{P}(L_i)$. Then $(L_1, L_2) \mapsto (Q_1, Q_2)$ is the section determined by the point $\Xi$.

Suppose that $\{c_P(\lambda)\}$ and $\{d_P(\lambda)\}$ are two $(G, M)$-families. Then

$$c_{\mathcal{P}}(\Lambda) = c_{P_1}(\lambda_1) d_{P_2}(\lambda_2),\ \mathcal{P} \in \mathcal{P}(\mathcal{M}),\ \Lambda \in i\mathfrak{a}^*_{\mathcal{M}},$$

is a $(\mathcal{G}, \mathcal{M})$-family, where

$$\mathcal{P} = (P_1, P_2), \quad P_1, P_2 \in \mathcal{P}(M),$$

and

$$\Lambda = (\lambda_1, \lambda_2), \quad \lambda_1, \lambda_2 \in i\mathfrak{a}^*_M.$$

Its restriction to $\mathfrak{b}$ is just

$$(cd)_P(\lambda) = c_P(\lambda) d_P(\lambda), \quad P \in \mathcal{P}(M),\ \lambda \in i\mathfrak{a}^*_M,$$

the product $(G, M)$-family. One checks that

$$\theta_{\mathcal{P}}(\lambda, \lambda) = 2^{(1/2)\dim(\mathfrak{a}^G_M)} \theta_P(\lambda),$$

if $\mathcal{P}$ corresponds to the point $(P, P)$. Corollary 7.7 becomes our corollary. □

Corollary 7.9 is reminiscent of earlier product formulae for $(G, M)$-families and in particular Proposition 4.10.14. It seems to be independent of this result, but it does imply Corollary 10.16 there which is a special case.

COROLLARY 7.10. *We have*

$$(cd)_M = \sum_{L \in \mathcal{L}(M)} d_M^L d_L.$$

Note that this is Corollary 4.10.16.

PROOF. Suppose that $\{c_P(\lambda)\}$ satisfies the condition of Corollary 7.8. The formula in Corollary 7.9 contains a sum over pairs $(L_1, L_2)$, with $L_i \in \mathcal{P}(M)$, such that $d_M^G(L_1, L_2) \neq 0$. We shall fix $L = L_1$ and use Corollary 7.8 with $\mathfrak{b} = \mathfrak{a}_{L_1}$ to interpret the remaining sums over $L_2$. Take $\xi$ to be the projection of $(-2H)$ onto the orthogonal complement of $\mathfrak{a}_{L_1}^G$ in $\mathfrak{a}_M^G$. Then

$$\xi + \mathfrak{b} = -2H + \mathfrak{a}_{L_1} = H_2 + \mathfrak{a}_{L_1}.$$

This intersects $\mathfrak{a}_{L_2}^G$ in the unique point $H_2$. But for a given $L_2$, the parabolic $Q_2$ is the unique element in $\mathcal{P}(L_2)$ such that $H_2$ belongs to $\mathfrak{a}_{Q_2}^+$. Combining Corollaries 7.8 and 7.9, we obtain our corollary. □

Here are some supplementary remarks on the Jacobians $d_M^G(\mathfrak{b}, L)$.

PROPOSITION 7.11. *We have*

$$d_M^G(\mathfrak{b}, L) = \sum_{L_1 \in \mathcal{L}(M_1)} d_M^{L_1}(M_1, L) d_{M_1}^G(\mathfrak{b}, L_1). \tag{7.11.1}$$

PROOF. Suppose that $M_1 \in \mathcal{L}(M)$ is fixed and that $\mathfrak{b}$ is a special subspace of $\mathfrak{a}_{M_1}$ (see 7.3 (4)). Suppose that $\{c_P(\lambda)\}$ is a $(G, M)$-family that satisfies the condition of Corollary 7.8. We can then apply Corollary 7.8 in two stages: first with $M_1$ as the base and then with $M$ itself. We obtain

$$\begin{aligned} c_{\mathfrak{b}} &= \sum_{L_1 \in \mathcal{L}(M_1)} d_{M_1}^G(\mathfrak{b}, L_1) c_{M_1}^{L_1} \\ &= \sum_{L_1 \in \mathcal{L}(M_1)} d_{M_1}^G(\mathfrak{b}, L_1) \sum_{L \in \mathcal{L}^{L_1}(M)} d_M^{L_1}(M_1, L) c_M^L. \end{aligned}$$

Let us agree to set $d_M^{L_1}(M_1, L) = 0$ if $L_1$ does not contain both $M_1$ and $L$. Then

$$c_{\mathfrak{b}} = \sum_{L \in \mathcal{L}(M)} \left( \sum_{L_1 \in \mathcal{L}(M_1)} d_M^{L_1}(M_1, L) d_{M_1}^G(\mathfrak{b}, L_1) \right) c_M^L.$$

On the other hand, direct application of Corollary 7.8 gives

$$c_{\mathfrak{b}} = \sum_{L \in \mathcal{L}(M)} d_M^G(\mathfrak{b}, L) c_M^L.$$

We can choose $\{c_P(\lambda)\}$ so as to compare the coefficients of these two expressions. Fix an element $L \in \mathcal{L}(M)$ with $\dim(\mathfrak{a}_M^L) = \dim(\mathfrak{b}^G)$, and set

$$c_P(\lambda) = \prod_{\alpha \in \Sigma_P^r} c_\alpha(\lambda(\alpha^\vee)), \quad \lambda \in i\mathfrak{a}_M^*, \ P \in \mathcal{P}(M),$$

where $\Sigma_P^r$ stands for the set of reduced roots of $(G, A_M)$, and

$$c_\alpha(z) = \begin{cases} e^z, \text{ if } \alpha \text{ vanishes on } \mathfrak{a}_L, \\ 1, \text{ otherwise.} \end{cases}$$

Then $\{c_P(\lambda)\}$ is a $(G,M)$-family which satisfies the condition of Corollary 7.8. It is easy to see that if $L'$ is any element in $\mathcal{L}(M)$ with $\dim(\mathfrak{a}_M^{L'}) = \dim(\mathfrak{b}^G)$, then $c_M^{L'}$ vanishes unless $L' = L$. It follows that $L$ gives the only nonvanishing summand in the two expansions for $c_{\mathfrak{b}}$. We obtain the proposition. □

There is a variant of (7.11.1) which we shall need.

PROPOSITION 7.12. *We have*

$$d_M^G(\mathfrak{b}, L) = \sum_{\mathfrak{b}_1 \in \mathcal{L}(\mathfrak{b})} d_M^{\mathfrak{b}_1}(\mathfrak{b}, M_1) d_{M_1}^G(\mathfrak{b}_1, L). \tag{7.12.1}$$

PROOF. Fix a special subspace $b \subset \mathfrak{a}_M$. Proposition 7.6 and its corollaries have versions for arbitrary $(\mathfrak{a}_G, b)$-families. They are all established by the same arguments. In particular, if $\{c_P(\lambda)\}$ and $\{d_P(\lambda)\}$ are $(G,M)$-families which both satisfy the conditions of Corollary 7.8, the remarks following Corollary 7.9 can be adapted to the resulting $(\mathfrak{a}_G, \mathfrak{b})$-families. The result is

$$(cd)_{\mathfrak{b}} = \sum_{\mathfrak{b}_1 \in \mathcal{L}(\mathfrak{b})} c_{\mathfrak{b}}^{\mathfrak{b}_1} d_{\mathfrak{b}_1}.$$

Apply Corollary 7.8 to the left-hand side of this. We obtain

$$\begin{aligned} (cd)_{\mathfrak{b}} &= \sum_{L \in \mathcal{L}(M)} d_M^G(\mathfrak{b}, L)(cd)_M^L \\ &= \sum_{L \in \mathcal{L}(M)} \sum_{M_1 \in \mathcal{L}^L(M)} d_M^G(\mathfrak{b}, L) c_M^{M_1} d_{M_1}^L. \end{aligned}$$

We can also apply Corollary 7.8 to the right-hand side. If $\mathfrak{b}_1$ is contained in $\mathfrak{a}_{M_1}$, we define $d_M^{\mathfrak{b}_1}(\mathfrak{b}, M_1)$ exactly as we defined $d_M^G(\mathfrak{b}, M_1)$, but with $\mathfrak{a}_G$ replaced by $\mathfrak{b}_1$. If $\mathfrak{b}_1$ is not contained in $\mathfrak{a}_{M_1}$, we simply set $d_M^{\mathfrak{b}_1}(\mathfrak{b}, M_1) = 0$. Note that if $G$ is replaced by $G_1 = M_{\mathfrak{b}_1}$, then $\mathfrak{b} + \mathfrak{a}_{G_1}$ becomes a special subspace of $\mathfrak{a}_M$, and one has

$$d_M^{\mathfrak{b}_1}(\mathfrak{b}, M_1) = d_M^{G_1}(\mathfrak{b} + \mathfrak{a}_{G_1}, M_1).$$

Applied in this context, Corollary 7.8 is easily converted to the formula

$$c_{\mathfrak{b}}^{\mathfrak{b}_1} = \sum_{M_1 \in \mathcal{L}(M)} d_M^{\mathfrak{b}_1}(\mathfrak{b}, M_1) c_M^{M_1}.$$

Therefore the right-hand side equals

$$\begin{aligned}\sum_{\mathfrak{b}_1 \in \mathcal{L}(\mathfrak{b})} \sum_{M_1 \in \mathcal{L}(M)} d_M^{\mathfrak{b}_1}(\mathfrak{b}, M_1) c_M^{M_1} d_{\mathfrak{b}_1} &= \sum_{\mathfrak{b}_1} \sum_{M_1} d_M^{\mathfrak{b}_1}(\mathfrak{b}, M_1) c_M^{M_1} \sum_{L \in \mathcal{L}(M_1)} d_{M_1}^G(\mathfrak{b}_1, L) d_{M_1}^L \\ &= \sum_{M_1 \in \mathcal{L}(M)} \sum_{L \in \mathcal{L}(M_1)} \left( \sum_{\mathfrak{b}_1 \in \mathcal{L}(\mathfrak{b})} d_M^{\mathfrak{b}_1}(\mathfrak{b}, M_1) d_{M_1}^G(\mathfrak{b}_1, L) \right) c_M^{M_1} d_{M_1}^L.\end{aligned}$$

Arguing as in the proof of Proposition 7.11, we choose $\{c_P(\lambda)\}$ and $\{d_P(\lambda)\}$ so as to isolate any given pair of coefficients. Equating the coefficients, one obtains the proposition. □

# 8 Descent

We want to establish descent formulae for our various distributions.

DEFINITION 8.1. (1) For any $\gamma \in M(F_S)$, recall (from Subsection 2.6) that $\gamma^{M_1}$ denotes the induced space in $M_1(F_S)$. If $\gamma$ is such that $M_{1,\gamma} = M_\gamma$, then $\gamma^{M_1}$ is just the $M_1(F_S)$-orbit of $\gamma$. In general, however, $\gamma^{M_1}$ is a finite union of $M_1(F_S)$-orbits $\{\gamma_i\}$ in $M_1(F_S)$.

We shall prove a formula for

$$I_{M_1}(\gamma^{M_1}, f) = \sum_i I_{M_1}(\gamma_i, f)$$

in terms of the distributions $\widehat{I}_M^L(\gamma, f_L)$.

(2) We shall in fact establish a more general result. Suppose that $\mathfrak{b}$ is a special subspace of $\mathfrak{a}_M$. In Subsection 7.3(10), we defined the Levi subgroup $M_{\mathfrak{b}} \in \mathcal{L}(M)$. If $\gamma_1$ belongs to $M_{\mathfrak{b}}(F_S)$, we can define the distribution $J_{\mathfrak{b}}(\gamma_1, f)$ on $\mathbb{H}_{ac}(G(F_S))$ exactly as in the special case that $\mathfrak{b} = \mathfrak{a}_M$, see [Ar88.1, (2.1),(6.5)].

We need only replace the volume $v_M(x)$ in [Ar88.1, (2.1)] by $v_{\mathfrak{b}}(x)$, the volume in $\mathfrak{b}^G$ of the convex hull of

$$\{H_Q(x)\,;\,Q \in \mathcal{P}(M_{\mathfrak{b}}),\ \mathfrak{a}_Q^+ \cap \mathfrak{b} \neq \varnothing\}.$$

(3) Similarly, copying the definition of $\phi_M$ from Section 4.15, we can introduce a map

$$\phi_{\mathfrak{b}} : \mathbb{H}_{\mathrm{ac}}(G(F_S)) \to \mathbb{I}_{\mathrm{ac}}(M_{\mathfrak{b}}(F_S)).$$

Some effort was required in [Ar88.1] and Chapter 4 to establish the existence and properties of $J_M$ and $\phi_M$. However, identical arguments suffice for arbitrary $\mathfrak{b}$; alternatively, one can deal with the general case directly from that of $M = M_{\mathfrak{b}}$ by applying Proposition 7.6. At any rate, we shall take for granted the obvious analogues of results and constructions that have hitherto been discussed only in the case that $\mathfrak{b}$ equals $\mathfrak{a}_M$.

(4) In particular, we have an invariant distribution $I_{\mathfrak{b}}(\gamma_1)$ on $\mathbb{H}_{\mathrm{ac}}(G(F_S))$. It is defined inductively by

$$I_{\mathfrak{b}}(\gamma_1,f) = J_{\mathfrak{b}}(\gamma_1,f) - \sum_{\mathfrak{b}_1 \in \mathcal{L}_0(\mathfrak{b})} \widehat{I}_{\mathfrak{b}}^{\mathfrak{b}_1}(\gamma_1, \phi_{\mathfrak{b}_1}(f)), \tag{8.1.1}$$

where

$$I_{\mathfrak{b}}^{\mathfrak{b}_1}(\gamma_1) = I_{\mathfrak{b}+\mathfrak{a}_{G_1}}^{G_1}(\gamma_1), \quad G_1 = M_{\mathfrak{b}_1}.$$

## 8.2 Seventh Induction Assumption

Included in the definition is the following induction assumption. *We assume that, for any $\mathfrak{b}_1 \in \mathcal{L}_0(\mathfrak{b})$, the distribution $I_{\mathfrak{b}}^{\mathfrak{b}_1}(\gamma_1)$ on $\mathbb{H}_{\mathrm{ac}}(G_1(F_S))$ is supported on characters.*

The next theorem will provide a formula which resolves this new induction hypothesis in terms of assumption 2.6.

The space $\mathfrak{b}$ is always contained in $\mathfrak{a}_{M_{\mathfrak{b}}}$. If the two spaces are the same, then $I_{\mathfrak{b}}(\gamma_1,f)$ is just equal to $I_{M_{\mathfrak{b}}}(\gamma_1,f)$. However, this need not always be so. For example, $M_{\mathfrak{b}}$ could be defined over a subfield $F_1$ of $F$, and $\mathfrak{b}$ could be the split component of $M_{\mathfrak{b}}$ over $F_1$. This might well be a proper subspace of the split component of $\mathfrak{a}_{M_{\mathfrak{b}}}$ of $M_{\mathfrak{b}}$ over $F$, in which case $I_{\mathfrak{b}}(\gamma_1,f)$ would not be equal to $I_{M_{\mathfrak{b}}}(\gamma_1,f)$.

If $\gamma$ belongs to $M(F_S)$, write $\gamma^{\mathfrak{b}} = \gamma^{M_{\mathfrak{b}}}$ for the induced class in $M_{\mathfrak{b}}(F_S)$, and set

$$I_{\mathfrak{b}}(\gamma^{\mathfrak{b}},f) = I_{\mathfrak{b}}(\gamma^{M_{\mathfrak{b}}},f).$$

THEOREM 8.3. *For $\gamma \in M(F_S)$ and $f \in \mathbb{H}_{\mathrm{ac}}(G(F_S))$, we have*

$$I_{\mathfrak{b}}(\gamma^{\mathfrak{b}},f) = \sum_{L\in\mathcal{L}(M)} d_M^G(\mathfrak{b},L)\widehat{I}_M^L(\gamma,f_L).$$

PROOF. Both sides depend only on the values of $f$ on

$$\{x \in G(F_S)\,;\, H_G(x) = H_G(\gamma)\}.$$

Since the restriction of $f$ to this subset coincides with that of some function in $\mathbb{H}(G(F_S))$, we can assume that $f$ itself belongs to $\mathbb{H}(G(F_S))$. We shall also assume for the moment that $\gamma \in M(F_S)$ is such that $M_\gamma = G_\gamma$. Then $\gamma^{\mathfrak{b}}$ equals $\gamma$ and $J_{\mathfrak{b}}(\gamma,f)$ equals (see Definition 8.1(2) for $v_{\mathfrak{b}}(x)$)

$$|D^G(\gamma)|^{1/2} \int_{G_\gamma(F_S)\backslash G(F_S)} f(x^{-1}\gamma x)v_{\mathfrak{b}}(x)\,\mathrm{d}x.$$

Applying Corollary 7.7 to the $(G,M)$-family

$$\{v_P(x) = e^{-\lambda(H_P(x))}\,;\, P \in \mathcal{P}(M)\},$$

we write

$$v_{\mathfrak{b}}(x) = \sum_{L\in\mathcal{L}(M)} d_M^G(\mathfrak{b},L)v_M^{L_Q}(x).$$

This allows us to make a standard change of variables in the integral over $G_\gamma(F_S)\backslash G(F_S)$, see [Ar88.1, (8.11)]. We find that $J_{\mathfrak{b}}(\gamma,f)$ equals

$$\sum_{L\in\mathcal{L}(M)} d_M^G(\mathfrak{b},L)J_M^L(\gamma,f_{Q_L}). \tag{8.3.1}$$

Our distribution $I_{\mathfrak{b}}(\gamma,f)$ equals the difference between (8.3.1) and the expression

$$\sum_{\mathfrak{b}_1\in\mathcal{L}_0(\mathfrak{b})} \widehat{I}_{\mathfrak{b}}^{\mathfrak{b}_1}(\gamma,\phi_{\mathfrak{b}_1}(f)). \tag{8.3.2}$$

We can assume inductively that the theorem holds for each of the distributions $I_{\mathfrak{b}}^{\mathfrak{b}_1}(\gamma)$. Then (8.3.2) may be written

$$\sum_{\mathfrak{b}_1\in\mathcal{L}_0(\mathfrak{b})}\ \sum_{M_1\in\mathcal{L}^{\mathfrak{b}_1}(M)} d_M^{\mathfrak{b}_1}(\mathfrak{b},M_1)\widehat{I}_M^{M_1}(\gamma,\phi_{\mathfrak{b}_1}(f)_{M_1}).$$

Now, $\phi_{\mathfrak{b}_1}(f)_{M_1}$ is a function in $\mathbb{I}_{\mathrm{ac}}(M_1(F_S))$. Its value at any representation $\tau_1 \in \Pi_{\mathrm{temp}}(M_1(F_S))$ equals

$$\mathrm{tr}(\mathcal{R}_{\mathfrak{b}_1}(\tau_1^{\mathfrak{b}_1}, Q_0) i_{Q_0}(\tau_1^{\mathfrak{b}_1}, f)).$$

Here $Q_0$ is a fixed element in $\mathcal{P}(G_1)$, where $G_1 = M_{\mathfrak{b}_1}$, and $\tau_1^{\mathfrak{b}_1}$ stands for the induced representation $\tau_1^{G_1}$; the operator $\mathcal{R}_{\mathfrak{b}_1}(\tau_1^{\mathfrak{b}_1}, Q_0)$ is obtained from the restriction to $\mathfrak{b}_1$ of the $(G, G_1)$-family

$$\mathcal{R}_Q(\nu, \tau_1^{\mathfrak{b}_1}, Q_0), \quad Q \in \mathcal{P}(G_1),\ \nu \in i\mathfrak{a}_{G_1}^*,$$

described in Section 4.12. It follows easily from Corollary 7.7 that

$$\phi_{\mathfrak{b}_1}(f)_{M_1} = \sum_{L \in \mathcal{L}(M_1)} d_{M_1}^G(\mathfrak{b}_1, L)\phi_{M_1}^L(f_{Q_L}).$$

See also Chapter 4, (11.8.1). Therefore (8.3.2) equals

$$\sum_{M_1 \in \mathcal{L}(M)} \sum_{L \in \mathcal{L}(M_1)} \sum_{\mathfrak{b}_1 \in \mathcal{L}_0(\mathfrak{b})} d_M^{\mathfrak{b}_1}(\mathfrak{b}, M) d_{M_1}^G(\mathfrak{b}_1, L)\widehat{\mathcal{L}}_M^{M_1}(\gamma, \phi_{M_1}^L(f_{Q_L})). \tag{8.3.3}$$

The section $L \mapsto Q_L$ is defined in (8.3.3) with respect to some point $\xi_1 \in \mathfrak{a}_{M_1}^{\mathfrak{b}_1}$ in general position, while in (8.3.1) it is defined with respect to a point $\xi \in \mathfrak{a}_M^{\mathfrak{b}}$. However, it turns out that the notation is consistent. For we need only consider elements $M_1$ such that $d_M^{\mathfrak{b}_1}(\mathfrak{b}, M) \neq 0$. This means that

$$\mathfrak{a}_M^{\mathfrak{b}_1} = \mathfrak{a}_{M_1}^{\mathfrak{b}_1} \oplus \mathfrak{b}^{\mathfrak{b}_1}.$$

Hence there is a natural isomorphism

$$\mathfrak{a}_M^{\mathfrak{b}} \simeq \mathfrak{a}_M/\mathfrak{b} \xrightarrow{\simeq} \mathfrak{a}_{M_1}/\mathfrak{b}_1 \simeq \mathfrak{a}_{M_1}^{\mathfrak{b}_1}.$$

We take $\xi_1$ to be the image of $\xi$. Then if $L$ is any element in $\mathcal{L}(M_1)$ with $d_{M_1}^G(\mathfrak{b}_1, L) \neq 0$, we have

$$\mathfrak{a}_{M_1}^G = \mathfrak{a}_L^G \oplus \mathfrak{b}_1^G,$$

and $\xi_1 + \mathfrak{b}_1$ and $\xi + \mathfrak{b}$ both intersect $\mathfrak{a}_L^G$ at the same point. Consequently for any given $L$, the parabolic $Q_L$ in (8.3.3) is the same as that in (8.3.1). In particular, $Q_L$ is independent of $\mathfrak{b}_1$. Thus the only part of the expression (8.3.3) which depends on $\mathfrak{b}_1$ is the sum

$$\sum_{\mathfrak{b}_1 \in \mathcal{L}_0(\mathfrak{b})} d_M^{\mathfrak{b}_1}(\mathfrak{b}, M_1) d_{M_1}^G(\mathfrak{b}_1, L).$$

This can be simplified. If $L \neq M_1$, we can replace $\mathcal{L}_0(\mathfrak{b})$ by $\mathcal{L}(\mathfrak{b})$, for the term corresponding to $\mathfrak{b}_1 = \mathfrak{a}_G$ vanishes. By (7.12.1), the sum is equal to $d_M^G(\mathfrak{b}, L)$. If

$L = M_1$, then

$$d^G_{M_1}(\mathfrak{b}_1, L) = d^G_{M_1}(\mathfrak{b}_1, M_1) = 0,$$

since $\mathfrak{b}_1 \neq \mathfrak{a}_G$, so in this case the summands are all zero. It follows that (8.3.3), which on the one hand equals the original expression (8.3.2), also equals

$$\sum_{L\in\mathcal{L}(M)} \sum_{\{M_1\in\mathcal{L}^L(M)\,;\,M_1\neq L\}} d^G_M(\mathfrak{b}, L)\widehat{I}^{M_1}_M(\gamma, \phi^L_{M_1}(f_{Q_L})).$$

This is easily combined with (8.3.1). From the inductive definition of $I^L_M(\gamma)$, we see that the difference between (8.3.1) and (8.3.2) equals

$$\sum_{L\in\mathcal{L}(M)} d^G_M(\mathfrak{b}, L)I^L_M(\gamma, f_{Q_L}).$$

Since $(f_{Q_L})_L$ equals $f_L$, this becomes

$$\sum_{L\in\mathcal{L}(M)} d^G_M(\mathfrak{b}, L)\widehat{I}^L_M(\gamma, f_L),$$

the required formula for $I_{\mathfrak{b}}(\gamma, f)$.

Now, suppose that $\gamma$ is an arbitrary element in $M(F_S)$. As in (2.11.1), we can write

$$I_{\mathfrak{b}}(\gamma^{\mathfrak{b}}, f) = \lim_{a\to e} \sum_{\mathfrak{b}_1\in\mathcal{L}(\mathfrak{b})} m^{\mathfrak{b}_1}_{\mathfrak{b}}(\gamma, a)I_{\mathfrak{b}_1}(a\gamma, f),$$

where $a$ approaches 1 through the regular points in $A_M(F_S)$. The theorem will be established by arguing as in the derivation of (7.12.1). For the function $m^{\mathfrak{b}_1}_{\mathfrak{b}}(\gamma, a)$ comes from a $(G, M)$-family

$$m_P(\lambda, \gamma, a), \quad P \in \mathcal{P}(M), \ \lambda \in i\mathfrak{a}^*_M,$$

which satisfies the condition of Corollary 7.8. See [Ar88.1, Lemma 5.1]. Moreover, we are assuming that $a \in A_M(F_S)$ is regular, so that $M_{a\gamma} = G_{a\gamma}$. Applying Corollary 7.8 and what we have just proved, we obtain

$$\begin{aligned}\sum_{\mathfrak{b}_1\in\mathcal{L}(\mathfrak{b})} m^{\mathfrak{b}_1}_{\mathfrak{b}}(\gamma, a)I_{\mathfrak{b}_1}(a\gamma, f) &= \sum_{\mathfrak{b}_1\in\mathcal{L}(\mathfrak{b})} \sum_{M_1\in\mathcal{L}(M)} d^{\mathfrak{b}_1}_M(\mathfrak{b}, M_1)m^{M_1}_M(\gamma, a)I_{\mathfrak{b}_1}(a\gamma, f)\\ &= \sum_{\mathfrak{b}_1}\sum_{M_1} d^{\mathfrak{b}_1}_M(\mathfrak{b}, M_1)m^{M_1}_M(\gamma, a) \sum_{L\in\mathcal{L}(M_1)} d^G_{M_1}(\mathfrak{b}_1, L)\widehat{I}^L_{M_1}(a\gamma, f_L)\\ &= \sum_{L\in\mathcal{L}(M)} d^G_M(\mathfrak{b}, L) \sum_{M_1\in\mathcal{L}^L(M)} m^{M_1}_M(\gamma, a)\widehat{I}^L_{M_1}(a\gamma, f_L).\end{aligned}$$

The last step follows from (7.12.1). But

$$\lim_{a\to e} \sum_{M_1\in\mathcal{L}^L(M)} m_M^{M_1}(\gamma, a)\widehat{I}^L_{M_1}(a\gamma, f_L) = \widehat{I}^L_M(\gamma, f_L),$$

by (2.9.1). Taking the limits in $a$ thus gives us

$$I_{\mathfrak{b}}(\gamma^{\mathfrak{b}}, f) = \sum_{L\in\mathcal{L}(M)} d_M^G(\mathfrak{b}, L)\widehat{I}^L_M(\gamma, f_L).$$

This completes the proof. □

We are of course interested in the special case that $\mathfrak{b} = \mathfrak{a}_{M_1}$, for some element $M_1 \in \mathcal{L}(M)$.

COROLLARY 8.4. *Given $\gamma \in M(F_S)$ and $f \in \mathbb{H}_{\mathrm{ac}}(G(F_S))$, we have*

$$I_{M_1}(\gamma^{M_1}, f) = \sum_{L\in\mathcal{L}(M)} d_M^G(M_1, L)\widehat{I}^L_M(\gamma, f_L).$$

COROLLARY 8.5. *Suppose that $\gamma \in M(F_S)$ is such that $M_\gamma = M_{1,\gamma}$. Then*

$$I_{M_1}(\gamma, f) = \sum_{L\in\mathcal{L}(M)} d_M^G(M_1, L)\widehat{I}^L_M(\gamma, f_L).$$

## 8.6 Descent Property

There is a similar descent property of $I_M(\tau, X, f)$. Once again, it is important to work in a slightly broader context. Suppose again that $\mathfrak{b}$ is a special subspace of $\mathfrak{a}_M$. If $\tau_1 \in \Pi(M_{\mathfrak{b}}(F_S))$ and $X_1 \in \mathfrak{a}_{M_1,S}$, we can define the distributions $J_{\mathfrak{b}}(\tau_1, X_1, f)$ on $\mathbb{H}_{\mathrm{ac}}(G(F_S))$ exactly the same way as in the special case that $\mathfrak{b} = \mathfrak{a}_M$. (See Sections 4.12, 14, 15.)

We can also define an invariant distribution $I_{\mathfrak{b}}(\tau_1, X_1)$ on $\mathbb{H}_{\mathrm{ac}}(G(F_S))$ inductively by setting

$$I_{\mathfrak{b}}(\tau_1, X_1, f) = J_{\mathfrak{b}}(\tau_1, X_1, f) - \sum_{\mathfrak{b}_1\in\mathcal{L}_0(\mathfrak{b})} \widehat{I}^{\mathfrak{b}_1}_{\mathfrak{b}}(\tau_1, X_1, \phi_{\mathfrak{b}_1}(f)). \quad (8.6.1)$$

## 8.7 Eighth Induction Assumption

Included in the definition is another induction assumption. Namely, *we assume that $I^{\mathfrak{b}_1}_{\mathfrak{b}}(\tau_1, X_1)$ is supported on characters*. This will be resolved in terms of our original induction hypothesis 2.6 by the next theorem, together with Theorem 6.2.

Suppose that $\tau \in \Pi(M(F_S))$ and $X \in \mathfrak{a}_{M,S}$. We shall write

$$J_{\mathfrak{b}}(\tau, X, f) = \int_{i\mathfrak{a}^*_{M,S}/i\mathfrak{b}^*_S} J_{\mathfrak{b}}(\tau^{\mathfrak{b}}_\lambda, h_{\mathfrak{b}}(X), f) e^{-\lambda(X)}\, d\lambda,$$

for any $f \in \mathbb{H}_{ac}(G(F_S))$. Here $h_{\mathfrak{b}}(X)$ is the projection of $X$ onto $\mathfrak{b}$. We often write $\tau^{\mathfrak{b}}_\lambda$ for the induced representation $\tau^{\mathfrak{b}}_\lambda = (\tau_\lambda)^{M_{\mathfrak{b}}}$.) The integral clearly depends only on the restriction of $f$ to $G(F_S)^Z$, $Z = h_G(X)$. Since this is compactly supported, we can always replace $f$ itself by a compactly supported function. It follows from standard estimates (Chapter 4, (20.3.5)) that the integral over $\lambda$ is absolutely convergent. Define an invariant distribution $I_{\mathfrak{b}}(\tau, X)$ on $\mathbb{H}_{ac}(G(F_S))$ inductively by

$$I_{\mathfrak{b}}(\tau, X, f) = J_{\mathfrak{b}}(\tau, X, f) - \sum_{\mathfrak{b}_1 \in \mathcal{L}_0(\mathfrak{b})} \widehat{I}^{\mathfrak{b}_1}_{\mathfrak{b}}(\tau, X, \phi_{\mathfrak{b}_1}(f)).$$

It then follows that

$$I_{\mathfrak{b}}(\tau, X, f) = \int_{i\mathfrak{a}^*_{M,S}/i\mathfrak{b}^*_S} I_{\mathfrak{b}}(\tau_\lambda, h_{\mathfrak{b}}(X), f) e^{-\lambda(X)}\, d\lambda,$$

with the integral converging absolutely.

THEOREM 8.8. *Given $\tau \in \Pi(M(F_S))$, $X \in \mathfrak{a}_{M,S}$, and $f \in \mathbb{H}_{ac}(G(F_S))$, we have*

$$I_{\mathfrak{b}}(\tau, X, f) = \sum_{L \in \mathcal{L}(M)} d^G_M(\mathfrak{b}, L) \widehat{I}^L_M(\tau, X, f_L).$$

PROOF. As above, we can assume that $f$ actually belongs to $\mathbb{H}(G(F_S))$. It also happens that we can restrict $\tau$, as we proceed to explain. For as in Proposition 3.5(2), we have

$$I_{\mathfrak{b}}(\tau, X, f) = \sum_{\mathfrak{p} \in \mathcal{P}(\mathfrak{b})} \omega_{\mathfrak{p}} I_{\mathfrak{b}}(\tau_{\varepsilon_{\mathfrak{p}}}, X, f) e^{-\varepsilon_{\mathfrak{p}}(X)},$$

where for each $\mathfrak{p}$, $\omega_{\mathfrak{p}} = \omega(\mathfrak{b}^+_{\mathfrak{p}}, 0)$, and $\varepsilon_{\mathfrak{p}}$ denotes a small regular point in the dual chamber $(\mathfrak{b}^*)^+_{\mathfrak{p}}$. Suppose that $L \in \mathcal{L}(M)$ is such that $d^G_M(\mathfrak{b}, L) \neq 0$. Then the canonical map

$$\mathfrak{b}^*/\mathfrak{a}^*_G \to \mathfrak{a}^*_M/\mathfrak{a}^*_L$$

is an isomorphism. Each chamber in the second space is, up to a set of measure 0, a union of images of chambers $(\mathfrak{b}^*)^+_{\mathfrak{p}}$. Moreover, for any small regular point $\varepsilon$ in $\mathfrak{a}^*_M$, the number

$$I_{M,\varepsilon}(\tau, X, f) = I_M(\tau_\varepsilon, X, f) e^{-\varepsilon(X)}$$

depends only on the chamber in $\mathfrak{a}^*_M$ that contains $\varepsilon$. Consequently

$$\sum_{\mathfrak{p}\in\mathcal{P}(\mathfrak{b})} \omega_{\mathfrak{p}} \widehat{I}_M^L(\tau_{\varepsilon_{\mathfrak{p}}}, X, f_L) e^{-\varepsilon_{\mathfrak{p}}(X)} = \sum_{R\in\mathcal{P}^L(M)} \omega_R \widehat{I}_{M,\varepsilon_R}^L(\tau, X, f_L)$$

$$= \widehat{I}_M^L(\tau, X, f_L).$$

It follows that if the theorem holds with $\tau$ replaced by $\tau_{\varepsilon_{\mathfrak{p}}}$, it then holds for $\tau$ itself. We may therefore assume that $\tau$ is in general position, as a point in some $\mathfrak{a}_M^*$-orbit in $\Pi(M(F_S))$.

The general position of $\tau$ implies that the function

$$J_{\mathfrak{b}}(\tau_\lambda, f) = \operatorname{tr}(\mathcal{R}_{\mathfrak{b}}(\tau_\lambda^{\mathfrak{b}}, Q_0) i_{Q_0}(\tau_\lambda^{\mathfrak{b}}, f))$$

is analytic for $\lambda \in i\mathfrak{a}_M^*$. Recall that $Q_0$ is a fixed element in $\mathcal{P}(M_{\mathfrak{b}})$, and $\mathcal{R}_{\mathfrak{b}}(\tau_\lambda^{\mathfrak{b}}, Q_0)$ is obtained from the restriction to $\mathfrak{b}$ of a $(G, M_{\mathfrak{b}})$-family

$$\mathcal{R}_Q(\nu, \tau_\lambda^{\mathfrak{b}}, Q_0), \quad Q \in \mathcal{P}(M_{\mathfrak{b}}), \ \nu \in i\mathfrak{a}_{M_{\mathfrak{b}}}^*.$$

As in Chapter 4, (11.8.1), we have

$$\operatorname{tr}(\mathcal{R}_{\mathfrak{b}}(\tau_\lambda^{\mathfrak{b}}, Q_0) i_{Q_0}(\tau_\lambda^{\mathfrak{b}}, f)) = \operatorname{tr}(\mathcal{R}_{\mathfrak{b}}(\tau_\lambda, P_0) i_{P_0}(\tau_\lambda, f))$$

for any fixed element $P_0 \in \mathcal{P}(M)$. It follows that

$$J_{\mathfrak{b}}(\tau, X, f) = \int_{i\mathfrak{a}_{M,S}^*/i\mathfrak{b}_S^*} \left( \int_{i\mathfrak{b}_S^*} J_{\mathfrak{b}}(\tau_{\lambda+\mu}, f) e^{-\mu(h_{\mathfrak{b}}(X))} \, d\mu \right) e^{-\lambda(X)} \, d\lambda$$

$$= \int_{i\mathfrak{a}_{M,S}^*} \operatorname{tr}(\mathcal{R}_{\mathfrak{b}}(\tau_\lambda, P_0) i_{P_0}(\tau_\lambda, f)) e^{-\lambda(X)} \, d\lambda.$$

If we apply Corollary 7.7 to the $(G, M)$-family

$$\mathcal{R}_P(\nu, \tau_\lambda, P_0), \quad P \in \mathcal{P}(M), \ \nu \in i\mathfrak{a}_M^*,$$

we find that $J_{\mathfrak{b}}(\tau, X, f)$ equals

$$\sum_{L\in\mathcal{L}(M)} d_M^G(\mathfrak{b}, L) \int_{i\mathfrak{a}_{M,S}^*} \operatorname{tr}(\mathcal{R}_M^{Q_L}(\tau_\lambda, P_0) i_{P_0}(\tau_\lambda, f)) e^{-\lambda(X)} \, d\lambda.$$

The argument used to prove Proposition 4.11.9 then allows us to write this last expression as

$$\sum_{L\in\mathcal{L}(M)} d_M^G(\mathfrak{b}, L) \int_{i\mathfrak{a}_{M,S}^*} J_M^L(\tau_\lambda, f_{Q_L}) e^{-\lambda(X)} \, d\lambda.$$

It follows that $J_{\mathfrak{b}}(\tau, X, f)$ equals

$$\sum_{L \in \mathcal{L}(M)} d_M^G(\mathfrak{b}, L) J_M^L(\tau, X, f_{Q_L}). \tag{8.8.1}$$

Our distribution $I_{\mathfrak{b}}(\tau, X, f)$ equals the difference between (8.8.1) and the expression

$$\sum_{\mathfrak{b}_1 \in \mathcal{L}_0(\mathfrak{b})} \widehat{I}_{\mathfrak{b}}^{\mathfrak{b}_1}(\tau, X, \phi_{\mathfrak{b}_1}(f)). \tag{8.8.2}$$

The proof is now identical to that of Theorem 8.3. Assuming inductively that Theorem 8.8 holds for the distribution $I_{\mathfrak{b}}^{\mathfrak{b}_1}(\tau, X)$, we are lead to an expansion of (8.8.2) into

$$\sum_{L \in \mathcal{L}(M)} \sum_{\{M_1 \in \mathcal{L}^L(M)\,;\, M_1 \neq L\}} d_M^G(\mathfrak{b}, L) \widehat{I}_M^{M_1}(\tau, X, \phi_{M_1}^L(f_{Q_L})).$$

It follows that the difference between (8.8.1) and (8.8.2) equals

$$\sum_{L \in \mathcal{L}(M)} d_M^G(\mathfrak{b}, L) \widehat{I}_M^L(\tau, X, f_L),$$

the required formula for $I_{\mathfrak{b}}(\tau, X, f)$. □

Consider the special case that $\mathfrak{b} = \mathfrak{a}_{M_1}$ for some element $M_1 \in \mathcal{L}(M)$. Then the distribution $I_{M_1}(\tau, X, f) = I_{\mathfrak{b}}(\tau, X, f)$ equals

$$\int_{i\mathfrak{a}_{M,S}^*/i\mathfrak{a}_{M_1,S}^*} I_{M_1}(\tau_\lambda, h_{M_1}(X), f) e^{-\lambda(X)}\, d\lambda,$$

an absolutely convergent integral.

COROLLARY 8.9. *Given $\tau \in \Pi(M(F_S))$, $X \in \mathfrak{a}_{M,S}$, and $f \in \mathbb{H}_{ac}(G(F_S))$, we have*

$$I_{M_1}(\tau, X, f) = \sum_{L \in \mathcal{L}(M)} d_M^G(M_1, L) \widehat{I}_M^L(\tau, X, f_L).$$

## 8.10 Proper Subspace

Suppose that $\mathfrak{b}$ is a *proper* subspace of $\mathfrak{a}_M$. Then $d_M^G(\mathfrak{b}, L)$ is nonzero only when $L \neq G$, in which case the distributions $\widehat{I}_M^L$ are all well defined. Strictly speaking, the two theorems are only valid for such $\mathfrak{b}$. However, until we complete the induction in the next sections, it will be understood that $\widehat{I}(f_G)$ really means $I(f)$, for any given invariant linear form $I$ on $\mathbb{H}_{\mathrm{ac}}(G(F_S))$. With this temporary abuse of notation, the theorem and corollaries of this section are all valid as stated.

## 9 Splitting

The splitting properties are essentially special cases of Theorem 8.3 and 8.8. However, they are important enough to discuss separately on their own. To state them, we take $S$ to be the disjoint union of two sets $S_1$ and $S_2$. We assume that each of $S$, $S_1$, and $S_2$ has the closure property.

PROPOSITION 9.1. *Suppose that*

$$\gamma = \gamma_1\gamma_2, \quad \gamma_i \in M(F_{S_i}),$$

*is an element in $M(F_S)$. Then for any function $f \in \mathbb{H}_{\mathrm{ac}}(G(F_S))$ of the form*

$$f = f_1 f_2, \quad f_i \in \mathbb{H}_{\mathrm{ac}}(G(F_{S_i})),$$

*we have*

$$I_M(\gamma, f) = \sum_{L_1, L_2 \in \mathcal{L}(M)} d_M^G(L_1, L_2) \widehat{I}_M^{L_1}(\gamma_1, f_{1,L_1}) \widehat{I}_M^{L_2}(\gamma_2, f_{2,L_2}).$$

PROOF. This is essentially a special case of Theorem 8.3. We say essentially because we must in fact replace $(G, M)$ by the pair $(\mathcal{G}, \mathcal{M}) = (G \times G, M \times M)$, in which the products are regarded as varieties over the ring $F \times F$. However, the definitions of Section 8 extend in a straightforward way to this setting. We take $\mathfrak{b}$ to be the space $\mathfrak{a}_M$, embedded diagonally in $\mathfrak{a}_{\mathcal{M}} = \mathfrak{a}_M \oplus \mathfrak{a}_M$. Notice that

$$G(F_S) = (G \times G)(F_{S_1} \times F_{S_2}) = \mathcal{G}((F \times F)_{S_1 \times S_2}).$$

It follows that

$$I_{\mathfrak{b}}(\gamma, f) = 2^{(1/2)\dim(\mathfrak{a}_M^G)} I_M(\gamma_1\gamma_2, f_1 f_2).$$

Obviously $M_{\mathfrak{b}} = M \times M = \mathcal{M}$, so that $\gamma_1\gamma_2 = (\gamma_1\gamma_2)^{\mathfrak{b}}$. As noted in the discussion prior to Corollary 7.9, $\mathcal{L}(M)$ is the set of pairs $\mathcal{L} = (L_1, L_2)$, $L_i \in \mathcal{L}(M)$. Clearly

$$\widehat{I}_{\mathcal{M}}^{\mathcal{L}}(\gamma, f_{\mathcal{L}}) = \widehat{I}_M^{L_1}(\gamma_1, f_{1,L_1}) \widehat{I}_M^{L_2}(\gamma_2, f_{2,L_2}).$$

As in the proof of Theorem 8.3, this really comes from the distribution $I_{\mathcal{M}}^{\mathcal{L}}(\gamma, f_{Q_{\mathcal{L}}})$, so there is no intrinsic difficulty with the notation $\widehat{I}_{\mathcal{M}}^{\mathcal{L}}$, even in the case $L_i = G$

where our induction hypothesis does not apply. Since

$$d^{\mathcal{G}}_{\mathcal{M}}(\mathfrak{b}, L) = 2^{(1/2)\dim(\mathfrak{a}^G_M)} d^G_M(L_1, L_2),$$

Theorem 8.3 gives the required formula for $I_M(\gamma,f)$. □

REMARK 9.2. (1) If we combine Proposition 9.1 with Corollary 8.4, we obtain the formula

$$I_M(\gamma,f) = \sum_{L \in \mathcal{L}(M)} \widehat{I}^L_M(\gamma_1,f_{1,L}) I_L(\gamma^L_2,f_2).$$

(2) According to the induction assumption of Section 2, the Fourier transform $\widehat{I}^{L_i}_M(\gamma_i)$ is defined if $L_i \subsetneq G$. However,

$$d^G_M(M,G) = d^G_M(G,M) = 1,$$

so there are terms in the formula of the proposition with $L_i = G$. For these terms it is understood that

$$\widehat{I}_G(\gamma_i,f_G) = I_G(\gamma_i,f),$$

as we agreed at the end of Section 8.

## 9.3 Splitting and Descent Combined

It is sometimes useful to combine the splitting and descent properties into one formula. Suppose that for each $v \in S$, $M_v$ is a Levi subgroup of $M$ which is defined over $F_v$. We can of course apply all our earlier definitions with $F$ replaced by $F_v$. In particular, we have the real vector space $\mathfrak{a}_{M_v}$, and the map

$$H_{M_v} : M_v(F_v) \to \mathfrak{a}_{M_v}.$$

We should point out that even if $M_v$ equals $M$, the spaces $\mathfrak{a}_{M_v}$ and $\mathfrak{a}_M$ need not be equal, for they are defined relative to the different fields $F_v$ and $F$.

Set $\mathfrak{M} = \prod_{v \in S} M_v$ and $\mathfrak{a}_{\mathfrak{M}} = \oplus_{v \in S} \mathfrak{a}_{M_v}$. If we think of $\mathfrak{M}$ as a Levi subgroup of $M$ defined over $F_S$, it will be clear how to extend our earlier definitions. For example, $\mathcal{L}(\mathfrak{M})$ will denote the set of

$$\mathfrak{L} = \prod_{v \in S} L_v, \quad L_v \in \mathcal{L}(M_v).$$

Given such an $\mathfrak{L}$, we can define the linear form

$$I^{\mathfrak{L}}_{\mathfrak{M}}(\gamma), \quad \gamma \in \mathfrak{M}(F_S),$$

on $\mathbb{H}_{\mathrm{ac}}(\mathfrak{L}(F_S))$, and the map $f \mapsto f_{\mathfrak{L}}$ from $\mathbb{H}_{\mathrm{ac}}(G(F_S))$ to $\mathbb{I}_{\mathrm{ac}}(\mathfrak{L}(F_S))$. We also have a number $d^G_{\mathfrak{M}}(M, \mathfrak{L})$. It is defined to be zero unless the natural map

$$\mathfrak{a}^M_{\mathfrak{M}} \oplus \mathfrak{a}^{\mathfrak{L}}_{\mathfrak{M}} \to \mathfrak{a}^G_{\mathfrak{M}}$$

is an isomorphism, in which case $d^G_{\mathfrak{M}}(M, \mathfrak{L})$ is the volume in $\mathfrak{a}^G_{\mathfrak{M}}$ of the parallelogram generated by orthonormal bases of $\mathfrak{a}^M_{\mathfrak{M}}$ and $\mathfrak{a}^{\mathfrak{L}}_{\mathfrak{M}}$.

COROLLARY 9.4. *Suppose that $\gamma = \prod_{v \in S} \gamma_v$ is a point in $\mathfrak{M}(F_S)$ and that*

$$\gamma^M = \prod_{v \in S} \gamma^M_v$$

*is the induced space in $M(F_S)$. Then*

$$I_M(\gamma^M, f) = \sum_{\mathfrak{L} \in \mathcal{L}(\mathfrak{M})} d^G_{\mathfrak{M}}(M, \mathfrak{L}) \widehat{I}^{\mathfrak{L}}_{\mathfrak{M}}(\gamma, f_{\mathfrak{L}}),$$

*for any function $f \in \mathbb{H}_{\mathrm{ac}}(G(F_S))$.*

PROOF. It is easy to see how to extend Theorem 8.3 in a formal way so that it includes Corollary 9.4 as well as Proposition 9.1 as special cases. Alternatively, the corollary follows by repeatedly applying Theorem 8.3 and Proposition 9.1 directly. □

REMARK 9.5. In the special case that $\gamma$ is regular, a similar formula was stated in [Ar81, Corollary 11.3]. However, the proof there does not apply in the generality claimed. For [Ar81] does not account for the fact that the space $\mathfrak{a}_M$ depends on the ground field over which $M$ is taken. Theorem 12.1 of [Ar81] is likewise affected, for it depends on this Corollary 11.3. As established in [Ar81], these results are only valid if $G$ is an inner twist of a split group. Sections 11 and 12 of [Ar81] have since been subsumed in other results and are no longer needed. For example, [Ar81, Theorem 12.1] can be replaced by the assertion that $\phi_M$ maps $\mathbb{H}_{\mathrm{ac}}(G(F_S))$ continuously to $\mathbb{I}_{\mathrm{ac}}(M(F_S))$. This was established as Theorem 4.20.3. It can also be proved quite simply by applying Corollary 7.7 directly to the $(G, M)$-family from which $\phi_M(f, \tau, X)$ is defined. However, the proof in Chapter 4 has the advantages of providing an obstruction, in terms of residues, for a function $\phi_M(f)$ to lie in $\mathbb{I}(M(F_S))$.

COROLLARY 9.6. *For each $v \in S$, set $M_v = M$, and suppose that the distributions*

$$I^{L_v}_{M_v}(\gamma_v), \quad L_v \in \mathcal{L}(M_v), \ \gamma_v \in M_v(F_v),$$

*are supported on characters. Then the corresponding distributions*

$$I^L_M(\gamma), \quad L \in \mathcal{L}(M), \ \gamma \in M(F_S),$$

*for $F_S$ are also supported on characters. In particular, the Induction Assumption* 2.6 *is valid for* $(G/F, S)$*, provided that it holds for* $(G/F_v, \{v\})$.

PROOF. We need only consider the case that $L = G$. Fix $\gamma \in M(F_S)$. We must show that $I_M(\gamma)$ annihilates the functions $f \in \mathbb{H}(G(F_S))$ such that $f_G = 0$. We leave the reader to check that any such function can be approximated by linear combinations of functions of the form

$$\prod_{v \in S} f_v, \quad f_v \in \mathbb{H}(G(F_v)),$$

in which $f_{w,G} = 0$ for some valuation $w$ in $S$. Corollary 9.4 tells us that $I_M(\gamma)$ vanishes on this latter function. □

## 9.7 Splitting Formula

The splitting formula for the dual distributions is similar. Let

$$\tau = \tau_1 \otimes \tau_2, \quad \tau_i \in \Pi(M(F_{S_i})),$$

be an arbitrary representation in $\Pi(M(F_S))$, and consider a point

$$X = (X_1, X_2), \quad X_i \in \mathfrak{a}_{M,S_i}.$$

For each $f \in \mathbb{H}_{\mathrm{ac}}(G(F_S))$, we shall write

$$J_M(\tau, X, f) = \int J_M(\tau_\lambda, X_1 + X_2, f) e^{-\lambda(X)} \, d\lambda$$

and

$$I_M(\tau, X, f) = \int I_M(\tau_\lambda, X_1 + X_2, f) e^{-\lambda(X)} \, d\lambda,$$

where each integral is taken over the direct sum of $i\mathfrak{a}^*_{M,S_1}$ and $i\mathfrak{a}^*_{M,S_2}$, modulo the diagonally embedded image of $i\mathfrak{a}^*_{M,S}$. Both integrals converge absolutely, and we have

$$I_M(\tau, X, f) = J_M(\tau, X, f) - \sum_{L \in \mathcal{L}_0(M)} \widehat{I}^L_M(\tau, X, \phi_L(f)).$$

Specializing Theorem 8.8, we obtain

PROPOSITION 9.8. *Let $\tau = \tau_1 \otimes \tau_2$ and $X = (X_1, X_2)$ be as above. Then for any function*

$$f = f_1 f_2, \quad f_i \in \mathbb{H}_{\mathrm{ac}}(G(F_{S_i})),$$

*we have*

$$I_M(\tau,X,f) = \sum_{L_1,L_2\in\mathcal{L}(M)} d_M^G(L_1,L_2)\widehat{I}_M^{L_1}(\tau_1,X_1,f_{1,L_1})\widehat{I}_M^{L_2}(\tau_2,X_2,f_{2,L_2}).$$

REMARK. Proposition 9.8 and also the results Theorem 8.8 and Corollary 8.9 of the last section have obvious analogues if $\tau$ is replaced by a standard representation $\rho \in \Sigma(M(F_S))$.

## 10 Local Vanishing Properties for GL($n$)

Let us look at an example. We shall show that, for GL($n$), the invariant distributions sometimes vanish. These vanishing results demonstrate how the descent formula of Section 8 can be usefully applied.

PROPOSITION 10.1. *Suppose that $G = \mathrm{GL}(n)$. Let $L$, $L_1$, and $L_2$ be Levi subgroups of $G$ over $F$, with $L \subset L_1$ and $L \subset L_2$, such that $d_L^G(L_1,L_2) \neq 0$. Then the natural map*

$$X(L_1)_F \oplus X(L_2)_F \to X(L)_F$$

*is surjective.*

PROOF. Fix an isomorphism

$$L \xrightarrow{\sim} \mathrm{GL}(n_1)\times\cdots\times\mathrm{GL}(n_r).$$

If

$$x \mapsto x_1\times\cdots\times x_r, \qquad x_i \in \mathrm{GL}(n_i,F),$$

is an arbitrary point in $L(F)$, set

$$\chi_i(x) = \det(x_i), \qquad 1 \leq i \leq r.$$

Then $\{\chi_i;\ 1 \leq i \leq r\}$ is a basis of $X(L)_F$. Once the isomorphism above is fixed, the group $L_1 \in \mathcal{L}(L)$ corresponds canonically to a partition of the set $\{1,\ldots,r\}$ into disjoint subsets $S_1,\ldots,S_p$. The characters

$$\prod_{i\in S_j}\chi_i, \qquad 1 \leq j \leq p,$$

form a basis of $X(L_1)_F$. Similarly, $L_2$ corresponds to a partition of $\{1, \dots, r\}$ into disjoint subsets $T_1, \dots, T_q$. We must show that each $\chi_i$ belongs to $X(L_1)_F \oplus X(L_2)_F$.

The nonvanishing of $d_L^G(L_1, L_2)$ is equivalent to the property that

$$\mathfrak{a}_{L_1}^* \oplus \mathfrak{a}_{L_2}^* \to \mathfrak{a}_L^*$$

is surjective, with one-dimensional kernel $\{(Z, -Z) \,;\, Z \in \mathfrak{a}_G^*\}$. The reader can check that this implies that (a) $p + q = r + 1$ and that (b) no proper nonempty subset of $\{1, \dots, r\}$ is a simultaneous union of sets $S_j$ or $T_k$. According to (a), one of the two partitions contains a set consisting of one element. To be definite, we can assume that $S_p = \{r\}$. Then the character $\chi_r$ belongs to $X(L_1)_F \oplus X(L_2)_F$. The element $r$ also belongs to a unique set $T_k$ and (b) implies that $T_k$ contains more than one element. In other words, $T_k' = T_k - \{r\}$ is not empty. We obtain two disjoint partitions $S_1, \dots, S_{p-1}$ and $T_1, \dots, T_k, \dots, T_q$ of the set $\{1 \dots, r-1\}$, which also satisfy (a) and (b). Since the character $\prod_{i \in T_k'} \chi_i$ belongs to $X(L_1)_F \oplus X(L_2)_F$, the proposition follows by induction on $r$. □

## 10.2 Conjugacy Classes

Let us recall some notions for the general linear group over a fixed field $F$. The characteristic polynomial identifies the semisimple conjugacy classes in GL$(n, F)$ with the polynomials in $F[X]$ of degree $n$ and nonzero constant term.

Regular semisimple conjugacy classes correspond to polynomials with distinct roots. If $\mathfrak{o}$ is a regular semisimple class, we define a partition

$$\mathfrak{p}(\mathfrak{o}) = (n_1, \dots, n_r), \quad n_1 \geq n_2 \geq \cdots \geq n_r,$$

of $n$ from the degrees of the irreducible factors of the characteristic polynomial.

A partition can also be defined for any Levi subgroup $M$ of GL$(n)$ defined over $F$. It is the unique partition

$$\mathfrak{p}(M) = (n_1, \dots, n_r), \quad n_1 \geq \cdots \geq n_r,$$

of $n$ such that $M$ is isomorphic to $\prod_{i=1}^r \mathrm{GL}(n_i)$. Note that $r$ is the dimension of the space $\mathfrak{a}_M$.

Partially order the partitions of $n$ by setting $\mathfrak{p}_1 \leq \mathfrak{p}_2$ whenever there are Levi subgroups $M_1 \subset M_2$ of GL$(n)$ such that $\mathfrak{p}_1 = \mathfrak{p}(M_1)$ and $\mathfrak{p}_2 = \mathfrak{p}(M_2)$. We have

LEMMA. (1) *For a given $\mathfrak{o}$ and $M$, $\mathfrak{p}(\mathfrak{o}) \leq \mathfrak{p}(M)$ if and only if $\mathfrak{o}$ intersects $M(F)$.*

(2) *Put $\mathfrak{p}(k) = (k, k, \dots, k)$ for a divisor $k$ of $n$. Then $\mathfrak{p}(k) \leq (n_1, \dots, n_r)$ if and only if $k$ divides each $n_i$.*

PROPOSITION 10.3. *Suppose that $M$ is a Levi subgroup of* $\mathrm{GL}(n)$ *and that* $\mathfrak{a}_M^{M_1} \cap \mathfrak{a}_M^{M_2} = \{0\}$ *for groups $M_1$ and $M_2$ in $\mathcal{L}(M)$ defined over $F$. Suppose also that $\mathfrak{p}(k) \leq \mathfrak{p}(M_1)$ and $\mathfrak{p}(k) \leq \mathfrak{p}(M_2)$ for some divisor $k$ of $n$. Then $\mathfrak{p}(k) \leq \mathfrak{p}(M)$.*

PROOF. The condition on $M_1$ and $M_2$ is equivalent to $\mathfrak{a}_M = \mathfrak{a}_{M_1} + \mathfrak{a}_{M_2}$. Now for $i = 1, 2$, let

$$\mathfrak{p}(M_i) = (n_{i1}, \ldots, n_{ir_i}).$$

Then $k$ divides each $n_{ij}$, and $M_i = \prod_{j=1}^{r_i} M_{ij}$, where $M_{ij}$ is isomorphic to $\mathrm{GL}(n_{ij})$. The intersection of $\mathfrak{a}_{M_1}$ and $\mathfrak{a}_{M_2}$ is a space of dimension at least one. Therefore

$$\dim \mathfrak{a}_M = \dim \mathfrak{a}_{M_1} + \dim \mathfrak{a}_{M_2} - \dim(\mathfrak{a}_{M_1} \cap \mathfrak{a}_{M_2}) \leq r_1 + r_2 - 1.$$

The group $M$ is the subgroup of $M_i$ defined by a subset of the simple roots of $M_i$ with respect to some ordering on a maximal split torus. It follows that

$$M = \prod_{j=1}^{r_i} (M \cap M_{ij}).$$

We claim that one of the groups $M_{ij}$ is contained in $M$. Assume the contrary, and suppose that $r_1 \leq r_2$. For each $j$, the subgroup $M \cap M_{2j}$ will be proper in $M_{2j}$. The length of $\mathfrak{p}(M)$ will be no less than $2r_2$. Since this must also equal $\dim \mathfrak{a}_M$, we obtain a contradiction. We have shown that $M_{2j} \subset M$ for some $j$. Therefore $M_{2j} \subset M_{1h}$ for some $h$. Let $n' = n - n_{2j}$. Then $k$ divides $n'$ and we have a partition $\mathfrak{p}'(d)$ of $n'$. There is clearly a unique subgroup $G'$ of $\mathrm{GL}(n)$ isomorphic to $\mathrm{GL}(n')$ such that $M_2$ is contained in $M_{2j}G'$. Then $M' = M \cap G'$, $M_1' = M_1 \cap G'$ and $M_2' = M_2 \cap G'$ are Levi subgroups of $G'$ such that $\mathfrak{a}_{M'}^{M_1'} \cap \mathfrak{a}_{M'}^{M_2'} = \{0\}$. The partition $\mathfrak{p}(M_1')$ is obtained by replacing $n_{1h}$ by $n_{1h} - n_{2j}$. Since $k$ divides $n_{1h} - n_{2j}$, we have $\mathfrak{p}'(k) \leq \mathfrak{p}(M_1')$. Similarly $\mathfrak{p}'(k) \leq \mathfrak{p}(M_2')$. It follows by induction on $n$ that $\mathfrak{p}'(k) \leq \mathfrak{p}(M')$. Since $\mathfrak{p}(M)$ is obtained from $\mathfrak{p}(M')$ by adjoining $n_{2j}$, we get $\mathfrak{p}(k) \leq \mathfrak{p}(M)$. □

Let $\gamma$ be an element of $M(F_S)$. The induced space $\gamma^G$ is defined in Definition 2.10.

PROPOSITION 10.4. *The induced space $\gamma^G$ can also be characterized analytically by the formula*

$$J_G(\gamma^G, f) = \lim_{a \to e} J_G(a\gamma, f), \quad f \in C_c^\infty(G(F_S)), \tag{10.4.1}$$

*where $a$ ranges over elements in $A_{M,\mathrm{reg}}(F_S)$ which are close to* 1.

PROOF. To establish this formula, note that the right-hand side equals

$$\lim_{a \to e} |D^G(a\gamma)|^{1/2} \int_{M(F_S)\backslash G(F_S)} \int_{M_\gamma(F_S)\backslash M(F_S)} f(x^{-1}m^{-1}a\gamma m x)\, dm\, dx.$$

If $P \in \mathcal{P}(M)$, this in turn equals

$$\begin{aligned}
&\lim_{a \to e} |D^M(\gamma)|^{1/2} \delta_P(a\gamma)^{1/2} \int_K \int_{N_P(F_S)} \int_{M_\gamma(F_S)\backslash M(F_S)} f(k^{-1}am^{-1}\gamma mnk)\, dm\, dn\, dk \\
&\quad = |D^M(\gamma)|^{1/2} \delta_P(\gamma)^{1/2} \int_K \int_{N_P(F_S)} \int_{M_\gamma(F_S)\backslash M(F_S)} f(k^{-1}m^{-1}\gamma mnk)\, dm\, dn\, dk,
\end{aligned}$$

by a standard change of variables formula. This last expression is obviously the integral of $f$ over the invariant measure on $\gamma^G$. It is therefore equal to the left-hand side of (10.4.1). □

## 10.5 Inner Twist

For the rest of this section, we shall assume that we have been given an inner twist

$$\eta : G \to G' = \mathrm{GL}(n).$$

This is an isomorphism over an algebraic closure $\overline{F}$ of $F$ such that for every $\sigma \in \mathrm{Gal}(\overline{F}/F)$, the map $\eta^{-1} \circ {}^{\sigma}\eta$ is equal to conjugation by an element in $G$. One knows (see, e.g., [We67]) that

$$G(F) \simeq \mathrm{GL}(n/d, D),$$

where $d$ is a divisor of $n$ and $D$ is a division algebra of degree $d^2$ over $F$.

We write $G'$ for the group $\mathrm{GL}(n)$. We shall show that our invariant distributions on $G'$ vanish on certain data related to $G$, in a sense that it depends only on the integer $d$. Suppose that $L$ is a Levi subgroup of $G'$ (defined over $F$). As in 10.2, write

$$\mathfrak{p}(L) = (n_1, \ldots, n_r), \quad n_1 \geq n_2 \geq \cdots \geq n_r,$$

for the unique partition of $n$ such that

$$L \simeq \mathrm{GL}(n_1) \times \cdots \times \mathrm{GL}(n_r).$$

We shall say that $L$ *comes from* $G$ if $d$ divides each of the integers $n_i$. This means that there is a Levi subgroup $M$ of $G$ such that $L = M'$. Suppose that $L_1 \subset L_2$ are two other Levi subgroups of $G'$ with $d_{L_1}^{G'}(M', L_2) \neq 0$. Then if $L_2$ comes from $G$, Proposition 10.3 asserts that $L_1$ also comes from $G$.

DEFINITION 10.6. (1) Recall that an element $\delta \in G'(F)$ is called *$F$-elliptic* if it lies in a maximal torus of $G'$ which is anisotropic over $F$, modulo $A_G$. Write $G'(F)_{\mathrm{ell}}$ for the set of such elements.

(2) By the theory of elementary divisors every conjugacy class in $G'(F)$ is induced from an elliptic class. In other words, for any $\delta \in G'(F)$ there is a Levi subgroup $L_1$ of $G'$, and an element $\gamma_1 \in L_1(F)_{\mathrm{ell}}$, such that $\delta$ belongs to the induced conjugacy class $\gamma_1^{G'}$. The pair $(L_1, \gamma_1)$ is uniquely determined by $\delta$ up to $G'(F)$-conjugacy.
(3) We shall say that $\delta$ *comes from* $G$ if the group $L_1$ comes from $G$. Write $G'(F)_G$ for the set of such elements.
(4) Observe that if $M'$ is a Levi subgroup of $G'$ which comes from $G$, we can define the subgroup $M'(F)_M$ of $M'(F)$.

## 10.7 Vanishing of Orbital Integrals

Suppose now that $F$ is a local field and that $S = \{v\}$, so that $F$ is $F_v = F_S$. Let $f'$ be a fixed function in $\mathbb{H}(G'(F))$ such that

$$I_{G'}(\zeta, f') = 0 \tag{10.7.1}$$

for any $G'$-regular element $\zeta \in G'(F)$ which does not belong to $G'(F)_G$.

PROPOSITION. *Suppose that $M'$ is a Levi subgroup of $G'$ which comes from $G$ and that $\delta$ belongs to $M'(F)$. Then $I_{M'}(\delta, f') = 0$, unless $\delta$ lies in $M'(F)_M$.*

REMARK. If $M' = G'$, the proposition is essentially a restatement of the definition of $f'$. It is of course the case that $M' \neq G'$ that is interesting.

PROOF. Assume that $I_{M'}(\delta, f') \neq 0$. Fix a pair

$$(L_1, \gamma_1), \quad \gamma_1 \in L_1(F)_{\mathrm{ell}}, \ \delta \in \gamma_1^{M'}.$$

We must show that $L_1$ comes from $M$.

The situation is made to order for our descent formula. For Corollary 8.4 immediately yields an expansion

$$I_{M'}(\delta, f') = \sum_{L_2 \in \mathcal{L}(L_1)} d_{L_1}^{G'}(M', L_2) \widehat{I}_{L_1}^{L_2}(\gamma_1, f'_{L_2}),$$

and hence the existence of some $L_2 \in \mathcal{L}(L_1)$ with

$$d_{L_1}^{G'}(M', L_2) \widehat{I}_{L_1}^{L_2}(\gamma_1, f'_{L_2}) \neq 0.$$

Now the distribution $\widehat{I}_{L_1}^{L_2}(\gamma_1, f'_{L_2})$ belongs to the closed linear span of the ordinary orbital integrals $\{\widehat{I}_{L_2}^{L_2}(\zeta, f'_{L_2})\}$, where $\zeta$ ranges over the $G$-regular points in $L_2(F)$. But $\widehat{I}_{L_1}^{L_2}(\gamma_1, f'_{L_2})$ does not vanish, so there exists such a $\zeta$ with

$$\widehat{I}_{L_2}^{L_2}(\zeta, f'_{L_2}) = I_G(\zeta, f') \neq 0.$$

It follows from the definition of $f'$ that $L_2$ comes from $G$. Applying Proposition 10.3, we see that $L_1$ also comes from $G$. This obviously implies that $L_1$ comes from $M$, our required condition. □

There is a parallel vanishing property for the distributions

$$I_{M'}(\tau, Y, f'), \quad \tau \in \Pi(M'(F)),\ Y \in \mathfrak{a}_{M',v}.$$

PROPOSITION 10.8. *Suppose that $M'$ is a Levi subgroup of $G'$ which comes from $G$ and that $L_1$ is a Levi subgroup of $M'$. Then*

$$I_{M'}(\tau, Y, f') = 0,$$

*for any $Y \in \mathfrak{a}_{M',v}$, and any induced representation*

$$\tau = \tau_{1,\lambda}^{M'}, \quad \lambda \in i\mathfrak{a}_{L,v}^*,\ \tau_1 \in \Pi(L_1(F)),$$

*unless $L_1$ comes from $M$.*

PROOF. The proof is similar to that of the last proposition. It is enough to show that if $L_1$ does not come from $M$, then the Fourier transform

$$I_{M'}(\tau_1, Y_1, f') = \int_{i\mathfrak{a}_{L_1,v}^*/i\mathfrak{a}_{M',v}^*} I_{M'}(\tau_{1,\lambda}^{M'}, Y_1, f')e^{-\lambda(Y_1)}\, d\lambda$$

vanishes, for every point $Y_1 \in \mathfrak{a}_{L_1,v}$ whose projection onto $\mathfrak{a}_{M',v}$ equals $Y$. The descent formula, Corollary 8.9, yields

$$I_{M'}(\tau_1, Y_1, f') = \sum_{L_2 \in \mathcal{L}(L_1)} d_{L_1}^{G'}(M', L_2)\widehat{I}_{L_1}^{L_2}(\tau_1, Y_1, f'_{L_2}).$$

The proposition then follows as above from Proposition 10.3. □

REMARK. Obviously, a similar vanishing property holds if $\tau$ and $\tau_1$ are replaced by standard representations $\rho \in \Sigma(M'(F))$ and $\rho_1 \in \Sigma(L_1(F))$.

## 11 Convex Polytopes

The aim of this section is to prove Proposition 7.6.

DEFINITION 11.1. (1) Let $\mathfrak{a}$ be a finite-dimensional Euclidean space. A *convex polytope* $\Pi$ in $\mathfrak{a}$ is the convex hull of a finite set of points. Fix such a $\Pi$. Let $\mathcal{F}(\Pi)$ denote the *set of closed faces* of $\Pi$. Then $\mathcal{F}(\Pi)$ is a partially ordered finite set whose elements are convex polytopes in their own right. The maximal element is just $\Pi$. The minimal elements form the subset $\mathcal{P}(\Pi)$ of *faces which are just points*. The faces in $\mathcal{P}(\Pi)$ are of course called the *vertices* of $\Pi$.

(2) Suppose that $F$ is a face in $\mathcal{F}(\Pi)$. The (open) *dual cone* $\mathfrak{a}_F^+$ is defined as follows. Choose a point $X_F$ in $F$ which does not lie on any proper subface of $F$. Form the cone generated by $\Pi - X_F$. Then $\mathfrak{a}_F^+$ is the relative interior of the corresponding dual cone. That is, $\mathfrak{a}_F^+$ is the intersection, over all points $X$ in the complement of $F$ in $\Pi$, of the half open spaces

$$\{Y \in \mathfrak{a};\ \langle Y, X - X_F\rangle < 0\}.$$

Let $\mathfrak{a}^F$ denote the subspace of $\mathfrak{a}$ spanned by $F - X_F$. Let $\mathfrak{a}_F$ be the orthogonal complement of $\mathfrak{a}^F$ in $\mathfrak{a}$. Then $\mathfrak{a}_F^+$ is an open convex cone in $\mathfrak{a}_F$ which is independent of $X_F$.

PROPOSITION 11.2. *The space $\mathfrak{a}$ is the disjoint union of the cones $\mathfrak{a}_F^+$.*

PROOF. The dual cones consist of cosets of $\mathfrak{a}_\Pi$. They are invariant under translation of $\Pi$. We may therefore assume that $\Pi$ contains the origin as an interior point. Let $\widehat{\Pi}$ be the *polar set* of $\Pi$. Thus $\widehat{\Pi}$ is the intersection, over all points $X \in \Pi$, of the closed half spaces

$$\{Y \in \mathfrak{a};\ \langle Y, X\rangle \leq 1\}.$$

Then $\widehat{\Pi}$ is another convex polytope, whose interior contains the origin. There is an incidence reversing bijection $F \mapsto \widehat{F}$ between the proper faces of $\Pi$ and of $\widehat{\Pi}$, and $\mathfrak{a}_F^+$ is just the cone generated by the relative interior of $\widehat{F}$. But any half line through the origin will intersect the relative interior of a unique proper face $\widehat{F}$. Thus $\mathfrak{a}$ is indeed a disjoint union of the cones $\mathfrak{a}_F^+$. □

## 11.3 Subspaces

Suppose that $\mathfrak{b}$ is a vector subspace of $\mathfrak{a}$. Let $\Pi_\mathfrak{b}$ be the projection of $\Pi$ onto $\mathfrak{b}$. Then $\Pi_\mathfrak{b}$ is also a convex polytope. We shall construct a section from $\Pi_\mathfrak{b}$ into $\Pi$. We must first fix a point $\xi$ in $\mathfrak{a}^\mathfrak{b}$, the orthogonal complement of $\mathfrak{b}$ in $\mathfrak{a}$, which is in general position. Let $\mathcal{F}(\Pi, \xi)$ denote the set of faces $F \in \mathcal{F}(\Pi)$ for which the set

$$\mathfrak{b}_{\xi,F} = (\xi + \mathfrak{b}) \cap \mathfrak{a}_F^+$$

is not empty. Then $(\xi + \mathfrak{b})$ is a disjoint union over $\mathcal{F}(\Pi, \xi)$ of the sets $\mathfrak{b}_{\xi,F}$. Define

$$\Pi(\xi) = \bigcup_{F \in \mathcal{F}(\Pi,\xi)} F.$$

The general position of $\xi$ implies that if $F$ belongs to $\mathcal{F}(\Pi, \xi)$ and if $F_1 \in \mathcal{F}(\Pi)$ is a face which is contained in $F$, then $F_1$ also belongs to $\mathcal{F}(\Pi, \xi)$. It follows that $\Pi(\xi)$ is a subcomplex of $\Pi$.

PROPOSITION 11.4. *The orthogonal projection of* $\mathfrak{a}$ *onto* $\mathfrak{b}$ *maps* $\Pi(\xi)$ *bijectively onto* $\Pi_{\mathfrak{b}}$.

PROOF. Let $\eta$ be a point in $\Pi_{\mathfrak{b}}$. The fiber at $\eta$ is the set

$$\Pi^\eta = \Pi \cap (\eta + \mathfrak{a}^{\mathfrak{b}}).$$

We must show that $\Pi^\eta$ intersects $\Pi(\xi)$ at precisely one point.

The faces of $\Pi(\xi)$ are the elements in $\mathcal{F}(\Pi, \xi)$. Observe that $\mathcal{F}(\Pi, \xi)$ is the subset of faces $F \in \mathcal{F}(\Pi)$ such that $\xi$ belongs to $(\mathfrak{a}_F^+ + \mathfrak{b})$. On the other hand, $\Pi^\eta$ is also a convex polytope, and its faces are of the form

$$F^\eta = F \cap \Pi^\eta, \quad F \in \mathcal{F}(\Pi).$$

Many of these intersections will be empty. Moreover, if $\eta$ is not in general position, different $F$ will give the same intersection. However, let us define $\mathcal{F}^\eta(\Pi)$ to be the set of elements $F \in \mathcal{F}(\Pi)$ such that $F^\eta$ contains a point $X_F^\eta$ in the relative interior of $F$. Any such $F$ will be minimal among those faces which have the same intersection with $\Pi^\eta$. Clearly $F \to F^\eta$ is a bijection from $\mathcal{F}^\eta(\Pi)$ onto the set of faces of $\Pi^\eta$.

Suppose that $F \in \mathcal{F}^\eta(\Pi)$ and that $F_1 \in \mathcal{F}(\Pi)$ is some other face such that $F_1^\eta = F^\eta$. Then $\mathfrak{a}^F \cap \mathfrak{a}^{\mathfrak{b}} = \mathfrak{a}^{F_1} \cap \mathfrak{a}^{\mathfrak{b}}$. Taking orthogonal complements, we obtain $\mathfrak{a}_F + \mathfrak{b} = \mathfrak{a}_{F_1} + \mathfrak{b}$. However, $F$ is minimal, so it is actually a face of $F_1$. This means that $\mathfrak{a}_{F_1}^+$ is contained in the closure of $\mathfrak{a}_F^+$. It follows easily that

$$\mathfrak{a}_F^+ + \mathfrak{b} \supseteq \mathfrak{a}_{F_1}^+ + \mathfrak{b}.$$

Thus in studying the intersection of $\Pi(\xi)$ with $\Pi^\eta$, we need only consider those faces of $\Pi(\xi)$ which belong to $\mathcal{F}^\eta(\Pi)$.

Suppose again that $F \in \mathcal{F}^\eta(\Pi)$. We shall define the dual cone $\mathfrak{a}_{F^\eta}^+$ of $F^\eta$. Set

$$C_F = \{t(X - X_F^\eta)\,;\, t \geq 0, X \in \Pi\}$$

and

$$C_{F^\eta} = \{t(X^\eta - X_F^\eta)\,;\, t \geq 0, X^\eta \in \Pi^\eta\}.$$

Then $C_F^\eta = C_F \cap \mathfrak{a}^{\mathfrak{b}}$. But $C_F$ and $\mathfrak{a}^{\mathfrak{b}}$ are both polyhedra cones. As is well known, the dual cone of their intersection equals the sum of their dual cones. It follows that the closure of $\mathfrak{a}_{F^\eta}^+$ equals the sum of the closure of $\mathfrak{a}_F^+$ with $\mathfrak{b}$. Taking the relative interior of these closed cones, we obtain

$$\mathfrak{a}_{F^\eta}^+ = \mathfrak{a}_F^+ + \mathfrak{b}.$$

We know that $\mathfrak{a}$ is the disjoint union of the cones $\mathfrak{a}_{F^\eta}^+$. We can therefore express $\mathfrak{a}$ as the disjoint union, over $F \in \mathcal{F}^\eta(\Pi)$, of the cones $\mathfrak{a}_F^+ + \mathfrak{b}$. In particular, $\xi$ lies in precisely one such cone. But $\xi$ is in general position, so we can assume that the cone in which it lies is open and corresponds to a vertex of $\Pi^\eta$. We have thus shown that there is precisely one face of $\Pi^\eta$ which meets $\Pi(\xi)$ and that this face is a vertex. In other words, $\Pi^\eta$ meets $\Pi(\xi)$ in precisely one point, as required. □

## 11.5 (G,M)-Families

Our purpose in discussing convex polytopes is for their connection with $(G, M)$-families. We aim to prove Proposition 7.6. Let us consider a typical example. For each $Q \in \mathcal{F}(M)$, let $\rho_Q \in \mathfrak{a}_Q^+$ be the usual vector defined by the square root of the modular function. Let $\Pi_M$ denote the convex hull of the finite set $\{\rho_P\,;\, P \in \mathcal{P}(M)\}$. Then $\Pi_M$ is a convex polytope which lies in $\mathfrak{a}_M^G$. There is an order preserving bijection $Q \mapsto \Pi_M^Q$ from $\mathcal{F}(M)$ onto the set of faces of $\Pi_M$. Moreover, the dual cone of $\Pi_M^Q$ is just the chamber $\mathfrak{a}_Q^+$. Thus the face $\Pi_M^Q$ and the chamber $\mathfrak{a}_Q^+$ are of complementary dimensions, and they intersect orthogonally at the point $\rho_Q$. Consider the $(G, M)$-family given by

$$c_P(\lambda) = e^{\lambda(\rho_P)}, \quad P \in \mathcal{P}(M),\ \lambda \in i\mathfrak{a}_M^*. \tag{11.5.1}$$

Then $c_M(\lambda)$ is just the integral of $e^{-\lambda(H)}$ over $\Pi_M$. More generally, suppose that $Q \in \mathcal{F}(M)$. Then $\Pi_M^Q$ lies in the affine space $\rho_Q + \mathfrak{a}_M^Q$ and inherits a Euclidean measure $\mathrm{d}H$ from that on $\mathfrak{a}_M^Q$. We have

$$c_M^Q(\lambda) = \int_{\Pi_M^Q} e^{-\lambda(H)}\,\mathrm{d}H, \quad \lambda \in \mathfrak{a}_{M,\mathbb{C}}^*. \tag{11.5.2}$$

In particular, $c_M^Q$ is just the volume in $\rho_Q + \mathfrak{a}_M^Q$ of the face $\Pi_M^Q$.

## 11.6 Special Subspaces

Now, as in Section 7, suppose that $\mathfrak{b}$ is a special subspace of $\mathfrak{a}_M$. Let $\Pi_{M,\mathfrak{b}}$ be the projection of $\Pi_M$ onto $\mathfrak{b}$. We claim that the $(\mathfrak{a}_G, \mathfrak{b})$-family associated to $\{c_P(\lambda)\}$ is also the one attached to the polytope $\Pi_{M,\mathfrak{b}}$. If $\mathfrak{p}$ is any element in $\mathcal{P}(\mathfrak{b})$ (see Definition 7.3(5)), let $Q$ be the unique element in $\mathcal{P}(M_\mathfrak{b})$ such that $\mathfrak{b}_\mathfrak{p}^+$ is contained in $\mathfrak{a}_Q^+$ and define $\rho_\mathfrak{p}$ to be the projection of $\rho_Q$ onto $\mathfrak{b}$. Then

$$c_{\mathfrak{p}}(\nu) = e^{\nu(\rho_{\mathfrak{p}})}, \quad \mathfrak{p} \in \mathcal{P}(\mathfrak{b}),\ \nu \in i\mathfrak{b}^*,$$

is the associated $(\mathfrak{a}_G, \mathfrak{b})$-family. On the other hand, $\Pi_{M,\mathfrak{b}}$ is the convex hull in $\mathfrak{b}$ of the set

$$\{\rho_{\mathfrak{p}}\,;\ \mathfrak{p} \in \mathcal{P}(\mathfrak{b})\}.$$

Indeed, it is trivial that $\Pi_{M,\mathfrak{b}}$ contains the convex hull. The converse is a minor extension of [Ar82.II, Lemma 3.1]. It is proved in the same way. Our claim, then, is justified. In particular, as in Section 4.10 we can write

$$c_{\mathfrak{b}}(\nu) = \int_{\Pi_{M,\mathfrak{b}}} e^{-\nu(\widetilde{H})}\, \mathrm{d}\widetilde{H}, \quad \nu \in i\mathfrak{b}^*, \tag{11.6.1}$$

where $\mathrm{d}\widetilde{H}$ is the Euclidean measure on $\mathfrak{b}$.

## 11.7 A Parametrization

We shall want to apply Proposition 11.4. As before, let $\xi$ be a point in $\mathfrak{a}_M^{\mathfrak{b}}$ in general position and write $\mathcal{F}(M, \xi)$ for the set of elements $Q \in \mathcal{F}(M)$ such that the set

$$\mathfrak{b}_{\xi,Q} = (\xi + \mathfrak{b}) \cap \mathfrak{a}_Q^+$$

is not empty. Then

$$\xi + \mathfrak{b} = \bigcup_{Q \in \mathcal{F}(M,\xi)} \mathfrak{b}_{\xi,Q}$$

is a decomposition of $\xi + \mathfrak{b}$ into a polyhedral complex. The vertices correspond to the parabolics $Q_L$ introduced in Section 7. The maximal cells correspond to the set

$$\mathcal{P}(M, \xi) = \mathcal{P}(M) \cap \mathcal{F}(M, \xi).$$

We note that $\mathcal{P}(M, \xi)$ is just the set of $P \in \mathcal{P}(M)$ which are contained in one of the parabolics $Q_L$. Of particular interest are the cells which are translates in $\xi + \mathfrak{b}$ of the chambers $\mathfrak{b}_{\mathfrak{p}}^+$ in $\mathfrak{b}$. Let us write $\mathcal{P}_{\mathrm{ext}}(M, \xi)$ for the subset of elements $P \in \mathcal{P}(M, \xi)$ such that the closure of $\mathfrak{a}_P^+$ intersects $\mathfrak{b}$ in an open set. This intersection must necessarily be the closure of a chamber $\mathfrak{b}_{\mathfrak{p}(P)}^+$ for a uniquely determined element $\mathfrak{p}(P)$ in $\mathcal{P}(\mathfrak{b})$.

PROPOSITION. *The map $P \mapsto \mathfrak{p}(P)$ is a bijection from $\mathcal{P}_{\mathrm{ext}}(M, \xi)$ onto $\mathcal{P}(\mathfrak{b})$.*

PROOF. Suppose that $\mathfrak{p}$ is an arbitrary element in $\mathcal{P}(\mathfrak{b})$. Let $Q$ be the unique element in $\mathcal{P}(M_{\mathfrak{b}})$ such that $\mathfrak{a}_Q^+$ contains $\mathfrak{a}_P^+$. Let $R$ be the unique element in $\mathcal{P}^{M_{\mathfrak{b}}}(M)$ such that $\xi$ belongs to $\mathfrak{a}_R^+$. Then $P = Q(R)$ is the unique element in $\mathcal{P}_{\mathrm{ext}}(M, \xi)$ with $\mathfrak{p}(P) = \mathfrak{p}$. □

Note that $\rho_{\mathfrak{p}(P)}$ is just the projection of $\rho_P$ onto $\mathfrak{b}$.

## 11.8 Cells in a Complex

We will use Proposition 11.4 to study the function

$$c_{\mathfrak{b}}(\nu), \quad \nu \in i\mathfrak{b}^*.$$

Observe that the maximal cells in the complex

$$\Pi_M(\xi) = \sum_{Q \in \mathcal{F}(M,\xi)} \Pi_M^Q$$

correspond to the parabolics $Q_L$, where $L$ ranges over the elements in $\mathcal{L}(M)$ with $d_M^G(\mathfrak{b}, L) \neq 0$. Let $\widetilde{\Pi}_M^{Q_L}$ be the projection of $\Pi_M^{Q_L}$ onto $\mathfrak{b}$. Then Proposition 11.4 asserts that $\Pi_{M,\mathfrak{b}}$ is the disjoint union of the sets $\widetilde{\Pi}_M^{Q_L}$, together with a set of measure 0. It follows from (11.6.1) that

$$c_{\mathfrak{b}}(\nu) = \int_{\Pi_{M,\mathfrak{b}}} e^{-\nu(\widetilde{H})}\, d\widetilde{H} = \sum_L \int_{\widetilde{\Pi}_M^{Q_L}} e^{-\nu(\widetilde{H})}\, d\widetilde{H}.$$

Fix $L$ for the moment and let $H \mapsto \widetilde{H}$ denote the orthogonal projection of $\mathfrak{a}_M^L$ onto $\mathfrak{b}^G$. We are assuming that $d_M^G(\mathfrak{b}, L) \neq 0$, so that this map is an isomorphism, and $d\widetilde{H} = d_M^G(\mathfrak{b}, L) dH$. Moreover, $e^{-\nu(H)} = e^{-\nu(\widetilde{H})}$, since $\nu$ belongs to $i\mathfrak{b}^*$. It follows that

$$\int_{\widetilde{\Pi}_M^{Q_L}} e^{-\nu(\widetilde{H})}\, d\widetilde{H} = d_M^G(\mathfrak{b}, L) \int_{\Pi_M^{Q_L}} e^{-\nu(H)}\, dH.$$

Combining these formulae with (11.5.2), we obtain

$$c_{\mathfrak{b}}(\nu) = \sum_{L \in \mathcal{L}(M)} d_M^G(\mathfrak{b}, L) c_M^{Q_L}(\nu), \quad \nu \in i\mathfrak{b}^*.$$

On the other hand, we have

$$c_{\mathfrak{b}}(\nu) = \sum_{\mathfrak{p} \in \mathcal{P}(\mathfrak{b})} e^{\nu(\rho_{\mathfrak{p}})} / \theta_{\mathfrak{p}}(\nu) = \sum_{P \in \mathcal{P}_{\mathrm{ext}}(M,\xi)} e^{\nu(\rho_P)} / \theta_{\mathfrak{p}(P)}(\nu)$$

from the correspondence between $\mathcal{P}_{\text{ext}}(M,\xi)$ and $\mathcal{P}(\mathfrak{b})$ of Proposition 11.7. It thus follows that

$$\sum_{L\in\mathcal{L}(M)} d_M^G(\mathfrak{b},L)c_M^{Q_L}(\nu) = \sum_{P\in\mathcal{P}_{\text{ext}}(M,\xi)} e^{\nu(\rho_P)}/\theta_{\mathfrak{p}(P)}(\nu), \tag{11.8.1}$$

for any point $\nu \in i\mathfrak{b}^*$ and for $\{c_P(\lambda)\}$ the $(G,M)$-family given by (11.5.1).

## *11.9 Proof of Proposition* 7.6

Suppose that $\{c_P(\lambda)\}$ is an arbitrary $(G,M)$-family. The expression

$$\sum_{L\in\mathcal{L}(M)} d_M^G(\mathfrak{b},L)c_M^{Q_L}(\lambda)$$

equals

$$\sum_{L\in\mathcal{L}(M)} d_M^G(\mathfrak{b},L) \sum_{\{P\in\mathcal{P}(M)\,;\,P\subset Q_L\}} c_P(\lambda)/\theta_{P\cap L}(\lambda).$$

Let $r_{P,\xi}(\lambda)$ denote the sum, over all elements $L\in\mathcal{F}(M)$ with $d_M^G(\mathfrak{b},L)\neq 0$ and with $Q_L\supset P$, of the terms

$$d_M^G(\mathfrak{b},L)/\theta_{P\cap L}(\lambda).$$

Then

$$\sum_{L\in\mathcal{L}(M)} d_M^G(\mathfrak{b},L)c_M^{Q_L}(\lambda) = \sum_{P\in\mathcal{P}(M,\xi)} c_P(\lambda)r_{P,\xi}(\lambda). \tag{11.9.1}$$

Set $\lambda$ equal to a point $\nu$ in $i\mathfrak{b}^*$, and for the moment take $\{c_P(\nu)\}$ to be the $(G,M)$-family defined by (11.5.1). Then we can combine (11.9.1) with (11.8.1). We obtain

$$\sum_{P\in\mathcal{P}_{\text{ext}}(M,\xi)} e^{\nu(\rho_P)}/\theta_{\mathfrak{p}(P)}(\nu) = \sum_{P\in\mathcal{P}(M,\xi)} e^{\nu(\rho_P)}r_{P,\xi}(\nu).$$

The functions $\{\theta_{\mathfrak{p}(P)}(\nu)^{-1}\}$ and $\{r_{P,\xi}(\nu)\}$ are all rational in $\nu$. Furthermore, by Proposition 11.4, the projection of the set

$$\{\rho_P\,;\,P\in\mathcal{P}(M,\xi)\}$$

onto $\mathfrak{b}$ is injective. Therefore the exponential functions

$$\nu \mapsto e^{\nu(\rho_P)}, \quad P \in \mathcal{P}(M, \xi),$$

are linearly independent over the field of rational functions. Setting the coefficients equal to 0, we find that

$$r_{P,\xi}(\nu) = \begin{cases} \theta_{\mathfrak{p}(P)}(\nu)^{-1}, & \text{if } P \in \mathcal{P}_{\text{ext}}(M, \xi), \\ 0, & \text{otherwise.} \end{cases}$$

Returning to the case that $\{c_P(\lambda)\}$ is arbitrary, we substitute the formula for $r_{P,\xi}(\nu)$ into the right-hand side of (11.9.1). We obtain

$$\begin{aligned} \sum_{L \in \mathcal{L}(M)} d_M^G(\mathfrak{b}, L) c_M^{Q_L}(\nu) &= \sum_{P \in \mathcal{P}_{\text{ext}}(M,\xi)} c_P(\nu)/\theta_{\mathfrak{p}(P)}(\nu) \\ &= \sum_{\mathfrak{p} \in \mathcal{P}(\mathfrak{b})} c_{\mathfrak{p}}(\nu)/\theta_{\mathfrak{p}}(\nu) = c_{\mathfrak{b}}(\nu). \end{aligned}$$

This completes the proof of Proposition 7.6.

# 12 The Invariant Trace Formula So Far

## *12.1 Review*

The purpose of the second part of this chapter, starting here, is to prove an explicit invariant trace formula. In the first part, Sections 1–11, we studied two families of invariant distributions. Now we shall exhibit these distributions as terms on the two sides of the invariant trace formula. The introduction of this chapter contains a general discussion of the problem. We now describe the formula in more detail, and review the progress so far.

Let $G$ be a connected reductive algebraic group over a number field $F$. Let $f$ be a function in the Hecke algebra of $G(\mathbb{A})$. We already have a "coarse" invariant trace formula,

$$\sum_{\mathfrak{o} \in \mathcal{O}} I_{\mathfrak{o}}(f) = \sum_{\chi \in \mathfrak{X}} I_{\chi}(f), \tag{12.1.1}$$

which was established in Chapter 4. This will be our starting point here. The terms on each side of (12.1.1) are invariant distributions, but as they stand, they are not explicit enough to be very useful. After reviewing the formula (12.1.1) later in this section, we shall study the two sides separately in Sections 13 and 14. These two sections are crucial. Building on earlier investigations of noninvariant distributions ([Ar82.II, Ar86]), later in this chapter we shall establish finer expansions for each side of (12.1.1). The resulting identity

$$\sum_{M} |W_0^M||W_0^G|^{-1} \sum_{\gamma \in (M(F))_{M,S}} a^M(S,\gamma) I_M(\gamma, f)$$
$$= \sum_{t \geq 0} \sum_{M} |W_0^M||W_0^G|^{-1} \int_{\Pi(M,t)} a^M(\tau_{1,\lambda}) I_M(\tau_{1,\lambda}, f)\, d\tau_{1,\lambda} \tag{12.1.2}$$

will be our explicit trace formula. The terms $I_M(\gamma,f)$ and $I_M(\tau_{1,\lambda},f)$ in (12.1.2) are essentially the invariant distributions studied in the first part of this chapter. The functions $a^M(S,\gamma)$ and $a^M(\tau_{1,\lambda})$ depend only on the Levi subgroup $M$, and are global in nature. They are strongly dependent on the discrete subgroup $M(F)$ of $M(\mathbb{A})$. More detailed descriptions of these objects, as well as the sets $(M(F))_{M,S}$ and $\Pi(M,t)$, are in Sections 13 and 14.

## 12.2 Induction

To define the invariant distributions $I_M(\gamma,f)$ and $I_M(\tau,X,f)$, we had to introduce an induction hypothesis in Subsection 2.6. This hypothesis remains in force. We shall finally settle this matter in Section 15. We shall show that the invariant distributions in the trace formula are all supported on characters. Using Theorem 6.2, we shall first establish in Proposition 15.2 that the distributions on the right-hand side of (12.1.2) have the required property. We shall then use formula (12.1.2) itself to deduce the same property of the distributions on the left (Theorem 15.1). This is a generalization of an argument introduced by Kazhdan in his Maryland lectures (see [DKV84, Ka86.1]). Theorems 6.2 and 15.1 are actually simple versions of a technique that can be applied generally. They provide a good introduction to the more complicated versions used for the comparison in Chapter 6.

## 12.3 Estimates

It is not known whether the right-hand side of (12.1.2) converges as a double integral over $t$ and $\tau$. It is a difficulty which originates with the Archimedean valuations of $F$. On the other hand, some result of this nature will definitely be required for many applications of the trace formula. In Section 16 we shall prove a weak estimate (Corollary 16.9) for the rate of convergence of the sum over $t$. It will be stated in terms of multipliers for the Archimedean part $\Pi_{v \in S_\infty} G(F_v)$ of $G(\mathbb{A})$. One would then hope that by varying the multipliers, one could separate the terms according to their Archimedean infinitesimal character. For comparison with inner forms, this is in fact what happens. One can use the estimate to eliminate problems caused by the Archimedean primes, as in Section 6.14. In general, Corollary 16.9 seems to be a natural device for isolating the contributions of a given infinitesimal character.

## 12.4 Simple Version

It is useful to have simple versions of the trace formula for functions

$$f = \prod_v f_v$$

that are suitably restricted. Since the terms in (12.1.2) are all invariant distributions, we will be able to impose conditions on $f$ strictly in terms of its orbital integrals. If at one place $v$ the semisimple orbital integrals of $f_v$ are supported on the elliptic set, then all the terms with $M \neq G$ on the (spectral) right-hand side of (12.1.2) vanish. If the same thing is true at two places, the terms with $M \neq G$ on the (geometric) left-hand side also vanish. These two assertions comprise Theorem 17.3. They are simple consequences of the descent and splitting formulae in Sections 8 and 9. We shall also see that in certain cases the remaining terms take a particularly simple form (Corollaries 17.5, 17.6, 17.8).

## 12.5 Case of GL($n$)

As with the first half of the chapter, we shall conclude (Section 18) by discussing the example of GL($n$). Groups related to GL($n$) by inner twisting are the simplest examples of general rank for which one can attempt a comparison of trace formulae. However, one must first establish some properties of the trace formula of GL($n$) itself. By imposing less stringent conditions than those of Section 17, we shall establish more delicate vanishing properties. The resulting formula for GL($n$) is then what should be compared with the trace formula for inner forms.

## 12.6 Hecke Spaces

The geometric side of the (noninvariant) trace formula was considered in Chapter 3 as a distribution on $C_c^\infty(G(\mathbb{A})^1)$. However, to deal with the other side of the trace formula, and to exploit the present knowledge of invariant harmonic analysis, we need to work with $K$-finite functions. In Section 1, we defined the Hecke spaces $\mathbb{H}(G(F_S))$ and $\mathbb{H}_{\mathrm{ac}}(G(F_S))$, where $S$ is any finite set of valuations of $F$ with the closure property. Recall that $\mathbb{H}_{\mathrm{ac}}(G(F_S))$ consists of the Hecke functions $f$ on $G(F_S)$ of "almost compact" support, in the sense that, for any $b \in C_c^\infty(\mathfrak{a}_{G,S})$, the function

$$f^b(x) = f(x)b(H_G(x)), \quad x \in G(F_S),$$

belongs to $\mathbb{H}(G(F_S))$. We say that $G$ is *unramified* at a place $v$ if $v$ is finite, $G$ is quasisplit at $v$ and $G$ splits over an unramified extension of $F_v$. Let $S_{\text{ram}}$ be the finite set of valuations of $F$ at which $G$ is ramified. By definition, $S_{\text{ram}}$ contains $S_\infty$, the set of Archimedean valuations of $F$. Suppose that $S$ contains $S_{\text{ram}}$. We can multiply any function on $G(F_S)$ with the characteristic function of $\prod_{v \notin S} K_v$, thereby identifying it with a function on $G(\mathbb{A})$. This allows us to define the adelic Hecke spaces

$$\mathbb{H}(G(\mathbb{A})) = \varinjlim_S \mathbb{H}(G(F_S))$$

and

$$\mathbb{H}_{\text{ac}}(G(\mathbb{A})) = \varinjlim_S \mathbb{H}_{\text{ac}}(G(F_S)).$$

Similarly, we can define the Hecke space

$$\mathbb{H}(G(\mathbb{A})^1) = \varinjlim_S \mathbb{H}(G(F_S)^1),$$

on $G(\mathbb{A})^1$. The terms in the trace formula are actually linear forms on $\mathbb{H}(G(\mathbb{A})^1)$. However, the restriction map $f \mapsto f^1$ sends $\mathbb{H}_{\text{ac}}(G(\mathbb{A}))$ to $\mathbb{H}(G(\mathbb{A})^1)$. We shall usually regard the terms as linear forms on $\mathbb{H}(G(\mathbb{A}))$ or $\mathbb{H}_{\text{ac}}(G(\mathbb{A}))$ that factor through this map.

DEFINITION 12.7. (1) In Section 4.19, we defined function spaces $\mathbb{I}(G(F_S))$ and $\mathbb{I}_{\text{ac}}(G(F_S))$ on

$$\Pi_{\text{temp}}(G(F_S)) \times \mathfrak{a}_{G,S}.$$

These spaces are sometimes referred to as *Paley-Wiener spaces*.

(2) Let $\Pi(G(\mathbb{A}))$ (respectively $\Pi_{\text{unit}}(G(\mathbb{A}))$, $\Pi_{\text{temp}}(G(\mathbb{A}))$) denote the set of equivalence classes of irreducible admissible (respectively unitarizable, tempered) representations of $G(\mathbb{A})$.

(3) When $S$ contains $S_{\text{ram}}$, we have $\mathfrak{a}_{G,S} = \mathfrak{a}_G$. We can identify a function $\varphi$ on

$$\Pi_{\text{temp}}(G(F_S)) \times \mathfrak{a}_G$$

with the function on $\Pi_{\text{temp}}(G(\mathbb{A})) \times \mathfrak{a}_G$ whose value at

$$(\pi, X), \quad \pi = \otimes_v \pi_v, \ X \in \mathfrak{a}_G,$$

equals

$$\varphi\left(\otimes_{v \in S} \pi_v, X\right) \cdot \prod_{v \notin S} \operatorname{tr} \pi_v(K_v), \quad \pi_v(K_v) = \int_{K_v} \pi_v(k_v)\, dk_v.$$

With this convention, we then define

$$\mathbb{I}(G(\mathbb{A})) = \varinjlim_{S} \mathbb{I}(G(F_S))$$

and

$$\mathbb{I}_{\rm ac}(G(\mathbb{A})) = \varinjlim_{S} \mathbb{I}_{\rm ac}(G(F_S)).$$

(4) Any of our definitions can be transferred from $G$ to a Levi subgroup $M \in \mathcal{L}$. In particular, we have spaces $\mathbb{I}(M(\mathbb{A}))$ and $\mathbb{I}_{\rm ac}(M(\mathbb{A}))$. It is easy to see that the maps $f \mapsto f_M$ and $f \mapsto \phi_M(f)$ , described in Subsections 1.4 and 4.20.1, respectively, extend to continuous maps from $\mathbb{H}_{\rm ac}(G(\mathbb{A}))$ to $\mathbb{I}_{\rm ac}(M(\mathbb{A}))$.
(5) Recall that $\mathcal{L}_0(M)$ denotes the set of Levi subgroups $L$ of $G$ with $M \subset L \subsetneq G$.
(6) Recall that a distribution attached to $G$ is *supported on characters* if it vanishes on every function $f$ such that $f_G = 0$, where $f_G(\pi) = \operatorname{tr} \pi(f)$.

## 12.8 Local Theory

The goal of the remainder of this chapter is to use the local theory developed thus far to study the trace formula.

Suppose that $M \in \mathcal{L}$ and that $S$ is a finite set of valuations of $F$ with the closure property. In Section 2, for each $\gamma \in M(F_S)$, we defined the invariant distribution

$$I_M(\gamma, f) = J_M(\gamma, f) - \sum_{L \in \mathcal{L}_0(M)} \widehat{I}^L_M(\gamma, \phi_L(f)).$$

In Section 3, for $\tau \in \Pi(M(F_S))$, $X \in \mathfrak{a}_{M,S}$, we defined another family of invariant distributions,

$$I_M(\tau, X, f) = J_M(\tau, X, f) - \sum_{L \in \mathcal{L}_0(M)} \widehat{I}^L_M(\tau, X, \phi_L(f)).$$

Both families are regarded as linear forms on $\mathbb{H}_{\rm ac}(G(F_S))$. These definitions were contingent on an induction hypothesis from Section 2 that we continue to carry. Recall we assumed that, for any $S$ and for any elements $M \in \mathcal{L}$ and $L \in \mathcal{L}_0(M)$, the distributions

$$I^L_M(\gamma), \quad \gamma \in M(F_S),$$

on $L(F_S)$ are all supported on characters. Then the distributions $I_M(\gamma)$, and, thanks to Theorem 6.2, also the distributions $I_M(\tau, X)$, are well defined. In Corollary 15.7, we shall complete the induction argument by showing that the condition holds when $L$ is replaced by $G$.

## 12.9 Lack of Symmetry

The distributions $I_M(\gamma)$ and $I_M(\tau, X)$ have many parallel properties. However, there is one essential difference between the two. If $\tau \in \Pi(M(\mathbb{A}))$ and $X \in \mathfrak{a}_M$, it is easy to see that $I_M(\tau, X)$ can be defined as a linear form on $\mathbb{H}(G(\mathbb{A}))$ or even $\mathbb{H}_{\text{ac}}(G(\mathbb{A}))$. This is a consequence of the original definition of $J_M(\tau, X)$ in terms of normalized intertwining operators, and in particular, the property $(\mathcal{R}_8)$ of Theorem 4.6.2. On the other hand, if $\gamma$ belongs to $M(\mathbb{A})$, there seems to be no simple way to define $I_M(\gamma)$ as a linear form on $\mathbb{H}(G(\mathbb{A}))$. This circumstance is responsible for a certain lack of symmetry in the trace formula. The terms on the geometric side depend on a suitably large finite set $S$ of valuations, while the terms on the spectral side do not.

## 12.10 A Subgroup

If $G(\mathbb{A})$ is replaced by $G(\mathbb{A})^1$, we can obviously define the sets $\Pi(G(\mathbb{A})^1)$, $\Pi_{\text{unit}}(G(\mathbb{A})^1)$ and $\Pi_{\text{temp}}(G(\mathbb{A})^1)$ as above. The terms on the spectral side of the trace formula will depend on elements $M \in \mathcal{L}$ and representations $\tau^1 \in \Pi_{\text{unit}}(M(\mathbb{A})^1)$. We shall generally identify a representation $\tau^1 \in \Pi_{\text{unit}}(M(\mathbb{A})^1)$ with the corresponding orbit

$$\{\tau_\lambda \,;\, \lambda \in i\mathfrak{a}_M^*\}$$

of $i\mathfrak{a}_M^*$ in $\Pi_{\text{unit}}(M(\mathbb{A}))$. More specifically, given $\tau^1$ on $M(\mathbb{A})^1$, we define $\tau$ to be an extension to $M(\mathbb{A})$. To be definite, one may assume the representation is extended trivially. Then

$$\tau_\lambda(m) = \tau(m)e^{\lambda(H_M(m))}.$$

With this convention, let us agree to write

$$J_M(\tau^1, f) = J_M(\tau_\lambda, 0, f), \quad \lambda \in i\mathfrak{a}_M^*,$$

and

$$I_M(\tau^1, f) = I_M(\tau_\lambda, 0, f),$$

for the values of the distributions at $X = 0$ and $f \in \mathbb{H}_{\text{ac}}(G(\mathbb{A}))$. The two terms on the right are independent of $\lambda$, and are therefore well defined functions of $\tau^1$. They also depend only on the restriction $f^1$ of $f$ to $G(\mathbb{A})^1$. This notation applies also to the map $f_G$. For if $\pi$ is an arbitrary representation in $\Pi(G(\mathbb{A}))$ and $X \in \mathfrak{a}_G$, we have

$$f_G(\pi, X) = J_G(\pi, X, f) = I_G(\pi, X, f).$$

Therefore if $\pi^1$ belongs to $\Pi_{\mathrm{unit}}(G(\mathbb{A})^1)$, it makes sense to write

$$f_G(\pi^1) = f_G(\pi_\lambda, 0) = \operatorname{tr} \pi^1(f^1), \quad \lambda \in i\mathfrak{a}_G^*, \ f \in \mathbb{H}_{\mathrm{ac}}(G(\mathbb{A})).$$

## 12.11 Non-Invariant Trace Formula

The first version of the noninvariant trace formula is the identity

$$\sum_{\mathfrak{o} \in \mathcal{O}} J_{\mathfrak{o}}(f^1) = J(f^1) = \sum_{\chi \in \mathfrak{X}} J_\chi(f^1), \quad f^1 \in C_c^\infty(G(\mathbb{A})^1), \tag{12.11.1}$$

discussed in Chapter 3, in which a certain linear form $J$ on $C_c^\infty(G(\mathbb{A})^1)$ is expanded in two different ways. The sets $\mathcal{O} = \mathcal{O}(G, F)$ and $\mathfrak{X} = \mathfrak{X}(G, F)$ parametrize orbit theoretic and representation theoretic data, respectively, but the corresponding terms are not given as explicitly as one would like.

Suppose that $J_*(f)$ stands for one of the summands in (12.11.1). We write $f$ for $f^1$ here to simplify the notation. Then $J_*$ is a linear form on $C_c^\infty(G(\mathbb{A})^1)$ which behaves in a predictable way under conjugation:

$$J_*(f^y) = \sum_{Q \in \mathcal{F}} |W_0^Q||W_0^G|^{-1} J_*^{M_Q}(f_{Q,y}), \quad y \in G(\mathbb{A}).$$

See Theorem 4.4.3 for details. Since we want to take $f$ to be in $\mathbb{H}_{\mathrm{ac}}(G(\mathbb{A}))$, we cannot use this formula. However, as in the proof of Proposition 4.14.3, we can transform it to an alternate formula

$$J_*(L_h f) = \sum_{Q \in \mathcal{F}} |W_0^{M_Q}||W_0^G|^{-1} J_*^{M_Q}(R_{Q,h} f), \tag{12.11.2}$$

which makes sense for functions $f \in \mathbb{H}_{\mathrm{ac}}(G(\mathbb{A}))$ and $h \in \mathbb{H}(G(\mathbb{A})^1)$.

DEFINITION. (1) Let $\mathcal{L}_0$ denote the set of Levi subgroups $L \in \mathcal{L}$ with $L \neq G$.
(2) Define a linear form $I_*(f) = I_*^G(f)$ on $\mathbb{H}_{\mathrm{ac}}(G(\mathbb{A}))$ inductively by setting

$$I_*(f) = J_*(f) - \sum_{M \in \mathcal{L}_0} |W_0^M||W_0^G|^{-1} \widehat{I}_*^M(\phi_M(f)). \tag{12.11.3}$$

## 12.12 Induction Assumption

The invariance of $I_*$ follows from (12.11.2) and the analogous formula for $\phi_M$, see equation (20.2.2) and Proposition 4.11 in Chapter 4. Implicit in the definition is the induction assumption that, *for any Levi subgroup $L \in \mathcal{L}_0$, the distribution $I_*^L$*

*is defined and is supported on characters.* This is what allows us to write $\widehat{I}^L_*$. This is the second induction assumption currently in effect. However, in the next two sections, we shall establish explicit formulae for $I_{\mathfrak{o}}$ and $I_\chi$ in terms of $I_M(\gamma)$ and $I_M(\pi)$, respectively. This will reduce this second assumption to the primary one, adopted in 2.6.

Substitute (12.11.3) for each of the terms (12.11.1). For $f \in \mathbb{H}_{\mathrm{ac}}(G(\mathbb{A}))$, the result is an identity

$$\sum_{\mathfrak{o}\in\mathcal{O}} I_{\mathfrak{o}}(f) = I(f) = \sum_{\chi\in\mathfrak{X}} I_\chi(f), \tag{12.12.1}$$

in which the invariant distribution

$$I(f) = J(f) - \sum_{M\in\mathcal{L}_0} |W_0^M||W_0^G|^{-1}\widehat{I}^M(\phi_M(f)), \tag{12.12.2}$$

is expanded in two ways, see Theorem 4.4.12. This is the first version of the invariant trace formula. In [Ar86] and [Ar82.II] there are more explicit formulae for the terms $J_{\mathfrak{o}}(f)$ and $J_\chi(f)$ in (12.11.1). These formulae will be reviewed in the next two sections. The purpose of the remainder of this chapter is to convert these formulae into explicit expansions of each side of the invariant formula (12.12.1). In the process, we will establish the required properties of local harmonic analysis.

# 13 The Geometric Side

We shall derive a finer expansion for the geometric left-hand side of (12.12.1). The result will be a sum of terms, indexed by orbits in $G(F)$, which separate naturally into local and global constituents. We shall first review the results of [Ar86], which provide a parallel expansion for the noninvariant distributions on the left-hand side of (12.11.1).

DEFINITION 13.1. (1) Recall that two elements in $G(F)$ are *equivalent* if their semisimple Jordan components are conjugate by $G(F)$. We let $\mathcal{O} = \mathcal{O}(G, F)$ denote the set of equivalence classes. We write $\gamma \sim \gamma'$ if $\gamma$, $\gamma'$ are equivalent. See also Section 3.2.3.

(2) The formulae in [Ar86] are stated in terms of another equivalence relation on $G(F)$. It depends on a finite set $S$ of valuations of $F$. Let $\sigma \in G(F)$ be a semisimple element, and let $U$ be a unipotent conjugacy class in $G_\sigma(F_S)$. The $(G, S)$-*equivalence classes* are defined to be the sets

$$G(F) \cap (\sigma U)^{G(F)} = \{\delta^{-1}\sigma u\delta \,;\, \delta \in G(F), u \in U \cap G(F)\}.$$

Any class $\mathfrak{o}$ in $\mathcal{O}$ breaks up into a finite set $\{\mathfrak{o}\}_{G,S}$ of $(G, S)$-equivalence classes.

The following theorem, Theorem 8.1 of [Ar85], defines the global coefficient for unipotent elements, as well as providing an expansion of the distribution which is furthest from the unramified case. Let $\mathcal{U}_M(F)$ denote the $F$-points of the variety of unipotent elements. Then $(\mathcal{U}_M(F))_{M,S}$ is just the $M(F_S)$-conjugacy classes in $\mathcal{U}_M(F)$.

THEOREM 13.2. *For any $S$ sufficiently large, there are uniquely determined coefficients*

$$a^M(S, u), \quad M \in \mathcal{L}, u \in (\mathcal{U}_M(F))_{M,S},$$

*such that*

$$J_{\text{unip}}(f) = \sum_{M \in \mathcal{L}} |W_0^M||W_0^G|^{-1} \sum_{u \in (\mathcal{U}_M(F))_{M,S}} a^M(S, u) J_M(u, f), \quad f \in C_c^\infty(G(F_S)^1).$$

*We have*

$$a^M(S, 1) = \text{vol}(M(F)\backslash M(\mathbb{A})^1).$$

The following is Theorem 8.1 of [Ar86]. It is one of the main results of that paper.

THEOREM. *For any $\mathfrak{o} \in \mathcal{O}$ and any $f^1 \in C_c^\infty(G(F_S)^1)$, there is an expansion*

$$J_\mathfrak{o}(f^1) = \sum_{M \in \mathcal{L}} |W_0^M||W_0^G|^{-1} \sum_{\gamma \in (M(F) \cap \mathfrak{o})_{M,S}} a^M(S, \gamma) J_M(\gamma, f^1). \tag{13.2.1}$$

Here $S$ is any finite set of valuations of $F$ which contains a certain set $S_\mathfrak{o}$, determined by $\mathfrak{o}$. The distributions $J_M(\gamma, f^1)$ are purely local, in the sense that they depend only on $\gamma$ as an element in $M(F_S)$. The functions $a^M(S, \gamma)$ carry the global information. They were defined by [Ar86, (8.1)] and [Ar85, Theorem 8.1] in the case that $S$ contains $S_\mathfrak{o}$. The more general definition is given below.

DEFINITION 13.3. (1) Suppose that $M \in \mathcal{L}$. A semisimple element $\sigma \in M(F)$ is said to be *$F$-elliptic* in $M$ if the split component of the center of $M_\sigma$ equals $A_M$.

(2) Suppose that $S$ is any finite set of valuations of $F$ that contains $S_\infty$. Let

$$K_v^M = K_v \cap M(F_v)$$

and set

$$K_S^M = \prod_{v \in S} K_v^M.$$

(3) Suppose that $\gamma$ is an element in $M(F)$ with semisimple Jordan component $\sigma$. Set $i^M(S,\sigma)$ equal to 1 if $\sigma$ is $F$-elliptic in $M$, and if, for every $v \notin S$, the set

$$\mathrm{Ad}(M(F_v))\sigma = \{m\sigma m^{-1}\,;\, M \in M(F_v)\}$$

intersects the compact set $K_v^M$. Otherwise, set $i^M(S,\sigma)$ equal to 0.

(4) Let $\{u;\, \sigma u \sim \gamma\}$ be the set of $u \in (\mathcal{U}_{G_\sigma})_{G_\sigma,S}$ such that $\sigma u$ is $(G,S)$-equivalent to $\gamma$.

(5) Let $i^G(\sigma) = G_{\sigma,+}(F)/G_\sigma(F)$. Here $G_{\sigma,+}$ denotes the centralizer, rather than the connected centralizer. The group $i^G(\sigma)$ acts transitively on $\{u\,;\, \sigma u \sim \gamma\}$.

(6) Define

$$a^M(S,\gamma) = i^M(S,\sigma)|i^M(\sigma)|^{-1} \sum_{\{u\,;\,\sigma u\sim\gamma\}} a^{M_\sigma}(S,u). \tag{13.3.1}$$

This definition matches the one in [Ar86] in the special case that $S$ contains $S_{\mathfrak{o}}$, where $\mathfrak{o}$ is the equivalence class in $\mathcal{O}$ which contains $\sigma$.

The second main result of [Ar86] is its Theorem 9.2.

THEOREM 13.4. *For any $f^1 \in C_\Delta^\infty(G(F_S)^1)$, there is an expansion*

$$J(f^1) = \sum_{M\in\mathcal{L}} |W_0^M||W_0^G|^{-1} \sum_{\gamma\in(M(F))_{M,S}} a^M(S,\gamma)J_M(\gamma,f^1), \tag{13.4.1}$$

Here $\Delta$ is a compact neighborhood in $G(\mathbb{A})^1$, and $S$ is any finite set of valuations of $F$ which contains a certain set $S_\Delta$ determined by $\Delta$. This latter set is large enough so that $\Delta$ is the product of a compact neighborhood in $G(F_{S_\Delta})^1$ with the characteristic function of

$$\prod_{v\notin S_\Delta} K_v^G.$$

By definition,

$$C_\Delta^\infty(G(F_S)^1) = C_\Delta^\infty(G(\mathbb{A})^1) \cap C_c^\infty(G(F_S)^1).$$

REMARK. Note that [Ar86] does not write down the general definition (13.3.1) for $a^M(S,\gamma)$. This is required for the expansion (13.4.1) to make sense.

PROPOSITION 13.5. *Suppose that $S \supset S_{\mathfrak{o}}$ is a finite set of valuations and $f \in \mathbb{H}_{\mathrm{ac}}(G(F_S))$. Then*

$$I_{\mathfrak{o}}(f) = \sum_{M\in\mathcal{L}} |W_0^M||W_0^G|^{-1} \sum_{\gamma\in(M(F)\cap\mathfrak{o})_{M,S}} a^M(S,\gamma)I_M(\gamma,f).$$

PROOF. By definition, $I_{\mathrm{o}}(f)$ equals the difference between $J_{\mathrm{o}}(f)$ and

$$\sum_{L\in\mathcal{L}_0} |W_0^L||W_0^G|^{-1}\widehat{I}^L_{\mathrm{o}}(\phi_L(f)).$$

We can assume inductively that, if $L \in \mathcal{L}_0$, the proposition holds for $I^L_{\mathrm{o}}$. Since $\phi_L$ maps $\mathbb{H}_{\mathrm{ac}}(G(F_S))$ to $\mathbb{I}_{\mathrm{ac}}(L(F_S))$, we obtain

$$\widehat{I}^L_{\mathrm{o}}(\phi_L(f)) = \sum_{M\in\mathcal{L}^L} |W_0^M||W_0^L|^{-1} \sum_{\gamma\in(M(F)\cap\mathrm{o})_{M,S}} a^M(S,\gamma)\widehat{I}^L_M(\gamma,\phi_L(f)).$$

This is valid whenever $S$ contains the finite set $S^L_{\mathrm{o}}$ associated with $L$. A look at the condition defining $S_{\mathrm{o}}$ on [Ar86, p. 203] reveals that $S_{\mathrm{o}}$ contains $S^L_{\mathrm{o}}$, so we can certainly take any $S \supset S_{\mathrm{o}}$. Combining this formula with (13.2.1), we write $I_{\mathrm{o}}(f)$ as

$$\sum_{M\in\mathcal{L}} |W_0^M||W_0^G|^{-1} \sum_{\gamma\in(M(F)\cap\mathrm{o})_{M,S}} a^M(S,\gamma)\left(J_M(\gamma,f) - \sum_{L\in\mathcal{L}_0(M)} \widehat{I}^L_M(\gamma,\phi_L(f))\right).$$

The expression in brackets on the right equals $I_M(\gamma,f)$, so we obtain the required formula for $I_{\mathrm{o}}(f)$. □

## 13.6 *Proof That Induction Assumption* 2.6 *Implies Half of Assumption* 12.12

PROOF. Assumption 2.6 asserts that, for any $L \in \mathcal{L}_0$, the distributions $I^L_M(\gamma)$ are all supported on characters. The last proposition provides an expansion for $I^L_{\mathrm{o}}$ in terms of the distributions $I^L_M(\gamma)$. Therefore $I^L_{\mathrm{o}}$ is also supported on characters. Thus half of Assumption 12.12 is included in the original assumption. In Section 14 we shall take care of the rest of Assumption 12.12. □

To be able to exploit the Proposition 13.5 effectively, we shall establish an important support property of the distributions $I_M(\gamma)$.

DEFINITION 13.7. Fix an element $M \in \mathcal{L}$, a finite set $S_1$ of valuations containing $S_{\mathrm{ram}}$, and a compact neighborhood $\Delta_1$ in $G(F_{S_1})$. Let $\mathbb{H}_{\Delta_1}(G(F_{S_1}))$ denote the set of functions in $\mathbb{H}(G(F_{S_1}))$ which are supported on $\Delta_1$.

PROPOSITION 13.8. *There is a compact subset $\Delta^M_1$ of $M(F_{S_1})^1$ such that, for any finite set $S \supset S_1$ and any $f$ in the image of $\mathbb{H}_{\Delta_1}(G(F_{S_1}))$ in $\mathbb{H}(G(F_S))$, the function of $\gamma \in M(F_S)^1$ given by*

$$\gamma \mapsto I_M(\gamma,f)$$

*is supported on the set*

$$\mathrm{Ad}(M(F_S))(\Delta_1^M K_S^M) = \{mcm^{-1}\,;\, m \in M(F_S), c \in \Delta_1^M K_S^M\}.$$

PROOF. Suppose that

$$\mathfrak{M}_1 = \prod_{v \in S_1} M_v$$

is a Levi subgroup of $M$ defined over $F_{S_1}$. That is, for each $v \in S_1$, $M_v$ is a Levi subgroup of $M$ which is defined over $F_v$. Let $M_v(F_v)'$ be the set of elements $\gamma_v \in M_v(F_v)$ whose semisimple component $\sigma_v$ satisfies the following two conditions:

(i) The connected centralizer $M_{\sigma_v}$ of $\sigma_v$ in $M$ is contained in $M_v$.
(ii) $\sigma_v$ is an $F_v$-elliptic point in $M_v$.

Set

$$\mathfrak{M}_1(F_{S_1})' = \prod_{v \in S_1} M_v(F_v)'.$$

Consider the restriction of the map

$$H_{\mathfrak{M}_1} = \oplus_{v \in S_1} H_{M_v} : \mathfrak{M}_1(F_{S_1}) \to \mathfrak{a}_{\mathfrak{M}_1} = \oplus_{v \in S_1} \mathfrak{a}_{M_v}$$

to $\mathfrak{M}_1(F_{S_1})'$. The map is certainly constant on the orbit of

$$\mathfrak{M}_1(F_{S_1}) = \prod_{v \in S_1} M_v(F_v).$$

The $F_v$-elliptic set in $M_v(F_v)$ has a set of representative which is compact modulo $A_{M_v}(F_v)$. It follows that $H_{\mathfrak{M}_1}$ is proper as a map on the space of $\mathfrak{M}_1(F_{S_1})$-orbits in $\mathfrak{M}_1(F_{S_1})'$. To prove the proposition, we shall combine this fact with the descent and splitting properties of $I_M(\gamma, f)$.

We may assume that

$$\Delta_1 = \prod_{v \in S_1} \Delta_v \quad \text{and} \quad f = \prod_{v \in S} f_v,$$

so that $f_v$ belongs to $\mathbb{H}_{\Delta_v}(G(F_v))$ if $v$ belongs to $S_1$, and $f_v$ equals the characteristic function of $K_v^G$ if $v$ belongs to the complement of $S_1$ in $S$. Suppose that

$$\gamma = \prod_{v \in S} \gamma_v$$

is an element in $M(F_S)^1$ such that $I_M(\gamma,f) \neq 0$. For each $v \in S_1$, let $\sigma_v$ be the semisimple part of $\gamma_v$, and let $A_{\sigma_v}$ be the split component of the center of $M_{\sigma_v}$. Set $M_v$ equal to the centralizer of $A_{\sigma_v}$ in $M$. Then $\gamma_v$ belongs to $M_v(F_v)'$. In other words, if

$$\mathfrak{M}_1 = \prod_{v \in S_1} M_v,$$

the element

$$\gamma_1 = \prod_{v \in S_1} \gamma_v$$

belongs to $\mathfrak{M}_1(F_{S_1})'$. If we were to replace $\gamma$ by an $M(F_{S_1})$-conjugate, $\mathfrak{M}_1$ would be similarly conjugated, but $I_M$ would remain nonzero. There are only finitely many $M(F_{S_1})$-orbits of Levi subgroups $\mathfrak{M}_1$ over $F_{S_1}$. It is therefore sufficient to fix $\mathfrak{M}_1$ and to consider only those elements $\gamma$ such that $\gamma_1$ belongs to $\mathfrak{M}_1(F_{S_1})'$.

For each valuation $w$ in $S - S_1$, we set $M_w = M$. We then define a Levi subgroup

$$\mathfrak{M} = \mathfrak{M}_1 \times \left( \prod_{w \in S - S_1} M_w \right) = \prod_{v \in S} M_v$$

of $M$ over $F_S$. Regarding $\gamma$ as an element in $\mathfrak{M}(F_S)$, we can form the induced class (see Subsection 2.6)

$$\gamma^M = \prod_{v \in S} \gamma_v^M.$$

But $M_{v,\gamma_v} = M_{\gamma_v}$ for each $v$, so $\gamma^M$ is just the $\mathfrak{M}(F_S)$-orbit of $\gamma$. Applying Corollary 9.4, we obtain

$$I_{\mathfrak{M}}(\gamma,f) = \sum_{\mathfrak{L} \in \mathcal{L}(\mathfrak{M})} d_{\mathfrak{M}}^G(M, \mathfrak{L}) \prod_{v \in S} \widehat{I}_{M_v}^{L_v}(\gamma_v, f_{v,L_v}) \neq 0.$$

Recalling the definition of the number $d_{\mathfrak{M}}^G(M, \mathfrak{L})$ in Subsection 9.3, we find that we can choose

$$\mathfrak{L} = \prod_{v \in S} L_v, \quad L_v \in \mathcal{L}(M_v),$$

so that the natural map $\mathfrak{a}_{\mathfrak{M}}^G \to \mathfrak{a}_M^G \oplus \mathfrak{a}_{\mathfrak{L}}^G$ is an isomorphism, and so that, for all $v \in S$,

$$\widehat{I}_{M_v}^{L_v}(\gamma_v, f_{v,L_v}) \neq 0. \qquad (13.8.1)$$

Suppose first that $w$ is a valuation in the complement of $S_1$ in $S$. Since $f_w$ is the characteristic function of $K_w^G$, Proposition 2.19 tells us that

$$\widehat{I}^{L_w}_{M_w}(\gamma_w, f_{w,L_w}) = I^{L_w}_M(\gamma_w, f_{w,Q_w}) = J^{L_w}_M(\gamma_w, f_{w,Q_w})$$

for any $Q_w \in \mathcal{P}(L_w)$. The function on the right is a weighted orbital integral. By [Ar88.1, Corollary 6.2], it is the integral with respect to a measure on the induced class $\gamma_w^G$. (See also the statement of Theorem 2.4 and Subsection 2.11.) Therefore the class $\gamma_w^G$ must intersect $K_w^G$. Combining the definition of the induced class $\gamma_w^G$ with the standard properties of the special maximal compact group $K_w$, we find that the $M(F_w)$-orbit of $\gamma_w$ intersects $K_w^M$. Notice in particular that $H_M(\gamma_w) = 0$.

We turn, finally, to the valuations in $S_1$. It remains for us to show that the $M(F_{S_1})$-orbit of $\gamma_1$ intersects a compact subset $\Delta_1^M$ of $M(F_{S_1})$ which depends only on $\Delta_1$. We are already assuming that $\gamma_1$ belongs to $\mathfrak{M}_1(F_{S_1})'$, so by the discussion above, we need only show that $H_{\mathfrak{M}_1}(\gamma_1)$ lies in a fixed compact subset of $\mathfrak{a}_{\mathfrak{M}_1}$. Set

$$\mathfrak{L}_1 = \prod_{v \in S_1} L_v.$$

It is clear that the natural map

$$\mathfrak{a}_{\mathfrak{M}_1} \to \mathfrak{a}_M \oplus \mathfrak{a}_{\mathfrak{L}_1}$$

is injective. But the image of $H_{\mathfrak{M}_1}(\gamma_1)$ in $\mathfrak{a}_M$ equals

$$H_M(\gamma_1) = H_M(\gamma) - \sum_{w \in S - S_1} H_M(\gamma_w) = 0,$$

since $\gamma$ belongs to $M(F_S)^1$. We have only to show that the image of $H_{\mathfrak{M}_1}(\gamma_1)$ in $\mathfrak{a}_{\mathfrak{L}_1}$, namely, the vector

$$H_{\mathfrak{L}_1}(\gamma_1) = \oplus_{v \in S_1} H_{L_v}(\gamma_v),$$

lies in a compact subset of $\mathfrak{a}_{\mathfrak{L}_1}$ which depends only on $\Delta_1$. For any $v \in S_1$, the distribution $\widehat{I}^{L_v}_{M_v}(\gamma_v, f_{v,L_v})$ depends only on the restriction of $f_v$ to the set

$$\{x_v \in G(F_v)\,;\, H_{L_v}(x_v) = H_{L_v}(\gamma_v)\}.$$

It follows from (13.8.1) that $H_{L_v}(\gamma_v)$ belongs to $H_{L_v}(\Delta_v)$, the image of the support of $f_v$. In other words, $H_{\mathfrak{L}_1}(\gamma_1)$ belongs to $\oplus_{v \in S_1} H_{L_v}(\Delta_v)$, a compact set which depends only on $\Delta_1$. This completes the proof of the proposition. □

DEFINITION 13.9. Suppose that $f$ belongs to $\mathbb{H}(G(\mathbb{A}))$.

(1) We shall write supp($f$) for the support of $f$.

(2) There exists a finite set $S$ of valuations of $f$, which contains $S_{\mathrm{ram}}$, such that $f$ is the image of a function in $\mathbb{H}(G(F_S))$. We shall write $V(f)$ for the minimal such set.
(3) If $S$ is any such set and $\gamma$ is a point in $(M(F))_{M,S}$, we shall understand $I_M(\gamma,f)$ to mean the value of the distribution $I_M(\gamma)$ at $f$, regarded as a function in $\mathbb{H}(G(F_S))$. Since we are thinking of $I_M(\gamma)$ as a local object, this convention is quite reasonable. It simply means that when $\gamma \in (M(F))_{M,S}$ parametrizes such a distribution, we should treat $\gamma$ as a point in $M(F_S)$ rather than $M(F)$.

THEOREM 13.10. *Suppose that $f \in \mathbb{H}(G(\mathbb{A}))$. Then*

$$I(f) = \sum_{M\in\mathcal{L}} |W_0^M||W_0^G|^{-1} \sum_{\gamma\in(M(F))_{M,S}} a^M(S,\gamma)I_M(\gamma,f),$$

*where $S$ is any finite set of valuations which is sufficiently large, in a sense that it depends only on* supp$(f)$ *and $V(f)$. The inner series can be taken over a finite subset of $(M(F))_{M,S}$ which also depends only on* supp$(f)$ *and $V(f)$.*

PROOF. By (12.12.1) and Proposition 13.5, we have

$$I(f) = \sum_{\mathfrak{o}\in\mathcal{O}} \sum_{M\in\mathcal{L}} |W_0^M||W_0^G|^{-1} \sum_{\gamma\in(M(F)\cap\mathfrak{o})_{M,S}} a^M(S,\gamma)I_M(\gamma,f),$$

where $S$ is any finite set of valuations that contains $S_{\mathfrak{o}}$. We shall use Proposition 13.8 to show that the sum over $\mathfrak{o}$ is finite.

Choose any finite set $S_1 \supset S_{\mathrm{ram}}$, and a compact neighborhood $\Delta_1$ in $G(F_{S_1})$, such that $f$ belongs to $\mathbb{H}_{\Delta_1}(G(F_{S_1}))$. Assume that $S$ contains $S_1$. Suppose that a class $\mathfrak{o}$ gives a nonzero contribution to the sum above. Then there is an $M \in \mathcal{L}$ and an element $\gamma \in (M(F) \cap \mathfrak{o})_{M,S}$ such that

$$a^M(S,\gamma)I_M(\gamma,f) \neq 0.$$

The nonvanishing of $a^M(S,\gamma)$ implies that, for each $v \in S$, the image of $\gamma$ in $M(F_v)$ lies in

$$\mathrm{Ad}(M(F_v))K_v^M.$$

The image of $\gamma$ in $M(F_S)$ then lies in $M(F_S)^1$, and therefore belongs to a set

$$\mathrm{Ad}(M(F_S))(\Delta_1^M K_S^M),$$

by Proposition 13.8. It follows that the $M(\mathbb{A})$-orbit of $\gamma$ meets the compact set $\Delta_1^M K^M$, and in particular that

$$\mathrm{Ad}(G(\mathbb{A}))\mathfrak{o} \cap \Delta_1^M K^M \neq \emptyset.$$

By [Ar86, Lemma 9.1], $\mathfrak{o}$ must belong to a finite subset $\mathcal{O}_1$ of $\mathcal{O}$. Since $\Delta_1^M$ depends only on $\Delta_1$, $\mathcal{O}_1$ clearly depends only on supp($f$) and $V(f)$. The required expansion for $I(f)$ then holds if $S$ is any finite set which contains the union of $S_1$ with the sets $S_{\mathfrak{o}}$, as $\mathfrak{o}$ ranges over $\mathcal{O}_1$. This establishes the first assertion of the theorem. The union over $\mathfrak{o} \in \mathcal{O}_1$ of the sets

$$(M(F) \cap \mathfrak{o})_{M,S}$$

is certainly a finite subset of $(M(F))_{M,S}$, so the second assertion also follows. □

# 14 The Spectral Side

## *14.1 Finer Expansion*

We shall derive a finer expansion for the right-hand side of (12.12.1). The result will be a sum of terms, indexed by irreducible representations, which separate naturally into local and global constituents. Again, there is a parallel expansion for the noninvariant distributions on the right-hand side of (12.11.1). It is provided by the results of [Ar82.II] and [CLL84]. However, these results are not immediately in the form we want, and it is necessary to review them in some detail.

DEFINITION 14.2. (1) The set $\mathfrak{X} = \mathfrak{X}(G, F)$ consists of cuspidal automorphic data. It was introduced in Section 3.3. It is the set of orbits

$$\chi = \{s(L, \theta)\,;\, s \in W_0\},$$

where $L$ is a Levi subgroup in $\mathcal{L}$, $\theta$ is an irreducible cuspidal automorphic representation of $L(\mathbb{A})^1$, and the pair $(L, \theta)$ is fixed by some element in the Weyl group $W_0^G$ of isomorphisms of $\mathfrak{a}_0$ induced from $G$.

The set $\mathfrak{X}$ has been used to describe the convergence of the spectral side, which is more delicate than that of the geometric side. However, for applications that involve a comparison of trace formulae, it is easier to handle the convergence by keeping track of Archimedean infinitesimal characters.

(2) Set

$$F_\infty = F_{S_\infty} = \prod_{v \in S_\infty} F_v.$$

(3) Regarding $G(F_\infty)$ as a real Lie group, we can define the abelian Lie algebra

$$\mathfrak{h} = i\mathfrak{h}_K \oplus \mathfrak{h}_0$$

as in [Ar82.I, Section3]. Then $\mathfrak{h}_0$ is the Lie algebra of a fixed maximal real split torus in $M_0(F_\infty)$ (the fixed minimal Levi subgroup), and $\mathfrak{h}_K$ is a fixed Cartan subalgebra of the centralizer of $\mathfrak{h}_0$ in

$$K_\infty = \prod_{v \in S_\infty} K_v.$$

(4) The complexification $\mathfrak{h}_\mathbb{C}$ is a Cartan subalgebra of the complex Lie algebra of $G(F_\infty)$. The real form $\mathfrak{h}$ is invariant under the complex Weyl group $W^G$ of $G(F_\infty)$. (By definition, $W^G$ equals $\mathrm{Ad}(\varepsilon)W$, where $\varepsilon$ is any element in $G(F_\infty)$ which normalizes $\mathfrak{h}_\mathbb{C}$, and $W$ is the complex Weyl group of $G(F_\infty)$ with respect to $\mathfrak{h}$.)

(5) Fix a Euclidean norm $\|\cdot\|$ on $\mathfrak{h}$ which is invariant under $W^G$. We shall also write $\|\cdot\|$ for the dual Hermitian norm on $\mathfrak{h}^*_\mathbb{C}$.

(6) To any representation $\tau \in \Pi(M(\mathbb{A}))$, we can associate the induced representation $\tau^G$ of $G(\mathbb{A})$. We multiply the representation by the factor $\delta_P^{1/2}$ before inducing, to preserve unitarity.

(7) Let $\nu_\tau$ denote the infinitesimal character of the Archimedean constituent of $\tau$. It is a $W$-orbit in $\mathfrak{h}^*_\mathbb{C}$. We shall actually be more concerned with the case that $\tau^1$ is a representation in $\Pi(M(\mathbb{A})^1)$. Then $\nu_{\tau^1}$ is determined a priori only as an orbit of $\mathfrak{a}^*_{M,\mathbb{C}}$ in $\mathfrak{h}^*_\mathbb{C}$. However, this orbit has a unique point of smallest norm in $\mathfrak{h}^*_\mathbb{C}$ (up to translation by $W$), and it is this point which we shall denote by $\nu_{\tau^1}$.

(8) If $t$ is a nonnegative real number, let $\Pi_{\mathrm{unit}}(M(\mathbb{A})^1, t)$ denote the set of representations $\tau^1 \in \Pi_{\mathrm{unit}}(M(\mathbb{A})^1)$ such that

$$\|\,\mathrm{Im}(\nu_{\tau^1})\| = t,$$

where $\mathrm{Im}(\nu_{\tau^1})$ is the imaginary part of $\nu_{\tau^1}$ relative to the real form $\mathfrak{h}^*$ of $\mathfrak{h}^*_\mathbb{C}$. We adopt similar notation when $M$ is replaced by a group $L \in \mathcal{L}$. In particular, if

$$\chi = \{s(L, \theta)\,;\, s \in W_0^G\}$$

is any class in $\mathfrak{X}$, we set $\nu_\chi = \nu_\theta$.

(9) Suppose that $L$ is a Levi subgroup in $\mathcal{L}$. Set

$$A_{L,\infty} = A_{L,\mathbb{Q}}(\mathbb{R}),$$

where $A_{L,\mathbb{Q}}$ is the split component of the center of the group obtained by restricting scalars from $F$ to $\mathbb{Q}$.

(10) Let

$$L^2_{\mathrm{disc},t}(L(F)A_{L,\infty}\backslash L(\mathbb{A})) \tag{14.2.1}$$

be the subspace of $L^2(L(F)A_{L,\infty}\backslash L(\mathbb{A}))$ which decomposes under $L(\mathbb{A})$ as a direct sum of representations in $\Pi_{\mathrm{unit}}(L(\mathbb{A}), t)$.

(11) For any group $Q$ in $\mathcal{P}(L)$ and point $\Lambda \in \mathfrak{a}^*_{L,\mathbb{C}}$, let

$$r_{Q,t}(\Lambda) : x \mapsto r_{Q,t}(\Lambda, x)$$

be the induced representation of $G(\mathbb{A})$ obtained from (14.2.1). Thus $r = r_{Q,t}$ acts by

$$(r(\Lambda, y)\psi)(g) = \psi(gy)e^{\langle \Lambda+\rho_Q, H_Q(gy)-H_Q(g)\rangle}.$$

(12) If $Q'$ is another group in $\mathcal{P}(L)$, the theory of Eisenstein series provides an intertwining operator $\mathcal{M}_{Q'|Q}(\Lambda)$ from $r_{Q,t}(\Lambda)$ to $r_{Q',t}(\Lambda)$.

PROPOSITION 14.3. *The representation $r_{Q,t}(\Lambda)$ is admissible.*

PROOF. The assertion is that the restriction of $r_{Q,t}(\Lambda)$ to $K$ contains any irreducible representation with only finite multiplicity. Since admissibility is preserved under parabolic induction, it is enough to show that the representation of $L(\mathbb{A})$ on (14.2.1) is admissible. To this end, we may assume that $L = G$. The assertion is then a consequence of Langlands' theory of Eisenstein series [La76, Chapter 7]. Indeed, one of the main results of [La76] is a decomposition

$$L^2_{\text{disc},t}(G(F)A_{G,\infty}\backslash G(\mathbb{A})) = \oplus_\chi L^2_{\text{disc},\chi}(G(F)A_{G,\infty}\backslash G(\mathbb{A})),$$

where $\chi$ ranges over the data in $\mathfrak{X}$ such that $\| \operatorname{Im}(\nu_\chi)\|$ equals $t$, and each corresponding summand is an admissible $G(\mathbb{A})$-module. On the other hand, the set of all $\chi$ whose associated cuspidal representations contain the restrictions of a given $K$-type have discrete infinitesimal characters. That is, the associated points $\{\nu_\chi\}$ form a discrete subset of $B + i\mathfrak{a}^*_M$, with $B$ a compact ball about the origin in $\mathfrak{a}^*_M$. It follows that there are only finitely many modules $L^2_{\text{disc},\chi}$ in the direct sum above, which contain a given $K$-type. The proposition follows. □

## 14.4 Induced Representations

Consider the representation $r_{Q,t}(\Lambda)$ of $G(\mathbb{A})$. Suppose that $s$ is an element in $W^G_0$ with representative $w$ in $G(F)$. If $y$ belongs to $G(\mathbb{A})$, we can define a linear map from the underlying Hilbert space of $r_{Q,t}(\Lambda)$ to that of $r_{sQ,t}(s\Lambda)$ by

$$(r_{Q,t}(s, \Lambda, y)\psi)(g) = \psi(w^{-1}gy)e^{\langle \Lambda+\rho_Q, H_Q(w^{-1}gy)-H_Q(g)\rangle}.$$

It satisfies

$$r_{Q,t}(s, \Lambda, y_1yy_2) = r_{sQ,t}(s\Lambda, y_1)r_{Q,t}(s, \Lambda, y)r_{Q,t}(\Lambda, y_2), \tag{14.4.1}$$

for any points $y_1$ and $y_2$ in $G(\mathbb{A})$. Indeed, with

$$a = \langle s\Lambda + \rho_{sQ}, H_{sQ}(gy_1) - H_{sQ}(g)\rangle,$$
$$b = \langle \Lambda + \rho_Q, H_Q(w^{-1}gy_1y) - H_Q(gy_1)\rangle,$$
$$c = \langle \Lambda + \rho_Q, H_Q(w^{-1}gy_1yy_2) - H_Q(w^{-1}gy_1y)\rangle,$$

we have

$$\begin{aligned} r(s\Lambda, y_1)r(s,\Lambda,y)r(\Lambda,y_2)\psi(g) &= r(s,\Lambda,y)r(\Lambda,y_2)\psi(gy_1)e^a \\ &= r(\Lambda,y_2)\psi(w^{-1}gy_1y)e^b e^a \\ &= \psi(w^{-1}gy_1yy_2)e^c e^b e^a. \end{aligned}$$

Note that $a+b+c = \langle \Lambda + \rho_Q, H_Q(w^{-1}gy_1yy_2) - H_Q(g)\rangle$. This map depends only on the image of $s$ in $W_0^G/W_0^{L_0}$. In particular, it is well defined for any element in $W^G(\mathfrak{a}_{L_0})$, the normalizer of $\mathfrak{a}_{L_0}$ in $W_0^G$.

Suppose that $s$ is an element in $W^G(\mathfrak{a}_{L_0})$ which fixes $\Lambda$. If $f$ is a function in $\mathbb{H}(G(\mathbb{A}))$, we write

$$r_{Q,t}(s,\Lambda,f^1) = \int_{G(\mathbb{A})^1} f(x) r_{Q,t}(s,\Lambda,x)\,\mathrm{d}x.$$

Then

$$\mathcal{M}_{Q|sQ}(\Lambda) r_{Q,t}(s,\Lambda,f^1)$$

is an operator of trace class on the underlying Hilbert space of $r_{Q,t}(\Lambda)$. According to (14.4.1), its trace is an invariant distribution which, by Proposition 14.3, can be written as a finite linear combination of irreducible characters

$$\operatorname{tr}\pi^1(f^1) = f_G(\pi^1), \quad \pi^1 \in \Pi_{\mathrm{unit}}(G(\mathbb{A})^1, t).$$

DEFINITION 14.5. (1) Consider the expression

$$\sum_{L\in\mathcal{L}} |W_0^L||W_0^G|^{-1} \sum_s |\det(s-1)_{\mathfrak{a}_L^G}|^{-1} \operatorname{tr}\left(\mathcal{M}_{Q|sQ}(0) r_{Q,t}(s,0,f^1)\right), \qquad (14.5.1)$$

where $Q$ stands for any element in $\mathcal{P}(L)$ and $s$ is summed over the Weyl group

$$W^G(\mathfrak{a}_L)_{\mathrm{reg}} = \{s \in W^G(\mathfrak{a}_L)\,;\ \det(s-1)_{\mathfrak{a}_L^G} \neq 0\}.$$

This is just the “discrete part” of the formula for

$$\sum_{\{\chi\in\mathfrak{X}\,;\,\|\mathrm{Im}(\nu_\chi)\|=t\}} J_\chi(f), \quad f \in \mathbb{H}(G(\mathbb{A})),$$

provided by [Ar82.II, Theorem 8.2].

(2) We can then rewrite (14.5.1) as

$$\sum_{\pi^1 \in \Pi_{\mathrm{unit}}(G(\mathbb{A})^1, t)} a^G_{\mathrm{disc}}(\pi^1) f_G(\pi^1), \tag{14.5.2}$$

a finite linear combination of irreducible characters. The complex-valued function

$$a_{\mathrm{disc}}(\pi^1) = a^G_{\mathrm{disc}}(\pi^1), \quad \pi^1 \in \Pi_{\mathrm{unit}}(G(\mathbb{A})^1, t),$$

which is defined by the equality of (14.5.1) and (14.5.2), is the primary global datum for the spectral side.

(3) It is convenient to work with a manageable subset of $\Pi_{\mathrm{unit}}(G(\mathbb{A})^1, t)$ which contains the support of $a^G_{\mathrm{disc}}(\pi^1)$. For $M \in \mathcal{L}$, $\tau^1 \in \Pi_{\mathrm{unit}}(M(\mathbb{A})^1, t)$ and $\lambda \in i\mathfrak{a}^*_M/i\mathfrak{a}^*_G$, we write $\tau$ for the trivial extension of $\tau^1$ to $M(\mathbb{A})$ and $\tau_\lambda$ for the representation given by

$$\tau_\lambda(x) = \tau(x) e^{\lambda(H_M(x))}.$$

The induced representation of $G$, $\tau^G_\lambda$, restricted to $G(\mathbb{A})^1$, denoted $(\tau^G_\lambda)^1$, has the same irreducible constituents as the induced representation $\tau^{G^1}_\lambda$. We let $\Pi_{\mathrm{disc}}(G^1, t)$ denote the subset of $\Pi_{\mathrm{unit}}(G(\mathbb{A})^1, t)$ consisting of the irreducible constituents of such $(\tau_\lambda)^{G^1}$, where $\tau_\lambda$ satisfies the following two conditions:

(i) $a^M_{\mathrm{disc}}(\tau^1) \neq 0$.
(ii) There is an element $s \in W^G(\mathfrak{a}_M)_{\mathrm{reg}}$ such that $s\tau_\lambda = \tau_\lambda$.

Any representation in $\Pi_{\mathrm{disc}}(G^1, t)$ is an irreducible constituent of an induced representation $r_{Q,t}(0)$, $Q \in \mathcal{F}$. From Proposition 14.3, we obtain

COROLLARY 14.6. *Suppose that $\Gamma$ is a finite subset of $\Pi(K)$. Then there are only finitely many representations $\pi^1 \in \Pi_{\mathrm{disc}}(G^1, t)$ whose restrictions to $K$ contain an element in $\Gamma$. In particular, there are only finitely many $\pi^1 \in \Pi(G(\mathbb{A})^1, t)$ which contain an element in $\Gamma$ and such that $a^G_{\mathrm{disc}}(\pi^1) \neq 0$.*

## 14.7 Global Ingredient

The remaining global ingredient is a function constructed from the global normalizing factors introduced in [Ar82.II, Section6]. We shall recall briefly how it is defined. Suppose that $M \in \mathcal{L}$ and that $\tau^1 = \otimes_v \tau_v$ belongs to $\Pi_{\mathrm{disc}}(M, t)$. Then $\tau^1$ is an irreducible constituent of some representations

$$r_{R,t}(0), \quad L \in \mathcal{L}^M, R \in \mathcal{P}^M(L).$$

If $P \in \mathcal{P}(M)$, we can form the induced representation

$$i_P^G(\tau_\lambda) = i_P(\tau_\lambda), \quad \lambda \in \mathfrak{a}^*_{M,\mathbb{C}}.$$

It is a subrepresentation of $r_{Q,t}(\lambda)$, where $Q$ is the group $P(R)$ in $\mathcal{P}(L)$ which is contained in $P$ and whose intersection with $M$ is $R$. If $P' \in \mathcal{P}(M)$ and $Q' = P'(R)$, the operator

$$\mathcal{M}_{P'|P}(\tau_\lambda) = \prod_v \mathcal{M}_{P'|P}(\tau_{v,\lambda}),$$

defined as an infinite product of unnormalized intertwining operators, is therefore equivalent to the restriction of $\mathcal{M}_{Q'|Q}(\lambda)$ to an invariant subspace. The theory of Eisenstein series tells us that the infinite product converges for certain $\lambda$ and can be analytically continued to an operator-valued function which is unitary when $\lambda \in i\mathfrak{a}^*_M$. But we also have the normalized intertwining operator

$$\mathcal{R}_{P'|P}(\tau_\lambda) = \prod_v \mathcal{R}_{P'|P}(\tau_{v,\lambda}) = \prod_v (m_{P'|P}(\tau_{v,\lambda})^{-1}\mathcal{M}_{P'|P}(\tau_{v,\lambda})),$$

described in Section 4.6. The infinite product reduces to a finite product at any smooth vector. It follows that the infinite product

$$m_{P'|P}(\tau_\lambda) = \prod_v m_{P'|P}(\tau_{v,\lambda})$$

of local normalizing factors converges for certain $\lambda$ and can be continued as a meromorphic function which is analytic for $\lambda \in i\mathfrak{a}^*_M$. Moreover,

$$m_{P''|P}(\tau_\lambda) = m_{P''|P'}(\tau_\lambda)m_{P'|P}(\tau_\lambda),$$

if $P''$ is a third element in $\mathcal{P}(M)$.

For a fixed $P' \in \mathcal{P}(M)$, we define the $(G, M)$-family

$$m_P(\nu, \tau_\lambda, P') = m_{P|P'}(\tau_\lambda)^{-1} m_{P|P'}(\tau_{\lambda+\nu}), \quad P \in \mathcal{P}(M),\ \nu \in i\mathfrak{a}^*_M.$$

Since

$$m_{P|P'}(\tau_{v,\lambda}) = \prod_{\alpha \in \Sigma^r_P \cap \Sigma^r_{\overline{P}'}} m_\alpha(\tau_v, \lambda(\alpha^\vee))$$

for each $v$ (Subsection 4.6.4), we have

$$m_{P|P'}(\tau_\lambda) = \prod_{\alpha \in \Sigma^r_P \cap \Sigma^r_{\overline{P}'}} m_\alpha(\tau, \lambda(\alpha^\vee)),$$

where $m_\alpha(\tau, z)$ equals an infinite product

$$\prod_v m_\alpha(\tau_v, z), \quad z \in \mathbb{C},$$

which converges in some half plane. Therefore the $(G, M)$-family is of the special sort considered in [Ar82.II, Section7]. In particular, if $L \in \mathcal{L}(M)$ and $Q \in \mathcal{P}(L)$, the number

$$m_M^L(\tau_\lambda) = \lim_{\nu \to 0} \sum_{\{P \in \mathcal{P}(M)\,;\, P \subset Q\}} m_P(\nu, \tau_\lambda, P')/\theta_P^Q(\nu)$$

can be expressed in terms of logarithmic derivatives

$$m_\alpha(\tau_\lambda, 0)^{-1} m'_\alpha(\tau_\lambda, 0), \quad \alpha \in \Sigma^r(L, A_M),$$

and is independent of $Q$ and $P'$ [Ar82.II, Proposition 7.5]. As a function of $\lambda \in i\mathfrak{a}_M^*$, it is a tempered distribution [Ar82.II, Lemma 8.4].

DEFINITION 14.8. (1) Let $M \in \mathcal{L}$ be a Levi subgroup. For each $M_1 \in \mathcal{L}^M$, that is, $M_1 \subset M$, set

$$\Pi_{M_1}(M, t) = \{\tau_{1,\lambda}\,;\, \tau_1^1 \in \Pi_{\mathrm{disc}}(M_1^1, t), \lambda \in i\mathfrak{a}_{M_1}^*/i\mathfrak{a}_M^*\}.$$

Recall that $\tau_1$ is an extension of $\tau_1^1$ to $M_1(\mathbb{A})$, and $\tau_{1,\lambda}$ is the twist of $\tau_1$ by $\exp(\lambda(H_M(\cdot))$.

Let $\Pi(M, t)$ be the disjoint union over all such $M_1$:

$$\Pi(M, t) = \bigcup_{M_1 \in \mathcal{L}^M} \Pi_{M_1}(M, t).$$

An arbitrary element of $\Pi(M, t)$ will then be denoted by $\tau_{1,\lambda}$.

(2) We define a measure $\mathrm{d}\tau_{1,\lambda}$ on $\Pi(M, t)$ by setting

$$\int_{\Pi(M,t)} \phi(\tau_{1,\lambda})\,\mathrm{d}\tau_{1,\lambda} = \sum_{M_1 \in \mathcal{L}^M} |W_0^{M_1}||W_0^M|^{-1} \sum_{\tau_1^1 \in \Pi_{\mathrm{disc}}(M_1^1, t)} \int_{i\mathfrak{a}_{M_1}^*/i\mathfrak{a}_M^*} \phi(\tau_{1,\lambda})\,\mathrm{d}\lambda,$$

for any suitable function $\phi$ on $\Pi(M, t)$.

(3) Let $\tau_{1,\lambda} \in \Pi_{M_1}(M, t)$. The global constituent of the spectral side is defined by setting

$$a^M(\tau_{1,\lambda}) = a_{\mathrm{disc}}^{M_1}(\tau_1^1) m_{M_1}^M(\tau_{1,\lambda}).$$

The representation $\tau_{1,\lambda}$ appears in the trace formula through its induced representation $(\tau^1_{1,\lambda})^{M^1}$. Consider first an arbitrary $\tau^1 \in \Pi_{\text{unit}}(M(\mathbb{A})^1)$. We have already agreed (see Subsection 12.12) to set

$$I_M(\tau^1, f) = I_M(\tau_\mu, 0, f),$$

where $\tau_\mu$ is any element of the associated $i\mathfrak{a}^*_M$-orbit of representations of $M(\mathbb{A})$ and $f \in \mathbb{H}(G(\mathbb{A}))$. For such $f$, we now set

$$I_M(\tau_{1,\lambda}, f) = I_M((\tau_{1,\lambda})^{M^1}, f),$$

where the distribution on the right is understood in the sense just described. It will be the local constituent of the spectral side of the trace formula.

## 14.9 An Expansion

The first step in the proof of Theorem 14.10 below will be to apply the expansion of $J_\chi$ provided by [Ar82.II, Theorem 8.2]. We will now state this expansion, applied to the distribution

$$J_t(f) = \sum_{\{\chi \in \mathfrak{X}\,;\, \|\text{Im}(\nu_\chi)\| = t\}} J_\chi(f).$$

It is the main result of [Ar82.II].

Recall that $J_\chi$ was discussed in Section 4.3. Let $M_1 \in \mathcal{L}$ and $L \in \mathcal{L}^{M_1}$. Let $Q \in \mathcal{P}(L)$ be of the form $P(R)$ where $R$ is a fixed parabolic subgroup of $M_1$ with Levi component $L$ and $P \in \mathcal{P}(M_1)$. Then

$$\left\{\mathcal{M}_{P_1}(\nu, \Lambda, Q) = \mathcal{M}_{P_1(R)|Q}(\Lambda)^{-1}\mathcal{M}_{P_1(R)|Q}(\Lambda + \nu)\,;\, P_1 \in \mathcal{P}(M_1), \nu \in i\mathfrak{a}^*_{M_1}\right\}$$

is a $(G, M_1)$-family. From this $(G, M_1)$-family, we obtain the operator

$$\mathcal{M}_{M_1}(\Lambda, Q) = \lim_{\nu \to 0} \sum_{P_1 \in \mathcal{P}(M_1)} \mathcal{M}_{P_1}(\nu, \Lambda, Q)/\theta_{P_1}(\nu).$$

THEOREM. *The distribution $J_t(f)$ is the sum over the Levi subgroups $M_1 \in \mathcal{L}$ and $L \in \mathcal{L}^{M_1}$, and over $s \in W^{M_1}(\mathfrak{a}_L)_{\text{reg}}$, of the product of*

$$|W^L_0||W^G_0|^{-1}|\det(s-1)_{\mathfrak{a}^{M_1}_L}|^{-1}$$

*with*

$$\int_{i\mathfrak{a}^*_{M_1}} \text{tr}\left(\mathcal{M}_{M_1}(\Lambda, Q)\mathcal{M}_{Q|sQ}(0) r_{Q,t}(s, \Lambda, f^1)\right) d\Lambda.$$

We can now state and prove the invariant spectral expansion.

THEOREM 14.10. *Suppose that $f \in \mathbb{H}(G(\mathbb{A}))$. Then*

$$I(f) = \sum_{t \geq 0} \sum_{M \in \mathcal{L}} |W_0^M||W_0^G|^{-1} \int_{\Pi(M,t)} a^M(\tau_{1,\lambda}) I_M(\tau_{1,\lambda}, f)\, d\tau_{1,\lambda},$$

*where the integral and outer sum each converge absolutely.*

PROOF. The trace of the operator

$$\mathcal{M}_{M_1}(\Lambda, Q)\mathcal{M}_{Q|sQ}(0) r_{Q,t}(s, \Lambda, f)$$

vanishes except on an invariant subspace on which the representation $r_{Q,t}(\Lambda)$ reduces to a sum of induced representations

$$i_P(\tau_{1,\Lambda}), \quad \tau_1^1 \in \Pi_{\mathrm{disc}}(M_1^1, t).$$

Here we continue to think of $\tau_{1,\Lambda}$ as a representative of the orbit associated to $\tau_1^1$. With this interpretation, the intertwining operator $\mathcal{M}_{P_1(R)|Q}(\Lambda)$ corresponds to a direct sum of operators

$$\mathcal{M}_{P_1|P}(\tau_{1,\Lambda}) = m_{P_1|P}(\tau_{1,\Lambda}) \mathcal{R}_{P_1|P}(\tau_{1,\Lambda}), \quad \tau_1^1 \in \Pi_{\mathrm{disc}}(M_1^1, t).$$

Therefore $\mathcal{M}_{M_1}(\Lambda, Q)$ corresponds to a direct sum of operators

$$\lim_{\nu \to 0} \sum_{P_1 \in \mathcal{P}(M_1)} m_{P_1}(\nu, \tau_{1,\Lambda}, P) \mathcal{R}_{P_1}(\nu, \tau_{1,\Lambda}, P)/\theta_{P_1}(\nu).$$

This last expression is obtained from a product of $(G, M)$-families. By Corollary 4.10.16, it equals

$$\sum_{M \in \mathcal{L}(M_1)} m_{M_1}^M(\tau_{1,\Lambda}) \mathcal{R}_M(\tau_{1,\Lambda}, P).$$

We now apply the definition of $a_{\mathrm{disc}}^G$. Given the observation above, we use the equality (14.5.1) and (14.5.2) (with $G$ replaced by $M_1$) to rewrite $J_t(f)$ as the sum over $M_1 \in \mathcal{L}$ and $M \in \mathcal{L}(M_1)$ of the product of $|W_0^{M_1}||W_0^G|^{-1}$ with

$$\sum_{\tau_1^1 \in \Pi_{\mathrm{disc}}(M_1^1, t)} \int_{i\mathfrak{a}_{M_1}^*} a_{\mathrm{disc}}^{M_1}(\tau_1^1) m_{M_1}^M(\tau_{1,\Lambda}) \operatorname{tr}(\mathcal{R}_M(\tau_{1,\Lambda}, P) i_P(\tau_{1,\Lambda}, f))\, \mathrm{d}\Lambda.$$

Observe that $m_{M_1}^M(\tau_{1,\Lambda})$ depends only on the projection $\lambda$ of $\Lambda$ onto $i\mathfrak{a}_{M_1}^*/i\mathfrak{a}_M^*$. Moreover, by the definition in Section 4.12, we have

$$\begin{aligned}\int_{i\mathfrak{a}_M^*} \operatorname{tr}(\mathcal{R}_M(\tau_{1,\Lambda+\mu}^M, P) i_P^G(\tau_{1,\Lambda+\mu}, f))\, \mathrm{d}\mu &= \int_{i\mathfrak{a}_M^*} J_M(\tau_{1,\Lambda+\mu}^M, f)\, \mathrm{d}\mu \\ &= J_M(\tau_{1,\Lambda}^M, 0, f) = J_M((\tau_{1,\lambda}^M)^1, f),\end{aligned}$$

if $P$ is any element in $\mathcal{P}(M)$. The last equality requires some explanation of the notation. Since $\tau_{1,\Lambda}^M$ is a representation of $M(\mathbb{A})$, we can associate with the orbit a representation of $M(\mathbb{A})^1$, with $\lambda$ a coset of $i\mathfrak{a}_M^*$ in $i\mathfrak{a}_{M_1}^*$. It is this representation which appears in the last displayed line. Decomposing the original integral over $\Lambda$ into a double integral of $(\lambda, \mu)$ in

$$(i\mathfrak{a}_{M_1}^*/i\mathfrak{a}_M^*) \times (i\mathfrak{a}_M^*),$$

we obtain

$$\begin{aligned} J_t(f) &= \sum_{M\in\mathcal{L}} \sum_{M_1\in\mathcal{L}^M} |W_0^{M_1}||W_0^G|^{-1} \\ &\quad \times \sum_{\tau_1^1\in\Pi_{\mathrm{disc}}(M_1^1,t)} \int_{i\mathfrak{a}_{M_1}^*/i\mathfrak{a}_M^*} a_{\mathrm{disc}}^{M_1}(\tau_1^1) m_{M_1}^M(\tau_{1,\lambda}) J_M(\tau_{1,\lambda}^M, f)\, d\lambda \\ &= \sum_{M\in\mathcal{L}} |W_0^M||W_0^G|^{-1} \sum_{M_1\in\mathcal{L}^M} \int_{\Pi_{M_1}(M,t)} a^M(\tau_{1,\lambda}) J_M(\tau_{1,\lambda}, f)\, d\tau_{1,\lambda} \\ &= \sum_{M\in\mathcal{L}} |W_0^M||W_0^G|^{-1} \int_{\Pi(M,t)} a^M(\tau_{1,\lambda}) J_M(\tau_{1,\lambda}, f)\, d\tau_{1,\lambda}. \end{aligned}$$

The convergence of the integral and the justification for our use of Fubini's theorem follow from the fact that $m_{M_1}^M(\tau_{1,\lambda})$ is tempered.

Set

$$I_t(f) = \sum_{\{\chi\in\mathfrak{X};\, \|\mathrm{Im}(\nu_\chi)\|=t\}} I_\chi(f).$$

Since the invariant $\chi$-expansion converges absolutely to $I(f)$, we have

$$I(f) = \sum_{t\geq 0} I_t(f), \tag{14.10.1}$$

the series converging absolutely. From the definition of $I_\chi(f)$, we obtain

$$I_t(f) = J_t(f) - \sum_{L\in\mathcal{L}_0} |W_0^L||W_0^G|^{-1} \widehat{I}_t^L(\phi_L(f)).$$

Assume inductively that

$$I_t^L(h) = \sum_{M\in\mathcal{L}^L} |W_0^M||W_0^L|^{-1} \int_{\Pi(M,t)} a^M(\tau_{1,\lambda}) I_M^L(\tau_{1,\lambda}, h)\, d\tau_{1,\lambda}$$

for any $L \in \mathcal{L}_0$ and any $h \in \mathbb{H}(L(\mathbb{A}))$. Combined with the formula above for $J_t(f)$, this tells us that $I_t(f)$ equals

$$\sum_{M\in\mathcal{L}} |W_0^M||W_0^G|^{-1} \int_{\Pi(M,t)} a^M(\tau_{1,\lambda}) \left( J_M(\tau_{1,\lambda}, f) - \sum_{L\in\mathcal{L}_0(M)} \widehat{I}_M^L(\tau_{1,\lambda}, \phi_L(f)) \right) \mathrm{d}\tau_{1,\lambda}.$$

It follows that

$$I_t(f) = \sum_{M\in\mathcal{L}} |W_0^M||W_0^G|^{-1} \int_{\Pi(M,t)} a^M(\tau_{1,\lambda}) I_M(\tau_{1,\lambda}, f)\, \mathrm{d}\tau_{1,\lambda}. \tag{14.10.2}$$

The theorem follows immediately from (14.10.1) and (14.10.2). □

## 14.11 Special Case

The definitions in this section have obvious analogues if the real number $t$ is replaced by a fixed datum $\chi \in \mathfrak{X}$. In particular, if $\|\operatorname{Im}(\nu_\chi)\| = t$, we have a subrepresentation $r_{Q,\chi}(\Lambda)$ of $r_{Q,t}(\Lambda)$. We shall sometimes write $\mathcal{A}^2_{Q,\chi}$ for the space of $K$-finite vectors in the underlying Hilbert space of $r_{Q,\chi}(\Lambda)$. The definitions also provide functions $a^{M_1}_{\mathrm{disc},\chi}$ and $a^M_\chi$ on respective subsets

$$\Pi_{\mathrm{disc}}(M_1^1, \chi) \subset \Pi_{\mathrm{disc}}(M_1^1, t), \quad M_1 \in \mathcal{L},$$

and

$$\Pi(M, \chi) \subset \Pi(M, t), \quad M \in \mathcal{L}.$$

The proof of Theorem 14.10 yields

COROLLARY 14.12. *Suppose that $f \in \mathbb{H}(G(\mathbb{A}))$ and $\chi \in \mathfrak{X}$. Then*

$$I_\chi(f) = \sum_{M\in\mathcal{L}} |W_0^M||W_0^G|^{-1} \int_{\Pi(M,\chi)} a^M_\chi(\tau_{1,\lambda}) I_M(\tau_{1,\lambda}, f)\, d\tau_{1,\lambda}.$$

## 14.13 Completion of Proof That Induction Assumption 2.6 Implies Induction Assumption 12.12

PROOF. For any element $L \in \mathcal{L}_0$, the corollary provides an expansion for $I^L_\chi$ in terms of the distributions

$$I^L_M(\tau_{1,\lambda}).$$

The distribution is defined in terms of the distributions $I^L_M(\tau_\mu, 0)$. But our Assumption 2.6 implies that the distributions $I^L_M(\tau_\mu, 0)$ are supported on characters. This is a consequence of Theorem 4.6.2. Therefore the distributions $I^L_\chi$ are also supported on characters. We have thus shown that the entire Assumption 12.12 is implied by the original one. □

## 15 Completion of the Induction Argument

We shall now show that *all the distributions which occur in the invariant trace formula are supported on characters.* These are local objects, so we shall not start off with the number field $F$ that has been fixed up until now.

THEOREM 15.1. *Let $F_1$ be a local field of characteristic* 0. *Let $G_1$ be a reductive group over $F_1$. Let $M_1$ be a Levi subgroup of $G_1$ over $F_1$. Then the distributions*

$$I_{M_1}(\gamma_1, f_1), \quad \gamma_1 \in M_1(F_1), f_1 \in \mathbb{H}(G_1(F_1)),$$

*are supported on characters.*

PROOF. This entire section deals with the proof of the theorem. Fix a positive integer $n_1$. Assume the theorem is valid for any $G_1$ defined over $F_1$ with $\dim_{F_1}(G_1) < n_1$. Having made this induction assumption, we fix $G_1$ and $F_1$ such that $\dim_{F_1}(G_1) = n_1$. If $L_1 \in \mathcal{L}_0(M_1)$, the distributions $I^{L_1}_{M_1}(\gamma_1)$ are (by Assumption 2.6) supported on characters. This assumption is what allowed us to define $I_{M_1}(\gamma_1)$ in the first place.

Let $f_1$ be a fixed function in $\mathbb{H}(G_1(F_1))$ such that $f_{1,G_1} = 0$. We must show that the distributions $I_{M_1}(\gamma_1)$ vanish on $f_1$. It is convenient to fix $M_1$ and to make an **additional induction assumption**. Assume that

$$I_{L_1}(\delta_1, f_1) = 0, \quad \delta_1 \in L_1(F_1), \tag{15.1.1}$$

for any $L_1 \in \mathcal{L}(M_1)$ with $L_1 \neq M_1$.

We must show that $I_{M_1}(\gamma_1, f_1)$ vanishes for each $\gamma_1 \in M_1(F_1)$.

If $\gamma_1$ is an arbitrary point in $M_1(F_1)$, we can write

$$\begin{aligned} I_{M_1}(\gamma_1, f_1) &= \lim_{a \to e} \sum_{L_1 \in \mathcal{L}(M_1)} m^{L_1}_{M_1}(\gamma_1, a) I_{L_1}(a\gamma_1, f_1) \\ &= \lim_{a \to e} I_{M_1}(a\gamma_1, f_1), \end{aligned}$$

by (15.1.1) and (2.9.1). Since $a$ stands for a small regular point in $A_{M_1}(F_1)$, we may assume without loss of generality that $G_{1,\gamma_1} = M_{1,\gamma_1}$. But now we can apply (2.12.1). This formula asserts that the function

$$\gamma \mapsto I_{M_1}(\gamma, f_1)$$

coincides with the orbital integral of a function on $M_1(F_1)$, for all points $\gamma$ whose semisimple part is close to that of $\gamma_1$.

It is known that the orbital integral of a function on $M_1(F_1)$ is completely determined by its values at the regular semisimple points. For $p$-adic $F_1$, this is [HC78, Theorem 10]. If $F_1$ is Archimedean, the result is due also to Harish-Chandra. The proof, which was never actually published, uses the Archimedean analogues of the techniques of [HC78]. In any case, it follows that if $I_{M_1}(\gamma, f_1)$ vanishes whenever $\gamma$ is $G_1$-regular, it vanishes for all $\gamma_1$. We may therefore assume that $\gamma_1$ itself is $G_1$-regular.

We can also assume that $\gamma_1$ is an $F_1$-elliptic point in $M_1(F_1)$. For $\gamma_1$ would otherwise belong to a proper Levi subgroup $M$ of $M_1$ defined over $F_1$, and we would be able to write

$$I_{M_1}(\gamma_1, f_1) = \sum_{L \in \mathcal{L}(M)} d_M^G(M_1, L) \widehat{I}_M^L(\gamma_1, f_{1,L}),$$

by the descent property of Corollary 8.5. Since $d_M^G(M_1, L) = 0$ unless $L$ is properly contained in $G$, the expression vanishes by our Induction Assumption 2.6. Thus it remains for us to show that $I_{M_1}(\gamma_1, f_1)$ vanishes when $\gamma_1$ is a fixed point in $M_1(F_1)$ which is $G_1$-regular and $F_1$-elliptic. For this basic case we shall use the global argument introduced by Kazhdan. See [DKV84] and [Ka86.1].

Suppose that $G$ is a reductive group over a number field $F$ such that $F_{v_1} \simeq F_1$ and $G_{v_1} = G_1$ for a valuation $v_1$ of $F$. Then

$$\dim_F(G) = \dim_{F_1}(G_1) = n_1.$$

It follows from Corollary 9.6 and our induction assumption on $n_1$ that, for any $S$, the distributions

$$I_M^L(\gamma), \quad M \in \mathcal{L}, \, L \in \mathcal{L}_0(M), \, \gamma \in M(F_S),$$

are all supported on characters. Therefore the group $G$, defined over $F$, satisfies the conditions of Section 12, and we can apply the results of Sections 13 and 14.

LEMMA 15.2. *Suppose that*

$$f = \prod_v f_v, \quad f_v \in \mathbb{H}(G(F_v)),$$

*is a function in* $\mathbb{H}(G(\mathbb{A}))$ *such that* $f_{v_1} = f_1$. *Then* $I(f) = 0$.

PROOF. Consider the spectral expansion

$$I(f) = \sum_{t \geq 0} \sum_{M \in \mathcal{L}} |W_0^M||W_0^G|^{-1} \int_{\Pi(M,t)} a^M(\tau_{1,\lambda}) I_M(\tau_{1,\lambda}, f)\, d\tau_{1,\lambda}$$

of Theorem 14.10. We shall show that the global distributions on $G(\mathbb{A})$,

$$I_M(\tau_{1,\lambda},f) = I_M(\tau_\mu,0,f), \quad \mu \in i\mathfrak{a}_M^*, \ M \in \mathcal{L}, \ \tau_{1,\lambda}^1 \in \Pi(M,t),$$

which occur on the right, vanish. As usual, $\tau_\mu$ is a representative of the $i\mathfrak{a}_M^*$-orbit of the representation of $M(\mathbb{A})$ induced from $\tau_{1,\lambda}^1$. (See Subsection 14.8.) In doing this, we will make essential use of the fact that $\tau$ is unitary.

It is clearly enough to establish the vanishing of the Fourier transform

$$I_M(\tau,\mathfrak{X},f) = \int I_M(\tau_\Lambda,f)e^{-\Lambda(\mathfrak{X})}\,d\Lambda,$$

where for a large finite set $S$ of valuations, $\mathfrak{X}$ belongs to the vector space of elements in $\oplus_{v\in S}\mathfrak{a}_{M,v}$ whose components sum to 0. The integral is over the imaginary dual vector space. According to the splitting formula, Proposition 9.8, we can write $I_M(\tau,\mathfrak{X},f)$ as a finite sum of products, over $v \in S$, of linear forms on the spaces $\mathbb{H}(L(F_v))$, $L \in \mathcal{L}(M)$. But if $L \in \mathcal{L}_0(M)$, our induction hypothesis, combined with Theorem 6.2, tells us that the distributions

$$\widehat{I}_M^L(\tau_{v_1},X_1,f_{1,L}), \quad \tau_{v_1} \in \Pi_{\mathrm{unit}}(M(F_{v_1})), \ X_1 \in \mathfrak{a}_{M,v_1},$$

are well defined. They must then vanish, since $f_{1,L} = 0$. It is therefore enough to show that the distributions

$$I_M(\tau_{v_1},X_1,f_1), \quad \tau_{v_1} \in \Pi_{\mathrm{unit}}(M(F_{v_1})), X_1 \in \mathfrak{a}_{M,v_1},$$

vanish. Recall that by abuse of notation, we denoted these distributions by $\widehat{I}_M(\tau_{v_1},X_1,f_{1,G})$ in the splitting formula.

The formula (3.8.1) gives an expansion for $I_M(\tau_{v_1},X_1,f_1)$ in terms of the local distributions associated with standard representations $\rho \in \Sigma(M(F_{v_1}))$. Only those $\rho$ with $\xi(\rho,\tau_{v_1}) \neq 0$ can occur in the expansion. (See Sections 4.9 and 4.12.) Since $\tau_{v_1}$ is unitarizable, this implies that $\rho$ has a unitary central character. It is sufficient to establish that, for any such $\rho$ and any such point $\lambda \in \mathfrak{a}_{M,\mathbb{C}}^*$ with a small real part, the local distributions

$$I_L(\rho_\lambda^L, h_L(X_1),f_1), \quad L \in \mathcal{L}(M), \ X_1 \in \mathfrak{a}_{M,v_1}, \tag{15.2.1}$$

all vanish. Since its central character is unitary, $\rho$ must be tempered or be induced from a proper parabolic subgroup of $M$. If $\rho$ is tempered,

$$I_L(\rho_\lambda^L, h_L(X_1),f_1) = \begin{cases} f_{1,G}(\rho_\lambda^G, h_G(X_1)), & \text{if } L = G,\\ 0, & \text{otherwise,}\end{cases}$$

by Proposition 3.4. But $f_{1,G} = 0$, so the local distribution vanishes also when $L = G$. In the other case,

$$\rho = \rho_1^M, \quad M_1 \subsetneq M, \ \rho_1 \in \Sigma(M_1(F_{v_1})),$$

and we can make use of the descent property of Corollary 8.9. We obtain an expression for a Fourier transform of (15.2.1) in terms of the local distributions

$$\widehat{I}^{M_2}_{M_1}(\rho_{1,\lambda}, Y_1, f_{1,M_1}), \quad M_2 \in \mathcal{L}_0(M_1), \ Y_1 \in \mathfrak{a}_{M_1,v_1}.$$

Since $M_2 \neq G$, the distributions are well defined and therefore vanish. Thus the distribution (15.2.1) vanishes in all cases. In other words, the spectral expansion reduces to 0, and $I(f)$ vanishes. This completes the proof of the proposition. □

## 15.3 *Choice of Global Objects*

We must decide how to choose the global objects $G$, $F$, and the place $v_1$ in order to prove Theorem 15.1. Our original element $\gamma_1$ in $M_1(F_1)$ belongs to a unique "maximal torus" $T_1 = G_{1,\gamma_1}$ in $M_1$, the connected centralizer of $\gamma_1$ in $G_1$. It is a torus in $M_1$ which is $F_1$-anisotropic modulo $A_{M_1}$. Let $E_1 \supset F_1$ be a finite Galois extension over which $G_1$ and $T_1$ split. Choose any number field $E$, with a valuation $w_1$, such that $E_{w_1} \simeq E_1$. The Galois group $\mathrm{Gal}(E_1/F_1)$ can be identified with the decomposition group of $E$ at $w_1$. It therefore acts on $E$. Let $F$ be the fixed field in $E$ of this group. Let $v_1$ be a valuation of $F$ which $w_1$ divides. Then $F_1 \simeq F_{v_1}$ and $\mathrm{Gal}(E_1/F_1) = \mathrm{Gal}(E/F)$. We can therefore use $G_1$ to twist the appropriate Chevalley group and maximal torus over $F$. We obtain a group $G$ and maximal torus $T$ defined over $F$, such that $G_1 = G_{v_1}$ and $T_1 = T_{v_1}$. Moreover, the construction is such that $M_1 = M_{v_1}$ and $\mathfrak{a}_{M_1} = \mathfrak{a}_M$, where $M$ is a Levi subgroup of $G$ which contains $T$ and is defined over $F$. It follows that

$$I_{M_1}(\gamma_1, f_1) = I_M(\gamma_1, f_1).$$

But the set $T(F)$ is dense in $T(F_{v_1})$. We can therefore approximate our local $G$-regular point $\gamma_1$ by global elements $\gamma \in T(F)$. Since $I_M(\gamma_1, f_1)$ is continuous in (regular) $\gamma_1$, we have only to show that $I_M(\gamma, f_1) = 0$.

PROPOSITION 15.4. *For any fixed $G$-regular element $\gamma$ in $T(F)$, we have* $I_M(\gamma, f_1) = 0$.

PROOF. We can use the trace formula to do this. We shall choose a suitable function

$$f = \prod_v f_v, \quad f_v \in \mathbb{H}(G(F_v)),$$

in $\mathbb{H}(G(\mathbb{A}))$ and apply Proposition 15.2. Observe first that $T$ is $F_{v_1}$-anisotropic modulo $A_M$. This means that $T$ is contained in no proper Levi subgroup of $M$ (relative to $F_{v_1}$). We can always replace $F$ by a finite extension in which $v_1$ splits completely. We may therefore assume that $T$ is also $F_{v_2}$-anisotropic modulo $A_M$, where $v_2$ is another valuation of $F$. Let $V = \{v_1, v_2, \dots, v_k\}$ be a large finite set of valuations of $F$ which contains $v_1$ and $v_2$, and outside of which $G$ and $T$ are unramified. At $v = v_1$, we have already been given our function $f_{v_1} = f_1$. If $v$ is any of the other valuations in $V$, let $f_v$ be any function which is supported on a very small open neighborhood of $\gamma$ in $G(F_v)$, and such that

$$\widehat{I}_M^M(\gamma, f_{v,M}) = I_G(\gamma, f_v) = 1.$$

If $v$ lies outside of $V$, let $f_v$ equal the characteristic function of $K_v^G$. Then $f = \prod_v f_v$ certainly belongs to $\mathbb{H}(G(\mathbb{A}))$. It follows from Proposition 15.2 and Theorem 13.10 that

$$\sum_{L \in \mathcal{L}} |W_0^L||W_0^G|^{-1} \sum_{\delta \in (L(F))_{L,S}} a^L(S, \delta) I_L(\delta, f) = 0. \tag{15.4.1}$$

Since $V = V(f)$, the shrinking of the support of the functions $f_{v_2}, \dots, f_{v_k}$ around $\gamma$ does not increase $V(f)$. Nor does it increase the support of $f$. It follows that in (15.4.1) the set $S$ may be chosen to be independent of $f$, and the sums over $\delta$ can be taken over finite sets which are also independent of $f$.

Suppose that $L \in \mathcal{L}$ and $\delta \in (L(F))_{L,S}$. We apply the splitting formula of Corollary 9.4 to $I_L(\delta, f)$. If $L \subset L_1 \subsetneq G$, we have

$$\widehat{I}_L^{L_1}(\delta, f_{v_1, L_1}) = 0,$$

by assumption. It follows that

$$I_L(\delta, f) = I_L(\delta, f_{v_1}) \cdot \prod_{v \neq v_1} \widehat{I}_L^L(\delta, f_{v,L}).$$

Now the function $f_{v_2}$ is supported on the $F_{v_2}$-anisotropic set in $M(F_{v_2})$. This means that $f_{v_2,L} = 0$ unless $L$ contains a conjugate of $M$. On the other hand, if $L$ contains a conjugate

$$wMw^{-1}, \quad w \in W_0,$$

of $M$, we can write

$$I_L(\delta, f_{v_1}) = I_{w^{-1}Lw}(w^{-1}\delta w, f_{v_1}),$$

by (2.14.1). If $M$ is properly contained in $w^{-1}Lw$, this vanishes by (15.1.1). Thus the contribution of $L$ to (15.4.1) vanishes unless $L$ is conjugate to $M$. Since the contributions from different conjugates of $M$ are equal, we obtain

$$\sum_{\delta \in (M(F))_{M,S}} a^M(S, \delta) \left( I_M(\delta, f_{v_1}) \prod_{v \neq v_1} I_G(\delta, f_v) \right) = 0. \tag{15.4.2}$$

Once again, $\delta$ can be summed over a finite set which is independent of how we shrink $f$.

The orbital integrals

$$I_G(\delta, f_{v_j}), \quad 2 \leq j \leq k,$$

vanish unless $\delta$ is close to the $G(F_{v_j})$-orbit of $\gamma$. In particular, the sum in (15.4.2) need only be taken over elements in $\delta$ which are regular semisimple. Consequently,

$$a^M(S, \delta) = |M_\delta(F)\backslash M(F, \delta)|^{-1} \operatorname{vol}(M_\delta(F)\backslash M_\delta(\mathbb{A})^1),$$

by [Ar86, Theorem 8.2]. Moreover, the $(M, S)$-equivalence classes of regular semisimple elements in $M(F)$ are just $M(F)$-orbits. It follows that

$$\sum_\delta c(\delta) I_M(\delta, f_{v_1}) = 0, \tag{15.4.3}$$

where $\delta$ is summed over those $M(F)$-orbits in $M(F)$ which are $G(F_{v_j})$-conjugate to $\gamma$ for $2 \leq j \leq k$, and which meet $K_v^G$ for $v$ outside of $V$, and where

$$c(\delta) = |M_\delta(F)\backslash M(F, \delta)|^{-1} \operatorname{vol}(M_\delta(F)\backslash M_\delta(\mathbb{A})^1) \cdot \prod_{v \in S-V} I_G(\delta, f_v).$$

The proof of the proposition will be complete once we show the following.

LEMMA 15.5. *Every such $\delta$ is also $G(F_{v_1})$-conjugate to $\gamma$.*

PROOF. As in [Ka86.1, Appendix], we use an argument from Galois cohomology. Note that $T$ is a maximal torus (in the usual sense) in $G$. The set of $G(F_v)$-conjugacy classes in $G(F_v)$ which are contained in the $G(\overline{F}_v)$-conjugacy class of $\gamma$ is known to be in bijective correspondence with a subset of

$$H^1(F_v, T) = H^1(\operatorname{Gal}(\overline{F}_v/F_v), T(\overline{F}_v)).$$

A similar assertion holds for $G(F)$-conjugacy classes. Let $E/F$ be a finite Galois extension which is unramified outside $V$ and over which $T$ splits. Then $H^1(F_v, T)$ equals $H^1(\operatorname{Gal}(E_w/F_v), T(E_w))$, and Tate-Nakayama theory (see [Ko86]) provides an isomorphism between this group and

$$\{\lambda^\vee \in X_*(T)\,;\, \operatorname{Norm}_{E_w/F_v}(\lambda^\vee) = 0\}/\{\lambda^\vee - \sigma\lambda^\vee\,;\, \lambda^\vee \in X_*(T), \sigma \in \operatorname{Gal}(E_w/F_v)\}, \tag{15.5.1}$$

and an isomorphism between

$$H^1(\operatorname{Gal}(E/F), T(\mathbb{A}_E)/T(E))$$

and

$$\{\lambda^\vee \in X_*(T)\,;\, \operatorname{Norm}_{E/F}(\lambda^\vee) = 0\}/\{\lambda^\vee - \sigma\lambda^\vee\,;\, \lambda^\vee \in X_*(T), \sigma \in \operatorname{Gal}(E/F)\}. \tag{15.5.2}$$

Here $w$ stands for a fixed valuation on $E$ which lies above a given $v$. Moreover, there is an exact sequence

$$H^1(\mathrm{Gal}(E/F), T(E)) \to \oplus_v H^1(\mathrm{Gal}(E_w/F_v), T(E_w)) \to H^1(\mathrm{Gal}(E/F), T(\mathbb{A}_E)/T(E)).$$

The first map is compatible with the embedding of $G(F)$-conjugacy classes into $\prod_v G(F_v)$, and the second arrow is given by the natural map

$$\oplus_v \lambda_v \mapsto \sum_v \lambda_v$$

from the direct sum of modules (15.5.1) into (15.5.2). Now, consider the conjugacy class of $\gamma$. Any $\delta$ which occurs in the sum (15.4.3) maps to an element $\oplus_v \lambda_v$ such that $\sum_v \lambda_v = 0$. If $v$ is one of the valuations $v_2, \dots, v_k$, then $\delta$ is $G(F_v)$-conjugate to $\gamma$, so that $\lambda_v = 0$. If $v$ lies outside $V$, then $\delta$ is $M(F_v)$-conjugate to an element in $K_v^G$. Since $(G, T)$ is unramified at $v$, we again have $\lambda_v = 0$ [Ko86, Proposition 7.1]. It follows that $\lambda_{v_1} = 0$. In other words, $\delta$ is $G(F_{v_1})$-conjugate to $\gamma$, completing the proof of the lemma and the proposition. □

## 15.6 *Proof of Theorem* 15.1

If $\delta$ is an element in $M(F)$ which is $G(F_{v_1})$-conjugate to $\gamma$, we have $\delta = y^{-1}\gamma y$, for some element $y \in M(F_{v_1})K_{v_1}$ which normalizes $M$. It follows from (2.14.1) that

$$I_M(\delta, f_{v_1}) = I_M(\gamma, f_1).$$

But for any $\delta$ which occurs in the sum (15.4.3), the number $c(\delta)$ is strictly positive. It follows from (15.4.3) that

$$I_M(\gamma, f_1) = 0.$$

As we noted earlier, this implies that

$$I_{M_1}(\gamma_1, f_1) = 0,$$

for our original point $\gamma_1 \in M_1(F_1)$. Theorem 15.1 is proved. □

COROLLARY 15.7. *Suppose that $G/F$ is, as usual, a connected reductive algebraic group over a number field $F$. Then for any $S$ and any $M \in \mathcal{L}$, the distributions*

$$I_M(\gamma), \quad \gamma \in M(F_S),$$

*are supported on characters.*

PROOF. This corollary follows immediately from Theorem 15.1 and Corollary 9.6. □

Corollary 15.7 justifies the Assumption 2.6. In particular, the distributions which occur in the invariant trace formula are all supported on characters. We have at last finished the extended induction argument, begun at the start of this chapter.

# 16 A Convergence Estimate

## *16.1 Trace Class Problem*

It is not known that the spectral expansion for $I(f)$ provided by Theorem 14.10 converges as a multiple integral over $t$, $M$, and $\pi$. The main obstruction is the trace class problem. This is essentially the question of showing that the operators

$$\oplus_{t\geq 0} r_{Q,t}(\Lambda, f), \quad Q \in \mathcal{F},\ f \in \mathbb{H}(G(\mathbb{A})),$$

are of trace class. We shall instead prove an estimate for the rate of convergence of the $\chi$-expansion. The estimate is an extension of some of the arguments used in the derivation of the trace formula. Although rather weak, it seems to be a natural tool for those applications which entail a comparison of trace formulae.

The estimate will be stated in terms of multipliers. See [Ar82.I].

DEFINITION 16.2. (1) Let $\mathcal{E}(\mathfrak{h})^W$ be the convolution algebra of compactly supported $W$-invariant distributions on $\mathfrak{h}$. (Recall that $\mathfrak{h}$ was defined in Subsection 14.2.)

(2) For $\alpha \in \mathcal{E}(\mathfrak{h})^W$ and $f \in \mathbb{H}(G(\mathbb{A}))$, let $f_\alpha$ be a new function in $\mathbb{H}(G(\mathbb{A}))$ such that

$$\pi(f_\alpha) = \hat{\alpha}(\nu_\pi)\pi(f), \quad \pi \in \Pi(G(\mathbb{A})).$$

(3) Similarly, for any function $\varphi$ in the global Paley-Wiener space $\mathbb{I}(G(\mathbb{A}))$ (see Section 12), there is another function $\varphi_\alpha \in \mathbb{I}(G(\mathbb{A}))$ such that

$$\varphi_\alpha(\pi) = \hat{\alpha}(\nu_\pi)\varphi(\pi), \quad \pi \in \Pi_{\text{temp}}(G(\mathbb{A})).$$

(4) We saw in Section 4.19.2 that the function $f_G$, for a local function $f$, could be interpreted as a function on $\Pi_{\text{temp}}(G(F_S))$ or $\Pi_{\text{temp}}(G(F_S)) \times \mathfrak{a}_{G,S}$. Similarly, $\varphi \in \mathbb{I}(G(\mathbb{A}))$ can be regarded as a function on $\Pi_{\text{temp}}(G(\mathbb{A}))$ or the product $\Pi_{\text{temp}}(G(\mathbb{A})) \times \mathfrak{a}_G$. The two interpretations are of course related by the Fourier transform

$$\varphi(\pi, X) = \int_{i\mathfrak{a}_G^*} \varphi(\pi_\lambda)e^{-\lambda(X)}\,\mathrm{d}\lambda, \quad \lambda \in \mathfrak{a}_G,$$

on $i\mathfrak{a}_G^*$.

(5) Suppose that $\alpha$ belongs to the subalgebra $C_c^\infty(\mathfrak{h})^W$. For $Z \in \mathfrak{a}_G$, let

$$\alpha_G(\pi, Z) = \int_{i\mathfrak{a}_G^*} \hat{\alpha}(\nu_\pi + \lambda)e^{-\lambda(Z)}\,d\lambda.$$

Then we have

$$\varphi_\alpha(\pi, X) = \int_{\mathfrak{a}_G} \varphi(\pi, Z)\alpha_G(\pi, X - Z)\,dZ. \tag{16.2.1}$$

Formula (16.2.1) is useful because it makes sense even if $\varphi$ belongs to the larger space $\mathbb{I}_{ac}(G(\mathbb{A}))$. For if $X$ remains within a compact set, the function

$$Z \mapsto \alpha_G(\pi, X - Z)$$

is supported on a fixed compact set. It follows that $\varphi \mapsto \varphi_\alpha$ extends to an action of $C_c^\infty(\mathfrak{h})^W$ on $\mathbb{I}_{ac}(G(\mathbb{A}))$ such that (16.2.1) holds.

(6) Similarly, $f \mapsto f_\alpha$ extends to an action of $C_c^\infty(\mathfrak{h})^W$ on $\mathbb{H}_{ac}(G(\mathbb{A}))$.

(7) Recall that if $f \in \mathbb{H}_{ac}(G(\mathbb{A}))$ and $X \in \mathfrak{a}_G$, then $f^X$ is the restriction of the function $f$ to

$$G(\mathbb{A})^X = \{x \in G(\mathbb{A})\,;\, H_G(x) = X\},$$

and

$$\pi(f^X) = \int_{G(\mathbb{A})^X} f(x)\pi(x)\,dx, \quad \pi \in \Pi(G(\mathbb{A})).$$

Then we have

$$\pi(f_\alpha^X) = \pi((f_\alpha)^X) = \int_{\mathfrak{a}_G} \pi(f^Z)\alpha_G(\pi, X - Z)\,dZ. \tag{16.2.2}$$

Setting $X = 0$, we obtain the formula

$$\pi(f_\alpha^1) = \int_{\mathfrak{a}_G} \pi(f^Z)\alpha_G(\pi, -Z)\,dZ, \quad \pi \in \Pi(G(\mathbb{A})), \tag{16.2.2$'$}$$

for the restriction $f_\alpha^1$ of $f_\alpha$ to $G(\mathbb{A})^1$.

(8) We do not want $f$ to be an arbitrary function in $\mathbb{H}_{ac}(G(\mathbb{A}))$. We must insist on some mild support and growth conditions on the functions $f^Z$ as $Z$ gets large. Fix a height function

$$\|x\| = \prod_v \|x_v\|_v, \quad x \in G(\mathbb{A}),$$

on $G(\mathbb{A})$ as in [Ar82.I, sections 2,3]. We shall say that a function $f \in \mathbb{H}_{ac}(G(\mathbb{A}))$ is *moderate* if there are positive numbers $c$ and $d$ such that $f$ is supported on

$$\{x \in G(\mathbb{A})\,;\, \log \|x\| \leq c(\|H_G(x)\| + 1)\},$$

and such that

$$\sup_{x \in G(\mathbb{A})}(|Df(x)| \exp\{-d\|H_G(x)\|\}) < \infty,$$

for any left invariant differential operator $D$ on $G(F_\infty)$. In a similar fashion, one can define the notion of a moderate function in $\mathbb{I}_{ac}(G(\mathbb{A}))$. A precise definition is in the appendix to [Ar88.3].

It is not hard to show that the map $f \mapsto f_G$ sends moderate functions in $\mathbb{H}_{ac}(G(\mathbb{A}))$ to moderate functions in $\mathbb{I}_{ac}(G(\mathbb{A}))$. Conversely, we have

PROPOSITION 16.3. *Suppose that $\Omega$ is a finite subset of $\Pi(K)$ and that $\varphi$ is a moderate function in $\mathbb{I}_{ac}(G(\mathbb{A}))_\Gamma$. Then there is a moderate function $f \in \mathbb{H}_{ac}(G(\mathbb{A}))_\Omega$ such that $f_G = \varphi$.*

This proposition can be regarded as a variant of the trace Paley-Wiener theorem. A proof is in the appendix to [Ar88.3].

DEFINITION 16.4. Write $C_N^\infty(\mathfrak{h})^W$, as usual, for the set of functions in $C_c^\infty(\mathfrak{h})^W$ which are supported on the ball of radius $N$.

PROPOSITION. *Suppose that $f$ is a moderate function in $\mathbb{H}_{ac}(G(\mathbb{A}))$. Then there is a positive number $c$ such that for any $\alpha \in C_N^\infty(\mathfrak{h})^W$, with $N > 0$, the function $f_\alpha$ is supported on*

$$\{x \in G(\mathbb{A})\,;\, \log \|x\| \leq c(\|H_G(x)\| + N + 1)\}.$$

PROOF. We can use the direct product decomposition $G(\mathbb{A}) = G(\mathbb{A})^1 \times A_{G,\infty}$ to identify each of the restricted functions $f^X$, $X \in \mathfrak{a}_G$, with a function in $\mathbb{H}(G(\mathbb{A})^1)$. The proposition then follows from [Ar82.I, Proposition 3.1] and the appropriate variant of (16.2.2). □

## 16.5 Convergence Estimate

We are now ready to state our convergence estimate. Fix a finite subset $\Gamma$ of $\Pi(K)$. For $L \in \mathcal{L}$ and $\chi \in \mathfrak{X}(G, F)$ (definition 14.2), a variant of definition 14.5 provides a set $\Pi_{disc}(L, \chi)$ of irreducible representations of $L(\mathbb{A})^1$. Let $\Pi_{disc}(L, \chi)_\Gamma$ be the subset of representations in $\Pi_{disc}(L, \chi)$ which contain representations in the restriction of $\Gamma$ to $K \cap L(\mathbb{A})$.

PROPOSITION. *Suppose that $\varphi$ is a moderate function in $\mathbb{I}_{\mathrm{ac}}(G(\mathbb{A}))_\Gamma$. Then there are positive numbers $C$ and $k$ such that, for any subset $\mathfrak{X}_1$ of $\mathfrak{X}(G,F)$ and any $\alpha \in C_N^\infty(\mathfrak{h})^W$ with $N > 0$, the expression*

$$\sum_{\chi\in\mathfrak{X}_1} |\widehat{I}_\chi(\varphi_\alpha)|$$

*is bounded by the supremum over $\chi \in \mathfrak{X}_1$, $L \in \mathcal{L}$, $\Lambda \in i\mathfrak{a}_L^*$, and $\sigma^1 \in \Pi_{\mathrm{disc}}(L,\chi)_\Gamma$ of*

$$Ce^{kN}|\hat{\alpha}(\nu_{\sigma^1} + \Lambda)|.$$

PROOF. By Proposition 16.3 there is a moderate function $f \in \mathbb{H}_{\mathrm{ac}}(G(\mathbb{A}))_\Gamma$ such that $f_G = \varphi$. Then

$$\widehat{I}_\chi(\varphi_\alpha) = I_\chi(f_\alpha) = I_\chi(f_\alpha^1),$$

for $\chi \in \mathfrak{X}$ and $\alpha \in C_N^\infty(\mathfrak{h})^W$. By Proposition 16.1 the function $f_\alpha^1$ is supported on a set

$$\{x \in G(\mathbb{A})^1 \,;\, \log\|x\| \leq c(1+N)\},$$

where the positive number $c$ depends only on $f$. We shall first estimate the sum $\sum_{\chi\in\mathfrak{X}_1} |J_\chi(f_\alpha)|$ of noninvariant distributions. We shall appeal to Proposition 2.2 and Lemma A.1 of [Ar82.I].

Fix a minimal parabolic subgroup $Q_0 \in \mathcal{P}(M_0)$ for $G$. Proposition 2.2 of [Ar82.I] applies to the distribution $J_\chi^T(f_\alpha)$, where $T$ is a point in $\mathfrak{a}_0$ such that the function

$$d_{Q_0}(T) = \min_{\alpha\in\Delta_{Q_0}} \{\alpha(T)\}$$

is suitably large. The assertion is that there is a positive number $C_0$ such that if

$$d_{Q_0}(T) > C_0 c(1+N), \tag{16.5.1}$$

and if $f_\alpha$ is as above, then $J_\chi^T(f_\alpha)$ equals an expression

$$\sum_{\{Q\in\mathcal{F}\,;\,Q\supset Q_0\}} \int_{i\mathfrak{a}_Q^*/i\mathfrak{a}_G^*} \Psi_{Q,\chi}^T(\Lambda, f_\alpha)\, d\Lambda.$$

Here

$$\Psi_{Q,\chi}^T(\Lambda, f_\alpha) = |\mathcal{P}(M_Q)|^{-1}\,\mathrm{tr}(\Omega_{Q|sQ,\chi}^T(s\Lambda) r_{Q,\chi}(s,\Lambda,f_\alpha^1)),$$

where $r_{Q,\chi}(s,\Lambda,f_\alpha^1)$ is the convolution operator from $\mathcal{A}^2_{Q,\chi}$ to $\mathcal{A}^2_{sQ,\chi}$, and $\Omega^T_{Q|sQ,\chi}(s\Lambda)$ is the linear map from $\mathcal{A}^2_{sQ,\chi}$ to $\mathcal{A}^2_{Q,\chi}$ such that for any pair of vectors $\phi \in \mathcal{A}^2_{Q,\chi}$ and $\phi_s \in \mathcal{A}^2_{sQ,\chi}$, the inner product

$$(\Omega^T_{Q|sQ,\chi}(s,\Lambda)\phi_s,\phi)$$

equals

$$\int_{G(F)A_{G,\infty}\backslash G(\mathbb{A})} \Lambda^T E_{sQ}(x,\phi_s,s\Lambda)\cdot\overline{\Lambda^T E_Q(x,\phi,\Lambda)}\,\mathrm{d}x.$$

Here $E_Q$ stands for the Eisenstein series associated with $Q$, and $\Lambda^T$ is the truncation operator. Recall that the truncation operator was defined in Section 3.9. Therefore

$$\sum_{\chi\in\mathfrak{X}_1}|J^T_\chi(f_\alpha)| \tag{16.5.2}$$

is bounded by

$$\sum_{\chi\in\mathfrak{X}_1}\sum_{Q\supset Q_0}|\mathcal{P}(M_Q)|^{-1}\int_{i\mathfrak{a}^*_Q/i\mathfrak{a}^*_G}\|\Omega^T_{Q|sQ,\chi}(s\Lambda)r_{Q,\chi}(s,\Lambda,f_\alpha^1)\|_1\,\mathrm{d}\Lambda,$$

where $\|\cdot\|_1$ denotes the trace class norm.

Suppose that $f$ is bi-invariant under an open compact subgroup $K_0$ of $G(\mathbb{A}_{\mathrm{fin}})$. According to [Ar82.I, Lemma A.1], there are positive numbers $C_{K_0}$ and $d_0$ such that

$$\sum_{\chi\in\mathfrak{X}_1}\sum_{Q\supset Q_0}|\mathcal{P}(M_Q)|^{-1}\int_{i\mathfrak{a}^*_Q/i\mathfrak{a}^*_G}\|r_{Q,\chi}(\Lambda,D^m)^{-1}_{K_0}\cdot\Omega^T_{Q|sQ,\chi}(s\Lambda)_{K_0}\|_1\,\mathrm{d}\Lambda$$

is bounded by

$$C_{K_0}(1+\|T\|)^{d_0}, \tag{16.5.3}$$

where $D^m$ is a certain left invariant differential operator on $G(F_\infty)^1$ and $(\cdot)_{K_0}$ denotes the restriction of a given operator to the space of $K_0$-invariant vectors.

In order to exploit this estimate, we note that

$$\|\Omega^T_{Q|sQ}(s\Lambda)r_{Q,\chi}(s,\Lambda,f_\alpha^1)\|_1$$

is no greater than

$$\|r_{Q,\chi}(\Lambda,D^m)^{-1}_{K_0}\cdot\Omega^T_{Q|sQ,\chi}(s\Lambda)_{K_0}\|_1\cdot\|r_{Q,\chi}(s,\Lambda,D^m f_\alpha^1)\|.$$

It follows that (16.5.2) is bounded by the product of (16.5.3) with

$$\sup_{\chi\in\mathfrak{X}_1} \sup_{Q\supset Q_0} \sup_{\Lambda\in i\mathfrak{a}_Q^*/i\mathfrak{a}_G^*} \|r_{Q,\chi}(s,\Lambda,D^m f_\alpha^1)\|. \tag{16.5.4}$$

Now $J_\chi^T(f_\alpha)$ is a polynomial in $T$ (Proposition 4.3.12), and $J_\chi(f_\alpha)$ is defined as its value at a fixed point $T_0$, see Subsection 4.3.14. We can certainly interpolate $J_\chi(f_\alpha)$ from the values of $J_\chi^T(f_\alpha)$ in which $T$ satisfies (16.5.1), see [Ar82.I, Lemma 5.2]. It follows that there is a number $C'_{K_0} > 0$, depending only on $K_0$, such that the original sum $\sum_{\chi\in\mathfrak{X}_1} |J_\chi(f_\alpha)|$ is bounded by the product of (16.5.4) with $C'_{K_0}(1+N)^{d_0}$.

Consider the expression (16.5.4). For a given $Q$, write

$$r_{Q,\chi}(s,\Lambda,D^m f_\alpha^1) = \oplus_{\sigma^1\in\Pi_{\mathrm{disc}}(M_Q^1,\chi)_\Gamma} r_{Q,\chi,\sigma^1}(s,\Lambda,D^m f_\alpha^1),$$

where $r_{Q,\chi,\sigma^1}$ denotes the representation induced from the isotypical component of $\sigma^1$. Then

$$\|r_{Q,\chi}(s,\Lambda,D^m f_\alpha^1) = \oplus_{\sigma^1\in\Pi_{\mathrm{disc}}(M_Q^1,\chi)_\Gamma} \|r_{Q,\chi,\sigma^1}(s,\Lambda,D^m f_\alpha^1)\|.$$

Since

$$D^m f_\alpha^1 = (D^m f)_\alpha^1,$$

the formula (16.2.2′) leads to an inequality

$$\begin{aligned}\|r_{Q,\chi,\sigma^1}(s,\Lambda,D^m f_\alpha^1)\| &\le \int_{\mathfrak{a}_G} \|r_{Q,\chi,\sigma^1}(s,\Lambda,D^m f^Z)\|\cdot|\alpha_G(\sigma_\Lambda^G,-Z)|\,\mathrm{d}Z\\ &\le \int_{\mathfrak{a}_G}\left(\int_{G(\mathbb{A})^Z} |(D^m f)(x)|\cdot\|r_{Q,\chi,\sigma^1}(s,\Lambda,x)\|\,\mathrm{d}x\right)\cdot|\alpha_G(\sigma_\Lambda^G,-Z)|\,\mathrm{d}Z.\end{aligned}$$

The operator $r_{Q,\chi,\sigma^1}(s,\Lambda,x)$ is unitary and has norm equal to 1. Observe also that the function $\alpha_G(\sigma_\Lambda,Z)$ vanishes unless $\|Z\|\le N$. It follows that

$$\|r_{Q,\chi,\sigma^1}(s,\Lambda,D^m f_\alpha^1)\| \le \left(\int_{G(\mathbb{A})_N} |D^m f(x)|\,\mathrm{d}x\right)\cdot \sup_{Z\in\mathfrak{a}_G}(|\alpha_G(\sigma_\Lambda^G,Z)|),$$

where

$$G(\mathbb{A})_N = \{x\in G(\mathbb{A})\,;\,\|H_G(x)\|\le N\}.$$

Since $f$ is moderate, the intersection of its support with $G(\mathbb{A})_N$ is contained in a set

$$\{x\in G(\mathbb{A})\,;\,\log\|x\|\le c(N+1)\},$$

whose volume depends exponentially on $N > 0$. Moreover, the supremum of $|Df(x)|$ on $G(\mathbb{A})_N$ is bounded by a function which also depends exponentially on $N$. It follows that

$$\int_{G(\mathbb{A})_N} |D^m f(x)|\, dx \leq C_0 e^{k_0 N},$$

for positive numbers $C_0$ and $k_0$ which are independent of $N$. On the other hand, we can write

$$\begin{aligned}\sup_{Z \in \mathfrak{a}_G} |\alpha_G(\sigma_\Lambda^G, Z)| &\leq \int_{i\mathfrak{a}_G^*} |\hat{\alpha}(\nu_{\sigma^1} + \Lambda + \mu)|\, d\mu \\ &\leq C_G \cdot \sup_{\mu \in i\mathfrak{a}_G^*} \left((1 + \|\Lambda + \mu\|^2)^{\dim \mathfrak{a}_G} \|\hat{\alpha}(\nu_{\sigma^1} + \Lambda + \mu)|\right),\end{aligned}$$

where

$$C_G = \int_{i\mathfrak{a}_G^*} (1 + \|\mu\|^2)^{-\dim \mathfrak{a}_G}\, d\mu.$$

Combining these facts, we see that the expression (16.5.4) is bounded by the product of $C_G C_0 e^{k_0 N}$ with the supremum over $\chi \in \mathfrak{X}_1$, $Q \supset Q_0$, $\Lambda \in i\mathfrak{a}_Q^*$, and $\sigma^1 \in \Pi_{\mathrm{disc}}(M_Q^1, \chi)_\Gamma$ of

$$(1 + \|\Lambda\|^2)^{\dim \mathfrak{a}_G} |\hat{\alpha}(\nu_{\sigma^1} + \Lambda)|.$$

We can now state an estimate for

$$\sum_{\chi \in \mathfrak{X}_1} |J_\chi(f_\alpha)|. \tag{16.5.5}$$

In order to remove the dependence on $Q_0$, we shall replace the supremum over $Q$ by one over $L \in \mathcal{L}$. Choose positive numbers $C_1'$ and $k_1'$ such that

$$C_{K_0}'(1 + N)^{d_0} C_G \cdot C_0 e^{k_0 N} \leq C_1' e^{k_1' N}.$$

Then (16.5.5) is bounded by the supremum over $\chi \in \mathfrak{X}_1$, $L \in \mathcal{L}$, $\Lambda \in i\mathfrak{a}_L^*$, and $\sigma^1 \in \Pi_{\mathrm{disc}}(L^1, \chi)_\Gamma$ of

$$C_1' e^{k_1' N} (1 + \|\Lambda\|^2)^{\dim \mathfrak{a}_G} |\hat{\alpha}(\nu_{\sigma^1} + \Lambda)|.$$

To remove the factors $(1 + \|\Lambda\|^2)$ from the estimate, we require a

LEMMA 16.6. *For any integer $m \geq 1$, we can choose a bi-invariant differential operator $z$ on $G(F_\infty)$, and multipliers $\alpha_1 \in C_c^M(\mathfrak{h})^W$ and $\alpha_2 \in C_c^\infty(\mathfrak{h})^W$ such that $f = (zf)_{\alpha_1} + f_{\alpha_2}$, for any function $f \in \mathbb{H}_{\mathrm{ac}}(G(\mathbb{A}))$.*

PROOF. This follows from a standard argument, see Proposition 3.4.1. For any $m$, one obtains a $W$-invariant differential operator $\zeta$ with constant coefficients on $\mathfrak{h}$, and a function $\alpha_1 \in C_c^m(\mathfrak{h})^W$ and $\alpha_2 \in C_c^\infty(\mathfrak{h})^W$, such that $\zeta\alpha_1 + \alpha_2$ is the Dirac measure at the origin in $\mathfrak{h}$. Let $z$ be the inverse image of $\zeta$ under the Harish-Chandra map. Then

$$f = f_{(\zeta\alpha_1+\alpha_2)} = (zf)_{\alpha_1} + f_{\alpha_2},$$

as required. □

## 16.7 Bounds

Returning to the proof of Proposition 16.5, we apply Lemma 16.6, with $m$ large, to our moderate function $f$. We see that (16.5.5) is bounded by

$$\sum_{\chi\in\mathfrak{X}_1} |J_\chi((zf)_{\alpha_1*\alpha}| + \sum_{\chi\in\mathfrak{X}_1} |J_\chi(f_{\alpha_2*\alpha})|.$$

Since the function $zf$ is also moderate, we can apply the estimate we have obtained to each of these sums. Notice that

$$\begin{aligned}
&\sup\nolimits_{\chi,L,\Lambda,\sigma^1} \big((1 + \|\Lambda\|^2)^{\dim \mathfrak{a}_G} |(\alpha_i * \alpha)^\wedge(\nu_{\sigma^1} + \Lambda)|\big)\\
&\leq \sup \big((1 + \|\Lambda\|^2)^{\dim \mathfrak{a}_G} |\hat{\alpha}_i(\nu_{\sigma^1} + \Lambda)| \cdot |\hat{\alpha}(\nu_{\sigma^1} + \Lambda)|\big)\\
&\leq \sup \big((1 + \|\Lambda\|^2)^{\dim \mathfrak{a}_G} |\hat{\alpha}_i(\nu_{\sigma^1} + \Lambda)|\big) \cdot \sup |\hat{\alpha}(\nu_{\sigma^1} + \Lambda)|\\
&\leq \sup \big((1 + \|\nu_{\sigma^1} + \Lambda\|^2)^{\dim \mathfrak{a}_G} |\hat{\alpha}_i(\nu_{\sigma^1} + \Lambda)|\big) \cdot \sup |\hat{\alpha}(\nu_{\sigma^1} + \lambda)|.
\end{aligned}$$

But the real parts of the points $\nu_{\sigma^1}$ lie in a fixed bounded set, and the functions $\hat{\alpha}_i$ decrease rapidly on cylinders (in a sense that depends on $m$). Therefore

$$\sup((1 + \|\nu_{\sigma^1} + \Lambda\|^2)^{\dim \mathfrak{a}_G} |\hat{\alpha}_i(\nu_{\sigma^1} + \Lambda)|) < \infty.$$

It follows that there are positive numbers $C_1$ and $k_1$ such that (16.5.5) is bounded by

$$C_1 e^{k_1 N} \sup\nolimits_{\chi\in\mathfrak{X}_1} \sup\nolimits_{L,\Lambda,\sigma^1}(|\hat{\alpha}(\nu_{\sigma^1} + \Lambda)|).$$

We must convert this into an estimate for

$$\sum_{\chi\in\mathfrak{X}_1} |\widehat{I}_\chi(\varphi_\alpha)|. \tag{16.7.1}$$

Suppose that $M \in \mathcal{L}_0$. It follows from Corollary 4.20.5 that the function $\phi_M(f)$ in $\mathbb{I}_{\mathrm{ac}}(M(\mathbb{A}))$ is also moderate. Since

$$\widehat{I}_\chi^M(\phi_M(f_\alpha)) = \widehat{I}_\chi^M(\phi_M(f)_\alpha),$$

we can apply the lemma inductively to $\phi_M(f_\alpha)$. We obtain positive $C_M$ and $k_M$, depending only on $f$, such that

$$\sum_{\chi \in \mathfrak{X}_1} |\widehat{I}_\chi^M(\phi_M(f_\alpha))|$$

is bounded by

$$C_M e^{k_M N} \sup_{\chi \in \mathfrak{X}_1} \sup_{L,\Lambda,\sigma^1}(|\hat{\alpha}(\nu_{\sigma^1} + \Lambda)|).$$

The required estimate for (16.7.1) then follows from the estimate for (16.5.5) and the formula

$$\widehat{I}_\chi(\varphi_\alpha) = J_\chi(f_\alpha) - \sum_{M \in \mathcal{L}_0} |W_0^M||W_0^G|^{-1}\widehat{I}_\chi^M(\phi_M(f_\alpha)).$$

Proposition 16.5 follows. □

We shall restate Proposition 16.5 in a simple form that is convenient for applications.

DEFINITION 16.8. Let $\mathfrak{h}_u^*$ denote the set of elements $\nu$ in the vector space $\mathfrak{h}_{\mathbb{C}}^*/i\mathfrak{a}_G^*$ such that $\overline{\nu} = s\nu$ for some element $s \in W$ of order 2. Here $\overline{\nu}$ stands for the conjugation of $\mathfrak{h}_{\mathbb{C}}^*$ relative to $\mathfrak{h}^*$. As is well known, the infinitesimal character $\nu_{\pi^1}$ of any unitary representation $\pi^1 \in \Pi_{\mathrm{unit}}(G(\mathbb{A})^1)$ belongs to $\mathfrak{h}_u^*$. Observe that if $r$ and $T$ are nonnegative real numbers, the set

$$\mathfrak{h}_u^*(r,T) = \{\nu \in \mathfrak{h}_u^* ;\ \|\mathrm{Re}(\nu)\| \leq r, \|\mathrm{Im}(\nu)\| \geq T\}$$

is invariant under $W$. (An element $\nu \in \mathfrak{h}_u^*$ is only a coset of $i\mathfrak{a}_G^*$ in $\mathfrak{h}_{\mathbb{C}}^*$, but $\|\nu\|$ is understood to be the minimum value of the norm on the coset.) Let $\mathfrak{h}^1$ be the orthogonal complement of $\mathfrak{a}_G$ in $\mathfrak{h}$. Then $\mathfrak{h}_u^*$ can be identified with a subset of the complex dual space of $\mathfrak{h}^1$.

COROLLARY 16.9. *Choose any function $f \in \mathbb{H}_{\mathrm{ac}}(G(\mathbb{A}))$. Then there are positive numbers $C$, $k$, and $r$ such that*

$$\sum_{t>T} |I_t(f_\alpha)| \leq Ce^{kN} \sup_{\nu \in \mathfrak{h}_u^*(r,T)}(|\hat{\alpha}(\nu)|)$$

*for any $T > 0$ and any $\alpha \in C_N^\infty(\mathfrak{h}^1)^W$, with $N > 0$.*

PROOF. Proposition 16.5 is stated for multipliers in $C_N^\infty(\mathfrak{h})^W$, but it is equally valid if $\alpha$ belongs to $C_N^\infty(\mathfrak{h}^1)^W$. To see this, apply the proposition to the sequence

$$\alpha_n(H+Z) = \alpha(H)\beta_n(Z), \quad H \in \mathfrak{h}^1,\, Z \in \mathfrak{a}_G,$$

in $C_c^\infty(\mathfrak{h})^W$, where $\beta_n \in C_c^\infty(\mathfrak{a}_G)$ approaches the Dirac measure at 1. The (upper) limits of each side of the resulting inequality give the analogous inequality for $\alpha$. Notice that $f_\alpha^1$ depends only on $f^1$, so that $f$ can indeed be an arbitrary function in $\mathbb{H}_{\mathrm{ac}}(G(\mathbb{A}))$.

We shall apply this version of Proposition 16.5 to the given $\alpha$, with $\varphi = f_G$, and with

$$\mathfrak{X}_T = \{\chi \in \mathfrak{X};\ \|\operatorname{Im}(\nu_\chi)\| > T\}.$$

Then

$$\sum_{t>T} |I_t(f_\alpha)| = \sum_{\chi \in \mathfrak{X}_T} |\widehat{I}_\chi(\varphi_\alpha)|.$$

Choose a finite subset $\Gamma$ of $\Pi(K)$ such that $\varphi$ belongs to $\mathbb{I}_{\mathrm{ac}}(G(\mathbb{A}))_\Gamma$. There is a positive number $r$ such that if $\pi$ is any representation in $\Pi_{\mathrm{unit}}(G(\mathbb{A}))$ whose $K$-spectrum meets $\Gamma$, the point $\nu_\pi$ belongs to

$$\{\nu \in \mathfrak{h}_u^*;\ \|\operatorname{Re}(\nu)\| \le r\}.$$

If $\chi$, $L$, $\Lambda$, and $\sigma^1$ are elements in $\mathfrak{X}_T$, $\mathcal{L}$, $i\mathfrak{a}_L^*$, and $\Pi_{\mathrm{disc}}(L^1, \chi)_\Gamma$, as in the inequality of Proposition 16.5, the points $\nu_{\sigma^1} + \Lambda$ then belong to $\mathfrak{h}_u^*(r, T)$. The corollary follows. □

REMARK 16.10. Suppose that $\mathfrak{h}^2$ is any vector subspace of $\mathfrak{h}$ which contains $\mathfrak{h}^1$. Then there will be an obvious variant of Corollary 16.9 for multipliers $\alpha \in C_N^\infty(\mathfrak{h}^2)^W$. For this, $f$ must again be taken to be a moderate function in $\mathbb{H}_{\mathrm{ac}}(G(\mathbb{A}))$.

## 17 Simpler Forms of the Trace Formula

### 17.1 The Full Trace Formula

The full trace formula is the identity, for any $f \in \mathbb{H}(G(\mathbb{A}))$,

$$\sum_{M \in \mathcal{L}} |W_0^M||W_0^G|^{-1} \sum_{\gamma \in (M(F))_{M,S}} a^M(S,\gamma) I_M(\gamma, f)$$

$$= \sum_{t \ge 0} \sum_{M \in \mathcal{L}} |W_0^M||W_0^G|^{-1} \int_{\Pi(M,t)} a^M(\tau_{1,\lambda}) I_M(\tau_{1,\lambda}, f)\, d\tau_{1,\lambda},$$

given by the two expansions for $I(f)$ in Theorems 13.10 and 14.10. In this section we shall investigate how the formula simplifies if conditions are imposed on $f$. The conditions will be invariant, in the sense that they depend only on the image of $f$ in $\mathbb{I}(G(\mathbb{A}))$. Equivalently, the conditions will depend only on the (invariant) orbital integrals of $f$.

DEFINITION 17.2. We shall say that a function $f \in \mathbb{H}(G(\mathbb{A}))$ is *discrete* at a valuation $v_1$ if $f$ is a finite sum of functions $\prod_v f_v, f_v \in \mathbb{H}(G(F_v))$, such that for all $M \in \mathcal{L}_0$, we have

$$f_{v_1,M} = 0.$$

This is implied by the vanishing of the orbital integral $I_G(\gamma_1,f_{v_1})$ for any $G$-regular element $\gamma_1 \in G(F_{v_1})$ which is not $F_{v_1}$-elliptic. For example, a pseudo-coefficient of a square-integrable representation is discrete, see [Ka86.1].

THEOREM 17.3. (1) *If $f$ is discrete at one place $v_1$, we have*

$$I(f) = \sum_{t\geq 0} \sum_{\pi^1\in\Pi_{\mathrm{disc}}(G^1,t)} a^G_{\mathrm{disc}}(\pi^1) I_G(\pi^1,f).$$

(2) *If $f$ is discrete at two places $v_1$ and $v_2$, we have*

$$I(f) = \sum_{\gamma\in(G(F))_{G,S}} a^G(S,\gamma) I_G(\gamma,f).$$

PROOF. We can assume that $f = \prod_v f_v$, with

$$f_{v_1,M} = 0$$

for all $M \in \mathcal{L}_0$. Part (1) will be a special case of the spectral expansion

$$I(f) = \sum_{t\geq 0} \sum_{M\in\mathcal{L}} |W_0^M||W_0^G|^{-1} \int_{\Pi(M,t)} a^M(\tau_{1,\lambda}) I_M(\tau_{1,\lambda},f)\, d\tau_{1,\lambda}.$$

The main step is to show that if $M \in \mathcal{L}_0$ and $\tau^1 \in \Pi_{\mathrm{unit}}(M(\mathbb{A})^1)$, then

$$I_M(\tau^1,f) = 0.$$

But this is very similar to the proof of Proposition 15.2. Using the splitting formula of Proposition 9.8, we reduce the problem to showing that

$$I_M(\tau_{v_1},X_1,f_{v_1}) = 0, \quad \tau_{v_1} \in \Pi_{\mathrm{unit}}(M(F_{v_1})),\ X_1 \in \mathfrak{a}_{M,v_1},\ M \in \mathcal{L}_0.$$

We then apply the expansion (3.8.1) into standard representations and the descent formula, Corollary 8.9. Since $\tau_{v_1}$ is unitarizable, the required vanishing formula follows as in Proposition 15.2. In particular, the terms with $M \neq G$ in the spectral expansion all vanish. Moreover,

$$\int_{\Pi(G,t)} a^G(\pi_{1,\lambda}) I_G(\pi_{1,\lambda}, f)\, d\pi_{1,\lambda}$$

$$= \sum_{M_1 \in \mathcal{L}} |W_0^{M_1}||W_0^G|^{-1} \sum_{\tau_1^1 \in \Pi_{\mathrm{disc}}(M_1^1,t)} \int_{i\mathfrak{a}_{M_1}^*/i\mathfrak{a}_G^*} a_{\mathrm{disc}}^{M_1}(\tau_1^1) m_{M_1}^G(\tau_{1,\lambda}) I_G(\tau_{1,\lambda}^G, f)\, d\lambda$$

$$= \sum_{\pi^1 \in \Pi_{\mathrm{disc}}(G^1,t)} a_{\mathrm{disc}}^G(\pi^1) I_G(\pi^1, f),$$

since

$$I_G(\tau_1^G, f) = \widehat{I}_{M_1}^{M_1}(\tau_1^1, f_{M_1}) = 0, \quad M_1 \neq G,\ \tau_1^1 \in \Pi_{\mathrm{unit}}(M_1(\mathbb{A})^1).$$

Part (1) follows.

Suppose that $f$ is also discrete at a second place $v_2$. Part (2) will be a special case of the geometric expansion

$$I(f) = \sum_{M \in \mathcal{L}} |W_0^M||W_0^G|^{-1} \sum_{\gamma \in (M(F))_{M,S}} a^M(S,\gamma) I_M(\gamma, f).$$

The set $S$ is large enough that it contains $v_1$ and $v_2$, and so that $f$ belongs to $\mathbb{H}(G(F_S))$. Write

$$f = f_1 f_2, \quad f_i \in \mathbb{H}(G(F_{S_i})),$$

where $S_1$ and $S_2$ are disjoint sets of valuations with the closure property, which contains $v_1$ and $v_2$, respectively, and whose union is $S$. From the splitting formula, Proposition 9.1, we obtain

$$I_M(\gamma, f) = \sum_{L_1, L_2 \in \mathcal{L}(M)} d_M^G(L_1, L_2) \widehat{I}_M^{L_1}(\gamma, f_{1,L_1}) \widehat{I}_M^{L_2}(\gamma, f_{2,L_2}).$$

The distributions on the right vanish unless $L_1 = L_2 = G$. Moreover, $d_M^G(G,G) = 0$ unless $M = G$. It follows that if $M \neq G$, the distribution $I_M(\gamma, f)$ equals 0, and the corresponding term in the geometric expansion vanishes. This gives (2). □

COROLLARY 17.4. *Suppose that $f$ is discrete at two places. Then*

$$\sum_{\gamma \in (G(F))_{G,S}} a^G(S,\gamma) I_G(\gamma, f) = \sum_{t \geq 0} \sum_{\pi^1 \in \Pi_{\mathrm{disc}}(G,t)} a_{\mathrm{disc}}^G(\pi^1) I_G(\pi^1, f).$$

Assume that $f \in \mathbb{H}(G(\mathbb{A}))$ is such that

$$f = \prod_v f_v, \quad f_v \in \mathbb{H}(G(F_v)).$$

With additional invariant restrictions on $f$, we shall be able to simplify the trace formula further.

COROLLARY 17.5. *Suppose there is a place $v_1$ such that for all $\pi_{v_1} \in \Pi_{\mathrm{unit}}(G(F_{v_1}))$, we have*

$$\operatorname{tr} \pi_{v_1}(f_{v_1}) = 0,$$

*whenever $\pi_{v_1}$ is a constituent of a (properly) induced representation*

$$\tau_{v_1}^G, \quad \tau_{v_1} \in \Pi_{\mathrm{unit}}(M(F_{v_1})), \quad M \in \mathcal{L}_0.$$

*Then*

$$I(f) = \sum_{t \geq 0} \operatorname{tr} r_{\mathrm{disc},t}(f),$$

*where $r_{\mathrm{disc},t}$ denotes the representation of $G(\mathbb{A}))$ on $L^2_{\mathrm{disc},t}(G(F)A_{G,\infty}\backslash G(\mathbb{A}))$ by right shifts.*

PROOF. If $M$ belongs to $\mathcal{L}_0$, the condition implies that

$$\operatorname{tr} \tau_{v_1}^G(f_{v_1}) = 0, \quad \tau_{v_1} \in \Pi_{\mathrm{temp}}(M(F_{v_1})),$$

so that $f_{v_1,M} = 0$. Therefore $f$ is discrete at $v_1$. Applying part (1) of Theorem 17.3, we obtain

$$\begin{aligned} I(f) &= \sum_{t \geq 0} \sum_{\pi^1 \in \Pi_{\mathrm{disc}}(G,t)} a^G_{\mathrm{disc}}(\pi^1) I_G(\pi^1, f) \\ &= \sum_{t \geq 0} \sum_{M \in \mathcal{L}} |W_0^M||W_0^G|^{-1} \sum_{s \in W(\mathfrak{a}_M)_{\mathrm{reg}}} |\det(s-1)_{\mathfrak{a}_M^G}|^{-1} \operatorname{tr}(\mathcal{M}_{Q|sQ}(0) r_{Q,t}(s,0,f)), \end{aligned}$$

in the notation of Section 14. Here $Q$ is any element in $\mathcal{P}(M)$. If $M \neq G$ then

$$\operatorname{tr}(\mathcal{M}_{Q|sQ}(0)\rho_{Q,t}(s,0,f))$$

is a linear combination of characters of unitary induced representations. It vanishes by assumption. If $M = G$ then

$$\operatorname{tr} \mathcal{M}_{Q|sQ}(0) r_{Q,t}(s,0,f) = \operatorname{tr} r_{G,t}(0,f) = \operatorname{tr} r_{\mathrm{disc},t}(f),$$

by definition. The corollary follows. □

COROLLARY 17.6. *Suppose there is a place $v_1$ such that*

$$I_G(\gamma_1, f_{v_1}) = 0$$

*for any element $\gamma_1 \in G(F_{v_1})$ which is not semisimple and $F_{v_1}$-elliptic. Suppose also that $f$ is discrete at another place $v_2$. Then*

$$I(f) = \sum_{\gamma \in \{G(F)_{\mathrm{ell}}\}} \mathrm{vol}(G(F,\gamma)A_{G,\infty}\backslash G(\mathbb{A},\gamma)) \int_{G(\mathbb{A},\gamma)\backslash G(\mathbb{A})} f(x^{-1}\gamma x)\, dx,$$

*where $\{G(F)_{\mathrm{ell}}\}$ denotes the set of $G(F)$-conjugacy classes of $F$-elliptic elements in $G(F)$, and $G(F,\gamma)$ and $G(\mathbb{A},\gamma)$ denote the centralizers of $\gamma$ in $G(F)$ and $G(\mathbb{A})$.*

PROOF. The conditions imply that $f$ is discrete at $v_1$ and $v_2$. We can therefore apply the formula

$$I(f) = \sum_{\gamma \in (G(F))_{G,S}} a^G(S,\gamma) I_G(\gamma, f)$$

of Theorem 17.3. If an element $\gamma \in G(F)$ is not $F$-elliptic, it is not $F_{v_1}$-elliptic, and $I_G(\gamma, f) = 0$. The corollary then follows from [Ar86, Theorem 8.2] and the definition of $I_G(\gamma, f)$. □

The conditions of Corollaries 17.5 and 17.6 sometimes arise naturally. For example, if $v_1$ is discrete, Kottwitz [Ko88] has introduced a simple function $f_{v_1}$ which satisfies the conditions of Corollary 17.6. Kottwitz also establishes a version of this corollary in [Ko88]. He imposes stronger conditions at $v_2$, but derives a formula without resorting to the invariant trace formula.

## 17.7 General Linear Group

For example, take $G = \mathrm{GL}(n)$. Suppose that $f$ is discrete at $v_1$. Any element $\gamma_1 \in G(F_{v_1})$ which is not $F_{v_1}$-elliptic belongs to a $G(F_{v_1})$-conjugacy class

$$\delta_1^G, \quad \delta_1 \in M(F_{v_1}),\ M \in \mathcal{L}_0.$$

Consequently,

$$I_G(\gamma_1, f_{v_1}) = \widehat{I}_M^M(\delta_1, f_{v_1,M}) = 0.$$

Therefore the first condition of Corollary 17.6 is satisfied. Moreover, it is known that any induced unitary representation

$$\tau_1^G, \quad \tau_1 \in \Pi_{\mathrm{unit}}(M(F_{v_1})),\ M \in \mathcal{L},$$

is irreducible ([Be84, Vo86]). Since

$$\operatorname{tr} \tau_1^G(f_{v_1}) = f_{v_1,M}(\tau_1) = 0, \quad M \in \mathcal{L}_0,$$

the condition of Corollary 17.5 also holds. Combining Corollaries 17.5 and 17.6, we obtain

COROLLARY 17.8. *Assume that $G = \mathrm{GL}(n)$ and that $f$ is discrete at two places $v_1$ and $v_2$. Then*

$$\sum_{\gamma \in \{G(F)_{\mathrm{ell}}\}} \mathrm{vol}(G(F,\gamma)A_{G,\infty}\backslash G(\mathbb{A},\gamma)) \int_{G(\mathbb{A},\gamma)\backslash G(\mathbb{A})} f(x^{-1}\gamma x)\, dx = \sum_{t \geq 0} \operatorname{tr} r_{\mathrm{disc},t}(f).$$

# 18 Global Vanishing Properties for GL($n$)

## 18.1 Simple Versions

The simple versions of the trace formula were obtained by placing rather severe restrictions on $f$. In many applications, one will need to prove that certain terms vanish for less severely restricted functions. We can illustrate this with the example of GL($n$), begun in Section 10.

Adopt the notation of Section 10. Then $G$ is an inner form of $G' = \mathrm{GL}(n)$ over a field $F$, and

$$\eta : G \to G' = \mathrm{GL}(n)$$

is a given inner twist. Let us write $\mathcal{L}'$ for the set of Levi subgroups of $G'$ which contain the group of diagonal matrices. For each $L \in \mathcal{L}'$, we have the partition

$$\mathfrak{p}(L) = (n_1, \ldots, n_r), \quad n_1 \geq n_2 \geq \cdots \geq n_r,$$

of $n$ such that

$$L \simeq \mathrm{GL}(n_1) \times \cdots \times \mathrm{GL}(n_r).$$

DEFINITION 18.2. Suppose that $\mathfrak{p}_1$ and $\mathfrak{p}_2$ are partitions of $n$. We shall write $\mathfrak{p}_1 \leq \mathfrak{p}_2$, as in Section 10 if there are groups $L_1 \subset L_2$ in $\mathcal{L}'$ such that $\mathfrak{p}_1 = \mathfrak{p}(L_1)$ and $\mathfrak{p}_2 = \mathfrak{p}(L_2)$.

Let $M_0$ be a minimal Levi subgroup of $G$ over $F$. Denote by $A_{M_0}$ the split component of the center of $M_0$. We shall assume that $\eta(M_0)$ is contained in a standard Levi subgroup of $G'$ and that the restriction of $\eta$ to $A_{M_0}$ is defined over $F$. Write $\mathcal{L}$ for the set of Levi subgroups of $G$ which contain $M_0$. Then the map

$$M \mapsto M' = \eta(M), \quad M \in \mathcal{L},$$

is an injection of $\mathcal{L}$ into $\mathcal{L}'$. The image of this map is easy to describe. Indeed,

$$G(F) = \mathrm{GL}(n/d, D)$$

where $d$ is a divisor of $n$, and $D$ is a division algebra of degree $d^2$ over $F$. The minimal group $M'$ in the image corresponds to the partition $\mathfrak{p}(d) = (d, \ldots, d)$. The other groups in the image correspond to partitions $(n_1, \ldots, n_r)$ such that $d$ divides each $n_i$. For each valuation $v$, we shall write $d_v$ for the order of the invariant of the division algebra at $v$. Then $d$ is the least common multiple of the integers $d_v$.

In Section 10 we describe the norm mapping $\gamma \mapsto \gamma'$ from conjugacy classes in $G(F)$ to (conjugacy classes in) $G'(F)$. It can be defined the same way for any element $M \in \mathcal{L}$. We also investigated certain functions on the local groups $G'(F_v)$. Let $f' = \prod_v f'_v$ be a fixed function in $\mathbb{H}(G'(\mathbb{A}))$ whose local constituents satisfy (10.7.1), that is, the orbital integrals of $f'_v$ vanish at the $G'$-regular elements which are not local norms. Recall that $(L(F))_{L,S}$ is defined in 13.1(2). In our case of inner forms of $\mathrm{GL}(n)$, this is the set of the conjugacy classes in $L(F)$.

PROPOSITION 18.3. *Suppose $L \in \mathcal{L}'$ and $\delta \in L(F)$. Embed $\delta$ in $(L(F))_{L,S}$, where $S \supset S_{\mathrm{ram}}$ is a large finite set of valuations. Then*

$$I_L(\delta, f') = 0,$$

*unless $L = M'$ and $\delta = \gamma'$, for elements $M \in \mathcal{L}$ and $\gamma \in M(F)$.*

PROOF. In the orbital integral, $\delta$ is to be considered as a point in $L(F_S)$. We must therefore regard $f' = \prod_{v \in S} f'_v$ as an element in $\mathbb{H}(G'(F_S))$. Assume that $I_L(\delta, f') \neq 0$. We must deduce that $L = M'$ and $\delta = \gamma'$.

Applying the splitting formula Corollary 9.4, we obtain

$$I_L(\delta, f') = \sum_{\{L_v\}} d(\{L_v\}) \widehat{I}_L^{L_v}(\delta, f'_{v,L_v}), \tag{18.3.1}$$

where the sum is taken over collections $\{L_v \in \mathcal{L}(L)\,;\, v \in S\}$, and $d(\{L_v\})$ is a number which vanishes unless

$$\mathfrak{a}_L^G = \oplus_{v \in S} \mathfrak{a}_L^{L_v}. \tag{18.3.2}$$

By assumption, the left-hand side of (18.3.1) is nonzero. Therefore there is a collection $\{L_v\}$ for which (18.3.2) holds and such that $\widehat{I}_L^{L_v}(\delta, f'_{v,L_v}) \neq 0$ for each $v \in S$. This implies that

$$\mathfrak{p}(d_v) \leq \mathfrak{p}(L_v), \quad v \in S.$$

Our first task is to show that $\mathfrak{p}(d) \leq \mathfrak{p}(L)$. Let $p$ be any rational prime. Let $p^k$ be the highest power of $p$ which divides $d$. Since $d$ is the least common multiple of $\{d_v\}$, there is a valuation $v \in S$ such that $p^k$ divides $d_v$. But the invariants of a central simple algebra add up to 0, so there must be a valuation $w \in S$, distinct from $v$, such that $p^k$ also divides $d_w$. It follows that $\mathfrak{p}(p^k) \leq \mathfrak{p}(L_v)$ and $\mathfrak{p}(p^k) \leq \mathfrak{p}(L_w)$. Since $\mathfrak{a}_L^{L_v} \cap \mathfrak{a}_L^{L_w} = \{0\}$, we can apply Proposition 10.3. The result is that $\mathfrak{p}(p^k) \leq \mathfrak{p}(L)$. In other words, the integer $p^k$ divides each of the numbers $n_1, \ldots, n_r$ which make up the partition $\mathfrak{p}(L)$. The same is therefore true of the integer $d$, so that $\mathfrak{p}(d) \leq \mathfrak{p}(L)$. In other words, $L = M'$ for an element $M \in \mathcal{L}$.

The final step is to apply Proposition 10.7. This vanishing result was stated only for local fields, but by the splitting formula, it extends immediately to $G'(F_S)$. Since $I_{M'}(\delta, f')$ does not vanish, and since $\delta$ belongs to $M'(F_S)$, the element $\delta$ must belong to a smaller set

$$M'(F_S)_M = \prod_{v \in S} M'(F_v)_M.$$

The set $M'(F_v)_M$ was introduced in Definition 10.6. Any element in $M'(F_v)_M$ comes from an element in $M(F_v)$. Since $S$ contains $S_{\mathrm{ram}}$, this implies that the global element $\delta$ comes everywhere from the inner form. This implies that $\delta$ comes globally from an element in $M(F)$. In other words, $\delta = \gamma'$, for some element $\gamma \in M(F)$. This completes the proof of the proposition. □

PROPOSITION 18.4. *Suppose that $L_1 \subset L$ are Levi subgroups in $\mathcal{L}'$ and that $S \supset S_{\mathrm{ram}}$ is a large finite set of valuations. Let $f'$ be as in the lines before Proposition* 18.3. *Then $I_L(\tau, Y, f') = 0$ for any $Y \in \mathfrak{a}_L$ and any induced representation $\tau = \tau_1^L$, $\tau_1 \in \Pi(L_1(F_S))$, unless both $L_1$ and $L$ are the images of elements in $\mathcal{L}$.*

PROOF. Suppose that $I_L(\tau, Y, f') \neq 0$. Using the splitting formula Proposition 9.8, we first argue as at the beginning of the proof of Proposition 18.3. This establishes that $L = M'$ for some element $M \in \mathcal{L}$. We then apply the local vanishing property, Proposition 10.8. This proves that $L_1 = M_1'$ for another $M_1 \in \mathcal{L}$. □

Proposition 18.3 and 18.4 are the first steps toward comparing the trace formulae of $G$ and $G'$. They assert that for functions $f'$ on $G'(\mathbb{A})$ as above, the distributions vanish at the data which do not come from $G$. The trace formula for $G'$ becomes

$$\sum_{M \in \mathcal{L}} |W_0^M||W_0^G|^{-1} \sum_{\gamma \in (M(F))_{M,S}} a^{M'}(S, \gamma') I_{M'}(\gamma', f')$$

$$= \sum_{t \geq 0} \sum_{M \in \mathcal{L}} |W_0^M||W_0^G|^{-1} \int_{\Pi(M',t)} a^{M'}(\tau_{1,\lambda}) I_{M'}(\tau_{1,\lambda}, f') \, d\tau_{1,\lambda}.$$

It is considerably harder to compare the terms which remain with the corresponding terms for $G$. This problem is the main topic of Chapter 6.

# Chapter 6
# Main Comparison

## 1 Notation and Conventions

### 1.1 *The Group*

Let $F$ be a number field. The results of this chapter also hold, we believe, for the function field of a smooth projective absolutely irreducible curve over a finite field, once the invariant trace formula—established for number fields in Chapters 3–5—is established for such a function field.

Let $G$ be a connected reductive linear algebraic group over $F$. Let $M_0$ be a fixed minimal Levi subgroup of $G$ over $F$. Let $\mathcal{L}$ denote the set of Levi subgroups of $G$ over $F$ that contain $M_0$. It is a finite set. Usually we will let $M$ denote an element of $\mathcal{L}$. Let $\mathcal{L}^M$ be the set of Levi subgroups over $F$ contained in $M$ (so that $\mathcal{L} = \mathcal{L}^G$), $\mathcal{L}(M)$ the set of Levi subgroups of $G$ over $F$ that contain $M$ (so that $\mathcal{L} = \mathcal{L}(M_0)$), $\mathcal{P}(M)$ (resp. $\mathcal{F}(M)$) the set of parabolic subgroups of $G$ over $F$ with Levi component (resp. containing) $M$, and $A_M$ the maximal split torus in the center of $M$.

### 1.2 *Hecke Algebra*

Write $F_v$ for the completion of $F$ at the place $v$. For each valuation $v$ of $F$, we fix a maximal compact subgroup $K_v$ of the group $G(F_v)$ of $F_v$-valued points of $G$. We choose $K_v$ to be hyperspecial for all $v$ where $G$ is unramified, namely, quasisplit over $F_v$ and split over an unramified extension of $F_v$. The group $G$ is unramified for almost all $v$. See [Ti79] for the definition of hyperspecial. If $G$ is defined over the ring $\mathcal{O}_v$ of integers of $F_v$, e.g., $G = \mathrm{GL}(n)$, then we can take $K_v = G(\mathcal{O}_v)$.

For each place $v$ of $F$, we write $\mathbb{H}(G(F_v))$ for the algebra of $K_v$-finite functions in the convolution algebra $C_c^\infty(G(F_v))$ of compactly supported smooth functions on



Y.Z. Flicker, *Arthur's Invariant Trace Formula and Comparison of Inner Forms*,
DOI 10.1007/978-3-319-31593-5_6

the group $G(F_v)$. It will be referred to as the *Hecke algebra* of $G(F_v)$. A choice of Haar measure $dx_v$ on $G(F_v)$ is implicit.

If $S$ is a finite set of places of $F$, put $F_S = \prod_v F_v$ and $G(F_S) = \prod_v G(F_v)$.

Denote by $\mathbb{A}$ the ring of adèles of $F$. Put $\mathbb{H}(G(F_S)) = \otimes_{v \in S}\mathbb{H}(G(F_v))$, the Hecke algebra of compactly supported smooth $K_S$-finite functions on $G(F_S)$, where $K_S = \prod_{v \in S} K_v$. Multiplying $f \in \mathbb{H}(G(F_S))$ by the characteristic function $1_{K^S}$ of $K^S = \prod_{v \notin S} K_v$, we obtain a $C_c^\infty$-function on $G(\mathbb{A})$, the group of $\mathbb{A}$-valued points of $G$. Put $\mathbb{H}(G(\mathbb{A}))$ for the union of all $\mathbb{H}(G(F_S)) \otimes 1_{K^S}$ over all $S$ such that $K_v$ is hyperspecial for all $v \notin S$.

We shall relate objects associated with an inner form $G$ of $G' = \mathrm{GL}(n)$ to analogous objects on $G'$. Let us recall the definition.

DEFINITION 1.3. The group $G$ (as in 1.1) is an *inner form* of $G'$ over $F$ if there is an isomorphism $\eta : G \to G'$ over a finite algebraic extension of $F$ such that, for every $\theta \in \mathrm{Gal}(\overline{F}/F)$, the composition $\eta_\theta = \eta^{-1} \circ \eta^\theta$ equals conjugation, $\mathrm{Int}(a_\theta)$, by an element $a_\theta$ in $G$. If $G' = \mathrm{GL}(n)$, the group $G$ is the multiplicative group of a central simple algebra over $F$ (see, e.g., [We67]).

We can choose $\eta$ such that $\eta(M_0)$ contains the standard (diagonal) minimal Levi subgroup of $G'$. Furthermore, we may assume that the restriction of $\eta$ to $A_{M_0}$ is defined over $F$.

## 1.4 Test Functions

Our aim is to compare the trace formula of $G$ with the trace formula of $G' = \mathrm{GL}(n)$ for matching test functions $f \in \mathbb{H}(G(\mathbb{A}))$ and $f' \in \mathbb{H}(G'(\mathbb{A}))$. It suffices to consider $f = \otimes_v f_v$ and $f' = \otimes_v f'_v$. We take $f_v = f'_v$ under the isomorphism $G(F_v) \simeq G'(F_v)$ for all places $v \notin S_{\mathrm{ram}} = \{$places where $G$ is not split$\}$. Denote the local correspondence of conjugacy classes by $\gamma \mapsto \gamma'$ (thus if $\gamma, \gamma'$ are semisimple, their characteristic polynomials are equal). More precisely, if $\{\gamma\}$ is a conjugacy class, namely, a $G(F)$-orbit in $G(F)$, then $\{\eta(\gamma)\}$ is a $G'(F)$-conjugacy class. It is this class that is represented by $\gamma'$. We write $\gamma'$ for an element in this class.

The local correspondence of test functions will be denoted by $f \mapsto f'$ (thus the orbital integral of $f'$ at $\gamma'$ which comes from $\gamma$ is equal to that of $f$ at $\gamma$, and it is 0 if $\gamma'$ is not in the image of $\gamma$). It is important to note that $f \in \mathbb{H}(G(F_S))$ is a function, and so is $f'$, but $f'$ is determined by $f$ only by matching its orbital integrals or character value. This $f'$ is in $\mathbb{H}(G'(F_S))$. In particular, $f'_{G'} : \pi \mapsto \mathrm{tr}\,\pi'(f')$ is uniquely determined by the orbital integrals of $f'$ on the regular set, by the local integrability of the character of $\pi'$. Thus $f$ determines uniquely $f'_{G'}$ in $\mathbb{I}(G'(F_S))$, and $f'$ is a representative in $\mathbb{H}(G'(F_S))$ of $f'_{G'}$. See Theorem 2.14.4 and Proposition 2.22.2 for the existence of $f'$ in the non-Archimedean case.

Sometimes, we will use $f'$, $\gamma'$ or $\pi'$ for general objects on $G'$, not necessarily coming from the local correspondence. This will not be mentioned explicitly unless the context makes the usage unclear.

DEFINITION 1.5. (1) Let $\Pi(G(F_S))$ be the set of equivalence classes of irreducible admissible representations of $G(F_S)$. The subset of tempered representations is denoted by $\Pi_{\text{temp}}(G(F_S))$ and those which are unitarizable by $\Pi_{\text{unit}}(G(F_S))$.

(2) For $M \in \mathcal{L}$ and $\tau \in \Pi(M(F_S))$, the normalized induced representation $i_P^G(\tau)$ is independent in the Grothendieck group of the choice of $P \in \mathcal{P}(M)$. We often denote it by $\tau^G$.

(3) For any valuation $v$ let $\mathcal{L}_v \supset \mathcal{L}$ be the set of Levi subgroups of $G$ defined over $F_v$ containing a fixed minimal one, $M_{0v}$.

(4) For $f \in \mathbb{H}(G(F_S))$, $M \in \mathcal{L}$, $P \in \mathcal{P}(M)$ and $\tau \in \Pi_{\text{temp}}(M(F_S))$, define

$$f_M(\tau) = \operatorname{tr} \tau^G(f) = \operatorname{tr} i_P^G(\tau, f).$$

The trace Paley-Wiener theorem ([CD84, BDK86, F95]) determines the image space

$$\mathbb{I}(G(F_S)) = \{f_G\, ; f \in \mathbb{H}(G(F_S))\} \subset \operatorname{Map}\left(\Pi_{\text{temp}}(G(F_S)), \mathbb{C}\right),$$

(see Section 5.1.8 for a discussion of the various results). Here $\mathbb{I}$ suggests "invariant." An element in the *Paley-Wiener space* $\mathbb{I}(G(F_S))$ defines a function $\pi \mapsto \varphi(\pi)$, $\pi \in \Pi_{\text{temp}}(G(F_S))$, $\varphi(\pi) = \operatorname{tr} \pi(f)$ for $f \in \mathbb{H}(G(F_S))$, with $\varphi = f_G$.

(5) A continuous linear map $I$ from $\mathbb{H}(G(F_S)$ to a topological vector space $\mathbb{V}$ is said to be *supported on characters* if it is 0 at any function $f$ with $f_G = 0$. As an example, the map $f \mapsto f_M$ from $\mathbb{H}(G(F_S))$ to $\mathbb{I}(M(F_S))$ has this property, since $f_M(\tau) = \operatorname{tr} \tau^G(f) = f_G(\tau^G)$.

(6) Suppose $I$ is supported on characters. Then there is a unique continuous map, $\widehat{I}$, from $\mathbb{I}(G(F_S))$ to $\mathbb{V}$, given by $\widehat{I}(f_G) = I(f)$, for $f \in \mathbb{H}(G(F_S))$.

For example, the map $f \mapsto f_M$ factors via a map $\varphi \mapsto \varphi_M$ from $\mathbb{I}(G(F_S))$ to $\mathbb{I}(M(F_S))$.

In Chapter 5 it was shown that the invariant distributions and maps obtained from the trace formula for $G$ are supported on characters. See Section 5.15.

We now collect various notation and definitions that we need about the correspondence. The local correspondence of representations is an injection $\pi \mapsto \pi'$ from $\Pi(G(F_S))$ into $\Pi(G'(F_S))$, dual to $\gamma \mapsto \gamma'$. It is determined by a character relation. The local correspondence was established for non-Archimedean fields in Chapter 2. For a discussion of the Archimedean case, see [BR10]. To describe it, we recall some definitions.

DEFINITION 1.6. (1) Write $\mathcal{L}'$ for the set of Levi subgroups of $G'$ which contain the group of diagonal matrices. The map $M \mapsto M'$ is an injection $\mathcal{L} \to \mathcal{L}'$.

(2) The centralizer of any element $\gamma$ in $G$ is connected. It will be denoted by $G_\gamma$.

(3) Let $\sigma \in G$ be semisimple. The centralizer $G_\sigma$ of $\sigma$ in $G$ is the group of units in a product of central simple algebras over $F$. Given a twisting $\eta : G \to G'$ as in 1.3 and a semisimple $\sigma$, there is an inner twist $\eta_\sigma : G_\sigma \to G'_{\sigma'}$ uniquely determined up to $G'(F)$-conjugacy. If $\mu \mapsto \mu_{\sigma'}$ is the associated map from

conjugacy classes in $G_\sigma(F)$ to conjugacy classes in $G'_{\sigma'}(F)$, we have

$$(\sigma\mu)' = \sigma'\mu_{\sigma'} \quad \text{for } \mu \in G_\sigma(F). \tag{1.6.1}$$

If $S$ is a finite set of valuations of $F$, similar observations apply when $\gamma$ and $\sigma$ are in $G(F_S) = \prod_{v\in S} G(F_v)$.

(4) If $v$ is a valuation, we attach a sign $e_v = e_v(G)$ to the group $G$. If $v$ is non-Archimedean and $r_v(G)$ is the $F_v$-split rank of $G$, then $e_v(G) = (-1)^{r_v(G)-r_v(G')} = (-1)^{r_v(G)-n}$. In particular, if $D_v$ is a division algebra of rank $m_v$ over $F_v$ and $G(F_v) = \mathrm{GL}(n_v, D_v)$, then $e_v(G) = (-1)^{n_v(m_v-1)}$. The last formula holds also if $v$ is Archimedean, then $m_v = 2$, and $D_v$ is the quaternion algebra. See [Ko83] for a definition of these signs in a more general context.

Let $S$ be a finite set of valuations. Put $e_S(G) = e_S = \prod_{v\in S} e_v$. Note that the value of $e_S$ is not changed if it is defined using $M \in \mathcal{L}$ instead of $G$.

(5) The representation $\pi \in \Pi_{\mathrm{temp}}(G(F_S))$ corresponds to $\pi' \in \Pi_{\mathrm{temp}}(G'(F_S))$ if their characters satisfy

$$\mathrm{ch}_\pi(\gamma) = e_S \cdot \mathrm{ch}_{\pi'}(\gamma') \tag{1.6.2}$$

for all $\gamma$ in the set $G_{\mathrm{reg}}(F_S)$ of regular elements in $G(F_S)$; see Theorem 2.13.8.

(6) Recall (Section 4.9.3) the set $\Sigma(G(F_S))$ of *standard representations*. It consists of all representations $\otimes_{v\in S}\tau_v^G$, induced from $\tau_v \in \Pi(M_v(F_v))$, $M_v \in \mathcal{L}_v$, where $\tau_v$ is the twist by a character of $A_{M_v}(F_v)$ of a tempered representations (thus $\tau_v$ is quasi-tempered).

Note that the character relations extend to an injection $\rho \mapsto \rho'$ from $\Sigma(G(F_S))$ to $\Sigma(G'(F_S))$. It does not extend to arbitrary representations $\pi \in \Pi(G(F_S))$.

(7) The discriminant of $\gamma \in G$ is $D(\gamma) = D^G(\gamma) = \det(I - \mathrm{Ad}(\sigma))|_{\mathfrak{g}/\mathfrak{g}_\sigma}$, where $\sigma$ is the semisimple part of $\gamma$, and $\mathfrak{g}$ , $\mathfrak{g}_\sigma$ are the Lie algebras of $G$ and $G_\sigma$. Since $D^G(\gamma)$ is not changed under field extension, we have $D^G(\gamma) = D^{G'}(\gamma')$. An element $\gamma$ is called *regular* if $D(\gamma) \neq 0$. For example, a unipotent $u \neq e$ in GL(2) is regular, as $\sigma = e$, $\mathfrak{g}_\sigma = \mathfrak{g}$, and $D(u) = 1$.

(8) Suppose that $\gamma = \prod_{v\in S}\gamma_v$ is a point in $G(F_S)$. Let

$$\gamma = \sigma u = \prod_{v\in S} \sigma_v u_v$$

be its Jordan decomposition. Since $G_{\sigma_v}$ is the multiplicative group of a product of central simple algebras, there exists a parabolic subgroup $P_{\sigma_v} = M_{\sigma_v}N_{\sigma_v}$ of $G_{\sigma_v}$ such that the unipotent element $u_v \in G_{\sigma_v}$ is contained in the Richardson orbit (the unique dense $P_{\sigma_v}$-orbit in $N_{\sigma_v}$) of $P_{\sigma_v}$. We shall write $A_{\sigma_v}$ for the $F_v$-split component of $M_{\sigma_v}$. It is a simple consequence of the definition of $P_{\sigma_v}$ that

$$\dim(G_{\gamma_v}) = \dim(M_{\sigma_v}).$$

(9) Define

$$e^G(\gamma) = \prod_{v \in S} e^G(\gamma_v)$$

where

$$e^G(\gamma_v) = e(G_{\sigma_v}) = e(M_{\sigma_v})$$

is the sign introduced in (4) above.

(10) Define the (normalized) *orbital integral* of any $f \in \mathbb{H}(G(F_S))$ by

$$I_G(\gamma, f) = |D^G(\gamma)|^{\frac{1}{2}} \int_{G_\gamma(F_S)\backslash G(F_S)} f(x^{-1}\gamma x)\, dx,$$

where $|D(\gamma)| = \prod_{v \in S} |D(\gamma_v)|_v$ and $G_\gamma(F_S) = \prod_{v \in S} G_{\gamma_v}(F_v)$.

(11) The distribution $I_G(\gamma, f)$ satisfies for a general $\gamma \in G(F_S)$ the equation

$$I_G(\gamma, f) = \lim_{a \to e} I_G(\sigma a, f), \quad a \in \prod_{v \in S} A_{\sigma_v}(F_v). \tag{1.6.3}$$

Suppose $G = \mathrm{GL}(2)$ and $\sigma_v = e$. If $\gamma = e$, this simply says that $f_v(e) = \lim f_v(a)$ as $a \to e$, $a \in A_{\sigma_v}(F_v)$ = center of $G(F_v)$. If $\gamma \neq e$, then $A_{\sigma_v}$ is conjugate to the diagonal subgroup, so $a \to e$ where $a$ is split regular in $G(F_v)$. For $u$ a regular unipotent element, we then obtain

$$I_G(u, f) = \lim_{a \to e} I_G(a, f).$$

Observe that if $a$ is a small regular point in $\prod_v A_{\sigma_v}(F_v)$, then

$$G_{\sigma a}(F_S) = \prod_{v \in S} M_{\sigma_v}(F_v).$$

(12) The distribution $I_G(\gamma)$ depends implicitly on a choice of a Haar measure on $G_{\sigma a}(F_S)$, as well as one on $G(F_S)$. We use the inner twist $\eta_\sigma$ to transfer the former to a Haar measure on

$$G'_{(\sigma a)'}(F_S) = \prod_{v \in S} M_{\sigma'_v}(F_v).$$

Combined with a fixed Haar measure on $G'(F_S)$, this $\eta_\sigma$ allows us to define the distribution $I_{G'}(\gamma')$ on $G'(F_S)$.

(13) We shall write $f \mapsto f'$ for the map of 1.4 from $\mathbb{H}(G(F_S))$ to $\mathbb{H}(G'(F_S))$ constructed in Chapter 2 by transferring orbital integrals.

LEMMA 1.7. *For all $\gamma \in G(F_S)$, we have*

$$I_G(\gamma,f) = e^G(\gamma)\widehat{I}_{G'}(\gamma',f'_{G'}), \tag{1.7.1}$$

PROOF. If $\gamma$ is semisimple, this follows from the definition of $f \mapsto f'$ and $D^G(\gamma) = D^{G'}(\gamma')$ (see part (7) above). If $\gamma$ is arbitrary, the formula follows from the regular semisimple case, the formula (1.6.3) and the definition of $e^G(\gamma)$. □

LEMMA 1.8. *Let $\pi \in \Pi_{\mathrm{temp}}(G(F_S))$. Let $\pi'$ be the image of $\pi$ under the local correspondence. For $f \in \mathbb{H}(G(F_S))$, we have*

$$\operatorname{tr}\pi(f) = e_S f'_{G'}(\pi'). \tag{1.8.1}$$

PROOF. Assume that $\gamma \in G(F_S)$ is semisimple regular. Then $e^G(\gamma) = 1$ and (1.7.1) becomes

$$I_G(\gamma,f) = \widehat{I}_{G'}(\gamma',f'_{G'}). \tag{1.8.2}$$

We use the Weyl integration formula, recalled in Subsection 2.18.2, with the Haar measure on the torus

$$G'_{\gamma'_v}(F_v), \quad v \in S,$$

obtained from that on $G_{\gamma_v}(F_v)$ by $\eta_{\gamma_v}$. Combining (1.8.2) with the character identity (1.6.2), the lemma follows from the Weyl integration formula. □

## 1.9 Vanishing Properties

Formulae (1.7.1) and (1.8.1) both come with supplementary vanishing properties. If $\zeta$ is an element in $G'(F_S)$ which is not of the form $\gamma'$ for $\gamma \in G(F_S)$, then $\widehat{I}_{G'}(\zeta,f'_{G'}) = 0$. See Theorem 2.14.4. If $\pi'$ is a representation in $\Pi_{\mathrm{temp}}(G'(F_S))$ which is not the image of any $\pi$, then $f'_{G'}(\pi') = 0$. See Section 2.22. In particular, the map $f \mapsto f'$ is supported on characters. It follows that if

$$I' : \mathbb{H}(G'(F_S)) \to \mathbb{V}'$$

is any map which is supported on characters, then the map $f \mapsto \widehat{I}'(f'_{G'})$ from $\mathbb{H}(G(F_S))$ to $\mathbb{V}'$ is also supported on characters. This will be applied without comment to the invariant distributions in the trace formula for $G'$.

## 1.10 Harish-Chandra Map

To discuss the measures, fix on the space $\mathfrak{a}_{M_0}$ a Euclidean norm $\|\cdot\|$ which is invariant under the Weyl group $W_0 = W_0^G$ of $(G, A_{M_0})$. For $M \in \mathcal{L}$, take the associated Euclidean measures on each of the spaces $\mathfrak{a}_M$. Recall that the Harish-Chandra map $H_M : M(\mathbb{A}) \to \mathfrak{a}_M$ is defined by

$$\langle H_M(x), \chi \rangle = \log |\chi(x)|, \qquad \chi \in \mathfrak{a}_M^* = \mathrm{Hom}(M(\mathbb{A}), \mathbb{R}).$$

Let $M(\mathbb{A})^1$ denote the kernel of the Harish-Chandra map. Then $M(\mathbb{A}) = M(\mathbb{A})^1 \times \mathfrak{a}_M$. The measure on $\mathfrak{a}_M$, together with an invariant measure on $M(\mathbb{A})$, provides an invariant measure on $M(\mathbb{A})^1$. To a $\chi' \in X(M')_F$ we can associate $\chi \in X(M)_F$ by $\chi(m) = \chi'(\eta(m))$ for $m \in M$. The mapping $\chi' \mapsto \chi$ is a surjection of $X(M')_F$ into $X(M)_F$. The dual mapping

$$\mathfrak{a}_M = \mathrm{Hom}(X(M)_F, \mathbb{R}) \to \mathrm{Hom}(X(M')_F, \mathbb{R}) = \mathfrak{a}_{M'}$$

is an isomorphism. We use it to identify the real vector spaces $\mathfrak{a}_M$ and $\mathfrak{a}_{M'}$. Using this identification we transfer the Euclidean measure from $\mathfrak{a}_M$ to $\mathfrak{a}_{M'}$. As above, we have a Haar measure on $M'(\mathbb{A})^1$, the kernel of $H_{M'}$, associated to a Haar measure on $M'(\mathbb{A})$.

## *1.11 Lie Algebras*

For each $M \in \mathcal{L}$, we have identified the spaces $\mathfrak{a}_{M'}$ and $\mathfrak{a}_M$. Denote the map $\mathfrak{a}_M \mapsto \mathfrak{a}_{M'}$ by $H \mapsto H'$. If $m \in M(F_S)$, we have

$$H_M(m)' = H_{M'}(\eta(m)). \tag{1.11.1}$$

Let $\tau \in \Pi(M(F_S))$ and $\lambda \in \mathfrak{a}_{M,\mathbb{C}}^*$. The representation given by

$$\tau_\lambda(m) = \tau(m) e^{\lambda(H_M(m))}$$

also belongs to $\Pi(M(F_S))$. Suppose that $\tau$ maps to $\tau'$ under the local correspondence. It is a consequence of (1.11.1) that the representation $\tau_\lambda$ maps to $\tau'_{\lambda'}$, where $\lambda \mapsto \lambda'$ via $\mathfrak{a}_{M,\mathbb{C}}^* \xrightarrow{\sim} \mathfrak{a}_{M',\mathbb{C}}^*$.

## *1.12 Closure Property*

Recall that in Section 4.1.4, we defined the *closure property* for a finite set $S$ of valuations. A set $S$ has this property if it contains an Archimedean valuation or all $v$ in $S$ have the same residual characteristic. An equivalent formulation is that, for each $M \in \mathcal{L}$, the subgroup

$$\mathfrak{a}_{M,S} = \{H_M(m)\,;\, m \in M(F_S)\}$$

of $\mathfrak{a}_M$ is closed. Assume $S$ has the closure property. We define

$$i\mathfrak{a}^*_{M,S} = i\mathfrak{a}^*_M / i\,\mathrm{Hom}(\mathfrak{a}_{M,S}, \mathbb{Z}).$$

The Euclidean measure on $i\mathfrak{a}^*_M$ dual to the measure on $\mathfrak{a}_M$ induces a measure $\mathrm{d}\lambda$ on $i\mathfrak{a}^*_{M,S}$.

DEFINITION 1.13. We can identify a function $\varphi$ in the Paley-Wiener space $\mathbb{I}(M(F_S))$ with the function on $\Pi_{\mathrm{temp}}(M(F_S)) \times \mathfrak{a}_{M,S}$ given by

$$\varphi(\tau, X) = \int_{i\mathfrak{a}^*_{M,S}} \varphi(\tau_\lambda) e^{-\lambda(X)}\,\mathrm{d}\lambda.$$

In a similar way we identify $\mathbb{I}(M'(F_S))$ with a space of functions on $\Pi_{\mathrm{temp}}(M'(F_S)) \times \mathfrak{a}_{M,S}$.

If $\varphi$ belongs to $\mathbb{I}(M(F_S))$, define

$$\varphi'(\tau', X') = e_S \varphi(\tau, X), \quad \tau \in \Pi_{\mathrm{temp}}(M(F_S)),\ X \in \mathfrak{a}_{M,S}. \tag{1.13.1}$$

By defining $\varphi'$ to be zero on points of $\Pi_{\mathrm{temp}}(M'(F_S)) \times \mathfrak{a}_{M,S}$ that are not in the image of any point in $\Pi_{\mathrm{temp}}(M(F_S)) \times \mathfrak{a}_{M,S}$, we obtain an element of $\mathbb{I}(M'(F_S))$.

LEMMA 1.14. *The definition* (1.13.1) *is compatible with our earlier Definition* 1.4.

PROOF. Indeed, suppose that $\varphi = h_M$ for $h \in \mathbb{H}(M(F_S))$. (See 1.5(4) for the definition of $h_M$.) Then

$$\begin{aligned}
\varphi'(\tau', X') &= e_S \varphi(\tau, X) \\
&= e_S \int_{i\mathfrak{a}^*_{M,S}} \mathrm{tr}\,\tau_\lambda(h) \cdot e^{-\lambda(X)}\,\mathrm{d}\lambda \\
&= e_S^2 \int_{i\mathfrak{a}^*_{M,S}} h'_{M'}((\tau_\lambda)') e^{-\lambda(X)}\,\mathrm{d}\lambda \\
&= \int_{i\mathfrak{a}^*_{M,S}} h'_{M'}(\tau'_{\lambda'}) e^{-\lambda'(X')}\,\mathrm{d}\lambda.
\end{aligned}$$

Since $\mathrm{d}\lambda$ equals $\mathrm{d}\lambda'$, we obtain

$$\varphi'(\tau', X') = \int_{i\mathfrak{a}^*_{M,S}} h'_{M'}(\tau'_{\lambda'}) e^{-\lambda'(X')}\,\mathrm{d}\lambda'.$$

This says precisely that $\varphi'$ is the function on $\Pi_{\mathrm{temp}}(M'(F_S)) \times \mathfrak{a}_{M,S}$ which is identified with $h'_{M'}$. □

### 1.15 Adele Spaces

We can define the spaces $\mathbb{H}(G(\mathbb{A}))$, $\mathbb{I}(G(\mathbb{A}))$ and the sets $\Pi(G(\mathbb{A}))$, $\Pi(G(\mathbb{A})^1)$, etc. Briefly, these function spaces are defined as limits over $S$ of the spaces $\mathbb{H}(G(F_S))$ and $\mathbb{I}(G(F_S))$. See Sections 5.12.6, 5.12.7 for a more detailed discussion. If $\pi$ belongs to $\Pi_{\text{temp}}(G(\mathbb{A}))$, the obvious analogue of (1.8.1) holds. The same holds if $\pi^1$ belongs to $\Pi_{\text{temp}}(G(\mathbb{A})^1)$.

## 2 Statement of Parts (i) and (ii) of the Main Comparison

### 2.1 Geometric Part

Arthur's invariant trace formula for $G$ is the equality $I_{\text{geom}}(f) = I_{\text{spec}}(f)$ of the geometric $I_{\text{geom}}(f)$ and spectral $I_{\text{spec}}(f)$ expansions of a distribution $I(f)$ on $G$. See Section 5.12 for a discussion of $I(f)$. The geometric side of the invariant trace formula for $G$ is the expansion

$$I_{\text{geom}}(f) = \sum_{M \in \mathcal{L}} \left(|W_0^M|/|W_0^G|\right) \sum_{\gamma \in (M(F))_{M,S}} a^M(S,\gamma) I_M(\gamma,f). \tag{2.1.1}$$

Recall that $(M(F))_{M,S}$ is the set of $(M,S)$-equivalence classes in $M(F)$. When $G$ is an inner form of $\mathrm{GL}(n)$, it is just the set of $M(F)$-orbits in $M(F)$. See Section 5.13.1 for the full definition.

### 2.2 Coefficients

At 5.(13.3.1) we defined the number $a^M(S,\gamma)$. Let us recall how it depends on $\gamma$. Let $\sigma$ be a semisimple element of $M(F)$. Let $i^M(S,\sigma)$ be 1 if $\sigma$ is $F$-elliptic in $M$ and the $M(F_v)$-orbit of $\sigma$ meets $K_v$ for every valuation $v \notin S$. Otherwise put $i^M(S,\sigma) = 0$. For an inner form $G$ of $G' = \mathrm{GL}(n)$, 5.(13.3.1) implies that, if $\gamma = \sigma u$ is the Jordan decomposition of $\gamma$, then

$$a^M(S,\gamma) = i^M(S,\sigma) a^{M_\sigma}(S,u). \tag{2.2.1}$$

In the notation of 5.(13.3.1), we have $\{u\,;\, \sigma u \sim \gamma\} = \{u\}$, the unipotent part of $\gamma$, and the reduction to (2.2.1) follows.

Our overall strategy will be to pull objects on $G'$ back to $G$, where they can be compared with the corresponding objects on $G$. We shall systematically denote objects on $G$ which have been obtained from $G'$ by a superscript $\mathcal{E}$, $\mathcal{E}$ standing for "endoscopic."

DEFINITION 2.3. (1) Define the linear form $I^{\mathcal{E}}$ on $\mathbb{H}(G(\mathbb{A}))$ using the linear form $I = I_{G'}$ on $\mathbb{H}(G'(\mathbb{A}))$ by

$$I^{\mathcal{E}}(f) = \widehat{I}(f'_{G'}). \tag{2.3.1}$$

We also have the distribution $\widehat{I}^{\mathcal{E}}$ on $\mathbb{I}(G(\mathbb{A}))$, defined by

$$\widehat{I}^{\mathcal{E}}(f_G) = I^{\mathcal{E}}(f),$$

as $\widehat{I}(f'_{G'})$ depends only on $f_G \in \mathbb{I}(G(\mathbb{A}))$.
(2) If $S$ is a large finite set of valuations, define

$$a^{M,\mathcal{E}}(S,\gamma) = a^{M'}(S,\gamma'). \tag{2.3.2}$$

## 2.4 Jordan Decomposition

Let $\gamma \in M(F)$ have Jordan decomposition $\sigma u$. Then $\sigma'$ is also semisimple and has the same characteristic polynomial as $\sigma$. In this case, $i^M(S,\sigma)$ equals $i^{M'}(S,\sigma')$. Using (2.2.1) for $M'$ and the definitions, we have

$$a^{M,\mathcal{E}}(S,\gamma) = i^M(S,\sigma)a^{M_\sigma,\mathcal{E}}(S,u). \tag{2.2.1$\mathcal{E}$}$$

## 2.5 Invariant Geometric Part

We will formally define $I^{\mathcal{E}}_M(\gamma,f)$ in Definition 5.8 below. For now, it is enough to say it is defined similarly to $I^{\mathcal{E}}(f)$.

**Proposition 2.5.1.** *For all $f \in \mathbb{H}(G(\mathbb{A}))$, we have*

$$I^{\mathcal{E}}(f) = \sum_{M\in\mathcal{L}} (|W_0^M|/|W_0^G|) \sum_{\gamma\in(M(F))_{M,S}} a^{M,\mathcal{E}}(S,\gamma)I^{\mathcal{E}}_M(\gamma,f). \tag{2.1.1$\mathcal{E}$}$$

PROOF. By (2.3.1) and (2.1.1) applied to $G'$,

$$I^{\mathcal{E}}(f) = \widehat{I}(f'_{G'}) = \sum_{L\in\mathcal{L}'} \left(|W_0^L|/|W_0^{G'}|\right) \sum_{\zeta\in(L(F))_{L,S}} a^L(S,\zeta)\widehat{I}_L(\zeta,f'_{G'}).$$

For each $L \in \mathcal{L}'$ and $\zeta \in L(F)$, the distribution $\widehat{I}_L(\zeta, f'_{G'})$ has the following global vanishing property. By Proposition 5.18.3, it is 0 unless $L = M'$ and $\zeta = \gamma'$ for some $M \in \mathcal{L}$ and $\gamma \in M(F)$. Now, $\gamma \mapsto \gamma'$ is an injection of $(M(F))_{M,S}$ into $(M'(F))_{M',S}$, since we have seen that the equivalence classes are just conjugacy classes. Hence, we obtain

$$I^{\mathcal{E}}(f) = \sum_{M \in \mathcal{L}} (|W_0^M|/|W_0^G|) \sum_{\gamma \in (M(F))_{M,S}} a^{M'}(S, \gamma') \widehat{I}_{M'}(\gamma', f'_{G'}).$$

Since $I_M^{\mathcal{E}}(\gamma, f) = \widehat{I}_{M'}(\gamma', f'_{G'})$ (see Corollary 5.9 below) and $a^{M,\mathcal{E}}(S, \gamma) = a^{M'}(S, \gamma')$ by (2.3.2), this equals (2.1.1$\mathcal{E}$), as required. □

REMARK 2.6. Given a function $f$ in $\mathbb{H}(G(\mathbb{A}))$, let $V = V(f)$ be the smallest set of valuations containing $S_{\text{ram}}$ such that $f$ lies in $\mathbb{H}(G(F_V)) \otimes 1_{K^V}$. By the results of Chapter 5 applied to each of $G$ and $G'$, (2.1.1) and (2.1.1$\mathcal{E}$) hold for any finite set $S$ of valuations of $F$ that is suitably large depending only on the support supp($f$) of $f$ and on $V(f)$. The sums over $\gamma$ in (2.1.1) and (2.1.1$\mathcal{E}$) can both be taken over a finite set that also depends only on supp($f$) and $V(f)$.

The following theorem is (part of) the primary result of this chapter.

THEOREM 2.7 (Geometric Part of Main Theorem).

(i) *Let $S$ be a finite set of valuations containing $S_{\text{ram}}$. Then for all $M \in \mathcal{L}$, $\gamma \in M(F_S)$, $f \in \mathbb{H}(G(F_S))$, we have*

$$I_M^{\mathcal{E}}(\gamma, f) = I_M(\gamma, f).$$

(ii) *Let $S$ be a sufficiently large, finite set of valuations. For every $\gamma \in M(F)$, we have*

$$a^{M,\mathcal{E}}(S, \gamma) = a^M(S, \gamma).$$

These are two out of four assertions comprising the main comparison theorem of the invariant trace formulae. The assertion (i) is local and (ii) is global. They imply a term by term identification of the geometric sides of the trace formulae of $G$ and $G'$. The other two assertions, (iii) and (iv), are the corresponding local and global spectral counterparts of (i) and (ii). The correspondence between automorphic representations follows from the resulting equality of spectral sides. See Section 2.25.

# 3 Statement of Parts (iii) and (iv) of the Main Theorem

## *3.1 Local and Global Constituents*

We will recall the definition of the global constituents $a^M(\tau)$ of the spectral side and the local constituents, the distributions $I_M(\tau)$. We shall then state the spectral parts

(iii) and (iv) of the Main Theorem; they are parallel to the geometric statements (i) and (ii), stated in the previous section.

DEFINITION 3.1.1. For every $t \geq 0$ put

$$I_t(f) = \sum_{M \in \mathcal{L}} (|W_0^M|/|W_0^G|) \int_{\Pi(M,t)} a^M(\tau_{1,\lambda}) I_M(\tau_{1,\lambda}, f)\, d\tau_{1,\lambda}. \tag{3.1.1}$$

The spectral side of the trace formula is the sum

$$I_{\text{spec}}(f) = \sum_{t \geq 0} I_t(f).$$

The trace formula asserts that $I_{\text{spec}}$ is equal to $I_{\text{geom}}$, namely, that $I_{\text{spec}}(f)$ is an expansion of $I_{\text{geom}}(f)$ in terms of the distributions

$$I_M(\tau_{1,\lambda}, f) = I_M(\tau_{1,\lambda}^M, 0, f)$$

whose definition is to be recalled.

The variable $t$, which ranges over the nonnegative real numbers, is required for convergence. We shall recall in a moment how it is used to keep track of the size of Archimedean infinitesimal characters. We will then review the definitions of $a^M(\tau)$ and $\Pi(M, t)$ from Section 5.14.

## 3.2 Cartan Subalgebra

Let $S_\infty$ denote the set of Archimedean valuations of $F$. Set $F_\infty = F_{S_\infty}$. Then $G'(F_\infty)$ can be regarded as a real Lie group. Let $\mathfrak{h}'_{\mathbb{C}}$ denote the standard Cartan subalgebra of its complex Lie algebra. Let $\mathfrak{h}' \subset \mathfrak{h}'_{\mathbb{C}}$ be the real form of $\mathfrak{h}'_{\mathbb{C}}$ associated to the split real form of $G'(F_\infty)$. Then $\mathfrak{h}'$ is invariant under the complex Weyl group $W'$ of $G'(F_\infty)$. By means of the inner twist $\eta$, we can identify $\mathfrak{h}'_{\mathbb{C}}$ with a Cartan subalgebra of the complex Lie algebra of the real Lie group $G(F_\infty)$. In this situation we set $\mathfrak{h} = \mathfrak{h}'$. Then $\mathfrak{h}$ is invariant under the complex Weyl group, $W$, of $G(F_\infty)$. It contains each of the real vector spaces $\mathfrak{a}_M$. It is convenient to fix a Euclidean norm $\|\cdot\|$ on $\mathfrak{h}$ which is invariant under $W$. We shall also write $\|\cdot\|$ for the dual Hermitian norm on $\mathfrak{h}^*_{\mathbb{C}}$.

## 3.3 Infinitesimal Characters

For $M \in \mathcal{L}$ and any representation $\tau \in \Pi(M(\mathbb{A}))$, we can associate the induced representation $\tau^G$ of $G(\mathbb{A})$. Let $\nu_\tau$ be the infinitesimal character of the Archimedean constituent of $\tau$. It is a $W$-orbit in $\mathfrak{h}_{\mathbb{C}}^*$. We shall be more concerned with the case that $\tau^1$ is a representation in $\Pi(M(\mathbb{A})^1)$. Then $\nu_{\tau^1}$ is a priori only determined as an orbit of $\mathfrak{a}_{M,\mathbb{C}}^*$ in $\mathfrak{h}_{\mathbb{C}}^*$. However, this orbit has a unique—up to translation by $W$—point of smallest norm in $\mathfrak{h}_{\mathbb{C}}^*$. This is the point we will denote by $\nu_{\tau^1}$ . If $t \geq 0$, define $\Pi(M(\mathbb{A})^1, t)$ to be the set of representations $\tau^1 \in \Pi(M(\mathbb{A})^1)$ such that

$$\|\mathrm{Im}(\nu_{\tau^1})\| = t,$$

where $\mathrm{Im}(\nu_{\tau^1})$ is the imaginary part of $\nu_{\tau^1}$ relative to the real form $\mathfrak{h}^*$ of $\mathfrak{h}_{\mathbb{C}}^*$.

DEFINITION 3.4. (1) The discrete constituents of (3.1.1) are defined in terms of a function

$$a_{\mathrm{disc}}^G(\tau^1), \quad \tau^1 \in \Pi(G(\mathbb{A})^1, t).$$

This function is defined by rewriting the expression

$$I_{\mathrm{disc},t}(f) =$$

$$\sum_{L \in \mathcal{L}} |W_0^L|/|W_0^G| \sum_{s \in W(\mathfrak{a}_L)_{\mathrm{reg}}} |\det(s-1)_{\mathfrak{a}_L^G}|^{-1} \operatorname{tr} \mathcal{M}(s,0) r_{Q,t}(0,f) \tag{3.4.1}$$

as

$$\sum_{\pi^1 \in \Pi(G(\mathbb{A})^1, t)} a_{\mathrm{disc}}^G(\pi^1) f_G(\pi^1), \tag{3.4.2}$$

a linear combination of characters. See Section 5.14 for a more detailed discussion. In particular, $Q$ is any element in $\mathcal{P}(L)$, and $r_{Q,t}$ is the induced representation of $G(\mathbb{A})^1$ obtained from the subrepresentation of $M(\mathbb{A})^1$ on $L^2(M(F)\backslash M(\mathbb{A})^1)$ which decomposes into a discrete sum of elements in $\Pi(M(\mathbb{A})^1, t)$. Here $\mathcal{M}(s, 0)$ is the global intertwining operator associated to an element in

$$W(\mathfrak{a}_L)_{\mathrm{reg}} = \{s \in W(\mathfrak{a}_L)\,;\, \det(s-1)_{\mathfrak{a}_L^G} \neq 0\},$$

and $W(\mathfrak{a}_L)$ denotes the Weyl group of $\mathfrak{a}_L$. For any function $f \in \mathbb{H}(G(\mathbb{A}))$, the sum in (3.4.2) can be taken over a finite subset of $\pi^1$ in $\Pi_{\mathrm{unit}}(G(\mathbb{A})^1, t)$. It is understood that

$$f_G(\pi^1) = I_G(\pi^1, f) = I_G(\pi_\lambda, 0, f).$$

(2) Suppose that $M_1 \in \mathcal{L}$. As in Section 5.14, we write $\Pi_{\mathrm{disc}}(M_1^1, t)$ for the subset of the set $\Pi_{\mathrm{unit}}(M_1(\mathbb{A})^1, t)$ consisting of irreducible constituents of induced representations

$$\sigma_\lambda^{M_1}, \quad L \in \mathcal{L}^{M_1}, \ \sigma^1 \in \Pi_{\mathrm{unit}}(L(\mathbb{A})^1, t), \ \lambda \in i\mathfrak{a}_L^*/i\mathfrak{a}_{M_1}^*,$$

where $\sigma_\lambda$ satisfies the following two conditions:

(i) $a_{\mathrm{disc}}^L(\sigma^1) \neq 0$.
(ii) There is an element $s \in W^{M_1}(\mathfrak{a}_L)_{\mathrm{reg}}$ such that $s\sigma_\lambda = \sigma_\lambda$.

(3) For any $M$, the set $\Pi(M, t)$ is the disjoint union over $M_1 \in \mathcal{L}^M$ of the sets

$$\Pi_{M_1}(M, t) = \{\tau_{1,\lambda}\,;\ \tau_1^1 \in \Pi_{\mathrm{disc}}(M_1^1, t), \lambda \in i\mathfrak{a}_{M_1}^*/i\mathfrak{a}_M^*\}.$$

(4) The global datum in (3.1.1) is the function

$$a^M(\tau_{1,\lambda}) = a_{\mathrm{disc}}^{M_1}(\tau_1^1) m_{M_1}^M(\tau_{1,\lambda}), \quad \tau_{1,\lambda} \in_{M_1} (M, t),$$

also introduced in Section 5.14. It can be defined for any representation

$$\tau = \tau_{1,\lambda}, \quad \tau_1^1 \in \Pi(M_1(\mathbb{A})^1), \ \lambda \in i\mathfrak{a}_{M_1}^*/i\mathfrak{a}_M^*,$$

but it vanishes unless $\tau_{1,\lambda}$ belongs to $\Pi_{M_1}(M, t)$ for some $t$. The function $m_{M_1}^M(\tau_{1,\lambda})$ is obtained from a $(G, M_1)$-family which is built out of the global normalizing factors. The measure in (3.1.1) is given by

$$d\tau_{1,\lambda} = |W_0^{M_1}|/|W_0^M|\, d\lambda, \quad \tau_{1,\lambda} \in \Pi_{M_1}(M, t).$$

(5) In our notation $\{\tau_{1,\lambda}\}$ stands for the orbit of $i\mathfrak{a}_{M_1}^*/i\mathfrak{a}_M$ in

$$\Pi_{\mathrm{unit}}(M_1(\mathbb{A}) \cap M(\mathbb{A})^1)$$

associated to a given $\tau_1^1 \in \Pi_{\mathrm{disc}}(M_1^1, t)$. But we shall often identify $\tau_{1,\lambda}$ with the induced representation $\tau_{1,\lambda}^M$ in $\Pi_{\mathrm{unit}}(M(\mathbb{A})^1)$. It is in this sense that the distribution $I_M(\tau_{1,\lambda}, f)$ in (3.1.1) is defined.

## *3.5 Convergence*

The function

$$I_M(\tau_{1,\lambda}, f) = I_M(\tau_{1,\lambda}^M, f), \quad \lambda \in i\mathfrak{a}_{M_1}^*/i\mathfrak{a}_M^*,$$

is rapidly decreasing. In fact, it extends to a meromorphic function in the complex domain which is rapidly decreasing on cylinders, as one sees directly from the definition of the distribution in 5.3. This property is implicit in the formula (3.1.1), and (6.1.1), and will be used later without comment.

Recall that the integral over $\Pi(M,t)$ in (3.1.1) converges absolutely, as does the sum $\sum_t I_t(f)$ (see Theorem 5.14.10). For proofs of absolute convergence in the context of the non-invariant spectral side of Arthur's trace formula, see [FLM11].

A similar expansion of course holds for $G'$. However, we would like to define functions which can be compared directly with $a^M(\tau)$.

Let us recall some numbers, initially introduced in Section 4.9

DEFINITION 3.6. (1) The numbers $\iota(\rho,\tau)$ and $\xi(\tau,\rho)$ describe the transformation formulae between standard and irreducible characters. Thus for $\rho \in \Sigma(M(F_S))$, put

$$\operatorname{tr}\rho = \sum_{\tau\in\Pi(M(F_S))} \iota(\rho,\tau)\operatorname{tr}\tau.$$

And, for $\tau \in \Pi(M(F_S))$, put

$$\operatorname{tr}\tau = \sum_{\rho\in\Sigma(M(F_S))} \xi(\tau,\rho)\operatorname{tr}\rho.$$

(2) Suppose that $\tau' \in \Pi(M'(F_S))$ and $\rho \in \Sigma(M(F_S))$. The numbers of (1) are not immediately defined for representations of different groups. However, we set

$$\xi(\tau',\rho) = e_S\xi(\tau',\rho').$$

For $\tau \in \Pi(M(F_S))$, we define

$$\delta(\tau',\tau) = \sum_{\rho\in\Sigma(M(F_S))} \xi(\tau',\rho)\iota(\rho,\tau).$$

If $G = G'$, we have

$$\delta(\tau',\tau) = \begin{cases} 1, & \tau' = \tau, \\ 0, & \text{otherwise.} \end{cases}$$

In general, the formula is not as simple.

Suppose $\tau' = \otimes_{v\in S}\tau'_v$ and $\tau = \otimes_{v\in S}\tau_v$. Then

$$\delta(\tau',\tau) = \prod_{v\in S}\delta(\tau'_v,\tau_v). \tag{3.6.1}$$

This is a consequence of the fact that $h'_{M'}(\rho')$ is $e_S\operatorname{tr}\rho(h)$ if $\rho'$ corresponds to $\rho$ (and $h'$ to $h$), and $h'_{M'}(\rho')$ is 0 if $\rho'$ does not correspond to some $\rho$.

PROPOSITION 3.7. *For $h \in \mathbb{H}(M(F_S))$ and $\tau' \in \Pi(M'(F_S))$, we have*

$$h'_{M'}(\tau') = \sum_{\tau \in \Pi(M(F_S))} \delta(\tau', \tau) \operatorname{tr} \tau(h).$$

PROOF. We have

$$\begin{aligned} h'_{M'}(\tau') &= \sum_{\rho' \in \Sigma(M'(F_S))} \xi(\tau', \rho') h'_{M'}(\rho') \quad \text{(by definition of the numbers } \xi(\tau', \rho')) \\ &= \sum_{\rho \in \Sigma(M(F_S))} \xi(\tau', \rho) \operatorname{tr} \rho(h) \quad \text{(by sentence above the statement of the proposition)} \\ &= \sum_{\rho \in \Sigma(M(F_S))} \sum_{\tau \in \Pi(M(F_S))} \xi(\tau', \rho) \iota(\rho, \tau) \operatorname{tr} \tau(h) \quad \text{(by definition of the numbers } \iota(\rho, \tau)) \\ &= \sum_{\tau \in \Pi(M(F_S))} \delta(\tau', \tau) \operatorname{tr} \tau(h). \quad \text{(by definition of the numbers } \delta(\tau', \tau)) \end{aligned}$$

□

## 3.8 *Adelic Statement*

Suppose that $\tau' = \otimes_v \tau'_v$ and $\tau = \otimes_v \tau_v$ are representations in $\Pi(M'(\mathbb{A}))$ and $\Pi(M(\mathbb{A}))$, respectively. Define

$$\delta(\tau', \tau) = \prod_v \delta(\tau'_v, \tau_v).$$

Almost all terms in the product are 0 or 1, so the product can be taken over a finite set. The adelic formulation is then included in the previous definitions, so it satisfies the formulae above. In particular, for $h \in \mathbb{H}(M(\mathbb{A}))$ and $\tau' \in \Pi(M'(\mathbb{A}))$, we have

$$h'_{M'}(\tau') = \sum_{\tau \in \Pi(M(\mathbb{A}))} \delta(\tau', \tau) \operatorname{tr} \tau(h).$$

Assume now that $\tau'^1$ and $\tau^1$ are members of $\Pi(M'(\mathbb{A})^1)$ and $\Pi(M(\mathbb{A})^1)$, respectively. We have identified these representations with orbits $\{\tau'_\eta\}$ and $\{\tau_\lambda\}$ in $\Pi(M'(\mathbb{A}))$ and $\Pi(M(\mathbb{A}))$. In this situation we fix $\eta$ and define

$$\delta(\tau'^1, \tau^1) = \sum_{\lambda \in \mathfrak{a}^*_{M,\mathbb{C}}} \delta(\tau'_\eta, \tau_\lambda).$$

There can be at most one nonzero summand on the right, and its value is independent of $\eta$.

LEMMA 3.9. *If $\tau_1^1$ is any representation in $\Pi(M_1(\mathbb{A})^1)$, the sum*

$$\sum_{\tau_1'^1 \in \Pi(M_1'(\mathbb{A})^1)} a_{\text{disc}}^{M_1'}(\tau_1'^1)\delta(\tau_1'^1, \tau_1^1)$$

*has only finitely many nonzero terms.*

PROOF. Extend $\tau_1^1$ to a representation $\tau_1$ in $\Pi(M_1(\mathbb{A}))$. Let $S$ be a finite set of valuations containing the valuations at which $G$ and $\tau_1$ are ramified. Let $\tau_1 = \tau_{1,S} \otimes \tau_1^S$ be the decomposition of $\tau_1$ corresponding to the decomposition

$$M_1(\mathbb{A}) = M_1(F_S)\left(\prod_{v \notin S} M_1(F_v)\right).$$

Consider the representations $\tau_1' = \tau_{1,S}' \otimes \tau_1'^S$ in $\Pi(M_1'(\mathbb{A}))$ such that the number

$$\delta(\tau_1', \tau_1) = \delta(\tau_{1,S}', \tau_{1,S}) \prod_{v \notin S} \delta(\tau_{1,v}', \tau_{1,v})$$

does not vanish. Now $\tau_1'^S$ is unramified. It is clear that there are only finitely many choices for $\tau_{1,S}'$. It follows from Proposition 5.14.6 that there are only finitely many such $\tau_1'$ with $a_{\text{disc}}^{M_1}(\tau_1'^1) \neq 0$. Therefore there are only finitely many nonzero summands in the series. □

DEFINITION 3.10. (1) Define

$$a_{\text{disc}}^{M_1,\mathcal{E}}(\tau_1^1) = \sum_{\tau_1'^1 \in \Pi(M_1'(\mathbb{A})^1)} a_{\text{disc}}^{M_1'}(\tau_1'^1)\delta(\tau_1'^1, \tau_1^1). \tag{3.10.1}$$

(2) For each

$$\tau_{1,\lambda}, \quad \tau_1^1 \in \Pi(M_1(\mathbb{A})^1), \quad \lambda \in \mathfrak{a}_{M_1,\mathbb{C}}^*/\mathfrak{a}_{M,\mathbb{C}}^*,$$

we define

$$a^{M,\mathcal{E}}(\tau_{1,\lambda}) = a_{\text{disc}}^{M_1,\mathcal{E}}(\tau_1^1) m_{M_1}^M(\tau_{1,\lambda}).$$

We now state parts (iii) and (iv) of the Main Theorem.

THEOREM 3.11 (Spectral Part of Main Theorem).

(iii) *Suppose that $S$ is a finite set of valuations which contains $S_{\text{ram}}$. Then for $\tau^1 \in \Pi_{\text{unit}}(M(\mathbb{A})^1)$ and $f \in \mathbb{H}(G(F_S))$, we have*

$$I_M^{\mathcal{E}}(\tau^1, f) = I_M(\tau^1, f).$$

(iv) *For any given*

$$\tau_{1,\lambda}, \quad \tau_1^1 \in \Pi(M_1(\mathbb{A})^1), \quad \lambda \in \mathfrak{a}^*_{M_1,\mathbb{C}}/\mathfrak{a}^*_{M,\mathbb{C}},$$

*we have*

$$a^{M,\mathcal{E}}(\tau_{1,\lambda}) = a^M(\tau_{1,\lambda}).$$

Parts (iii) and (iv) consist of a local and global assertions parallel to parts (i) and (ii) introduced in the previous section. This completes the statement of the Main Theorem. They imply a term by term identification of the trace formulae of $G$ and $G'$. Part (iv) of the Theorem 3.11 will provide the correspondence between automorphic representations.

## 3.12 Proof by Induction

The Main Theorems 2.7 and 3.11 will be proved by induction. We shall prove in Section 10 that the local assertion (iii) of the Main Theorem 3.11 can be proved from the Induction Assumption (i) of the Main Theorem 2.7.

The global assertion (iv) requires an additional induction assumption. *We assume that for $M_1 \in \mathcal{L}$ with $M_1 \neq G$ and $\tau_1 \in \Pi(M_1(\mathbb{A})^1)$ that*

$$a^{M_1,\mathcal{E}}_{\text{disc}}(\tau_1) = a^{M_1}_{\text{disc}}(\tau_1).$$

In particular, $a^{M_1,\mathcal{E}}_{\text{disc}}(\tau_1)$ vanishes unless $\tau_1 \in \Pi_{\text{disc}}(M_1, t)$ for some $t$, and necessarily must be unitary. If $\tau_1$ is unitary, $M_1 \subset M$ and $\lambda \in i\mathfrak{a}^*_{M_1}/i\mathfrak{a}^*_M$, then $\tau = \tau_{1,\lambda}$ is well defined. In addition, if $\tau_1 \in \Pi_{\text{disc}}(M_1, t)$, the function $m^M_{M_1}(\tau_{1,\lambda})$ is defined (in the following Section 4). Then the function $a^{M,\mathcal{E}}(\tau)$ is well defined, and if $M_1 \neq G$, we have

$$a^{M,\mathcal{E}}(\tau) = a^M(\tau).$$

# 4 Normalization Factors and the Trace Formula

## 4.1 Normalizing Factors

The invariant trace formula of Chapter 5 depends on a normalization of the intertwining operators between induced representations. Since we shall compare the trace formulae for $G$ and $G'$, we will want to choose the normalizations for the two groups in a compatible way.

Let $S$ be a finite set of valuations with the closure property. Fix a Levi subgroup $M \in \mathcal{L}$ and a representations $\tau = \otimes_{v\in S}\tau_v$ in $\Pi(M(F_S))$. Associated to parabolic subgroups $P$ and $Q$ in $\mathcal{P}(M)$, there are intertwining operators

$$\mathcal{M}_{Q|P}(\tau_\lambda), \quad \lambda \in \mathfrak{a}^*_{M,\mathbb{C}},$$

between the induced representations $i_P^G(\tau_\lambda)$ and $i_Q^G(\tau_\lambda)$. These operators are defined by integrals over $N_Q(F_S) \cap N_{\overline{P}}(F_S)$, and so depends on a choice of a Haar measure on this group. In order to put the trace formula into invariant form, it is necessary to define meromorphic scalar valued functions

$$m_{Q|P}(\tau_\lambda) = \prod_{\alpha \in \Sigma_Q \cap \Sigma_{\overline{P}}} m_\alpha(\tau, \lambda(\alpha^\vee)), \quad \lambda \in \mathfrak{a}^*_{M,\mathbb{C}} \tag{4.1.1}$$

so that the normalized intertwining operators

$$\mathcal{R}_{Q|P}(\tau_\lambda) = m_{Q|P}(\tau_\lambda)^{-1}\mathcal{M}_{Q|P}(\tau_\lambda)$$

satisfy the conditions of Theorem 4.6.2. Here $\Sigma_P$ denotes the set of roots $\alpha$ of $(P, A_M)$. For each $\alpha$, the product

$$m_\alpha(\tau, s) = \prod_{v \in S} m_\alpha(\tau_v, s), \quad s \in \mathbb{C},$$

is a meromorphic function of one complex variable. In Chapter 4, Sections 6-8, we saw that such normalizing factors could be chosen for any group. However, to show that this can be done in a compatible way for $G$ and $G'$, we must use the more precise results of Shahidi [Sha84].

## 4.2 Explicit Expression

The intertwining operators and the normalizing factors are given by products over $v \in S$, so we can work with a given valuation. For the moment, then, we shall suppose that $S$ consists of a single valuation $v$. Let $\psi_v$ be a fixed nontrivial additive character on $F_v$. We shall consider first the special case that $G = G' = \mathrm{GL}(n)$. To define the normalizing factors, it is enough to define the functions $m_\alpha(\tau, s)$ for any root $\alpha$ of $(G, A_M)$ and any $\tau \in \Pi(M(F_v))$. There is an isomorphism

$$M \xrightarrow{\sim} \prod_{i=1}^{r} \mathrm{GL}(n_i),$$

where $(n_1, \ldots, n_r)$ is a partition of $n$. The root $\alpha$ is associated to an ordered pair $(p, q)$ of distinct integers between 1 and $r$, and $\tau$ corresponds to a representation

$$\tau_1 \times \cdots \times \tau_r, \quad \tau_i \in \Pi(\mathrm{GL}(n_i, F_v)).$$

If $\tau$ is tempered, define

$$m_\alpha(\tau, s) = L(s, \tau_p \otimes \tilde{\tau}_q)\varepsilon(s, \tau_p \otimes \tilde{\tau}_q, \psi_v)^{-1} L(s+1, \tau_p \otimes \tilde{\tau}_q)^{-1}. \tag{4.2.1}$$

Then $m_\alpha(\tau, s)$ is a meromorphic function of $s$ with the property that

$$m_\alpha(\tau_\lambda, s) = m_\alpha(\tau, \lambda(\alpha^\vee) + s), \quad \lambda \in i\mathfrak{a}_M^*.$$

By meromorphic continuation in $\lambda$ the definition can then be extended to standard representations. Finally, if $\tau$ is an arbitrary representation in $\Pi(M(F_v))$, let $\rho$ be the standard representation of which $\tau$ is the Langlands quotient, and set

$$m_\alpha(\tau, s) = m_\alpha(\rho, s).$$

It follows from the results of Shahidi [Sha84] that there are Haar measures, depending on $\psi_v$, on the groups $N_Q(F_v) \cap N_{\overline{P}}(F_v)$, such that the normalizing factors (4.1.1) have all the right properties. See also Section 4.8.

## 4.3 General Case

We return to the general case, with $G$ an inner form over $F$ of $G' = \mathrm{GL}(n)$. We fix the normalizing factors as above for $G'$. We shall show that the normalizing factors for $G$ can be defined in terms of those for $G'$. Suppose that $\alpha$ is a root of $(G, A_M)$. Then $\alpha'$ is a root of $(G', A_{M'})$. There is an isomorphism

$$M' \xrightarrow{\sim} \prod_{i=1}^{r} \mathrm{GL}(n_i).$$

As above, $\alpha'$ corresponds to a pair $(p, q)$. Given the representation $\tau \in \Pi(M(F_v))$, we set

$$m_\alpha(\tau, s) = m_{\alpha'}(\tau', s), \quad s \in \mathbb{C}, \tag{4.3.1}$$

where $\tau'$ corresponds to $\tau$. Then $m_\alpha(\tau, s)$ is a meromorphic function of $s$.

PROPOSITION 4.4. *The normalizing factors defined by* (4.1.1) *and* (4.3.1) *satisfy all the properties of Theorem* 4.6.2.

PROOF. Many of the required properties follow from standard properties of the operators $\mathcal{M}_{Q|P}(\tau)$ and the general form (4.1.1) of the normalizing factors. There is, in fact, only one condition to verify. We must show that

$$m_{P|\overline{P}}(\tau_\lambda) m_{\overline{P}|P}(\tau_\lambda) = \mu_M(\tau_\lambda)^{-1}, \quad P \in \mathcal{P}(M), \quad \lambda \in \mathfrak{a}_{M,\mathbb{C}}^*, \tag{4.4.1}$$

for any $\tau \in \Pi_{\text{temp}}(M(F_v))$. Here $\mu_M(\tau_\lambda)$ denotes Harish-Chandra's $\mu$-function. As explained in Section 4.6, all of the required properties of the normalized operators will follow from (4.4.1).

Note that $G$ is obtained from a central simple algebra. The definition (4.3.1) is

$$m_\alpha(\tau, s) = m_{\alpha'}(\tau', s)$$

where $\tau \mapsto \tau'$. Therefore

$$m_{Q|P}(\tau_\lambda) = m_{Q'|P'}(\tau'_{\lambda'})$$

by (4.1.1). Since (4.4.1) is true for $G'$, we have only to show that

$$\mu_M(\tau) = \mu_{M'}(\tau') \tag{4.4.2}$$

for any $\tau \in \Pi_{\text{temp}}(M(F_v))$. Let $\Pi_{\text{disc}}(M(F_v))$ be the subset of representations in $\Pi_{\text{temp}}(M(F_v))$ which are square-integrable modulo $A_M(F_v)$. Any $\tau \in \Pi_{\text{temp}}(M(F_v))$ is obtained by induction from a representation $\tau_1 \in \Pi_{\text{disc}}(M_1(F_v))$, where $M_1$ is a Levi component of a parabolic subgroup of $M$ over $F_v$. This follows in the non-Archimedean case from Proposition 2.19.2. By definition

$$\mu_M(\tau) = \mu_{M_1}(\tau_1).$$

Therefore if we are willing to assume that $M$ is defined only over $F_v$, we need only prove (4.4.2) for $\tau \in \Pi_{\text{disc}}(M(F_v))$. We shall do so by comparing the Plancherel formulae for $G$ and $G'$.

Let $f$ be a function in $\mathbb{H}(G(F_v))$ whose character vanishes on any irreducible tempered representation which is not equivalent to some

$$i_P(\tau), \quad P \in \mathcal{P}(M), \quad \tau \in \Pi_{\text{disc}}(M(F_v)).$$

The Plancherel formula for $G$ ([HC76, HCpl, Wl95]) provides a number $\gamma_M$ such that

$$f(1) = \gamma_M \int_{\Pi_{\text{disc}}(M(F_v))} d^M(\tau)\mu_M(\tau)f_M(\tau)\,\mathrm{d}\tau.$$

Here $d^M(\tau)$ is the formal degree of $\tau$. It is not to be confused with $\mathrm{d}\tau$. The latter stands for the measure on $\Pi_{\text{disc}}(M(F_v))$ which is obtained from our Haar measure on $i\mathfrak{a}^*_{M,v}$ and the free action

$$\tau \mapsto \tau_\lambda, \quad \lambda \in i\mathfrak{a}^*_{M,v}.$$

By properties of matching test functions $f$ and $f'$ (see Lemma 1.7), we have

$$f(1) = I_G(1,f) = e_v I_{G'}(1,f').$$

The sign $e_v$ is defined in Subsection 1.6(4). Moreover, $f'$ vanishes for any tempered representation of $G'(F_v)$ which is not equivalent to one of the form

$$\pi' = i_{P'}(\tau'), \quad P \in \mathcal{P}(M),\ \tau \in \Pi_{\mathrm{disc}}(M(F_v)).$$

But by the properties of the correspondence $\tau \mapsto \tau'$, we have

$$f'(\pi') = f'_{M'}(\tau') = e_v^{-1} f_M(\tau).$$

Combined with the Plancherel formula for $G'$, these observations tell us that

$$f(1) = \gamma_{M'} \int_{\Pi_{\mathrm{disc}}(M(F_v))} d^{M'}(\tau')\mu_{M'}(\tau') f_M(\tau)\, d\tau.$$

We choose Haar measures on the groups $N_Q(F_v) \cap N_{\overline{P}}(F_v)$ to match those on $N_{Q'}(F_v) \cap N_{\overline{P}'}(F_v)$ under the isomorphism $\eta$. The numbers $\gamma_M$ and $\gamma_{M'}$ are defined in terms of certain integrals on these groups and are therefore equal. By varying $f$, and taking note of the trace Paley-Wiener theorems ([BDK86, F95, CD84]), we see that

$$d^{M'}(\tau')\mu_{M'}(\tau') = d^M(\tau)\mu_M(\tau), \quad \tau \in \Pi_{\mathrm{disc}}(M(F_v)).$$

In the special case that $M$ equals $G$, the functions $\mu_G$ and $\mu_{G'}$ both equal 1. It follows that $d^{G'}(\tau')$ equals $d^G(\tau)$. The same formula of course holds for arbitrary $M$. We therefore obtain the formula (4.4.2) for any representation $\tau \in \Pi_{\mathrm{disc}}(M(F_v))$. This establishes the proposition. □

REMARK. If the valuation $v$ is Archimedean, Proposition 4.4 is essentially a special case of the general results of Section 4.7.

## *4.5 The Trace Formula*

We have thus defined the local normalizing factors for $G$. They depend on the additive characters $\psi_v$. We assume that each $\psi_v$ is the local component of a fixed nontrivial additive character $\psi$ of $\mathbb{A}/F$. This allows us to build global normalizing factors from infinite products of local ones.

Having chosen normalizations for the intertwining operators, we can write down the full trace formula for $G$ as in Chapter 5. It may be regarded as an identity

$$I(f) = \sum_{M \in \mathcal{L}} |W_0^M||W_0^G|^{-1} \sum_{\gamma \in (M(F))_{M,S}} a^M(S,\gamma) I_M(\gamma, f)$$

$$= \sum_{t} \sum_{M \in \mathcal{L}} |W_0^M||W_0^G|^{-1} \int_{\Pi(M,t)} a^M(\tau_{1,\lambda}) I_M(\tau_{1,\lambda}, f) \, d\tau_{1,\lambda},$$

in which a certain linear functional $I$ on $\mathbb{H}(G(\mathbb{A}))$ is expressed in two different ways. Both sides break up into constituents which are of either a local or a global nature. We shall discuss these separately in the following sections.

## 4.6 Comparison of Normalizing Factors

We next discuss some further questions related to the comparison of normalizing factors.

Suppose that $\tau^1 \in \Pi_{\text{disc}}(M, t)$. Then $\tau^1$ can be identified with an orbit $\{\tau_\lambda\}$ of $i\mathfrak{a}_M^*$ in $\Pi_{\text{unit}}(M(\mathbb{A}))$. The global normalizing factors are functions

$$m_{P_1|P_2}(\tau_\lambda) = \prod_{\alpha \in \Sigma_{P_1} \cap \Sigma_{\overline{P}_2}} m_\alpha(\tau^1, \lambda(\alpha^\vee)), \quad P_1, P_2 \in \mathcal{P}(M),$$

which are meromorphic in $\lambda \in \mathfrak{a}_{M,\mathbb{C}}^*$ and regular for imaginary $\lambda$. They satisfy

$$m_{P_1|P_3}(\tau_\lambda) = m_{P_1|P_2}(\tau_\lambda) m_{P_2|P_3}(\tau_\lambda). \tag{4.6.1}$$

(See 5.14.7.) The global factors $m_\alpha$ can be obtained from the local normalizing factors defined in Subsection 4.1. If $\tau^1$ is identified with a unitary representation $\otimes_v \tau_v$ of $M(\mathbb{A})$, they are related by an infinite product

$$m_\alpha(\tau, \lambda(\alpha^\vee)) = \prod_v m_\alpha(\tau_v, \lambda(\alpha^\vee)),$$

which converges in some right half plane. Suppose that $\tau' = \otimes_v \tau'_v$ is the image of $\tau$. The formula (4.3.1) for the local normalizing factors can be written

$$m_\alpha(\tau_v, s) = m_{\alpha'}(\tau'_v, s). \tag{4.6.2}$$

Observe that if the representation $\tau'$ is not automorphic, then $m_{\alpha'}(\tau', s)$ may not be defined for all $s$. It is defined in general for $s$ in some right half plane.

## 4.7 Equality of Normalizing Factors

The functions which occur in the trace formula are built out of the $(G, M)$-family

$$m_P(\nu, \tau_\lambda, P_0) = m_{P|P_0}(\tau_\lambda)^{-1} m_{P|P_0}(\tau_{\lambda+\nu}), \quad P \in \mathcal{P}(M),\ \nu \in i\mathfrak{a}_M^*. \tag{4.7.1}$$

The associated functions $m_M^L(\tau_\lambda)$, $L \in \mathcal{L}(M)$, are analytic in $\lambda \in i\mathfrak{a}_M^*$ and are independent of $P_0$.

PROPOSITION. *For each $L \in \mathcal{L}(M)$, we have*

$$m_M^L(\tau_\lambda) = m_{M'}^{L'}(\tau'_{\lambda'}),$$

*where $\tau'$ corresponds to $\tau$.*

PROOF. Since the $(G, M)$-family (4.7.1) is defined as a product of functions indexed by roots, we can apply Lemma 7.1 of [Ar82.II]. We obtain

$$m_M^L(\tau_\lambda) = \sum_\Phi \mathrm{vol}\left(\mathfrak{a}_M^L/\mathbb{Z}(\Phi^\vee)\right) \prod_{\alpha\in\Phi} m_\alpha(\tau, \lambda(\alpha^\vee))^{-1} m_\alpha^{(1)}(\tau, \lambda(\alpha^\vee)),$$

where $\Phi$ is taken over all subsets of the roots of $(G, A_M)$ for which

$$\Phi^\vee = \{\alpha^\vee\,;\, \alpha \in \Phi\}$$

is a basis of $\mathfrak{a}_M^L$, and $m_\alpha^{(1)}$ denotes the derivative with respect to the second variable. Observe that the map $X \mapsto X'$ from $\mathfrak{a}_M^L$ to $\mathfrak{a}_{M'}^{L'}$ sends any "coroot" $\alpha^\vee$ to the corresponding "coroot" $(\alpha')^\vee$. We have

$$\mathrm{vol}\left(\mathfrak{a}_M^L/\mathbb{Z}(\Phi^\vee)\right) = \mathrm{vol}\left(\mathfrak{a}_{M'}^{L'}/\mathbb{Z}((\Phi')^\vee)\right).$$

Combined with (4.6.2) and the fact that $\lambda(\alpha^\vee) = \lambda'((\alpha')^\vee)$, this gives

$$\prod_{\alpha\in\Phi} m_\alpha(\tau, \lambda(\alpha^\vee))^{-1} m_\alpha^{(1)}(\tau, \lambda(\alpha^\vee)) = \prod_{\alpha'\in\Phi'} m_{\alpha'}(\tau', \lambda'((\alpha')^\vee))^{-1} m_{\alpha'}^{(1)}(\tau', \lambda'(\alpha')^\vee).$$

Applying Lemma 7.1 of [Ar82.II] to the function

$$m_{M'}^{L'}(\tau'_{\lambda'}),$$

we obtain the required formula. □

## 4.8 A Variant

It is actually a slight variant of Proposition 4.7 that we will need to use. If $\varphi'^1$ is any representation in $\Pi_{\mathrm{disc}}(M', t)$, set

$$m_{P_1|P_2}(\varphi'_\lambda) = \prod_{\alpha \in \Sigma_{P_1} \cap \Sigma_{\overline{P}_2}} m_{\alpha'}(\varphi'^1, \lambda'((\alpha')^\vee)),$$

and let

$$m_P(\nu, \varphi'_\lambda, P_0) = m_{P|P_0}(\varphi'_\lambda)^{-1} m_{P|P_0}(\varphi'_{\lambda+\nu}), \quad P \in \mathcal{P}(M),\ \nu \in i\mathfrak{a}^*_M,$$

be the associated $(G, M)$-family. The proof of Proposition 4.7 gives

COROLLARY. *For each $L \in \mathcal{L}(M)$, we have*

$$m^L_M(\varphi'_\lambda) = m^{L'}_{M'}(\tau'_{\lambda'}).$$

Suppose that $\tau^1$ is as above and that

$$\tau_\lambda = \otimes_v \tau_{v,\lambda}, \quad \tau_v \in \Pi_{\mathrm{unit}}(M(F_v)).$$

We can write

$$m_{P_1|P_2}(\tau_\lambda) = \prod_v m_{P_1|P_2}(\tau_{v,\lambda}),$$

the right-hand side being defined by analytic continuation. The local factors do not satisfy the product formula (4.6.1). However, it is important to consider certain quotients of local factors which do satisfy this formula. The functions $m^L_M(\tau_\lambda, \rho_\lambda)$ discussed in Section 6 below arise from examples of this sort. More examples are provided by coupling representations of $G$ and $G'$.

Suppose that $\varphi' = \otimes_v \varphi'_v$ and $\tau = \otimes_v \tau_v$ are arbitrary representations in $\Pi(M'(\mathbb{A}))$ and $\Pi(M(\mathbb{A}))$, respectively. Assume that the number

$$\delta(\varphi', \tau) = \prod_v \delta(\varphi'_v, \tau_v)$$

defined in Definition 3.6(2) does not vanish. For each root $\alpha$ of $(G, A_M)$ and each $v$, set

$$\tilde{m}_\alpha(\varphi'_v, \tau_v, s) = m_{\alpha'}(\varphi'_v, s)^{-1} m_\alpha(\tau_v, s).$$

This function is constructed out of local $L$-functions. If $\varphi'_v$ and $\tau_v$ are unramified, the representations $\varphi'_v$ must be the one corresponding to $\tau_v$. It follows from the product formula above that in this case

$$\tilde{m}_\alpha(\varphi'_v, \tau_v, s) = 1.$$

We can therefore define

$$\tilde{m}_\alpha(\varphi', \tau, s) = \prod_v \tilde{m}_\alpha(\varphi'_v, \tau, s)$$

as a product over a finite set $S$. We can then set

$$\tilde{m}_{P_1|P_2}(\varphi'_\lambda, \tau_\lambda) = \prod_{\alpha \in \Sigma_{P_1} \cap \Sigma_{\overline{P}_2}} \tilde{m}_\alpha(\varphi', \tau, \lambda(\alpha^\vee)).$$

Now for any given $v$, we have

$$\delta(\varphi'_v, \tau_v) = \sum_{\rho_v \in \Sigma(M(F_v))} e_v \xi(\varphi'_v, \rho'_v)\iota(\rho_v, \tau_v) \neq 0,$$

so we may choose $\rho_v$ and $\rho'_v$ so that $\xi(\varphi'_v, \rho'_v)\iota(\rho_v, \tau_v)$ does not vanish. Since

$$m_\alpha(\rho_v, s) = m_{\alpha'}(\rho'_v, s),$$

we can write $\tilde{m}_\alpha(\varphi'_v, \tau_v, s)$ as the product of

$$(m_{\alpha'}(\rho'_v, s)/m_{\alpha'}(\varphi'_v, s))^{-1}$$

and

$$m_\alpha(\rho_v, s)^{-1} m_\alpha(\tau_v, s).$$

By Proposition 4.9.10, these are rational functions of $s$ if $v$ is Archimedean and of $q_v^{-s}$ if $v$ is non-Archimedean of residual order $q_v$. The same is therefore true of $\tilde{m}_\alpha(\varphi'_v, \tau_v, s)$. Another consequence of the Proposition 4.9.10 is the formula

$$\tilde{m}_\alpha(\varphi'_v, \tau_v, s)\tilde{m}_{-\alpha}(\varphi'_v, \tau_v, s) = 1.$$

In other words,

$$\tilde{m}_\alpha(\varphi', \tau_v, s)\tilde{m}_{-\alpha}(\varphi', \tau, s) = 1.$$

From this we easily see that

$$\tilde{m}_{P_1|P_3}(\varphi'_\lambda, \tau_\lambda) = \tilde{m}_{P_1,P_2}(\varphi'_\lambda, \tau_\lambda)\tilde{m}_{P_2|P_3}(\varphi'_\lambda, \tau_\lambda),$$

for any $P_1$, $P_2$, and $P_3$ in $\mathcal{P}(M)$.

We define a $(G, M)$-family

$$\tilde{m}_P(\nu, \varphi'_\lambda, \tau_\lambda, P_0) = \tilde{m}_{P|P_0}(\varphi'_\lambda, \tau_\lambda)^{-1}\tilde{m}_{P|P_0}(\varphi'_{\lambda+\nu}, \tau_{\lambda+\nu}), \quad P \in \mathcal{P}(M),\ \nu \in i\mathfrak{a}^*_M.$$

PROPOSITION 4.9. (a) *Take* $\varphi'^1$ *and* $\tau^1$ *as above. Then for each* $L \in \mathcal{L}(M)$, $\tilde{m}_M^L(\varphi'_\lambda, \tau_\lambda)$ *is a rational function of the variables* (6.2.1) *of Section* 6 *below, which is independent of* $P_0$. *In* (6.2.1), $S$ *is understood to be any finite set of valuations outside of which* $\varphi'$ *and* $\tau$ *are unramified.*

(b) *Suppose in addition that* $\varphi'^1 \in \Pi_{\text{disc}}(M', t)$ *and* $\tau^1 \in \Pi_{\text{disc}}(M, t)$. *Then*

$$\tilde{m}_{P_1|P_2}(\varphi'_\lambda, \tau_\lambda) = m_{P_1|P_2}(\varphi'_\lambda)^{-1} m_{P_1|P_2}(\tau_\lambda).$$

*In particular, for each* $L \in \mathcal{L}(M)$ *the function* $\tilde{m}_M^L(\varphi'_\lambda, \tau_\lambda)$ *is regular for* $\lambda \in i\mathfrak{a}_M^*$. *Moreover,*

$$m_M^L(\tau_\lambda) = \sum_{L_1 \in \mathcal{L}^L(M)} m_M^{L_1}(\varphi'_\lambda) \tilde{m}_{L_1}^L(\varphi'_\lambda, \tau_\lambda).$$

*We have written* $\varphi'_\lambda$ *and* $\tau_\lambda$ *when in the last formula we really mean the induced representations to* $L'_1$ *and* $L_1$, *respectively.*

PROOF. Everything but the last assertion of the proposition follows from the discussion above. Notice that under the conditions of (b), we have a decomposition

$$m_P(\nu, \tau_\lambda, P_0) = \tilde{m}_P(\nu, \varphi'_\lambda, \tau_\lambda, P_0) m_P(\nu, \varphi'_\lambda, P_0)$$

into a product of $(G, M)$-families. The last assertion therefore follows from Corollary 4.10.16. □

## 4.10 Similar Families

It is clear that there are other $(G, M)$-families which are similar to those just discussed. For example, suppose that $\rho = \otimes_v \rho_v$ is a standard representation in $\Sigma(M(\mathbb{A}))$ such that the number

$$\xi(\varphi', \rho) = \prod_v \xi(\varphi'_v, \rho_v)$$

does not vanish. Then replacing $\tau$ by $\rho$ in the discussion above, we can define a $(G, M)$-family

$$\tilde{m}_P(\nu, \varphi'_\lambda, \rho_\lambda, P_0) = \tilde{m}_{P|P_0}(\varphi'_\lambda, \rho_\lambda)^{-1} \tilde{m}_{P|P_0}(\varphi'_{\lambda+\nu}, \rho_{\lambda+\nu}).$$

For another example, let $\sigma'$ and $\varphi'$ be representations in $\Pi(M'(\mathbb{A}))$ such that for some $\tau \in \Pi(M(\mathbb{A}))$, the numbers $\delta(\sigma', \tau)$ and $\delta(\varphi', \tau)$ are nonzero. Then

$$\tilde{m}_P(\nu, \sigma'_\lambda, \varphi'_\lambda, P_0) = \tilde{m}_P(\nu, \sigma'_\lambda, \tau_\lambda, P_0) / \tilde{m}_P(\nu, \varphi'_\lambda, \tau_\lambda, P_0)$$

is a $(G,M)$-family which is independent of $\tau$. It satisfies an obvious analogue of Proposition 4.9. Notice that Corollary 4.10.16 provides additional expansion formulae

$$\tilde{m}_M^L(\varphi'_\lambda, \rho_\lambda) = \sum_{L_1 \in \mathcal{L}^L(M)} \tilde{m}_M^{L_1}(\varphi'_\lambda, \tau_\lambda)\tilde{m}_{L_1}^L(\tau_\lambda, \rho'_\lambda) \qquad (4.10.1)$$

and

$$\tilde{m}_M^L(\sigma'_\lambda, \tau_\lambda) = \sum_{L_1 \in \mathcal{L}^L(M)} \tilde{m}_M^{L_1}(\sigma'_\lambda, \varphi'_\lambda)\tilde{m}_{L_1}^L(\varphi'_\lambda, \tau_\lambda). \qquad (4.10.2)$$

Finally, suppose that $\varphi'$, $\rho$, and $\tau$ are *arbitrary* representations in $\Pi(M'(\mathbb{A}))$, $\Sigma(M(\mathbb{A}))$ and $\Pi(M(\mathbb{A}))$, respectively. The functions $m_M^L(\tau_\lambda, \rho_\lambda)$ used in Section 6 below are obtained from the $(G,M)$-family

$$m_P(\nu, \tau_\lambda, \rho_\lambda, P_0) = \xi(\tau, \rho)\tilde{m}_P(\nu, \tau_\lambda, \rho_\lambda, P_0), \quad P \in \mathcal{P}(M),\ \nu \in i\mathfrak{a}_M^*.$$

In a similar fashion, we define

$$m_P(\nu, \varphi'_\lambda, \tau_\lambda, P_0) = \delta(\varphi', \tau)\tilde{m}_P(\nu, \varphi'_\lambda, \tau_\lambda, P_0)$$

and

$$m_P(\nu, \varphi'_\lambda, \rho_\lambda, P_0) = \xi(\varphi', \rho)\tilde{m}_P(\nu, \varphi'_\lambda, \rho_\lambda, P_0)$$

for $P \in \mathcal{P}(M)$ and $\nu \in i\mathfrak{a}_M^*$. These two new $(G,M)$-families (as well as the earlier ones) satisfy versions of Proposition 4.7. We shall comment explicitly only on the case of the latter one.

PROPOSITION 4.11. *For each $L \in \mathcal{L}(M)$, we have*

$$m_M^L(\varphi'_\lambda, \rho_\lambda) = m_{M'}^{L'}(\varphi'_{\lambda'}, \rho'_{\lambda'}),$$

*with $\rho \mapsto \rho'$ corresponding representations.*

PROOF. By definition,

$$m_M^L(\varphi'_\lambda, \rho_\lambda) = \xi(\varphi', \rho)\tilde{m}_M^L(\varphi'_\lambda, \rho_\lambda).$$

Arguing as in Proposition 4.7, we see that

$$\tilde{m}_M^L(\varphi'_\lambda, \rho_\lambda) = \tilde{m}_M^L(\varphi'_{\lambda'}, \rho'_{\lambda'}),$$

for $\rho \mapsto \rho'$ corresponding and $\xi(\varphi', \rho') \neq 0$. Moreover,

$$\xi(\varphi', \rho) = \xi(\varphi', \rho'),$$

since $e_S = 1$ for any large finite set $S$ of valuations. Therefore

$$m_M^L(\varphi'_\lambda, \rho_\lambda) = \xi(\varphi', \rho')\tilde{m}_{M'}^{L'}(\varphi'_{\lambda'}, \rho'_{\lambda'}).$$

This is

$$m_{M'}^{L'}(\varphi'_{\lambda'}, \rho'_{\lambda'}),$$

as required. □

# 5 The Distributions $I_M(\gamma)$ and $I_M^{\mathcal{E}}(\gamma)$

The geometric side of the trace formula is a sum of terms which are indexed by orbits. These terms can be separated naturally into local and global constituents. We shall look first at the local constituents.

## 5.1 *Near Singular Elements*

Let $M \in \mathcal{L}$. Let $S$ be a finite set of valuations of $F$ with the closure property. Recall that, for $\gamma \in M(F_S)$ and $f \in \mathbb{H}(G(F_S))$, we have defined in Section 5.2 the (geometric) invariant distributions

$$I_M(\gamma, f).$$

They depend only on the $M(F_S)$-orbit of $\gamma$. If $M_\gamma = G_\gamma$, we saw that they can be defined in terms of weighted orbital integrals. For general $\gamma$, the functions $m_M^L(\gamma, a)$ allow us to define

$$I_M(\gamma, f) = \lim_{a \to e} \sum_{L \in \mathcal{L}(M)} m_M^L(\gamma, a) I_L(a\gamma, f). \tag{5.1.1}$$

Here we use $a \in A_M(F_S)$ taking small regular values. See Section 5.2 and in particular 5.(2.9.1).

We shall recall some properties of these distributions.

DEFINITION 5.2. Suppose that $\sigma$ is a semisimple element in $M(F_S)$. Consider two functions $\phi_1$ and $\phi_2$ which are defined on an open subset $\Xi$ of $\sigma M_\sigma(F_S)$ whose closure contains an $M_\sigma(F_S)$-invariant neighborhood of $\sigma$. For $\gamma \in \Xi$, we write

$$\phi_1(\gamma) \overset{(M,\sigma)}{\sim} \phi_2(\gamma)$$

if the difference is an orbital integral on $M(F_S)$ for $\gamma$ near $\sigma$. That is, if there is a function $h \in C_c^\infty(M(F_S))$ and a neighborhood $U$ of $\sigma$ in $M(F_S)$ such that

$$\phi_1(\gamma) - \phi_2(\gamma) = I_M^M(\gamma, h), \quad \gamma \in \Xi \cap U.$$

PROPOSITION 5.3. *Suppose that $G_\sigma = M_\sigma$. For $\gamma \in M_\sigma(F_S)$ and $f \in \mathbb{H}(G(F_S))$, we have*

$$I_M(\gamma,f) \overset{(M,\sigma)}{\sim} 0. \tag{5.3.1}$$

PROOF. This is 5.(2.12.1). □

Let us now recall the descent and splitting, from Sections 5.8 and 5.9.

Let $M_1 \in \mathcal{L}^M$. Suppose $\gamma \in M_1(F_S)$ satisfies $M_{1,\gamma} = M_\gamma$. The descent property says that

$$I_M(\gamma,f) = \sum_{L \in \mathcal{L}(M_1)} d_{M_1}^G(M,L)\widehat{I}_{M_1}^L(\gamma,f_L). \tag{5.3.2}$$

The number $d_{M_1}^G(M,L)$ was introduced in Section 5.7. The key property is that it vanishes unless the map

$$\mathfrak{a}_{M_1}^M \oplus \mathfrak{a}_{M_1}^L \to \mathfrak{a}_{M_1}^G$$

is an isomorphism.

Assume now that $S$ is the disjoint union of $S_1$ and $S_2$. Let $f = f_1 f_2$ and $\gamma = \gamma_1\gamma_2$ be the corresponding decompositions. The splitting property says that

$$I_M(\gamma,f) = \sum_{L_1,L_2 \in \mathcal{L}(M)} d_M^G(L_1,L_2)\widehat{I}_M^{L_1}(\gamma_1,f_{1,L_1})\widehat{I}_M^{L_2}(\gamma_2,f_{2,L_2}). \tag{5.3.3}$$

We now recall some properties that apply to particular fields. In particular, a differential equation and a germ expansion. They were first discussed in Section 5.2.

PROPOSITION 5.4. *Assume that $S = \{v\}$, so that $F_S = F_v$ is a local field.*

(1) *Assume that $F_v$ is Archimedean. Let $z$ be an element of the center $\mathcal{Z}(G(F_v))$ of the universal enveloping algebra of the complexified Lie algebra of $G(F_v)$. Let $z_L$ denote its image under the natural map from $\mathcal{Z}(G(F_v))$ to $\mathcal{Z}(L(F_v))$. There is an associated linear differential operator, $\partial_M^L(\gamma, z_L)$, on $M(F_v) \cap L_{\mathrm{reg}}$. It is invariant under conjugation by $M(F_v)$. For $\gamma \in M(F_v) \cap G_{\mathrm{reg}}$, the differential equation is*

$$I_M(\gamma, zf) = \sum_{L \in \mathcal{L}(M)} \partial_M^L(\gamma, z_L)I_L(\gamma,f). \tag{5.4.1}$$

(2) *Assume now that $F_v$ is non-Archimedean. Let $\sigma \in M(F_v)$ be semisimple. The set of conjugacy classes of unipotent elements in $L_\sigma(F_v)$ is denoted by $(\mathcal{U}_{L_\sigma}(F_v))$.*

*For $\gamma \in \sigma M_\sigma(F_v) \cap G_{\text{reg}}$, the germ expansion provides $(M,\sigma)$-equivalence classes $g^L_M(\gamma,\delta)$ of functions on the L regular elements $\gamma \in \sigma M_\sigma(F_v)$, satisfying*

$$I_M(\gamma,f) \overset{(M,\sigma)}{\sim} \sum_{L\in\mathcal{L}(M)} \sum_{\delta\in\sigma(\mathcal{U}_{L_\sigma}(F_v))} g^L_M(\gamma,\delta) I_L(\delta,f). \tag{5.4.2}$$

Consider now the linear forms $I_{M'}(\gamma')$ on $\mathbb{H}(G'(F_S))$. We want to describe a vanishing property for these distributions. Recall that there is an injection $M \mapsto M'$ from $\mathcal{L}$ into $\mathcal{L}'$.

PROPOSITION 5.5. *Suppose that $f$ is a function in $\mathbb{H}(G(F_S))$. Then*

$$I_{M'}(\delta,f') = 0, \quad \textit{for } \delta \in M'(F_S) \smallsetminus \{\gamma' \,;\, \gamma \in M(F_S)\}. \tag{5.5.1}$$

*In other words, $I_{M'}(\delta,f')$ vanishes if $\delta \in M'(F_S)$ does not come from $M(F_S)$.*

PROOF. This is Proposition 5.10.8 □

DEFINITION 5.6. Let $\gamma$ be an element of $M(F_S)$ such that $M_\gamma = G_\gamma$. For $f \in \mathbb{H}(G(F_S))$ define

$$I^{\mathcal{E}}_M(\gamma,f) = e^M(\gamma)\widehat{I}_{M'}(\gamma',f'_{G'}). \tag{5.6.1}$$

Suppose $M = G$, so that any element of $G$ is of this form. Formula (1.7.1) becomes

$$I^{\mathcal{E}}_G(\gamma,f) = I_G(\gamma,f). \tag{5.6.2}$$

Assume now that $\gamma$ is an arbitrary element in $M(F_S)$. For any small regular point $a \in A_M(F_S)$, we have $M_{a\gamma} = G_{a\gamma}$. Consequently, for any $L \in \mathcal{L}(M)$, the distribution $I^{\mathcal{E}}_L(a\gamma,f)$ is defined.

PROPOSITION 5.7. *The expression*

$$\sum_{L\in\mathcal{L}(M)} m^L_M(\gamma,a) I^{\mathcal{E}}_L(a\gamma,f)$$

*extends to a continuous function of $a \in A_M(F_S)$ in a neighborhood of the identity.*

PROOF. Let $\gamma = \sigma u$ be the Jordan decomposition of $\gamma$. Recall that $\mu \mapsto \mu_{\sigma'}$ denotes the map from the conjugacy classes of $M_\sigma$ to the conjugacy classes of its quasisplit form $M'_{\sigma'}$. By (1.6.1),

$$(a\gamma)' = \sigma' u_{\sigma'} a_{\sigma'} = \gamma' a_{\sigma'}.$$

According to (5.6.1),

$$I_L^{\mathcal{E}}(a\gamma,f) = e^L(a\gamma)\widehat{I}_{L'}((a\gamma)',f'_{G'}) = e^M(\gamma)\widehat{I}_{L'}(\gamma' a_{\sigma'},f'_{G'}).$$

We will relate the functions $m_M^L(\gamma,a)$ and $m_{M'}^{L'}(\gamma',a_{\sigma'})$.

Suppose that $\sigma = \prod_{v\in S}\sigma_v$, $u = \prod_{v\in S}u_v$ and $a = \prod_{v\in S}a_v$. Then $M_\sigma = \prod_{v\in S}M_{\sigma_v}$. The $(G,M)$-family

$$m_P(\nu,\gamma,a) = \prod_{v\in S}\prod_{\beta} m_\beta\left(\frac{1}{2}\nu,u_v,a_v\right), \quad P\in\mathcal{P}(M),\ \nu\in i\mathfrak{a}_M^*$$

was defined in [Ar88.1]. The $\beta$ are the roots of $(P_{\sigma_v},A_{M_{\sigma_v}})$. For each such $\beta$, we have

$$m_\beta\left(\frac{1}{2}\nu,u_v,a_v\right) = |\beta(a_v)-\beta(a_v)^{-1}|_v^{\rho(\beta,u_v)\nu(\beta^\vee)}.$$

The real numbers $\rho(\beta,u_v)$ defined there depend only on the geometric conjugacy class of $u_v\in M_{\sigma_v}$. In particular,

$$\rho(\beta,u_v) = \rho(\beta',(u_v)_{\sigma_v'}),$$

where $\beta'$ is the root of $G'$ associated to $\beta$.

The function $m_M^L(\gamma,a)$ is obtained from the family of $m_P(\nu,\gamma,a)$. It follows that

$$m_{M'}^{L'}(\gamma',a_{\sigma'}) = m_M^L(\gamma,a).$$

We have

$$\begin{aligned}\sum_{L\in\mathcal{L}(M)} m_M^L(\gamma,a)I_L^{\mathcal{E}}(a\gamma,f) &= e^M(\gamma)\sum_{L\in\mathcal{L}(M)} m_M^L(\gamma,a)\widehat{I}_{L'}(\gamma' a_{\sigma'},f'_{G'})\\ &= e^M(\gamma)\sum_{L\in\mathcal{L}(M)} m_{M'}^{L'}(\gamma',a_{\sigma'})\widehat{I}_{L'}(\gamma' a_{\sigma'},f'_{G'}).\end{aligned}$$

By 5.(2.11.1), this function, defined for regular elements $a'\in A_{M'}(F_S)$, extends to a continuous function around $a'=e$. Therefore the original expression extends to a continuous function around $a=e$. □

The proposition justifies the following:

DEFINITION 5.8. If $\gamma$ is any element in $M(F_S)$, we define

$$I_M^{\mathcal{E}}(\gamma,f) = \lim_{a\to e}\sum_{L\in\mathcal{L}(M)} m_M^L(\gamma,a)I_L^{\mathcal{E}}(a\gamma,f), \qquad a\in A_M(F_S). \tag{5.1.1$\mathcal{E}$}$$

In other words, we set

$$I_M^{\mathcal{E}}(\gamma,f) = e^M(\gamma)\widehat{I}_{M'}\left(\gamma',f'_{G'}\right). \tag{5.8.1}$$

Then $I_M^{\mathcal{E}}(\gamma)$ is an invariant distribution on $\mathbb{H}(G(F_S))$. Comparing $I_M^{\mathcal{E}}(\gamma)$ with $I_M(\gamma)$ is one of the main tasks in this chapter.

COROLLARY 5.9. *Suppose that $S$ contains all the Archimedean and ramified places and that $\gamma \in M(F_S)$. Then*

$$I_M^{\mathcal{E}}(\gamma,f) = \widehat{I}_{M'}(\gamma',f'_{G'}).$$

PROOF. For such $S$ and $\gamma$, we have $e^M(\gamma) = 1$. □

Our purpose is to prove that $I_M^{\mathcal{E}}(\gamma,f) = I_M(\gamma,f)$. To that end, we list some properties that $I_M^{\mathcal{E}}(\gamma,f)$ shares with $I_M(\gamma,f)$.

PROPOSITION 5.10. (1) *Suppose that $\sigma$ is a semisimple element in $M(F_S)$ such that $G_\sigma = M_\sigma$. Then for any $\gamma \in M_\sigma(F_S)$, we have*

$$I_M^{\mathcal{E}}(\gamma,f) \overset{(M,\sigma)}{\sim} 0. \tag{5.3.1$\mathcal{E}$}$$

(2) *For $\gamma \in M_1(F_S)$ with $M_{1,\gamma} = M_\gamma$ there is a descent property*

$$I_M^{\mathcal{E}}(\gamma,f) = \sum_{L\in\mathcal{L}(M_1)} d_{M_1}^G(M,L)\widehat{I}_{M_1}^{L,\mathcal{E}}(\gamma,f_L). \tag{5.3.2$\mathcal{E}$}$$

(3) *For $\gamma = \gamma_1\gamma_2$ and $f = f_1f_2$, we have the splitting property*

$$I_M^{\mathcal{E}}(\gamma,f) = \sum_{L_1,L_2\in\mathcal{L}(M)} d_M^G(L_1,L_2)\widehat{I}_M^{L_1,\mathcal{E}}(\gamma_1,f_{1,L_1})\widehat{I}_M^{L_2,\mathcal{E}}(\gamma_2,f_{2,L_2}). \tag{5.3.3$\mathcal{E}$}$$

PROOF. (1) This follows from the characterization of orbital integrals on $M(F_S)$ (Theorem 2.14.4 or [AC89, Proposition I.3.1]), the vanishing property (5.5.1), and property (5.3.1) applied to $G'$.

(2), (3) Assume $\gamma$ is regular. We may use the corresponding properties for $G'$ and the fact that

$$d_{M_1}^G(M,L) = d_{M_1'}^{G'}(M',L').$$

If $\gamma$ is a general element, we need (5.1.1$\mathcal{E}$) and 5.(7.11.1). The proof is similar to Theorem 5.8.3. □

Later, we will establish the analogues of (5.4.1) and (5.4.2) (the differential equation and germ expansion, respectively).

PROPOSITION 5.11. *Let $\gamma \in M(F_S)$ be $G$-regular and semisimple. Suppose that for any such $\gamma$ the function $f \in \mathbb{H}(G(F_S))$ satisfies*

$$I_M(\gamma,f) = I_M^{\mathcal{E}}(\gamma,f).$$

*Then this equality holds for every element in $M(F_S)$.*

PROOF. Suppose $\delta \in M(F_S)$ has semisimple component $\sigma$, with $G_\sigma = M_\sigma$. The orbital integral at $\delta$ of any function on $M(F_S)$ is completely determined by its orbital integrals at elements $\gamma \in \sigma M_\sigma(F_S)$ which are in general position and near to $\sigma$. Then (5.3.1) and (5.3.1$\mathcal{E}$) imply that

$$I_M(\delta,f) = I_M^{\mathcal{E}}(\delta,f).$$

Now, let $\gamma$ be any element of $M(F_S)$. Let $a$ be a small point in general position in $A_M(F_S)$. Then $\delta = \gamma a$ is as above, and

$$I_M(\gamma a,f) = I_M^{\mathcal{E}}(\gamma a,f).$$

Using definitions (5.1.1) and (5.1.1$\mathcal{E}$) gives

$$I_M(\gamma,f) - I_M^{\mathcal{E}}(\gamma,f) = \lim_{a\to e} \sum_{L\in\mathcal{L}(M)} m_M^L(\gamma,a)(I_L(\gamma a,f) - I_L^{\mathcal{E}}(\gamma a,f)) = 0,$$

as required. □

## 6 The Distributions $I_M(\tau,X)$ and $I_M^{\mathcal{E}}(\tau,X)$

### 6.1 Spectral Side

We shall now consider the spectral side of the trace formula. As on the geometric side, the terms have a local and global component. We will begin by discussing the local properties of the spectral terms.

Let $S$ denote a finite set of valuations of $F$ with the closure property. Let $M$ be an element of $\mathcal{L}$. The local terms of the spectral side are related to the distributions, defined for $\tau \in \Pi(M(F_S))$, $X \in \mathfrak{a}_{M,S}$, and $f \in \mathbb{H}(G(F_S))$ and denoted

$$I_M(\tau,X,f),$$

introduced in Section 5.3. These distributions are also defined for standard representations $\rho \in \Sigma(M(F_S))$. The distributions are related by an expansion, given by 5.(3.8.1), which states,

$$I_M(\tau,X,f) =$$

$$\sum_{P\in\mathcal{P}(M)} \omega_P \sum_{L\in\mathcal{L}(M)} \sum_{\rho\in\Sigma(M(F_S))} \int_{\varepsilon_P + i\mathfrak{a}_{M,S}^*/i\mathfrak{a}_{L,S}^*} m_M^L(\tau_\lambda,\rho_\lambda) I_L(\rho_\lambda^L, h_L(X),f) e^{-\lambda(X)}\, d\lambda. \tag{6.1.1}$$

Recall that $\varepsilon_P$ is a small point in $(\mathfrak{a}_P^*)^+$. The number $\omega_P$ is given by

$$\omega_P = \mathrm{vol}\{X \in \mathfrak{a}_P^+ ;\ \|X\| \le 1\}/\ \mathrm{vol}\{X \in \mathfrak{a}_P ;\ \|X\| \le 1\}.$$

For any $L$, the vector $h_L(X)$ is the projection of $X$ onto $\mathfrak{a}_L$.

## 6.2 Normalizing Factor

The function $m^L_M(\tau_\lambda, \rho_\lambda)$ was defined in Section 4.12. For now, it is enough to recall that it is obtained from a $(G, M)$-family arising from the local normalizing factors. It is a rational function of the variables

$$\{\lambda(\alpha^\vee), q_v^{-\lambda(\alpha^\vee)}; \alpha \in \Sigma(G, A_M), v \in S \smallsetminus S_\infty\}, \tag{6.2.1}$$

where $\Sigma(G, A_M)$ denotes the set of roots of $(G, A_M)$ and $q_v$ denotes the residual cardinality of the local field $F_v$ for $v \in S \smallsetminus S_\infty$.

## 6.3 A Distribution

Suppose that $\tau \in \Pi_{\text{unit}}(M(F_S))$. Then (6.1.1) simplifies to

$$I_M(\tau, X, f) =$$

$$\sum_{L \in \mathcal{L}(M)} \sum_{\rho \in \Sigma(M(F_S))} \int_{\varepsilon_M + i\mathfrak{a}^*_{M,S}/i\mathfrak{a}^*_{L,S}} m^L_M(\tau_\lambda, \rho_\lambda) I_L(\rho^L_\lambda, h_L(X), f) e^{-\lambda(X)} \, d\lambda,$$

where $\varepsilon_M$ is a small regular point in $\mathfrak{a}^*_M$.

## 6.4 Local Correspondence

Recall that the local correspondence provides a map $\rho \mapsto \rho'$ from $\Sigma(M(F_S))$ to $\Sigma(M'(F_S))$ such that, for any standard representation $\rho \in \Sigma(M(F_S))$ and any Hecke function $h \in \mathbb{H}(M(F_S))$, we have

$$\operatorname{tr} \rho(h) = e_S h'_{M'}(\rho').$$

There is also the map $\tau \mapsto \tau'$ from $\Pi(M(F_S))$ into $\Pi(M'(F_S))$. If $\tau$ is nontempered, this map does not give a simple character identity.

DEFINITION. Let $\rho \in \Sigma(M(F_S))$. Let $\rho' \in \Sigma(M'(F_S))$ be the corresponding representation. Define

$$I^{\mathcal{E}}_M(\rho, X, f) = e_S \widehat{I}_{M'}(\rho', X', f'_{G'}).$$

For each $L \in \mathcal{L}(M)$, $\rho \in \Sigma(M(F_S))$, and $\lambda \in \mathfrak{a}_{M,\mathbb{C}}^*$, there is the induced representation $\rho_\lambda^L$ of $L(F_S)$. It belongs to $\Sigma(L(F_S))$. If $\lambda$ is in general position, then $\rho_\lambda^L$ is irreducible. (See 4.9.3 for the definition of $\lambda$ regular.)

DEFINITION 6.5. Let $\tau \in \Pi(M(F_S))$. Define (by just adding $\mathcal{E}$ to $I_M$ and $I_L$ in (6.1.1))

$$I_M^{\mathcal{E}}(\tau, X, f) =$$

$$\sum_{P \in \mathcal{P}(M)} \omega_P \sum_{L \in \mathcal{L}(M)} \sum_{\rho \in \Sigma(M(F_S))} \int_{\varepsilon_P + i\mathfrak{a}_{M,S}^*/i\mathfrak{a}_{L,S}^*} m_M^L(\tau_\lambda, \rho_\lambda) I_L^{\mathcal{E}}(\rho_\lambda^L, h_L(X), f) e^{-\lambda(X)}\, \mathrm{d}\lambda. \tag{6.5.1}$$

## 6.6 *Orbits of Representations*

We will employ new notation to describe the local components of the trace formula. We identify a representation $\tau^1 \in \Pi(M(F_S)^1)$ with the orbit $\{\tau_\lambda\,;\, \lambda \in \mathfrak{a}_{M,\mathbb{C}}^*\}$ of $\mathfrak{a}_{M,\mathbb{C}}^*$ in $\Pi(M(F_S))$, where $\tau$ is a lift of $\tau^1$ to $M(F_S) = M(F_S)^1 \times A(\mathbb{R}^0) = M(F_S)^1 \times \mathfrak{a}_M$, say with 1 on $A(\mathbb{R}^0)$. If $\tau^1$ is unitary, we identify it with the smaller orbit $\{\tau_\lambda\,;\, \lambda \in i\mathfrak{a}_M^*\}$ of $i\mathfrak{a}_M^*$ in $\Pi_{\mathrm{unit}}(M(F_S))$. We will use the same conventions for representations in $\Pi(M(\mathbb{A})^1)$ and $\Pi_{\mathrm{unit}}(M(\mathbb{A})^1)$.

DEFINITION 6.7. Let $\tau^1 \in \Pi_{\mathrm{unit}}(M(F_S)^1)$. Recall (Section 5.14) that we set

$$I_M(\tau^1, f) = I_M(\tau_\lambda, 0, f).$$

We now set

$$I_M^{\mathcal{E}}(\tau^1, f) = I_M^{\mathcal{E}}(\tau_\lambda, 0, f).$$

We have seen that the first expression describes the local spectral terms of the trace formula of $G$. The second expression is closely related to the analogous terms for $G'$. Both expressions are independent of $S$ if $S$ is sufficiently large, so are defined for $\tau^1 \in \Pi_{\mathrm{unit}}(M(\mathbb{A})^1)$.

# 7 Geometric Induction Assumption

## 7.1 *Induction Assumption*

We make an **induction assumption** that will be kept until the end of this chapter. *We assume that assertion (i) and (ii) of the main theorem hold if $G$ is replaced by* *any* $G_1$ *satisfying* $\dim G_1 < \dim G$, *where* $G_1$ *is a product of varieties satisfying the same conditions as* $G$. In particular, the induction assumption applies if $G$ is replaced by a proper Levi subgroup $M \in \mathcal{L}$ of $G$. More generally, if $M \in \mathcal{L}$ and $\sigma$ is a semisimple element in $M(F)$, $M_\sigma$ satisfies the same conditions as $G$. And, unless $M = G$ and $\sigma$ is an element in $A_G(F)$, we have $\dim M_\sigma < \dim G$.

We now give some immediate consequences of the induction assumption.

PROPOSITION 7.2. *Let* $S$ *be a finite set of valuations containing* $S_{\mathrm{ram}}$. *Let* $M_1 \in \mathcal{L}^M$ *be a Levi subgroup with* $M_1 \neq M$. *If* $\gamma \in M_1(F_S) \cap G_{\mathrm{reg}}$, *then we have*

$$I_M^{\mathcal{E}}(\gamma,f) = I_M(\gamma,f). \tag{7.2.1}$$

PROOF. By (5.3.2) and (5.3.2$\mathcal{E}$), we have

$$I_M^{\mathcal{E}}(\gamma,f) - I_M(\gamma,f) = \sum_{L \in \mathcal{L}(M_1)} d_{M_1}^G(M,L)(\widehat{I}_{M_1}^{L,\mathcal{E}}(\gamma,f_L) - \widehat{I}_{M_1}^L(\gamma,f_L)).$$

By definition, the numbers $d_{M_1}^G(M,L)$ vanish unless the map

$$\mathfrak{a}_{M_1}^M \oplus \mathfrak{a}_{M_1}^L \to \mathfrak{a}_{M_1}^G$$

is an isomorphism. Suppose now that $L = G$. Since $M_1 \neq M$ the map is not an isomorphism and the number vanishes. Assume now that $L \neq G$. The induction assumption gives

$$\widehat{I}_{M_1}^{L,\mathcal{E}}(\gamma,f_L) - \widehat{I}_{M_1}^L(\gamma,f_L) = 0.$$

The proposition follows. □

PROPOSITION 7.3. *Take* $S$ *to be the disjoint union of* $S_0$ *and* $S_1$, *where* $S_0$ *contains* $S_{\mathrm{ram}}$ *and* $S_1$ *consists of a single unramified valuation. Suppose that* $f = f_0 f_1$ *and* $\gamma = \gamma_0\gamma_1$ *are the corresponding decompositions. Then the local assertion* (i) *of the Main Theorem 2.7 implies* $I_M^{\mathcal{E}}(\gamma_1,f_1)$ *equals* $I_M(\gamma_1,f_1)$.

PROOF. The splitting properties (5.3.3) and (5.3.3$\mathcal{E}$) imply that $I_M^{\mathcal{E}}(\gamma,f) - I_M(\gamma,f)$ equals

$$\sum_{L_0,L_1 \in \mathcal{L}(M)} d_M^G(L_0,L_1)(\widehat{I}_M^{L_0,\mathcal{E}}(\gamma_0,f_{0,L_0})\widehat{I}_M^{L_1,\mathcal{E}}(\gamma_1,f_{1,L_1}) - \widehat{I}_M^{L_0}(\gamma_0,f_{0,L_0})\widehat{I}_M^{L_1}(\gamma_1,f_{1,L_1})).$$

If neither $L_0$ nor $L_1$ equals $G$, the Induction Assumption 7.1 implies

$$\widehat{I}_M^{L_0,\mathcal{E}}(\gamma_0,f_{0,L_0})\widehat{I}_M^{L_1,\mathcal{E}}(\gamma_1,f_{1,L_1}) - \widehat{I}_M^{L_0}(\gamma_0,f_{0,L_0})\widehat{I}_M^{L_1}(\gamma_1,f_{1,L_1}) = 0.$$

Suppose one of $L_0$ or $L_1$ equals $G$. In this case, the number $d_M^G(L_0,L_1)$ vanishes unless the other Levi subgroup equals $M$. According to the definitions in

Section 5.7.4,

$$d_M^G(G, M) = d_M^G(M, G) = 1.$$

Combining these facts gives

$$I_M^{\mathcal{E}}(\gamma, f) - I_M(\gamma, f) = \sum_{i=0}^{1} (I_M^{\mathcal{E}}(\gamma_i, f_i) - I_M(\gamma_i, f_i)) \prod_{j \neq i} \widehat{I}_M^M(\gamma_j, f_{j,M}). \tag{7.3.1}$$

Part (i) of the Main Theorem 2.7 implies the left side of (7.3.1) vanishes. Part (i) also implies that the summand with $i = 0$ on the right side vanishes, which gives the desired equality. □

We can deduce global statements from the induction assumption. Indeed, the induction assumption implies most cases of the global assertion (ii) of the Main Theorem 2.7.

PROPOSITION 7.4. *Let $M \in \mathcal{L}$. Let $\gamma = \sigma u$ be an element of $M(F)$. If $M = G$, assume $\sigma \notin A_G(F)$. If $S$ is a sufficiently large finite set of valuations, we have*

$$a^{M,\mathcal{E}}(S, \gamma) = a^M(S, \gamma).$$

PROOF. The assumptions give $\dim M_\sigma < \dim G$. Thus we may apply the global part of the induction hypothesis to $M_\sigma$. The conclusion then follows from (2.2.1) and (2.2.1$\mathcal{E}$). □

PROPOSITION 7.5. *For $f \in \mathbb{H}(G(F_S))$, we have*

$$I^{\mathcal{E}}(f) - I(f) = \sum_{M \in \mathcal{L}} |W_0^M|/|W_0^G| \sum_{\gamma \in (M(F))_{M,S}} a^M(S, \gamma)(I_M^{\mathcal{E}}(\gamma, f) - I_M(\gamma, f))$$
$$+ \sum_{\xi \in A_G(F)} \sum_{u \in (\mathcal{U}_G(F))_{G,S}} (a^{G,\mathcal{E}}(S, u) - a^G(S, u)) I_G(\xi u, f).$$

PROOF. This follows from (2.1.1), (2.1.1$\mathcal{E}$) and (5.6.2). □

## 8 The Numbers $\varepsilon_M(S)$

In this section we will derive some consequences of the local assertion (i) of the Main Theorem 2.7. The assertion only applies if $G$ is unramified at every valuation outside $S$. It is natural to consider more general $S$. For example, if $v$ is any valuation (in $S_{\mathrm{ram}}$), we can ask if $I_M^{\mathcal{E}}(\gamma_v)$ equals $I_M(\gamma_v)$. This of course holds for $v \notin S_{\mathrm{ram}}$, where $G_v \simeq G'_v$.

The following theorem gives some results for more general $S$.

THEOREM 8.1. *Let $M \in \mathcal{L}$. Let $S$ be a finite set of valuations with the closure property. If $S$ contains $S_{\mathrm{ram}}$, we further assume that, for any $L \in \mathcal{L}(M)$, $\gamma \in L(F_S)$ and $f \in \mathbb{H}(G(F_S))$, we have*

$$I_L^{\mathcal{E}}(\gamma,f) = I_L(\gamma,f).$$

*Let $L \in \mathcal{L}(M)$. Then there are unique complex numbers*

$$\varepsilon_L(S) = \varepsilon_L^G(S)$$

*such that, for $\gamma \in M(F_S)$ and $f \in \mathbb{H}(G(F_S))$, we have*

$$I_M^{\mathcal{E}}(\gamma,f) = \sum_{L\in\mathcal{L}(M)} \widehat{I}_M^L(\gamma, \varepsilon_L(S)f_L). \tag{8.1.1}$$

*These numbers have a descent property. For $M_1 \subset M$ we have*

$$\varepsilon_M(S) = \sum_{L\in\mathcal{L}(M_1)} d_{M_1}^G(M,L)\varepsilon_{M_1}^L(S). \tag{8.1.2}$$

*They also satisfy a splitting property. For any disjoint union $S = S_1 \cup S_2$, we have*

$$\varepsilon_M(S) = \sum_{L_1,L_2\in\mathcal{L}(M)} d_M^G(L_1,L_2)\varepsilon_M^{L_1}(S_1)\varepsilon_M^{L_2}(S_2). \tag{8.1.3}$$

PROOF. The uniqueness of $\varepsilon_L(S)$ is clear from (8.1.1). The proof will be by induction. If $M = G$, then the theorem holds with $\varepsilon_G(S) = 1$. Fix $M \neq G$. Assume inductively that the theorem is valid whenever $M$ is replaced by any $L \in \mathcal{L}(M)$ with $L \neq M$. In particular, we assume for such $L$ that the numbers $\varepsilon_L(S)$ are all defined.

We need the following lemma:

LEMMA 8.2. *Assume we are in the setting of the theorem, but that $L \supsetneq M$. For $\gamma \in M(F_S)$ and $f \in \mathbb{H}(G(F_S))$, define*

$$\varepsilon_M(\gamma,f) = I_M^{\mathcal{E}}(\gamma,f) - \sum_{L\supsetneq M} \widehat{I}_M^L(\gamma, \varepsilon_L(S)f_L).$$

*This function has descent and splitting properties identical to* (5.3.2) *and* (5.3.3).

PROOF. We will deduce these properties for $\varepsilon_M(\gamma,f)$ by using the corresponding properties for $I_M^{\mathcal{E}}(\gamma,f)$, $\widehat{I}_M^L(\gamma,f_L)$, and $\varepsilon_L(S)$.

We start by proving the descent property. Let $M_1$, $\gamma$ be as in (5.3.2). By (5.3.2) and (5.3.2$\mathcal{E}$), we can write $\varepsilon_M(\gamma,f)$ as the difference between

$$\sum_{L\in\mathcal{L}(M_1)} d_{M_1}^G(M,L)\widehat{I}_{M_1}^{L,\mathcal{E}}(\gamma,f_L)$$

and

$$\sum_{L \supsetneq M} \sum_{L_1 \in \mathcal{L}^L(M_1)} d_{M_1}^L(M, L_1) \widehat{I}_{M_1}^{L_1}(\gamma, f_{L_1}) \varepsilon_L(S). \tag{8.2.1}$$

Consider (8.2.1). By only taking terms for which $d_{M_1}^L(M, L_1) \neq 0$, we can write this expression as

$$\sum_{L_1 \supsetneq M_1} \sum_{L \in \mathcal{L}(L_1)} d_{M_1}^L(M, L_1) \widehat{I}_{M_1}^{L_1}(\gamma, f_{L_1}) \varepsilon_L(S). \tag{8.2.2}$$

Indeed, recall that $d_{M_1}^L(M, L_1)$ is nonzero if and only if

$$\mathfrak{a}_{M_1}^M \oplus \mathfrak{a}_{M_1}^{L_1} \to \mathfrak{a}_{M_1}^L$$

is an isomorphism. This implies that the nonzero contributions in (8.2.1) and (8.2.2) are identical. For a nonzero term, the element $L$ in the sum will be strictly larger than $M$, since it comes from a term in (8.2.1). The induction assumption then says that $\varepsilon_L(S)$ satisfies the descent property (8.1.2). Applying the descent property to $\varepsilon_L(S)$ and then applying 5.(7.11.1) allows us to write

$$\sum_{L \in \mathcal{L}(L_1)} d_{M_1}^L(M, L_1) \varepsilon_L(S) = \sum_{L \in \mathcal{L}(L_1)} d_{M_1}^G(M, L) \varepsilon_{L_1}^L(S).$$

Applying this to (8.2.1) shows that (8.2.1) is equal to

$$\sum_{L_1 \supsetneq M_1} \sum_{L \in \mathcal{L}(L_1)} d_{M_1}^G(M, L) \widehat{I}_{M_1}^{L_1}(\gamma, \varepsilon_{L_1}^L(S) f_{L_1}).$$

Substituting this expression for (8.2.1) into the expansion of $\varepsilon_M(\gamma, f)$ gives

$$\varepsilon_M(\gamma, f) = \sum_{L \in \mathcal{L}(M_1)} d_{M_1}^G(M, L) \left( \widehat{I}_{M_1}^{L,\mathcal{E}}(\gamma, f_L) - \sum_{\{L_1\, ;\, M_1 \subsetneq L_1 \subset L\}} \widehat{I}_{M_1}^{L_1}(\gamma, \varepsilon_{L_1}^L(S) f_{L_1}) \right).$$

Rewriting gives

$$\varepsilon_M(\gamma, f) = \sum_{L \in \mathcal{L}(M_1)} d_{M_1}^G(M, L) \widehat{\varepsilon}_{M_1}^L(\gamma, f_L), \tag{8.1.2*}$$

which is the descent property.

Let $\gamma = \gamma_1 \gamma_2$, $f = f_1 f_2$, as in (5.3.3). To prove the splitting property, we apply (5.3.3), (5.3.3$\mathcal{E}$), and (8.1.3) to the formula for $\varepsilon_M(\gamma, f)$. The result is

$$\varepsilon_M(\gamma, f) = \sum_{L_1, L_2 \in \mathcal{L}(M)} d_M^G(L_1, L_2) \widehat{\varepsilon}_M^{L_1}(\gamma_1, f_{1,L_1}) \widehat{\varepsilon}_M^{L_2}(\gamma_2, f_{2,L_2}). \tag{8.1.3*}$$

□

REMARK. In order to prove the lemma, it suffices to assume for $\gamma \in L(F_S)$, $f \in \mathbb{H}(G(F_S))$, $S \supset S_{\text{ram}}$, $L \in \mathcal{L}(M)$, $L \neq M$, that $I_L^{\mathcal{E}}(\gamma,f) = I_L(\gamma,f)$.

## 8.3 *Proof of Theorem* 8.1

We now resume the proof of Theorem 8.1. As mentioned, we only need to prove the existence of $\varepsilon_M(S)$. We will use (decreasing) induction on the size of $S$. If $S \supset S_{\text{ram}}$, take $\varepsilon_G(S) = 1$ and $\varepsilon_L(S) = 0$ for $L \neq G$. If not, assume inductively the theorem holds for a given $S$. In particular, the numbers $\varepsilon_M(S)$ are defined for all $M$. We need to show, for any $\gamma \in M(F_S)$, that

$$\varepsilon_M(\gamma,f) = \varepsilon_M(S)\widehat{I}_M^M(\gamma,f_M).$$

Let $S = S_1 \cup S_2$ be a disjoint union. We need to prove the theorem for the nonempty $S_1$ and $S_2$.

Let $\gamma = \gamma_1\gamma_2, f = f_1f_2$. Let

$$c_0 = \sum_{L_1,L_2 \in \mathcal{L}_0(M)} d_M^G(L_1,L_2)\varepsilon_M^{L_1}(S_1)\varepsilon_M^{L_2}(S_2).$$

Recall that $\mathcal{L}_0(M)$ denotes the Levi subgroups $M \subset L \subsetneq G$. Applying the splitting property (8.1.3*) gives

$$\varepsilon_M(\gamma,f) = \varepsilon_M(\gamma_1,f_1)\widehat{I}_M^M(\gamma_2,f_{2,M}) + \widehat{I}_M^M(\gamma_1,f_{1,M})\varepsilon_M(\gamma_2,f_2) + c_0\widehat{I}_M^M(\gamma_1,f_{1,M})\widehat{I}_M^M(\gamma_2,f_{2,M}). \tag{8.3.1}$$

By induction, $\varepsilon_M(\gamma,f) = \varepsilon_M(S)\widehat{I}_M^M(\gamma,f_M)$. Let $\gamma_2$ and $f_2$ be such that $\widehat{I}_M^M(\gamma_2,f_{2,M}) \neq 0$.

Let $\gamma_1$ be such that $\widehat{I}_M^M(\gamma_1,f_{1,M})$ vanishes. We claim that $\varepsilon_M(\gamma_1,f_1)$ vanishes. The splitting property implies

$$\varepsilon_M(\gamma,f) = \varepsilon_M(S)\widehat{I}_M^M(\gamma_1,f_{1,M})\widehat{I}_M^M(\gamma_2,f_{2,M}) = 0.$$

Then from (8.3.1) we get

$$\varepsilon_M(\gamma_1,f_1)\widehat{I}_M^M(\gamma_2,f_{2,M}) = 0,$$

which implies the claim.

It follows that there is a function $\varepsilon_M(S_1,\gamma_1)$ of $\gamma_1$ such that

$$\varepsilon_M(\gamma_1,f_1) = \varepsilon_M(S_1,\gamma_1)\widehat{I}_M^M(\gamma_1,f_{1,M}).$$

Similarly,

$$\varepsilon_M(\gamma_2, f_2) = \varepsilon_M(S_2, \gamma_2)\widehat{I}_M^M(\gamma_2, f_{2,M}),$$

for a function $\varepsilon_M(S_2, \gamma_2)$. Plugging everything into (8.3.1) gives

$$\varepsilon_M(S_1, \gamma_1) + \varepsilon_M(S_2, \gamma_2) + c_0 = \varepsilon_M(S).$$

This equation implies that $\varepsilon_M(S_1, \gamma_1)$ and $\varepsilon_M(S_2, \gamma_2)$ are independent of $\gamma_1$ and $\gamma_2$. That is, there is a number $\varepsilon_M(S_1)$ such that

$$\varepsilon_M(\gamma_1, f_1) = \varepsilon_M(S_1)\widehat{I}_M^M(\gamma_1, f_{1,M}).$$

There is a similar equation for $\varepsilon_M(S_2)$. This gives an inductive definition of $\varepsilon_M(S)$.

The properties (8.1.2) and (8.1.3) follow from (8.1.2*) and (8.1.3*). □

COROLLARY 8.4. *Suppose that S contains $S_{\mathrm{ram}}$ or consists of a single unramified valuation. Then*

$$\varepsilon_M(S) = \begin{cases} 1, & M = G \\ 0, & M \neq G. \end{cases}$$

PROOF. Clear. □

It is probable that this corollary is true for any $S$. In that direction, we have the following proposition:

PROPOSITION 8.5. *Let $L \in \mathcal{L}(M)$. Let $S$ be a finite set of valuations with the closure property. If $S$ contains $S_{\mathrm{ram}}$, assume for $\gamma \in L(F_S)$ and $f \in \mathbb{H}(G(F_S))$ that*

$$I_L^{\mathcal{E}}(\gamma, f) = I_L(\gamma, f).$$

*We have*

$$\varepsilon_M(S) = \begin{cases} 1, & M = G, \\ 0, & M \neq G. \end{cases}$$

PROOF. We are in the setting of Theorem 8.1. In particular, the numbers $\varepsilon_L(S)$ are defined for all $L \in \mathcal{L}(M)$ and $S$ with the closure property. We saw above that if $M = G$, we can take $\varepsilon_M(S) = 1$, so assume that $M \subsetneq G$. We will use another inductive argument. Specifically, we assume for $M \subsetneq L \subsetneq G$ that $\varepsilon_M^L(S) = 0$. Suppose that $M$ is not minimal. In this case, the descent property (8.1.2) implies $\varepsilon_M(S) = 0$. Assume then that $M$ is minimal. The splitting property (8.1.3) and induction on the size of $S$ show that

$$\varepsilon_M(S) = \sum_{v \in S} \varepsilon_M(v).$$

So, it suffices to show that $\varepsilon_M(v)$ vanishes.

Recall that $G(F) = A^{\times}(F)$ for a simple algebra $A$ of degree $n$ over $F$. We will need some facts from the theory of such algebras. For any $v$, $A$ has an invariant $i_v \in \mathbb{Q}/\mathbb{Z}$ whose order $d_v$ divides $n$. The number $\varepsilon_M(v)$ depends only on the pair $(G(F_v), M(F_v))$. Fix $n$ and $M$ minimal over $F_v$. We can define

$$\varepsilon(i_v) = \varepsilon_M(v),$$

a complex number depending only on $i_v \in \mathbb{Z}/n\mathbb{Z}$. The theory tells us that there is a unique simple algebra $A$ over $F$ attached to a finite set $\{i_v \in \mathbb{Z}/n\mathbb{Z}\,;\ v \in S\}$ such that $\sum i_v = 0$. If $S \supset S_{\mathrm{ram}}$, then

$$\sum_{v\in S} \varepsilon(i_v) = \varepsilon_M(S) = 0.$$

It follows that the $\varepsilon(i_v) = 0$. □

# 9 Comparison of Germs

We are motivated to define a subspace of $\mathbb{H}(G(F_S))$ by the induction assumption of the last section.

DEFINITION 9.1. (1) Recall that $G$ is the group of units of a central simple algebra. Let $S_G$ denote the set of finite places at which $G$ is not split, namely, ramifies.

(2) Consider the functions $f = \otimes_{v\in S} f_v$ of $\mathbb{H}(G(F_S))$, $f_v \in \mathbb{H}(G(F_v))$, with the $f_v$ satisfying the following property. For each $v \in S_G \cap S$, $\xi_v \in A_G(F_v)$ and $e \neq u_v \in \mathcal{U}_G(F_v)$, the orbital integral of $f_v$ vanishes at $\xi_v u_v$. Then define $\mathbb{H}(G(F_S))^0$ to be the subspace of $\mathbb{H}(G(F_S))$ spanned by such functions.

(3) Orbital integrals are invariant and define distributions in the space $\mathbb{I}(G(F_S))$. Define a subspace $\mathbb{I}(G(F_S))^0$ of $\mathbb{I}(G(F_S))$ in a similar way, as the span of $f_G$ for $f$ in $\mathbb{H}(G(F_S))^0$.

(4) We also define the spaces $\mathbb{H}(G(\mathbb{A}))^0$ and $\mathbb{I}(G(\mathbb{A}))^0$ in an analogous way, using $1_{K^S}$ to extend members of $\mathbb{H}(G(F_S))^0$ and $\mathbb{I}(G(F_S))^0$.

The goal of this section is to prove the following proposition.

PROPOSITION 9.2. *If $v$ is a non-Archimedean valuation of $F$ and $f \in \mathbb{H}(G(F_v))^0$, then $I_M^{\mathcal{E}}(\gamma,f)$ and $I_M(\gamma,f)$ have the same germ expansion.*

We will begin by showing that part (i) of the Main Theorem 2.7 implies a germ identity. This will allow us to use the induction assumption.

The germs for $G(F_v)$ and $G'(F_v)$ belong to different equivalence classes. However, they can still be compared directly. Let $\sigma \in M(F_v)$ be semisimple and let $\sigma'$ be its image. Take $f \in \mathbb{H}(G(F_v))$. Consider the germ expansion about $\sigma'$ for

$$\widehat{I}_{M'}(\zeta, f'_{G'}).$$

Recall that there is a bijection between $\mathcal{L}(M)$ and $\mathcal{L}(M')$, given by $L \mapsto L'$. Let $\zeta \in \sigma' M'_{\sigma'}(F_v) \cap G'_{\text{reg}}$. The bijection allows us to write

$$\widehat{I}_{M'}(\zeta, f'_{G'}) \overset{(M',\sigma')}{\sim} \sum_{L \in \mathcal{L}(M)} \sum_{\eta \in \sigma'(\mathcal{U}_{L'_{\sigma'}}(F_v))} g_{M'}^{L'}(\zeta, \eta) \widehat{I}_{L'}(\eta, f'_{G'}).$$

Recall that $(\mathcal{U}_{L_\sigma}(F_v))$ denotes the set of conjugacy classes of unipotent elements in $L_\sigma(F_v)$ and similarly for $L'_{\sigma'}$. Formula (5.5.1) says that $\widehat{I}_{G'}(\zeta, f'_{G'}) = 0$ unless there is a $\delta \in \sigma(\mathcal{U}_{L_\sigma}(F_v))$ such that $\zeta = \delta'$. We can then rewrite the germ expansion as

$$\widehat{I}_{M'}(\zeta, f'_{G'}) \overset{(M',\sigma')}{\sim} \sum_{L \in \mathcal{L}(M)} \sum_{\delta \in \sigma(\mathcal{U}_{L_\sigma}(F_v))} g_{M'}^{L'}(\zeta, \delta') \widehat{I}_{L'}(\delta', f'_{G'}). \tag{9.2.1}$$

By (5.5.1) the function $I_{M'}(\zeta, f')$ vanishes unless $\zeta = \gamma'$ for $\gamma \in M(F_v) \cap G_{\text{reg}}$.

PROPOSITION 9.3. *For each $L$ and $\delta$ there is a function $g_{M'}^{L'}(\zeta, \delta')$ within the $(M', \sigma')$-equivalence class which has the same property. That is, the function vanishes unless $\zeta = \gamma'$ as above.*

PROOF. We may assume the proposition inductively for $L \neq G$. Fix $\delta_1 \in \sigma(\mathcal{U}_{G_\sigma}(F_v))$. Choose $f_1 \in \mathbb{H}(G(F_v))$ such that

$$I_G(\delta, f_1) = \begin{cases} e^G(\delta_1), & \delta = \delta_1, \\ 0, & \text{otherwise.} \end{cases}$$

By (1.7.1),

$$\widehat{I}_{G'}(\delta', f'_{1,G'}) = \begin{cases} 1, & \delta = \delta_1, \\ 0, & \text{otherwise.} \end{cases}$$

Substituting $f_1$ into the expansion (9.2.1) and applying the induction assumption justifies the proposition. □

The orbital integral of a function on $M'(F_v)$ that vanishes at all $\zeta$ in $M'(F_v)$ which is not conjugate to some $\gamma'$ is equal to an orbital integral of a function on $M(F_v)$ which is zero at all orbits not of $\gamma$. See Theorem 2.14.4. Each germ

$$g_{M'}^{L'}(\gamma', \delta'), \quad \gamma \in \sigma M_\sigma(F_v) \cap G_{\text{reg}},$$

can then be regarded as an $(M, \sigma)$-equivalence class. We will use this identification to compare the germs for $G$ and $G'$. For $\gamma \in \sigma M_\sigma(F_v) \cap G_{\text{reg}}$, the expansion above becomes

$$\widehat{I}_M(\gamma', f'_{G'}) \overset{(M,\sigma)}{\sim} \sum_{L\in\mathcal{L}(M)} \sum_{\delta\in\sigma(\mathcal{U}_{L_\sigma}(F_v))} g_{M'}^{L'}(\gamma', \delta')\widehat{I}_{L'}(\delta', f'_{G'}).$$

Applying (5.6.1) to the left side of the equation gives

$$I_M^{\mathcal{E}}(\gamma, f) \overset{(M,\sigma)}{\sim} \sum_{L\in\mathcal{L}(M)} \sum_{\delta\in\sigma(\mathcal{U}_{L_\sigma}(F_v))} g_{M'}^{L'}(\gamma', \delta')\widehat{I}_{L'}(\delta', f'_{G'}). \tag{9.3.1}$$

PROPOSITION 9.4. *Suppose that part* (i) *of the Main Theorem* 2.7 *holds for* $G$. *Let* $v$ *be a non-Archimedean valuation of* $F$. *For each* $u \in (\mathcal{U}_G(F_v))$ *and each* $\gamma \in M(F_v) \cap G_{\mathrm{reg}}$, *we have*

$$g_{M'}^{G'}(\gamma', u') \overset{(M,e)}{\sim} e^G(u) g_M^G(\gamma, u).$$

PROOF. Since we are assuming part (i) of the Main Theorem, $G$ and $M$ satisfy the conditions of Proposition 8.5. Let $\gamma \in M(F_v), f \in \mathbb{H}(G(F_v))$. Proposition 8.5 and Theorem 8.1 together imply that

$$I_M^{\mathcal{E}}(\gamma, f) = I_M(\gamma, f).$$

Using (5.8.1) we obtain

$$\widehat{I}_{L'}(u', f'_{G'}) = e^L(u)^{-1} I_L^{\mathcal{E}}(u, f) = e^L(u)^{-1} I_L(u, f).$$

Applying (9.3.1) gives

$$I_M(\gamma, f) \overset{(M,e)}{\sim} \sum_{L\in\mathcal{L}(M)} \sum_{u\in(\mathcal{U}_L(F_v))} g_{M'}^{L'}(\gamma', u') e^L(u)^{-1} I_L(u, f).$$

On the other hand, applying (5.4.2) gives

$$I_M(\gamma, f) \overset{(M,e)}{\sim} \sum_{L\in\mathcal{L}(M)} \sum_{u\in(\mathcal{U}_L(F_v))} g_M^L(\gamma, u) I_L(u, f).$$

For $L \subsetneq G$, we assume inductively that

$$e^L(u)^{-1} g_{M'}^{L'}(\gamma', u') \overset{(M,e)}{\sim} g_M^L(\gamma, u).$$

Taking the difference of the two expansions for $I_M(\gamma, f)$ shows that

$$\sum_{u\in(\mathcal{U}_G(F_v))} (e^G(u)^{-1} g_{M'}^{G'}(\gamma', u') - g_M^G(\gamma, u)) I_G(u, f) \overset{(M,e)}{\sim} 0.$$

This formula holds in the space of $(M, e)$-equivalence classes of germs of functions. Thus, for any $f$ and $u \in (\mathcal{U}_G(F_v))$, we obtain

$$e^G(u)^{-1} g_{M'}^{G'}(\gamma', u') - g_M^G(\gamma, u) \overset{(M,e)}{\sim} 0,$$

as desired. □

We continue to carry the induction assumption that (i) of the Main Theorem 2.7 holds for proper Levi subgroups of $G$. We will eventually use Proposition 9.4 to deduce the equality of most of the germs. However, there is a pair of germs we can compare using local methods. That is, without resorting to the Main Theorem and the global argument of its proof.

PROPOSITION 9.5. *Let $v$ be a non-Archimedean valuation of $F$. For $\gamma \in M(F_v) \cap G_{\mathrm{reg}}$, we have*

$$g_{M'}^{G'}(\gamma', e) \overset{(M,e)}{\sim} e_v g_M^G(\gamma, e).$$

PROOF. Recall that $G$ is the multiplicative group of a central simple algebra. The local correspondence provides an injection of $\Pi_{\mathrm{temp}}(G(F_v))$ into $\Pi_{\mathrm{temp}}(G'(F_v))$. Denote the correspondence by $\pi \mapsto \pi'$. For $f \in \mathbb{H}(G(F_v))$, the correspondence satisfies the formula

$$f_G(\pi) = e_v f'_{G'}(\pi')$$

In addition, if $\gamma \in G(F_v)_{\mathrm{reg}}$, we have the character identity

$$\mathrm{ch}_\pi(\gamma) = e_v \, \mathrm{ch}_{\pi'}(\gamma').$$

Any cuspidal representation in $\Pi(G'(F_v))$ is of the form $\pi'$, for a unique cuspidal representation $\pi$ in $\Pi(G(F_v))$ (see [BZ76] for the definition of cuspidal). That $\pi$ corresponding to a cuspidal $\pi'$ is cuspidal can be deduced from the character identity here and theorems of Deligne and Casselman [Cas77]. See Proposition 2.7.3. Fix a corresponding pair $\pi$ and $\pi'$.

Let $f$ be a matrix coefficient of the cuspidal $\pi$. It satisfies $\mathrm{tr}\,\pi(f) \neq 0$. Since $\pi$ is a cuspidal representation, $f$ is a member of $\mathbb{H}(G(F_v))$. Let $\gamma$ be an element of $M(F_v) \cap G_{\mathrm{reg}}$. By [Ar87],

$$I_M(\gamma, f) = (-1)^{\dim(A_M/A_G)} \, \mathrm{vol}(G_\gamma(F_v)/A_M(F_v))^{-1} \cdot \mathrm{tr}\,\pi(f) \cdot |D_M^G(\gamma)|_v^{\frac{1}{2}} \, \mathrm{ch}_\pi(\gamma).$$

Note that if $\gamma$ is not $F_v$-elliptic in $M$, then the $F_v$-split component of $G_\gamma$ is larger than $A_M$. In this case the equation vanishes.

By definition, the function $f'_{G'}$ is only defined as a member of $\mathbb{I}(G'(F_v))$. However, we can represent it as a matrix coefficient of $\pi'$. For any $G$-regular element $\gamma \in M(F_v)$, we have

$$\begin{aligned} I_M(\gamma,f) &= (-1)^{\dim(A_M/A_G)} \operatorname{vol}(G_\gamma(F_v)/A_M(F_v))^{-1} \cdot \operatorname{tr}\pi(f) \cdot |D^G(\gamma)|_v^{\frac{1}{2}} \operatorname{ch}_\pi(\gamma) \\ &= (-1)^{\dim(A_{M'}/A_{G'})} \operatorname{vol}(G'_{\gamma'}(F_v)/A_{M'}(F_v))^{-1} \cdot \operatorname{tr}\pi'(f') \cdot |D^{G'}(\gamma')|_v^{\frac{1}{2}} \operatorname{ch}_{\pi'}(\gamma') \\ &= I_{M'}(\gamma',f'). \end{aligned}$$

For such $\gamma$, we have $I_{M'}(\gamma',f') = I_M^{\mathcal{E}}(\gamma,f)$. By Proposition 5.11, this equality holds for all of $M(F_v)$. Setting $\gamma = 1$ and applying (5.8.1) gives

$$I_M(e,f) = I_M^{\mathcal{E}}(e,f) = e_v I_{M'}(e,f').$$

We need the following lemma.

LEMMA 9.6. *If $u \neq e$ is a unipotent element in $M(F_v)$, then $I_M(u,f) = 0$.*

PROOF. Since $G$ comes from a central simple algebra, there exists a proper Levi subgroup $M_1 \subsetneq M$ and $u_1 \in \mathcal{U}_{M_1}(F_v)$ such that $u$ can be represented as an induced unipotent conjugacy class, $u_1^M$. Applying the descent formula Corollary 5.8.4 gives

$$I_M(u,f) = I_M(u_1^M,f) = \sum_{L\in\mathcal{L}(M_1)} d_{M_1}^G(M,L)\widehat{I}_{M_1}^L(u_1,f_L).$$

Since $f$ is a cusp form on $G(F_v)$, $f_L = 0$ for proper Levi subgroups $L \subsetneq G$. If $L = G$, the number $d_{M_1}^G(M,L)$ vanishes. Thus $I_M(u,f)$ vanishes. □

The same argument applied to $G'$ implies that $I_{M'}(u',f')$ vanishes.

## 9.7 Near the Identity

Let $\gamma$ be a $G$-regular element of $M(F_v)$ that is close to $e$. Then

$$I_M(\gamma,f) \overset{(M,e)}{\sim} \sum_{L\in\mathcal{L}(M)} \sum_{u\in(\mathcal{U}_L(F_v))} g_M^L(\gamma,u)I_L(u,f).$$

Since $I_M^{\mathcal{E}}(\gamma,f) = I_{M'}(\gamma',f')$, (9.3.1) gives

$$I_{M'}(\gamma',f') \overset{(M,e)}{\sim} \sum_{L\in\mathcal{L}(M)} \sum_{u\in(\mathcal{U}_L(F_v))} g_{M'}^{L'}(\gamma',u')I_{L'}(u',f').$$

The left-hand sides are equal, and substituting the formulae obtained above and using Lemma 9.6 allows us to write

$$\sum_{L\in\mathcal{L}(M)} e_v g_M^L(\gamma,e) I_{L'}(e,f') \overset{(M,e)}{\sim} \sum_{L\in\mathcal{L}(M)} g_{M'}^{L'}(\gamma',e) I_{L'}(e,f').$$

Assume the result inductively for $L \subsetneq G$. Namely, assume

$$e_v g_M^L(\gamma,e) \overset{(M,e)}{\sim} g_{M'}^{L'}(\gamma',e).$$

We see then that

$$e_v g_M^G(\gamma,e) I_{G'}(e,f') \overset{(M,e)}{\sim} g_{M'}^{G'}(\gamma',e) I_{G'}(e,f').$$

Since $I_{G'}(e,f') \neq 0$, we obtain Proposition 9.5. □

PROPOSITION 9.8. *Suppose that $v$ is a non-Archimedean valuation of $F$. Let $\sigma$ be a semisimple element of $M(F_v)$. Assume further that $\mathfrak{a}_{M_\sigma} = \mathfrak{a}_M$. Then for $\gamma \in \sigma M_\sigma(F_v) \cap G_{\mathrm{reg}}$ and $f \in \mathbb{H}(G(F_v))^0$, we have*

$$I_M^{\mathcal{E}}(\gamma,f) \overset{(M,\sigma)}{\sim} \sum_{L\in\mathcal{L}(M)} \sum_{\delta\in\sigma(\mathcal{U}_{L_\sigma}(F_v))} g_M^L(\gamma,\delta) I_L^{\mathcal{E}}(\delta,f). \tag{5.4.2$\mathcal{E}$}$$

PROOF. It is known that the germs depend only on the unipotent part of $\delta$. Let us make this more precise. Write $\delta = \sigma u$ and $\gamma = \sigma\mu$ for $u \in \mathcal{U}_{L_\sigma}(F_v)$ and $\mu \in \mathcal{U}_{M_\sigma(F_v)}$. By [Ar88.1, Lemma 9.2]

$$g_M^L(\gamma,\delta) = \begin{cases} g_{M_\sigma}^{L_\sigma}(\mu,u), & \text{if } \mathfrak{a}_{L_\sigma} = \mathfrak{a}_L, \\ 0, & \text{otherwise.} \end{cases} \tag{9.8.1}$$

This formula will allows us to only consider groups with dimension smaller than $G$, where we can apply the induction assumption that (i) of the Main Theorem 2.7 holds for such groups.

By (9.3.1), we have

$$I_M^{\mathcal{E}}(\gamma,f) \overset{(M,\sigma)}{\sim} \sum_{L\in\mathcal{L}(M)} \sum_{\delta\in\sigma(\mathcal{U}_{L_\sigma}(F_v))} g_{M'}^{L'}(\gamma',\delta') \widehat{I}_{L'}(\delta',f'_{G'}). \tag{9.8.2}$$

Let $\tau = \sigma'$. Applying (9.8.1) to $(G',\tau)$ and using (1.6.1) gives

$$g_{M'}^{L'}(\gamma',\delta') = g_{M'}^{L'}(\tau\mu_\tau,\tau u_\tau) = \begin{cases} g_{M'_\tau}^{L'_\tau}(\mu_\tau,u_\tau), & \text{if } \mathfrak{a}_{L_\sigma} = \mathfrak{a}_L, \\ 0, & \text{otherwise,} \end{cases}$$

for $\gamma$ and $\delta$ as in (9.8.1). That is, the sum in (9.8.2) need only be taken over $L \in \mathcal{L}(M)$ satisfying $\mathfrak{a}_{L_\sigma} = \mathfrak{a}_L$. Consider such an $L$, further satisfying $L_\sigma \neq G$. The induction assumption says that part (i) of the Main Theorem 2.7 holds for $L_\sigma$. Proposition 9.4 applied to $L_\sigma$ gives

$$g_{M'_\tau}^{L'_\tau}(\mu_\tau, u_\tau) \overset{(M_\sigma,e)}{\sim} e^{L_\sigma}(u) g_{M_\sigma}^{L_\sigma}(\mu, u).$$

Consider the property of germs given in [Ar88.1, Proposition 10.2]. We note that any group in $\mathcal{L}^{L_\sigma}(M_\sigma)$ is of the form $L_{1\sigma}$ for a unique $L_1 \in \mathcal{L}^L(M)$. Set $[u_1^{L_\sigma} : u]$ is 0 or 1, depending on whether or not the induced conjugacy class $u_1^{L_\sigma}$ equals $u$. The property says that

$$g_{M_\sigma}^{L_\sigma}(\mu, u) = \sum_{u_1 \in (\mathcal{U}_{L_{1\sigma}}(F_v))} g_{M_\sigma}^{L_{1\sigma}}(\mu, u_1)[u_1^{L_\sigma} : u].$$

Assume $[u_1^{L_\sigma} : u] = 1$. Set $\delta_1 = \sigma u_1$. Note that this is the Jordan decomposition of $\delta_1$. We have

$$e^{L_\sigma}(u) = e^{L_{1\sigma}}(u_1) = e^{L_1}(\delta_1).$$

In addition, (9.8.1) gives

$$g_{M_\sigma}^{L_{1\sigma}}(\mu, u_1) = g_M^{L_1}(\gamma, \delta_1).$$

Taken together, these facts imply that the inner sum in (9.8.2) is equal to

$$\sum_{\delta_1 \in \sigma(\mathcal{U}_{L_{1\sigma}}(F_v))} e^{L_1}(\delta_1) g_M^{L_1}(\gamma, \delta_1) \widehat{I}_{L'}((\delta_1^L)', f'_{G'}). \tag{9.8.3}$$

The stated equality was established for $L \in \mathcal{L}(M)$ satisfying $\mathfrak{a}_{L_\sigma} = \mathfrak{a}_L$ and $L_\sigma \neq G$. Assume now that $L$ is such that $\mathfrak{a}_{L_\sigma} \neq \mathfrak{a}_L$. By (9.8.1),

$$g_{M'}^{L'}(\gamma', \delta') = 0$$

and

$$g_M^{L_1}(\gamma, \delta_1) = 0.$$

Items (9.8.2) and (9.8.3) then vanish and in particular they remain equal.

We consider the final case, when $L_\sigma = G$. If this is so, then $L = G$ and $\sigma$ is central. Expression (9.8.2) becomes

$$\sum_{\delta \in \sigma(\mathcal{U}_G(F_v))} g_{M'}^{G'}(\gamma', \delta') \widehat{I}_{G'}(\delta', f'_{G'}). \tag{9.8.2*}$$

Furthermore (9.8.3) equals

$$\sum_{\delta \in \sigma(\mathcal{U}_G(F_v))} e^G(\delta) g_M^G(\gamma, \delta) \widehat{I}_{G'}(\delta', f'_{G'}). \tag{9.8.3*}$$

We are assuming that $f$ belongs to $\mathbb{H}(G(F_v))^0$. If $G$ splits at $v$, this puts no restrictions on $f$. In this case $G(F_v) = G'(F_v)$ and (9.8.2*) equals (9.8.3*). If $G$ does not split at $v$, and $\delta \neq \sigma$, then

$$\widehat{I}_{G'}(\delta', f') = 0.$$

Since $\sigma$ is central, Proposition 9.5 implies

$$g_{M'}^{G'}(\delta', \sigma') = g_{M'}^{G'}(\gamma', e) = e_v g_M^G(\gamma, e) = e^G(\sigma) g_M^G(\gamma, \sigma).$$

Accordingly, (9.8.2*) equals (9.8.3*) in this case too.

We have shown that (9.8.2) equals (9.8.3) for any $L \in \mathcal{L}(M)$. Equivalently, $I_M^{\mathcal{E}}(\gamma, f)$ is $(M, \sigma)$-equivalent to the sum over $L \in \mathcal{L}(M)$ in (9.8.3). Definition 5.8 says

$$e^{L_1}(\delta_1) \widehat{I}_{L'}((\delta_1^L)', f') = I_{L_1}^{\mathcal{E}}(\delta_1, f).$$

Therefore (9.3.1) is

$$I_M^{\mathcal{E}}(\gamma, f) \overset{(M,\sigma)}{\sim} \sum_{L_1 \in \mathcal{L}(M)} \sum_{\delta_1 \in \sigma(\mathcal{U}_{L_{1\sigma}}(F_v))} g_M^{L_1}(\gamma, \delta_1) I_{L_1}^{\mathcal{E}}(\delta_1, f),$$

as desired. □

## 10 Comparison of $I_M^{\mathcal{E}}(\pi, X, f)$ and $I_M(\pi, X, f)$

The following theorem is the main result of this section.

THEOREM 10.1. *Part* (i) *of the Main Theorem* 2.7 *implies that* $I_M(\tau, X, f)$ *equals* $I_M^{\mathcal{E}}(\tau, X, f)$. *In particular, the local assertion* (iii) *of the Main Theorem* 3.11 *follows from the local assertion* (i).

The main tool will be the constructions introduced in Chapter 5.

### *10.2 Function Spaces*

Fix a finite set $S$ of valuations of $F$ with the closure property. In Sections 4.19 and 5.4, we defined function spaces

$$\widetilde{\mathbb{H}}_{\mathrm{ac}}(G(F_S)) \supset \mathbb{H}_{\mathrm{ac}}(G(F_S)) \supset \mathbb{H}(G(F_S))$$

and

$$\widetilde{\mathbb{I}}_{\mathrm{ac}}(G(F_S)) \supset \mathbb{I}_{\mathrm{ac}}(G(F_S)) \supset \mathbb{I}(G(F_S)).$$

The spaces were defined so that those in the second row are the images under the invariant Fourier transform of the corresponding space in the first row. That is, under $f \mapsto f_G$. We recall that the spaces in the second row consist of functions $\varphi(\pi, X)$ on $\Pi_{\mathrm{temp}}(G(F_S)) \times \mathfrak{a}_{G,S}$ with different conditions on the second variable.

The conditions defining $\mathbb{I}(G(F_S))$ require $\varphi(\pi, X)$ to be smooth and compactly supported in $X$. The space $\mathbb{I}_{\mathrm{ac}}(G(F_S))$ relaxes the support condition, while $\widetilde{\mathbb{I}}_{\mathrm{ac}}(G(F_S))$ also relaxes the smoothness condition. The invariant distributions on $\mathbb{H}(G(F_S))$ we have described extend naturally to distributions on $\widetilde{\mathbb{H}}_{\mathrm{ac}}(G(F_S))$. Consequently, their Fourier transforms extend to $\widetilde{\mathbb{I}}_{\mathrm{ac}}(G(F_S))$.

## 10.3 Relative Maps

Recall that, for a pair of Levi subgroups $M \subset L$ in $\mathcal{L}$, we have defined maps $\theta_M^L$ and ${}^c\theta_M^L$, from $\widetilde{\mathbb{H}}_{\mathrm{ac}}(L(F_S))$ to $\widetilde{\mathbb{I}}_{\mathrm{ac}}(M(F_S))$. See (4.15.1) and (4.15.2) of Chapter 5 for the definitions. We showed in Proposition 5.4.17 that, for $\gamma \in M(F_S)$ and $f \in \widetilde{\mathbb{H}}_{\mathrm{ac}}(G(F_S))$, they satisfy

$$\sum_{L \in \mathcal{L}(M)} \hat{\theta}_M^L({}^c\theta_L(f)) = \sum_{L \in \mathcal{L}(M)} {}^c\hat{\theta}_M^L(\theta_L(f)) = 0, \tag{10.3.1}$$

$$I_M(\gamma, f) = \sum_{L \in \mathcal{L}(M)} {}^c\widehat{I}_M^L(\gamma, \theta_L(f)), \tag{10.3.2}$$

and

$${}^cI_M(\gamma, f) = \sum_{L \in \mathcal{L}(M)} \widehat{I}_M^L(\gamma, {}^c\theta_L(f)). \tag{10.3.3}$$

Recall that ${}^cI_M^L(\gamma)$ is an invariant distribution on $\widetilde{\mathbb{H}}_{\mathrm{ac}}(L(F_S))$ which depends only on the $M(F_S)$-orbit of $\gamma$. See equation 5.(4.9.1) for the definition. The distribution has a key property. Namely, if $f$ belongs to $\mathbb{H}(G(F_S))$, then ${}^cI_M(\gamma, f)$ is compactly supported in $\gamma$ in the space of $M(F_S)$-orbits in $M(F_S)$. See Proposition 5.4.10.

As before, we will pull back the analogous objects on $G'$ to $G$.

DEFINITION 10.4. (1) Let $\gamma \in M(F_S)$ be $G$-regular. Let $f \in \widetilde{\mathbb{H}}_{ac}(G(F_S))$. We define

$$^cI^{\mathcal{E}}_M(\gamma,f) = {}^c\widehat{I}_{M'}(\gamma',f'_{G'}).$$

(2) Let $f \in \widetilde{\mathbb{H}}_{ac}(G(F_S))$, $\tau \in \Pi_{\mathrm{temp}}(M(F_S))$ and $X \in \mathfrak{a}_{M,S}$. We set

$$\theta^{\mathcal{E}}_M(f,\tau,X) = e_S\hat{\theta}_{M'}(f'_{G'},\tau',X')$$

and

$$^c\theta^{\mathcal{E}}_M(f,\tau,X) = e_S{}^c\hat{\theta}_{M'}(f'_{G'},\tau',X').$$

PROPOSITION 10.5. *The functions $\theta^{\mathcal{E}}_M(f)$ and $^c\theta^{\mathcal{E}}_M(f)$ of $(\tau,X)$, defined by the expressions in* 10.4, *both belong to $\widetilde{\mathbb{I}}(M(F_S))$. Moreover, we have*

$$\sum_{L\in\mathcal{L}(M)} \hat{\theta}^{L,\mathcal{E}}_M({}^c\theta^{\mathcal{E}}_L(f)) = \sum_{L\in\mathcal{L}(M)} {}^c\hat{\theta}^{L,\mathcal{E}}_M(\theta^{\mathcal{E}}_L(f)) = 0, \tag{10.3.1$\mathcal{E}$}$$

$$I^{\mathcal{E}}_M(\gamma,f) = \sum_{L\in\mathcal{L}(M)} {}^c\widehat{I}^{L,\mathcal{E}}_M(\gamma,\theta^{\mathcal{E}}_L(f)), \tag{10.3.2$\mathcal{E}$}$$

*and*

$$^cI^{\mathcal{E}}_M(\gamma,f) = \sum_{L\in\mathcal{L}(M)} \widehat{I}^{L,\mathcal{E}}_M(\gamma,{}^c\theta^{\mathcal{E}}_L(f)). \tag{10.3.3$\mathcal{E}$}$$

PROOF. Let $B$ be a small ball in $\mathfrak{a}_P$ centered at the origin. Let $\nu_P$ be a point in the chamber $(\mathfrak{a}^*_P)^+$ which is far from the walls. Set

$$\omega_P(X) = \mathrm{vol}(\mathfrak{a}^+_P \cap B)/\,\mathrm{vol}(B).$$

For any $\tau' \in \Pi_{\mathrm{temp}}(M'(F_S))$, Proposition 5.4.16 gives

$$\hat{\theta}_{M'}(f'_{G'},\tau',X') = \sum_{P\in\mathcal{P}(M)} \omega_P(X)e^{-\nu'_P(X')}\widehat{I}_{M'}(\tau'_{\nu'_P},X',f'_{G'}).$$

The vanishing property Proposition 5.10.8 implies that $\hat{\theta}_{M'}(f'_{G'},\tau',X')$ vanishes if $\tau'$ does not come from some $\tau$. It follows from the definitions that $\theta^{\mathcal{E}}_M(f)$ belongs to $\widetilde{\mathbb{I}}_{ac}(M(F_S))$.

We make an observation before proceeding. By (1.13.1) and the definitions above, for any $f \in \widetilde{\mathbb{H}}_{ac}(G(F_S))$, we have

$$\theta_G^{\mathcal{E}}(f)'(\pi', X') = e_S \theta_G^{\mathcal{E}}(f, \pi, X)$$
$$= \hat{\theta}_{G'}(f'_{G'}, \pi', X').$$

Said differently,

$$\theta_G^{\mathcal{E}}(f)' = \hat{\theta}_{G'}(f'_{G'}).$$

This formula is rather trivial, for the maps are defined in Chapter 5 so that $\theta_G(f) = {}^c\theta_G(f) = f_G$.

With this observation we can proceed to the discussion of ${}^c\theta_M^{\mathcal{E}}(f)$. The corresponding formulae with $G$ replaced by $L \in \mathcal{L}$ do not hold. There is a partial result. Let $I'$ be an invariant distribution on $\widetilde{\mathbb{H}}_{\mathrm{ac}}(L'(F_S))$ that is supported on characters. Then

$$\widehat{I'}(\theta_L^{\mathcal{E}}(f)') = \widehat{I'}(\hat{\theta}_{L'}(f'_{G'})).$$

The linear form

$$I'(h) = {}^c\theta_{M'}^{L'}(h, \tau', X')$$

on $\widetilde{\mathbb{H}}_{\mathrm{ac}}(L'(F_S))$ satisfies the two conditions above. Consequently,

$$\sum_{L \in \mathcal{L}(M)} {}^c\hat{\theta}_M^{L,\mathcal{E}}(\theta_L^{\mathcal{E}}(f), \tau, X) = e_S \sum_{L \in \mathcal{L}(M)} {}^c\hat{\theta}_{M'}^{L'}(\theta_L^{\mathcal{E}}(f)', \tau', X')$$
$$= e_S \sum_{L \in \mathcal{L}(M)} {}^c\hat{\theta}_{M'}^{L'}(\hat{\theta}_{L'}(f'_{G'}), \tau', X').$$

This vanishes by (10.3.1) applied to $G'$. Since $\theta_G^{\mathcal{E}}(f) = f_G$, we obtain

$${}^c\theta_M^{\mathcal{E}}(f, \tau, X) = - \sum_{L \in \mathcal{L}_0(M)} {}^c\hat{\theta}_M^{L,\mathcal{E}}(\theta_L^{\mathcal{E}}(f), \tau, X).$$

It follows inductively that ${}^c\theta_M^{\mathcal{E}}(f)$ belongs to $\widetilde{\mathbb{I}}_{\mathrm{ac}}(M(F_S))$. Half of (10.3.1$\mathcal{E}$) also follows. The rest of the statement follows similarly. □

THEOREM 10.6. *Fix $M \in \mathcal{L}$. Let $S$ be a finite set of valuations with the closure property. If $S \supset S_{\mathrm{ram}}$, assume in addition that for $f \in \mathbb{H}(G(F_S))$, $L \in \mathcal{L}(M)$ and $\gamma \in L(F_S)$, we have*

$$I_L^{\mathcal{E}}(\gamma,f) = I_L(\gamma,f).$$

*Then for any $f \in \widetilde{\mathbb{H}}_{\mathrm{ac}}(G(F_S))$ and $X \in \mathfrak{a}_{M,S}$, we have*

(1)

$$\theta_M^{\mathcal{E}}(f) = \theta_M(f),$$

(2)

$$^c\theta_M^{\mathcal{E}}(f) = {}^c\theta_M(f),$$

(3)

$$I_M^{\mathcal{E}}(\rho,X,f) = I_M(\rho,X,f), \quad \textit{for all } \rho \in \Sigma(M(F_S)),$$

*and*

(4)

$$I_M^{\mathcal{E}}(\tau,X,f) = I_M(\tau,X,f), \quad \textit{for all } \tau \in \Pi(M(F_S)).$$

PROOF. By the induction assumption of (i) of the Main Theorem 2.7, if $L \subset L_1 \subsetneq G$, $S \supset S_{\mathrm{ram}}$ and $\gamma \in L(F_S)$, we have

$$I_L^{L_1,\mathcal{E}}(\gamma) = I_L^{L_1}(\gamma).$$

We can then assume inductively that the four formulae hold if $G$ is replaced by such an $L_1$. We shall further assume inductively that the four formulae hold for $G$, but with $M$ replaced by any Levi subgroup $L \supsetneq M$.

It suffices to prove the theorem for $f \in \mathbb{H}(G(F_S))$. Indeed, the restriction of a function in $\widetilde{\mathbb{H}}_{\mathrm{ac}}(G(F_S))$ to any fixed set

$$G(F_S)^Z = \{x \in G(F_S)\,;\, H_G(x) = Z\}, \quad Z \in \mathfrak{a}_{G,S},$$

coincides with the restriction of a function in $\mathbb{H}(G(F_S))$.

We need a lemma.

LEMMA 10.7. *Assume the situation of the theorem. For $\tau \in \Pi_{\mathrm{temp}}(M(F_S))$, the difference*

$$^c\theta_M^{\mathcal{E}}(f,\tau,X) - {}^c\theta_M(f,\tau,X)$$

*is compactly supported in $X$.*

PROOF. Let $\gamma \in M(F_S)$ be $G$-regular. The hypotheses allow us to use Theorem 8.1. In particular, we can use the expansion (8.1.1). Applying successively the descent properties (8.1.2) and (5.3.2) allows us to write (8.1.1) as

$$I_M^{\mathcal{E}}(\gamma,f) = \sum_{L_1\in\mathcal{L}(M)} \varepsilon_M^{L_1}(S) I_{L_1}(\gamma,f). \tag{10.7.1}$$

We expect a similar formula to hold for ${}^cI_M^{\mathcal{E}}(\gamma,f)$. Consider the expression

$${}^cI_M^{\mathcal{E}}(\gamma,f) - \sum_{L_1\in\mathcal{L}(M)} \varepsilon_M^{L_1}(S)\cdot {}^cI_{L_1}(\gamma,f). \tag{10.7.2}$$

By (10.3.3) and (10.3.3$\mathcal{E}$), we can write this as the sum of

$$\widehat{I}_M^M(\gamma, {}^c\theta_M^{\mathcal{E}}(f) - {}^c\theta_M(f)) \tag{10.7.3}$$

and

$$\sum_{L\supsetneq M} \left\{ \widehat{I}_M^{L,\mathcal{E}}(\gamma, {}^c\theta_L^{\mathcal{E}}(f)) - \sum_{L_1\in\mathcal{L}^L(M)} \varepsilon_M^{L_1}(S)\widehat{I}_{L_1}^L(\gamma, {}^c\theta_L(f)) \right\}. \tag{10.7.4}$$

If $L \neq M$, our induction hypothesis and (10.7.1) (with $G$ replaced by $L$) imply

$$\sum_{L_1\in\mathcal{L}^L(M)} \varepsilon_M^{L_1}(S)\widehat{I}_{L_1}^L(\gamma, {}^c\theta_L(f)) = \widehat{I}_M^{L,\mathcal{E}}(\gamma, {}^c\theta_L(f)) = \widehat{I}_M^{L,\mathcal{E}}(\gamma, {}^c\theta_L^{\mathcal{E}}(f)).$$

Thus (10.7.4) vanishes. We get that (10.7.2) equals (10.7.3).

Since $f$ is a member of $\mathbb{H}(G(F_S))$, the function (10.7.2) has bounded support as a function of $\gamma$ in the space of $M(F_S)$-orbits in $M(F_S)$. The same then holds for (10.7.3). For $X \in \mathfrak{a}_{M,S}$, (10.7.3) is the orbital integral in

$$\{\gamma \in M(F_S)\,;\, H_M(\gamma) = X\}$$

of a function defined on

$$M(F_S)^X = \{x \in M(F_S)\,;\, H_M(x) = X\}.$$

The tempered characters of this function are the expressions of the lemma. □

DEFINITION 10.8. Recall the following definition, (5.4.1) of Chapter 5.

(1) For each $\tau \in \Pi_{\text{temp}}(M(F_S))$ put

$${}^c\theta_M(f,\tau_\lambda) = \int_{\mathfrak{a}_{M,S}} {}^c\theta_M(f,\tau,X)e^{\lambda(X)}\,\mathrm{d}X.$$

It is a meromorphic function of $\lambda \in \mathfrak{a}_{M,\mathbb{C}}^*$.

The definition extends to standard representations $\rho \in \Sigma(M(F_S))$ by analytic continuation.

(2) For $\rho \in \Sigma(M(F_S))$, define

$$^cI_M(\rho, X, f) = \sum_{L \in \mathcal{L}(M)} \widehat{I}^L_M(\rho, X, {}^c\theta_L(f)).$$

DEFINITION 10.9. *For each $\tau \in \Pi_{\text{temp}}(M(F_S))$ put, for $\lambda \in \mathfrak{a}^*_{M,\mathbb{C}}$,*

$$^c\theta^{\mathcal{E}}_M(f, \tau_\lambda) = \int_{\mathfrak{a}_{M,S}} {}^c\theta^{\mathcal{E}}_M(f, \tau, X) e^{\lambda(X)}\, dX.$$

LEMMA 10.10. *If $\tau'$ is the image of $\tau$, we have*

$$^c\theta^{\mathcal{E}}_M(f, \tau_\lambda) = e_S \cdot {}^c\hat{\theta}_{M'}(f'_{G'}, \tau'_{\lambda'}).$$

PROOF. Since $\mathrm{d}X = \mathrm{d}X'$,

$$\begin{aligned} ^c\theta^{\mathcal{E}}_M(f, \tau_\lambda) &= \int_{\mathfrak{a}_{M,S}} e_S \cdot {}^c\hat{\theta}_{M'}(f'_{G'}, \tau', X') e^{\lambda(X)}\, \mathrm{d}X \\ &= \int_{\mathfrak{a}_{M',S}} e_S \cdot {}^c\hat{\theta}_{M'}(f', \tau', X') e^{\lambda'(X')}\, \mathrm{d}X' \\ &= e_S {}^c\hat{\theta}_{M'}(f'_{G'}, \tau'_{\lambda'}). \end{aligned}$$

□

These formulae extend by analytic continuation to $\rho \in \Sigma(M(F_S))$.
Recall Proposition 5.5.8.

PROPOSITION 10.11. *The distribution $^cI_M(\rho, X, f)$ can be computed inductively from the function $^c\theta_M(f, \rho_\lambda)$. Namely, if $\beta \in C^\infty_c(\mathfrak{a}_{M,S})$ is a test function approaching the Dirac measure at the origin, and $X \in \mathfrak{a}_{M,S}$ is a point at which $^cI_M(\rho, X, f)$ is smooth, then*

$$^cI_M(\rho, X, f) = \lim_\beta \sum_{P \in \mathcal{P}(M)} \omega_P \int_{\varepsilon_P + i\mathfrak{a}^*_{M,S}} \widehat{\beta}(\lambda) \cdot {}^c\theta_M(f, \rho_\lambda) e^{-\lambda(X)}\, d\lambda. \tag{10.11.1}$$

If $\mathfrak{a}_{M,S}$ is discrete, $X$ can be any point and $\beta$ may be removed from (10.11.1).

DEFINITION 10.12. *For any representation $\rho \in \Sigma(M(F_S))$, define*

$$^cI^{\mathcal{E}}_M(\rho, X, f) = \sum_{L \in \mathcal{L}(M)} \widehat{I}^{L,\mathcal{E}}_M(\rho, X, {}^c\theta^{\mathcal{E}}_L(f)).$$

LEMMA 10.13. *For any smooth point $X$, we have*

$$ {}^cI_M^{\mathcal{E}}(\rho, X, f) = \lim_\beta \sum_{P \in \mathcal{P}(M)} \omega_P \int_{\varepsilon_P + i\mathfrak{a}_{M,S}^*} \widehat{\beta}(\lambda) \cdot {}^c\theta_M^{\mathcal{E}}(f, \rho_\lambda) e^{-\lambda(X)}\, d\lambda. \qquad (10.11.1\mathcal{E}) $$

PROOF. We have

$$ \begin{aligned} {}^cI_M^{\mathcal{E}}(\rho, X, f) &= \sum_{L \in \mathcal{L}(M)} e_S \widehat{I}_{M'}^{L'}(\rho', X', {}^c\theta_L^{\mathcal{E}}(f)') \\ &= e_S \sum_{L \in \mathcal{L}(M)} \widehat{I}_{M'}^{L'}(\rho', X', {}^c\hat{\theta}_{L'}(f'_{G'})) \\ &= e_S \cdot {}^c\widehat{I}_{M'}(\rho', X', f'_{G'}). \end{aligned} $$

Apply (10.11.1) to the last expression, with $f$, $\rho$, and $X$ replaced by $f'_{G'}$, $\rho'$ and $X'$. We obtain

$$ \lim_\beta \sum_{P \in \mathcal{P}(M)} \omega_P \int_{\varepsilon_P + i\mathfrak{a}_{M,S}^*} \widehat{\beta}(\lambda') e_S \cdot {}^c\hat{\theta}_{M'}(f'_{G'}, \rho'_{\lambda'}) e^{-\lambda'(X')}\, d\lambda'. $$

We substitute the formula above for ${}^c\theta_M^{\mathcal{E}}(f, \rho_\lambda)$. Since $d\lambda' = d\lambda$, the lemma follows. □

LEMMA 10.14. *We have*

$$ e^{-\mu(X)} \cdot {}^cI_M(\tau_\mu, X, f) = \lim_\beta \int_{\mu + i\mathfrak{a}_{M,S}^*} \widehat{\beta}(\lambda) {}^c\theta_M(f, \tau_\lambda) e^{-\lambda(X)}\, d\lambda. $$

PROOF. We will apply (10.11.1) with $\rho = \tau_\mu$, for $\tau \in \Pi_{\mathrm{temp}}(M(F_S))$ and $\mu \in \mathfrak{a}_M^*$. We take $\mu$ to be in general position. Then the contours of integration on the right side of (10.11.1) can be deformed to $i\mathfrak{a}_{M,S}^*$. Thus

$$ \begin{aligned} {}^cI_M(\tau_\mu, X, f) &= \lim_\beta \int_{i\mathfrak{a}_{M,S}^*} \widehat{\beta}(\lambda) \cdot {}^c\theta_M(f, \tau_{\mu+\lambda}) e^{-\lambda(X)}\, d\lambda \\ &= e^{\mu(X)} \lim_\beta \int_{\mu + i\mathfrak{a}_{M,S}^*} \widehat{\beta}(\lambda - \mu) \cdot {}^c\theta_M(f, \tau_\lambda) e^{-\lambda(X)}\, d\lambda. \end{aligned} $$

We chose $\beta$ to approach the Dirac measure at the origin. But

$$ \lambda \mapsto \widehat{\beta}(\lambda - \mu) $$

is the Fourier-Laplace transform of a function

$$ X \mapsto e^{-\mu(X)} \beta(X) $$

which also approaches the Dirac measure at the origin. We can therefore replace $\widehat{\beta}(\lambda - \mu)$ by $\widehat{\beta}(\lambda)$, and the lemma follows. □

LEMMA 10.15. *For any $\tau \in \Pi_{\mathrm{temp}}(M(F_S))$ and $\mu \in \mathfrak{a}_M^*$, we have*

$$ {}^cI_M^{\mathcal{E}}(\tau_\mu, X, f) = {}^cI_M(\tau_\mu, X, f). $$

PROOF. A similar formula to that of the last lemma arises from (10.11.1$\mathcal{E}$). Taking the difference of the two formulae shows that

$$ \begin{aligned} & e^{-\mu(X)}({}^cI_M^{\mathcal{E}}(\tau_\mu, X, f) - {}^cI_M(\tau_\mu, X, f)) \\ &= \lim_\beta \int_{\mu + i\mathfrak{a}_{M,S}^*} \widehat{\beta}(\lambda)({}^c\theta_M^{\mathcal{E}}(f, \tau_\lambda) - {}^c\theta_M(f, \tau_\lambda))e^{-\lambda(X)}\, \mathrm{d}\lambda. \end{aligned} \tag{10.15.1} $$

Note that

$$ \lambda \mapsto {}^c\theta_M^{\mathcal{E}}(f, \tau_\lambda) - {}^c\theta_M(f, \tau_\lambda) $$

is the Fourier transform of the compactly supported function

$$ X \mapsto {}^c\theta_M^{\mathcal{E}}(f, \tau, X) - {}^c\theta_M(f, \tau, X). $$

Thus it is an entire function. The integral over $\mu + i\mathfrak{a}_{M,S}^*$ can be deformed to any translate of $i\mathfrak{a}_{M,S}^*$. Consequently, the left side of (10.15.1) is independent of $\mu$ for almost all $\mu$ and $X$. We saw at equation (4.11.2) of Chapter 5 that the value of the left side of (10.15.1) at any $\mu$ and $X$ can be expressed in terms of its values at points in general position. In particular, the expression is independent of $\mu$, for any choice of $\mu$.

Proposition 5.4.12 says that

$$ \sum_{P \in \mathcal{P}(M)} \omega_P(X)e^{-\nu_P(X)} \cdot {}^cI_M(\tau_{\nu_P}, X, f) = 0. $$

Applying the formula for $G'$ gives

$$ \sum_{P \in \mathcal{P}(M)} \omega_P(X)e^{-\nu_P(X)} \cdot {}^cI_M^{\mathcal{E}}(\tau_{\nu_P}, X, f) = 0. $$

Since (10.15.1) is independent of $\mu$, we can express its value at $\mu$ by using these sums. Consequently, it vanishes for any $\mu$. □

## 10.16 Proof of Theorem 10.6 continued

We resume the proof of Theorem 10.6. We start with part (2). We apply the last lemma with $\mu = 0$. Proposition 5.4.16 asserts that, for $\tau \in \Pi_{temp}(M(F_S))$, we have

$$^{c}I_M(\tau, X, f) = {}^{c}\theta_M(f, \tau, X).$$

Applying the result to $G'$ gives

$$^{c}I_M^{\mathcal{E}}(\tau, X, f) = {}^{c}\theta_M^{\mathcal{E}}(f, \tau, X).$$

Combining the formulae gives part (2) of the theorem.

## 10.17 Part (1)

Part (1) of the theorem, that $\theta_M^{\mathcal{E}}(f) = \theta_M(f)$, follows from (2), the induction hypothesis, and (10.3.1) and (10.3.1$\mathcal{E}$).

## 10.18 Part (3)

Part (3) of the theorem requires a comparison of (10.11.1) with (10.11.1$\mathcal{E}$). Part (2) of the theorem implies that for $\tau \in \Pi_{temp}(M(F_S))$ and $\lambda \in \mathfrak{a}_{M,\mathbb{C}}^*$, we have

$$^{c}\theta_M^{\mathcal{E}}(f, \tau_\lambda) = {}^{c}\theta_M(f, \tau_\lambda).$$

By analytic continuation, the formula is valid for standard representations $\rho \in \Sigma(M(F_S))$. The right sides of (10.11.1) and (10.11.1$\mathcal{E}$) are then equal. Consider the equality of left sides

$$^{c}I_M^{\mathcal{E}}(\rho, X, f) = {}^{c}I_M(\rho, X, f).$$

For $L \in \mathcal{L}(M)$ with $L \neq G$, the induction assumption and part (2) give

$$\widehat{I}_M^{L,\mathcal{E}}(\rho, X, {}^{c}\theta_L^{\mathcal{E}}(f)) = \widehat{I}_M^{L}(\rho, X, {}^{c}\theta_L(f)).$$

We get

$$I_M^{\mathcal{E}}(\rho, X, f) = I_M(\rho, X, f),$$

part (3) of the theorem.

## 10.19 *Part* (4)

Lastly, we prove part (4). Let $\tau \in \Pi(M(F_S))$. Recall that at (6.5.1) we defined $I_M^{\mathcal{E}}(\tau, X, f)$ as an expansion in terms of the distributions

$$I_L^{\mathcal{E}}(\rho_\lambda, h_L(X), f).$$

We also saw at (6.1.1) that $I_M(\tau, X, f)$ satisfies a similar expansion in terms of the distributions

$$I_L(\rho_\lambda, h_L(X), f).$$

Part (4) follows from part (3), with $M$ replaced by $L$. □

COROLLARY 10.20. *Under the assumptions of Theorem* 10.6, *for every* $f \in \widetilde{\mathbb{H}}_{\mathrm{ac}}(G(F_S))$ *and any* $G$*-regular element* $\gamma \in M(F_S)$, *we have*

$$^cI_M^{\mathcal{E}}(\gamma, f) = \sum_{L \in \mathcal{L}(M)} \varepsilon_M^L(S)\, {}^cI_L(\gamma, f).$$

PROOF. In the proof of Theorem 10.6, we established the equality of (10.7.2) with (10.7.3). By part (2), (10.7.3) vanishes. This gives the corollary. □

PROPOSITION 10.21. *The local assertion* (*iii*) *of the Main Theorem* 3.11 *follows from Theorem* 10.6.

PROOF. We cannot assume the equality of $I_L^{\mathcal{E}}(\gamma)$ and $I_L(\gamma)$. However, we still have the original induction assumption. Namely, if $L \subset L_1 \subsetneq G$, then $I_L^{L_1, \mathcal{E}}(\gamma)$ and $I_L^{L_1}(\gamma)$ are equal. Suppose that $\tau \in \Pi_{\mathrm{unit}}(M(F_S))$. We must show for $f \in \mathbb{H}(G(F_S))$ that

$$I_M^{\mathcal{E}}(\tau, 0, f) = I_M(\tau, 0, f).$$

We noted that these distributions have the same expansions in terms of the distributions associated to standard representations $\rho \in \Sigma(M(F_S))$; see (6.1.1) and (6.5.1). Furthermore, only $\rho$ with $\xi(\tau, \rho) \neq 0$ can occur in the expansions. Since $\tau$ is unitary, any such $\rho$ has a unitary central character. It suffices to prove that for such a $\rho$, a point $\lambda \in \mathfrak{a}_{M,\mathbb{C}}^*$ with small real part, $L \in \mathcal{L}(M)$ and $X \in \mathfrak{a}_M$ that

$$I_L^{\mathcal{E}}(\rho_\lambda, h_L(X), f) = I_L(\rho_\lambda, h_L(X), f). \tag{10.21.1}$$

We will use the splitting and descent formulae for $I_L(\rho_\lambda, h_L(X), f)$ and $I_L^{\mathcal{E}}(\rho_\lambda, h_L(X), f)$. The splitting property allows us to reduce to the case when $S$ is a single valuation $v$. So, assume we are in this case. Since $\rho$ has unitary central character, $\rho$ must be tempered or induced from a proper parabolic subgroup. If $\rho$ is tempered, then the proof of Lemma 5.3.4 implies

$$I_L^{\mathcal{E}}(\rho_\lambda, h_L(X), f) = I_L(\rho_\lambda, h_L(X), f) = \begin{cases} 0, & L \neq G \\ f_G(\rho_\lambda^G, h_G(X)), & L = G. \end{cases}$$

Otherwise, there is a proper Levi subgroup $M_1 \subsetneq M$ and $\rho_1 \in \Sigma(M_1(F_v))$ satisfying

$$\rho = \rho_1^M.$$

We will use the descent property of each side of (10.21.1). The descent formula shows that we only need to establish that, for $L_1 \in \mathcal{L}(M)$, $L_1 \neq G$ and $X_1 \in \mathfrak{a}_{M_1}$, we have

$$\widehat{I}_{M_1}^{L_1,\mathcal{E}}(\rho_{1,\lambda}, X_1, f_{L_1}) = \widehat{I}_{M_1}^{L_1}(\rho_{1,\lambda}, X_1, f_{L_1}).$$

Since we are assuming the equality of $I_L^{L_1,\mathcal{E}}(\gamma)$ and $I_L^{L_1}(\gamma)$, the formula follows from Theorem 10.6, applied to $(L_1, M_1)$. Thus (10.21.1) holds, and so too does part (iii) of the Main Theorem 3.11. □

## 11 A Formula for $I_t^{\mathcal{E}}(f)$

We continue with a discussion of the global assertion (iv) of the Main Theorem 3.11, having shown that (iii) follows from (i). The goal of this section is an expansion for $I^{\mathcal{E}}(f)$ which is dual to that of Proposition 2.5.1. We first will derive an inversion formula for $I_M^{\mathcal{E}}(\tau, f)$.

LEMMA 11.1. *Suppose that $\tau \in \Pi_{\mathrm{unit}}(M'(\mathbb{A})^1)$. Then for any small point $\varepsilon_M \in \mathfrak{a}_M^*$ in general position, the distribution*

$$I_M^{\mathcal{E}}(\tau, f) = \widehat{I}_{M'}(\tau, f'_{G'})$$

*equals*

$$\sum_{L \in \mathcal{L}(M)} \sum_{\varphi \in \Pi(M(\mathbb{A})^1)} \int_{\varepsilon_M + i\mathfrak{a}_M^*/i\mathfrak{a}_L^*} m_M^L(\tau_\lambda, \varphi_\lambda) I_L^{\mathcal{E}}(\varphi_\lambda, f)\, d\lambda. \tag{11.1.1}$$

**Remark 2.** Only those $\varphi$ with $\delta(\tau, \varphi) \neq 0$ will contribute to (11.1.1). Since $\tau$ is unitary, such $\varphi$ will have a unitary central character. Thus $\varphi$ can be identified with an orbit $\{\varphi_\lambda\}$ of $i\mathfrak{a}_M^*$ in $\Pi(M(\mathbb{A}))$.

PROOF. By relabeling the outer summation index $L$ by $L_1$, we can insert the following expression—see definitions 6.5 and 6.7 and equation (6.5.1)—for $I_{L_1}^{\mathcal{E}}(\varphi_\lambda, f)$ into the formula:

$$\sum_{Q\in\mathcal{P}(L_1)} \omega_Q \sum_{L\in\mathcal{L}(L_1)} \sum_{\rho\in\Sigma(M(\mathbb{A})^1)} \int_{\varepsilon_Q+i\mathfrak{a}^*_{L_1}/i\mathfrak{a}^*_L} m^L_{L_1}(\varphi_{\lambda+\mu},\rho_{\lambda+\mu}) I^{\mathcal{E}}_L(\rho_{\lambda+\mu},f)\,\mathrm{d}\mu.$$

The point $\varepsilon_Q$ can be taken to be small relative to the point $\varepsilon_M$. Since $\lambda\in\varepsilon_M+i\mathfrak{a}^*_M$, we can deform the contour of integrating in $\mu$ to $i\mathfrak{a}^*_{L_1}/i\mathfrak{a}^*_L$. Thus (11.1.1) is given by

$$\sum_{M\subset L_1\subset L} \sum_{\varphi\in\Pi(M(\mathbb{A})^1)} \sum_{\rho\in\Sigma(M(\mathbb{A})^1)} \int_{\varepsilon_M+i\mathfrak{a}^*_M/i\mathfrak{a}^*_L} m^{L_1}_M(\tau_\lambda,\varphi_\lambda) m^L_{L_1}(\varphi_\lambda,\rho_\lambda) I^{\mathcal{E}}_L(\rho_\lambda,f)\,\mathrm{d}\lambda.$$

We can move the sums over $L_1$ and $\varphi$ inside the integral. Their contributions are given by

$$\sum_{\varphi} \sum_{\{L_1\,;\,M\subset L_1\subset L\}} m^{L_1}_M(\tau_\lambda,\varphi_\lambda) m^L_{L_1}(\varphi_\lambda,\rho_\lambda).$$

This expression comes from a product of two $(G,M)$-families. By 4.11 it equals

$$\tilde{m}^L_M(\tau_\lambda,\rho_\lambda)\left(\sum_{\varphi\in\Pi(M(\mathbb{A})^1)} \delta(\tau,\varphi)\xi(\varphi,\rho)\right).$$

But

$$\sum_{\varphi\in\Pi(M(\mathbb{A})^1)} \delta(\tau,\varphi)\xi(\varphi,\rho) = \sum_{\rho_1\in\Sigma(M(\mathbb{A})^1)} \xi(\tau,\rho_1) \sum_{\varphi\in\Pi(M(\mathbb{A})^1)} \iota(\rho_1,\varphi)\xi(\varphi,\rho) = \xi(\tau,\rho).$$

The product of this expression with $\tilde{m}^L_M(\tau_\lambda,\rho_\lambda)$ is $m^L_M(\tau_\lambda,\rho_\lambda)$. So, we can write (11.1.1) as

$$\sum_{L\in\mathcal{L}(M)} \sum_{\rho\in\Sigma(M(\mathbb{A})^1)} \int_{\varepsilon_M+i\mathfrak{a}^*_M/i\mathfrak{a}^*_L} m^L_M(\tau_\lambda,\rho_\lambda) I^{\mathcal{E}}_L(\rho_\lambda,f)\,\mathrm{d}\lambda. \tag{11.1.2}$$

Consider the expression

$$\sum_{\rho\in\Sigma(M(\mathbb{A})^1)} m^L_M(\tau_\lambda,\rho_\lambda) I^{\mathcal{E}}_L(\rho_\lambda,f).$$

Proposition 4.11 says this equals

$$\sum_{\rho} m^{L'}_{M'}(\tau_{\lambda'},\rho'_{\lambda'}) I^{\mathcal{E}}_L(\rho_\lambda,f).$$

But

$$I_L^{\mathcal{E}}(\rho_\lambda,f) = \widehat{I}_{L'}(\rho'_{\lambda'},f'_{G'}),$$

so the expression equals

$$\sum_{\rho} m_{M'}^{L'}(\tau_{\lambda'},\rho'_{\lambda'})\widehat{I}_{L'}(\rho'_{\lambda'},f'_{G'}).$$

If $\rho' \in \Sigma(M'(\mathbb{A})^1)$ is not the image of a $\rho$, then by 5.10.8 we have

$$I_{L'}(\rho'_{\lambda'},f') = 0.$$

The sum in the expression under consideration may be taken over all of $\Sigma(M'(\mathbb{A})^1)$, and we get

$$\sum_{\rho'} m_{M'}^{L'}(\tau_{\lambda'},\rho'_{\lambda'})\widehat{I}_{L'}(\rho'_{\lambda'},f'_{G'}).$$

We substitute this expression into (11.1.2). Noting that $\mathrm{d}\lambda = \mathrm{d}\lambda'$, we get

$$\sum_{L\in\mathcal{L}(M)}\ \sum_{\rho'\in\Sigma(M'(\mathbb{A})^1)} \int_{\varepsilon_M+i\mathfrak{a}_M^*/i\mathfrak{a}_L^*} m_{M'}^{L'}(\tau_{\lambda'},\rho'_{\lambda'})\widehat{I}_{L'}(\rho'_{\lambda'},f'_{G'})\,\mathrm{d}\lambda.$$

Because $\tau$ is unitary, we can apply the formula (6.1.1). Consequently, (11.1.2) equals

$$\widehat{I}_{M'}(\tau,f').$$

□

DEFINITION 11.2. Equation (2.3.1) defined $I^{\mathcal{E}}$ as the pullback to $G$ of the distribution $I$ on $G'$. Let

$$I_t^{\mathcal{E}}(f) = \widehat{I}_t(f'_{G'}) = \sum_{L\in\mathcal{L}'} |W_0^L|/|W_0^{G'}| \int_{\Pi(L,t)} a^L(\tau_{1,\lambda})\widehat{I}_L(\tau_{1,\lambda},f'_{G'})\,\mathrm{d}\tau_{1,\lambda}.$$

Applying (3.1.1) to $G'$, we obtain

$$I^{\mathcal{E}}(f) = \sum_{t\geq 0} I_t^{\mathcal{E}}(f).$$

By Proposition 5.18.4, $\widehat{I}_L(\tau,f'_{G'})$ vanishes unless $L$ is the image $L = M'$ of an element $M$ of $\mathcal{L}$. We can therefore rewrite the equation as

$$I_t^{\mathcal{E}}(f) = \sum_{M\in\mathcal{L}} |W_0^M|/|W_0^G| \int_{\Pi(M',t)} a^{M'}(\tau_{1,\lambda})\widehat{I}_{M'}(\tau_{1,\lambda},f'_{G'})\,\mathrm{d}\tau_{1,\lambda}. \tag{11.2.1}$$

We shall transform this into a formula involving the functions $a^{M,\mathcal{E}}$ and $I_M^{\mathcal{E}}$.

DEFINITION 11.3. For $M_1 \in \mathcal{L}^M$, $L \in \mathcal{L}^{M_1}$, $\sigma^1 \in \Pi(L(\mathbb{A})^1, t)$ and $\lambda \in i\mathfrak{a}_L^*/i\mathfrak{a}_M^*$, we have the induced representations $\sigma_\lambda^{M_1}$. We then let $\Pi_{\mathrm{disc}}^{\mathcal{E}}(M_1^1, t)$ be the subset of $\Pi(M_1(\mathbb{A})^1, t)$ consisting of irreducible constituents of such induced representations $\sigma_\lambda^{M_1}$ in which $\sigma_\lambda$ satisfies

(1) $a_{\mathrm{disc}}^{L,\mathcal{E}}(\sigma^1) \neq 0$ and
(2) there exists an element $s \in W^{M_1}(\mathfrak{a}_L)_{\mathrm{reg}}$ such that $s\sigma_\lambda = \sigma_\lambda$.

Comparing this with the earlier definition, 3.4(2), of $\Pi_{\mathrm{disc}}(M_1^1, t)$, the only difference is the additional $\mathcal{E}$. If $M_1 \neq G$, the induction assumption implies that $\Pi_{\mathrm{disc}}^{\mathcal{E}}(M_1^1, t)$ equals $\Pi_{\mathrm{disc}}(M_1^1, t)$.

Recalling $\Pi(M, t)$, see 3.4(3), we make the following definition.

DEFINITION 11.4. For $M_1 \in \mathcal{L}^M$, we set

$$\Pi_{M_1}^{\mathcal{E}}(M, t) = \{\tau_{1,\lambda}\,;\ \tau_1^1 \in \Pi_{\mathrm{disc}}^{\mathcal{E}}(M_1, t),\ \lambda \in i\mathfrak{a}_{M_1}^*/i\mathfrak{a}_M^*\}.$$

We then let

$$\Pi^{\mathcal{E}}(M, t) = \bigcup_{M_1 \in \mathcal{L}^M} \Pi_{M_1}^{\mathcal{E}}(M, t),$$

noting that the union is disjoint.

PROPOSITION 11.5. *Suppose that $t \geq 0$ and $f \in \mathbb{H}(G(\mathbb{A}))$. Then*

$$I_t^{\mathcal{E}}(f) = \sum_{M \in \mathcal{L}} |W_0^M|/|W_0^G| \int_{\Pi^{\mathcal{E}}(M,t)} a^{M,\mathcal{E}}(\varphi) I_M^{\mathcal{E}}(\varphi, f)\, d\varphi. \qquad (2.1.1\mathcal{E})$$

PROOF. It follows from the formulae above that

$$I_t^{\mathcal{E}}(f) = \sum_{\{M_1, M \in \mathcal{L}\,;\, M \supset M_1\}} |W_0^{M_1}|/|W_0^G| \sum_{\tau \in \Pi_{\mathrm{disc}}(M_1', t)} \int_{i\mathfrak{a}_{M_1}^*/i\mathfrak{a}_M^*} a_{\mathrm{disc}}^{M_1'}(\tau) m_{M_1'}^{M'}(\tau_{\lambda'}) \widehat{I}_{M'}(\tau_{\lambda'}, f_{G'}')\, \mathrm{d}\lambda'.$$

Substituting the expansion from Lemma 11.1 gives that $\widehat{I}_{M'}(\tau_{\lambda'}, f_{G'}')$ is

$$\sum_{L \in \mathcal{L}(M)} \sum_{\varphi_1 \in \Pi(M_1(\mathbb{A})^1, t)} \int_{\varepsilon_M + i\mathfrak{a}_M^*/i\mathfrak{a}_L^*} m_{M_1}^M(\tau_\lambda) m_M^L(\tau_{\lambda+\mu}, \varphi_{1,\lambda+\mu}) I_L^{\mathcal{E}}(\varphi_{1,\lambda+\mu}, f)\, \mathrm{d}\mu.$$

Initially, the sum should be over $\varphi \in \Pi(M(\mathbb{A})^1)$. Because $\lambda$ is in general position, the sum reduces to one over $\Pi(M_1(\mathbb{A})^1)$. Since $\|\,\mathrm{Im}(\nu_\lambda)\| = t$, the summand vanishes unless $\varphi_1 \in \Pi(M_1(\mathbb{A})^1, t)$.

Since $\mathrm{d}\lambda' = \mathrm{d}\lambda$, we can insert this expression into the formula for $I_t^{\mathcal{E}}(f)$, allowing us to write $I_t^{\mathcal{E}}(f)$ as the sum over $\{M_1, L\,;\, M_1 \subset L\}$ and $\varphi_1 \in \Pi(M_1(\mathbb{A})^1, t)$ of

$$\sum_{\{M\,;\,M_1\subset M\subset L\}} \int_{\varepsilon_M + i\mathfrak{a}_{M_1}^*/i\mathfrak{a}_L^*} \sum_{\tau\in\Pi_{\mathrm{disc}}(M_1',t)} a_{\mathrm{disc}}^{M_1'}(\tau) m_{M_1}^M(\tau_\lambda) m_M^L(\tau_\lambda, \varphi_{1,\lambda}) I_L^{\mathcal{E}}(\varphi_{1,\lambda}, f)\, \mathrm{d}\lambda. \tag{11.5.1}$$

We need a lemma.

LEMMA 11.6. *We have that* (11.5.1) *equals*

$$\sum_{\{L_1\,;\,M_1\subset L_1\subset L\}} \int_{\varepsilon_{L_1} + i\mathfrak{a}_{M_1}^*/i\mathfrak{a}_L^*} a_{\mathrm{disc}}^{M_1,\mathcal{E}}(\varphi_1) m_{M_1}^{L_1}(\sigma_\lambda) \tilde{m}_{L_1}^L(\sigma_\lambda, \varphi_{1,\lambda}) I_L^{\mathcal{E}}(\varphi_{1,\lambda}, f)\, d\lambda. \tag{11.6.1}$$

PROOF. The summand in (11.5.1) corresponding to a given $\tau$ vanishes unless $\delta(\tau, \varphi_1) \neq 0$. Fix $\sigma$. For another $\tau$, we can write

$$m_{M_1}^M(\tau_\lambda) = \sum_{\{L_1\,;\,M_1\subset L_1\subset M\}} m_{M_1}^{L_1}(\sigma_\lambda) \tilde{m}_{L_1}^M(\sigma_\lambda, \tau_\lambda).$$

We substitute this expression in (11.5.1) and take the sum over $L_1$ outside the integral. As in Lemma 4.9 the function $\tilde{m}_{L_1}^M(\sigma_\lambda, \tau_\lambda)$ is slowly increasing and regular for $\lambda$ in a cylinder about the imaginary space $i\mathfrak{a}_{M_1}^*$. Let $\varepsilon_{L_1}$ be a small regular point in $\mathfrak{a}_{L_1}^*$, depending only on $L_1$. The previous comment allows us to deform the contour of integration in (11.5.1) from $\varepsilon_M + i\mathfrak{a}_{M_1}^*/i\mathfrak{a}_L^*$ to $\varepsilon_{L_1} + i\mathfrak{a}_{M_1}^*/i\mathfrak{a}_L^*$. We may then bring the sum over $M$ in (11.5.1) inside the integral and the sum over $L_1$ and $\tau$. By (4.10.2) we have

$$\sum_{\{M\,;\,L_1\subset M\subset L\}} \tilde{m}_{L_1}^M(\sigma_\lambda, \tau_\lambda) m_M^L(\tau_\lambda, \varphi_{1,\lambda}) = \delta(\tau, \varphi_1) \tilde{m}_{L_1}^L(\sigma_\lambda, \varphi_{1,\lambda}).$$

Since $\varphi_1 \in \Pi(M_1(\mathbb{A})^1, t)$, we have

$$\sum_{\tau\in\Pi_{\mathrm{disc}}(M_1',t)} a_{\mathrm{disc}}^{M_1'}(\tau)\delta(\tau, \varphi_1) = a_{\mathrm{disc}}^{M_1,\mathcal{E}}(\varphi_1).$$

The lemma follows. □

LEMMA 11.7. (1) *If* $M_1 = G$ *then* (11.6.1) *simplifies to*

$$\sum_{\pi\in\Pi_{\mathrm{disc}}^{\mathcal{E}}(G,t)} a_{\mathrm{disc}}^{\mathcal{E}}(\pi) I_G^{\mathcal{E}}(\pi, f). \tag{11.7.1}$$

(2) *If* $M_1 \neq G$ *then* (11.6.1) *is given by*

$$\int_{i\mathfrak{a}^*_{M_1}/i\mathfrak{a}^*_L} a^{M_1,\mathcal{E}}_{\mathrm{disc}}(\varphi_1) m^L_{M_1}(\varphi_{1,\lambda}) I^{\mathcal{E}}_L(\varphi_{1,\lambda},f)\, d\lambda.$$

PROOF. (1) This is clear.

(2) In this case, we may use our induction assumption (part (iv) of the Main Theorem 3.11) that $a^{M_1,\mathcal{E}}_{\mathrm{disc}}(\varphi_1)$ equals $a^{M_1}_{\mathrm{disc}}(\varphi_1)$. We can assume then that $\varphi_1 \in \Pi_{\mathrm{disc}}(M_1,t)$. Lemma 4.9 tells us that $\tilde{m}^L_{L_1}(\sigma_\lambda,\varphi_{1,\lambda})$ is analytic for $\lambda$ in a cylinder about the imaginary space $i\mathfrak{a}^*_{M_1}$. We saw in Section 10 that

$$I^{\mathcal{E}}_L(\varphi_{1,\lambda},f) = I_L(\varphi_{1,\lambda},f).$$

Furthermore, Proposition 5.3.7 says that the function on the right is analytic for $\lambda$ near $i\mathfrak{a}^*_{M_1}$. Consequently, the contour of integration in (11.6.1) can be deformed from $\varepsilon_{L_1} + i\mathfrak{a}^*_{M_1}/i\mathfrak{a}^*_L$ to $i\mathfrak{a}^*_{M_1}/i\mathfrak{a}^*_L$. Taking the sum over $L_1$ in (11.6.1) inside the integral and applying Lemma 4.9 gives

$$\sum_{\{L_1\,;\,M_1\subset L_1\subset L\}} m^{L_1}_{M_1}(\sigma_\lambda)\tilde{m}^L_{L_1}(\sigma_\lambda,\varphi_{1,\lambda}) = m^L_{M_1}(\varphi_{1,\lambda}).$$

We obtain (2). □

Returning to the main proof, we see that $I^{\mathcal{E}}_t(f)$ is the sum of (11.7.1) and

$$\sum_L \sum_{\{M_1\subset L;\, M_1\neq G\}} |W^{M_1}_0|/|W^G_0| \sum_{\varphi_1\in\Pi_{\mathrm{disc}}(M_1,t)} \int_{i\mathfrak{a}^*_{M_1}/i\mathfrak{a}^*_L} a^{M_1,\mathcal{E}}_{\mathrm{disc}}(\varphi_1) m^L_{M_1}(\varphi_{1,\lambda}) I^{\mathcal{E}}_L(\varphi_{1,\lambda},f)\, \mathrm{d}\lambda.$$

By definition 3.10(2), this is

$$\sum_{L\in\mathcal{L}} |W^L_0||W^G_0|^{-1} \int_{\Pi^{\mathcal{E}}(L,t)} a^{L,\mathcal{E}}(\varphi) I^{\mathcal{E}}_L(\varphi,f)\, \mathrm{d}\varphi,$$

the formula of the proposition. □

PROPOSITION 11.8. *Suppose that* $t \geq 0$ *and* $f \in \mathbb{H}(G(\mathbb{A}))$. *Let* $f^1$ *denote the restriction of* $f$ *to* $G(\mathbb{A})^1$. *Then*

$$I^{\mathcal{E}}_t(f) - I_t(f) = \sum_{\pi\in\Pi(G(\mathbb{A})^1,t)} (a^{\mathcal{E}}_{\mathrm{disc}}(\pi) - a_{\mathrm{disc}}(\pi))\, \mathrm{tr}\,\pi(f^1).$$

PROOF. We will use (3.1.1) and (2.1.1$\mathcal{E}$) to expand the difference on the left. Let $M_1 \subset M \subset G$ with $M_1 \neq G$. Then Induction Assumption (iv) of Section 3 implies that the sets $\Pi^{\mathcal{E}}_{M_1}(M,t)$ and $\Pi_{M_1}(M,t)$ are the same, and for $\varphi \in \Pi_{M_1}(M,t)$ that

$$a^{M,\mathcal{E}}(\varphi) = a^M(\varphi).$$

These numbers vanish unless $\varphi$ is unitary. We saw in Section 10 that if $\varphi$ is unitary then $I_M^{\mathcal{E}}(\varphi,f) = I_M(\varphi,f)$. Hence the only contribution to the expansion of $I_t^{\mathcal{E}}(f) - I_f(f)$ comes from $M_1 = M = G$. But if $\pi \in \Pi(G(\mathbb{A})^1)$, then Theorem 10.6(4) implies that

$$I_G^{\mathcal{E}}(\pi,f) = I_G(\pi,f) = \operatorname{tr}\pi(f^1).$$

The proposition follows. □

## 11.9 Parallel Discussions

We have seen two parallel discussions, one dealing with the geometric sides of the two trace formulae, the other dealing with the spectral sides. We note a few similarities between the results. Aside from the duality of the statements comprising the Main Theorem, one compares Proposition 2.5.1 and Proposition 11.5, as well as Propositions 7.5 and 11.8. We will begin in the next section a study of the geometric side which has no parallel on the spectral side. The discussion will allow us to take advantage of the two different formulae for $I^{\mathcal{E}}(f) - I(f)$ given by Propositions 7.5 and 11.8.

# 12 The Map $\varepsilon_M$

The trace formula simplifies if at two places the orbital integrals of $f_v$ are supported on the elliptic set. We will use a similar idea, but will remove restrictions on $f$.

DEFINITION 12.1. (1) Let $M \in \mathcal{L}$. Consider functions $f \in \mathbb{H}(G(\mathbb{A}))$ of the form $f = \prod_v f_v, f_v \in \mathbb{H}(G(F_v))$, satisfying the following property. There are two finite places $v_1$ and $v_2$ where $G$ is unramified such that, unless $L \in \mathcal{L}$ contains a conjugate of $M$, the components $f_{v_i}$ $(i = 1, 2)$ satisfy

$$f_{v_i,L} = 0.$$

Then define $\mathbb{H}(G(\mathbb{A}), M)$ to be the subspace of $\mathbb{H}(G(\mathbb{A}))$ spanned by such $f$.

(2) If $S$ is a finite set of places containing $S_{\text{ram}}$ and at least two other places, define $\mathbb{H}(G(F_S), M)$ in the same way. Note that $\mathbb{H}(G(F_S), M)$ is a subspace of $\mathbb{H}(G(\mathbb{A}), M)$.

## 12.2 A New Induction Assumption

Let us fix $M \in \mathcal{L}$ with $M \neq G$. We have the induction assumption that the Main Theorem holds if $G$ is replaced by a proper Levi subgroup. We will make a **new induction assumption**. Namely, we assume, for $S$ containing $S_{\mathrm{ram}}$, $M \subsetneq L \in \mathcal{L}$, $\gamma \in L(F_S)$, and $f \in \mathbb{H}(G(F_S))$, that

$$I_L^{\mathcal{E}}(\gamma,f) = I_L(\gamma,f).$$

We will complete the proof of part (i) of the Main Theorem 2.7, that $I_M^{\mathcal{E}}(\gamma,f) = I_M(\gamma,f)$, in Section 16. The subgroup $M$ will remain fixed and the new induction assumption will be in force until part (i) is completed. The base of the induction is $M$ maximal and proper. In this case $L = G$, and the required equality is (5.6.2).

PROPOSITION 12.3. *Let $S$ be a finite set of valuations of $F$ that is sufficiently large, depending only on* supp$(f)$ *and* $V(f)$. *For* $f \in \mathbb{H}(G(\mathbb{A}), M)$, *the distribution*

$$I^{\mathcal{E}}(f) - I(f)$$

*is equal to the sum of*

$$|W(\mathfrak{a}_M)|^{-1} \sum_{\gamma \in (M(F))_{M,S}} a^M(S,\gamma)(I_M^{\mathcal{E}}(\gamma,f) - I_M(\gamma,f))$$

*and*

$$\sum_{\xi \in A_G(F)} \sum_{u \in (\mathcal{U}_G(F))_{G,S}} (a^{\mathcal{E}}(S,u) - a(S,u)) I_G(\xi u,f).$$

PROOF. Suppose $S$ is the disjoint union of $S_0 \supset S_{\mathrm{ram}}$ and $S_1$, $S_2$, where $S_i = \{v_i\}$ for unramified valuations $v_i$. We can assume that

$$f = \prod_{i=0}^{2} f_i$$

for $f_i \in \mathbb{H}(G(F_{S_i}))$, where for $L \in \mathcal{L}$ and $i = 1, 2$, we have

$$f_{i,L} = 0$$

unless $L$ contains a conjugate of $M$. For any $L \in \mathcal{L}$ and $\gamma \in L(F)$, two applications of (7.3.1) gives

$$I_L^{\mathcal{E}}(\gamma,f) - I_L(\gamma,f) = \sum_{i=0}^{2} (I_L^{\mathcal{E}}(\gamma,f_i) - I_L(\gamma,f_i)) \prod_{j \neq i} \widehat{I}_L^L(\gamma,f_{j,L}).$$

The product, and thus the expansion, vanishes unless $L$ contains a conjugate of $M$. For $w \in W_0$, 5.(2.14.1) implies

$$I_{wLw^{-1}}(w\gamma w^{-1}, f) = I_L(\gamma, f).$$

A similar assertion holds for $I_L^{\mathcal{E}}(\gamma, f)$. Unless $L$ is conjugate to $M$, the new induction assumption implies $I_L^{\mathcal{E}}(\gamma, f) = I_L(\gamma, f)$. The number of $L$ conjugate to $M$ is

$$|W_0^M||W_0^G|^{-1}|W(\mathfrak{a}_M)|^{-1}.$$

The proposition then follows from Proposition 7.5. □

We would like to prove that $I^{\mathcal{E}}(f)$ equals $I(f)$. The following proposition will allow us to apply the trace formula for $M$ to the difference in Proposition 12.3. Recall that the spaces $\mathbb{H}(G(F_S))^0$ were defined in Section 9.

PROPOSITION 12.4. *For every $S$ containing $S_{\mathrm{ram}}$ and $f \in \mathbb{H}(G(F_S))^0$, the function*

$$\gamma \mapsto I_M^{\mathcal{E}}(\gamma, f) - I_M(\gamma, f)$$

*is the orbital integral of a function on $M(F_S)$.*

We will actually consider the more general situation in which $S$ is a finite set of valuations with the closure property. The new induction assumption means that Theorem 8.1 applies for $L \in \mathcal{L}(M)$ with $L \neq M$. In particular, the numbers $\varepsilon_L(S)$ (see Theorem 8.1) are defined, and for all $\gamma \in L(F_S)$, and $f \in \mathbb{H}(G(F_S))$, we have

$$I_L^{\mathcal{E}}(\gamma, f) = \sum_{L_1 \in \mathcal{L}(L)} \widehat{I}_L^{L_1}(\gamma, \varepsilon_{L_1}(S) f_{L_1}).$$

PROPOSITION 12.5. *Let $S$ be a finite set of valuations with the closure property. Then for all $L \in \mathcal{L}(M)$, there is a unique map,*

$$\varepsilon_L : \mathbb{H}(G(F_S))^0 \to \mathbb{I}_{\mathrm{ac}}(L(F_S)),$$

*such that, for all $\gamma \in M(F_S)$ and $f \in \mathbb{H}(G(F_S))^0$, we have*

$$I_M^{\mathcal{E}}(\gamma, f) = \sum_{L \in \mathcal{L}(M)} \widehat{I}_M^L(\gamma, \varepsilon_L(f)). \tag{12.5.1}$$

*These maps have the following descent property. For $M_1 \subset M$, we have*

$$\varepsilon_M(f)_{M_1} = \sum_L d_{M_1}^G(M, L) \widehat{\varepsilon}_{M_1}^L(f_L). \tag{12.5.2}$$

*They also satisfy a splitting property. For $f = f_1 f_2$ as in (5.3.3), we have*

$$\varepsilon_M(f) = \sum_{L_1, L_2} d_M^G(L_1, L_2) \widehat{\varepsilon}_M^{L_1}(f_{1,L_1}) \widehat{\varepsilon}_M^{L_2}(f_{2,L_2}). \tag{12.5.3}$$

REMARK 12.6. (1) A function in $\mathbb{I}_{\mathrm{ac}}(M(F_S))$ is uniquely determined by its orbital integrals. Thus the uniqueness of $\varepsilon_L$ follows inductively from (12.5.1). Equation (12.5.1) implies that $\varepsilon_M$ is supported on characters. This justifies the usage of $\widehat{\varepsilon}_M(f_G) = \varepsilon_M(f)$, appearing in (12.5.2) and (12.5.3).

(2) We can think of the proposition as a weak version of part (i) of the Main Theorem 2.7. To see this, assume for $\gamma \in M(F_S)$ and $f \in \mathbb{H}(G(F_S))$ that

$$I_M^{\mathcal{E}}(\gamma, f) = I_M(\gamma, f)$$

when $S$ contains $S_{\mathrm{ram}}$. In this case, Theorem 8.1 holds for $M$, and for any $S$ we have

$$I_M^{\mathcal{E}}(\gamma, f) = \sum_{L \in \mathcal{L}(M)} \widehat{I}_M^L(\gamma, \varepsilon_L(S) f_L).$$

Proposition 12.5 follows from this equation, setting $\varepsilon_M(f) = \varepsilon_M(S) f_M$.

(3) The new induction assumption of Subsection 12.2 allows us to apply (2) to $L$ properly containing $M$. In particular, we can assume for $L \supsetneq M$ that

$$\varepsilon_L(f) = \varepsilon_L(S) f_L. \tag{12.6.1}$$

We can rewrite (12.5.1) as

$$\widehat{I}_M^M(\gamma, \varepsilon_M(f)) = I_M^{\mathcal{E}}(\gamma, f) - \sum_{L \supsetneq M} \widehat{I}_M^L(\gamma, \varepsilon_L(S) f_L). \tag{12.5.1*}$$

Let

$$\varepsilon_{M,M_1}(S) = \sum_{L \in \mathcal{L}(M_1)} d_{M_1}^G(M, L) \varepsilon_{M_1}^L(S).$$

If $M_1 \subsetneq M$, then formula (12.5.2) can be written as

$$\varepsilon_M(f)_{M_1} = \varepsilon_{M,M_1}(S) f_{M_1}. \tag{12.5.2*}$$

Set

$$\tilde{d}_M^G(S_1, S_2) = \sum_{\{L_1, L_2 \in \mathcal{L}(M)\,;\, L_1, L_2 \neq M\}} d_M^G(L_1, L_2) \varepsilon_M^{L_1}(S_1) \varepsilon_M^{L_2}(S_2).$$

Then formula (12.5.3) is

$$\varepsilon_M(f) = \varepsilon_M(f_1)f_{2,M} + f_{1,M}\varepsilon_M(f_2) + \tilde{d}_M^G(S_1, S_2)f_M. \tag{12.5.3*}$$

(4) Suppose $S$ contains $S_{\text{ram}}$ or is a single unramified valuation. Then $\varepsilon_L(S)$ vanishes unless $M \subsetneq L \subsetneq G$. In this case, for any $f \in \mathbb{H}(G(F_S))$, (12.5.1*) is

$$I_M^{\mathcal{E}}(\gamma, f) - I_M(\gamma, f) = \widehat{I}_M^M(\gamma, \varepsilon_M(f)). \tag{12.5.1**}$$

This is how $\varepsilon_M(f)$ will be applied to Proposition 12.3. Note that $\varepsilon_{M_1}^L(S)$ vanishes if $M_1 \subsetneq L \subsetneq G$. If $M_1 \subseteq M$, then (12.5.2*) becomes

$$\varepsilon_M(f)_{M_1} = 0. \tag{12.5.2**}$$

(5) Assume $S_0$ contains $S_{\text{ram}}$ and $S_1 = \{v\}$ consists of an unramified valuation. Let $S$ be the disjoint union of $S_0$ and $S_1$. If $f = f_0 f_1$, (12.5.3*) becomes

$$\varepsilon_M(f) = \varepsilon_M(f_0)f_{1,M} + f_{0,M}\varepsilon_M(f_1). \tag{12.5.3**}$$

Let $f_1$ be the characteristic function of $K_v$. Equation (12.5.1**) implies

$$\varepsilon_M(f_1) = 0.$$

Consequently, $\varepsilon_M$ can be extended to a map from $\mathbb{H}(G(\mathbb{A}))^0$ to $\mathbb{I}_{\text{ac}}(M(\mathbb{A}))$.

## 12.7 *Proof of Proposition* 12.5

PROOF. We will now begin the proof of Proposition 12.5. It will be completed in the next section. Fix $S$ and $f$. Recall that in Lemma 8.2, for $\gamma \in M(F_S)$ and $f \in \mathbb{H}(G(F_S))$, we set

$$\varepsilon_M(\gamma, f) = I_M^{\mathcal{E}}(\gamma, f) - \sum_{L \supsetneq M} \widehat{I}_M^L(\gamma, \varepsilon_L(S)f_L).$$

This is equal to the right side of (12.5.1*). The existence of $\varepsilon_M$ is equivalent to $\varepsilon_M(\gamma, f)$ being an orbital integral in $\gamma$ of a function in $\mathbb{H}_{\text{ac}}(M(F_S))$. We will first examine the properties (12.5.2) and (12.5.3).

Lemma 8.2 holds under the new induction assumption we made in Subsection 12.2. Properties (12.5.2) and (12.5.3) will follow from (8.1.2*) and (8.1.3*), once we have proved the existence of $\varepsilon_M(f)$. Formula (8.1.3*) provides a formula for $\varepsilon_M(\gamma, f)$ in terms of functions defined for a valuation $v \in S$. This reduces the proof of the existence of $\varepsilon_M(f)$ to the case $S = \{v\}$. We will treat the real and $p$-adic cases separately in the next section. We will now give some properties of $\varepsilon_M(\gamma, f)$.

LEMMA 12.8. *If* $\gamma \in M(F_S)$, *then*

$$\varepsilon_M(\gamma, f) = \lim_{a \to e} \varepsilon_M(a\gamma, f).$$

PROOF. We apply formulae (5.1.1) and (5.1.1$\mathcal{E}$) to $\varepsilon_M(\gamma, f)$. Then $\varepsilon_M(\gamma, f)$ equals the limit, as $a$ approaches 1 through regular elements in $A_M(F_S)$, of

$$\sum_{L_1 \supset M} m_M^{L_1}(\gamma, a) \left( I_{L_1}^{\mathcal{E}}(a\gamma, f) - \sum_{\{L \in \mathcal{L}(L_1)\,;\, L \neq M\}} \widehat{I}_{L_1}^{L}(a\gamma, \varepsilon_L(f)) \right)$$

If $L_1 \neq M$, the sum on the right is over $\mathcal{L}(L_1)$, and the expression in parentheses vanishes.

If $L_1 = M$, the expression in parentheses is $\varepsilon_M(a\gamma, f)$. The lemma follows. □

## 12.9 Limit Case

The formula is consistent with $\varepsilon_M(\gamma, f)$ being the orbital integral of a function on $M(F_S)$. The lemma allows us to only consider $\gamma$ such that $M_\gamma = G_\gamma$. Properties (5.3.1) and (5.3.1$\mathcal{E}$) tell us that we only need to consider $\gamma \in M(F_S)$ that are $G$-regular. We only need to prove for $\gamma \in M(F_S) \cap G_{\text{reg}}$ that

$$\gamma \mapsto \varepsilon_M(\gamma, f)$$

equals the orbital integral in $\gamma$ of a function in $\mathbb{H}_{\text{ac}}(M(F_S))$. We continue to assume that $S = \{v\}$.

Recall that we discussed the distributions ${}^cI_M(\gamma)$ and ${}^cI_M^{\mathcal{E}}(\gamma)$ in Section 10.

DEFINITION 12.10. Let $\gamma \in M(F_S) \cap G_{\text{reg}}$. Set

$${}^c\varepsilon_M(\gamma, f) = {}^cI_M^{\mathcal{E}}(\gamma, f) - \sum_{L \in \mathcal{L}_0(M)} \varepsilon_M^L(S)\, {}^cI_L(\gamma, f).$$

The advantage of ${}^c\varepsilon_M(\gamma, f)$ is that, according to Lemma 5.4.10, it has bounded support as a function of $\gamma$ in the space of $M(F_S)$-orbits in $M(F_S)$.

There is a formula relating $\varepsilon_M(\gamma, f)$ and ${}^c\varepsilon_M(\gamma, f)$.

PROPOSITION 12.11. *Suppose that* $\gamma \in M(F_S) \cap G_{\text{reg}}$. *Then*

$$\varepsilon_M(\gamma, f) - {}^c\varepsilon_M(\gamma, f) = \widehat{I}_M^M(\gamma, \theta_M^{\mathcal{E}}(f) - \theta_M(f)) = \widehat{I}_M^M(\gamma, {}^c\theta_M(f) - {}^c\theta_M^{\mathcal{E}}(f)).$$

PROOF. By definition,

$$\varepsilon_M(\gamma, f) = I_M^{\mathcal{E}}(\gamma, f) - \sum_{L \supsetneq M} \widehat{I}_M^L(\gamma, \varepsilon_L(S) f_L).$$

Applying the descent properties (5.3.2) and (8.1.2) gives

$$\varepsilon_M(\gamma,f) = I_M^{\mathcal{E}}(\gamma,f) - \sum_{L_1 \in \mathcal{L}_0(M)} \varepsilon_M^{L_1}(S) I_{L_1}(\gamma,f).$$

By (10.3.2) and (10.3.2$\mathcal{E}$),

$$\varepsilon_M(\gamma,f) - {}^c\varepsilon_M(\gamma,f) = (I_M^{\mathcal{E}}(\gamma,f) - {}^cI_M^{\mathcal{E}}(\gamma,f)) - \sum_{L_1 \in \mathcal{L}_0(M)} \varepsilon_M^{L_1}(S)(I_{L_1}(\gamma,f) - {}^cI_{L_1}(\gamma,f))$$

$$= \sum_{L \in \mathcal{L}_0(M)} \left( {}^c\widehat{I}_M^{L,\mathcal{E}}(\gamma, \theta_L^{\mathcal{E}}(f)) - \sum_{L_1 \in \mathcal{L}^L(M)} \varepsilon_M^{L_1}(S) \cdot {}^c\widehat{I}_{L_1}^L(\gamma, \theta_L(f)) \right).$$

Consider a summand corresponding to $L \supsetneq M$. Using the induction assumption in Subsection 12.2, we can apply Theorem 10.6 and Corollary 10.20. But then such a summand vanishes, since

$${}^c\widehat{I}_M^{L,\mathcal{E}}(\gamma, \theta_L^{\mathcal{E}}(f)) = {}^c\widehat{I}_M^{L,\mathcal{E}}(\gamma, \theta_L(f)) = \sum_{L_1 \in \mathcal{L}^L(M)} \varepsilon_M^{L_1}(S) \cdot {}^c\widehat{I}_{L_1}^L(\gamma, \theta_L(f)).$$

If $L = M$, the summand is

$$\widehat{I}_M^M(\gamma, \theta_M^{\mathcal{E}}(f) - \theta_M(f)).$$

The first identity of the proposition follows. Equations (10.3.1) and (10.3.1$\mathcal{E}$), Theorem 10.6, and the induction assumption of Subsection 12.2 imply that

$$\theta_M^{\mathcal{E}}(f) - \theta_M(f) = {}^c\theta_M(f) - {}^c\theta_M^{\mathcal{E}}(f).$$

The second identity of the proposition is a consequence of this formula. □

# 13 Cancellation of Singularities

## *13.1 Proof of Proposition* 12.5 *continued*

This section continues the proof of Proposition 12.5. More specifically, we need to show for a valuation $v$ of $F$, a function $f \in \mathbb{H}(G(F_v))^0$, and $\gamma \in M(F_v) \cap G_{\mathrm{reg}}$ that the function

$$\gamma \mapsto \varepsilon_M(\gamma,f)$$

is the orbital integral of a function in $\mathbb{H}_{\mathrm{ac}}(M(F_v))$.

PROPOSITION. *Suppose that $v$ is non-Archimedean. Fix $f \in \mathbb{H}(G(F_v))^0$. Let $\sigma \in M(F_v)$ be semisimple. Then for $\gamma \in \sigma M_\sigma(F_v) \cap G_{\mathrm{reg}}$, we have*

$$\varepsilon_M(\gamma,f) \overset{(M,\sigma)}{\sim} 0. \tag{13.1.1}$$

*That is, $\varepsilon_M(\gamma,f)$ is the orbital integral of a function $\varepsilon_M(f)$ in $\mathbb{H}_{\mathrm{ac}}(M(F_v))$.*

PROOF. Assume that $\mathfrak{a}_{M_\sigma} \neq \mathfrak{a}_M$. Then $\sigma M_\sigma(F_v)$ is contained in a proper Levi subgroup $M_1 \subsetneq M$. We saw in the previous section that if $\gamma \in M_1(F_v) \cap G_{\mathrm{reg}}$, then

$$\varepsilon_M(\gamma,f) = \varepsilon_{M,M_1}(v)\widehat{I}^{M_1}_{M_1}(\gamma,f_{M_1}).$$

In particular, the function on the right is an orbital integral on $M(F_v)$, and (13.1.1) holds in this case.

Assume now that $\mathfrak{a}_{M_\sigma} = \mathfrak{a}_M$. By (5.4.2) and (5.4.2$\mathcal{E}$) of Proposition 9.8, the function

$$\varepsilon_M(\gamma,f) = I^{\mathcal{E}}_M(\gamma,f) - \sum_{L_1 \supsetneq M} \widehat{I}^{L_1}_M(\gamma, \varepsilon_{L_1}(f))$$

is $(M,\sigma)$-equivalent to

$$\sum_{L\in\mathcal{L}(M)} \sum_{\delta\in\sigma(\mathcal{U}_{L_\sigma}(F_v))} g^L_M(\gamma,\delta)\left(I^{\mathcal{E}}_L(\delta,f) - \sum_{\{L_1\in\mathcal{L}(L)\,;\,L_1\neq M\}} \widehat{I}^{L_1}_L(\delta,\varepsilon_{L_1}(f))\right).$$

We can assume that $L \supsetneq M$, since the functions $g^M_M(\gamma,\delta)$ are $(M,\sigma)$-equivalent to 0. If $L \supsetneq M$, the sum over $L_1$ is actually over $\mathcal{L}(L)$. Thus the expression in parentheses vanishes, and (13.1.1) holds in general.

By Proposition 12.11, $\varepsilon_M(\gamma,f)$ equals the sum of ${}^c\varepsilon_M(\gamma,f)$ and the invariant orbital integral in $\gamma$ of ${}^c\theta_M(f) - {}^c\theta^{\mathcal{E}}_M(f)$. The functions ${}^c\theta_M(f)$ and ${}^c\theta^{\mathcal{E}}_M(f)$ are elements of $\widetilde{\mathbb{I}}_{\mathrm{ac}}(M(F_v))$. Recall that in 5.4.2 we mentioned that if $v$ is non-Archimedean, then $\widetilde{\mathbb{I}}_{\mathrm{ac}}(M(F_v))$ and $\mathbb{I}_{\mathrm{ac}}(M(F_v))$ coincide. In particular, for $\gamma \in \sigma M_\sigma(F_v)$, we have

$$\widehat{I}^M_M(\gamma, {}^c\theta_M(f) - {}^c\theta^{\mathcal{E}}_M(f)) \overset{(M,\sigma)}{\sim} 0.$$

For $\gamma \in \sigma M_\sigma(F_v) \cap G_{\mathrm{reg}}$, (13.1.1) implies that

$${}^c\varepsilon_M(\gamma,f) \overset{(M,\sigma)}{\sim} 0.$$

By definition, then, ${}^c\varepsilon_M(\gamma,f)$ equals the orbital integral of a function in $\mathbb{H}(M(F_v))$ for any regular element $\gamma \in \sigma M_\sigma(F_v)$ close to $\sigma$. By Proposition 5.4.10, ${}^c\varepsilon_M(\gamma,f)$ has bounded support as a function of $\gamma$ in the space of $M(F_v)$-orbits in $M(F_v)$. A partition of unity argument shows that only finitely many $\sigma$ are needed, so that ${}^{c}\varepsilon_M(\gamma, f)$ is everywhere equal to an orbital integral. That is, there is a function ${}^{c}\varepsilon_M(f) \in \mathbb{I}(M(F_v))$ such that

$$ {}^{c}\varepsilon_M(\gamma, f) = \widehat{I}^M_M(\gamma, {}^{c}\varepsilon_M(f)). $$

Therefore

$$ \varepsilon_M(f) = {}^{c}\varepsilon_M(f) + {}^{c}\theta_M(f) - {}^{c}\theta^{\mathcal{E}}_M(f) $$

is the required function. □

PROPOSITION 13.2. *Suppose that $v$ is Archimedean. Let $\sigma$ be a semisimple element in $M(F_v)$. Regard $G(F_v)$ as a real Lie group. Let $f \in \mathbb{H}(G(F_v))^0$ be a fixed function. Then* (13.1.1) *holds.*

PROOF. Note that, in this case, $\mathbb{H}(G(F_v))^0 = \mathbb{H}(G(F_v))$. The main step in proving the existence of $\varepsilon_M(f)$ is showing that $\varepsilon_M(\gamma, f)$ behaves like the orbital integral of a Schwarz function on $M(F_v)$. Let $T$ be a maximal torus in $G$ defined over $F_v$. So, $T$ is the centralizer in $G$ of $\gamma \in G_{\mathrm{reg}}(F_v)$. We will prove that the restriction of $\varepsilon_M(\gamma, f)$ to $T(F_v)$ satisfies two conditions. The first condition is that any derivative of $\varepsilon_M(\gamma, f)$ is locally bounded on $T(F_v)$. The second is that the function has appropriate behavior across the singular hyperspaces of $T(F_v)$. These will be two out of three conditions of Theorem 4.7 of [She79] that are equivalent to $\varepsilon_M(\gamma, f))$ being an orbital integral of a Schwarz function on $M(F_v)$. For the third condition, see Subsection 13.7 below.

## 13.3 First Condition

We begin the discussion of the first condition by considering the differential equation satisfied by $\varepsilon_M(\gamma, f)$. Set $\mathcal{Z}_v = \mathcal{Z}(G(F_v))$. Let $z \in \mathcal{Z}_v$. By (5.4.1) and its $\mathcal{E}$-analogue (see [AC89], Lemma 2.4.4), the function

$$ \varepsilon_M(\gamma, zf) = I^{\mathcal{E}}_M(\gamma, zf) - \sum_{L_1 \supsetneq M} \widehat{I}^{L_1}_M(\gamma, (zf)_{L_1})\varepsilon_{L_1}(v) $$

is equal to

$$ \sum_{L \in \mathcal{L}(M)} \partial^L_M(\gamma, z_L) \left( I^{\mathcal{E}}_L(\gamma, f) - \sum_{\{L_1 \in \mathcal{L}(L)\,;\, L_1 \neq M\}} \widehat{I}^{L_1}_L(\gamma, f_{L_1})\varepsilon_{L_1}(v) \right). $$

If $L \neq M$, the sum over $L_1$ is over all of $\mathcal{L}(L)$, and the expression in parentheses vanishes. If $L = M$, the expression in parentheses is $\varepsilon_M(\gamma, f)$. Let $\partial(h_T(z))$ be the invariant differential operator on $T(F_v)$ obtained from $z$ by the Harish-Chandra map. Lemma 12.4 of [Ar88.1] gives

$$\partial_M^M(\gamma, z_M) = \partial(h_T(z)).$$

It follows that for $\gamma \in T_{\mathrm{reg}}(F_v)$, we have

$$\varepsilon_M(\gamma, zf) = \partial(h_T(z))\varepsilon_M(\gamma, f). \tag{13.3.1}$$

This differential equation can be combined with a technique of Harish-Chandra ([HC57]) to establish that the derivatives of $\varepsilon_M(\gamma, f)$ are locally bounded. We will sketch the argument.

## 13.4 A Bound

Let $\Omega$ be a compact subset of $T(F_v)$. Set $\Omega_{\mathrm{reg}} = \Omega \cap T_{\mathrm{reg}}(F_v)$. Recall 5.(2.18.1). It says that, if $\partial(u)$ is an invariant differential operator on $T(F_v)$, there are positive numbers $c(f)$ and $q$ such that, for all $\gamma \in \Omega_{\mathrm{reg}}$, we have

$$|\partial(u)I_M(\gamma, f)| \le c(f)|D^G(\gamma)|^{-q}.$$

The numbers depend on $\partial(u)$, but $q$ is independent of $f$. A similar assertion holds for $I_M^{\mathcal{E}}(\gamma, f)$ and therefore also for $\varepsilon_M(\gamma, f)$. Suppose $\partial(u_1), \dots, \partial(u_n)$ are generators over $\partial(h_T(\mathcal{Z}_v))$ of the module of differential operators on $T(F_v)$ with constant coefficients. Then there exists $z_i \in \mathcal{Z}_v$ such that any $\partial(u)$ can be written

$$\partial(u) = \partial(u_1)h_T(z_1) + \cdots + \partial(u_n)h_T(z_n).$$

Applying the differential equation (13.3.1) gives

$$\partial(u)\varepsilon_M(\gamma, f) = \sum_{i=1}^{n} \partial(u_i)\varepsilon_M(\gamma, z_i f).$$

It follows that, for $\gamma \in T_{\mathrm{reg}}(F_v)$, there is a number $q$, independent of $\partial(u)$, such that

$$|\partial(u)\varepsilon_M(\gamma, f)| \le c(f)|D^G(\gamma)|^{-q}.$$

We will use this inequality to show that $\partial(u)\varepsilon_M(\gamma, f)$ is bounded on $\Omega_{\mathrm{reg}}$. The result is a consequence of the following lemma of Langlands' lecture "Cancellation of singularities at the real places" (I.A.S., p. 21–22). See [LW13].

LEMMA 13.5. *Let $\lambda_1, \dots, \lambda_k$ be a finite set of linear forms on $\mathbb{R}^m$. Let $\phi$ be a smooth function on the set*

$$B' = \left\{ \xi \in \mathbb{R}^m;\ \|\xi\| \le 1,\quad \prod_{i=1}^{k} \lambda_i(\xi) \neq 0 \right\}.$$

*Suppose for any differential operator $\Delta$ with constant coefficients on $\mathbb{R}^m$ that, for $\xi \in B'$, we have*

$$|\Delta\phi(\xi)| \leq c_\Delta \left|\prod_{i=1}^{k} \lambda_i(\xi)\right|^{-q},$$

*for $q$ independent of $\Delta$. Then $\Delta\phi$ is bounded on $B'$.*

This completes the discussion of the first condition stated in the first paragraph of the proof of Proposition 13.2.

Let us now discuss the second condition of 13.2, namely, the behavior of $\varepsilon_M(\gamma, f)$ across the singular hyperplanes. Recall the following definition:

DEFINITION 13.6. A semisimple element $\sigma \in T(F_v)$ is *semiregular* if the derived group of $G_\sigma$ is three dimensional.

The condition is equivalent to requiring that for such $\sigma$ and $G$-regular $\gamma$ in $T(F_v)$ near $\sigma$ that $\varepsilon_M(\gamma, f)$ be $(M, \sigma)$-equivalent to 0. If $M_\sigma = G_\sigma$ this fact follows from (5.3.1) and (5.3.1$\mathcal{E}$).

Suppose that $\sigma \in T(F_v)$ is semiregular with $M_\sigma = G_\sigma$. Then $\sigma$ lies on a hyperspace in $T(F_v)$ defined by a real root $\beta$ relative to the action of $T(F_v)$ on the Lie algebra of $G(F_v)$. The coroot $\beta^\vee$ belong to the Lie algebra of $T(F_v)$. For $r \in \mathbb{R}$ set

$$\gamma_r = \sigma \exp(r\beta^\vee).$$

It suffices to show that for any invariant differential operator $\partial(u)$, the function $\partial(u)\varepsilon_M(\gamma_r, f)$ is smooth at $r = 0$. Associated to $T$ and $\beta$, we have a Cayley transform

$$C : T_0 \to T_{01}.$$

This is an inner automorphism on $G_\sigma$ which maps $T_0$ to a torus $T_{01}$ in $G_\sigma$ which is $F_v$-anisotropic modulo the center of $G_\sigma$. Let $M_1 \in \mathcal{L}(M)$ be the Levi subgroup such that

$$A_{M_1} = A_M \cap T_{01}.$$

Then $C\beta$ is a noncompact imaginary root of $M_1(F_v)$. For $s \in \mathbb{R}$ small and nonzero,

$$\delta_s = \sigma \exp(sC\beta^\vee)$$

is a $G$-regular point in

$$T_1(F_v) = \sigma T_{01}(F_v).$$

Let $w_\beta$ be the reflection about $\beta^\vee$. Then

$$\partial(u_1) = \partial(Cw_\beta u - Cu)$$

is an invariant differential operator on $T_1(F_v)$. For $\gamma \in T_{\mathrm{reg}}(F_v)$, let

$$I_M^\beta(\gamma,f) = I_M(\gamma,f) + \|\beta_M^\vee\| \log(|\gamma^\beta - \gamma^{-\beta}|) I_{M_1}(\gamma,f).$$

Then 5.(2.17.1) implies that

$$\lim_{r\to 0+} (\partial(u) I_M^\beta(\gamma_r,f) - \partial(u) I_M^\beta(\gamma_{-r},f)) = n_\beta \lim_{s\to 0} \partial(u_1) I_{M_1}(\delta_s,f).$$

Here $n_\beta$ is the cosine of the angle between $\beta^\vee$ and $\mathfrak{a}_M$, and $\|\beta_M^\vee\|$ denotes the norm of the projection $\beta_M^\vee$ of $\beta^\vee$ onto $\mathfrak{a}_M$. We can apply the same formula to $G'$. The objects discussed can be mapped to corresponding objects on $G'$. For $\gamma \in T_{\mathrm{reg}}(F_v)$, let

$$I_M^{\mathcal{E},\beta}(\gamma,f) = I_M^{\mathcal{E}}(\gamma,f) + \|\beta_M^\vee\| \log(|\gamma^\beta - \gamma^{-\beta}|) I_{M_1}^{\mathcal{E}}(\gamma,f).$$

After transferring the objects to $G'$, we see that

$$\lim_{r\to 0+} (\partial(u) I_M^{\mathcal{E},\beta}(\gamma_r,f) - \partial(u) I_M^{\mathcal{E},\beta}(\gamma_{-r},f)) = n_\beta \lim_{s\to 0} \partial(u_1) I_{M_1}^{\mathcal{E},\beta}(\delta_s,f).$$

Now

$$\lim_{r\to 0+} (\partial(u)\varepsilon_M(\gamma_r,f) - \partial(u)\varepsilon_M(\gamma_{-r},f)) \tag{13.6.1}$$

equals the jump at $r = 0$ of

$$\partial(u) I_M^{\mathcal{E}}(\gamma_r,f) - \sum_{L_1 \supsetneq M} \partial(u) \widehat{I}_M^{L_1}(\gamma_r, \varepsilon_{L_1}(f)).$$

Suppose $L_1$ does not contain $M_1$. Then $\sigma$ is regular in $L_1$, and the summand will be smooth at $r = 0$. We may therefore take the sum over $L_1 \in \mathcal{L}(M_1)$. The induction hypothesis implies that

$$\|\beta_M^\vee\| \log(|\gamma^\beta - \gamma^{-\beta}|) \left( I_{M_1}^{\mathcal{E}}(\gamma,f) - \sum_{L_1 \in \mathcal{L}(M_1)} \widehat{I}_{M_1}^{L_1}(\gamma, \varepsilon_{L_1}(f)) \right) = 0.$$

Therefore (13.6.1) equals the jump at $r = 0$ of the expression

$$\partial(u) I_M^{\mathcal{E},\beta}(\gamma_r,f) - \sum_{L_1 \supset M_1} \partial(u) \widehat{I}_M^{L_1,\beta}(\gamma_r, \varepsilon_{L_1}(f)).$$

Applying the two formulae, we can write (13.6.1) as

$$n_\beta \lim_{s\to 0} \partial(u_1)\left( I^{\mathcal{E}}_{M_1}(\delta_s,f) - \sum_{L_1\in\mathcal{L}(M_1)} \widehat{I}^{L_1}_{M_1}(\delta_s,\varepsilon_{L_1}(f))\right).$$

Since $M_1 \supsetneq M$, the term in parentheses vanishes by the induction assumption. Thus the function

$$\partial(u)\varepsilon_M(\gamma_r,f)$$

is smooth at $r=0$. Equivalently, for $\gamma \in T(F_v)\cap G_{\mathrm{reg}}$, we have

$$\varepsilon_M(\gamma,f) \overset{(M,\sigma)}{\sim} 0.$$

This was the second condition of 13.2.

## 13.7 Descent Properties

We have verified two of the three conditions of Theorem 4.7 of [She79] that are equivalent to $\varepsilon_M(\gamma,f)$ being an orbital integral of a Schwarz function on $M(F_v)$. Rather than work directly with the Schwarz space, we will proceed inductively as in [She79].

Recall the descent properties (5.3.2), (5.3.2$\mathcal{E}$), (8.1.2*), and (12.5.2*). They are local properties. If $S=\{v\}$, they hold if $M_1 \in \mathcal{L}_v$, rather than just $\mathcal{L}$. Suppose that $M_1 \subsetneq M$ is an element of $\mathcal{L}_v$. The induction assumption of Subsection 12.2 allows us to apply (8.1.2*). For $L \subsetneq G$, we are assuming that

$$\widehat{\varepsilon}^{L}_{M_1}(\gamma,f_L) = \widehat{I}^{M_1}_{M_1}(\gamma,\widehat{\varepsilon}^{L}_{M_1}(f_L)) = \varepsilon^{L}_{M_1}(v)\widehat{I}^{M_1}_{M_1}(\gamma,f_{M_1}).$$

Let

$$\varepsilon_{M,M_1}(v) = \sum_{L\in\mathcal{L}(M_1)} d^G_{M_1}(M,L)\varepsilon^L_{M_1}(v).$$

Then for $\gamma\in M_1(F_v)\cap G_{\mathrm{reg}}$, we have

$$\varepsilon_M(\gamma,f) = \varepsilon_{M,M_1}(v)\widehat{I}^{M_1}_{M_1}(\gamma,f_{M_1}).$$

LEMMA 13.8. *The number $\varepsilon_{M,M_1}(v)$ is independent of $M_1$.*

PROOF. It suffices to show that, if $M_1' \in \mathcal{L}_v$ satisfies $M_1'\subset M_1$, then

$$\varepsilon_{M,M_1'}(v) = \varepsilon_{M,M_1}(v).$$

We can assume, for $L \subsetneq G$, that $\varepsilon^L_{M_1}(v)$ satisfies the descent property (8.1.2). This implies that

$$\varepsilon_{M,M_1}(v) = \sum_{L' \in \mathcal{L}(M'_1)} \varepsilon^{L'}_{M'_1}(v) \sum_{\{L \in \mathcal{L}(M)\,;\, L \supset L'\}} d^L_{M'_1}(M_1, L') d^G_{M_1}(M, L).$$

By 5.(7.11.1), this equals

$$\sum_{L' \in \mathcal{L}(M'_1)} \varepsilon^{L'}_{M'_1}(v) d^G_{M'_1}(M, L') = \varepsilon_{M,M'_1}(v),$$

as required. □

## 13.9 Reduction

The previous lemma justifies writing

$$\varepsilon_M(v) = \varepsilon_{M,M_1}(v).$$

For any $\gamma \in M_1(F_v) \cap G_{\mathrm{reg}}$, we have shown that

$$\varepsilon_M(\gamma, f) = \varepsilon_{M,M_1}(v) \widehat{I}^{M_1}_{M_1}(\gamma, f_{M_1}) = \varepsilon_M(v) \widehat{I}^M_M(\gamma, f_M).$$

If $M$ is not minimal over $F_v$, we have seen how to define $\varepsilon_M(v)$. We are still unable to apply Theorem 8.1 to $M$. Set

$$\varepsilon'_M(v) = \begin{cases} \varepsilon_M(v), & \text{if } M \text{ is not minimal over } F_v \\ 0, & \text{if } M \text{ is minimal over } F_v. \end{cases}$$

Define

$$\varepsilon'_M(\gamma, f) = \varepsilon_M(\gamma, f) - \varepsilon'_M(v) \widehat{I}^M_M(\gamma, f_M).$$

Since $\widehat{I}^M_M(\gamma, f_M)$ is just the orbital integral of a function in $\mathbb{H}(M(F_v))$, the function $\gamma \mapsto \varepsilon'_M(\gamma, f)$ satisfies the same two conditions established for $\varepsilon_M(\gamma, f)$. Observe that if $M_1 \subsetneq M$ is as above and $\gamma \in M_1(F_v) \cap G_{\mathrm{reg}}$, then

$$\varepsilon'_M(\gamma, f) = 0.$$

## 13.10 Elliptic Case

Suppose $T$ is an elliptic maximal torus of $M$ over $F_v$. That is, $T$ is not contained in any proper Levi subgroup $M_1 \subsetneq M$. In this case, there is a locally constant function

$$c : T_{\mathrm{reg}}(F_v) \to \{z \in \mathbb{C}\,;\ |z| = 1\}$$

such that the function

$$\widetilde{\varepsilon}_M(\gamma,f) = c(\gamma)\varepsilon'_M(\gamma,f),$$

defined initially for $\gamma \in T_{\mathrm{reg}}(F_v)$, extends to a smooth function on $T(F_v)$. Furthermore, $\widetilde{\varepsilon}_M(\gamma,f)$ is skew-symmetric under the Weyl group

$$W^M(T) = N_T(M)/Z_T(M).$$

This property follows from the two conditions and the vanishing property above.

## 13.11 Orthogonal Basis

Observe that for $X \in \mathfrak{a}_M$, the set

$$T(F_v)^X = M(F_v)^X \cap T(F_v) = \{x \in M(F_v)\,;\ H_M(x) = X\} \cap T(F_v)$$

is compact.

For $M_1 \in \mathcal{L}_v$, $M_1 \subsetneq M$ and $\tau_1 \in \Pi_{\mathrm{temp}}(M_1(F_v))$, we have the induced representation $\tau_1^M$. Let us write $\Pi_{\mathrm{disc}}(M(F_v))$ for the set of representations in $\Pi_{\mathrm{temp}}(M(F_v))$ which are not of this form.

The function $c(\gamma)$ has an additional property. Namely, the set of functions of $\gamma \in T(F_v)$ given by

$$\Phi_\tau(\gamma) = c(\gamma)|D^M(\gamma)|^{1/2}\,\mathrm{ch}_\tau(\gamma),$$

for $\tau$ ranging over a set of representatives of $i\mathfrak{a}_M^*$-orbits in $\Pi_{\mathrm{disc}}(M(F_v))$, forms an orthogonal basis of a Hilbert space of functions on $T(F_v)$ which includes $\widetilde{\varepsilon}_M(\cdot,f)$. Indeed, each $\Phi_\tau$ is a multiple of the skew-symmetrization of a *one-dimensional* character on $T(F_v)$. This is a well-known result of Harish-Chandra. For $\tau \in \Pi_{\mathrm{disc}}(M(F_v))$, define

$$\varepsilon'_M(f,\tau,X) = |W^M(T)|^{-1}\int_{T(F_v)^X} \widetilde{\varepsilon}_M(\gamma,f)\overline{\Phi_\tau(\gamma)}\,d\gamma. \tag{13.11.1}$$

If $\tau \in \Pi_{\mathrm{temp}}(M(F_v))$ does not belong to $\Pi_{\mathrm{disc}}(M(F_v))$, we set $\varepsilon'_M(f,\tau,X) = 0$.

LEMMA 13.12. *The function $\varepsilon'_M(f,\tau,X)$ is a Schwarz function of $X$.*

PROOF. Since the function is smooth, it suffices to show that for any invariant differential operator $\Delta$ on $\mathfrak{a}_M$, the function of $X \in \mathfrak{a}_M$ given by

$$\Delta\varepsilon'_M(f,\tau,X)$$

is rapidly decreasing. Observe that $\varepsilon'_M(f,\tau,X)$ equals the integral over the set of $M(F_v)$-orbits in $M(F_v)^X$ of the product of $\varepsilon'_M(\gamma,f)$ with $|D^M(\gamma)|^{1/2}\Theta_\tau(\gamma)$.

Proposition 12.11 says that for any $G$-regular $\gamma \in M(F_v)^X$, we can write $\varepsilon'_M(\gamma,f)$ as the sum

$${}^c\varepsilon_M(\gamma,f) - \varepsilon'_M(v)\widehat{I}^M_M(\gamma,f_M) + \widehat{I}^M_M(\gamma,{}^c\theta_M(f) - {}^c\theta^{\mathcal{E}}_M(f)).$$

The first difference has bounded support on the set of $M(F_v)$-orbits in $M(F_v)^X$ and vanishes if $X$ lies outside a compact set.

The integral of the product of the last function with $|D^M(\gamma)|^{1/2}\Theta_\tau(\gamma)$ equals

$${}^c\theta_M(f,\tau,X) - {}^c\theta^{\mathcal{E}}_M(f,\tau,X).$$

It follows that for any point $X \in \mathfrak{a}_M$ outside a fixed compact set, we have

$$\varepsilon'_M(f,\tau,X) = {}^c\theta_M(f,\tau,X) - {}^c\theta^{\mathcal{E}}_M(f,\tau,X).$$

Now, Corollary 5.5.6 implies that $\Delta{}^c\theta_M(f,\tau,X)$ is a rapidly decreasing function of $X \in \mathfrak{a}_M$. A similar assertion applies to $\Delta{}^c\theta^{\mathcal{E}}_M(f,\tau,X)$. Consequently, $\Delta\varepsilon'_M(f,\tau,X)$ is rapidly decreasing, and $\varepsilon'_M(f,\tau,X)$ is a Schwarz function of $X$. □

LEMMA 13.13. *As a function of $(\tau,X)$, our $\varepsilon'_M(f,\tau,X)$ belongs to the space $\mathbb{I}_{\mathrm{ac}}(M(F_v))$.*

PROOF. There are two properties to establish. We need to show that the function is $(K_v \cap M(F_v))$-finite. That is, we need to find a finite subset $\Gamma_M$ of $\Pi(K_v \cap M(F_v))$ such that $\varepsilon'_M(f,\tau,X)$ vanishes unless the restriction of $\tau$ to $(K_v \cap M(F_v))$ contains a representation in $\Gamma_M$.

We also need to show that for $X \in \mathfrak{a}_M$, the function $\varepsilon'_M(f,\tau,X)$ belongs to the Paley-Wiener space in the natural coordinates (taken modulo $i\mathfrak{a}^*_M$) on $\Pi_{\mathrm{temp}}(M(F_v))$. This holds, since the Paley-Wiener requirement is trivial unless $\tau$ is properly induced, in which case $\varepsilon'_M(f,\tau,X)$ is 0.

We will use the differential equation (13.3.1) to establish the first property.

For $\tau \in \Pi_{\mathrm{temp}}(M(F_v))$, set

$$\varepsilon'_M(f,\tau) = \int_{\mathfrak{a}_M} \varepsilon'_M(f,\tau,X)\,dX.$$

Let $\tau(z_M)$ denote the infinitesimal character of a representation evaluated at $z_M$. The differential equation, definition (13.11.1), and the fact that $\varepsilon'_M(f, \tau, X)$ is a Schwarz function of $X$ together imply that for $\tau \in \Pi_{\text{temp}}(M(F_v))$, we have

$$\varepsilon'_M(zf, \tau) = \tau(z_M)\varepsilon'_M(f, \tau).$$

Thus, as a function of $f$, the linear form $\varepsilon'_M(f, \tau)$ is an invariant eigendistribution of $\mathcal{Z}_v$. For $\tau$ in general position and $P \in \mathcal{P}(M)$, there is a smooth function $c(\tau)$ on $\Pi_{\text{temp}}(M(F_v))$ such that

$$\varepsilon_M(f, \tau) = c(\tau) \operatorname{tr} i_P(\tau, f).$$

Since $f$ is $K_v$-finite there is a finite subset $\Gamma$ of $\Pi(K_v)$ such that $\operatorname{tr} i_P(\tau, f)$ vanishes unless the $K_v$-spectrum of $i_P(\tau)$ meets $\Gamma$. The first property then holds if we take $\Gamma_M$ to be the set of irreducible constituents of restrictions of elements in $\Gamma$ to $K_v \cap M(F_v)$. This proves that the function

$$\varepsilon'_M(f) : (\tau, X) \mapsto \varepsilon'_M(f, \tau, X)$$

belongs to $\mathbb{I}_{\text{ac}}(M(F_v))$. □

## 13.14 Fourier Inversion

For any $\gamma \in M(F_v)$, the orbital integral $\widehat{I}^M_M(\gamma, \varepsilon'_M(f))$ is defined. Applying Fourier inversion on $T(F_v)^X$ to (13.11.1), one sees that

$$\widehat{I}^M_M(\gamma, \varepsilon'_M(f)) = \varepsilon'_M(\gamma, f).$$

Define

$$\varepsilon_M(f) = \varepsilon'_M(f) + \varepsilon'_M(v) f_M.$$

Since $f_M$ belongs to $\mathbb{I}(M(F_v))$, the function $\varepsilon_M(f)$ belongs to $\mathbb{I}_{\text{ac}}(M(F_v))$. The formula

$$\widehat{I}^M_M(\gamma, \varepsilon_M(f)) = \varepsilon_M(\gamma, f)$$

follows from the definitions and the analogous formula for $\varepsilon'_M(f)$. We have then defined the map $\varepsilon_M$ when $S$ consists of one Archimedean valuation. This was the final step, so the proof of Proposition 13.2, and thus Proposition 12.5, is complete.

□

□

COROLLARY 13.15. *Let S be a finite set of valuations with the closure property. For* $f \in \mathbb{H}(G(F_S))^0$ *there is a function* ${}^c\varepsilon_M(f) \in \widetilde{\mathbb{I}}(M(F_S))$ *such that*

$$\varepsilon_M(f) = {}^c\varepsilon_M(f) + {}^c\theta_M(f) - {}^c\theta_M^{\mathcal{E}}(f).$$

*In particular, for* $\tau \in \Pi(M(F_S))$, *the function* $\varepsilon_M(f, \tau, X)$ *is a Schwarz function of* $X \in \mathfrak{a}_{M,S}$.

PROOF. By Proposition 12.11,

$${}^c\varepsilon_M(f) = \varepsilon_M(f) - {}^c\theta_M(f) + {}^c\theta_M^{\mathcal{E}}(f)$$

is a function in $\widetilde{\mathbb{I}}_{\mathrm{ac}}(M(F_S))$ whose orbital integral at $\gamma \in M(F_S) \cap G_{\mathrm{reg}}$ equals ${}^c\varepsilon_M(\gamma,f)$. It follows inductively from Proposition 5.4.10 that ${}^c\varepsilon(\gamma,f)$ vanishes if $X = H_M(\gamma)$ lies outside a compact set. Since ${}^c\varepsilon_M(f, \tau, X)$ equals the integral of the normalized character of $\tau$ against ${}^c\varepsilon_M(\gamma,f)$, this function also vanishes if $X$ is large. Therefore ${}^c\varepsilon_M(f)$ belongs to $\widetilde{\mathbb{I}}(M(F_S))$. The second assertion of the corollary follows from Corollary 5.5.6, as we saw above in the special case that $S$ consists of one Archimedean prime. □

## 13.16 Moderate Functions

As noted in Section 12, $\varepsilon_M$ extends to a map from $\mathbb{H}(G(\mathbb{A}))^0$ to $\mathbb{I}_{\mathrm{ac}}(M(\mathbb{A}))$. In 4.19.3 we introduced a space of *moderate functions*, which lies between $\mathbb{I}(M(\mathbb{A}))$ and $\mathbb{I}_{\mathrm{ac}}(M(\mathbb{A}))$. (See also section 5.11.) Without repeating the definition, let us say that for a function $\phi \in \mathbb{I}_{\mathrm{ac}}(M(\mathbb{A}))$ to be moderate, it must satisfy a weak growth condition and an equally weak support condition.

COROLLARY 13.17. *For each* $f \in \mathbb{H}(G(\mathbb{A}))^0$, *the function* $\varepsilon_M(f)$ *is a moderate function.*

PROOF. We can assume $f$ is of the form $\prod_v f_v$. Let

$$\varepsilon_{M,v}(f) = \varepsilon_M(f_v) \prod_{w \neq v} f_{w,M}.$$

By (12.5.3*), there is a number $d_0$ such that

$$\varepsilon_M(f) = \sum_v \varepsilon_{M,v}(f) + d_0 f_M.$$

Almost all functions $\varepsilon_{M,v}(f)$ vanish. The function $f_M$ belongs to $\mathbb{I}(M(\mathbb{A}))$ and is moderate. It is therefore enough to fix a valuation $v$ and prove that for a fixed function $f \in \mathbb{H}(G(F_v))^0$, the function $\varepsilon_M(f)$ is a moderate function in $\mathbb{I}_{\mathrm{ac}}(M(F_v))$.

For $M_1$ a Levi subgroup of $M$ over $F_v$, $\tau_1 \in \Pi_{\mathrm{temp}}(M_1(F_v))$, and $X_1 \in \mathfrak{a}_{M_1,v}$ whose projection onto $\mathfrak{a}_{M,v}$ equals $X$, we can define a function

$$X_1 \mapsto \int_{i\mathfrak{a}^*_{M_1,v}/i\mathfrak{a}^*_{M,v}} \varepsilon_M(f, \tau^M_{1,\Lambda}, X)e^{-\Lambda(X_1)}\, \mathrm{d}\Lambda.$$

We must establish the two conditions (growth and support) for such functions. If $M_1 \subsetneq M$, formula (12.5.2*) implies that the function is compactly supported. If $M_1 = M$, Corollary 13.15 asserts that the function belongs to the Schwarz space. In either case, the growth and support conditions hold. □

We will show that $\varepsilon_M$ behaves nicely under multipliers. Let $\mathfrak{h}^1$ be the orthogonal complement of $\mathfrak{a}_G$ in the space $\mathfrak{h}$ discussed in Subsection 3.2.

DEFINITION 13.18. Let $\mathcal{E}(\mathfrak{h}^1)^W$ be the convolution algebra of compactly supported, $W$-invariant distributions on $\mathfrak{h}^1$.

Multipliers are attached to elements $\alpha \in \mathcal{E}(\mathfrak{h}^1)^W$.

Recall that there is an action $f \mapsto f_\alpha$ of the algebra $\mathcal{E}(\mathfrak{h}^1)^W$ on $\mathbb{H}(G(\mathbb{A}))$ such that for $\tau \in \Pi(M(\mathbb{A}))$ we have

$$i_P(\tau, f_\alpha) = \widehat{\alpha}(\nu_\tau) i_P(\tau, f).$$

There is a compatible action $\phi \mapsto \phi_\alpha$ of $\mathcal{E}(\mathfrak{h}^1)^W$ on $\mathbb{I}_{\mathrm{ac}}(M(\mathbb{A}))$ which for any $\tau \in \Pi_{\mathrm{temp}}(M(\mathbb{A}))$ and $X \in \mathfrak{a}_M$ is given by

$$\phi_\alpha(\tau, X) = \int_{\mathfrak{a}^G_M} \phi(\tau, Y) \int_{i\mathfrak{a}^*_M/i\mathfrak{a}^*_G} \widehat{\alpha}(\nu_\tau + \mu)e^{-\mu(X-Y)}\, \mathrm{d}\mu\, \mathrm{d}Y.$$

The reader can check that there is a natural map $\alpha \mapsto \alpha'$ from $\mathcal{E}(\mathfrak{h}^1)^W$ to $\mathcal{E}(\mathfrak{h}^1 \cap \mathfrak{h}')^{W'}$ such that for $f \in \mathbb{H}(G(\mathbb{A}))$, we have

$$(f_\alpha)' = f'_{\alpha'}.$$

COROLLARY 13.19. *For $f \in \mathbb{H}(G(\mathbb{A}))^0$ and $\alpha \in \mathcal{E}(\mathfrak{h}^1)^W$, we have*

$$\varepsilon_M(f_\alpha) = \varepsilon_M(f)_\alpha.$$

PROOF. Fix a function of the form

$$f_0 = \prod_{v \text{ finite}} f_v, \quad f_v \in \mathbb{H}(G(F_v)),$$

with the following property. For any $f_\infty \in \mathbb{H}(G(F_\infty))$, the function $f = f_\infty f_0$ belongs to $\mathbb{H}(G(\mathbb{A}))^0$. We shall vary $f_\infty$. Suppose that $\tau = \tau_\infty \otimes \tau_0$ is a representation in $\Pi_{\mathrm{temp}}(M(\mathbb{A}))$. Let

$$\varepsilon_M(f, \tau) = \int_{\mathfrak{a}_M} \varepsilon_M(f, \tau, X)\, \mathrm{d}X.$$

Using the differential equation (13.3.1), we can argue as above to show that for $P \in \mathcal{P}(M)$ and a scalar $c(\tau, f_0)$ independent of $f_\infty$ we have

$$\varepsilon_M(f, \tau) = c(\tau, f_0) \operatorname{tr} i_P(\tau_\infty, f_\infty).$$

It follows that for $\alpha \in \mathcal{E}(\mathfrak{h}^1)^W$ we have

$$\begin{aligned}\varepsilon_M(f_\alpha, \tau) &= c(\tau, f_0) \operatorname{tr} i_P(\tau_\infty, (f_\infty)_\alpha) \\ &= \widehat{\alpha}(\nu_\tau) c(\tau, f_0) \operatorname{tr} i_P(\tau_\infty, f_\infty) \\ &= \widehat{\alpha}(\nu_\tau) \varepsilon_M(f, \tau).\end{aligned}$$

We obtain

$$\begin{aligned}\varepsilon_M(f_\alpha, \tau, X) &= \int_{i\mathfrak{a}_M^*} \varepsilon_M(f_\alpha, \tau_\lambda) e^{-\lambda(X)}\, \mathrm{d}\lambda \\ &= \int_{\mathfrak{a}_M^G} \varepsilon_M(f_\alpha, \tau, Y) \int_{i\mathfrak{a}_M^*/i\mathfrak{a}_G^*} \widehat{\alpha}(\nu_\tau + \mu) e^{-\mu(X-Y)}\, \mathrm{d}\mu\, \mathrm{d}Y,\end{aligned}$$

as required. □

## 14 Separation by Infinitesimal Character

We can now apply the map $\varepsilon_M$ to the formula for $I^{\mathcal{E}}(f) - I(f)$ in Proposition 12.3.

DEFINITION 14.1. Let $\mathbb{H}(G(\mathbb{A}), M)^0$ be the space of functions $f$ in

$$\mathbb{H}(G(\mathbb{A}), M) \cap \mathbb{H}(G(\mathbb{A}))^0$$

that vanish at any element in $G(\mathbb{A})$ whose component at each finite place $v$ belongs to $A_G(F_v)$.

Combined with the earlier definition of $\mathbb{H}(G(\mathbb{A}))^0$, the condition ensures that the orbital integral of $f$ will vanish at any element of the form $\xi u$, for $\xi \in A_G(F)$ and $u \in \mathcal{U}_G(F)$. Note that the function $f$ will still remain in $\mathbb{H}(G(\mathbb{A}), M)^0$ if it is modified at the Archimedean places.

PROPOSITION 14.2. *Suppose that $f \in \mathbb{H}(G(\mathbb{A}), M)^0$. Let $I^M$ be the analogue of $I$ for $M$. Then*

$$I^{\mathcal{E}}(f) - I(f) = |W(\mathfrak{a}_M)|^{-1} \widehat{I}^M(\varepsilon_M(f)).$$

PROOF. Consider the formula for $I^{\mathcal{E}}(f) - I(f)$ provided by Proposition 12.3. The conditions on $f$ imply that the second term in the formula vanishes. If $S$ is a large finite set of valuations, (12.5.1**) implies that the first term equals

$$|W(\mathfrak{a}_M)|^{-1} \sum_{\gamma \in (M(F))_{M,S}} a^M(S,\gamma)\widehat{I}^M_M(\gamma, \varepsilon_M(f)).$$

LEMMA 14.3. *The function $\varepsilon_M(f)$ lies in $\mathbb{I}_{\mathrm{ac}}(M(\mathbb{A}))$. Furthermore, it is cuspidal at two places. In other words, $\varepsilon_M(f)$ is a finite sum of functions $\prod_v \varphi_v$ in $\mathbb{I}_{\mathrm{ac}}(M(\mathbb{A}))$ such that for two unramified places $v_1$ and $v_2$ and any $M_1 \subsetneq M$, we have*

$$\varphi_{v_i,M_1} = 0, \quad i = 1, 2.$$

PROOF. This follows from (12.5.2*), (12.5.3*), and the fact that $f \in \mathbb{H}(G(\mathbb{A}), M)$. □

Applying the splitting property (5.3.3) to $M$, we find that, for any such $M_1$ and $\gamma \in M_1(F)$,

$$\widehat{I}^M_{M_1}(\gamma, \varepsilon_M(f)) = 0.$$

Since all terms in the sum over $M_1$ are zero except $M_1 = M$, it follows from (2.1.1), applied to $M$, that

$$\begin{aligned}\sum_{\gamma \in (M(F))_{M,S}} a^M(S,\gamma)\widehat{I}^M_M(\gamma, \varepsilon_M(f)) &= \sum_{M_1 \in \mathcal{L}^M} |W_0^{M_1}|/|W_0^M| \sum_{\gamma \in (M_1(F))_{M_1,S}} a^{M_1}(S,\gamma)\widehat{I}^M_{M_1}(\gamma, \varepsilon_M(f)) \\ &= \widehat{I}^M(\varepsilon_M(f)).\end{aligned}$$

The proposition follows. □

## 14.4 Spectral Expansion

Fix $f \in \mathbb{H}(G(\mathbb{A}), M)^0$. Combined with the expansions $I^{\mathcal{E}} = \sum_t I^{\mathcal{E}}_t$ and $I = \sum_t I_t$, Proposition 14.2 yields the formula

$$\sum_{t \geq 0} \left( I^{\mathcal{E}}_t(f) - I_t(f) - |W(\mathfrak{a}_M)|^{-1}\widehat{I}^M_t(\varepsilon_M(f)) \right) = 0. \qquad (14.4.1)$$

We are going to apply the spectral expansion of the distributions on the left. We will then try to deduce what remains of (iv) of the Main Theorem 3.11. As noted in Subsection 3.5, our control over the convergence is very weak. In this section we shall simply isolate the terms in (14.4.1) according to their Archimedean infinitesimal character.

We shall use an argument based on multipliers. We have associated the real vector space $\mathfrak{h}^1$ to the real Lie group $G(F_\infty)^1$; see Subsection 13.18. It is convenient to work with a subset of the complex dual space $\mathfrak{h}^*_{\mathbb{C}}/\mathfrak{a}^*_{G,\mathbb{C}}$ of $\mathfrak{h}^1$ which contains the infinitesimal character of any unitary representation of $G(F_\infty)^1$.

DEFINITION 14.5. Let $\mathfrak{h}^*_u$ ("$u$" for unitary) denote the set of points $\nu$ in $\mathfrak{h}^*_{\mathbb{C}}/i\mathfrak{a}^*_G$ such that $\overline{\nu} = -s\nu$ for some element $s \in W$ of order 2. Here $\overline{\nu}$ denotes the conjugation on $\mathfrak{h}^*_{\mathbb{C}}$ relative to $\mathfrak{h}^*$.

The Archimedean infinitesimal character $\nu_{\pi^1}$ associated to $\pi^1 \in \Pi_{\mathrm{unit}}(G(\mathbb{A})^1)$ belongs to the subset

$$(\mathfrak{h}')^*_u = \mathfrak{h}^*_u \cap (\mathfrak{h}' \cap \mathfrak{h}^1)^*_{\mathbb{C}}$$

of $\mathfrak{h}^*_u$. For nonnegative numbers $r$ and $T$, the set

$$\mathfrak{h}^*_u(r,T) = \{\nu \in \mathfrak{h}^*_u;\ \|\mathrm{Re}\,\nu\| \leq r,\ \|\operatorname{Im}\nu\| \geq T\}$$

is invarient under $W$. Note that an element $\nu \in \mathfrak{h}^*_u$ is a coset of $i\mathfrak{a}^*_G$ in $\mathfrak{h}^*_{\mathbb{C}}$. For $\|\nu\|$ we take the minimum value of the norm on the coset.

The multipliers enter through an estimate from Chapter 5. The result applies to moderate functions in $\mathbb{I}_{\mathrm{ac}}(G(\mathbb{A}))$.

LEMMA 14.6. *Suppose that $\phi$ is a moderate function. Then there exist positive numbers $C$, $k$, and $r$ such that for any $T > 0$ and $\alpha \in C^\infty_N(\mathfrak{h}^1)^W$ with $N > 0$ the inequality*

$$\sum_{t>T} |\widehat{I}_t(\phi_\alpha)| \leq Ce^{kN} \sum_{\nu\in\mathfrak{h}^*_u(r,T)} (|\widehat{\alpha}(\nu)|) \tag{14.6.1}$$

*holds.*

PROOF. This is Corollary 5.17.5. □

Let $f \in \mathbb{H}(G(\mathbb{A}), M)^0$. It follows easily from the definitions that, if $\alpha \in \mathcal{E}(\mathfrak{h}^1)^W$, the function $f_\alpha$ also belongs to $\mathbb{H}(G(\mathbb{A}), M)^0$. In particular, $f_\alpha$ satisfies (14.4.1).

LEMMA 14.7. *For $T \geq 0$ the expression*

$$\left|\sum_{t\leq T} \left(I^{\mathcal{E}}_t(f_\alpha) - I_t(f_\alpha) - |W(\mathfrak{a}_M)|^{-1}\widehat{I}^M_t(\varepsilon_M(f_\alpha))\right)\right| \tag{14.7.1}$$

*is bounded by*

$$Ce^{kN}\,\sup\nolimits_{\nu\in\mathfrak{h}^*_u(r,T)}(|\widehat{\alpha}(\nu)|). \tag{14.7.2}$$

PROOF. Expression (14.7.1) is bounded by

$$\sum_{t>T} \left(|I_t(f_\alpha)| + |I^{\mathcal{E}}_t(f_\alpha)| + |W(\mathfrak{a}_M)|^{-1}|\widehat{I}^M_t(\varepsilon_M(f_\alpha))|\right).$$

By Corollary 13.19 we can write

$$I_t(f_\alpha) = \widehat{I}_t(f_{G,\alpha}),$$

$$I_t^{\mathcal{E}}(f_\alpha) = \widehat{I}_t((f_\alpha)'_{G'}) = \widehat{I}_t(f'_{G',\alpha'})$$

and

$$\varepsilon_M(f_\alpha) = \varepsilon_M(f)_\alpha.$$

Consequently, (14.7.1) is bounded by

$$\sum_{t>T} |\widehat{I}_t^G(f_{G,\alpha})| + \sum_{t>T} |\widehat{I}_t^{G'}(f'_{G',\alpha'})| + |W(\mathfrak{a}_M)|^{-1} \sum_{t>T} |\widehat{I}_t^M(\varepsilon_M(f)_\alpha)|.$$

The functions $f_G \in \mathbb{I}(G(\mathbb{A}))$ and $f'_{G'} \in \mathbb{I}(G'(\mathbb{A}))$ are both moderate. By Corollary 13.17, $\varepsilon_M(f) \in \mathbb{I}_{\mathrm{ac}}(M(\mathbb{A}))$ is also moderate. We can apply (14.6.1) to these three functions. Observe that $(\mathfrak{h}')^*_u(r, T)$, the analogue of the set defined above for $G'$, is contained in $\mathfrak{h}^*_u(r, T)$. Applying the estimate gives the following statement. There are positive numbers $C$, $k$, and $r$ such that, for any $\alpha \in C_N^\infty(\mathfrak{h}^1)^W$ with $N > 0$ and for any $T > 0$, the expression (14.7.1) is bounded by (14.7.2). □

## 14.8 A Useful Function

Fix $\nu_1$ in $\mathfrak{h}^*_u$. Enlarging $r$ in (14.7.2) if necessary, we may assume that $\nu_1$ belongs to the cylinder

$$\mathfrak{h}^*_u(r) = \mathfrak{h}^*_u(r, 0).$$

PROPOSITION. *There is a function $\alpha_1 \in C_c^\infty(\mathfrak{h}^1)^W$ such that $\widehat{\alpha}_1$ maps $\mathfrak{h}^*_u(r)$ to the unit interval and such that the inverse image of* 1 *under $\widehat{\alpha}_1$ is the finite set*

$$W \cdot \nu_1 = \{s\nu_1 \,;\, s \in W\}.$$

PROOF. For $\alpha \in C_c^\infty(\mathfrak{h}^1)^W$, we have the function of $\nu \in \mathfrak{h}^*_u$ given by

$$\widehat{\alpha} : \nu \mapsto \widehat{\alpha}(\nu).$$

Consider the space of such functions on $\mathfrak{h}^*_u$. The definition of $\mathfrak{h}^*_u$ ensures that the real and imaginary parts of any such function also belong to the space. We can therefore find a function $\alpha_0 \in C_c^\infty(\mathfrak{h}^1)^W$, with $\widehat{\alpha}_0(\nu_1) \neq 0$, such that $\widehat{\alpha}_0$ is real valued on $\mathfrak{h}^*_u$.

Let $p_0 = 1, p_1, \ldots, p_m$ be a set of generators of the algebra of $W$-invariant polynomials on $\mathfrak{h}_{\mathbb{C}}^*/\mathfrak{a}_{G,\mathbb{C}}^*$. We can assume that each $p_i$ is real valued on $\mathfrak{h}_u^*$. Since $\widehat{\alpha}_0$ is rapidly decreasing at infinity on $\mathfrak{h}_u^*(r)$, the function

$$\beta(\nu) = (p_0(\nu)\widehat{\alpha}_0(\nu), \ldots, p_m(\nu)\widehat{\alpha}_0(\nu))$$

maps $\mathfrak{h}_u^*(r)$ continuously to a compact rectangle $[a,b]^{m+1}$ in $\mathbb{R}^{m+1}$. Set

$$s = (s_0, \ldots, s_m) = \beta(\nu_1).$$

Then

$$\beta^{-1}(\{s\}) = W \cdot \nu_1.$$

For each $i$, let

$$q_i : [a,b] \to [0,1]$$

be a real polynomial such that $q_i^{-1}(\{1\}) = \{s_i\}$. Since $s_0 \neq 0$, we can assume $q_0$ has no constant term. Consequently,

$$\widehat{\alpha}_1(\nu) = \prod_{i=0}^{m} q_i(p_i(\nu)\widehat{\alpha}_0(\nu))$$

is the Fourier-Laplace transform of a function $\alpha_1 \in C_c^\infty(\mathfrak{h}^1)^W$. It satisfies the requirements of the proposition. □

## 14.9 Convolution

Fix $\alpha_1$ as in the last proposition. It belongs to $C_{N_1}^\infty(\mathfrak{h}^1)^W$ for some $N_1 > 0$. If $r$ and $k$ are as in (14.7.2), choose $T > 0$ so that for all $\nu \in \mathfrak{h}_u^*(r, T)$, we have

$$|\widehat{\alpha}_1(\nu)| \leq e^{-2kN_1}.$$

This is possible, because $\widehat{\alpha}_1$ is rapidly decreasing on $\mathfrak{h}_u^*(r)$. Set

$$\alpha_m = \underbrace{\alpha_1 * \cdots * \alpha_1}_{m}.$$

Then $\alpha_m \in C_{mN_1}^\infty(\mathfrak{h}^1)^W$ and

$$\widehat{\alpha}_m(\nu) = \widehat{\alpha}_1(\nu)^m.$$

LEMMA 14.10. *The expression*

$$\sum_{t\le T}\Big(I_t^{\mathcal{E}}(f_{\alpha_m})-I_t(f_{\alpha_m})-|W(\mathfrak{a}_M)|^{-1}\widehat{I}_t^M(\varepsilon_M(f)_{\alpha_m})\Big) \tag{14.10.1}$$

*is bounded in absolute value by*

$$Ce^{-kN_1 m}.$$

*Consequently,* (14.10.1) *approaches* 0 *as m approaches* $\infty$.

PROOF. Take $\alpha=\alpha_m$ above. □

## 14.11 Separation

This assertion is a significant improvement over the formula (14.4.1), since the sum in (14.10.1) can be taken over a finite set which is independent of $m$. This will allow us to take the expansions of the terms in (14.10.1) and study the limit as $m$ approaches $\infty$.

Apply Proposition 11.8 to the function $f_{\alpha_m}$. Since

$$\operatorname{tr}\pi((f_{\alpha_m})^1)=\operatorname{tr}\pi(f^1)\widehat{\alpha}(\nu_\pi)^m,$$

we see that

$$\sum_{t\le T}(I_t^{\mathcal{E}}(f_{\alpha_m})-I_t(f_{\alpha_m}))$$

equals

$$\sum_{t\le T}\ \sum_{\pi^1\in\Pi(G(\mathbb{A})^1,t)}(a_{\mathrm{disc}}^{\mathcal{E}}(\pi^1)-a_{\mathrm{disc}}(\pi^1))\operatorname{tr}\pi^1(f^1)\widehat{\alpha}(\nu_{\pi^1})^m. \tag{14.11.1}$$

We now expand $I_t^M(\varepsilon_M(f)_{\alpha_m})$. The function $\varepsilon_M(f)_{\alpha_m}$ is a finite sum of functions which are cuspidal at two places. By Theorem 5.17.3 we have

$$\widehat{I}_t^M(\varepsilon_M(f)_{\alpha_m})=\sum_{\tau^1\in\Pi_{\mathrm{disc}}(M,t)}a_{\mathrm{disc}}^M(\tau^1)\widehat{I}_M^M(\tau^1,\varepsilon_M(f)_{\alpha_m}).$$

Before proceeding, we need an additional result.

PROPOSITION 14.12. *Suppose that* $\tau^1\in\Pi_{\mathrm{unit}}(M(\mathbb{A})^1)$. *Then there is a Schwarz function*

$$\lambda\mapsto\varepsilon_M(f^1,\tau^1,\lambda)$$

*of $\lambda \in i\mathfrak{a}_M^*/i\mathfrak{a}_G^*$ such that for any $\alpha \in C_c^\infty(\mathfrak{h}^1)^W$, we have*

$$\widehat{I}_M^M(\tau^1, \varepsilon_M(f)_\alpha) = \int_{i\mathfrak{a}_M^*/i\mathfrak{a}_G^*} \varepsilon_M(f^1, \tau^1, \lambda)\widehat{\alpha}(\nu_{\tau^1} + \lambda)\, \mathrm{d}\lambda.$$

PROOF. As an element of $\mathbb{I}_{\mathrm{ac}}(M(\mathbb{A}))$, the function $\varepsilon_M(f)_\alpha$ is initially defined on $\Pi_{\mathrm{temp}}(M(\mathbb{A})) \times \mathfrak{a}_M$. But, it may be naturally extended to a function on $\Pi(M(\mathbb{A})) \times \mathfrak{a}_M$. (This is a reflection of the fact that a function in $\mathbb{H}_{\mathrm{ac}}(M(\mathbb{A}))$ is compactly supported on any set $M(\mathbb{A})^X$ and can therefore be integrated against a nontempered character.) Identify $\tau^1$ with an orbit $\{\tau_\lambda\}$ of $i\mathfrak{a}_M^*$ in $\Pi_{\mathrm{unit}}(M(\mathbb{A}))$. By definition, $\widehat{I}_M^M(\tau, \varepsilon_M(f)_\alpha)$ is the value of $\varepsilon_M(f)_\alpha$ at $(\tau_\lambda, 0)$. Let

$$\alpha_M(\tau_\lambda, -Y) = \int_{i\mathfrak{a}_M^*/i\mathfrak{a}_G^*} \widehat{\alpha}(\nu_\tau + \lambda + \mu)e^{\mu(Y)}\, \mathrm{d}\mu.$$

By 5.(16.2.1), for any $\lambda$, we have

$$\widehat{I}_M^M(\tau, \varepsilon_M(f)_\alpha) = \int_{\mathfrak{a}_M^G} \varepsilon_M(f, \tau_\lambda, Y)\alpha_M(\tau_\lambda, -Y)\, \mathrm{d}Y.$$

Since $\alpha_M(\tau_\lambda, -Y)$ is compactly supported as a function of $Y \in \mathfrak{a}_M^G$, the integral over $\mathfrak{a}_M^G$ converges.

If we can show that

$$\varepsilon_M(f, \tau, X)$$

is a Schwarz function of $X \in \mathfrak{a}_M$, the proposition will follow from Fourier inversion. Here we will use the unitarity of $\tau$.

Assume that $f$ is of the form $\prod_v f_v$. Let

$$\varepsilon_{M,v}(f) = \varepsilon_M(f_v) \prod_{w \neq v} f_{w,M}.$$

By (12.5.3*), there is a number $d_0$ such that

$$\varepsilon_M(f) = \sum_v \varepsilon_{M,v}(f) + d_0 f_M.$$

Since

$$f_M(\tau_{\lambda+\mu})$$

is a Schwarz function of $\mu \in i\mathfrak{a}_M^*$,

$$f_M(\tau_\lambda, X) = \int_{i\mathfrak{a}_M^*} f_M(\tau_{\lambda+\mu})e^{-\mu(X)}\, \mathrm{d}\mu$$

is a Schwarz function of $X \in \mathfrak{a}_M$. This leaves the function $\varepsilon_{M,v}(f)$. Almost all of them vanish, so we need to show that, for a fixed $v$, $\varepsilon_{M,v}(f, \tau_\lambda, X)$ is a Schwarz function.

Fix $\lambda$. For $\tau_v \in \Pi_{\text{unit}}(M(F_v))$, we can write

$$\tau_\lambda = \tau_v \otimes \tau^v,$$

where

$$\tau^v = \otimes_{w \neq v} \tau_w.$$

Recall the decomposition

$$\operatorname{tr} \tau_v = \sum_{\rho_v \in \Sigma(M(F_v))} \xi(\tau_v, \rho_v) \operatorname{tr} \rho_v$$

into standard representations. Then

$$\varepsilon_{M,v}(f, \tau_\lambda, X) = \sum_{\rho_v \in \Sigma(M(F_v))} \xi(\tau_v, \rho_v) \varepsilon_{M,v}(f, \rho_v \otimes \tau^v, X).$$

Since $\tau_v$ is unitarizable, any $\rho_v$ with $\xi(\tau_v, \rho_v) \neq 0$ is either tempered or induced from a proper parabolic subgroup. If $\rho_v$ is properly induced, it follows from (12.5.2*) that there is a proper Levi subgroup $M_1 \subsetneq M$ such that

$$\varepsilon_{M,v}(f, \rho_v \otimes \tau^v, X) = \varepsilon_{M,M_1}(v) \int_{i\mathfrak{a}_M^*} f_M((\rho_v \otimes \tau^v)_\mu) e^{-\mu(X)}\, d\mu.$$

This is a Schwartz function of $X \in \mathfrak{a}_M$. Suppose now that $\rho_v$ is tempered. Then Corollary 13.15 ensures that $\varepsilon_M(f_v, \rho_v, X_v)$ is a Schwartz function of $X_v \in \mathfrak{a}_{M,v}$. It follows that $\varepsilon_{M,v}(f, \rho_v \otimes \tau^v, X)$ is a Schwartz function of $X \in \mathfrak{a}_M$.

This establishes that $\varepsilon_{M,v}(f, \tau_\lambda, X)$ is a Schwartz function. The proposition follows. □

## 14.13 Application of a Formula

Apply the proposition to the formula for $\widehat{I}_t^M(\varepsilon_M(f)_{\alpha_m})$ above. We see that

$$\sum_{t \leq T} |W(\mathfrak{a}_M)|^{-1} \widehat{I}_t^M(\varepsilon_M(f)_{\alpha_m})$$

equals

$$|W(\mathfrak{a}_M)|^{-1} \sum_{t \leq T} \sum_{\tau^1 \in \Pi_{\mathrm{disc}}(M^1, t)} a^M_{\mathrm{disc}}(\tau^1) \int_{i\mathfrak{a}^*_M / i\mathfrak{a}^*_G} \varepsilon_M(f^1, \tau^1, \lambda) \widehat{\alpha}_1(\nu_{\tau^1} + \lambda)^m \, \mathrm{d}\lambda. \tag{14.13.1}$$

We have shown that (14.10.1) equals the difference between (14.11.1) and (14.13.1). Consequently, this difference approaches 0 as $m$ approaches $\infty$. In each of the expressions (14.11.1) and (14.13.1), the sums over $t$ and $\pi$ are finite. We first apply the dominated convergence theorem to (14.13.1). Since $\varepsilon_M(f^1, \tau^1, \lambda)$ is a Schwartz function on $i\mathfrak{a}^*_M / i\mathfrak{a}^*_G$, and

$$0 \leq \widehat{\alpha}_1(\nu_{\tau^1} + \lambda) < 1, \tag{14.13.2}$$

except possibly at a finite number of $\lambda$, we see that (14.13.1) approaches 0 as $m$ approaches $\infty$. The same is then true of (14.11.1).

We next consider the terms in (14.11.1). If $\pi^1$ does not belong to the set

$$\Pi_{\nu_1}(G(\mathbb{A})^1) = \{\pi^1 \in \Pi(G(\mathbb{A})^1) \, ; \, \nu_{\pi^1} \in W(\nu_1)\},$$

the inequality (14.13.2) holds. Consequently, the corresponding term in (14.11.1) approaches 0 as $m$ approaches $\infty$.

On the other hand, if $\pi^1$ belongs to $\Pi_{\nu_1}(G(\mathbb{A})^1)$, the term equals

$$(a^{\mathcal{E}}_{\mathrm{disc}}(\pi^1) - a_{\mathrm{disc}}(\pi^1)) \operatorname{tr} \pi^1(f^1).$$

We can assume that $\| \mathrm{Im}(\nu_1) \| \leq T$. This ensures that all such terms will be included in (14.11.1). Letting $m$ approach $\infty$, we obtain the following important result:

PROPOSITION 14.14. *For each $f \in \mathbb{H}(G(\mathbb{A}), M)^0$ and $\nu_1 \in \mathfrak{h}^1_u$, we have*

$$\sum_{\pi^1 \in \Pi_{\nu_1}(G(\mathbb{A})^1)} (a^{\mathcal{E}}_{\mathrm{disc}}(\pi^1) - a_{\mathrm{disc}}(\pi^1)) \operatorname{tr} \pi^1(f^1) = 0.$$

## 15 Elimination of Restrictions on $f$

Proposition 14.14 will finally allows us to prove the equality of $I^{\mathcal{E}}(f)$ and $I(f)$, for certain functions $f$. However, for this to be effective, we first need to extend the formula of Proposition 14.14 to a larger class of functions.

DEFINITION 15.1. (1) As in the last section, fix $\nu_1 \in \mathfrak{h}^*_u$.
(2) We let $\mathbb{A}_{\mathrm{fin}}$ denote the finite adèles. Let

$$K_1 = \prod_{v \text{ finite}} K_{1,v}$$

be an open, compact subgroup of $G(\mathbb{A}_{\text{fin}})$.

(3) Write $\Pi_{\nu_1,K_1}(G(\mathbb{A})^1)$ for the set of representations $\pi^1 \in \Pi(G(\mathbb{A})^1)$ such that $\nu_{\pi^1} \in W \cdot \nu_1$, and such that $\pi^1$ contains a nonzero $K_1$-fixed vector.

(4) Put

$$a^{\mathcal{E}}_{\text{disc}}(\pi^1) = \sum_{\pi'^1 \in \Pi(G'(\mathbb{A})^1)} a^{G'}_{\text{disc}}(\pi'^1)\delta(\pi'^1, \pi^1).$$

LEMMA 15.2. (1) *There are only finitely many* $\pi^1 \in \Pi_{\nu_1,K_1}(G(\mathbb{A})^1)$ *such that* $a_{\text{disc}}(\pi^1) \neq 0$.

(2) *There are only finitely many* $\pi^1 \in \Pi_{\nu_1,K_1}(G(\mathbb{A})^1)$ *with* $a^{\mathcal{E}}_{\text{disc}}(\pi^1) \neq 0$.

PROOF. (1) See Lemma 5.14.6.

(2) There is an open compact subgroup $K_1'$ of $G'(\mathbb{A}_{\text{fin}})$ such that if $\delta(\pi'^1, \pi^1) \neq 0$ for some $\pi^1 \in \Pi_{\nu_1,K_1}(G(\mathbb{A})^1)$, then $\pi'^1 \in \Pi_{\nu_1',K_1'}(G'(\mathbb{A})^1)$. But there are only finitely many $\pi'^1 \in \Pi_{\nu_1',K_1'}(G'(\mathbb{A})^1)$ with $a^{G'}_{\text{disc}}(\pi'^1) \neq 0$, by Lemma 5.14.6. □

DEFINITION 15.3. (1) Write $\mathbb{H}(G(\mathbb{A}), M)_{K_1}$ for the subspace of $\mathbb{H}(G(\mathbb{A}), M)$ consisting of the functions which are bi-invariant under $K_1$.

(2) Set

$$\mathbb{H}(G(\mathbb{A}), M)^0_{K_1} = \mathbb{H}(G(\mathbb{A}), M)_{K_1} \cap \mathbb{H}(G(\mathbb{A}), M)^0.$$

(3) Let

$$\Sigma_{\nu_1,K_1}(G(\mathbb{A})^1)$$

be the set of representations $\rho^1 \in \Sigma(G(\mathbb{A})^1)$ such that $\nu_{\rho^1} \in W \cdot \nu_1$ and such that $\rho^1$ contains a $K_1$-fixed vector.

## 15.4 Finite Sum

Proposition 14.14 tells us that for $f \in \mathbb{H}(G(\mathbb{A}), M)^0_{K_1}$, we have

$$\sum_{\pi^1 \in \Pi_{\nu_1,K_1}(G(\mathbb{A})^1)} (a^{\mathcal{E}}_{\text{disc}}(\pi^1) - a_{\text{disc}}(\pi^1)) \operatorname{tr} \pi^1(f^1) = 0.$$

The sum can be taken over a finite set which depends only on $(\nu_1, K_1)$. We can write

$$\operatorname{tr} \pi^1(f^1) = \sum_{\rho^1 \in \Sigma_{\nu_1,K_1}(G(\mathbb{A})^1)} \xi(\pi^1, \rho^1) \operatorname{tr} \rho^1(f^1).$$

Let

$$A(\rho^1) = \sum_{\pi^1 \in \Pi_{v_1,K_1}(G(\mathbb{A})^1)} (a^{\mathcal{E}}_{\mathrm{disc}}(\pi^1) - a_{\mathrm{disc}}(\pi^1))\xi(\pi^1, \rho^1). \tag{15.4.1}$$

Then for $f \in \mathbb{H}(G(\mathbb{A}), M)^0_{K_1}$, we have

$$\sum_{\rho^1 \in \Sigma_{v_1,K_1}(G(\mathbb{A})^1)} A(\rho^1) \operatorname{tr} \rho^1(f^1) = 0. \tag{15.4.2}$$

## *15.5 An Expansion*

The goal of this section is the following proposition:

PROPOSITION. *The identity* (15.4.2) *holds if $f$ belongs to* $\mathbb{H}(G(\mathbb{A}), M)_{K_1} \supset \mathbb{H}(G(\mathbb{A}), N)^0_{K_1}$.

We start the proof. It is completed in Subsection 15.7. Recall that $G$ is the group of units of a central simple algebra. We will use an approximation argument. Let $v$ be a fixed valuation from the exceptional set $S_G$. Write $\kappa = K_{1,v}$. Then $\mathbb{H}(G(F_v))^0_\kappa$ is the space of compactly supported functions $f_v$ on $G(F_v)$ which are bi-invariant under $\kappa$ and such that for any $\xi \in A_G(F_v)$ and $1 \neq u \in \mathcal{U}_G(F_v)$, we have

$$I_G(\xi u, f_v) = 0.$$

We write $\mathbb{A}^v$ for the ring of adèles which are 0 at $v$.

PROPOSITION 15.6. *Suppose that $f^v$ is a smooth, compactly supported function on $G(\mathbb{A}^v)$ such that* (15.4.2) *holds for any function $f = f^v f_v$, $f_v \in \mathbb{H}(G(F_v))^0_\kappa$. Then* (15.4.2) *holds for $f = f^v f_v$ with $f_v \in \mathbb{H}(G(F_v))_\kappa$.*

PROOF. Let $C$ be a fixed finite subset of the lattice

$$\mathfrak{a}_{G,v} = \{H_G(x)\,;\, x \in G(F_v)\}.$$

Write $\mathbb{H}_v$ for the subspace of $\mathbb{H}(G(F_v))_\kappa$ consisting of the functions which are supported on

$$\{x \in G(F_v)\,;\, H_G(x) \in C\}.$$

Put

$$\mathbb{H}^0_v = \mathbb{H}_v \cap \mathbb{H}(G(F_v))^0_\kappa.$$

Let $I$ be the finite set of pairs $i = (\xi, u)$, where $\xi \in A_G(F_v)/(A_G(F_v) \cap \kappa)$ satisfies $H_G(\xi) \in C$ and $u$ ranges over the nontrivial unipotent conjugacy classes in $G(F_v)$. Set $J_i = I_G(\xi u)$. Then $\{J_i\,;\, i \in I\}$ is a linearly independent set of linear forms on $\mathbb{H}_v$ whose kernel is $\mathbb{H}^0_v$. Choose elements $\{f^j_v\,;\, j \in I\}$ in $\mathbb{H}_v$ such that

$$J_i(f_v^j) = \begin{cases} 1, & i = j, \\ 0, & i \neq j. \end{cases}$$

Let $f_v$ denote an element of $\mathbb{H}_v$. The map

$$f_v \mapsto \overline{f}_v = f_v - \sum_{i \in I} J_i(f_v) f_v^i$$

is a projection of $\mathbb{H}_v$ onto $\mathbb{H}_v^0$. By assumption,

$$\sum_{\rho^1} A(\rho^1) \operatorname{tr} \rho^1(f^v \overline{f}_v) = 0.$$

Let

$$\beta^i = \sum_{\rho^1} A(\rho^1) \operatorname{tr} \rho^1(f^v f_v^i).$$

We obtain

$$\sum_{\rho^1} A(\rho^1) \operatorname{tr} \rho^1(f^v f_v) = \sum_i \beta^i J_i(f_v). \tag{15.6.1}$$

We must show that (15.6.1) vanishes.

Let $L \in \mathcal{L}_v$. Denote by

$$\Pi_{\mathrm{disc}}(L(F_v))_\kappa$$

the set of representations in $\Pi_{\mathrm{temp}}(L(F_v))$ which are square-integrable modulo the center and which have a nonzero $\kappa \cap L(F_v)$-fixed vector. Let $\mathfrak{o}$ be an orbit of the compact group

$$i\mathfrak{a}_{L,v}^* = i(\mathfrak{a}_L^* / \operatorname{Hom}(\mathfrak{a}_{L,v}, \mathbb{Z}))$$

in $\Pi_{\mathrm{disc}}(L(F_v))_\kappa$. There are only finitely many pairs $(L, \mathfrak{o})$.

For any such orbit, let $W_\mathfrak{o}$ be the stabilizer of $\mathfrak{o}$ in $W(\mathfrak{a}_L)$. For $\rho^v \in \Sigma_{v_1,K_1}(G(\mathbb{A}^v))$, $P \in \mathcal{P}(L)$, $\sigma \in \mathfrak{o}$ and $\mu \in \mathfrak{a}_L^*$, we can consider representations of the form

$$\rho^v \otimes i_P(\sigma_\mu).$$

Let $\Sigma_\mathfrak{o}$ be the set of $\rho^1 \in \Sigma_{v_1,K_1}(G(\mathbb{A})^1)$ which are restrictions to $G(\mathbb{A})^1$ of representations of this form. The point $\sigma_\mu$ in $\mathfrak{o} \times \mathfrak{a}_L^*$ is uniquely determined as a $W_\mathfrak{o}$-orbit, modulo translation by $i\mathfrak{a}_{G,v}^*$ in $\mathfrak{o}$. We will write $X_{\rho^1} = \sigma_\mu$.

Two sets $\Sigma_\mathfrak{o}$ and $\Sigma_{\mathfrak{o}'}$ are either equal or disjoint, depending on whether or not $\mathfrak{o}$ and $\mathfrak{o}'$ are $W_0$-conjugate or not. The set $\Sigma_{v_1,K_1}(G(\mathbb{A})^1)$ is a union of sets $\Sigma_\mathfrak{o}$. Therefore it is enough to show that, for each $\mathfrak{o}$ and $f_v \in \mathbb{H}_v$, the number

$$\sum_{\rho^1 \in \Sigma_{\mathfrak{o}}} A(\rho^1) \operatorname{tr} \rho^1(f^v f_v) \tag{15.6.2}$$

vanishes.

Fix $(L, \mathfrak{o})$ and $f_v \in \mathbb{H}_v$. Let $\mathbb{I}_{\mathfrak{o}}$ be the space of functions

$$\phi : \mathfrak{o} \to \mathbb{C}$$

which satisfy the following three conditions:

(i) For $\sigma \in \mathfrak{o}$, $\phi(\sigma_\Lambda)$ is a finite Fourier series in $\Lambda \in i\mathfrak{a}^*_{L,v}$.
(ii) For $\sigma \in \mathfrak{o}$ and $w \in W_{\mathfrak{o}}$, we have $\phi(w\sigma) = \phi(\sigma)$.
(iii) For $\lambda \in i\mathfrak{a}^*_{G,v}$, we have $\phi(\sigma_\lambda) = \phi(\sigma)$.

The second and third conditions ensure that the numbers $\phi(X_{\rho^1})$ are well defined.

For each $\phi \in \mathbb{I}_{\mathfrak{o}}$, by the trace Paley-Wiener theorem for $G(F_v)$, [BDK86] or [F95], there is a function $f_v^\phi \in \mathbb{H}_v$ such that for $\rho^1 \in \Sigma_{\nu_1, K_1}(G(\mathbb{A})^1)$ the number

$$\operatorname{tr} \rho^1(f^v f_v^\phi)$$

is zero unless $\rho^1 \in \Sigma_{\mathfrak{o}}$, in which case

$$\operatorname{tr} \rho^1(f^v f_v^\phi) = \operatorname{tr} \rho^1(f^v f_v) \phi(X_{\rho^1}).$$

We replace $f_v$ by $f_v^\phi$ in (15.6.1). Let

$$c_{\rho^1} = A(\rho^1) \operatorname{tr} \rho^1(f^v f_v).$$

The left side of (15.6.1) becomes

$$\sum_{\rho^1 \in \Sigma_{\mathfrak{o}}} c_{\rho^1} \phi(X_{\rho^1}).$$

To evaluate the right side, we use the fact that every unipotent class in $G(F_v)$ is induced. For $u \in (\mathcal{U}_G(F_v))$ there is a $W_0$-orbit $\mathcal{L}(u)$ in $\mathcal{L}_v$ such that for each $L_1 \in \mathcal{L}(u)$ and $Q \in \mathcal{P}(L_1)$ the intersection $u \cap N_Q$ is dense in $N_Q$. Let $\chi_\sigma(\xi)$ denote the central character of $\sigma$ at $\xi$. Let $\mu_L^u(\sigma)$ be the number which vanishes unless there is an element $L_1 \in \mathcal{L}(u)$ containing $L$, in which case $\mu_L^u(\sigma)$ is the Plancherel density associated to the Levi subgroup $L$ of $L_1$. For $P \in \mathcal{P}(L)$, define

$$B_i(\sigma) = \chi_\sigma(\xi)^{-1} \mu_L^u(\sigma) \operatorname{tr} i_P(\sigma, f_v).$$

It is a smooth function on $\mathfrak{o}$. If $i = (\xi, u)$, put

$$J_i(f_v^\phi) = \int_{\mathfrak{o}} B_i(\sigma) \phi(\sigma)\, d\sigma.$$

Notice that, since $u \neq 1$, $B_i(\sigma) = 0$ if $L = G$. Let

$$B(\sigma) = \sum_i \beta^i B_i(\sigma).$$

It is a smooth function on $\mathfrak{o}$, vanishing if $L = G$. Equation (15.6.1) becomes

$$\sum_{\rho^1 \in \Sigma_\mathfrak{o}} c_{\rho^1}\phi(X_{\rho^1}) = \int_\mathfrak{o} B(\sigma)\phi(\sigma)\,d\sigma. \tag{15.6.3}$$

The last step is to show that each side of (15.6.3) is zero. We can assume $L \neq G$, so that

$$\mathfrak{o}^1 = \mathfrak{o}/i\mathfrak{a}^*_{G,v}$$

is a compact torus of positive dimension. On the right side of (15.6.3), we have a distribution on $\mathfrak{o}^1$ which is a smooth function. On the left side, we have a finite sum of point distributions on the complexification of $\mathfrak{o}^1$. Since the points $\{X_{\rho^1}\}$ are defined as $W_\mathfrak{o}$-orbits, and $B$ is symmetric under $W_\mathfrak{o}$, we do not need to assume that $\phi$ is symmetric under $W_\mathfrak{o}$. It can be any finite Fourier series on $\mathfrak{o}^1$. Its Fourier transform can be any compactly supported function on the dual lattice. Consider the Fourier transform of each side of (15.6.3) as a distribution on the dual lattice. The left-hand side is a finite sum of exponentials. The right-hand side is a rapidly decreasing function. Thus both sides vanish.

Having shown that each side of (15.6.3) is zero, we take $\phi = 1$. We obtain

$$\sum_{\rho^1 \in \Sigma_\mathfrak{o}} c_{\rho^1} = 0.$$

The expression on the left is (15.6.2), so the proof of the proposition is complete. □

## 15.7 *Proof of Proposition 5.5*

We now complete the proof of Proposition 15.5. We apply Proposition 15.6 to each place in $S_G$. It follows inductively that (15.4.2) holds for any function in $\mathbb{H}(G(\mathbb{A}), M)_{K_1}$ which vanishes on $G(F_\infty)A_G(\mathbb{A}_{\text{fin}})$. We will remove this last restriction.

We are free to modify an arbitrary function $f \in \mathbb{H}(G(\mathbb{A}), M)_{K_1}$ outside a finite set $S$ of valuations. Choose any unramified place $w$ outside of $S$ such that $K_{1,w}$ equals $K_w$, the standard maximal compact subgroup. Let $h$ be a variable function in $\mathbb{H}(G(F_w))_{K_w}$. Define a function of $x \in G(\mathbb{A})$ by

$$f^h(x) = f(x)h(x_w).$$

Evaluate the left-hand side of (15.4.2) on the function $f^h(x)$. The expression vanishes if $h$ is zero on $A_G(F_w)$, so as a linear form in $h$, it may be expressed in terms of the Plancherel density. On one hand, the Plancherel density is a continuous function on the unramified representations in $\Pi_{\mathrm{temp}}(G(F_w))$. On the other hand, the sum in (15.4.2) may be taken over a finite set. It follows that the linear form vanishes on any $h$. Therefore (15.4.2) holds for any function $f \in \mathbb{H}(G(\mathbb{A}), M)_{K_1}$. This is the claim of Proposition 15.5. □

## 15.8 Sum Over Irreducible Representations

Let us translate (15.4.2) back into a sum over irreducible representations. Let $f$ be an element in $\mathbb{H}(G(\mathbb{A}), M)_{K_1}$. We substitute (15.4.1) back into (15.4.2). We obtain

$$\sum_{\pi^1 \in \Pi_{\nu_1, K_1}(G(\mathbb{A})^1)} (a^{\mathcal{E}}_{\mathrm{disc}}(\pi^1) - a_{\mathrm{disc}}(\pi^1)) \operatorname{tr} \pi^1(f^1) = 0. \tag{15.8.1}$$

If $\pi^1$ belongs to the complement of $\Pi_{\nu_1, K_1}(G(\mathbb{A})^1)$ in $\Pi_{\nu_1}(G(\mathbb{A})^1)$, then $\operatorname{tr} \pi^1(f^1)$ equals 0, so we can take the sum over the larger set. Any function in $\mathbb{H}(G(\mathbb{A}), M)$ belongs to $\mathbb{H}(G(\mathbb{A}), M)_{K_1}$, for some $K_1$. It follows that for $f \in \mathbb{H}(G(\mathbb{A}), M)$, we have

$$\sum_{\pi^1 \in \Pi_{\nu_1}(G(\mathbb{A})^1)} (a^{\mathcal{E}}_{\mathrm{disc}}(\pi^1) - a_{\mathrm{disc}}(\pi^1)) \operatorname{tr} \pi^1(f^1) = 0. \tag{15.8.2}$$

PROPOSITION 15.9. *For any $f \in \mathbb{H}(G(\mathbb{A}), M)$, we have*

$$I^{\mathcal{E}}(f) = I(f).$$

PROOF. Let $t$ be a nonnegative real number. For any $f \in \mathbb{H}(G(\mathbb{A}), M)$ by Proposition 11.8, we have

$$I^{\mathcal{E}}_t(f) - I_t(f) = \sum_{\{\nu_1 \,;\, \|\operatorname{Im}(\nu_1)\| = t\}} \sum_{\pi^1 \in \Pi_{\nu_1}(G(\mathbb{A})^1)} (a^{\mathcal{E}}_{\mathrm{disc}}(\pi^1) - a_{\mathrm{disc}}(\pi^1)) \operatorname{tr} \pi^1(f^1).$$

This is 0, by (15.8.2). We therefore obtain

$$I^{\mathcal{E}}(f) = \sum_t I^{\mathcal{E}}_t(f) = \sum_t I_t(f) = I(f),$$

as required. □

# 16 Completion of the Proof of the Main Theorem

Having established Proposition 15.9, we return to the geometric sides of the trace formulae. At last we will prove that $I_M^{\mathcal{E}}(\gamma)$ equals $I_M(\gamma)$.

PROPOSITION 16.1. *Suppose* $f \in \mathbb{H}(G(\mathbb{A}), M)$. *Let* $S \supset S_{\text{ram}}$ *be a large finite set of valuations depending only on* supp($f$) *and* $V(f)$. *The sum of the expressions*

$$|W(\mathfrak{a}_M)|^{-1} \sum_{\gamma \in (M(F))_{M,S}} a^M(S, \gamma)(I_M^{\mathcal{E}}(\gamma,f) - I_M(\gamma,f)) \tag{16.1.1}$$

*and*

$$\sum_{\xi \in A_G(F)} \sum_{u \in (\mathcal{U}_G(F))_{G,S}} (a^{\mathcal{E}}(S,u) - a(S,u))I_G(\xi u,f) \tag{16.1.2}$$

*vanishes. The sums in* (16.1.1) *and* (16.1.2) *can be taken over finite sets, depending only on* supp($f$) *and* $V(f)$.

PROOF. This follows from Proposition 12.3 and Proposition 15.9. □

## 16.2 Sets of Places

Assume that $S$ is the disjoint union of a finite set $S_0 \supset S_{\text{ram}}$ with further sets

$$S_i = \{v_i\}, \quad 1 \leq i \leq k,$$

where $v_i$ are valuations at which $G$ splits. For $f_i \in \mathbb{H}(G(F_{S_i}))$ let

$$f = \prod_{i=0}^{k} f_i.$$

It follows inductively from (7.3.1) that for $\gamma \in M(F)$, we have

$$I_M^{\mathcal{E}}(\gamma,f) - I_M(\gamma,f) = \sum_{i=0}^{k} (I_M^{\mathcal{E}}(\gamma,f_i) - I_M(\gamma,f_i)) \prod_{j \neq i} \widehat{I}_M^M(\gamma,f_{j,M}). \tag{16.2.1}$$

We will take $\gamma \in M(F)$ to be regular and elliptic at $v_1$ and $v_2$. This means that the torus $M_\gamma/A_M$ is anisotropic at $v_1$ and $v_2$. We will use $v_1$ and $v_2$ to isolate the contributions from $\gamma$ to (16.1.1). Indeed, for $i = 1,2$, we can choose $f_i$ to be supported on a very small neighborhood of $\gamma$ in $G(F_S)$, such that

$$\widehat{I}_M^M(\gamma, f_{i,M}) = I_G(\gamma, f_i) = 1.$$

Then $f_i$ will be supported on the $F_v$-elliptic set in $M$, and the function $f$ above will belong to $\mathbb{H}(G(F_S), M)$.

## 16.3 Splitting Formula Used

Apply the splitting formula (16.2.1) to the terms in (16.1.1). Shrinking the function $f_1$ and $f_2$ around $\gamma$ does not increase the support of $f$ or the set $V(f)$. Therefore the set $S$ may be chosen independently of $f_1$ and $f_2$, and the sums in (16.1.1) and (16.1.2) may be taken over fixed finite sets.

It follows that $f_1$ and $f_2$ may be chosen so that (16.1.2) vanishes, so that the only contributions to (16.1.1) come from conjugates of $\gamma$. But for $w \in W(\mathfrak{a}_M)$, we have

$$I_M^{\mathcal{E}}(w\gamma w^{-1}, f) - I_M(w\gamma w^{-1}, f) = I_M^{\mathcal{E}}(\gamma, f) - I_M(\gamma, f).$$

Thus we need only consider the summand in (16.1.1) corresponding to $\gamma$. Moreover, $\gamma$ is semisimple. By Theorem 8.2 of [Ar86], if $S$ is sufficiently large in a sense depending only on $\gamma$, we have

$$a^M(S, \gamma) = \mathrm{vol}(M_\gamma(F)\backslash M_\gamma(\mathbb{A})^1).$$

In particular, this number is nonzero. Hence, for $\gamma, f_1$, and $f_2$ as above, we have

$$\sum_{i=0}^{k} (I_M^{\mathcal{E}}(\gamma, f_i) - I_M(\gamma, f_i)) \prod_{j \neq i} \widehat{I}_M^M(\gamma, f_{j,M}) = 0. \tag{16.3.1}$$

## 16.4 Special Set of Places

Suppose that $V$ is a finite set of valuations of $F$ that contains $S_{\mathrm{ram}}$ or consists of one valuation where $G$ splits. We can arrange that $V$ equals one of the sets $S_i$ above, with $i \neq 1, 2$. Choose $\gamma \in M(F)$ as in (16.3.1). Let $f_V$ be a function in $\mathbb{H}(G(F_V))$. Suppose first that

$$\widehat{I}_M^M(\gamma, f_{V,M}) = I_G(\gamma, f_V) = 0.$$

Then the only contribution to (16.3.1) will be the summand corresponding to $V = S_i$. For the sets $S_j$ other than $S_1$, $S_2$, and $S_i$, choose $f_j$ to be any function such that

$$\widehat{I}_M^M(\gamma, f_{j,M}) \neq 0.$$

The left-hand side of (16.3.1) becomes a nonzero multiple of

$$I_M^{\mathcal{E}}(\gamma, f_V) - I_M(\gamma, f_V).$$

We conclude that this distribution vanishes for any $f_V$ whose orbital integral vanishes at $\gamma$. It follows that there is a number, $\varepsilon_M(\gamma)$, such that for any $f_V \in \mathbb{H}(G(F_V))$, we have

$$I_M^{\mathcal{E}}(\gamma, f_V) - I_M(\gamma, f_V) = \varepsilon_M(\gamma) I_G(\gamma, f_V). \qquad (16.4.1)$$

## 16.5 Augmented Set

Let $V^+$ be the union of $V$ with the valuations $v_1$ and $v_2$ above. Write $U_{V^+}(M)$ for the set of elements of the form

$$\prod_{v \in V^+} \gamma_v, \qquad \gamma_v \in M(F_v) \cap G_{\mathrm{reg}},$$

such that $\gamma_{v_i}$ is $F_{v_i}$-elliptic in $M(F_{v_i})$ for $i = 1, 2$. Then $U_{V^+}(M)$ is open in

$$M(F_{V^+}) = \prod_{v \in V^+} M(F_v).$$

The set $M(F)$ is dense in $M(F_{V^+})$, so the intersection of $M(F)$ with $U_{V^+}(M)$ is dense in $U_{V^+}(M)$. It follows that we can approximate any $G$-regular element $\gamma_V \in M(F_V)$ by elements $\gamma$ which occur in (16.4.1). Since $I_M^{\mathcal{E}}(\gamma_V, f_V)$, $I_M(\gamma_V, f_V)$, and $I_G(\gamma_V, f_V)$ are smooth on $M(F_V) \cap G_{\mathrm{reg}}$, we see that $\varepsilon_M$ extends to a smooth function on this space. Furthermore, for $\gamma_V \in M(F_V) \cap G_{\mathrm{reg}}$, we have

$$I_M^{\mathcal{E}}(\gamma_V, f_V) - I_M(\gamma_V, f_V) = \varepsilon_M(\gamma_V) I_G(\gamma_V, f_V). \qquad (16.5.1)$$

LEMMA 16.6. *We have* $\varepsilon_M(\gamma_V) = 0$.

PROOF. Consider first the case that $V$ consists of one valuation $v$ where $G$ splits. Take $f_V = f_v$ to be the characteristic function of $K_v$. Then the left-hand side of (16.5.1) vanishes. On the other hand, if $\gamma_v$ belongs to $K_v$, the orbital integral $I_G(\gamma_v, f_v)$ does not vanish. It follows that $\varepsilon_M(\gamma_v) = 0$ for any such $\gamma_v$. Since $G$ splits completely at $v$, the left-hand side of (16.5.1) vanishes if $f_v$ is any $K_v$-bi-invariant function in $\mathbb{H}(G(F_v))$. For a given $\gamma_v$ we can always choose such an $f_v$ so that $I_G(\gamma_v, f_v) \neq 0$. It follows that $\varepsilon_M(\gamma_v) = 0$ in this case, for all $\gamma_v$.

Now take $V = S_0$ to be any finite set which contains $S_{\mathrm{ram}}$. Let

$$S = \bigcup_{i=0}^{k} S_i = S_0 \cup \{v_1, \ldots, v_k\},$$

as at the beginning of the argument. Choose $\gamma \in M(F)$ as in (16.3.1), and let $\gamma_i$ be the image of $\gamma$ in $M(F_{S_i})$. We substitute the formula (16.5.1), with $V$ replaced by $S_i$, into (16.3.1). Choosing the functions $f_i$ appropriately, we find that

$$\sum_{i=0}^{k} \varepsilon_M(\gamma_i) = 0.$$

We are free to drop any of the terms in this sum corresponding to unramified valuations at which $\gamma$ is integral. This means that we can take $\gamma$ to be any $G$-regular element in $M(F)$ which is elliptic at $v_1$ and $v_2$ and which is integral outside $S$. Suppose that $G$ splits at each of the places $v_1, \ldots, v_k$. Then for each $1 \leq i \leq k$, we have

$$\varepsilon_M(\gamma_i) = 0.$$

It follows for any such $\gamma$, we have

$$\varepsilon_M(\gamma_0) = 0.$$

For $k$ sufficiently large, the set of $\gamma \in M(F)$ which are integral outside of $S$, and which are elliptic at $v_1$ and $v_2$, projects onto a dense subset of $M(F_{S_0}) = M(F_V)$. It follows that for all $\gamma_V \in M(F_V) \cap G_{\mathrm{reg}}$, we have

$$\varepsilon_M(\gamma_V) = 0.$$

□

## 16.7 *Proof of Main Theorem* 2.7(i)

We will now complete the proof of assertion (i) of the Main Theorem 2.7. Let $V$ be a finite set of valuations containing $S_{\mathrm{ram}}$. Let $\gamma$ be a $G$-regular element in $M(F_V)$. For $f \in \mathbb{H}(G(F_V))$ we have established the formula

$$I_M^{\mathcal{E}}(\gamma, f) = I_M(\gamma, f).$$

It follows from Proposition 5.11 that the formula holds for any element $\gamma \in M(F_V)$. We have finished the induction argument begun in Section 12, where we first fixed $M$. In other words, the formula holds for any $M \in \mathcal{L}$. This completes the proof of part (i) of the Main Theorem 2.7.

## 16.8 Proof of Main Theorem 2.7(ii)

We move on to part (ii). We proved that the global assertion (ii) of the Main Theorem 2.7 was a consequence of the induction hypothesis of Section 7 unless $M = G$ and

$$\gamma = \xi u, \quad \xi \in A_G(F), u \in (\mathcal{U}_G(F))_{G,S},$$

for any large finite set $S$. To deal with this case, we return to the discussion at the beginning of this section, with $M$ a minimal element in $\mathcal{L}$. Then $\mathbb{H}(G(\mathbb{A}), M)$ equals $\mathbb{H}(G(\mathbb{A}))$. Since we have established the local assertion (i) of the Main Theorem 2.7, the expression (16.1.1) vanishes. Therefore so does (16.1.2). Now, $G$ is such that $(\mathcal{U}_G(F))_{G,S}$ equals $(\mathcal{U}_G(F))$, the set of unipotent conjugacy classes in $G$ defined over $F$. It follows that, for each $f \in \mathbb{H}(G(\mathbb{A}))$, we have

$$\sum_{\xi \in A_G(F)} \sum_{u \in (\mathcal{U}_G(F))} (a^{\mathcal{E}}(S,u) - a(S,u)) I_G(\xi u, f) = 0.$$

Fix any $u_1$ in $(\mathcal{U}_G(F))$. Choose $f \in \mathbb{H}(G(\mathbb{A}))$ such that

$$I_G(\xi u, f) = \begin{cases} 1, \text{ if } (\xi, u) = (1, u_1), \\ 0, \text{ otherwise.} \end{cases}$$

We see then that $a^{\mathcal{E}}(S, u_1)$ equals $a(S, u_1)$. This finishes the remaining case of the global assertion (ii) of the Main Theorem 2.7. □

## 16.9 Proof of Main Theorem 3.11(iii)

We proved the local assertion (iii) of the Main Theorem 3.11 in Section 10. The induction hypothesis of Subsection 3.12 reduces the global assertion (iv) of the Main Theorem 3.11 to proving the equality of $a^{\mathcal{E}}_{\text{disc}}(\pi^1)$ and $a_{\text{disc}}(\pi^1)$, for $\pi^1 \in \Pi(G(\mathbb{A})^1)$. Any such $\pi^1$ belongs to a set $\Pi_{\nu_1,K_1}(G(\mathbb{A})^1)$, so we shall fix $\nu_1$ and $K_1$. Since we have established (i) and (ii), we can apply (15.8.1) with any $M \in \mathcal{L}$. Take $M$ to be minimal, and note that $\mathbb{H}(G(\mathbb{A}), M)_{K_1}$ equals $\mathbb{H}(G(\mathbb{A}))_{K_1}$. For $f \in \mathbb{H}(G(\mathbb{A}))_{K_1}$, we obtain

$$\sum_{\pi^1 \in \Pi_{\nu_1,K_1}(G(\mathbb{A})^1)} (a^{\mathcal{E}}_{\text{disc}}(\pi^1) - a_{\text{disc}}(\pi^1)) \operatorname{tr} \pi^1(f^1) = 0.$$

The sum may be taken over a finite set. However, the set of linear forms defined on $\mathbb{H}(G(\mathbb{A}))_{K_1}$ by

$$f \mapsto \operatorname{tr} \pi^1(f^1),$$

parametrized by $\Pi_{\nu_1,K_1}(G(\mathbb{A})^1)$, is linearly independent. This follows from the linear independence of Archimedean characters and the non-Archimedean trace Paley-Wiener theorem [BDK86, F95]. It follows that for $\pi^1 \in \Pi_{\nu_1,K_1}(G(\mathbb{A})^1)$, we have

$$a_{\text{disc}}^{\mathcal{E}}(\pi^1) = a_{\text{disc}}(\pi^1).$$

This completes the proof of (iv) of the Main Theorem 3.11. □

## *16.10 Proof of Main Theorem* 3.11(iv)

It is the global assertion (iv) of the Main Theorem 3.11 that is relevant to the comparison of automorphic representations. It tells us that for $\pi^1 \in \Pi(G(\mathbb{A})^1)$, we have

$$a_{\text{disc}}^{G,\mathcal{E}}(\pi^1) = a_{\text{disc}}^{G}(\pi^1).$$

Recall that $I_{\text{disc},t}(f)$ is the linear combination of characters on $G(\mathbb{A})^1$ given explicitly by the expression (3.4.1). Then

$$\begin{aligned} I_{\text{disc},t}(f) &= \sum_{\pi^1 \in \Pi(G(\mathbb{A})^1,t)} a_{\text{disc}}^{G}(\pi^1) I_G(\pi^1, f) \\ &= \sum_{\pi^1 \in \Pi(G(\mathbb{A})^1,t)} a_{\text{disc}}^{G,\mathcal{E}}(\pi^1) I_G^{\mathcal{E}}(\pi^1, f). \end{aligned}$$

It follows from the definition (3.10.1) and the case $M = G$ of Lemma 11.1 that this last expression equals

$$\sum_{\pi'^1 \in \Pi(G'(\mathbb{A})^1,t)} a_{\text{disc}}^{G'}(\pi'^1) \widehat{I}_{G'}(\pi'^1, f'_{G'}).$$

This is equal to $\widehat{I}_{\text{disc},t}(f'_{G'})$. Then (iii) and (iv) of the Main Theorem 3.11 provide an identity

$$I_{\text{disc},t}(f) = \widehat{I}_{\text{disc},t}(f'_{G'}) \tag{16.10.1}$$

between the discrete parts of the trace formulae for $G$ and $G'$.

## 16.11 Modulo the Center

Instead of using characters on $G(\mathbb{A})^1$, it is sometimes more convenient to deal with characters on $G(\mathbb{A})$ which are equivariant with respect to a subgroup of the center. Let $G_{\mathbb{Q}}$ be obtained from $G$ by restricting scalars from $F$ to $\mathbb{Q}$. Let $A_{G_{\mathbb{Q}}}$ be the $\mathbb{Q}$-split component of the center. We can take

$$A_{G,\infty} = A_{G_{\mathbb{Q}}}(\mathbb{R})^0.$$

Then $A_{G,\infty}$ is a subgroup of $\prod_{v \in S_\infty} A_G(F_v)$. The Harish-Chandra map

$$H_G : A_{G,\infty} \to \mathfrak{a}_G$$

is an isomorphism, which we use to pull back the Haar measure on $\mathfrak{a}_G$ to a Haar measure on $A_{G,\infty}$. For $a \in A_{G,\infty}$ and $x \in G(\mathbb{A})$, define

$$f_a(x) = f(ax).$$

If $\mu \in i\mathfrak{a}_G^*$, define

$$I_{\text{disc},t,\mu}(f) = \int_{A_{G,\infty}} I_{\text{disc},t}(f_a) e^{\mu(H_G(a))} \, \mathrm{d}a.$$

This transforms the characters on $G(\mathbb{A})^1$ which occur in $I_{\text{disc},t}$ into $\mu$-equivariant characters on $G(\mathbb{A})$. We can repeat the same construction for $G'(\mathbb{A})$. Since

$$(f_a)' = f'_{a'},$$
$$e^{\mu(H_G(a))} = e^{\mu'(H_{G'}(a'))},$$

and

$$\mathrm{d}a' = \mathrm{d}a,$$

we obtain

$$I_{\text{disc},t,\mu}(f) = I_{\text{disc},t,\mu'}(f'), \tag{16.11.1}$$

from identity (16.10.1).

# Bibliography

[Ar74] J. Arthur, The Selberg trace formula for groups of $F$-rank one. Ann. Math. **100**, 326–385 (1974)

[Ar76] J. Arthur, The characters of discrete series as orbital integrals. Invent. Math. **32**, 205–261 (1976)

[Ar79] J. Arthur, Eisenstein series and the trace formula, in *Automorphic Forms, Representations and L-Functions. Proceedings of Symposia in Pure Mathematics*, vol. 33, Part I (American Mathematical Society, Providence, 1979)

[Ar81] J. Arthur, The trace formula in invariant form. Ann. Math. **114**, 1–74 (1981)

[Ar82.I] J. Arthur, On a family of distributions obtained from Eisenstein series. I. Applications of the Paley-Wiener theorem. Am. J. Math. **104**, 1243–1288 (1982)

[Ar82.II] J. Arthur, On a family of distributions obtained from Eisenstein series. II: Explicit formulas. Am. J. Math. **104**, 1289–1336 (1982)

[Ar83] J. Arthur, A Paley-Wiener theorem for real reductive groups. Acta Math. **150**, 1–89 (1983)

[Ar85] J. Arthur, A measure on the unipotent variety. Can. J. Math. **37**, 1237–1274 (1985)

[Ar86] J. Arthur, On a family of distributions obtained from orbits. Can. J. Math. **38**, 179–214 (1986)

[Ar87] J. Arthur, The characters of cuspidal representations as weighted orbital integrals. Proc. Indian Acad. Sci. (Math. Sci.) **97**, 3–19 (1987)

[Ar88.1] J. Arthur, The local behavior of weighted orbital integrals. Duke Math. J. **56**, 223–293 (1988)

[Ar88.2] J. Arthur, The invariant trace formula I. Local theory. J. Am. Math. Soc. **1**, 323–383 (1988)

[Ar88.3] J. Arthur, The invariant trace formula II. Global theory. J. Am. Math. Soc. **1**, 501–554 (1988)

[Ar89] J. Arthur, Intertwining operators and residues I. Weighted characters. J. Funct. Anal. **84**, 19–84 (1989)

[Ar93] J. Arthur, On elliptic tempered characters. Acta Math. **171**, 73–138 (1993)

[Ar05] J. Arthur, An introduction to the trace formula, in *Harmonic Analysis, the Trace Formula, and Shimura Varieties. Clay Mathematics Proceedings*, vol. 4 (American Mathematical Society, Providence, 2005), pp. 1–263

[AC89] J. Arthur, L. Clozel, *Simple Algebras, Base Change, and the Advanced Theory of the Trace Formula*. Annals of Mathematics Studies, vol. 120 (Princeton University Press, Princeton, 1989)

[AFOO] A. Aydoğdu, Y.Z. Flicker, E. Özkan, R. Öztürk, Explicit forms of the trace formula for GL(2). 55 pp. (2013, Preprint); J. Anal. Math. (2016, to appear)



Y.Z. Flicker, *Arthur's Invariant Trace Formula and Comparison of Inner Forms*, DOI 10.1007/978-3-319-31593-5

[Ba05] A. Badulescu, Un théorème de finitude dans le spectre automorphe pour les formes intérieures de $GL_n$ sur un corps global. Bull. Lond. Math. Soc. **37**, 651–657 (2005)
[Ba08] A. Badulescu, Global Jacquet-Langlands correspondence, multiplicity one and classification of automorphic representations. Invent. Math. **172**, 383–438 (2008)
[BR10] A. Badulescu, D. Renard, Unitary dual of GL($n$) at Archimedean places and global Jacquet-Langlands correspondence. Compos. Math. **146**, 1115–1164 (2010)
[BS] E. van den Ban, S. Souaifi, A comparison of Paley-Wiener theorems for real reductive Lie groups. J. Reine Angew. Math. **695**, 99–150 (2014)
[Be84] J. Bernstein, P-invariant distributions on GL($N$) and the classification of unitary representations of GL($N$) (non-Archimedean case), in *Lecture Notes in Mathematics*, vol. 1041 (Springer, New York, 1984), pp. 50–102
[BD84] J. Bernstein, P. Deligne, Le "centre" de Bernstein, in *Représentations des Groupes Réductifs sur un Corps Local* (Hermann, Paris, 1984), pp. 1–32
[BDK86] J. Bernstein, P. Deligne, D. Kazhdan, Trace Paley-Wiener theorem for reductive p-adic groups. J. d'Anal. Math. **47**, 180–192 (1986)
[BZ76] J. Bernstein, A.V. Zelevinsky, Representations of the group GL($n, F$), where $F$ is a non-Archimedean local field. Russ. Math. Surv. **31**(3), 1–68 (1976)
[BZ77] J. Bernstein, A.V. Zelevinsky, Induced representations of reductive $p$-adic groups I. Ann. Sci. ENS **10**, 441–472 (1977)
[Bo62] A. Borel, *Ensembles fondamentaux pour les groupes arithmétiques*. Colloque sur la Théorie des Groupes Algébriques (Bruxelles) (Librairie Universitaire/Gauthier-Villars, Louvain/Paris, 1962), pp. 23–40
[Bo76] A. Borel, Admissible representations of a semi-simple group over a local field with vectors fixed under an Iwahori subgroup. Invent. Math. **35**, 233–259 (1976)
[BH78] A. Borel, G. Harder, Existence of discrete cocompact subgroups of reductive groups over local fields. J. Reine Angew. Math. **298**, 53–64 (1978)
[BHC62] A. Borel, Harish-Chandra, Arithmetic subgroups of algebraic groups. Ann. Math. **75**, 484–535 (1962)
[BJ79] A. Borel, H. Jacquet, Automorphic forms and automorphic representations, in *Proceedings of Symposia in Pure Mathematics*, vol. 33, Part 1 (American Mathematical Society, Providence, 1979), pp. 189–202
[BW80] A. Borel, N. Wallach, *Continuous Cohomology, Discrete Subgroups, and Representations of Reductive Groups*. Annals of Mathematics Studies, vol. 94 (Princeton University Press, Princeton, 1980)
[Bu98] D. Bump, *Automorphic Forms and Representations* (Cambridge University Press, Cambridge, 1998)
[BF83] C. Bushnell, A. Frölich, *Gauss Sums and p-Adic Division Algebras*. Springer Lecture Notes, vol. 987 (Springer, New York, 1983)
[Car79] P. Cartier, Representations of $p$-adic groups: A survey, in *Proceedings of Symposia in Pure Mathematics*, vol. 33, Part I (American Mathematical Society, Providence, 1979), pp. 111–155
[Cas77] W. Casselman, Characters and Jacquet modules. Math. Ann. **230**, 101–105 (1977)
[Cas] W. Casselman, Introduction to the theory of admissible representations of $p$-adic reductive groups. Unpublished
[CF68] J. Cassels, A. Frohlich, *Algebraic Number Theory* (Academic, London, 1968)
[CLL84] L. Clozel, J.P. Labesse, R.P. Langlands, *Morning Seminar on the Trace Formula*. Lecture Notes (Institute for Advanced Study, Princeton, 1984)
[CD84] L. Clozel, P. Delorme, Le Théorème de Paley-Wiener invariant pour les groupes de Lie réductifs. Invent. Math. **77**, 427–453 (1984)
[CD85] L. Clozel, P. Delorme, Sur le théorème de Paley-Wiener invariant pour les groupes de Lie réductifs réels. CR Acad. Sci. Paris **300**, 331–333 (1985)
[CD90] L. Clozel, P. Delorme, Le théorème de Paley-Wiener invariant pour les groupes de Lie réductif *II*. Ann. Sci. École Norm. Sup. **23**, 193–228 (1990)
[Co74] L. Cohn, *Analytic Theory of Harish-Chandra's c-Function*. Lecture Notes in Mathematics, vol. 429 (Springer, New York/Berlin, 1974)

[De76] P. Deligne, Le support du caractère d'une représentation supercuspidale. C. R. Acad. Sci. Paris Sér. A-B **283**(Part A), 155–157 (1976)
[DK] P. Deligne, D. Kazhdan, On representations of local division algebras. Unpublished manuscript
[DKV84] P. Deligne, D. Kazhdan, M.-F. Vignéras, Représentations des algèbres centrales simples p-adiques, in *Représentations des Groupes Réductifs sur un Corps Local* (Hermann, Paris, 1984), pp. 33–117
[Dm05] P. Delorme, Sur le théorème de Paley-Wiener d'Arthur. Ann. Math. **162**, 987–1029 (2005)
[Dr77] V. Drinfeld, Elliptic modules II. Math. USSR Sb. **31**, 159–170 (1977)
[FLM11] T. Finis, E. Lapid, W. Müller, On the spectral side of Arthur's trace formula-absolute convergence. Ann. Math. **174**, 173–195 (2011)
[Ft79] D. Flath, Decomposition of representations into tensor products, in *Proceedings of Symposia in Pure Mathematics*, vol. 33, Part 1 (American Mathematical Society, Providence, 1979)
[F85] Y. Flicker, On twisted liftings. Trans. Am. Math. Soc. **290**, 161–178 (1985)
[F87.1] Y. Flicker, Rigidity for automorphic forms. J. Anal. Math. **49**, 135–202 (1987)
[F87.2] Y. Flicker, Stable base change for spherical functions. Nagoya Math. J. **106**, 121–142 (1987)
[F88] Y. Flicker, On endo-lifting. Compos. Math. **67**, 271–300 (1988)
[F90.1] Y. Flicker, Regular trace formula and base change for GL($n$). Ann. Inst. Fourier **40**, 1–30 (1990)
[F90.2] Y. Flicker, Transfer of orbital integrals and division algebras. J. Ramanujan Math. Soc. **5**, 107–122 (1990)
[F95] Y. Flicker, Bernstein isomorphism and good forms, in *Proceedings of Symposia in Pure Mathematics*, vol. 58, Part II (American Mathematical Society, Providence, 1995), pp. 171–196
[F06] Y. Flicker, *Automorphic Representations of Low Rank Groups* (World Scientific, Hackensack, 2006)
[F11] Y. Flicker, The tame algebra. J. Lie Theory **21**, 469–489 (2011)
[F13] Y. Flicker, *Drinfeld Moduli Schemes and Automorphic Forms. The Theory of Elliptic Modules with Applications*. Springer Briefs in Mathematics (Springer, New York, 2013), vi + 154 pp. ISBN:978-1-4614-5887-6; 978-1-4614-5888-3
[F14] Y. Flicker, The trace formula for GL(2) over a function field. Doc. Math. **19**, 1–62 (2014)
[F15] Y. Flicker, Counting Galois representations with at most one, unipotent, monodromy. Am. J. Math. **137**, 1–25 (2015)
[F16] Y. Flicker, Invariant distributions are supported on characters. Preprint
[FK87.1] Y. Flicker, D. Kazhdan, Geometric Ramanujan conjecture and Drinfeld reciprocity law, in *Proceedings of Selberg Symposium*, Oslo (1987)
[FK87.2] Y. Flicker, D. Kazhdan, Metaplectic correspondence. Publ. Math. IHES **64**, 53–110 (1987)
[FK88] Y. Flicker, D. Kazhdan, A simple trace formula. J. d'Anal. Math. **50**, 189–200 (1988)
[Fu97] K. Fujiwara, Rigid geometry, Lefschetz-Verdier trace formula and Deligne's conjecture. Invent. Math. **127**(3), 489–533 (1997)
[GK75] I. Gelfand, D. Kajdan, Representations of the group GL($n, K$) where $K$ is a local field, in *Lie Groups and Their Representations (Proc. Summer School, Bolyai János Math. Soc., Budapest, 1971)* (Halsted, New York, 1975), pp. 95–118
[GJ72] R. Godement, H. Jacquet, *Zeta Functions of Simple Algebras*. Springer Lecture Notes, vol. 260 (New York, Springer, 1972)
[HC57] Harish-Chandra, Fourier transforms on a semisimple Lie algebra I. Am. J. Math. **79**, 193–257 (1957)
[HC66] Harish-Chandra, Discrete series for semisimple Lie groups II. Acta Math. **116**, 1–111 (1966)
[HC68] Harish-Chandra, Notes by J.G.M. Mars, *Automorphic Forms on Semisimple Lie Groups*. Springer Lecture Notes, vol. 62 (Springer, New York, 1968)
[HC70] Harish-Chandra, Notes by G. van Dijk, *Harmonic Analysis on Reductive p-Adic Groups*. Lecture Notes in Mathematics, vol. 162 (Springer, New York, 1970)

[HC73] Harish-Chandra, Harmonic analysis on reductive $p$-adic groups, in *Harmonic Analysis on Homogeneous Spaces. Proceedings of Symposia in Pure Mathematics*, vol. 26 (American Mathematical Society, Providence, 1973)

[HC75] Harish-Chandra, Harmonic analysis on real reductive groups. I. The theory of the constant term. J. Funct. Anal. **19**, 104–204 (1975)

[HC76] Harish-Chandra, Harmonic analysis on real reductive groups. III. The Maass-Selberg relations and the Plancherel formula. Ann. Math. **104**, 117–201 (1976)

[HC78] Harish-Chandra, Admissible invariant distributions on reductive $p$-adic groups. Queen's Papers in Pure Appl. Math. **48**, 281–341 (1978)

[HCpl] Harish-Chandra, The Plancherel formula for reductive $p$-adic groups. in *Collected Papers*, vol. IV (Springer, New York, 1984), pp. 353–367

[Ho63] L. Hörmander, *Linear Partial Differential Operators* (Springer, New York, 1963)

[Ho74] R. Howe, The Fourier transform and germs of characters. Case of $\mathrm{GL}(n)$ over a $p$-adic field. Math. Ann. **208**, 305–322 (1974)

[Hu75] J. Humphreys, *Linear Algebraic Groups*. Graduate Texts in Mathematics (Springer, New York, 1975)

[JL70] H. Jacquet, R.P. Langlands, *Automorphic Forms on* GL(2). Springer Lecture Notes, vol. 114 (Springer, New York, 1970)

[JPSS83] H. Jacquet, I. Piatetskii-Shapiro, J. Shalika, Rankin-Selberg convolutions. Am. J. Math. **105**, 367–464 (1983)

[JS81] H. Jacquet, J. Shalika, On Euler products and the classification of automorphic forms II. Am. J. Math. **103**, 777–815 (1981)

[Ka86.1] D. Kazhdan, Cuspidal geometry on $p$-adic groups. J. Anal. Math. **47**, 1–36 (1986)

[Ka86.2] D. Kazhdan, Representations of groups over close local fields. J. Anal. Math. **47**, 175–179 (1986)

[KS75] A. Knapp, E.M. Stein, Singular integrals and the principal series, IV. Proc. Natl. Acad. Sci. U.S.A. **72**, 4622–4624 (1975)

[KS80] A. Knapp, E.M. Stein, Intertwining operators for semisimple groups, II. Invent. Math. **60**, 9–84 (1980)

[KZ77] A. Knapp, G. Zuckerman, Classification theorems for representations of semisimple Lie groups, in *Lecture Notes in Mathematics*, vol. 587 (Springer, New York, 1977), pp. 138–159

[Kn65] M. Kneser, Galois-Kohomologie halbeinfacher algebraischer Gruppen über $p$-adischen Körpern, I. Math. Z. **88**, 40–47 (1965); II, Math. Z. **89**, 250–272 (1965)

[Ko80] R. Kottwitz, Orbital integrals on $\mathrm{GL}_3$. Am. J. Math. **102**, 327–384 (1980)

[Ko83] R. Kottwitz, Sign changes in harmonic analysis on reductive groups. Trans. Am. Math. Soc. **278**, 289–297 (1983)

[Ko86] R. Kottwitz, Stable trace formula: Elliptic singular terms. Math. Ann. **275**, 365–399 (1986)

[Ko88] R. Kottwitz, Tamagawa numbers. Ann. Math. **127**, 629–646 (1988)

[LW13] J.-P. Labesse, J.-L. Waldspurger, *La formule des traces tordue d'après le Friday Morning Seminar*. CRM Monograph Series, vol. 31 (American Mathematical Society, Providence, 2013), xxvi + 234 pp.

[La66] R.P. Langlands, Eisenstein series, in *Algebraic Groups and Discontinuous Subgroups. Proceedings of Symposia in Pure Mathematics*, vol. 9 (American Mathematical Society, Providence, 1966)

[La76] R.P. Langlands, *On the Functional Equations Satisfied by Eisenstein Series*. Springer Lecture Notes, vol. 544 (Springer, New York, 1976)

[La88] R.P. Langlands, On the classification of irreducible representations of real algebraic groups, in *Mathematical Surveys and Monographs*, vol. 31 (American Mathematical Society, Providence, 1988)

[Le96] B. Lemaire, Intégrabilité locale des charactères-distributions de $\mathrm{GL}_N(F)$ où $F$ est un corps non-archimédien de charactéristique quelconque. Compos. Math. **100**, 41–75 (1996)

[MW89] C. Moeglin, J.-L. Waldspurger, Le spectre résiduel de $\mathrm{GL}(n)$. Ann. Sci. École Norm. Sup. (4) **22**, 605–674 (1989)

[MW95] C. Moeglin, J.-L. Waldspurger, *Spectral Decomposition and Eisenstein Series: Une Paraphrase de l'Écriture* (Cambridge University Press, Cambridge, 1995)
[MT62] G.D. Mostow, T. Tamagawa, On the compactness of arithmetically defined homogeneous spaces. Ann. Math. **76**, 446–463 (1962)
[MP96] A. Moy, G. Prasad, Jacquet functors and unrefined minimal K-types. Comment. Math. Helv. **71**, 98–121 (1996)
[PS79] I. Piatetski-Shapiro, Multiplicity one theorems, in *Proceedings of Symposia in Pure Mathematics*, vol. 33, Part 1 (American Mathematical Society, Providence, 1979)
[PR93] V. Platonov, A. Rapinchuk, *Algebraic Groups and Number Theory*. Pure and Applied Mathematics, vol. 139 (Academic, Boston, 1993), ISBN-13:978-0125581806
[Ra72] R. Rao, Orbital integrals in reductive groups. Ann. Math. **96**, 505–510 (1972)
[Ri99] L. Ribes, *Introduction to Profinite Groups and Galois Cohomology* (Queen's University, Kingston, 1999)
[Se56] A. Selberg, Harmonic analysis and discontinuous groups in weakly symmetric Riemannian spaces with applications to Dirichlet series. J. Indian Math. Soc. **20**, 47–87 (1956)
[Se62] A. Selberg, Discontinuous groups and harmonic analysis. Proc. Int. Congr. Math. 177–189 (1962)
[Sr71] J.-P. Serre, Cohomologie des groupes discrets, in *Prospects in Mathematics*. Annals of Mathematics Studies, vol. 70 (Princeton University Press, Princeton, 1971)
[Sr94] J.-P. Serre, *Cohomologie Galoisienne*. Lecture Notes in Mathematics, vol. 5 (Springer, New York, 1994), IX + 181 pp. ISBN:978-3-540-58002-7
[Sha81] F. Shahidi, On certain $L$-functions. Am. J. Math. **103**, 297–355 (1981)
[Sha83] F. Shahidi, Local coefficients and normalization of intertwining operators for $\mathrm{GL}(n)$. Comput. Math. **48**, 271–295 (1983)
[Sha84] F. Shahidi, Fourier transforms of intertwining operators and Plancherel measures for $\mathrm{GL}(n)$. Am. J. Math. **106**, 67–111 (1984)
[Sha85] F. Shahidi, Local coefficients as Artin factors for real groups. Duke Math. J. **52**, 973–1007 (1985)
[Shal72] J. Shalika, A theorem on semisimple $\mathcal{P}$-adic groups. Ann. Math. **95**, 226–242 (1972)
[Shal74] J. Shalika, The multiplicity one theorem for $\mathrm{GL}_n$. Ann. Math. **100**, 171–193 (1974)
[She79] D. Shelstad, Characters and inner forms of a quasisplit group over $\mathbb{R}$. Compos. Math. **39**, 11–45 (1979)
[Si80] A. Silberger, *Introduction to Harmonic Analysis on Reductive p-adic Groups*. Mathematical Notes (Princeton University Press, Princeton, 1980)
[Si81] A. Silberger, Discrete series and classification for $p$-adic groups, I. Am. J. Math. **103**, 1241–1321 (1981)
[SS70] T. Springer, R. Steinberg, Conjugacy classes, in *Seminar on Algebraic Groups and Related Finite Groups*. Lecture Notes in Mathematics, vol. 131 (1970), pp. 167–266
[St74] R. Steinberg, *Conjugacy Classes in Algebraic Groups*. Lecture Notes in Mathematics, vol. 366 (Springer, Berlin, 1974)
[Tc90] M. Tadic, Induced representations of $\mathrm{GL}(n, A)$ for a $p$-adic division algebra $A$. J. Reine Agnew. Math. **405**, 48–77 (1990)
[Ta66] J. Tate, The cohomology groups of tori in finite Galois extensions of number fields. Nagoya Math. J. **27**, 709–719 (1966)
[Ta79] J. Tate, Number theoretic background, in *Proceedings of Symposia in Pure Mathematics*, vol. 33, Part 2 (American Mathematical Society, Providence, 1979)
[Ti79] J. Tits, Reductive groups over local field, in *Proceedings of Symposia in Pure Mathematics*, vol. 33, Part 1 (American Mathematical Society, Providence, 1979)
[Va07] Y. Varshavsky, Lefschetz-Verdier trace formula and a generalization of a theorem of Fujiwara. Geom. Funct. Anal. **17**, 271–319 (2007)
[Vi82] M.-F. Vignéras, Caractérization des intégrales orbitales sur un groupe réductif $p$-adique. J. Fac. Sci. Univ. Tokyo **28**, 945–961 (1982)
[Vo81] D. Vogan, *Representations of Real Reductive Groups* (Birkhaüser, Bolton/Base/Stuttgart, 1981)

[Vo82] D. Vogan, Irreducible characters of semisimple Lie groups. IV. Character-multiplicity duality. Duke Math. J. **49**, 943–1073 (1982)

[Vo86] D. Vogan, The unitary dual of GL($n$) over an Archimedean field. Invent. Math. **83**, 449–505 (1986)

[Wl95] J.-L. Waldspurger, La formule de Plancherel pour les groupes p-adiques (d'après Harish-Chandra). J. Inst. Math. Jussieu **2**, 235–333 (2003)

[Wa75] N. Wallach, On Harish-Chandra's generalized C-functions. Am. J. Math. **97**, 386–403 (1975)

[We67] A. Weil, *Basic Number Theory*. Grundlehren der mathematischen Wissenschaften, vol. 144 (Springer, New York, 1967), ISBN:978-3-662-00048-9

[Ze80] A. Zelevinsky, Induced representations of reductive $p$-adic groups II. Ann. Sci. ENS **13**, 165–210 (1980)

# Symbol Index


$A(G')$, 31
$A(\mathbb{R}^0)$, 142
$A(\rho^1)$, 536
$A^-$, 71, 81
$A_0$, 218
$A_M$, 11, 215
$A_M$-orthogonal, 261
$A_P$, 11, 215
$A_P(\mathbb{R}^0)$, 143
$A_{\mathfrak{M}}$, 253
$A_\theta$, 71
$A_c^\infty(G(F))$, 70, 73
$A_{L,\infty}$, 406
$A_P$, 139
$\mathbb{A}$, 141
$\mathbb{A}^\times$, 142
$\mathbb{A}^u$, 29, 130
$\mathbb{A}^v$, 536
$\mathbb{A}_{\text{fin}}$, 534
$\mathbb{A}_f$, 94
$\mathbb{A}_f$, 141
$\mathbb{A}$, 10
$\mathcal{A}_\rho$, 295
$\mathcal{A}^2_{Q,\chi}$, 415
$\underline{\mathrm{Ad}}(g)X$, 38
$\overline{A}_c^\infty(G(F))$, 70, 73
$(\mathfrak{a}_0^*)^+$, 255
$(\mathfrak{a}_G, \mathfrak{b})$-family, 359
$(\mathfrak{a}_G, \mathfrak{b}_1)$-family, 359
$(\mathfrak{a}_P^Q)^*$, 222
$(\mathfrak{a}_{P_1}^{P_2})^*$, 140
$(\mathfrak{a}_P^*)^+$, 143
$\alpha_G(\pi, Z)$, 424
$\mathfrak{a}^F$, 384
$\mathfrak{a}_F$, 384
$\mathfrak{a}_M$, 11, 215, 322
$\mathfrak{a}_M^*$, 11, 322
$\mathfrak{a}_M^L$, 215
$\mathfrak{a}_M^{\mathfrak{b}}$, 358
$\mathfrak{a}_P$, 11
$\mathfrak{a}_P^*$, 11, 216
$\mathfrak{a}_P^+$, 216
$\mathfrak{a}_P^Q$, 222
$\mathfrak{a}_{\mathfrak{M}}$, 253
$\mathfrak{a}_{G,S}$, 216, 322
$\mathfrak{a}_{G,S}^*$, 217
$\mathfrak{a}_{G,S}^\vee$, 217
$\mathfrak{a}_{M,S}$, 12, 288, 447, 448
$\mathfrak{a}_{M,S}^*$, 288
$\mathfrak{a}_{M,S}^\vee$, 288
$\mathfrak{a}_{M,\mathbb{C}}^*$, 11
$\mathfrak{a}_{M_\beta}$, 241
$\mathfrak{a}_{P,\mathbb{C}}^*$, 11
$\mathfrak{a}_{P_1}^{P_2}$, 140
$\mathfrak{a}_P$, 139
$\mathfrak{a}_P^+$, 143
$\mathfrak{a}_P^*$, 139
$\mathfrak{a}_{\mathfrak{M},S}$, 292
$\mathfrak{a}_{\mathfrak{M},S}^*$, 292
$\mathfrak{a}_{\mathfrak{M},S}^\vee$, 292
$\mathfrak{a}_{\mathfrak{M},\mathbb{C}}^*/i\mathfrak{a}_{\mathfrak{M},S}^\vee$, 292
$\| a \|$, 142
$a^G(\gamma)$, 318
$a^G(\pi)$, 318
$a^M(S, \gamma)$, 18, 449
$a^M(S, u)$, 398
$a^M(\tau_{1,\lambda})$, 411
$a^M(\tau_{1,\lambda}))$, 22




Y.Z. Flicker, *Arthur's Invariant Trace Formula and Comparison of Inner Forms*,
DOI 10.1007/978-3-319-31593-5

$a^{M,\mathcal{E}}(S,\gamma)$, 450
$a^{M,\mathcal{E}}(\tau_{1,\lambda})$, 457
$a_\Lambda = a(\Lambda)$, 292
$a_\chi^M$, 415
$a_{\text{disc}}^G(\pi)$, 21
$a_{\text{disc}}^G(\tau^1)$, 453
$a_{\text{disc}}^{\mathcal{E}}(\pi^1)$, 535
$a_{\text{disc}}^{M_1,\mathcal{E}}(\tau_1^1)$, 457
$a_{\text{disc},\chi}^{M_1}$, 415

$B(T/F)$, 79
$\mathbb{B}$, 51
$\mathfrak{B}_P(\tau)_\chi$, 153
$(\mathfrak{b}_\mathfrak{q}, \mathfrak{b})$-family, 359
$\mathfrak{b}$, 358
$\mathfrak{b}^G$, 358
$\mathfrak{b}_\mathfrak{p}^+$, 358
$\mathfrak{b}_{\xi,F}$, 384

$C' : R(G) \to R(G')$, 112
$C : R(G') \to R(G)$, 112
$C_F$, 385
$C_F^\eta$, 385
$C_N^\infty(\mathfrak{a}_{\mathfrak{M},S})$, 308
$C_\Delta^\infty(G(F_S)^1)$, 399
$C_c^\infty(G(F_S))$, 12
$C_c^\infty(G(F_v), Z_0(F_v), \omega)$, 55
$C_c^\infty(G(\mathbb{A}))$, 55
$C_c^\infty(G,\omega)$, 44
$C_c^\infty(\mathfrak{g})$, 67
$C_{F^\eta}$, 385
$C_c(G(\mathbb{A})^1)$, 145
$C_c^\infty(G(\mathbb{A})^1)$, 146
$C_c^m(G(\mathbb{A})^1)$, 146
$C_c^m(G(\mathbb{A})^1)^L$, 146
$\mid C \mid (\pi')$, 115
$\overline{C}_c^\infty(G(F))$, 34, 73
$(cd)_P(\lambda)$, 264
$\chi^S$, 12
$\chi_s$, 185
$\langle \chi, \chi_1 \rangle$, 97
$|\cdot|$, 11
$\text{ch}_\pi(\gamma)$, 444
$\{c(L)\}$, 235
$c(\zeta)$, 155
$c_L(\lambda)$, 358
$c_M$, 15, 263, 358
$c_M(\lambda)$, 262, 358, 386
$c_M(\nu)$, 15
$c_M^Q$, 386
$c_M^Q(\lambda)$, 358, 386
$c_P(\lambda)$, 258, 357, 386
$c_P(\nu)$, 14
$c'_Q$, 263
$c'_Q(\lambda)$, 258
$c_Q(\lambda)$, 358
$c_R(\lambda)$, 358
$c_R^Q(\lambda)$, 266
$c_\mathfrak{b}$, 359
$c_\mathfrak{b}(\lambda)$, 359
$c_\mathfrak{p}(\lambda)$, 359
$c_\mathfrak{p}(\nu)$, 387

$D$-lift, 116
$D$-lifts, 10
$D^G(\gamma)$, 444
$D^M(m)$, 280
$\Delta_P$, 216
$\Delta_P^Q$, 222
$\Delta_P^\vee$, 216
$\text{DS}(G(\mathbb{A}))$, 116
$\text{DS}(m,t,\omega)$, 119
$\mid D(\gamma) \mid$, 445
$\Delta$, 399
$\Delta(g)$, 44
$\Delta(x)$, 106
$\Delta_0$, 139
$\Delta_{P_1}^{P_2}$, 140
$\Delta_P$, 139
$\Delta_Q^\Lambda$, 169
$\Delta_\mathfrak{p}$, 358
$\Delta_\mathfrak{p}^\vee$, 359
$\widehat{\Delta}_P^Q$, 222
$\widehat{\Delta}_{P_1}^{P_2}$, 140
$\delta(\tau',\tau)$, 455
$\delta_P(a)$, 106
$\delta_P(m)$, 34
$\delta_P(p)$, 143
$\mathrm{d}\tau_{1,\lambda}$, 23, 411, 454
$d$-lift, 9, 112
$d(O(x))$, 37
$d(P',P)$, 238
$d_M^Q(\mathfrak{b},L)$, 360
$d_v$-lift, 116
$d_{Q_0}(T)$, 426

$E(X)$, 111
$E(x,\phi_\zeta)$, 153
$E_\Gamma v$, 297
$E_{P_2}(x,\phi_\zeta)$, 155
$E_P(x,\phi,\zeta)$, 206
$\mathcal{E}(\mathfrak{h}^1)^W$, 525

$\mathcal{E}(\mathfrak{h})^W$, 423
$\mathcal{E}_\Omega$, 293
$\epsilon$-close, 136
$\epsilon_Q^P(\Lambda)$, 168
$e^G(\gamma)$, 445
$e^G(\gamma_v)$, 445
$e_S(G)$, 444
$e_v(G)$, 444

$F$, 10
$F$-elliptic, 17, 381, 398
$F$-forms, 79
$F$-rank of $G$, 60
$F(P_1, P_2)$, 193
$F^\eta$, 385
$F^1(x, T)$, 170
$F^{P_1}(x, T)$, 170
$F_S$, 10, 216
$F_\infty$, 405
$F_\infty^\times$, 117
$F_P$, 154
$F_{\text{good}}(G(F))$, 74
$F_{\text{tr}}(G(F))$, 74
$\mathbb{F}$, 51
$\mathcal{F}(M)$, 215, 441
$\mathcal{F}^L(M)$, 215
$\mathcal{F}(M)$, 11, 321
$\mathcal{F}(M, \xi)$, 387
$\mathcal{F}(\mathfrak{b})$, 359
$\mathcal{F}(\Pi)$, 383
$\mathcal{F}(\Pi, \xi)$, 384
$\|f\|_{X,Y}$, 146
$f \mapsto f'$, 442
$f^*(h)$, 93
$f^0$, 53
$f^1$, 395
$f^S$, 94
$f^Z$, 289
$f^b(x)$, 309
$f^y(x)$, 231
$f_G$, 308
$f_G(\pi, X)$, 322
$f_G(\pi, Z)$, 308
$f_G(\pi^1)$, 453
$f_L(\tau)$, 340
$f_M$, 16, 34
$f_M(\tau)$, 443
$f_M(\tau, X)$, 322
$f_N(m)$, 34, 106
$f_P$, 322
$f_Q(m)$, 229
$f_S$, 94
$f_\alpha$, 423
$f_t$, 72
$f_v^0$, 55
$f_{EP}$, 52
$f_{Q,y}(m)$, 232, 286
$f_\omega(x)$, 159
$f_{i,\Gamma}(m')$, 297

$(G(F))$, 318
$(G(F))_{G,S}$, 17
$(G, M)$-family, 14, 262
$(G, S)$-equivalence classes, 397
$(G, S)$-equivalent, 17
$G$, 10, 139
$G'$-cuspidal, 8, 89, 90
$G'(F)_G$, 382
$G'(F)_{\text{ell}}$, 381
$G(F)$-domain $D$, 68
$G(F)_\sigma$, 51
$G(F_S)$, 216
$G(F_S)^Z$, 12, 289, 325
$G(F_S, N)$, 307
$G(F_\infty)^1$, 145
$G(F_v)^1$, 145
$G^{\text{ss}}$, 52
$G_X$, 67
$G_\gamma$, 11
$G_\gamma(F_S)$, 445
$G_\gamma^0$, 52
$G_x$, 33
$G_t$, 144
$\Gamma_Q'(H, X)$, 224
$\mathbb{G}_m$, 142
$\Gamma(x)$, 164
$\Gamma_\Omega(\sigma_\Lambda^M, P_0)$, 299
$\widetilde{\Gamma}_Q(H, X)$, 225
$\gamma \mapsto \gamma'$, 442
$\gamma \sim \gamma'$, 397
$\gamma^{M_1}$, 365
$\gamma^{\mathfrak{b}}$, 366
$\mathfrak{g}$, 444
$\mathfrak{g}_\sigma$, 444
$\mathfrak{g}$, 32
$GG_\lambda$, 71

$H(R, \gamma)$, 146
$H(\gamma)$, 146
$H(\gamma)^0$, 146
$H^1(F, A)$, 58
$H_G$, 216
$H_M$, 11
$H_P$, 326
$H_P(w_s^{-1})$, 220

$H_P(x)$, 219
$H^i_e(G(F), V)$, 54
$H_{P_v}$, 326
$H_P(x)$, 142
$H_{\mathfrak{M}_1}$, 401
$\mathbb{H}(G(F_S))$, 307, 442, 447
$\mathbb{H}(G(F_S))^0$, 483
$\mathbb{H}(G(F_S))_\Gamma$, 307
$\mathbb{H}(G(F_S), M)$, 507
$\mathbb{H}(G(F_v))$, 441
$\mathbb{H}(G(\mathbb{A}))^0$, 483
$\mathbb{H}(G(\mathbb{A}), M)$, 507
$\mathbb{H}(G(\mathbb{A}), M)^0$, 526
$\mathbb{H}(G(\mathbb{A}), M)_{K_1}$, 535
$\mathbb{H}(G(\mathbb{A}), M)^0_{K_1}$, 535
$\mathbb{H}_N(G(F_S))_\Gamma$, 307
$\mathbb{H}_{\mathrm{ac}}(G(F_S))$, 310
$\mathbb{H}(G(\mathbb{A}))$, 12, 393
$\mathbb{H}(G(\mathbb{A})^1)$, 393
$\mathbb{H}^\Phi(G(F_S))$, 339
$\mathbb{H}_{\mathrm{ac}}(G(F_S))$, 12
$\mathbb{H}'_{\mathrm{ac}}(G(F_S))$, 325
$\mathbb{H}_{\mathrm{ac}}(G(F_S))$, 323
$\mathbb{H}_{\mathrm{ac}}(G(F_S))_\Gamma$, 322
$\mathbb{H}_{\mathrm{ac}}(G(\mathbb{A}))$, 393
$\mathbb{H}_{\Delta_1}(G(F_{S_1}))$, 400
$\mathcal{H}_P(\tau)$, 152
$\mathcal{H}_P(\tau)_{\chi,K_0}$, 153
$\mathcal{H}_P(\tau)_{\chi,K_0,W}$, 153
$\mathcal{H}_P(\tau)_\chi$, 153
$\mathcal{H}^0_P(\tau)$, 152
$\widetilde{\mathbb{H}}(G(F_S))$, 339
$\widetilde{\mathbb{H}}(L(F_S))$, 340
$\widetilde{\mathbb{H}}_{\mathrm{ac}}(L(F_S))$, 340
$\mathbb{H}_{\mathrm{ac}}(G(\mathbb{A}))$, 12
$\mathfrak{h}$, 405
$\mathfrak{h}_0$, 406
$\mathfrak{h}_K$, 406
$\mathfrak{h}^*_u$, 431
$\mathfrak{h}_\mathbb{C}$, 406
$h(H)$, 114
$h(\pi)$, 114
$h_L(X)$, 475
$h_M$, 304
$h_\mathfrak{L}$, 304
$h_\mathfrak{b}(X)$, 371

$I(S, f, x, y)$, 159
$I(g, f)$, 44
$I^G(x, f)$, 33
$I^L$, 236
$I^\mathcal{E}(f)$, 450
$I_*(f)$, 396
$I_G(\gamma, f)$, 318
$I_G(\gamma)$, 18
$I_G(\gamma, f)$, 445
$I_G(\pi, f)$, 318
$I_M(\gamma)$, 19
$I_M(\gamma, f)$, 329, 469
$I_M(\tau, X, f)$, 23, 334, 474
$I_M(\tau, \mathfrak{X}, f)$, 418
$I_M(\tau^1, f)$, 395
$I^L_M(\tau, X)$, 334
$I^\mathcal{E}_M(\gamma, f)$, 472
$I^\mathcal{E}_M(\rho, X, f)$, 475
$I^\mathcal{E}_M(\tau, X, f)$, 476
$I^\mathcal{E}_M(\tau^1, f)$, 476
$I^L_\chi$, 237
$I^M_\gamma(f)$, 280
$I^L_\mathfrak{o}$, 237
$I_{\mathrm{geom}}(f)$, 449
$I_{\mathrm{spec}}(f)$, 452
$I_t(f)$, 414, 452
$I_{L_1}(\gamma^{L_1}, f)$, 330
$I_{M,\mu}(\tau, X, f)$, 335
$I_{M,\varepsilon}(\tau, X, f)$, 371
$I_{M_1}(\gamma^{M_1}, f)$, 365
$I_P(\tau_\zeta)$, 153
$I_\mathfrak{b}(\gamma^\mathfrak{b}, f)$, 366
$I_\mathfrak{b}(\gamma_1, f)$, 366
$I_\mathfrak{b}(\tau, X, f)$, 371
$I_\mathfrak{b}(\tau_1, X_1, f)$, 370
$I^{\mathfrak{b}_1}_\mathfrak{b}(\gamma_1)$, 366
$\mathbb{I}(G(F_S))$, 309, 443
$\mathbb{I}(G(F_S))^0$, 483
$\mathbb{I}(G(F_S))_\Gamma$, 309
$\mathbb{I}(G(\mathbb{A}))^0$, 483
$\mathbb{I}_N(G(F_S))_\Gamma$, 308
$\mathbb{I}_\mathfrak{o}$, 538
$\mathbb{I}_{\mathrm{ac}}(G(F_S))$, 310
$\mathbb{I}_{\mathrm{ac}}(G(F_S))_\Gamma$, 310
$\mathbb{I}(G(F_S))$, 16
$\mathbb{I}(G(\mathbb{A}))$, 394
$\mathbb{I}(M(\mathbb{A}))$, 394
$\mathbb{I}^\Phi(G(F_S))$, 339
$\mathbb{I}_{\mathrm{ac}}(G(F_S))$, 16
$\mathbb{I}'_{\mathrm{ac}}(G(F_S))$, 325
$\mathbb{I}_{\mathrm{ac}}(G(F_S))$, 323
$\mathbb{I}_{\mathrm{ac}}(G(\mathbb{A}))$, 394
$\mathbb{I}_{\mathrm{ac}}(M(\mathbb{A}))$, 394
$\mathrm{Im}(\nu_{\tau^1})$, 406
$\mathrm{Int}(g)x$, 38
$\mathrm{Irr}(G(F))$, 6
$\mathrm{Irr}(G)$, 11
$\widehat{I}$, 236
$\widehat{I}(f_G) = I(f)$, 443

$\widetilde{I}^{\mathcal{E}}(f_G)$, 450
$\widetilde{I}^L_M(\gamma, \phi_L(f))$, 329
$\widetilde{I}^G(x,f)$, 33
$\widetilde{\mathbb{I}}(G(F_S))$, 339
$\widetilde{\mathbb{I}}(L(F_S))$, 340
$\mathbb{I}_{\rm ac}(L(F_S))$, 340
${}^M I(f^M)$, 80
${}^c I_M(\gamma,f)$, 343
${}^c I_M(\rho,X,f)$, 496
${}^c I_M(\tau,X,f)$, 343, 352
${}^c I_{M,\mu}(\tau,X,f)$, 343
$\iota(\rho,\pi)$, 254
$\iota(\rho,\tau)$, 455
$\mathrm{ind}_M^G\,\tau$, 69
$\mathrm{inv}\,A$, 4, 85
$i^G(\sigma)$, 18
$i^M(S,\sigma)$, 17, 399, 449
$i_M^G\tau$, 69
$i_M^{G*}$, 76
$i_P(\tau,X)^*$, 155
$i_P(\tau,h)^*$, 155
$i_P^G(\tau)$, 443
$i_P^G(\tau_\lambda)$, 13
$i_P^G(\tau_\lambda,f)$, 219
$i_{P'}(\tau_\lambda,f)$, 349
$i_{P'}^G(\tau,f^Z)$, 289

$J(\omega)$, 67
$J_M(\gamma,f)$, 19, 327
$J_M(\rho,X)$, 290
$J_M(\rho_\lambda,f)$, 275
$J_M(\sigma_\Lambda,f)$, 274
$J_M(\tau,X,f)$, 15, 288
$J_M(\tau^1,f)$, 395
$J_M(\tau_\lambda)$, 15
$J_M(\tau_\lambda,f)$, 274, 352
$J_M(\tau_\lambda,f^Z)$, 289
$J_M^L(\gamma,f)$, 281
$J_M^L(\tau,f)$, 282
$J_P(\tau)$, 65
$J_\chi^L(f)$, 230
$J_\chi^{L,T}$, 223
$J_\chi^{L,T}(f)$, 230
$J_{\mathfrak{o}}^L(f)$, 230
$J_{\mathfrak{o}}^T(f)$, 224
$J_{\mathfrak{o}}^{L,T}$, 223
$J_{\mathfrak{o}}^{L,T}(f)$, 230
$J_c^\infty(G(F))$, 70, 73
$J_t(f)$, 412
$J_{M,\gamma}^L(f)$, 281
$J_{M,\gamma}^{M_Q}(f)$, 281
$J_{M,\mu}(\tau,X,f)$, 336
$J_{M,\tau}^L(f)$, 283
$J_{P,\mathfrak{o}}(x,y)$, 177
$J_\chi^T(f)$, 201
$J_{\mathfrak{b}}(\gamma_1,f)$, 365
$J_{\mathfrak{b}}(\tau_1,X_1,f)$, 370
$J_{\mathfrak{o}_i}^{L,T}$, 223
$J_{\mathfrak{o}}^T(f)$, 177
$J^\infty(G(F))$, 34
$J_{L_1}(\gamma^{L_1},f)$, 330
$J_{M,\tau}^{M_Q}(f_Q)$, 283
$J_{\mathfrak{b}}(\tau,X,f)$, 371
$j_{\mathfrak{o}}^T(x,f)$, 177

$K$, 11, 141, 218
$K$-finite function, 157
$K_S^M$, 398
$K_v$-finite, 12, 441, 523
$K_{P,\chi}(x,y)$, 162, 229
$K_{P,\mathfrak{o}}(x,y)$, 150, 224
$K_P(x,y)$, 150
$K_\chi(x,y)$, 162
$K_{\mathfrak{o}}(x,y)$, 150
$K_f(x,y)$, 146
$k_P(x)$, 219
$k_\chi^T(x,f)$, 165
$k_{\mathfrak{o}}^T(x,f)$, 165

$(L_hf)(x)$, 286
$(L_yf)(x)$, 286
$(L_{Q,h}f)(m)$, 286
$(\Lambda^{T,P_1}\phi)(x)$, 190
$(\Lambda^T\phi)(x)$, 183
$L$ comes from $G$, 381
$L$-group, 247
$L$-vector, 103
$L(G(F)\backslash G(\mathbb{A}))$, 55, 88
$L(G(F)\backslash G(\mathbb{A}))$, 7
$L(\pi',r)$, 115
$L(s,\eta)$, 246
$L(w)$, 239
$L^-$, 81
$L^2(G(F)\backslash G(\mathbb{A}))$, 88
$L^2(G(F)\backslash G(\mathbb{A}))$, 7
$L^2(N_P(\mathbb{A})M_P(F)\backslash G(\mathbb{A})^1)_\chi$, 229
$L^2_{\rm disc,t}(M(F)A_{M,\infty}\backslash M(\mathbb{A}))$, 20
$L^2(M(F)\backslash M(\mathbb{A})^1)$, 150
$L^2(M(F)\backslash M(\mathbb{A})^1)_\chi$, 152
$L_0(G(F)\backslash G(\mathbb{A}))$, 88
$L_0 = L_0(G(F)\backslash G(\mathbb{A}))$, 57
$L_0 = L_0(G(F)\backslash G(\mathbb{A}))$, 7
$L_c(G(F)\backslash G(\mathbb{A}))$, 88

$L_c = L_c(G(F)\backslash G(\mathbb{A}))$, 7
$L_d = L_d(G(F)\backslash G(\mathbb{A}))$, 88
$L_d = L_d(G(F)\backslash G(\mathbb{A}))$, 7
$L_w$, 75
$L^2_{\text{cusp}}(M(F)\backslash M(\mathbb{A})^1)$, 150
$\Lambda(z)$, 293
$\Lambda_\Omega$, 293
$\mathcal{L}$, 117, 441
$\mathcal{L}(M)$, 215, 441
$\mathcal{L}^L(M)$, 215
$\mathcal{L}^M$, 441
$\mathcal{L}$, 11
$\mathcal{L}(M)$, 11, 321
$\mathcal{L}(\mathfrak{b})$, 359
$\mathcal{L}_0(M)$, 11
$\Lambda_1^{T,P_1}$, 192
$\Lambda_2^{T,P_1}$, 192
$\widehat{L}_{\mathcal{P}}$, 154
$\lambda \in \mathfrak{a}^*_{M,\mathbb{C}}$, 13
$\lambda_Q$, 258
$l(\pi)$, 111

$(M(s,\tau_\zeta)\phi_\zeta)(x)$, 153
$(M,\rho)$, 151
$(M,c)$, 146
$M$, 11
$M(F_S)^X$, 495
$M(\mathbb{A})^1$, 447
$M(\mathbb{A})^1$, 11
$M_0$, 11, 218
$M_P$, 11, 215
$M_\lambda$, 71
$M_\theta$, 71
$M_v(F_v)'$, 401
$M_w$, 75
$M_P$, 139
$M_{\mathfrak{b}}$, 359
$\mathcal{M}(s,0)$, 453
$\mathcal{M}(s,\tau,\zeta)$, 206
$\mathcal{M}(s,\zeta)$, 206
$\mathcal{M}_P^T(\tau)_\chi$, 202
$\mathfrak{M}$, 278
$\mathfrak{M}(F_S)$, 253
$\mathfrak{M}_1$, 401
$\mathfrak{M}_1(F_{S_1})$, 401
$\mathfrak{M}_1(F_{S_1})'$, 401
$\mathrm{MW}(\tau',r)$, 122
$\mu$-function, 257
$\mu(\Gamma\backslash G(F))$, 52
$\mu_{\mathcal{O}}(\phi)$, 67
$m(\pi)$, 57
$m_M^L(\gamma,a)$, 327
$m_M^L(\tau_\lambda)$, 22
$m_M^L(\tau_\lambda,\rho_\lambda,R')$, 276
$m_M^Q(\tau_\lambda,\rho_\lambda,P')$, 276
$m_P(\nu,\gamma,a)$, 472
$m_P(\nu,\tau_\lambda,P')$, 22
$m_P(\nu,\tau_\lambda,P_0)$, 464
$m_P(\nu,\tau_\lambda,\rho_\lambda,P')$, 275
$m_P(x)$, 219
$m_R^Q(\nu,\tau_\lambda,\rho_\lambda,P')$, 276
$m_{M,\mu}^L(\tau,X,J_L(f))$, 290
$m_{M_1}^M(\tau_{1,\lambda})$, 454

$N(M(F),G(F))$, 72
$N(M,G)$, 80
$N_P$, 215
$N_I$, 188
$N_I^-$, 188
$N_{P_1}^{P_2}$, 140
$N_P$, 139
$\mathcal{N}$, 305
$\mathfrak{N}$, 67
$\overline{N}$, 72
$\mathfrak{n}^I(F)'$, 188
$\mathfrak{n}_P$, 143
$\nabla(f)$, 294
$\nu_{\mathfrak{L}}$, 304
$\nu_\tau$, 113, 406
$\nu_{\tau^1}$, 406, 453
$n$-admissible, 130
$n(A)$, 145
$n_0$-admissible element, 135
$n_0$-admissible function, 135
$n_P(x)$, 219
$n_P(A_1)$, 145

$O(x)$, 37
$O_\gamma(f)$, 52
$\Omega = (\mathcal{E}_\Omega, \Lambda_\Omega)$, 293
$\mathcal{O}$, 146, 222
$\mathcal{O}^L$, 223
$\mathcal{O}_v$, 141
$\mathcal{O} = \mathcal{O}(G,F)$, 397
$\Omega(\mathfrak{a}_0;P)$, 183
$\Omega(\mathfrak{a}_1;P)$, 181, 207
$\Omega(\mathfrak{a}_1,\mathfrak{a}_2)$, 142
$\mathfrak{o}$, 180
$\omega(c,X)$, 338
$\omega_P$, 288, 335, 474
$\omega_P(X)$, 340
$\pi(f\,\mathrm{d}x)$, 56

$P$, 11
$P_\beta$, 241
$P_\lambda$, 71
$P_t$, 71
$P_0$, 139
$P_S$, 161
$P_\Lambda$, 169
$\Pi_{\text{temp}}(M(F_S))$, 240
$\Psi$, 305
$\mathcal{P}(M)$, 215, 441
$\mathcal{P}^L(M)$, 215
$\mathcal{P}_\chi$, 151
$\mathcal{P}(M)$, 11, 321
$\mathcal{P}(M, \xi)$, 387
$\mathcal{P}(\mathfrak{b})$, 358
$\mathcal{P}(\Pi)$, 383
$\mathcal{P}_{\text{ext}}(M, \xi)$, 387
$\overline{P}$, 72
$\Phi^G(x, f)$, 33
$\Pi(M)$, 152
$\Pi(G(F))$, 69
$\Pi(G(F_S))$, 443
$\Pi(G(\mathbb{A})^1)$, 395
$\Pi(G)$, 11
$\Pi(M(\mathbb{A})^1, t)$, 453
$\Pi(M, \chi)$, 415
$\Pi(M, t)$, 22, 411, 454
$\Pi(\xi)$, 384
$\Pi^\eta$, 385
$\Pi^{\mathcal{E}}(M, t)$, 504
$\Pi^{P_2}(M)$, 152
$\Pi_M$, 386
$\Pi_M(\xi)$, 388
$\Pi_M^Q$, 386
$\Pi_{M,\mathfrak{b}}$, 386
$\Pi_{M_1}(M, t)$, 22, 411, 454
$\Pi_{M_1}^{\mathcal{E}}(M, t)$, 504
$\Pi_{\mathfrak{b}}$, 384
$\Pi_{\text{disc}}(G^1, t)$, 22
$\Pi_{\text{temp}}(M(F_S))$, 16
$\Pi_{\text{unit}}(M(\mathbb{A})^1, t)$, 20
$\Pi_{\nu_1, K_1}(G(\mathbb{A})^1)$, 535
$\Pi_{\text{disc}}(G^1, t)$, 409
$\Pi_{\text{disc}}(L(F_v))_\kappa$, 537
$\Pi_{\text{disc}}(M_1^1, \chi)$, 415
$\Pi_{\text{disc}}(M_1^1, t)$, 453
$\Pi_{\text{disc}}^{\mathcal{E}}(M_1^1, t)$, 504
$\Pi_{\text{temp}}(G(F_S))$, 443
$\Pi_{\text{temp}}(G(\mathbb{A})^1)$, 395
$\Pi_{\text{unit}}(G(F_S))$, 443
$\Pi_{\text{unit}}(G(\mathbb{A})^1)$, 395
$\Pi_{\text{unit}}(M(\mathbb{A})^1, t)$, 406
$\Psi_2(\Lambda, x)$, 209
$\Psi_{Q,\chi}^T(\Lambda, f_\alpha)$, 426
$\varepsilon_P$, 335
$\widehat{\Pi}$, 384
$\widetilde{\Phi}^G(x, f)$, 33
$\widetilde{\Pi}_M^{Q_L}$, 388
(PW(2)), 74
$\hat{\psi}^M(m)$, 152
$\mathfrak{p}(M)$, 379
$\mathfrak{p}(\mathfrak{o})$, 379
$\mathfrak{p}(k)$, 379
$\phi'(i)$, 236
$\phi(\pi, Z)$, 16
$\phi_M$, 323
$\phi_M(f)$, 16, 291
$\phi_M(f, \sigma_\Lambda^M)$, 292
$\phi_M(f, \tau, X)$, 291
$\phi_M(f, \tau_\lambda)$, 349, 351
$\phi_M^L$, 236
$\phi_M^L(f, \tau)$, 273
$\phi_P^L(\lambda, f, \tau, P_0)$, 271
$\phi_{M,\mu}(f)$, 311
$\phi_{M,\mu}(f, \tau, X)$, 311
$\phi_{M,\mu}^b(f, \sigma, \mathfrak{X})$, 312
$\phi_{M,\nu_P}(f, \tau, X)$, 341
$\phi_{P_1,P_2}(y)$, 187
$\phi_Q^P(\Lambda, H)$, 168
$\phi_{\mathfrak{b}}$, 366
$\phi_\zeta(x)$, 153
$\pi'$ comes from $\pi$, 8, 89
$\pi'$ comes locally from $G(\mathbb{A})$ , 116
$\pi_v'$ comes from $\pi_v$, 8, 89
$\pi(u(\sigma', r), s)$, 113
$\pi(u(\sigma, r), s)$, 113
$\pi^S$, 94
$\pi_0^u$, 29
$\pi_1 << \pi_2$, 112
$\pi_1 < \pi$, 110
$\pi_S$, 94
$\pi_v$ corresponds to $\pi_v'$, 89
$\pi_v'$ comes from $\pi_v$, 116
$\pi_\zeta$, 152
$\prod U$, 115
$\psi(\Lambda)$, 293
$\psi_k$, 239
$\psi_2(x)$, 209
$\varphi(\pi_\lambda, Z)$, 308
$\varphi(\sigma, \mathfrak{X})$, 309
$\varphi(\tau, X)$, 448
$\varphi^b(\pi, Z)$, 310
$\varphi_\alpha$, 423
$\varpi_\alpha$, 185

${}^{c}\phi_M(f, \tau, X)$, 341
PPW(1)), 74
$P\Pi(M(F_S))$, 240

$Q(R)$, 215
$Q : \mathrm{Irr}(G) \rightarrow \mathrm{Irr}(G')$, 112
$q(G)$, 52, 54
$q_{\tau,\Gamma}(\lambda)$, 348

$(R_y f)(x)$, 286
$(R_{Q,h} f)(m)$, 286
$R^*(G(F))$, 74
$R_I(G')$, 31
$R_I(G(F))$, 69
$R_{\mathbb{Z}}(G(F))$, 69
$R_{\mathfrak{L}}(\mu, \mathcal{N})$, 305
$R_{Q,h}$, 16
$\mathbb{R}^0$, 142
$\mathbb{R}^{\times}_{>0}$, 142
$\mathcal{R}_M(\rho_\lambda, P')$, 275
$\mathcal{R}_M(\tau_\lambda, P')$, 274
$\mathcal{R}^Q_M(\sigma^M_\Lambda, P)$, 296
$\mathcal{R}_P(\lambda, \tau, P_0)$, 270
$\mathcal{R}_P(\nu, \rho_\lambda, P')$, 275
$\mathcal{R}_P(\nu, \tau_\lambda, P')$, 274
$\mathcal{R}_M(\tau_\lambda, P_0)$, 15
$\mathcal{R}_P(\nu, \tau_\lambda, P_0)$, 15
$\mathcal{R}_{\mathfrak{b}_1}(\tau_1^{\mathfrak{b}_1}, Q_0)$, 368
$\mathrm{Res}_\Omega\, \psi(\Lambda_0)$, 293
$\mathrm{Res}_\Omega\, \psi(\Lambda_\Omega + \lambda)$, 305
$\overline{\rho}$, 253
$\rho(f_{i,\Gamma})$, 297
$\rho_{P,t}(\Lambda)$, 21
$\rho_P$, 143
$\rho_{\mathfrak{p}(P)}$, 388
$r(f\,\mathrm{d}x)$, 56
$r^{G*}_M$, 76
$r_P$, 150
$r_P(f)$, 150
$r_{Q,\chi}(\Lambda)$, 415
$r_{Q,\chi}(s, \Lambda, D^m f^1_\alpha)$, 428
$r_{Q,t}$, 453
$r_{Q,t}(\Lambda)$, 407
$r_{Q,t}(s, \Lambda, f^1)$, 408
$r_S(f)$, 161
$r_\chi(f)$, 162
$r_{\mathrm{cusp}}$, 214
$r_x(X)$, 155
$\mathrm{R}(R_h f)(x)$, 286
$\mathrm{rr}_0(f)$, 57

$S$, 10
$S_G$, 483
$S_\infty$, 12, 94, 216
$S_\Delta$, 399
$S_e(G(F))$, 69
$S_{\mathfrak{o}}$, 398
$S_{\mathrm{ram}}$, 12, 393
$\Sigma(G(F_S))$, 253
$\Sigma_P(G, T)$, 247
$\Sigma^r_P$, 241
$\mathbb{S}$, 52
$\mathcal{S}$, 164
$\mathcal{S}$, 158
$\mathcal{S}^{P_1}(T_0, T, \omega)$, 170
$\mathcal{S}^{P_1}(T_0, \omega)$, 169
$\Sigma$, 147
$\Sigma(\gamma)$, 147
$\Sigma(G(F_S))$, 444
$\Sigma(\beta)$, 358
$\Sigma^r_P$, 364
$\Sigma_{\mathfrak{o}}$, 537
$\Sigma_{\nu_1,K_1}(G(\mathbb{A})^1)$, 535
$\mathrm{sgn}_\sigma(g)$, 52
$\mathrm{supp}(\pi)$, 111
$\mathrm{supp}(f)$, 403
$\sigma^M(m)$, 279
$\sigma^{\mathfrak{M}}$, 279
$\sigma^2_1(H)$, 166
$\sigma^{P_2}_{P_1}(H)$, 166
$\overset{(M,\sigma)}{\sim}$, 330

$T_0$, 220
$T_P(\nu, \tau_\lambda)$, 276
$T_\nabla(A(\rho))$, 295
${}^{x}T(\phi)$, 67
$(\tau^G_\lambda)^1$, 409
$\hat{\tau}^{P_2}_{P_1} = \hat{\tau}^2_1$, 163
$\hat{\tau}_P$, 163
$\hat{\theta}$, 324
$\hat{\theta}(f_G)$, 16
$\tau \in \Pi(M(F_S))$, 13
$\tau^1$, 395
$\tau^G$, 443
$\tau_1 \times \cdots \times \tau_r$, 110
$\tau^Q_P$, 222
$\tau_\lambda(m)$, 13, 218, 395, 447
$\tau^G_\lambda$, 219, 322
$\tau^{\mathfrak{b}}_\lambda$, 371
$\tau_u$, 113
$\tau_{1,\lambda}$, 411

$\tau_{P_1}^{P_2} = \tau_1^2$, 163
$\tau_P$, 163
$\tau_{v,\lambda_v}(m_v)$, 278
$\theta_M(f)$, 345
$\theta_P(\nu)$, 15
$\theta_Q^R(\Lambda)$, 226
$\theta_{\Gamma_L}(k_L)$, 296
$\theta_{\mathfrak{p}}(\lambda)$, 359
$\widehat{\tau}^{Q/P}(H)$, 225
$\widehat{\tau}_P^Q$, 222
$\widehat{\theta}_Q^R(\Lambda)$, 226
${}^c\theta_M(f)$, 345
${}^c\theta_M(f, \tau_\lambda)$, 350, 351, 495
${}^c\theta_M^{\mathcal{E}}(f, \tau_\lambda)$, 496

$U(L)$, 235
$U(L)'$, 235
$U_\infty$, 37
$U_d$, 37
$\mathcal{U}(\mathfrak{g}(F_\infty)^1 \otimes \mathbb{C})$, 145, 155
$\mathcal{U}$, 17
$\{u\,;\,\sigma u \sim \gamma\}$, 18
$u(\sigma', r)$, 113
$u(\sigma, r)$, 113

$V(M)$, 236
$V(f)$, 404
$V_P(\tau)$, 13, 219
$\varepsilon_L$, 509
${}^c\varepsilon_M(\gamma, f)$, 512
$v(x, T)$, 180
$v_M(x)$, 18, 326
$v_P(\lambda, x)$, 18, 267, 327
$v_Q'(k, y)$, 232
$v_{M_v}(\lambda, x_v)$, 326
$v_{M_v}(x_v)$, 326

$W(M(F), G(F))$, 72
$W(M, G)$, 80
$W(M, L)$, 75
$W(\mathfrak{a}_L)$, 453
$W(\psi_v)$, 88
$W^G(\mathfrak{a}_L)_{\mathrm{reg}}$, 408
$W^G(\mathfrak{a}_M)$, 21
$W^G(\mathfrak{a}_M)_{\mathrm{reg}}$, 21
$W_0^G$, 11
$W_M$, 75

$X(G(F))$, 74
$X(M)_F$, 11, 215
$X(M_P)$, 139
$X(\pi)$, 111
$X^L$, 146
$X^{\mathrm{reg}}$, 35
$X_F^\eta$, 385
$\Xi^M(m)$, 279
$\mathfrak{X}$, 151, 222
$\mathfrak{X}(G)$, 214
$\mathfrak{X}^L$, 223
$\mathfrak{X} = \mathfrak{X}(G, F)$, 405
$\mathfrak{X}_T$, 432
$\xi(\pi, \rho)$, 254
$\xi(\tau', \rho)$, 455
$\xi(\tau, \rho)$, 455
$\xi_v^0$, 56
$x$-regular, 132

$Z$, 141
$Z(\tau', r)$, 112
$Z(\tau, r)$, 113
$Z_u(\tau', r)$, 112
$\mathfrak{z}$, 141
$z(\sigma)$, 113
$z(\tau)$, 110, 113
$Z\mathcal{Z}(G)$, 255

# Subject Index


additional induction assumption, 416
adjacent parabolic subgroups $P$ and $P'$ in $\mathcal{P}(M)$, 14
admissible $A(\sigma_\Lambda^M)$, 296
admissible complex representation, 56
admissible reflection, 101
almost compact support, 310
almost compactly supported functions, 12
Arthur's invariant trace formula, 10, 449
associated parabolic subgroup, 142
automorphic $(\pi, V)$, 56
automorphic $G(\mathbb{A})$-module, 8, 88, 108

block equivalence, 253, 256
Borel subgroup, 79

central exponent of $\pi$, 81
chambers of $\mathfrak{a}$, 145
character of $\pi$, 68
character relation, 6
closure property, 12, 217, 447
coinvariant, $M(F)$-module $\pi_N$ of $N(F)$, 71
Combinatorial Lemma, 76
continuous spectrum, 7, 88
convex polytope $\Pi$, 383
coroot, 139
cuspidal $G(\mathbb{A})$-module, 7, 88, 93, 135
cuspidal function, 57, 58
cuspidal representation, 151
cuspidal support, 101
cuspidal support of $\pi$, 109

decay (of central exponent), 81
Deligne-Kazhdan correspondence, 90
descent formula, 20
discrete function, 31, 76, 129, 433
discrete series, 5, 82
discrete series $\pi$, 87
discrete spectrum, 7, 88
distribution, 235
distribution $T$, 67
distribution on $G(F_S)$ induced from $\nabla$, 296
division algebra splits, 8
dominates a linear form, 74
dual cone $\mathfrak{a}_F^+$, 384
dual lattice, 1

eighth induction assumption, 370
Eisenstein series, 153
elementary operations, 111
elliptic $G(F_v)$-module, 93
elliptic (regular) element $\gamma$, 146
elliptic regular $x$, 32
elliptic semisimple $x$, 32
equivalent elements $\gamma$ and $\gamma'$, 146
equivalent elements in $G(F)$, 397
equivalent pairs, 151
essentially tempered, 65
Euler-Poincaré function, 52

fifth induction assumption, 345
first induction assumption, 329
Fourier transform, 1, 67, 551–553




Y.Z. Flicker, *Arthur's Invariant Trace Formula and Comparison of Inner Forms*,
DOI 10.1007/978-3-319-31593-5

Fourier transform $\widehat{T}$, 67
fourth induction assumption, 342
Frobenius reciprocity, 71

Galois cohomology, 63
generic $\pi$, 7, 89
Geometric Lemma, 75
Geometric Part of Main Theorem, 451
germ expansion, 34, 41
germs, 40
global Hecke algebra, 12
global intertwining operator, 153
global normalizing factor, 22
global Whittaker model, 89
good form, 31, 74
Grothendieck group, 5, 10, 66, 69

Harish-Chandra map $H_M$, 447
Harish-Chandra's $\mu$-function, 461
Hecke algebra, 12, 273
Hecke space $\mathbb{H}(G(F_S))$, 321
height function, 144
Hermitian $\pi$, 114

induced orbit $\gamma^G$, 330
induced space $\gamma^{L_1}$, 330
induction assumption, 396, 476
inner form of $G'$, 442
inner forms, 80
inner twisting, 79
integral Grothendieck group, 69
intertwining operator, 13, 14, 238
invariant distribution, 38, 67, 231, 287
invariant linear form, 13, 18, 23, 318, 319
invariant orbital integral, 514
invariant trace formula, 237
Iwasawa decomposition, 142

Langlands classification, 109
Langlands quotient, 109
legnth $l(X)$, 111
Levi $S$-subgroup, 278
Levi decomposition, 11
lift of $G(\mathbb{A})$ module, 8
lift of $G(F_v)$-module $\pi_v$, 107
lift of irreducible $G(\mathbb{A})$-module $\pi = \otimes_v \pi_v$, 89
line generated by $\tau$, 111
linked segments, 111

matching function, 91
matching functions, 80
matrix coefficient of $\pi$, 65
moderate function, 425
moderate function $\varphi \in \mathbb{I}_{ac}(G(F_S))$, 310
moderate function $f \in \mathbb{H}_{ac}(G(F_S))$, 310
multiplicity one theorem, 89, 90, 117
multiplicty free representation, 101
multisegment, 111

new induction assumption, 508
nilpotent distribution $\mu$, 68
non-degenerate $\pi$, 88
non-degenerate $G'(\mathbb{A})$-module, 7
norm map, 80
normalized coefficient, 27
normalized intertwining operator, 240
normalized orbital integral, 445
normalizedly induced, 65, 69
normalizing factor, 240
normalizing factors, 14

orbital integral, 2, 18, 128, 132, 280, 320, 417, 421, 442, 474, 484, 509, 513, 514
orbital integral of $f$, 33, 52

Paley-Wiener space $\mathbb{I}(G(F_S))$, 322, 443
Paley-Wiener spaces, 393
parabolic subgroup, 3, 14, 65, 72, 73, 141, 170, 183
Plancherel formula, 461
Poisson summation formula, 1, 317
polar set of $\Pi$, 384
primitive element, 144
product formula, 142
proper Levi subgroup, 84, 99, 124, 487
pseudo-coefficient, 27, 54, 92

quasi-correspond, 107
quasi-square-integrable support of $\pi$, 109
quasisplit $F$-group, 79

rank of $G$, 32
rapidly decreasing function, 186
regular $X$, 69
regular conjugacy class, 5, 6, 59, 60, 78, 79, 86, 87, 96

regular element, 444
regular function, 132
regular in $X$, 35
regular representation $r$, 145
regular-discrete function, 28
relevant $\pi$, 87, 103
relevant $G(F)$-module $\pi$, 5
residue datum for $\mathfrak{L}$, 293
Richardson orbit, 444
right regular representation, 2, 214
rigid $\pi$, 111
rigidity one theorem, 117
rigidity theorem, 7, 10, 89, 95, 108, 120–122, 135
root, 139
root $\beta$ of $\mathfrak{b}$, 358
root of $(G, A_{\mathfrak{M}})$, 253

second induction assumption, 334
segment, 103
segment of length $r$ with end $\nu^{r-1}\tau'$, 110
Selberg trace formula, v, 2, 3, 126, 128, 129, 146, 317
semiregular semisimple element, 517
semisimple orbital integral, 392
semisimple regular $x$, 32
set of roots of $(P, A)$ $\Sigma$, 147
seventh induction assumption, 366
sixth induction assumption, 345
slowly increasing cuspidal form, 183
smooth complex representation, 56
smooth function, 12
special $\mathfrak{b}$, 358
Spectral Part of Main Theorem, 457
Spherical, 130
split, division algebra, 89
splitting formula, 20
square-integrable $\pi$, 65
square-integrable $G(F)$-module, 5–7, 25, 29
stable function, 79
stably conjugate, 57, 79
standard $G(F)$-module, 66
standard $P$, 32
standard decomposition of $\pi$, 112
standard Levi subgroup, 5, 86, 109
standard parabolic subgroup, 5, 71, 81, 82, 110, 139, 151, 169, 183, 207, 229, 252
standard representations, 109, 253, 444
strong multiplicity one theorem, 89, 117
submersive, 39
submersive continuous map, 39
suitable regular $T$, 165
support of $T$, 67
support of the distribution $D$, 38
supported on characters, 324, 443
supported on characters (linear form), 16

Tate-Nakayama theory, 63
tempered $\pi$, 65, 87
tempered $G(F)$-module $\pi$, 5, 6, 31
third induction assumption, 342
trace form, 31, 74
trace formula, 223
trace identity, 82
Trace Paley-Wiener theorem, 74
truncation operator, 183

unitarizable $\pi$, 7, 56, 89
unitarizable space, 65
unitary, 88
unitary $\pi$, 7, 89
unnormalized induced representation, 69
unramified, 56
unramified $\chi$, 206, 212
unramified $\pi_u$, 131
unramified $\pi_v$, 7, 89
unramified $G$ at a place $v$, 393
unramified class $\mathfrak{o}$, 180

vanishing of distribution $T$, 67
vertices of $\Pi$, 383

weight factor $v(x, T)$, 180
weighted character, 15
weighted orbital integral, 15, 18, 19, 280, 325, 327, 403
well-positioned Levi subgroup over $F_S$, 253
well-positioned relative to $M$, 217
Weyl group, 142, 247
Weyl integration formula, 30, 97
Whittaker model, 7, 88, 96, 108

Zelevinsky involution, 115
Zelevinsky length, 110

MIX
Papier aus verantwortungsvollen Quellen
Paper from responsible sources
FSC® C105338

If you have any concerns about our products,
you can contact us on
**ProductSafety@springernature.com**

In case Publisher is established outside the EU,
the EU authorized representative is:
**Springer Nature Customer Service Center GmbH**
**Europaplatz 3, 69115 Heidelberg, Germany**

Printed by Libri Plureos GmbH
in Hamburg, Germany